Physics

Ken Dobson
David Grace
David Lovett

Collins Educational
An imprint of HarperCollins*Publishers*

Published by Collins Educational
An imprint of HarperCollins*Publishers* Ltd
77–85 Fulham Palace Road
London W6 8JB

First published 1997
ISBN 0 00 322328 0

British Library Cataloguing in Publication Data:
A catalogue record for this book is available from the British Library.

Edited by Pat Winter and John Evans
Design by Glynis Edwards
Cover design by Michael Faulkner
Illustrations by Barking Dog, Tom Cross, Jeremy Gower, Peter Harper,
Illustrated Arts, Pantek and TTP
Picture research by Marilyn Rawlings

Printed and bound by Rotolito Lombarda, Pioltello (Italy)

CONTENTS

1 How to use this book

To the student

THIS BOOK GIVES you a thorough coverage of the core material for advanced physics courses. It also includes the main optional topics of examination syllabuses. It has been written to convey the excitement and immediacy of modern physics and its uses in real-life situations, in the hope that you will develop an enthusiasm for physics, and gain the basic tools for a possible future physics career.

Physics in context

Physics is a subject of enormous breadth. It gives basic explanations of how the stars evolve, planets move and also the nature of subatomic particles. Physics enables us to alter our surroundings – to build bridges, launch satellites and make delicate instruments for microsurgery. It has given us the Internet and strong, lightweight sports equipment; it extends into every area of our life.

Each century has had its own discoveries and development of ideas in physics: at the time the ideas seemed extraordinary but they soon became commonplace. For instance, Einstein's theory of relativity astonished a world familiar only with the classical mechanics of Newton. Yet now we see Einstein's theory as a set of useful ideas that emerge naturally from studying light and the motion of objects.

Physics, then, can be described as having two aspects. First it is a body of information containing the rules that govern the natural world and help us to appreciate our surroundings. Secondly, physics

Mapping the world of physics

THE NATURAL WORLD

Ordinary matter –
solids, liquids, gases

Condensed matter –
physical electronics, magnetism

Atoms and nuclei

Deep matter –
fundamental particles

Electromagnetic radiation

The Earth –
structure, atmosphere,
hydrosphere, biosphere

Planets and stars

The cosmos

THE MADE WORLD

Constructions

Transport

Communications and information

Energy sources and control

Medical physics

ORGANISING PRINCIPLES

Concepts and laws

Dynamics and kinematics

Energy and mass

Waves and particles

Fields

Thermodynamics

Waves and oscillations

Quantum physics

Relativity

Symmetry

TOOLS

Of description and analysis:
language and mathematics

Of investigation: equipment,
strategies and techniques

SKILLS

Manipulative

Investigational

Information retrieval

Information technology (IT)

Interpersonal

PHYSICS AND SOCIETY

The nature of scientific activity

How discoveries are made

The history of physics

A map helps a traveller to picture a journey in advance as well as to follow the route. We know, too, that more than one type of map can serve the same purpose. Similarly with physics: this map is just one way in which we can sub-divide the body of physics information and activities. The map also corresponds to how this book has been planned, so it should help you to identify the landmarks on the journey through your studies.

In the map, we divide the world into the **natural** and the **made worlds**. As observations tell us, these worlds overlap and interact: burning fuels for energy affects the atmosphere and ultimately the climate of the Earth, for example. The **organising principles** – the concepts and laws that physi-cists have developed – apply to both worlds.

provides the tools for the many human activities – of engineers, astronomers, electronics designers, medical researchers, the business managers of physics projects, and others – that allow us to alter and construct the material world to suit our needs and to pursue our wish to discover the unknown.

Doing physics

Physics begins with **phenomena** – things that happen, that we observe. Physicists attempt to make sense of these things by developing **concepts** – sets of ideas or organising principles, such as force and energy – and **laws**, like Faraday's laws of electromagnetism.

As a student of physics, you will grasp one of the most important aspects of doing physics, which is to create a **model**. A model organises phenomena, ideas and theory into a structure that makes sense; it also allows us to make **predictions** – and real **practical devices** that actually work. For example, a wave model of light helps us understand the colours in a thin film of oil. It also helped astronomers to make better telescopes. The wave model was then linked with models of charge and current to predict the existence of radio waves, and led to radio and TV.

A model need not be complete – or even 'true' – to be useful. This century, the wave model of light had to be extended to include a particle model, which led to a better understanding of light and also of atoms and electrons, and then to the invention of lasers.

Tools

As a student tackling modern physics, you need the right tools to explore phenomena and models. **Mathematics** provides some of the most useful tools for making progress in physics. There are also the tools you use in the laboratory, such as meters, oscilloscopes, power supplies and data loggers.

Skills

Mathematics involves both knowledge and **skills –** the ability to choose the appropriate tools. There are other skills that enable you to learn and operate successfully. They include:

- the skills of **manipulation**, to use equipment safely and accurately in the laboratory,

- the skills of **investigation,** to design, carry out, analyse and evaluate experiments,

- the **information retrieval** skills, to gather information from books, magazines and other data sources – and from listening,

- the skills of **information technology** (IT), to collect, order, analyse or present data using computers and other electronic systems,

- the **interpersonal skills**, to communicate with others, both in writing and in speech, and to work effectively as part of a team.

Using this book

The 'meat' of the book is in Chapters 2 to 19. They describe and explain the physics you will need to cover the **subject core** for physics. Whether you study the other chapters will depend on the particular syllabus you are following.

Each chapter begins with an **Opener**, describing an often familiar example of the way that an aspect of the physics in the chapter is used.

A brief overview describes the physics developed in the chapter. Key words and ideas are printed in **bold**, and also appear in the index at the end of the book.

2 Moving in space and time

THE FIRE, POLICE and ambulance services all need to respond swiftly to calls, and this often means driving as fast as possible through heavy traffic to the scene of an emergency. In built-up areas, an ambulance service has to respond to 50 per cent of emergencies within 8 minutes and 95 per cent within 14 minutes of the call.

The West Midlands Ambulance Service receives 850 emergency calls every 24 hours, so it is essential that the closest of their 64 ambulance crews responds to a call. They are helped by the first satellite tracking and mapping link to be used by an ambulance service anywhere in the world.

Each ambulance has a receiver–computer system on board, which continually monitors radio signals from geostationary satellites. The signals pinpoint an ambulance's position to within a few metres, and this information is relayed by radio to the control centre. There, a real-time mapping system displays on screen all the ambulance positions and accidents as they happen. Then, for each incident, the staff at the centre can direct the best placed ambulance to go to the scene of the emergency by the fastest route.

Introduction

Movement is one of the most fundamental topics in the study of

The ideas in this chapter

The ideas of **force** and **gravity** are fundamental to understanding the behaviour of objects on Earth. Objects will only move if a force is acting on them. Gravity is a force which acts on every object. For

Feature boxes describe modern applications of science or historical developments.

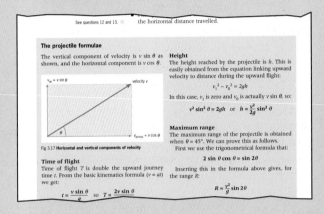

SATELLITES FOR SURVEYING

SATELLITES ARE NOW used to survey areas of land larger than a few square kilometres. The first question a map-maker needs to answer is: Where precisely am I measuring *from*? To get an accurate position on the ground, the surveying equipment needs to communicate with at least four satellites (Figs 2.8 and 2.9). Three produce a position 'fix', and the fourth provides an extra check.

For this, the positions of the satellites themselves need to be accurately known, so they are constantly monitored from fixed ground stations. The satellites emit positioning signals at ultra-high frequency radio waves. From the time delay for each satellite, its precise position is known.

Ionised layers that arise and disappear in the atmosphere slightly alter the radio wave speed. As for the varying satellite speeds, corrections are made to ensure that surveying data are accurate.

By 1994, Global Positioning Systems were working to accuracies of about 2 metres, and the accuracy was expected to improve to 10 centimetres within a few years.

GPS satellites

fixed monitor station

surveyor ship

Extension passages, which have a coloured tint over them, take ideas to a deeper level.

See questions 12 and 13. ■ the horizontal distance travelled.

The projectile formulae

The vertical component of velocity is $v \sin \theta$ as shown, and the horizontal component is $v \cos \theta$.

$v_{up} = v \sin \theta$

velocity v

θ

$v_{across} = v \cos \theta$

Fig 3.17 **Horizontal and vertical components of velocity**

Time of flight
Time of flight T is double the upward journey time t. From the basic kinematics formula ($v = at$) we get:

$$t = \frac{v \sin \theta}{g} \quad \text{so} \quad T = \frac{2v \sin \theta}{g}$$

Height
The height reached by the projectile is h. This is easily obtained from the equation linking upward velocity to distance during the upward flight:

$$v_1^2 - v_0^2 = 2gh$$

In this case, v_1 is zero and v_0 is actually $v \sin \theta$, so:

$$v^2 \sin^2 \theta = 2gh \quad \text{or} \quad h = \frac{v^2}{2g} \sin^2 \theta$$

Maximum range
The maximum range of the projectile is obtained when $\theta = 45°$. We can prove this as follows.
First we use the trigonometrical formula that:

$$2 \sin \theta \cos \theta = \sin 2\theta$$

Inserting this in the formula above gives, for the range R:

$$R = \frac{v^2}{g} \sin 2\theta$$

Keeping track of progress

At the end of the text for each chapter is a **Summary** of the main points that you should have understood and learned in your study.

There are **Self-test questions** in the margin, for you to check that you have really understood what you have read.

SUMMARY

This chapter should help you to understand a number of ideas and acquire several numerical skills. Having studied it, you should be able to:

■ Understand that distance can be measured in units of time (seconds) as well as in units of distance such as metres and kilometres.

■ Know that the speed of light, c, is fundamental to the measurement of distances.

■ Know that measured motion is always relative to an agreed frame of reference.

■ Analyse distance–time and speed–time graphs.

■ Use the equations of motion of an object moving in a straight line with uniform acceleration.

■ Know the difference between vector and scalar quantities.

■ Use vector diagrams and/or trigonometry to analyse vector motion.

■ Understand the basic principles of triangulation.

You should know the following **equations of motion**.

■ For **motion at a constant speed** (or an average speed):

distance = (average) speed × time

$x = vt$

■ For **uniform (steady) acceleration**:

(a) acceleration = $\dfrac{\text{change in velocity}}{\text{time taken to change}}$

$a = \dfrac{v_1 - v_0}{t}$ or $a = \dfrac{\Delta v}{\Delta t}$

which is often written as $v_1 = v_0 + at$.

(b) distance covered = average speed × time

$x = \dfrac{v_1 + v_0}{2} t$

which is equivalent to: $x = v_0 t + \frac{1}{2}at^2$.

When t is unknown, use $v_0^2 - v_0^2 = 2ax$.

(c) For non-uniform motion, we need to deal in small changes:

$\Delta x = v\Delta t$, or $v = \dfrac{dx}{dt}$ in calculus notation,

$\Delta v = a\Delta t$, or $a = \dfrac{dv}{dt}$ in calculus notation.

QUESTIONS

1 Model aircraft can be controlled 'in real time' by moving joysticks on a ground-based radio transmitter. Why wouldn't a control system like this be much use in controlling the fly-past of a space probe near Jupiter? (Assume equally good signal reception both ways.)

2 Calculate the following:

a) The distance in km of an aircraft from an airport whose radar measures it to be at a radar time-distance of 1.67 ms.

b) The time it would take a radio message to reach Mars from Earth when they are at their closest separation of 7.83×10^7 km.

3

a) The vertical probe radar from a mapping satellite crosses from the sea to a cliff. It records a difference in light-flight time of 1.17 microseconds as it does so. Show that the cliff is 175 m high.

b) To what accuracy must the timer in the satellite be able to measure time if it is to map height differences of 0.3 m?

4 Astronomers measure the distances of the nearer stars using a base line such that AM as in Fig 2.Q6 is 499 light-seconds long, which is the radius of the Earth's orbit round the Sun. The angle they measure that is equivalent to angle ACB is called the *parallax* of the star.

The parallax angles for two stars are given in the table. Calculate their distances from Earth in light years.

Star	Parallax (')
Sirius	1.05×10^{-4}
Alpha Centauri	2.07×10^{-4}

5 A radar speed detector shows a frequency change of 8 kHz when monitoring the speed of an aircraft. The frequency of the radar beam used is 10^{10} Hz. What is the relative speed between aircraft and detector?

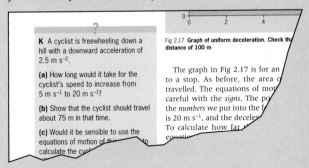

0   2   4

K A cyclist is freewheeling down a hill with a downward acceleration of 2.5 m s^{-2}.

(a) How long would it take for the cyclist's speed to increase from 5 m s^{-1} to 20 m s^{-1}?

(b) Show that the cyclist should travel about 75 m in that time.

(c) Would it be sensible to use the equations of motion of this to calculate the cyc...

Fig 2.17 **Graph of uniform deceleration.** Check th... distance of 100 m

The graph in Fig 2.17 is for an ... to a stop. As before, the area o... travelled. The equations of mot... careful with the *signs*. The po... the *numbers* we put into the f... is 20 m s^{-1}, and the decel... To calculate how far t... equatio...

Near the end of each chapter there are more **Questions**. Some are very simple, to check whether or not you have grasped the main ideas; others are more searching and test deeper understanding – these include questions set in examinations. Often, there are 'learning' questions, which lead you through proofs of formulae or develop important concepts in an alternative way to that in the text.

Questions are linked to the text by **Marginal notes** in the text.

See question 7. ■

The area of the triangle is $\frac{1}{2}(v_1 - v ...$ distance x travelled in time t is:

Most chapters have an **Assignment**. Typically, this is a self-contained extended exercise with information about an application of physics and can include real data to analyse. Questions range from simple comprehension to more challenging ones where you manipulate information in the Assignment.

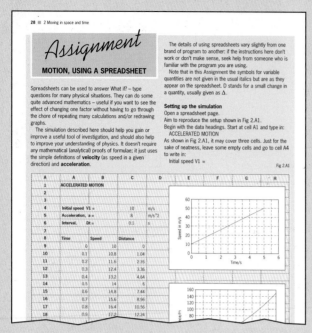

Each chapter or chapter group ends with a **Map**, a diagram of the main concepts covered, to help you see how they interlink. (Even better, construct your own map when you complete a chapter.)

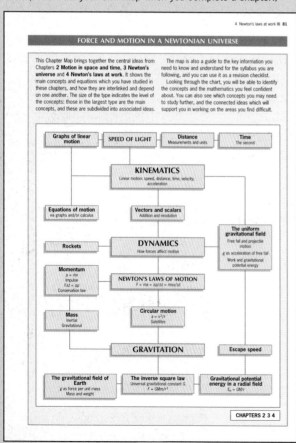

Cross-references and reminders

Ideas from one topic in this book are often relevant in another topic, so the text includes page number cross-references. Sometimes, though, it is more helpful to repeat brief definitions as short **Reminders** – the ticked boxes in the margin.

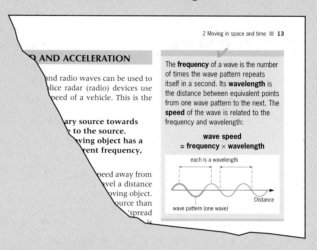

Mathematics

This book assumes that you have a basic knowledge of mathematics, but not that you study mathematics to advanced level. This means that any mathematics is developed fully from first principles in the text. **Examples** help you through the working. Appendix 2 gives an overview of the mathematics you need to do physics at advanced level – both as you learn it and in examinations.

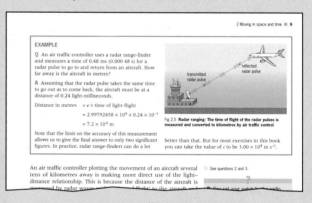

Practical work and investigations

This is not a practical book: experiments are mentioned in the text but not described in full. However, Appendix 3 gives advice on practical and investigative work, together with other learning activities, and lists the key experiments that normally form part of an advanced course. There is also a list of ideas for student investigations.

FORCES, MOVEMENT AND MATERIALS

THIS SECTION IS about the physics of everyday objects. We look at the forces that make objects move, stretch and break, and at the way that objects resist these effects. The underlying ideas are based on the theories and laws that Sir Isaac Newton first set out in the seventeenth century. He defined his laws so precisely that we still use his definitions to describe and predict the motion of everything – from the molecules in a gas, to space probes.

Everyday objects are made of ordinary matter formed into structures. By 'ordinary matter' we mean the atoms and molecules as they exist in bulk, as gases, liquids and solids. (In later sections we deal with subatomic matter that obeys other, quantum rules.)

To learn about the structures of ordinary matter, we look at natural things such as gases, crystals and metals. Then we see the way that their properties, and their behaviour when forces are applied, decide the structures we make from them, such as transport vehicles and buildings.

We also look at a fundamental idea that was unknown in Newton's day – the concept of energy. This idea runs through most chapters in the book, so it is important enough to have a short item on its own. The ideas of force and energy are linked in the very useful concept of a field, developed here in terms of the gravitational field.

Most of the key ideas you meet in this section will reappear – so it's well worth learning them thoroughly.

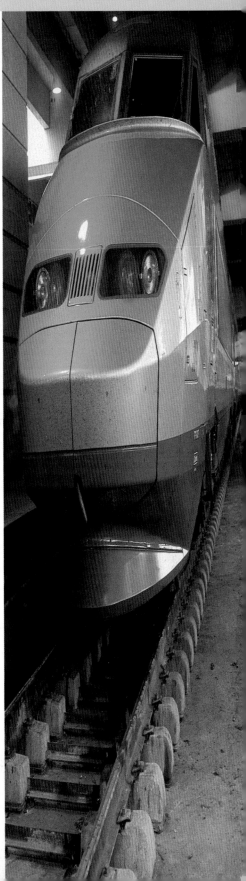

2 Moving in space and time

THE FIRE, POLICE and ambulance services all need to respond swiftly to calls, and this often means driving as fast as possible through heavy traffic to the scene of an emergency. In built-up areas, an ambulance service has to respond to 50 per cent of emergencies within 8 minutes and 95 per cent within 14 minutes of the call.

The West Midlands Ambulance Service receives 850 emergency calls every 24 hours, so it is essential that the closest of their 64 ambulance crews responds to a call. They are helped by the first satellite tracking and mapping link to be used by an ambulance service anywhere in the world.

Each ambulance has a receiver–computer system on board, which continually monitors radio signals from geostationary satellites. The signals pinpoint an ambulance's position to within a few metres, and this information is relayed by radio to the control centre. There, a real-time mapping system displays on screen all the ambulance positions and accidents as they happen. Then, for each incident, the staff at the centre can direct the best placed ambulance to go to the scene of the emergency by the fastest route.

Introduction

Movement is one of the most fundamental topics in the study of physics, with applications essential to our everyday lives. For example, in transport, drivers and pilots need to know how fast cars, trains and aircraft are moving, so that they can control them – while planners need to know this so that they can make timetables.

At an atomic level, knowing about the movement of particles has led to explanations of the behaviour of solids, liquids and gases.

But motion is very complicated. On a typical car journey, for example, the driver may make thousands of changes in speed and direction. Similarly, particles of gas are invisible, so how can we even think of following their movements?

Historically, the accurate study of motion began with trying to predict the movements of the Sun, the Moon, the planets and the stars. At the time, people believed that the Earth was at the centre of the Universe and everything else moved round the Earth in some way. People also believed that the stars and the planets had an effect on their lives and behaviour, and so needed accurate data to help them make *astrological* predictions.

People didn't see the need to explain the movement of everyday objects – smoke rose and apples fell from trees because it was natural for them to do so. Also, speed was not the important feature of everyday life that it is today. People moved slowly – on foot, on horseback or on sailing ships. Clocks were rare, and measured time no more accurately than to the nearest minute.

In the seventeenth century, many of these aspects of life began to change. First, there was a great increase in trade worldwide. As a result, there was a need to improve the skills of navigation. It was far more difficult to navigate a ship across an ocean than it has been simply to sail along to coast from port to port. The positions and movement of the Sun, the planets and the stars now needed to play a far more important role – as an aid to navigation, rather than for fortune-telling.

At the same time, the use of gunpowder was making the technology of war more sophisticated, with the development of cannons. The Italian physicist Galileo Galilei (1564–1642) studied the behaviour of moving objects and so helped to explain the movement of cannon balls. This further explained how cannons could be fired more accurately.

The result of these studies of motion in the seventeenth century was that scientists had very clear ideas of how to measure the quantities involved in motion: distance, time, speed, velocity and acceleration.

The study of motion on its own is called **kinematics**. When forces are taken into account, the study becomes known as **dynamics**. Some of the world's most well-known physicists – Galileo, Newton and Einstein in particular – did their greatest work on the motion of objects.

The study of motion has led to the development of accurate measuring instruments. Examples are telescopes, chronometers and satellites for ocean navigation, radar, and lasers for surveying. Studying motion has also made physicists think deeply about the nature of space and time, and has led to theories about gravity, special and general relativity, and the origin and expansion of the Universe. Some of these more advanced ideas are dealt with in later chapters.

In this chapter, before we look closely at the movement of objects, we first see how the necessary measurements are made, and how the units of measurement have been established. In particular, we shall review the use of light as the basis of many measuring techniques today, and measurements of distances ranging from the small scale to the scale of the Universe.

Fig 2.1 **A nineteenth-century navigator uses a sextant to measure the angle of altitude of the Moon. From this he could work out the latitude of his ship**

1 TIME USED TO MEASURE DISTANCE

In moving, an object changes its position in space – so we need to be able to measure *distance*. Moving takes time – so *time* also has to be measured. In practice, we are often more concerned with time than distance, as these common statements show:

House for sale. Five minutes from the station.

Paris is just 40 minutes from London, by regular air shuttle.

The signposts of footpaths occasionally tell you the time it will take you to walk somewhere, rather than the distance, as in Fig 2.2. You can see that in each of these examples an assumption is being made about the speed of travel.

Fig 2.2 **The signpost tells the hikers that they are over 2 hours' walk from Nancroix**

The letters in the word laser stand for light **a**mplification by **s**timulated **e**mission of **r**adiation, and those in radar for **r**adio **d**irection **a**nd **r**anging.

2 MEASURING DISTANCES WITH ELECTROMAGNETIC WAVES

Using **time units** to measure distance may seem strange, even 'unscientific'. Yet time is the basis of most modern ways of measuring distance. Air traffic controllers use the time it takes radar pulses to travel to and from an aircraft to chart its position in a crowded flight pattern (Fig 2.5). Surveyors use the time-of-flight of a pulse of laser light to measure distances accurately (Fig 2.7). Cartographers making accurate maps now use radar (Fig 2.8) or laser beams.

On the scale of distances in the Universe (Fig 2.3), astronomers also use radar to establish the precise positions of planets. As in Table 2.1, they measure far vaster distances in the **time units** based on the **speed of light** – light-seconds and light-years:

$$\text{time unit (in seconds)} = \frac{\text{distance of an object (in metres)}}{\text{speed of light (in metres per second)}}$$

Fig. 2.3 **The distance of Venus from the Earth has been measured accurately using radar. This measurement (see page 11) is the first step in all astronomical distance measuring**

Table 2.1 **Some distances in units of length and in time units**

Object	Average distance from Earth/metre	Time units
Geostationary satellite	4.2×10^7	0.1401 s
Car from radar speed trap	100	3.335×10^{-7} s
Moon	3.844×10^8	1.282 s
Sun	1.496×10^{11}	499.0 s
Sirius	8.2×10^{16}	8.7 y
Andromeda galaxy	2.1×10^{22}	2.2×10^6 y

All these techniques depend on the measuring signal having a constant speed. Many of the applications rely on the fact that electromagnetic waves are easily reflected, especially by metals. Both radio waves and light waves are electromagnetic waves of constant speed (when travelling in a vacuum). Light is such a fundamental and useful measuring tool that the SI unit of distance, the **metre**, is defined in terms of the speed of light:

Fig 2.4 **The Andromeda galaxy, a spiral galaxy and the nearest to our own Milky Way Galaxy. Both are composed of countless stars, cooler objects, dust and gas. Its name is from the constellation in which it can be seen with the naked eye**

See question 1. ■

?

A How long ago did the light leave the Andromeda galaxy shown in Fig 2.4?

A metre is the distance travelled by light in a vacuum in the fraction 1/299 792 458 of a second.

Where does the number 299 792 458 come from? To explain, we need to be aware that the metre is one of the seven base units that have been agreed on by scientists internationally, while other units are defined in terms of these seven. (Another base unit relevant here is the second, defined in terms of the vibrations of a caesium-133 atom.)

In 1983, it was decided that the distance light travels in a vacuum in this very tiny fraction of a second would be the new definition of a metre. This means that light is defined as travelling at a speed of 299 792 458 metres per second, the speed also of all electromagnetic waves in a vacuum. The speed is given the symbol c, and we can say that:

distance = c × time of light-flight
  (m) (m s^{-1}) (s)

This is a use of the familiar formula:

distance = speed × time

$$x = v \times t$$

EXAMPLE

Q An air traffic controller uses a radar range-finder and measures a time of 0.48 ms (0.000 48 s) for a radar pulse to go to and return from an aircraft. How far away is the aircraft in metres?

A Assuming that the radar pulse takes the same time to go out as to come back, the aircraft must be at a distance of 0.24 light-milliseconds.

Distance in metres $= c \times$ time of light-flight

$$= 2.99792458 \times 10^8 \times 0.24 \times 10^{-3}$$

$$= 7.2 \times 10^4 \text{ m}$$

Note that the limit on the accuracy of this measurement allows us to give the final answer to only two significant figures. In practice, radar range-finders can do a lot

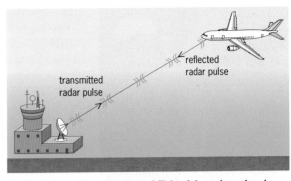

Fig 2.5 **Radar ranging: The time of flight of the radar pulses is measured and converted to kilometres by air traffic control**

better than that. But for most exercises in this book you can take the value of c to be 3.00×10^8 m s^{-1}.

An air traffic controller plotting the movement of an aircraft several tens of kilometres away is making more direct use of the light–distance relationship. This is because the distance of the aircraft is measured by radar waves with a 'time of flight' to the aircraft and back that is accurately measured using a quartz clock.

Mapping with light

Triangulation

Making modern maps of places on Earth involves the surveying technique called **triangulation**. This is carried out either on the ground or by using survey satellites. The technique relies on two of the properties of light:

- Light beams travel in straight lines
- Light travels at a known speed

Triangulation is a technique that was used, and probably invented, by the Ancient Egyptians. Simple triangulation needs a **base line** of carefully measured length (see Fig 2.6). Two distant objects (X and Y) are chosen. The angle each object makes with the base line is measured from each end of the base line. Then the distances to the objects are found using simple trigonometry. Nowadays a laser

See questions 2 and 3.

B You set your watch by the radio time signal. If the broadcasting aerial is 50 km away, how 'slow' is your watch because of the finite speed of light? Should this bother you?

C Outline some advantages and disadvantages of measuring distances such as the following using light (or radar): **(a)** from Earth to the Moon, **(b)** the thickness of a layer of paint on a car body, **(c)** the width of a room.

D Look at Fig 2.6. Give a reason why the objects at X and Y were used for surveying. Why was the pylon at Z not used?

See question 6.

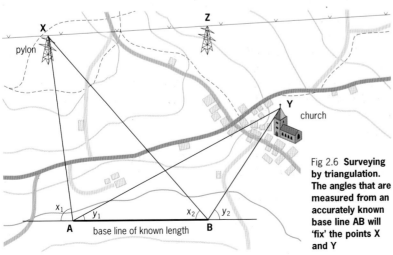

Fig 2.6 **Surveying by triangulation. The angles that are measured from an accurately known base line AB will 'fix' the points X and Y**

surveying instrument (Fig 2.7) measures a distance in light-seconds and converts this measurement into metres.

The accuracy of maps made using triangulation depends on very reliable clocks. Angles are also measured with great accuracy, but cannot be measured as exactly as light-times. However, a long base line increases accuracy, and accurate maps are based on lines many kilometres long.

Fig 2.7 **Surveyors use a laser instrument to measure distances. The instrument calculates a distance from the time a pulse travels to and from an object**

SATELLITES FOR SURVEYING

SATELLITES ARE NOW used to survey areas of land larger than a few square kilometres. The first question a map-maker needs to answer is: Where precisely am I measuring *from*? To get an accurate position on the ground, the surveying equipment needs to communicate with at least four satellites (Figs 2.8 and 2.9). Three produce a position 'fix', and the fourth provides an extra check.

For this, the positions of the satellites themselves need to be accurately known, so they are constantly monitored from fixed ground stations. The satellites emit positioning signals at ultra-high frequency radio waves. From the time delay for each satellite, its precise position is known.

Ionised layers that arise and disappear in the atmosphere slightly alter the radio wave speed. As for the varying satellite speeds, corrections are made to ensure that surveying data are accurate.

By 1994, Global Positioning Systems were working to accuracies of about 2 metres, and the accuracy was expected to improve to 10 centimetres within a few years.

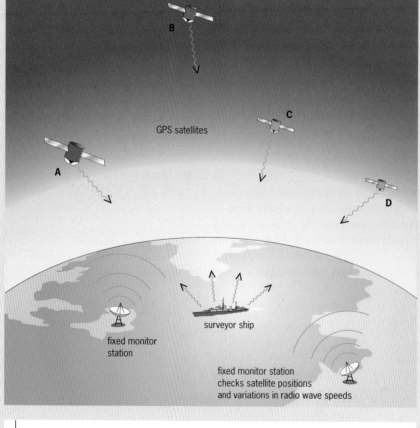

Fig 2.8 **The map-maker is surveying from a ship, and the ship's precise position is first worked out using data from four satellites**

Fig 2.9 **The first satellite signal, A, arrives, and signals B, C and D follow at different time intervals because they travel different distances**

LIGHT AND ASTRONOMICAL DISTANCES

CAPTAIN COOK EARNED his reputation as a great navigator and a careful, accurate maker of naval charts. In 1768 the British Admiralty sent him on a voyage to the far side of the world in a vast project – to measure the dimensions of the Solar System.

It had proved too difficult to measure the distance to the Sun and the planets using parallax, that is, by measuring the positions relative to each other of these objects in the sky from different places on the Earth. The angles needed to be measured very accurately, and the observations had to be made at exactly the same time at different places. Without reliable, accurate clocks, this was impossible.

Edmund Halley, of Halley's comet fame, suggested another method: to observe and time the transit of the planet Venus across the disc of the Sun from different places on the Earth. Transits of Venus are rare astronomical events, but were predicted for 1761 and 1769, with the following transit not due until 1874. Observations made in 1761 were not adequate to give

a distance to the Sun. So it was all the more important not to lose the opportunity in 1769. Expeditions were set up to observe from Canada and Norway in the north, and from Tahiti in the South Seas, which is where Captain Cook went.

The relative distances of the planets and the Sun from Earth were well known at the time (Fig 2.10), but the scale of the Solar System was not known with any accuracy. Observers viewed the transit of Venus from places on the Earth separated along a north–south line of known distance. By measuring the time it took for Venus to move across the Sun's disc, the observers could determine the position with some accuracy of the apparent paths of Venus, as shown in Fig 2.11. They could then work out the distance AB separating the paths, and hence the angle subtended by AB. From this angle they calculated the distance from Earth to the Sun. It was then a simple matter to calculate the distances to all the planets in the Solar System, a truly remarkable achievement.

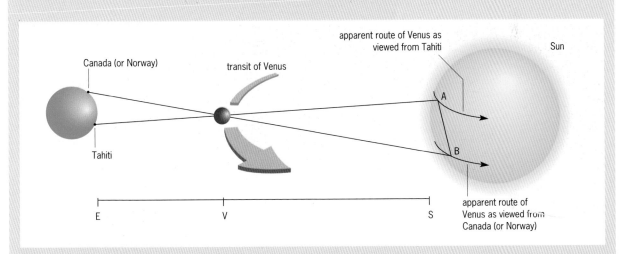

Fig 2.10 **How observations were made of the transit of Venus in 1769. The *ratio* of the distances of the Sun (ES) and Venus (EV) from Earth were known, and the distance was known between the latitudes of Canada and Tahiti**

Fig 2.11 **Observers measured the angle subtended by AB and could then calculate the distance to the Sun**

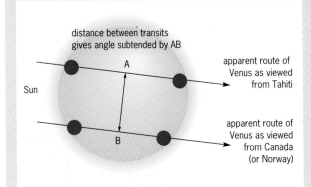

The astronomical unit

This enterprise laid down the main base line for the measurement of all astronomical distances – the radius of the Earth's orbit. This distance is called the astronomical unit or AU. It is 1.496×10^{11} m (499.0 light-seconds). As much by luck as by good judgement, the 1769 results were found to be accurate to within about 1 per cent of the modern value. (See Chapter 27 for the modern version of this measurement.)

How far away are the stars?

Stars are too far away for their distances to be measured using a base line drawn on Earth. But astronomers noticed that some stars appeared to move relative to other stars as the Earth moved round the Sun. They concluded that this effect was the same as you see when you move in front of a window: a point on the window frame appears to move against the background of more distant objects. This effect is called **parallax**, and for stars it is called stellar parallax (see Fig 2.12).

See question 4. ■

During the nineteenth century, the parallaxes of thousands of stars were measured, and hence their distances from Earth were calculated. But many stars showed no detectable parallax. They were too far away for the annual motion of the Earth across a distance of 3×10^{11} metres to make any measurable difference to their apparent position. In particular there was a class of bright cloudy objects that showed no parallax, and astronomers concluded that these objects (called nebulae, the Latin for 'clouds') must be on the fringes of the Universe. Measurements on this scale, which helped reveal the expansion of the Universe, are described in Chapter 27.

Fig 2.12 **Stellar parallax, due to the motion of Earth round the Sun**

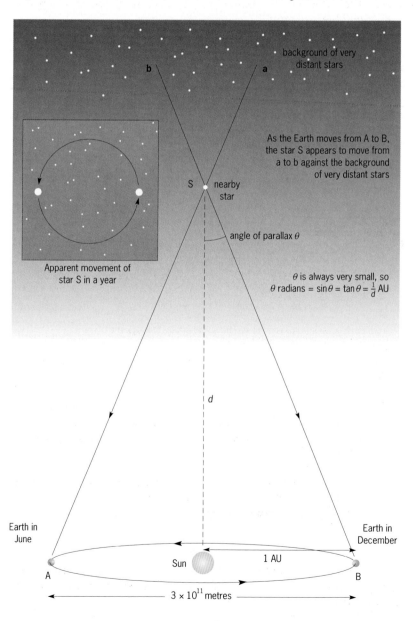

background of very distant stars

As the Earth moves from A to B, the star S appears to move from a to b against the background of very distant stars

S nearby star

angle of parallax θ

θ is always very small, so
θ radians $= \sin\theta = \tan\theta = \frac{1}{d}$ AU

Apparent movement of star S in a year

d

Earth in June

Earth in December

Sun

1 AU

A

B

3×10^{11} metres

3 KINEMATICS: SPEED AND ACCELERATION

Electromagnetic waves such as light and radio waves can be used to measure *speed* as well as distance. Police radar (radio) devices use the **Doppler effect** to measure the speed of a vehicle. This is the effect:

> **A signal is directed from a stationary source towards an object that is moving relative to the source.**
> **Then, the signal reflected from the moving object has a different wavelength, and hence a different frequency, from the source signal.**

For example, when an object is moving at constant speed away from the source of a signal, each wave (or pulse) has to travel a distance further than the previous wave to catch up with the moving object. A reflected wave then has a longer journey back to the source than the previous wave. This makes the reflected waves more 'spread out' – their wavelength is longer, and hence their frequency is lower, than the source frequency. The reverse happens when the reflecting object is moving towards the source.

Fig 2.13 illustrates the Doppler effect as described above for a radar signal. In (a), the wavelength of the reflected signal is longer, and the frequency lower, than of the source signal, and in (b), with the car moving towards the source, the wavelength is shorter, and the frequency higher. The radar device used by the police measures the *difference* in frequency between the transmitted and the received signal and displays the speed of the vehicle on a screen in miles per hour (see Fig 2.14).

The **frequency** of a wave is the number of times the wave pattern repeats itself in a second. Its **wavelength** is the distance between equivalent points from one wave pattern to the next. The **speed** of the wave is related to the frequency and wavelength:

wave speed = frequency × wavelength

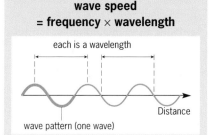

each is a wavelength

wave pattern (one wave)

Distance

E Look at Fig 2.13(a) and suggest a reading on the radar device for waves reflected from the vehicle.

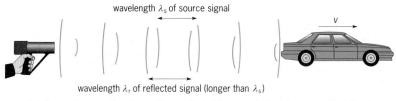

wavelength λ_s of source signal

v

wavelength λ_r of reflected signal (longer than λ_s)

(a) Car moving away from transmitter: reflected *frequency* is less than transmitted signal

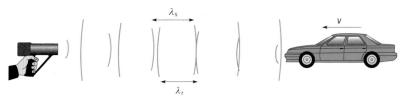

λ_s

v

λ_r

(b) Car moving towards transmitter: reflected *frequency* is greater than transmitted signal

Fig 2.13 **The Doppler effect – the effect on radar waves directed at a vehicle** (a) **moving away from the source,** (b) **moving towards it. (Remember that the speed of all signals are always c, the speed of light)**

Fig 2.14 **The display on the instrument indicates whether the vehicle is speeding or not**

In the simple case when only the reflecting object is moving (such as when the speed of a vehicle is being recorded), the following formula is the basis for calculating the speed *v* of the object:

$$\Delta f = \frac{2fv}{c}$$

where Δf is the change in frequency of the radar signal and *c* the speed of light. See the Appendix for a full derivation of this formula for the Doppler effect.

See question 5.

EXAMPLE

Q A police radar speed detector instrument uses a frequency of 10^{10} Hz. It measures a Doppler frequency change of 600 Hz when it is pointed at a moving car.

a) What speed is shown by the detector?

b) This may not be the true speed of the car. Explain why.

A

a) Using the simple Doppler formula, $\Delta f = 2fv/c$, we get:

$$600 = 10^{10} \times \frac{2v}{c}$$

So:

$$v = \frac{600 \times 3.00 \times 10^8}{2 \times 10^{10}}$$
$$= 9 \text{ m s}^{-1}$$

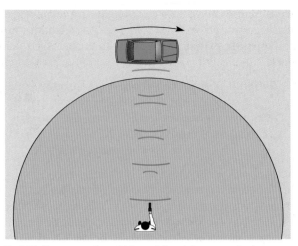

Fig 2.15 **Movement of a vehicle at right angles to a radar beam**

b) This will be the true speed only if the direction of the radar beam is the same as the actual direction of motion of the car. At any other angle, the true speed will be greater than the measured speed. For example, if the car were moving at right angles to the beam, the speed reading would be zero, as shown in Fig 2.15 (the transmitted and reflected frequencies are the same).

Speed from distance and time

The simplest way to measure speed in the laboratory and in everyday life is to measure how far an object travels in a measured time:

$$\text{speed} = \frac{\text{distance moved}}{\text{time taken}}$$

in m s^{-1} or km h^{-1}, for example.

Short times can be measured using an electronic clock stopped and started by a light gate, say. Speed in longer journeys may need only the accuracy of a stopwatch (to the nearest minute, second or tenth of a second). Distances are measured by metre rules, tape measures or from maps.

Acceleration

The police are not usually interested in measuring the *acceleration* of the car – although car manufacturers are. Acceleration of a car is a measure of how quickly its velocity changes with time:

$$\text{acceleration} = \frac{\text{change in velocity}}{\text{time for velocity to change}}$$

in m s^{-2}.

Expressed as a formula using standard symbols:

$$a = \frac{v_1 - v_0}{t} \qquad [1]$$

where v_0 is the *initial* velocity (at the start of timing) and v_1 is the velocity t seconds later.

4 THE EQUATIONS OF MOTION

Real objects tend to move in quite complicated ways. It is better to start with the simplest possible cases and build in the complications later. We start by considering objects that move *steadily*, in a *straight line*. We deal here with the motion of objects that move at constant speeds or with constant (or *uniform*) accelerations. This is of course a very common kind of motion – for example, falling objects tend to move with a constant acceleration of 9.8 m s^{-2}. We can describe this kind of motion by using graphs and by the *equations of motion.*

Graphs of motion

The equations of motion (kinematic equations) can be derived most easily by analysing the graphs that illustrate the motion.

Moving at a constant speed (zero acceleration)

Graphs A and B in Fig 2.16(a) illustrate an object moving at a steady speed. **Graph A** is a **distance–time** graph. It is a straight line with a constant slope, showing that the object covers the *same* extra distance Δx in equal extra intervals of time Δt.

Speed is defined as the rate at which distance changes with respect to time. In Graph A, the object covers 5 m in any given second: its speed is 5 *metres per second* (5 m s^{-1}). In general:

$$\text{speed } v = \frac{\Delta x}{\Delta t}$$

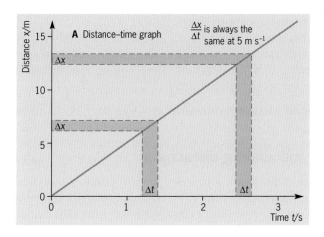
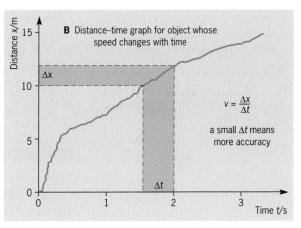

Fig 2.16(a) **Graphs for a moving object**

The quantity $\Delta x/\Delta t$ is the slope (or tangent) of the graph. This ratio will always give us the speed, even if the graph is not a straight line, but to get an accurate result we have to make Δt very small if, as in **Graph B**, the **speed is changing with time**.

But in Graph A, and in Graph C of Fig 2.16(b), $\Delta x/\Delta t$ doesn't change with time, and so we can use the simpler equation:

$$v = \frac{x}{t} \text{ or } x = vt \qquad [2]$$

In everyday situations, such as making a long journey by car, it makes sense to consider v as an *average* speed. Drivers make an experienced guess that they can manage an average speed of, say, 100 km per hour, and so will cover 500 km in five hours' driving.

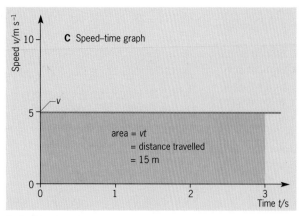

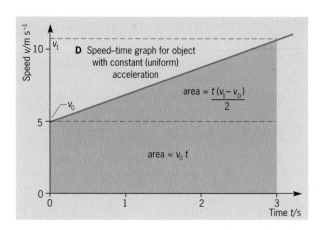

Fig 2.16(b) **Graphs for a moving object**

The idea of making values like Δt very small is the basis of *calculus*. The equations of motion may be derived using calculus, as shown on page 24.

?

F Estimate your average speed for your journey to school or college.

G Think about the journey you make going to school or college. Sketch a rough distance–time graph of this journey. Label the key features and the time and place at which you are travelling at the greatest speed.

H How long would a car take to increase its speed from 8 m s^{-1} to 24 m s^{-1} at a constant acceleration of 3.2 m s^{-2}?

Graph C is a **speed–time** graph for movement of the object in Graph A. It shows a constant value for v. It also gives us an insight into a very useful rule:

> **The area between the graph line and the time axis tells us the distance travelled during any interval of time.**

So the coloured area on the graph is the quantity vt (in this example, 15 metres, the distance travelled in 3 seconds at a speed of 5 m s^{-1}).

This rule works whatever the shape of the graph, however, and is useful in analysing non-uniform motion, as shown in Graph B.

Moving with uniform acceleration

Graph D in Fig 2.16(b) is a **speed–time** graph for an object whose speed increases steadily with time. In other words, it is moving with a **constant (uniform) acceleration**.

The speed at the start of timing is v_0, and after t seconds it has become v_1. We often want to know what the speed v_1 of an object will be after a period of acceleration. Rearranging the defining equation 1 above for acceleration gives:

$$\textbf{speed after acceleration } v_1 = v_0 + at \qquad\qquad [3]$$

Equations for calculating distance x

We can use Graph D to obtain a formula for the distance travelled by an accelerating object in a time t. To do this we apply the rule that the distance travelled in time t equals the relevant area under the graph.

Graph D consists of a rectangle topped by a triangle. The area of the rectangle is $v_0 t$, where v_0 is the speed when $t = 0$ (the *initial* speed).

See question 7. ■

The area of the triangle is $\frac{1}{2}(v_1 - v_0)t$, so the equation for the total distance x travelled in time t is:

$$\text{distance} = \text{total area under graph}$$

$$x = v_0 t + \tfrac{1}{2}(v_1 - v_0)t \qquad\qquad [4]$$

or: $\qquad\qquad x = \tfrac{1}{2}(v_0 + v_1)t$

which is simply: $\quad$ distance = average speed × time

This formula is useful if we happen to know the values of v_0 and v_1. In practice, we often know *either* v_0 *or* v_1, and the acceleration a.

When final speed v_1 is not known

We have seen that a is defined in terms of v_0, v_1 and t, as in equation 1:

$$a = \frac{v_1 - v_0}{t}$$

which can be rearranged to give:

$$v_1 - v_0 = at$$

Substituting for $(v_1 - v_0)$ in equation 4 above gives:

$$x = v_0 t + \tfrac{1}{2}at^2 \qquad [5]$$

When acceleration *a* is not known

Another way of looking at this situation is by using *average* speed. Where the acceleration is constant, we can assume an overall speed that is the *average* of the initial and final speeds. Then:

$$\text{distance} = \text{average speed} \times \text{time}$$

or:

$$x = \tfrac{1}{2}(v_0 + v_1)t \qquad [6]$$

Graph D in Fig 2.16(b) shows this as well. Check it by simplifying equation 4, the formula for the sum of the areas of the rectangle and the triangle.

When time *t* is not known

Finally, it would be useful to have a formula that allows us to make calculations when we have no information about the time during which speed (velocity) changes occur. This formula can be obtained by substituting for t. Rearranging equation 1 gives:

$$t = \frac{v_1 - v_0}{a}$$

Inserting this expression for t into equation 5 gives:

$$x = v_0\left(\frac{v_1 - v_0}{a}\right) + \tfrac{1}{2}a\left(\frac{v_1 - v_0}{a}\right)^2$$

Check yourself that this simplifies to:

$$v_1^2 - v_0^2 = 2ax \qquad [7]$$

See question 8, 9 and 10.

I (a) A small boy lets go of a tyre and it rolls down a hill. It accelerates at 3 m s^{-2}. How far does it travel in 8 s?

(b) The boy tries to run alongside it. His maximum speed is 4 m s^{-1}. How far does he run before the tyre runs away from him?

J Sketch the speed–time graph you would expect for an object thrown vertically upwards. Label its main features.

EXAMPLE

Q A hawk is hovering above a field at a height of 50 metres. It sees a mouse directly below it and dives vertically with an acceleration of 9 m s^{-2}.

a) At what speed will it be travelling just before it reaches the ground?

b) How long does it take to reach the ground?

A

a) We need to calculate the final speed v_1. We know the distance, the initial velocity and the acceleration. As we don't know the time of travel, equation 7 applies here: $v_1^2 - v_0^2 = 2ax$. The initial velocity v_0 of the hawk, relative to the ground, is zero, so:

$$v_1^2 - 0 = 2 \times 9 \times 50 = 900$$

and

$$v_1 = 30 \text{ m s}^{-1}$$

b) Any equation that includes t (except for equation 2, zero acceleration) could be used here, but equation 3, $v_1 = v_0 + at$, is simplest:

$$30 = 0 + (9 \times t)$$

giving

$$t = 3.3 \text{ s}$$

Note that the answer can only be approximate. The physics is likely to be a simplification of the real situation – hawks don't fly that precisely!

Deceleration

Deceleration is a slowing down: it is a *negative* acceleration.

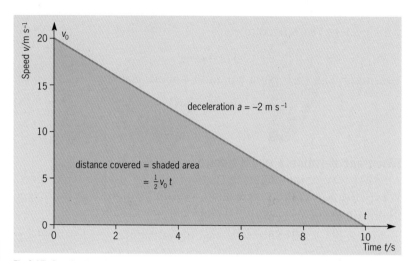

Fig 2.17 **Graph of uniform deceleration. Check that the area under the graph represents a distance of 100 m**

The graph in Fig 2.17 is for an object that decelerates uniformly to a stop. As before, the area of the triangle gives the distance travelled. The equations of motion still apply, but you have to be careful with the *signs*. The positive and negative signs are part of the *numbers* we put into the formulae. Suppose that the initial speed is 20 m s^{-1}, and the deceleration (negative acceleration) is –2 m s^{-2}. To calculate how far the object will travel in 10 seconds, using equation 5 we write:

$$\text{distance travelled} = v_0 t + \tfrac{1}{2}at^2$$
$$= (20 \times 10) + [0.5 \times (-2) \times 10^2]$$
$$= 200 - 100$$
$$= 100 \text{ m}$$

?

K A cyclist is freewheeling down a hill with a downward acceleration of 2.5 m s^{-2}.

(a) How long would it take for the cyclist's speed to increase from 5 m s^{-1} to 20 m s^{-1}?

(b) Show that the cyclist should travel about 75 m in that time.

(c) Would it be sensible to use the equations of motion of this section to calculate the cyclist's final speed if the hill was a kilometre long? Justify your answer.

See questions 11 and 12. ■

5 FRAMES OF REFERENCE

How fast are you travelling at the moment?

The answer may seem to depend on whether you are reading this book on a train that is travelling at 140 km h^{-1}, a plane travelling at 1000 km h^{-1} or just sitting at your desk. *But in fact, as the question stands, it is meaningless.*

Even if you are at your desk, both you and the desk are spinning with the Earth at a speed of just over 1000 km h^{-1} (if you are at the latitude of London). The Earth itself is moving in an orbit round the Sun at a speed of 107 000 km h^{-1}. The Sun, and its Solar System, are orbiting round the centre of the Galaxy at a speed of almost 1 000 000 km h^{-1}. You are not, of course, aware of any of these movements. The coffee in your cup is unshaken by these astronomical speeds. The question at the start of this section makes sense only when you add words such as 'relative to the room' or 'relative to the Earth' or 'relative to the Galactic centre'.

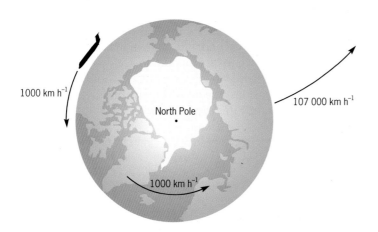

Fig 2.18 **How fast would you say you are travelling by plane?**

This is the **simple principle of relativity**, first stated by Galileo in the seventeenth century. His ideas of relativity were developed much further in the early twentieth century by Albert Einstein, as we shall see later.

Much of the time we take our measurements relative to the Earth, which we assume to be still and our normal '**frame of reference**'. This phrase simply means that the measurements we make are with reference to our static selves and the fixed objects that surround us. So, when we measure in an experiment that a dynamics trolley is moving at 0.55 m s^{-1}, we take it for granted that this motion is relative to the walls, floor and laboratory bench. We don't keep asking ourselves, 'relative to what?'. But we have some interesting problems to solve when our personal reference frame is moving relative to the Earth, as when it is a ship at sea, an aircraft in flight or a canoe on, say, a tidal river.

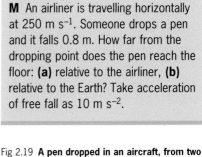

?

L Someone says, 'The ultimate fixed point must be the centre of the Universe. Everything that moves must have a real velocity relative to that.' Do you agree with this statement? Is it likely to be of any practical use?

Moving frames of reference

The equations of motion developed on pages 14–17 apply just as well inside an aircraft moving at speed relative to the Earth as they do for an object on the 'fixed' Earth itself.

If you drop a pen to the floor of an aircraft it will seem to you to move in exactly the same way as it does in your classroom on Earth. You don't need to take into account the fact that in the three-tenths of a second it takes to fall, the pen also moves forward a distance of a hundred metres or so. You too are moving forward at the same speed, so relative to you the pen falls straight downwards. You and the pen share the same reference frame – the aircraft. For both you and the pen, you could use the equation $v = at$ to calculate the speed of the pen when it hits the floor.

?

M An airliner is travelling horizontally at 250 m s^{-1}. Someone drops a pen and it falls 0.8 m. How far from the dropping point does the pen reach the floor: **(a)** relative to the airliner, **(b)** relative to the Earth? Take acceleration of free fall as 10 m s^{-2}.

Fig 2.19 **A pen dropped in an aircraft, from two frames of reference. When viewed from a different frame of reference, a straight line becomes a parabola**

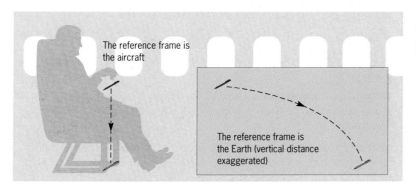

The reference frame is the aircraft

The reference frame is the Earth (vertical distance exaggerated)

Galileo, who used a sailing ship as his example of a moving frame, was aware of relative motion. He needed this idea to explain why the Earth could move without our noticing its movement. But it seems to have been as hard for people in Galileo's time to understand that the Earth could possibly move without leaving behind everything on it – sea, ships, the air and flying birds – as it is for people today to grasp Einstein's theories of relativity!

EXAMPLE

Q A canoe club plans a return trip on a river that flows at an average speed of 1.5 km h^{-1}. The canoeists will first paddle their canoes upstream for 6 km, then return. The leader estimates that the group can paddle at an average speed of 4.5 km h^{-1} on still water.

a) How long should the leader expect the trip to last, ignoring any stops for rest or refreshment?

b) What is the group's (expected) average speed for the journey?

A The leader must understand that the frame of reference is provided by the fixed Earth, with its map distance of 6 km each way.

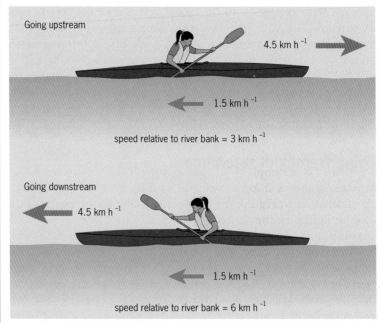

Fig 2.20 **Canoeing up and down the river**

a)
Going upstream: The canoe speed relative to the water is 4.5 km h^{-1}, but relative to the river bank it is only 3 km h^{-1} (4.5 − 1.5).

$$t = x/v = 6/3 = 2 \text{ hours}$$

So the group will take 2 hours of paddling time to reach their up-river destination.

Returning downstream: The distance is still 6 km, but now the speed relative to the bank is 6 km h^{-1} (4.5 + 1.5). They should cover the distance in just 1 hour.

b) Therefore the total journey takes 3 hours. The canoeists travel 12 km at an average speed (relative to the solid Earth) of 4 km h^{-1}.

See question 13.

6 VECTORS

When 2 + 2 does not equal 4: the importance of vectors

As shown in Fig 2.21, if you move 2 metres to your right, and then 2 metres forwards, you have travelled 4 metres but are only 2.8 metres from where you started. This is an example of where 'how far you have travelled' is not the same as 'how far you are from where you started'.

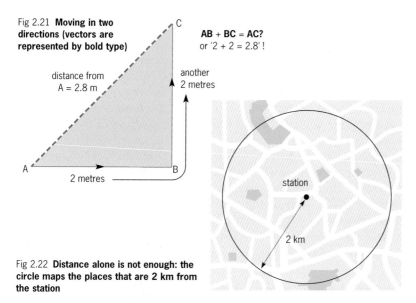

Fig 2.21 **Moving in two directions (vectors are represented by bold type)**

AB + BC = AC?
or '2 + 2 = 2.8' !

distance from A = 2.8 m

another 2 metres

A   2 metres   B

Fig 2.22 **Distance alone is not enough: the circle maps the places that are 2 km from the station**

station

2 km

Table 2.2 **Some common vectors and scalars in physics**

Scalars	Vectors
frequency	displacement
speed	velocity
mass	acceleration
density	momentum
energy	force
charge	current
resistance	electric field

This is a fairly obvious example of the fact that some quantities in physics don't mean much unless you also specify a *direction* to go with them. For example, it is not very useful simply to tell friends that your house is 2 km from the station. This information isn't clear enough: as in Fig 2.22, to find your house they would have to search along the circumference of a circle 2 km in radius!

Quantities that need both a number and a direction to define them properly are called **vector quantities** (or **vectors**). Those that are useful without a direction being specified are called **scalar quantities** (or **scalars**). Table 2.2 lists some common vectors and scalars. Velocity is a vector; it refers to speed-in-a-given-direction. Speed is useful, but only to describe the 'number' part of velocity. It is useful to know, for example, that the *speed* of a satellite in a circular orbit is constant, even though its *velocity* is constantly changing as the direction of travel changes. You will read more about satellite motion in Chapter 4.

?

N Are the following quantities vectors or scalars? Explain your answers. **(a)** Pressure, **(b)** time period of oscillation, **(c)** volume.

Putting vectors together

When we add vectors we have to take account of direction as well as size. If the directions are opposite and in the same straight line, we call one direction positive and the other negative, using the same convention as in graphs:

Going right is positive, going left is negative.

(In the canoeing example earlier, we used 'upstream' and 'downstream' to define directions.)

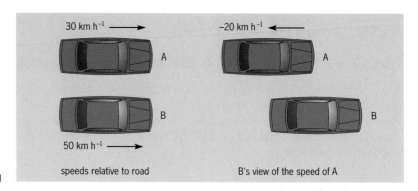

Fig 2.23 **Relativity on the road**

30 km h^{-1} ⟶   A

−20 km h^{-1} ⟵   A

B

B

50 km h^{-1} ⟶

speeds relative to road

B's view of the speed of A

See question 14. ■

?

O A child passenger in a train rolls a ball along the floor at a speed of 5 m s^{-1}. The train is travelling at a speed of 35 m s^{-1}. **(a)** What other fact do you need to know to find out the speed of the ball relative to the rail track? **(b)** What is the range of possible values of this relative speed?

Think of two cars travelling in the same direction along a road, one at 30 km h^{-1}, the other at 50 km h^{-1}, as in Fig 2.23. It is obvious that the *speed* of one car relative to the other is 20 km h^{-1}. But if you were in the slow car, you would see the faster car overtake you at 20 km h^{-1}. From your (moving) frame of reference, the faster car has a relative velocity of +20 km h^{-1}. An observer in the faster car would see your car apparently moving backwards with a relative velocity of −20 km h^{-1}.

If the cars were on a collision course, travelling in opposite directions, their relative speeds would be 80 km h^{-1}.

Again, think of throwing something from a moving vehicle. If it is thrown forwards with a speed of 10 km h^{-1} from a car moving with a speed of 30 km h^{-1}, it will be travelling at the even more dangerous speed of 40 km h^{-1} relative to a bystander.

Things moving at an angle to each other

But how do we tackle predictions of movement when the important vectors are not in the same line? This is a problem navigators deal with on water and in the air. Their vehicles may move in a medium that itself may be moving. There are usually tidal currents or winds. Suppose the canoeists in the Example on page 20 had to cross a wide estuary with a strong current of 3.0 km h^{-1}, as shown in Fig 2.24. How would this current affect their motion?

As Fig 2.24 shows, for every 4.5 units of distance they travel across the water, the current moves them downstream by 3 units. Relative to the banks of the estuary, they move along a path (AC) at a speed we can calculate to be 5.4 km h^{-1}, using Pythagoras' theorem. This path will *not* take them to their destination!

It would make more sense for the canoeists to head upstream at such an angle that the *combination* of their paddling velocity and the current velocity would lie on the path they want to move along. This combination is called the **resultant** velocity.

See question 15. ■

At what angle to the straight-across direction AX should they point the bows of the canoes? Fig 2.25 shows the situation. When the diagram is drawn to scale, it is possible to measure the angles and the value of the resultant. But it is usually quicker to use simple trigonometry to calculate the angle θ:

$$\sin \theta = 3/4.5 = 0.667$$

So: $$\theta = 42°$$

The canoeists would travel along the line AX at a speed given by:

$$\tan \theta = 3/\text{speed}$$

So the speed is 3.35 km h^{-1}.

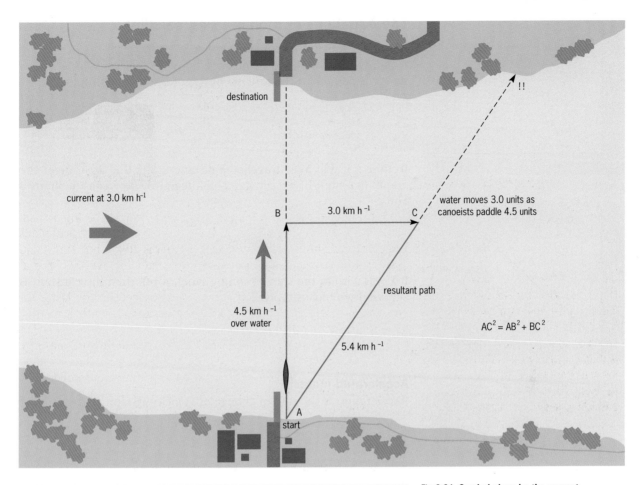

destination

current at 3.0 km h⁻¹

$$3.0 \text{ km h}^{-1}$$

B

water moves 3.0 units as
canoeists paddle 4.5 units

4.5 km h⁻¹
over water

resultant path

C

5.4 km h⁻¹

$$AC^2 = AB^2 + BC^2$$

!!

A
start

Fig 2.24 **Carried along by the current**

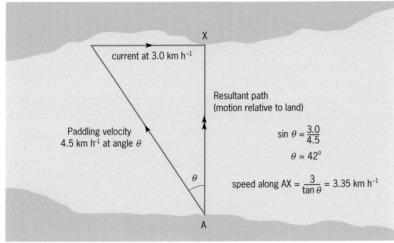

X

current at 3.0 km h⁻¹

Resultant path
(motion relative to land)

Paddling velocity
4.5 km h⁻¹ at angle θ

$$\sin \theta = \frac{3.0}{4.5}$$

$$\theta = 42°$$

$$\text{speed along AX} = \frac{3}{\tan \theta} = 3.35 \text{ km h}^{-1}$$

θ

A

Fig 2.25

?

P (a) An aircraft flies due north at an air speed of 500 km h⁻¹. It crosses
the high altitude jet stream (of air) which is moving due east with a speed of
100 km h⁻¹. What is the aircraft's **velocity** relative to the ground?

(b) The jet stream was discovered by military aircraft in the Second World
War when it was found that Atlantic crossings east–west took much longer
than those in the opposite direction. The aircraft has an air speed of about
400 km h⁻¹. What difference in flight times would you expect for
transatlantic journeys of 6000 km?

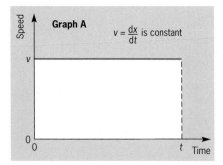

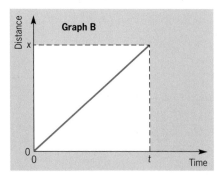

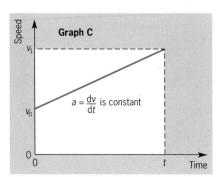

The equations of motion using calculus

An object moving at constant speed

As shown in Graph A, an object is moving at a constant speed v, which is defined as the rate of change of distance with time:

$$v = \frac{dx}{dt}$$

or:

$$dx = v \, dt$$

In time t, it will have travelled a distance x. As the 'area under the graph' in Graph B, we can get x by integrating between the limits 0 and t:

$$\int dx = v \int dt$$

giving:

$$x = vt + \text{constant} \qquad [1]$$

If x was zero at the start of timing when $t = 0$, then the constant is zero and we have simply:

$$x = vt \qquad [2]$$

Accelerated motion

Acceleration is the rate of change of velocity (Graph C), ie:

$$\frac{dv}{dt} = a$$

and so:

$$dv = a \, dt$$

This last equation can be integrated to relate velocity, acceleration and time:

$$\int dv = a \int dt$$

$$v = at + c \qquad [3]$$

where c is a constant of integration. The physical meaning of c is that it is the velocity at time $t = 0$, so it is v_0. Also v, the velocity at time t, is v_1. So we now have the equation:

$$v_1 = v_0 + at \qquad [4]$$

The equation relating distance to velocity, time and acceleration may be obtained by integrating equation 3, first putting v_0 for c:

$$v = at + v_0$$

ie:

$$\frac{dx}{dt} = at + v_0 \ \left(\text{since } v = \frac{dx}{dt}\right)$$

So:

$$\int dx = a \int t \, dt + v_0 \int dt$$

$$x = a\frac{t^2}{2} + v_0 t + C$$

C is another constant of integration which represents the value of x at time $t = 0$. This is usually taken to be zero in everyday calculations, and the equation simplifies to equation 5 on page 17, namely $x = v_0 t + \frac{1}{2}at^2$.

Equation 7, $v_1^2 - v_0^2 = 2ax$, may be obtained by substitution as shown on page 17.

SUMMARY

This chapter should help you to understand a number of ideas and acquire several numerical skills. Having studied it, you should be able to:

■ Understand that distance can be measured in units of time (seconds) as well as in units of distance such as metres and kilometres.

■ Know that the speed of light, c, is fundamental to the measurement of distances.

■ Know that measured motion is always relative to an agreed frame of reference.

■ Analyse distance–time and speed–time graphs.

■ Use the equations of motion of an object moving in a straight line with uniform acceleration.

■ Know the difference between vector and scalar quantities.

■ Use vector diagrams and/or trigonometry to analyse vector motion.

■ Understand the basic principles of triangulation.

You should know the following **equations of motion**.

■ For **motion at a constant speed** (or an average speed):

$$\text{distance} = (\text{average}) \text{ speed} \times \text{time}$$

$$x = vt$$

■ For **uniform (steady) acceleration**:

(a) $\text{acceleration} = \dfrac{\text{change in velocity}}{\text{time taken to change}}$

$$a = \frac{v_1 - v_0}{t} \quad \text{or} \quad a = \frac{\Delta v}{\Delta t}$$

which is often written as $v_1 = v_0 + at$.

(b) distance covered = average speed × time

$$x = \frac{v_1 + v_0}{2} t$$

which is equivalent to: $x = v_0 t + \frac{1}{2}at^2$.

When t is unknown, use $v_0^2 - v_0^2 = 2ax$.

(c) For non-uniform motion, we need to deal in small changes:

$$\Delta x = v\Delta t, \quad \text{or} \quad v = \frac{\mathrm{d}x}{\mathrm{d}t} \text{ in calculus notation,}$$

$$\Delta v = a\Delta t, \quad \text{or} \quad a = \frac{\mathrm{d}v}{\mathrm{d}t} \text{ in calculus notation.}$$

QUESTIONS

1 Model aircraft can be controlled 'in real time' by moving joysticks on a ground-based radio transmitter. Why wouldn't a control system like this be much use in controlling the fly-past of a space probe near Jupiter? (Assume equally good signal reception both ways.)

2 Calculate the following:

a) The distance in km of an aircraft from an airport whose radar measures it to be at a radar time-distance of 1.67 ms.

b) The time it would take a radio message to reach Mars from Earth when they are at their closest separation of 7.83×10^7 km.

3

a) The vertical probe radar from a mapping satellite crosses from the sea to a cliff. It records a difference in light-flight time of 1.17 microseconds as it does so. Show that the cliff is 175 m high.

b) To what accuracy must the timer in the satellite be able to measure time if it is to map height differences of 0.3 m?

4 Astronomers measure the distances of the nearer stars using a base line such that AM as in Fig 2,Q6 is 499 light-seconds long, which is the radius of the Earth's orbit round the Sun. The angle they measure that is equivalent to angle ACB is called the *parallax* of the star.

The parallax angles for two stars are given in the table. Calculate their distances from Earth in light years.

Star	Parallax (°)
Sirius	1.05×10^{-4}
Alpha Centauri	2.07×10^{-4}

5 A radar speed detector shows a frequency change of 8 kHz when monitoring the speed of an aircraft. The frequency of the radar beam used is 10^{10} Hz. What is the relative speed between aircraft and detector?

6 Look at Fig 2.Q6. Surveyors use base line AB which is 500.0 m long to establish the position of communications tower C.

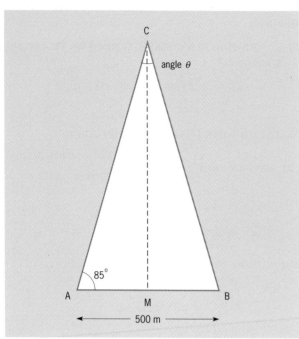

Fig 2.Q6

a) Use the relationship:

tan 85.00° = CM/250

to show that the distance CM is 2858 m.

b) If the communications tower C was a lot further away, the angles MAC and MBC would be close to 90°, and the angle ACB would tend to zero. Show that when θ = 0.20° the distance CM is given quite accurately by using the formula:

tan 0.10 = 250/CM

and by the formula:

sin 0.10 = 250/CM.

Explain why this is so.

7 Draw speed/time sketch graphs (freehand without numbers on the axes) to illustrate the following examples of objects in motion.

a) A car accelerating from rest to a steady speed.

b) A stone dropped from the roof of a tall building.

c) A tennis ball thrown vertically upwards, from the time it leaves your hand until you catch it again.

d) A tennis ball dropped from a height of about three metres which bounces once.

8 A student wants to find the distance below ground level of the water surface in a well. She uses a stopwatch and finds that the sound of the splash from a small pebble she drops into the well arrives 4.7 s after she lets the pebble go.

a) Use this result and the value of the acceleration of free fall (9.8 m s^{-1}) to calculate a value for the depth of the well.

b) Give two factors that might affect the accuracy of this value. Explain which one of them is likely to have the greater effect on the accuracy.

9 Plot the following data for a space probe which is moving in a straight line away from Earth.

Time (s)	0	100	200	300	400	500	600
Speed (m s^{-1})	5374	5329	5283	5238	5193	5147	5102

a) What does the shape of the graph tell you about the motion of the space probe?

b) Calculate (graphically or otherwise) the deceleration of the spacecraft.

10 Select the appropriate equation/s of motion and the data given at the end of the question and use them to answer the following questions.

a) How far would an aircraft travel in 4 hours at an average speed of 600 m s^{-1} against a head wind of 50 m s^{-1}?

b) What is the expected driving time for travelling from London to Edinburgh at an average speed of 100 km h^{-1}?

c) How long would a spacecraft take to reach the nearest star, Proxima Centauri, from Earth if the craft travelled at 0.2c?
(The speed of light c = 3.00 × 10^8 m s^{-1}.)

d) A spacecraft leaves Earth orbit (where its speed relative to the centre of the Earth was zero) and accelerates at 1.5 g to a steady speed (relative to Earth) of 20 000 m s^{-1}.

  (i) How long does the acceleration phase take?

  (ii) How far from its original position has the spacecraft moved in this time?

  (iii) How long will the spacecraft take to reach Mars? Road distance from London to Edinburgh = 600 km
Speed of light c = 3 × 10^8 m s^{-1}
Distance from Earth to Proxima Centauri = 4.3 light-years
Length of travel path between Earth and Mars = 5.0 × 10^8 m.

11 A car is travelling at 25 m s^{-1} when it is braked. It stops in a time of 12 s.

a) Show that its acceleration is –2.1 m s^{-2}.

b) How far does it travel after the brakes are applied?

12 A car is travelling at the town speed limit of 13 m s^{-1} when the driver sees a child run out into the road in front of the car and fall down. The driver has a reaction time (thinking time) of 0.4 s and the brakes can slow the car at a constant rate of 2.6 m s^{-2}. The car stopped just a few centimetres from the child. How far in front of the car was the child when the driver saw it?

13 Look back at the Example on page 20.

a) With the given values of river and paddling speeds, will any return journey be done at an average speed of 4 km h^{-1}? Try this with two different values of journey distance.

b) Prove the answer to part (a) algebraically.

14 Two aircraft are approaching each other directly on the same straight line. One has a speed of 200 km h^{-1}, the other a speed of 180 km h^{-1}. What is their relative speed?

15 The yacht Blue Star is sailing in a tidal stream of 3 knots (nautical miles per hour) in the direction shown in Fig 2.Q5. The speed of the yacht relative to the water is 6 knots. In which direction, and at what speed, will the yacht move relative to the fixed buoy marked A?

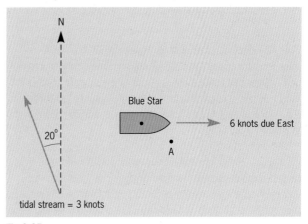

Fig 2.Q5

See page 81 for the CHAPTER MAP that covers
Chapters 2, 3 and 4

Assignment

MOTION, USING A SPREADSHEET

Spreadsheets can be used to answer What if? – type questions for many physical situations. They can do some quite advanced mathematics – useful if you want to see the effect of changing one factor without having to go through the chore of repeating many calculations and/or redrawing graphs.

The simulation described here should help you gain or improve a useful tool of investigation, and should also help to improve your understanding of physics. It doesn't require any mathematical (analytical) proofs of formulae; it just uses the simple definitions of **velocity** (as speed in a given direction) and **acceleration**.

The details of using spreadsheets vary slightly from one brand of program to another: if the instructions here don't work or don't make sense, seek help from someone who is familiar with the program you are using.

Note that in this Assignment the symbols for variable quantities are not given in the usual italics but are as they appear on the spreadsheet. D stands for a small change in a quantity, usually given as Δ.

Setting up the simulation

Open a spreadsheet page.

Aim to reproduce the setup shown in Fig 2.A1.

Begin with the data headings. Start at cell A1 and type in:

    ACCELERATED MOTION

As shown in Fig 2.A1, it may cover three cells. Just for the sake of neatness, leave some empty cells and go to cell A4 to write in:

    Initial speed V1 =

Fig 2.A1

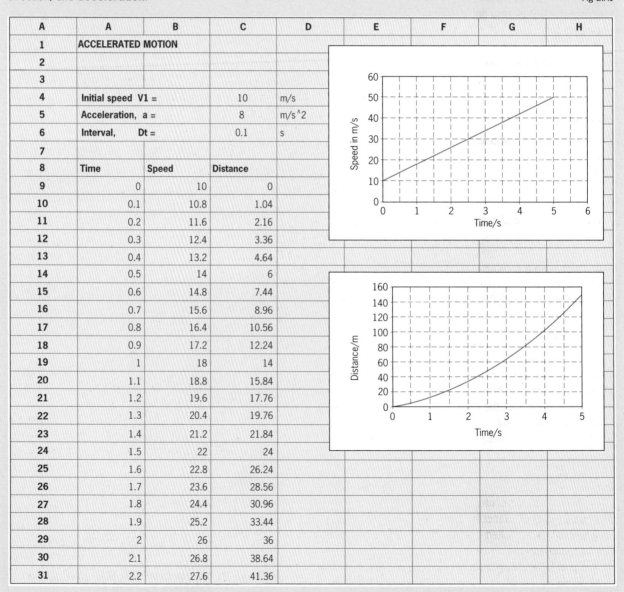

A	A	B	C	D	E	F	G	H
1	ACCELERATED MOTION							
2								
3								
4	Initial speed V1 =		10	m/s				
5	Acceleration, a =		8	m/s^2				
6	Interval, Dt =		0.1	s				
7								
8	Time	Speed	Distance					
9	0	10	0					
10	0.1	10.8	1.04					
11	0.2	11.6	2.16					
12	0.3	12.4	3.36					
13	0.4	13.2	4.64					
14	0.5	14	6					
15	0.6	14.8	7.44					
16	0.7	15.6	8.96					
17	0.8	16.4	10.56					
18	0.9	17.2	12.24					
19	1	18	14					
20	1.1	18.8	15.84					
21	1.2	19.6	17.76					
22	1.3	20.4	19.76					
23	1.4	21.2	21.84					
24	1.5	22	24					
25	1.6	22.8	26.24					
26	1.7	23.6	28.56					
27	1.8	24.4	30.96					
28	1.9	25.2	33.44					
29	2	26	36					
30	2.1	26.8	38.64					
31	2.2	27.6	41.36					

Next, type in cell A5:
   Acceleration a =
and in cell A6:
   Interval, Dt =
Add the units in the cells: D4   D5   D6
                              m/s   m/s^2 s
Head the columns: A8   B8   C8
                    Time Speed Distance

The algorithm
Choose any initial (starting) *speed* and any *acceleration* you like for an object, and the spreadsheet will work out its speed at any time. It will also calculate the total distance travelled in that time. It will make calculations for every fraction of a second, decided by Interval, Dt =.

To begin, choose Dt to be 0.1 s. So, in cell C6, type:
   0.1
Now set up the *time* column by filling it from A8 to A59 with numbers starting at zero and increasing by 0.1. This is best done by a formula. In cell A9, put:
   0
In cell A10, insert this formula:
   =A9+$C$6

Then COPY this formula into cells A11 through to A59. Use **Help** or read the manual if you don't know how to do this.

Choose the value of acceleration, eg in cell C5, type:
   8
Choose the value of initial speed. For example, in cell C4, type:
   10
Fill cell B9 with your chosen (and alterable) value of V1. Type in:
   =$C$4
The $ signs are a spreadsheet convention to lock the values to the specified cell that holds that value throughout the calculations. In cell C9, put the value zero:
   0

Calculating speed
The spreadsheet will calculate the speed by using the definition of **acceleration as change in speed per unit time**. Thus we expect the change of speed in a short time interval Dt to be simply:
                    acceleration × Dt

In cell B10, type the algorithm above in the form:
   =B9+($C$5*$C$6)

This means: Add the contents of cell B9 (initial speed) to the change in speed (acceleration × time interval).

   When you have finished typing and have ticked the OK box (or its equivalent), the value 10.8 should appear in cell B10.
   You now have to COPY the formula down to the end of the column (to cell B59). Do this. The COPY facility will

automatically change the cell location each time, so adding the extra speed (aDt) to each newly calculated value of the speed. Look at a few of the cells in the column to see what this means. If you have done all this correctly you should have values as shown in Fig 2.A1.
   Now use the graph facility to produce a graph of speed against time, See Fig 2.A1. Ensure that the program plots an 'XY' graph. Insert the graph into a convenient place in the spreadsheet so that you can monitor what happens when you change parameters. For example, you can change the value of acceleration from 8 m s^{-2} to some other value. Try some negative ones.

Distance and the distance–time graph
Before going any further, return to the original values as shown in Fig 2.A1.
   The extra distance covered in a small time Dt is:

average speed in the time interval × interval Dt

The average speed is calculated for the second time interval (represented by the row A10 to C10) by the algorithm:
   $\frac{1}{2}$ × [speed at end of previous interval (in cell B9, 10m/s)]
   PLUS
   speed at end of present interval (in B10, 10.8 m/s).

To find the extra distance, this value for the overall average must be multiplied by Dt. To find the total distance covered, the extra distance has to be added to the previously calculated distance (which is of course shown in the previous cell each time).

So type into cell C10:
   =C9+0.5*(B9 + B10)*$C$6
Check that this produces a value of 1.04 in cell C10. Then COPY the formula as before to fill the column (C10 to C59). Then graph it and display the graph.
   As before, investigate the effect of changing the initial values in cells C4, C5 and C6.
   Having worked through this, you can now answer the following questions.

1
a) Explain the shape of the distance–time plot when the acceleration is **(i)** positive, **(ii)** negative.
b) Explain why the distance–time plot is not a straight line.
c) Explain the differences between the distance–time plots for **(i)** positive and **(ii)** negative values of acceleration.

2 Use the worksheet to obtain values in the following circumstances:
a) The speed at a time 60 s after starting the motion, with an initial speed of zero and an acceleration of 2.5 m s^{-2}.
b) The distance covered in the first 30 s in the situation in **a)**.
c) How long it will take for a vehicle with an acceleration of 1.5 m s^{-2}, starting from rest, to cover a distance of 300 m.

The German long-jumper Heike Drechsler in action

SUCCESS IN LONG-JUMPING comes from a combination of the right body build, hard training and natural skill. But, ultimately, the length of the jump depends on the jumper keeping in the air for as long as possible while moving forward as fast as possible.

A good long-jumper hits the take-off board at a speed of about 10 m s^{-1}. Maximum distance is achieved when this speed is shared equally between forward motion and upwards motion, that is, when taking off at an angle of 45° to the ground.

In order to try to reach this 45° angle, what the jumper actually does is to leap vertically upwards as high as possible while at the same time moving forwards as fast as possible. A good jumper can raise the centre of mass of his or her body by about a metre. (Men can raise it by about 1.2 m; women about 1.0 m.)

Our knowledge of gravity tells us that to reach a height of 1.2 m and fall back again needs a time in the air of almost exactly 1 s. At a speed of 10 m s^{-1} this means that the body would move a distance of 10 m forwards – that is, the maximum possible long-jump distance would be 10 m. The current world record is 8.96 m (Ivan Pedroso of Cuba on 29 July 1995), so there is still some way to go!

Interestingly, the record of 8.90 m set at altitude by Bob Beaman in Mexico in 1968 was a huge improvement on the previous record, and stayed unbroken for 23 years. This suggests that the present record may be broken at altitude, and also how difficult it will be to reach the theoretical maximum.

The ideas in this chapter

The ideas of **force** and **gravity** are fundamental to understanding the behaviour of objects on Earth. Objects will only move if a force is acting on them. Gravity is a force which acts on every object. For example, if you throw a ball into the air, the force of your throw makes it move upwards. The force of gravity then causes the ball to fall back downwards.

These ideas were not fully understood by scientists until the time of Isaac Newton, over three hundred years ago. In this chapter we explain Newton's theories, and then in Chapter 4 we look at some of the ways in which an understanding of forces has been used in science and engineering to transform the world we live in.

1 NEWTON'S LAWS

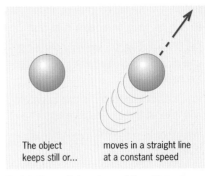

The object keeps still or... moves in a straight line at a constant speed

Fig 3.1 **Newton's imaginary, (almost) empty universe**

Suppose you roll a ball along the ground. It soon begins to slow down and eventually stops. Right from the time of Aristotle, over 300 years BC, up until Newton's time, scientists believed that this was the natural way for objects to behave. Newton showed that this belief, which had been accepted for two thousand years, was wrong. He showed that there was a force acting – the force of **friction** – which causes the ball to slow down. Without this force, the ball would carry on, without slowing down.

Newton first of all considered an imaginary universe with just one object in it, as in Fig 3.1. This object will stay still (be 'at rest') or, if it is moving, carry on moving in a straight line at a steady speed, for ever.

It is this simple idea that led Newton to develop the fundamental laws of physics. It is a part of Newton's **first law of motion** – see the box below. But of course the real Universe is not this simple. If it were, how could we tell if the object was moving or not? Or if it moved in a straight line? And who or where would 'we' be to ask the questions or make sense of the answers? When you do try to answer these questions you will be following closely in the footsteps of Albert Einstein – see Chapter 25.

But let us ignore these questions for now, and return to Newton. He went on to put something else in his imaginary universe: a **force**. If the object changes its motion in any way, by starting to move, by accelerating or slowing down, or by changing its direction of motion, **then a force must have acted on it**. This effect produced by a force defines what a force *is*, in Newton's world, and this idea completes his first law.

But where would this force come from? There must be something else in Newton's primitive universe, a 'force-causing' object or system of some kind. So far, all this seems fairly obvious – to us at least. But then Newton went on to say something really unexpected. He said that if the first object was acted upon by a force, then the other 'force-causing' object must also have a force acting on it. Forces can

?

A You throw a ball across a field. Draw a diagram to show the ball half-way in its flight. Use labelled arrows to show any forces acting on the ball. Explain briefly the origin or cause of the force or forces you have drawn.

See question 3.

Newton's laws of motion

Newton defined three laws of motion. Law 1 defines a force. Law 2 contains a set of complex related ideas, introducing the new ideas of **momentum** (ie **mass** is now involved) and the **rate of change** of momentum with time. This law allows us to measure the size of a force. Law 3 seems very simple but is also very far-reaching, since hidden in it is probably the most fundamental law of all physics – the law of the **conservation of momentum**.

First law

A body at rest will stay at rest, and a body moving in a straight line will continue to move in the same direction and at the same speed, unless an external (unbalanced) force acts on it.

Second law

The effect of a force is to change the speed and/or direction of motion of a body. When the force changes the momentum of the body, the rate of change of momentum is proportional to the size of the force causing the change, and is in the same direction as the force.

Law 2 leads to the equations:

$$F = ma$$

and
$$Ft = m\Delta P \quad \text{(see page 34).}$$

Third law

Forces occur in pairs: when a body is acted upon by a force, there must be another body which also has a force acting on it. The forces are equal in size but act in opposite directions.

B A book lying on a table is at rest.

(a) Draw a diagram showing the sizes and directions of the forces acting on the book.

(b) One of these forces is due to gravity and, like all forces, must be one of a pair. What is the other force and on what does it act?

C A rope is tied to a large box and pulled. You know that a force is acting on the box but observe that it doesn't move. What can you deduce from this observation?

D What pairs of forces are involved in the following situations?

(a) A tennis ball is hit by a tennis racket.

(b) A ball is dropped but has not yet reached the ground.

(c) You start moving by taking a step forwards.

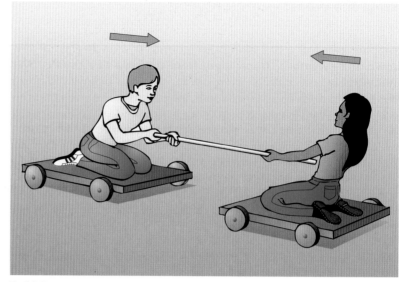
Fig 3.2 **Newton's third law. If one pulls or pushes, both move**

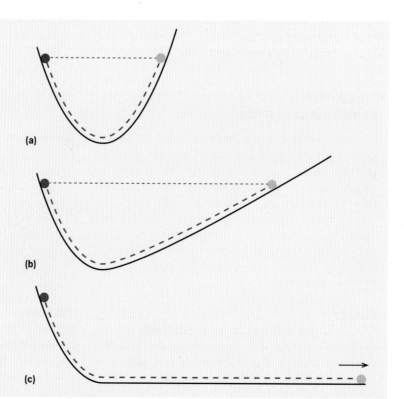

only exist in pairs ('force' and 'antiforce'). The pair of forces must be equal in size, but act in opposite directions (see Fig 3.2). This idea was expressed in the third law of motion.

Newton's imaginary universe was not just a science fiction story. It was based on observations made by Newton and many other scientists (see Galileo's ideas in Fig 3.3 for example). Although the basic idea might seem obvious to us, we must realise that at the time it was much more difficult to measure movement than it is now. It was a world without fuel-driven engines that move cars and trains and raise aircraft from the ground. There were no speedometers, and measurements of distance and time were, by our standards, very inaccurate.

See questions 2 and 4.

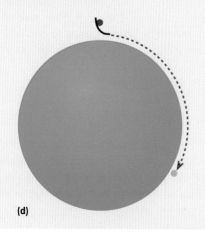
Fig 3.3 **Another scientist who studied motion was Galileo (1564–1642). These 'thought-experiments' show a ball and a slope with no frictional forces operating. In** (a) **and** (b) **the ball would roll down the slope and roll back up the other side to the same height. In** (c) **the ball would continue rolling for ever.** (d) **The Earth is round, which suggested to Galileo that 'unforced' motion is naturally circular. Newton disagreed with this idea, and showed it to be incorrect**

Using the planets for measuring

There was, though, a class of objects whose movements had been studied with great care for thousands of years – the planets. It was vital for navigators to know about the motion and positions of the planets and stars, especially in Newton's day, which was a time when trade by sea was developing rapidly.

The planets moved in 'empty' space, with effectively no friction, and so measurements of distance and time from astronomy were the most accurate quantities that Newton could use. Indeed, all measurements of time were based on the daily and annual motion of the Earth around the Sun, and of the Moon around the Earth.

Newton is famous for solving problems posed by the motion of these astronomical objects. To achieve his results, he defined the fundamental ideas of force, mass and momentum as expressed in his three laws of motion; and he put forward a law of gravity which explained the motion of the Moon and planets. He also developed a new branch of mathematics, the **calculus**, which allowed him to make calculations more easily.

Newton was not the only scientist in the seventeenth century. He and his fellow 'natural philosophers', as scientists were then known, have been credited with starting the first **scientific revolution**. They proposed that theories were worthless unless matched to observation – to what really happens. In physics, this meant that theories of the motion of bodies must include mathematical formulae that could accurately describe – and predict – this motion.

Let us now look at the fundamental ideas that had to be made clear before motion could be understood, for instance the motion of such things as planets, or of more familiar objects such as ships or horses.

2 MOMENTUM

First of all, how could 'motion' be quantified, that is, be given number values? What aspects of motion could be predicted?

To answer these questions, Newton had the idea of 'quantity of motion'. This is what we now call **momentum**. Newton realised that there were two aspects to this: speed in a given direction (which is **velocity**), and mass.

Momentum (P) is the product **mass** × **velocity**, or:

$$P = mv$$

A vector is a quantity with magnitude and also direction. Velocity is a vector, so momentum is also a vector. It has the same direction as the velocity of the mass.

Newton's first law of motion defines force simply and qualitatively (page 31). His second law extends the first law to give the relationship between a force and its effect on the motion of a body. A force changes the motion of a body, so the **rate of change** of momentum is proportional to the force. Force equals the rate of change of momentum:

$$\text{Force} = \frac{\textbf{change in momentum}}{\textbf{time}}$$

$$F = \frac{\Delta P}{t} \text{ (in newtons)}$$

where the symbol ΔP stands for the change in momentum.

See questions 16, 18 and 19. ▪

This force formula can be made more practically useful in two ways, as follows.

Impulse

Rearranging the formula gives:

$$Ft = \Delta P$$

The quantity Ft, the product of force and the time for which it acts, is called the **impulse**. Thus we have:

Impulse = change in momentum

This idea is particularly useful when we have situations where the force is variable, when it acts for a short time or when both force and time are hard to measure independently. This is the case in collisions, for example. There is more about the use of this way of looking at changes in motion in Chapter 4.

Change in momentum due to a constant force

When we are dealing with well-behaved forces that stay constant over the time interval t, we can define an initial velocity v_0 and a final velocity v_1 (see Chapter 2, page 14) such that the change in momentum is:

$$\Delta P = \frac{mv_1 - mv_0}{t}$$

Therefore:

$$F = \frac{mv_1 - mv_0}{t}$$

$$= m\frac{v_1 - v_0}{t}$$

$$F = ma$$

This is the well-known relationship:

Force = mass × acceleration

Note that this formula assumes that mass m does not change.

EXAMPLE

Q The engines of an airliner exert a force of 120 kN during take-off. The mass of the airliner is 40 tonnes (1 tonne = 1000 kg). Calculate **a)** the acceleration produced by the engines, **b)** the minimum length of runway needed if the speed required for take-off is 80 m s^{-1}.

A

a) Using the formula $F = ma$ gives:

$$a = F/m = (1.2 \times 10^5)/(4 \times 10^4)$$
$$= 3 \text{ m s}^{-2}$$

b) We use the equation of motion $v_1^2 = v_0^2 + 2ax$. Rearranging this, knowing that $v_0 = 0$:

$$x = \frac{v_1^2}{2a} = \frac{3600}{6} = 600 \text{ m}$$

Fig 3.4 **The airliner taking off**

Mass changes also affect momentum

The simple formula $F = ma$ (or $F = m\,\mathrm{d}v/\mathrm{d}t$) assumes that mass stays constant while the force is acting. But this is not true in many cases, such as rocketry, where the force is produced by expelling material from the back of the rocket. See Chapter 4 for more about this. Also, when objects are travelling at very high speeds, the mass changes with speed as a result of relativistic effects (see Chapters 25 and 26). So, using the calculus notation, a more accurate formulation of the second law relationship is:

$$F = \frac{\mathrm{d}(mv)}{\mathrm{d}t} = \frac{\mathrm{d}P}{\mathrm{d}t}$$

3 INERTIAL MASS AND GRAVITATIONAL MASS

It was hard for Newton to define 'mass', other than as 'quantity of matter'. It certainly was not the same as **weight** (see page 38)! Finally, Newton defined mass in two ways. His first definition was equivalent to this:

> **Mass is what a body has which makes it hard to accelerate.**

This means that we can work out the mass of a body by seeing how much it accelerates when we apply a standard force to it, then apply the formula $m = F/a$. The word **inertia** means reluctance to move, so this kind of mass is called **inertial mass**.

But Newton had also produced another entirely novel theory, the **theory of gravity**. In this theory:

> **Any object with mass produces a force of attraction which acts on other masses.**

So mass can be measured by measuring the size of the force-pair exerted between two objects. This is essentially what we do when we measure the force of gravity on an object to find its 'weight', using a spring balance, say. The result we get from this kind of measurement is called **gravitational mass**.

Are these two kinds of mass equivalent?

There is no reason, in Newtonian physics, why these measurements should give the same result for a given object, but they always do. This equivalence has been tested by experiment to be exact to 1 part in 10^{12}. This might seem a rather trivial point to worry about. After all, mass is mass is mass, isn't it? But it was thinking about this very equivalence that Einstein hit on his general theory of relativity.

G A force of 12 N is applied to an object. Calculate its mass if it changes speed (without changing direction) from $5\ \mathrm{m\ s^{-1}}$ to $25\ \mathrm{m\ s^{-1}}$ in 4 s.

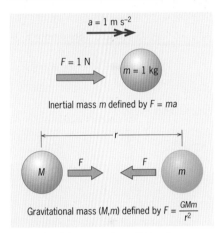

Fig 3.5 **Inertial mass and gravitational mass. Are they the same?**

See question 5.

H A net force of 15 N acts on an object and it accelerates at $5\ \mathrm{m\ s^{-2}}$.

(a) What is the mass of the object?

(b) What force would be needed to accelerate the object by $9.8\ \mathrm{m\ s^{-2}}$?

I Our Universe is expanding, meaning that all the very large masses in it (such as galaxies) are moving away from each other. One theory states that this expansion may be partly due to the gravitational mass of an object being very slightly different from its inertial mass. Which mass would have to be the larger in order to produce an expanding univers? Justify your answer.

J Outline the principle of any experiment you can think of that would test the theory that gravitational mass and inertial mass are equivalent.

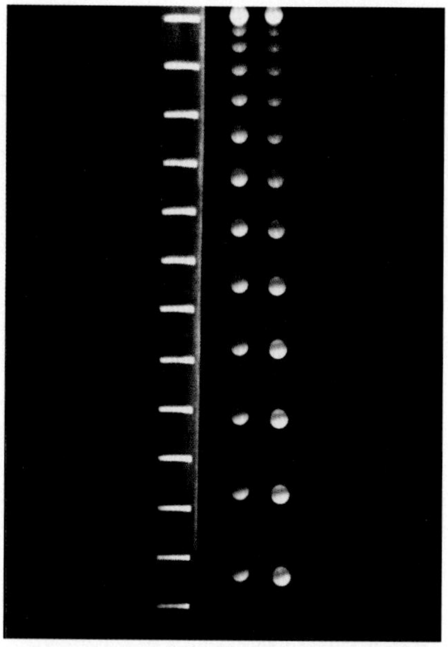

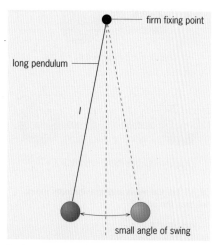

Fig 3.6 **In a vacuum, all objects accelerate at the same rate. Left: golf ball; right: table tennis ball**

See question 6. ■

Fig 3.7 **A simple pendulum. The period *T* is the time it takes for the pendulum bob to swing across and back to its starting point. At least 20 swings should be timed in order to get an accurate value**

?

K The simple pendulum was once the most accurate device for controlling clocks.

(a) Calculate the length of a simple pendulum that would swing with a period of exactly 1 second.

(b) Eighteenth-century navigators needed very accurate clocks to help locate their position at sea. Why were pendulum clocks not much use for this purpose?

4 THE FORCE OF GRAVITY AND THE ACCELERATION OF FREE FALL

Newton realised that objects fall because the Earth exerts a force on them – the **force of gravity**. This force causes an object to accelerate towards the ground. It was already well known that under the right conditions all objects, whatever their mass, fell with the same value of acceleration. We call this the **acceleration of free fall**. The most important condition is that the objects must be falling perfectly freely, with no friction or other opposing force to slow them down. The acceleration in air of a small lead ball is different from that of a light paper ball. This is because the objects are falling through a resistive medium, air. But in a vacuum all objects accelerate at the same rate.

A simple laboratory experiment shows that the acceleration of free fall is approximately 9.8 m s^{-2}. This value is given the symbol g. More accurate experiments show that g varies over the Earth's surface. Some values are given in Table 3.1.

Table 3.1 **Values of *g* at different places on the Earth's surface**

Location	London	Calcutta	Tokyo	Sydney	North Pole
Acceleration of free fall/m s^{-2}	9.812	9.788	9.798	9.797	9.832

There are several reasons for this variation:

● The Earth is not a perfect sphere, and so different places on the Earth are at different distances from the centre.

● The Earth's crust is not uniform in density, and so there is more or less mass under different places.

● The Earth is spinning, and some places (eg on the Equator) are spinning with a greater speed than others.

Why these differences should change the acceleration of free fall will become clear as you work through this chapter.

Measuring *g*

Accurate measurements of g use methods based on measuring the period of a swinging pendulum, as in Fig 3.7. The period T of a simple pendulum is given by the formula:

$$T = 2\pi\sqrt{\frac{l}{g}}$$

where l is the length of the pendulum. See page 115 for more about the simple pendulum. For most calculations at A-level, a value for g of 9.8 m s^{-2} is accurate enough.

Why do all objects fall with the same acceleration?

We learn from careful experiments that the gravitational force (due to the Earth) on an object of mass m accelerates the object with the local value of g. By Newton's second law, the force F must be such that:

$$F = mg$$

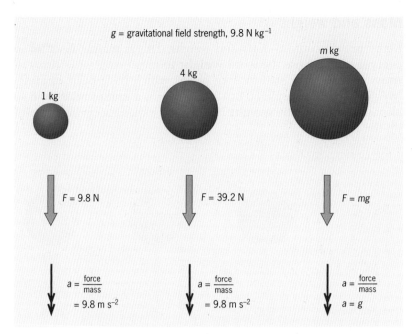

g = gravitational field strength, 9.8 N kg^{-1}

1 kg

4 kg

m kg

F = 9.8 N

F = 39.2 N

$F = mg$

$a = \dfrac{\text{force}}{\text{mass}}$

= 9.8 m s^{-2}

$a = \dfrac{\text{force}}{\text{mass}}$

= 9.8 m s^{-2}

$a = \dfrac{\text{force}}{\text{mass}}$

$a = g$

Fig 3.8 **In a gravitational field, all objects at a given place have the same acceleration**

Now, g is always the same at a particular place on Earth, whatever the mass of the object. This means that F must be proportional to m. If the mass doubles, for example, the force must double. (Einstein explains *why* gravity works in this way in his general theory of relativity.)

5 THE STRENGTH OF A GRAVITATIONAL FIELD

We use the quantity **gravitational field strength** to help define the effect of a gravitational field on a mass. At GCSE, you came across the idea of a magnetic or electric field to describe the pattern in space made by these types of force. We can also think of the space around any mass (such as the Earth) as having a **gravitational field**. There is more about this in Chapter 4.

The *strength* of the gravitational field is measured by the force it exerts on any object placed in the field, per unit of mass. As explained before, the value of the field strength is measured to be numerically the same as the acceleration of free fall, g. So, near the surface of the Earth, the value is approximately 9.8 newtons per kilogram.

■ See question 17.

By the definition of force:

Gravitational force on mass = mass × acceleration of free fall

$$F = mg$$

By our definition of gravitational field strength as force per unit mass:

Gravitational field strength $= \dfrac{\text{gravitational force}}{\text{mass}} = \dfrac{mg}{m} = g$

This means that sometimes you will see g given in units of N kg^{-1}, which relates to gravitational field strength, and sometimes in units of m s^{-2}, when we are thinking about the acceleration of free fall.

L Explain why the quantity g is sometimes quoted as 9.8 m s^{-2} and sometimes as 9.8 N kg^{-1}.

M The gravitational field strength at the surface of Mars is 3.8 N kg^{-1}. How long will it take a mass of 2 kg to reach a speed of 20 m s^{-1} in free fall?

Mass and weight

Many people confuse mass and weight. This is mainly because in everyday life we use the word **weight** to mean what physicists call **mass**. Potatoes are sold in 2 kg packs, which have been carefully 'weighed' by the suppliers. But a kilogram is a unit of mass (and so also is a 'pound'). The **weight** is what you experience when you lift up the pack.

Fig 3.9 **The distinction between mass and weight. Mass is the quantity of matter in a body (unit: kilogram). Weight is the force which acts on the mass in a gravitational field (unit: newton). Here, the three bodies all have the same mass, but very different weights**

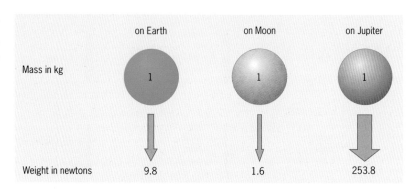

See questions 6 and 9. ■

The simplest definition of weight is as follows:

The weight of an object is the gravitational force exerted on the object by the mass of the Earth.

This force depends on the mass of the object, as explained above. Close to the Earth's surface, a mass of 1 kg is attracted by a force of approximately 9.8 N. A mass of 2 kg feels heavier than a mass of 1 kg because the gravitational force on it is twice as great: 19.6 N.

This force varies with distance from the centre of the Earth. But even at the height of a typical Earth satellite (say, 100 km above the surface), the force per kilogram is only slightly less at 9.5 N kg^{-1}. This is of course the force that keeps the satellite orbiting the Earth (see Chapter 4).

Weight is only one of the **pair of forces** involved. The other force is the pull of the object on the Earth, in accordance with Newton's third law (see Fig 3.10).

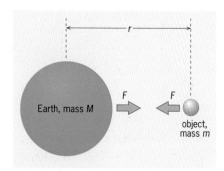

Fig 3.10 **Forces always occur in pairs. Just as the object is pulled by the Earth, so the Earth is pulled by the object**

6 MOVING IN A UNIFORM GRAVITATIONAL FIELD

In this section we deal with movement near the Earth's surface, so that the field strength is reasonably constant at 9.8 N kg^{-1}.

EXAMPLE

Q A small coin is dropped down a well. The splash is heard 2.3 s later. How deep is the well? (Assume that sound travels quickly enough for any delaying effect to be ignored.)

A The appropriate formula, from the kinematics formulae on pages 15–17, must involve acceleration (in this case g), the initial speed (in this case zero), the distance moved x, and the time t taken to fall:

$$x = v_0 t + \tfrac{1}{2}at^2$$

Inserting the given values: $x = 0 + \tfrac{1}{2} \times 9.8 \times (2.3)^2$

$$= 26 \text{ m}$$

See questions 1, 7 and 10. ■

Terminal speed

How fast would the coin in this example be travelling just before it hit the water? The final speed can be calculated from:

$$v_1 = v_0 + at$$
$$= 0 + 9.8 \times 2.3$$
$$= 23 \text{ m s}^{-1}$$

If a table-tennis ball had been dropped into the well instead of a coin, the time taken for it to reach the water would have been a lot longer. This is because, as an object speeds up in a medium (here air), the resistive (frictional) force opposing it increases (see Fig 3.11). When the opposing force is equal to the accelerating force, the *net* force on the object is zero. The object stops accelerating and carries on at a steady speed. In the case of a falling object this is called its **terminal speed** (or **terminal velocity**).

A steel ball has a terminal speed of about 40 m s^{-1}, compared with 5 m s^{-1} for a table-tennis ball. If you fell out of an aeroplane you would reach a terminal speed of about 70 m s^{-1}. A parachute increases the air resistance, producing a terminal speed of just a few metres per second, depending on its design.

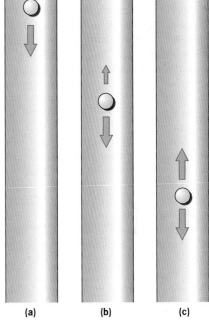

Fig 3.11 **Forces on a table-tennis ball as it falls down a well.**

(a) **At the top the ball is stationary, so the air resistance is zero.**

(b) **As the ball moves down the well, air resistance increases, slowing down the rate of acceleration.**

(c) **The air resistance exactly balances the weight of the ball, so the ball is no longer accelerating. The speed of the ball at this time is its terminal speed (or terminal velocity)**

(a)   (b)   (c)

F_f = frictional forces

$F_g = mg$

Fig 3.12 **The sky divers have reached terminal velocity**

N A small steel ball is placed just above the surface of a jar of oil, about 0.5 m tall, and then let go. Describe the motion of the ball and illustrate your answer with a sketch of the speed–time graph you would expect.

O Identify the force-pairs in the following circumstances (there might be more than one):

(a) the Moon moving in its orbit;

(b) a stone just after it has been dropped into a well;

(c) a 'free fall' parachutist who has reached terminal speed.

UP, UP AND AWAY

MOST PEOPLE THINK that birds flap their wings to stop them from falling out of the sky, and early human pioneers of flight tried to reproduce this action. They designed aircraft with flapping wings, always with disastrous results!

What actually happens – for both birds and aircraft – is that their wing shape, together with their forward motion, creates an upward force called **lift**. Lift is a pressure effect caused by the air moving faster over the top of the wing than it does underneath. This depends on wing design. In both birds and aircraft, the upper part of the wing is raised while the underside is flatter, so that the air passing over the upper part has further to travel and so moves more quickly. The result of this is that the air pressure above the wing is less than the pressure underneath, and so the wing is pushed upwards.

When the forward thrust of birds and aircraft is reduced, they either fall directly out of the sky or glide forwards and downwards, with the force of gravity moving them 'downhill'. Many birds are very well designed for this – seagulls, eagles and buzzards, for example, are very efficient gliders. But if all they did was glide, they would eventually and inevitably reach the ground.

Aircraft have engines to provide forward thrust, but how do birds keep moving forward? This is where

flapping their wings comes in. In fact, the movement of a bird's wing is a kind of swimming stroke. The wing is angled to push against the air as it moves backwards, and then tilts itself on the forward half of the stroke to reduce drag. The net result is a forward thrust. All this happens very quickly and has been studied by high-speed ciné photography.

The best analysis has come from studying film of birds flying through a cloud of small soap bubbles. The soap bubbles show air patterns (eddies) left behind the bird, rather like the way water bubbles show a ship's wake.

Up and down

EXAMPLE

Q The diagram shows a velocity–time graph for a ball thrown vertically upwards with an initial speed of 14.7 m s^{-1}. **a)** How high will it go? **b)** What is the total distance it travels? **c)** What is its time of flight? (Air resistance can be ignored.)

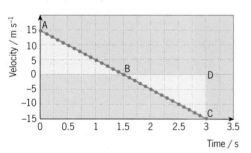

Fig 3.13 **Graph of velocity against time for a ball thrown vertically upwards with an initial speed of 14.7 m s^{-1}**

A

a) In this problem, direction is important, and we must be aware of the significance of the *signs* of vector quantities. We use the convention that *up* is positive and *down* is negative. The height, distance x, is given by the area under the graph for the portion A to B. The appropriate equation is:

$$v_1^2 = v_0^2 + 2ax$$

At the top of its flight, the ball has zero velocity, with $a = -9.8$ m s^{-2}.

We get: $0 = (14.7)^2 - (2 \times 9.8 \times x)$
giving: $19.6 \times x = (14.7)^2$
The height the ball reached: $x = 11.0$ m

b) The 'total distance travelled' can mean two things. Strictly, we could interpret it as the **displacement** of the ball from where it started. This distance is given by the sum of the two areas OAB and BCD. These areas are equal in size but opposite in sign, so add up to zero. This tells us no more than that the ball ended up in the same place as it started! How does the algebra cope with this? The final velocity is -14.7 m s^{-1}. So, using the same formula for the whole journey:

$$(-14.7)^2 = (14.7)^2 - (2 \times 9.8 \times x)$$

which also results in $x = 0$.

Taking the obvious alternative as the **non-vector** (or **scalar**) sum of the up-and-down distances, the ball has moved through a distance of 22 m.

c) What is the ball's time of flight – the time it was in the air? We can work this out using the formula $v_1 = v_0 + at$, which can be rearranged to give:

$$t = \frac{v_1 - v_0}{a} = \frac{-14.7 - 14.7}{-9.8}$$
$$= 3 \text{ s}$$

P A boy running at 3 m s⁻¹ throws a
ball vertically upwards with an initial
speed of 5 m s⁻¹.

(a) If he carries on running at the
same speed, will the ball return to
him? (Ignore air resistance.)

(b) How long will the ball be in the air?

(c) How far has the boy run in this time?

Q This would be a good time to
make use of a spreadsheet to
investigate motion in a uniform
gravitational field: see the details
in the Assignment for Chapter 2,
page 28, or the Assignment at
the end of this chapter.

See questions 11, 14 and 15.

7 PROJECTILES

A projectile is an object that has been propelled by a force. After you
kick a football in the air or lob a tennis ball with a racket, the ball
has a velocity whose direction is in between the horizontal and the
vertical, as shown in Fig 3.14. It moves both upwards and sideways.
The tennis ball in the diagram has a velocity v in the direction
shown, when $v \sin \theta$ is the **vertical component** and $v \cos \theta$ is
horizontal component.

But the force of gravity acts downwards – it can only affect the
up-and-down motion. Assuming air resistance is too small to worry
about (though it usually isn't!), the speed in the sideways direction
stays the same (Figs 3.15 and 3.16). That is, in a simple, uniform
gravitational field the horizontal and vertical components of velocity
are independent of each other.

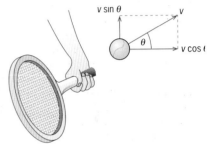

Fig 3.14 **The velocity of a tennis ball**

Fig 3.16 **A ball falling in a semi-parabolic curve**

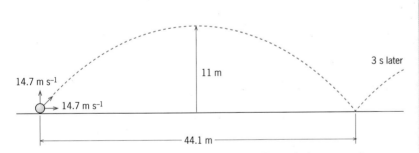

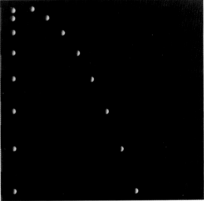

Fig 3.15 **Vertical and horizontal motion are independent. The vertical movement is
affected by gravity. The horizontal speed stays the same until the ball hits the ground**

EXAMPLE

Suppose the ball of the last Example was kicked so that it started
upwards with a speed of 14.7 m s⁻¹, and also moved sideways
with the same speed. We have calculated that its time of flight
would be 3 s. This is independent of whatever sideways motion it
has. So, after 3 s, the ball hits the ground.

Q How far did it move sideways before it bounced?

A It moved for 3 s at a speed of 14.7 m s⁻¹, so it must have travelled
a distance of 3 × 14.7 m, namely 44.1 m.

R A tennis ball is lobbed at an angle
of 50° to the ground with a speed of
16 m s⁻¹. It is hit from a point very
close to the ground at the base-line.

(a) What is its vertical speed?

(b) How long will it be in the air before
reaching the ground again?

(c) A tennis court has a length of
23.8 m from one base-line to the other.
Will the tennis ball land in court?

The Extension box gives a more general formula for solving projectile problems, but it is usually easy enough to work out results by following these steps:

● First find the separate vertical and horizontal velocities by simple vector theory (finding components).

● Then calculate the time of flight, t, as above, using the formula $v_1 = v_0 + at$ (simplified to $v = gt$).

● Finally, use t and the horizontal component of velocity to find the horizontal distance travelled.

See questions 12 and 13. ■

The projectile formulae

The vertical component of velocity is $v \sin \theta$ as shown, and the horizontal component is $v \cos \theta$.

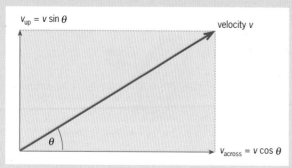

Fig 3.17 **Horizontal and vertical components of velocity**

Time of flight

Time of flight T is double the upward journey time t. From the basic kinematics formula ($v = at$) we get:

$$t = \frac{v \sin \theta}{g} \quad \text{so} \quad T = \frac{2v \sin \theta}{g}$$

Range

The horizontal **range** R of the projectile is how far it travels in time T at speed $v \cos \theta$. We use the formula:

$$\textbf{Distance = velocity} \times \textbf{time}$$

$$R = Tv \cos \theta = \frac{2v^2 \sin \theta \cos \theta}{g}$$

Height

The height reached by the projectile is h. This is easily obtained from the equation linking upward velocity to distance during the upward flight:

$$v_1^2 - v_0^2 = 2gh$$

In this case, v_1 is zero and v_0 is actually $v \sin \theta$, so:

$$v^2 \sin^2 \theta = 2gh \quad \text{or} \quad h = \frac{v^2}{2g} \sin^2 \theta$$

Maximum range

The maximum range of the projectile is obtained when $\theta = 45°$. We can prove this as follows.

First we use the trigonometrical formula that:

$$2 \sin \theta \cos \theta = \sin 2\theta$$

Inserting this in the formula above gives, for the range R:

$$R = \frac{v^2}{g} \sin 2\theta$$

For a given value of v, the greatest value of the range R occurs when $\sin 2\theta$ is a maximum. The maximum value for the sine of an angle is 1, which occurs for an angle of 90°. Thus the maximum value of R occurs when $2\theta = 90°$, or $\theta = 45°$. **This also means that the horizontal speed equals the vertical speed.**

S A circus performer is fired from a cannon, hoping to land in a net 25 m away. The cannon is pointing at an angle of 45° to the horizontal. At what speed must the person leave the cannon to be sure of landing in the net?

T (Ignore air resistance in this question.) A batsman hits a baseball so that it moves off with a velocity of 24 m s^{-1} at an angle of 30° to the horizontal.

(a) Show that the ball should reach the ground at about 51 m away (at least).

(b) A fielder is 20 m away from the expected landing point. He can run at a speed of 8 m s^{-1}. Is he likely to be in a position to catch the ball? Justify your decision, and explain what effect air resistance might have on the situation.

8 GRAVITY ON THE WAY TO THE MOON

The most direct way of getting an object to the Moon might *seem* to be to wait until the Moon is directly overhead and then project the object vertically upwards, really fast. *How* fast we work out in the next chapter – the speed would have to be at least 11 km s^{-1}.

But it is not that simple. By the time the object reached the Moon's orbit, the Moon would have moved on. So we should have calculated when to throw the object upwards so that it reached the Moon's orbit just when the Moon happened to be there. We would also have to do some calculating to allow for the fact that the object was thrown from a spinning Earth and so would have some sideways speed as well. The path of our object moving in a gravitational field will, in general, be an ellipse. Space dynamics is not simple! (More about this in Chapter 4.)

On its way to the Moon the object would be acted on by a force due to the Earth's gravity. This acts towards Earth and would decelerate the object. Table 3.2 shows some actual data from the Apollo 11 Moon Mission of July 1969, one of a series of projects in which a space capsule was sent from Earth to the Moon.

Speed at start and end of time interval/m s^{-1}	Distance from Earth's centre at start and end of each interval/10^6 m	Acceleration in each 600 s interval/m s^{-2}	Mean distance r from Earth's centre during each time interval/10^6 m
5374	26.3	–0.45	27.65
5102	29.0		
3633	54.4	–0.12	55.4
3560	56.4		
2619	95.7	–0.04	96.45
2594	97.2		
1796	169.9	–0.01	170.4
1788	170.9		

Table 3.2 **Data from Apollo 11 Moon Mission. The time interval for each pair of speed readings is 600 s**

The acceleration of the spacecraft is found by calculating the change in speed during each time interval, and dividing that by the size of the interval, 600 s. Table 3.2 shows not only that the spacecraft is slowing down on its way to the Moon, but also that the deceleration gets less as the craft gets further away from the Earth. The deceleration is due to the Earth's pull on the spacecraft, so this force must be decreasing, as shown in Fig 3.18. As explained earlier, the deceleration of the spacecraft equals the value of the Earth's gravitational field strength g at that point in space.

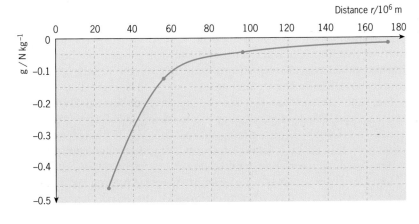

Fig 3.18 **Graph of gravity against distance for the Apollo 11 data in Table 3.2. It shows that the value of g decreases as distance from the Earth's centre increases**

The inverse square law

The plot of the graph in Fig 3.18 shows that the measured value of g falls off rapidly with distance. It also assumes that the spacecraft kept to the straight line between Earth and the Moon. This is only roughly true, as shown in Fig 3.19.

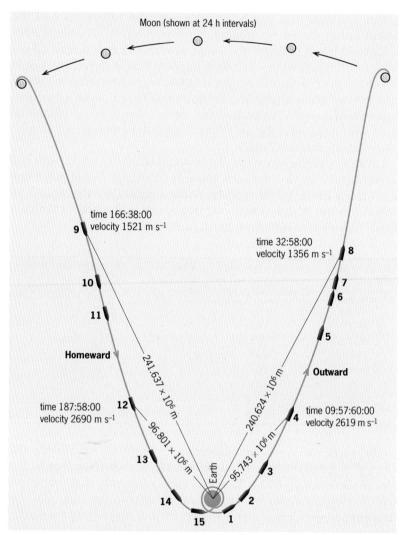

Moon (shown at 24 h intervals)

time 166:38:00
velocity 1521 m s^{-1}
9

time 32:58:00
velocity 1356 m s^{-1} 8

10

7
6

11

5

Homeward

241.637 × 10^6 m

240.624 × 10^6 m

Outward

time 187:58:00
velocity 2690 m s^{-1} 12

96.801 × 10^6 m

time 09:57:60:00
velocity 2619 m s^{-1}
4

13

95.743 × 10^6 m

Earth

3

14

2

15   1

Fig 3.19 **The path of Apollo 11**

Back in the seventeenth century, Newton proposed that the strength of the Earth's gravitational field should **vary inversely as the square of the distance from its centre**. Newton could not check this idea with spacecraft data, but he was able to confirm that the motion of the Moon in its orbit agreed with his theory to an accuracy of about 0.5 per cent, and that the motion of the planets in their orbits was also consistent with this inverse square law. Read more about this in the next chapter.

Fig 3.20 shows that the idea of an inverse square law makes sense, as the 'effect' of gravity has to spread out over larger and larger areas. In an inverse square law of force, the size of the force on any object is reduced by a factor of 4 when the distance doubles, by 9 when the distance trebles, and so on.

Fig 3.21 shows the values of g from Table 3.2 plotted against $1/r^2$, where r is the distance from the centre of the Earth. It is nearly a

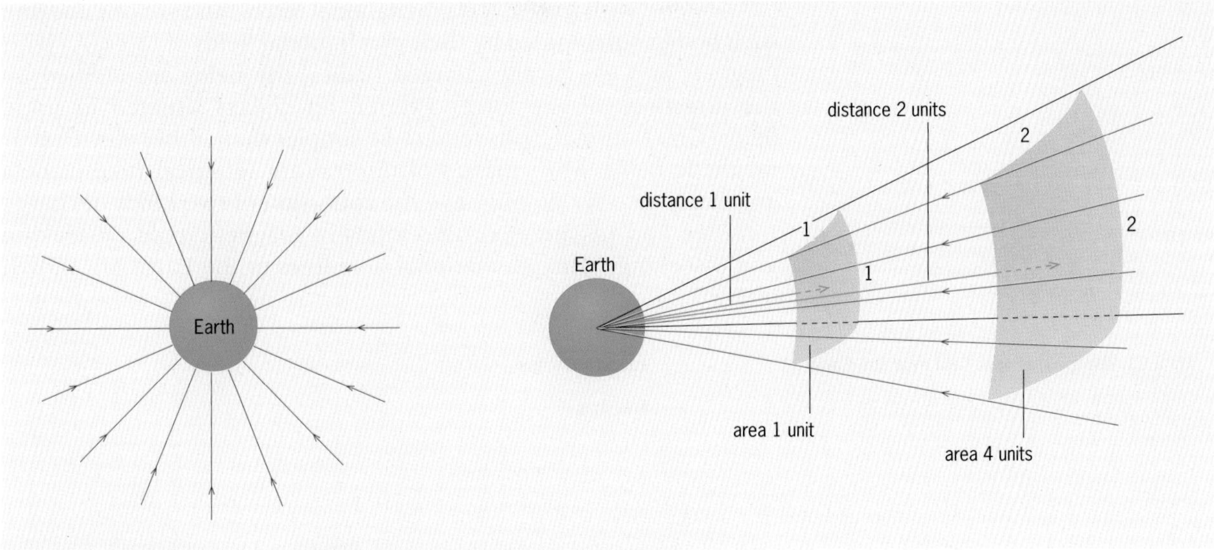

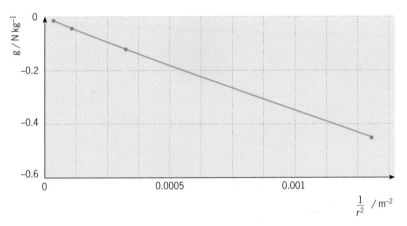

Fig 3.20 **Field lines radiating from a sphere, suggesting that an inverse square law is plausible. At double the distance, a given set of force lines are spread over four times the area**

Fig 3.21 **Graph of g against 1/r² for the data in** Table 3.2. **It is nearly a straight line, supporting the theory of an inverse square law**

See question 20.

straight line, supporting the idea that g is proportional to $1/r^2$. The data are not ideal, since the spacecraft was not moving along a straight line joining the centres of the Earth and the Moon.

9 VISUALISING A FIELD

You can get a picture of a magnetic field by sprinkling iron filings on a sheet of paper placed on top of a magnet. Each small piece of iron acts as a kind of probe, lining itself up with the direction of the magnetic force exerted on it at that point. The field lines formed show the direction of the force field at any point (Fig 3.22). We can do similar experiments to show the shapes of electric fields (see Chapter 10).

We cannot do these experiments to show a gravitational field because the forces involved are so weak. However, we can visualise it in the same way.

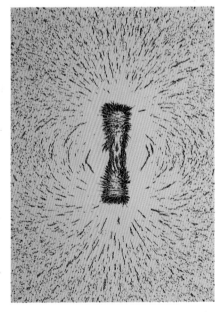

Fig 3.22 **Iron filings showing the field around a magnet**

All bodies with mass exert gravitational forces, and we can imagine such bodies surrounded by their gravitational fields of force. As with magnets and electrically charged bodies, the fields are defined by the directions of the forces exerted on objects situated in these fields. Gravitational fields tend to be simpler than either electrical or magnetic fields. Most masses of interest are spherical (like planets and stars), and we do not have the confusion of two kinds of charge (positive and negative) or two kinds of magnetic pole (N-seeking and S-seeking). The gravitational field lines of the Earth are shown in Fig 3.23.

Fig 3.23 **The gravitational field near the Earth**

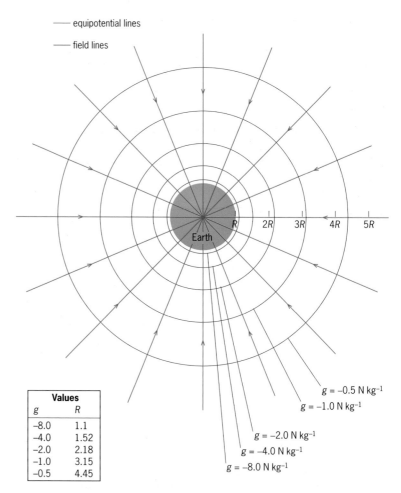

Fig 3.24 **The Earth's gravitational field is distorted near the surface by large features such as mountains**

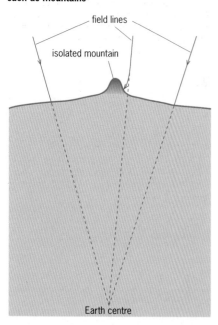

Values	
g	R
−8.0	1.1
−4.0	1.52
−2.0	2.18
−1.0	3.15
−0.5	4.45

If it is looked at in detail, the simple field of Fig 3.23 is distorted near large mountains, or particularly dense regions in the Earth's crust, as shown in Fig 3.24.

See question 8. ■

A plumb-line shows the direction of the Earth's gravitational field at any point. On a perfectly still, uniform and spherical Earth the plumb-line points directly towards the centre of the Earth in one direction, and towards the zenith (a point directly overhead) in the other. But near a mountain, the plumb-line veers away slightly from this line and towards the mountain.

We often picture the strength of a field by drawing lines close together for strong fields and further apart for weaker fields. So, as in Fig 3.21, drawing lines radially out from Earth represents the geometry of the inverse-square law neatly and accurately. When the distance from the Earth's centre is doubled, a bundle of lines is spread over an area four times as great.

?

U The Earth's radius is 6400 km. How far apart are two places on Earth if the directions of the radial gravitational field at those places differed by 1 degree?

10 NEWTON'S LAW OF GRAVITATION

Newton proposed a law of gravitation which allows us to work out the magnitude of the force of mutual attraction between two masses. Imagine two objects of mass M and m respectively, separated by a distance r between their centres. They attract each other with a force F given by:

$$f = G\frac{Mm}{r^2}$$

G ('big G') is called the universal gravitational constant (or constant of gravitation).

The currently accepted value of $G = 6.672\ 59 \times 10^{-11}$ N m^2 kg^{-2}. For most calculations we take $G = 6.67 \times 10^{-11}$ N m^2 kg^{-2}.

See questions 21 and 22.

?

V Use the gravitation formula to check that the units for G are N m^2 kg^{-2}.

Measuring G

It is difficult to measure G, the universal gravitational constant. The force between two 1 kg masses 1 m apart is very small, just 6.67×10^{-11} N. The first accurate measurement was made by the English scientist Henry Cavendish in 1798. (He used a version of the apparatus that the French scientist Charles Coulomb used to measure the much larger force between two electrically charged bodies.)

Fig 3.25 **Apparatus to measure the value of the Universal Gravitational Constant, G. (a) Principle of the Cavendish torsion balance. The masses M and m attract each other. (b) Boys' apparatus. The deflection of AB is measured using a beam of light, so the apparatus can be very small. The photo shows modern apparatus for measuring G using the principle seen in (a)**

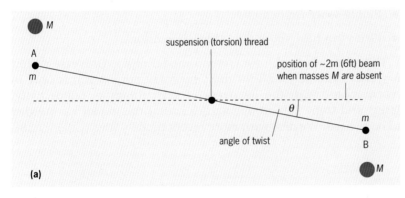

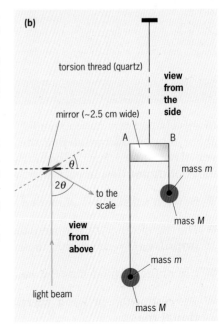

Fig 3.25(a) shows the type of **torsion balance** that Cavendish used in later experiments. The gravitational force between two pairs of masses is measured by the twisting (torsion) of a fine wire. Cavendish's apparatus was large, as he wanted to make the forces and movements involved as large as possible. But this produced many errors – including errors due to air currents and the effect of thermal expansion.

Later experimenters worked with much smaller apparatus. In 1895, C. V. Boys used a beam (AB) an inch (2.54 cm) long, compared with Cavendish's 6 ft version. Boys' torsion thread was made of very thin quartz, which he made by firing melted quartz from a crossbow! He used a mirror and a beam of light to measure the twisting angle.

Once we have measured G, we can calculate the masses of bodies such as the Sun, the Earth, the Moon and the planets. For example, we know that the force between the Earth and a 1 kg mass is approximately 9.81 N. Rearranging the Newtonian formula gives:

$$\text{Mass of Earth, } M = \frac{Fr^2}{Gm} = \frac{9.81 \times (6.38 \times 10^6)^2}{6.67 \times 10^{-11} \times 1.00}\ \text{kg}$$

$$= 5.99 \times 10^{24}\ \text{kg}$$

So what exactly *is* a force?

Newton was worried by his theory of gravity. It worked – in that it gave the right results – but he could not see how the force of gravity could act between objects separated by completely empty space. 'Action at a distance', with nothing to carry the action, did not seem to be scientific. What carried the pull? In the end he gave up and wrote '*Hypotheses non fingo*' – 'I don't give explanations.'

This remained the main problem with gravity for a long time – but as long as a theory gives the right results, physicists are usually happy enough to live with it!

A hundred years after Newton, the forces of gravity, electricity and magnetism were linked by a common inverse square model (see Chapters 14 and 27). Then, early in the nineteenth century, Michael Faraday (1791–1867), discoverer of many effects in electromagnetism, proposed the idea of the **field**, as described above. With these two ideas, everything that happened to bodies could be explained and described very neatly using mathematics.

But the 'force carrier gap' still existed in these nineteenth-century field theories. Modern physics has filled the gap, using some rather strange new ideas which we deal with later. For the record, they can be briefly summarised as:

● Forces are carried through space by certain **particles** – photons, gravitons, gluons.

● Empty space isn't empty!

● Fields can have more than three dimensions.

The point about the Newtonian model of the Universe is that it works extremely well for most everyday purposes. It is only in recent years, with the discovery of such things as radioactivity, lasers and some sub-atomic particles, that the model has needed to be revised. We deal with more recent models later (Chapter 28).

SUMMARY

In this chapter you have extended your knowledge about motion by considering the effect of forces: you have moved from **kinematics** to the beginnings of **dynamics**. In particular, you have learnt:

■ Newton's three laws of motion.

■ The significance of **momentum** ($P = mv$) and **impulse** ($Ft = \Delta P$) within these laws.

■ Use of the relationships $F = ma$ and $Ft = \Delta P$.

■ The distinction between **inertial** and **gravitational mass**.

■ The acceleration of free fall, g, and its relationship to the strength of a gravitational field.

■ The distinction between **mass** and **weight.**

■ How to tackle problems involving motion in a uniform gravitational field (projectile motion).

■ Gravitation follows the **inverse square law**.

■ The concept of a **field** of force.

■ How the universal gravitational constant G is measured

QUESTIONS

In all explanations, you are expected to use the correct physical terms – *force, mass, velocity, acceleration, momentum,* etc.

Take g to be 9.8 m s^{-2} or 9.8 N kg^{-1};
$G = 6.67 \times 10^{-11}$ N m^2 kg^{-2}.

Introductory questions

These questions are designed to make you think about the physical principles underlying the topics in this chapter.

1 A skier slides down a snow slope which is at a constant angle of 30°, starting from rest. The slope is perfectly smooth and the acceleration of the skier down the slope is given by $g \sin \theta$.

a) How fast is the skier travelling after 6 seconds?

b) How far has she travelled in this time?

c) If the snow is perfectly smooth, how could the skier stop if she needed to do so as quickly as possible?

2 A friend of yours who is not a student of physics refuses to accept that 'forces must always occur in pairs'. He gives some examples to disprove this idea:

a) A stone falling from a cliff.

b) A nail being driven into a piece of wood by a hammer.

c) A car coming to a stop when it is braked.

Your friend says it is obvious that in each of these cases there is only one object that clearly has a force exerted on it, and that is the only force there is.

d) Use your understanding of Newtonian physics to explain why he is wrong in each case.

3 Newton's first law of motion (see page 39) was also thought of by Galileo, over 40 years earlier. However, Galileo gave the Moon as one example of an object that will naturally carry on moving for ever, as no force is acting on it.

Do you agree with Galileo about the motion of the Moon? Justify your answer.

4 In naval battles during the Napoleonic Wars (at the turn of the eighteenth century) the ships were wooden and carried as many as 50 guns (cannons which fired large iron balls) on each side of the ship. It was sometimes considered dangerous to fire all the guns on one side of the ship simultaneously (in a 'broadside'). It was more usual for each gun to be fired in turn, one after the other in a 'rolling broadside'.

Explain the possible danger of a full broadside to the ship that fired it, and why a rolling broadside would be safer.

5
a) Explain the differences between *inertial mass, gravitational mass* and *weight*.

b) Which of these three quantities would be different for a fixed very large number of iron atoms on the Moon compared with exactly the same number on Earth?

6 A cannon ball was fired horizontally from the top of a sea cliff. After a second or so it was about 150 metres from the cliff and about 10 metres nearer the sea.

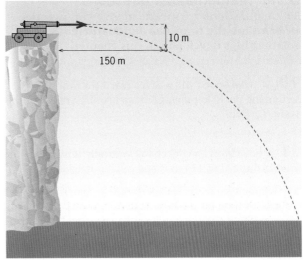

Fig 3.Q6

a) Copy the diagram which shows the cannon ball at that time, and mark on it the force/s then acting on the ball.

b) Does the force of gravity affect the sideways motion of the cannon ball? Explain your answer.

7

a) What force is exerted on a mass of 4.80 kg at a place where the gravitational field strength is 5.83 N kg^{-1}?

b) With what acceleration will the mass move, if it is free to do so?

c) With what acceleration would a mass of 2000 tonnes move at the same place in the gravitational field?

Use your answers to parts a), b) and c) to explain why the acceleration of free fall is equivalent to the gravitational field strength at any point.

8

a) Sketch diagrams to illustrate the gravitational fields:
   (i) of a spherical mass,
   (ii) of a spherical mass equal in size to the one in (i) but of twice the density,
   (iii) between two spheres of equal mass and density separated by a distance equal to twice their diameters,
   (iv) in the space on the surface of the Earth approximately the size of your school.

b) Explain the features of your diagram which distinguish between the four cases you have illustrated.

9 *There is no force, however great, can pull a string, however fine, into a perfect horizontal line.*

Why not?

Some mathematical questions
These questions are meant to give you practice in handling the formulae and mathematical relationships developed in this chapter.

10 At what speed will a diver reach the water in a swimming pool after diving from a board 10 m above the water surface?

11 A boy throws a cricket ball vertically upwards with an initial speed of 12 m s^{-1} and catches it again.

a) Before doing any calculation, sketch a graph of velocity against time for the ball. Take the upward direction as positive and the downward as negative.

b) How long does it take the ball to reach its maximum height?

c) How high does the ball go?

12 A cricketer standing on the boundary throws a cricket ball at the wicket-keeper who is 70 m away. He can throw the ball at a maximum speed of 24 m s^{-1}. Is it possible for him to throw the ball so that it reaches the wicket-keeper without bouncing? Justify your answer.

13 A tennis court is 23.8 metres long. A player hits the ball on her base-line from effectively ground level, at an angle of 60° to the ground, intending to lob her opponent. (A lob is a stroke which makes the ball go very high – but it must land in the court). Show that the maximum speed with which the ball can be hit if it just lands on her opponent's base-line is just over 16 m s^{-1}.

14 A refrigerator falls from the balcony of a tall block of flats into a garden. The balcony is 40 m above the ground. The refrigerator is comparatively undamaged as it buries itself 45 cm into the soft ground.

a) Ignoring air resistance, at what speed does the refrigerator hit the ground?

b) What is the average value of the deceleration of the refrigerator as it buries itself?

15 A hot-air balloon is travelling horizontally just 20 m above the ground when the balloonist decides to gain height. To achieve this, 20 kg of sand is released from the balloon. The mass of the balloon is now 400 kg.

a) How long will it take the balloon to reach a vertical speed of 5 m s^{-1}?

b) How high is the balloon now?

c) Another sandbag is now dropped from the balloon.
   (i) At what speed will the bag hit the ground?
   (ii) How long will it take to reach the ground?

16 A bullet of mass 4 g leaves the barrel of a rifle at a speed of 300 m s^{-1}. The barrel is 0.74 m long.

a) What is the average force exerted (i) on the bullet, (ii) on the rifle.

b) The rifle has a mass of 1.5 kg, and the inexperienced user holds it loosely so that it is 0.02 m away from his shoulder when it is fired. Estimate the speed at which the rifle butt hits the user's shoulder.

17 The table shows the value of the gravitational field strength g for some planets in the Solar System. An astronaut has a mass of 75 kg. Copy the table and complete it to show how much the astronaut would weigh on each planet.

Planet:	Mercury	Venus	Earth	Mars	Jupiter
g/N kg^{-1}	3.6	0.87	9.8	3.7	26
Weight of astronaut					

18 An electron is accelerated from rest in an electric field so that it increases its speed by 2.5×10^5 m s^{-1} in a distance of 10 cm. The mass of an electron is 9.1×10^{-31} kg.

a) Calculate the average acceleration of the electron.

b) What average force was exerted on the electron?

c) Electrons are accelerated in TV tubes. In designing a TV tube, do the manufacturers need to take into account the gravitational force exerted on an electron? Justify your answer.

19 A builder is employed to rebuild a chimney stack on a house. He uses a wooden box attached to a single pulley system to lift the new bricks to the roof. When he has finished he loads the box as full as he can with old bricks, having taken the precaution to tie the rope firmly at ground level as shown in the diagram. The box is now 12 m above the ground.

Fig 3.Q19

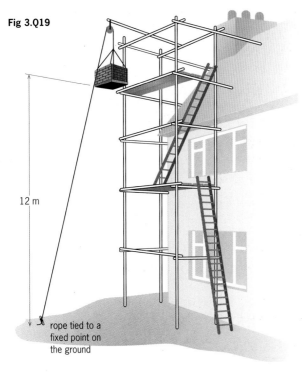

12 m

rope tied to a
fixed point on
the ground

He climbs down to the ground and unties the rope. Unfortunately the box of bricks has a mass of 75 kg, while the builder has a mass of only 55 kg.

Assuming that the builder keeps holding on to the rope, describe in as much mathematical detail as necessary what happens next. Ignore friction.

20 The following set of data was obtained by monitoring a planetary space probe as it moved directly away from Earth. The distance is measured from the centre of the Earth.

Distance/10^6 m	8.2	10.2	12.2	14.2	16.2	20.2
speed/m s^{-1}	10 210	9208	8471	7901	7424	6789
speed/m s^{-1} 100 s later	9620	8820	8201	7703	7248	6691

Calculate the deceleration of the space probe at each distance. Then plot a suitable graph to check that the data confirms the inverse square law of gravity.

21
a) Calculate the force of gravitational attraction between the Earth and the Moon.

b) Hence find the acceleration of the Moon towards the Earth.

c) Does the Earth accelerate towards the Moon?

d) Are the Moon and the Earth getting closer together?

Data: GM for the Earth is 4×10^{14} N m^2 kg^{-1};
Earth–Moon distance = 60 times the Earth's radius;
g at the Earth's surface = 9.8 N kg^{-1};
mass of Earth 5.98×10^{24} kg;
mass of Moon = 7.3×10^{22} kg.

22 Estimate the gravitational force of attraction between you and the person sitting next to you in the laboratory or classroom. State clearly what assumptions you make.

See page 81 for the CHAPTER MAP that covers
Chapters 2, 3 and 4

Assignment

ELECTRICITY VERSUS GRAVITY; HOW HIGH CAN A MOUNTAIN BE?

Ordinary matter – a rock for instance – is made up of crystals that are held together simply by interlocking with each other or are 'glued' together by some kind of matrix or glassy material. Large single crystals are rare, but where they exist they are often stronger than the same substance in a poly-crystalline form.

The force that holds the particles in crystals together is an electric force. Many of the crystals that make up rocks on the Earth's surface are ionic crystals and, for these, the electric force is a force of attraction between two oppositely charged particles (ions).

The charge on an ion is small, usually one or a few electronic charges. Each charge is of size e, where e is 1.6×10^{-19} C. But because the ions are very close together, the force between them is large when compared with the force of gravity between them, or with the force of gravity acting on an ion due to the mass of the Earth.

But suppose we have an *extremely* large crystal that is many kilometres tall. What forces would be acting on a layer of ions at the base? Would a crystal be strong enough to with-stand the forces due to its own weight?

Think of building a pyramid with dough. Dough gently flows under pressure: when a dough pyramid is large enough its base will spread. This is because the forces between its particles are quite small when compared with those between the particles in rocks, or even in a solid like ice. But then ice can flow – glaciers do this. A glacier moves 'under its own weight'. It is of course the effect of gravity acting on

the large bulk of the ice which tends to push the particles apart so that the binding is weak enough for the glacier to flow or possibly to fracture.

In principle, this is what could happen to our mountain which we can think of as a very large crystal (though in fact mountains are made of very many crystals). The questions below lead you through a very simple model to produce a very rough estimate of the maximum height a mountain could be, assuming that it would start slipping apart at the base if the force of gravity happened to be larger than the electric force holding its base together.

1 The force between a pair of ions

We take a common but fairly weak material, common salt (NaCl).

a) The density of common salt is 2170 kg m^{-3}. What is the volume of 1 mole of this substance? A mole of salt has a mass of 0.0585 kg.

b) A mole of any substance has 6×10^{23} particles in it. How many pairs of Na$^+$ and Cl$^-$ ions are there in the volume you worked out in a)?

c) Show that the volume occupied by one pair of ions is about 5×10^{-29} m^3.

d) Picture the ions as being on opposite sides of a cube of volume 5×10^{-29} m^3 (Fig 3.A2). Calculate the value of d, the length of the side of this cube.

e) The ions each carry a charge e (see the introduction). The electric force between them is given by $F = ke^2/d^2$ where k is the electric force constant and has value 9×10^9 N m^2 C^{-2}. Show that the electric force of attraction between the ions is about 6×10^{-19} N.

Fig 3.A2 **A pair of ions positioned on opposite faces of a cube**

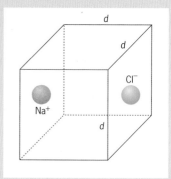

2 The force due to gravity

The masses of sodium and chlorine ions are 3.8×10^{-26} kg and 5.9×10^{-26} kg respectively. G is 6.7×10^{-11} N m^2 kg^{-2}.

a) What is the gravitational force of attraction between the ions?

b) How many times larger than this is the electric force?

Fig.3.A1 **A glacier: the dirt and rocks mark the flow lines in the ice**

3 The sliding mountain

a) To represent matter at the base of our gigantic crystal mountain, imagine a cuboid crystal of cross-sectional area A square metres, as in Fig 3.A3. To pull this crystal apart will need a force large enough to overcome the electric attraction of all the ion pairs in such an area.

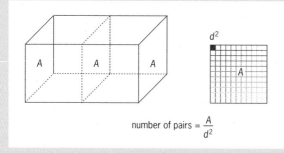

$$\text{number of pairs} = \frac{A}{d^2}$$

Fig 3.A3 **Part of a crystal containing many pairs of ions**

(i) Write down an expression for this force in terms of your answer to **1 e)** above, and A and d. **(ii)** Show that the sideways *'pressure'* needed to pull the crystal apart is about 5 Pa. This is an estimate of the crystal's tensile strength.

Salt is not a very strong material, but, even so, our calculation for the forces in a salt crystal, which you would have to overcome to pull it apart, is a great underestimate. Also, rocks tend to be made of stronger materials than salt. A typical strong rock might have a tensile strength of as much as 200 MPa (and would possibly be stronger in compression). This is the stress at which the rock would be pulled apart.

b) The pressure at the base of a rock column of height h is $\rho h g$ where ρ is its density – about 2600 kg m^{-3} for rock in the Earth's crust. Calculate a value for h at which a mountain would produce a pressure at its base equal to the pressure needed to pull the rock apart.

c) Mount Everest is the highest mountain on Earth, at just under 9 km high. The highest mountain on Mars is about 25 km high. Suggest some reasons for the difference.

ENERGY

ENERGY is one of the most fundamental ideas in physics. There are few chapters in this book in which the word does not appear. These two pages give a summary of how physicists today use the concept of energy.

Defining energy

This is the simplest definition of energy:

Energy is the ability of a system to do work.

Work is done when a force moves something. The idea of work is dealt with on page 58.

A **system** consists of at least two parts. Examples:

- A mixture of a **fuel** and **oxygen** can do work via an internal combustion engine.

- A **mass** at a height above the **Earth**'s surface can do work via a machine when the mass is allowed to move downwards under the force of gravity.

- A **moving mass** can do work via a machine when slowing down relative to the **machine**. For example, a moving mass of water can turn a stationary turbine wheel and generate electricity. (This will not happen if the turbine is floating along at the same speed as the water.)

The **quantity of energy** is defined by the **state** of the system, for example, how far apart masses or charges are, and how much mass or charge is involved. In atomic physics we often call 'state' an **energy level**.

We can rarely measure the total energy in a system. What we measure is an energy *change* as a system moves from one state to another, or an energy *difference* between one system and another.

Using energy

It is impossible to use all the energy change of a system to do work. This is sometimes for practical or design reasons – such as inefficiency due to friction or work done in moving parts of a machine other than a load. But there is also a more fundamental reason for inefficiency which applies to **heat engines** which rely on thermal transfer, like steam turbines and car engines. There is more about this in Chapter 15.

Measuring energy

The unit for both energy and work is the **joule** (J). A joule is the work done when a force of 1 **newton** (N) moves something through a distance of 1 metre: $1 \text{ J} = 1 \text{ N m}$. Energy is measured in the same unit since it is defined as the capacity to do work.

Forms of energy?

In elementary work it is sometimes helpful to think of different *forms* of energy, such as *chemical* energy, *nuclear* energy, *electrical* energy and *heat* energy. We then talk about 'converting energy from one form to another'.

This approach to energy is less useful – and can be misleading – in advanced studies. For example, if we take a closer look at 'heat energy' we find that when a solid object is heated ('gets hotter') its particles vibrate more. The particles in a solid are held together in fixed positions by forces between them (bonds) and the particles vibrate about these positions. This means that the particles have a changing mix of kinetic and potential energy. In a gas, the extra energy that makes the gas hotter is spread amongst the particles as kinetic energy. For a body at a fixed temperature the sum of the particles' kinetic and potential energies is constant. This is called **internal energy**. Many textbooks still use the term *heat* or *thermal energy* to describe it, but the term is unhelpful.

These are the few *fundamental* forms of energy:

- **Potential energy**: the energy of a system in which a body is in a force field of some kind which exerts a force on it. Gravity fields exert forces on mass itself, giving rise to **gravitational potential energy**. Electric fields act on charges and give rise to *electrical potential energy*. Nuclear forces act on nucleons and they have potential energy often called *binding energy*. These kinds of potential energy are fully dealt with in the relevant chapters.

- **Kinetic energy** is the energy of a mass moving relative to the observer, measured by the work it could do by being brought to rest (ie to the same speed as the observer, usually at rest on Earth).

- **Radiation energy** describes the energy carried by electromagnetic waves, which are a combination of moving electric and magnetic fields. The energy is transferred by photons, each carrying its quantum of energy at the speed of light.

All other 'forms' of energy can be described more simply and fundamentally as combinations of these three basic types.

'Transferring energy' is more useful than 'converting' it

In most practical situations, energy becomes important when it is being **transferred** from one system to another. It is therefore sounder and more useful to have a good understanding of **energy transfer processes**.

Electrical transfer or electrical energy?

Instead of talking about the 'electrical energy' in a wire, it is better to think of the wire as a vehicle for the transfer of energy by an **electrical transfer** process. Think of a battery connected to a torch bulb, see Fig E.1. The chemicals in the battery contain charged particles (ions) which are separated and so produce an electrical potential difference and hence a force which can move electrons around the circuit. When electrons move, the force does work, pushing them against other electrons, atoms and ions in a conductor to give them extra kinetic and potential energy. The filament resists the flow of electrons, the work done raises the internal energy and so the temperature of the filament rises. In time, a balance is achieved in the filament – the work done electrically equals the radiation energy (infrared, light) emitted.

The battery will 'run down', as chemicals change and the total electrical potential difference between the ions decreases. Note that the current of electrons in the wire connecting the battery to the bulb does not 'carry a load of energy' to the bulb. Although the electrons move randomly at high speeds they move along the wire with a very *small* drift speed. The kinetic energy they have due to this slow drift is also very small. The job of the electrons is to transmit a force, just like the chain in a bicycle. This is an electric force and so can do work, for example, on a machine such as an electric motor, or by interacting with charged particles in a resistor and making them move (increasing internal energy or 'heat').

Heating and working

Energy can move from one object to another via thermal processes: **conduction, convection, change of state** or **radiation**. These are random processes, because they involve the random motion of particles in the substances. Compare this with *working*, which is an ordered process where a clearly directed force acts on something.

Strictly speaking, *heating* is a process in which energy is transferred from a hot object to a cooler one via the processes described above. But of course you can heat something by a non-random process – think of sawing though a a piece of metal. This is *working* (you are using a directed force), but the metal and saw get hot. In the same way, the current in a bulb filament does work to make it hot. Using 'heating' in this way is an example of physicists stealing an ordinary word from the dictionary, giving it a special meaning and so causing some confusion to students.

To summarise, we can increase the **internal energy** of an object by supplying energy via **work** and also by supplying it via **random thermal processes due to temperature differences**. Physicists shorten this last phrase to 'heating'. There is more about this in Chapter 15 which deals with the interrelation of heating and working in what we call **thermodynamics**.

The law of conservation of energy

This states that in any closed system or set of systems the total energy remains constant. It is also stated as:

Energy can neither be created nor destroyed.

It is one of the most fundamental laws of physics. It means that when a system loses a quantity of energy, other systems must together gain an exactly equal quantity of energy.

The law has been extended to include the idea that mass and energy are equivalent (see Chapter 26) so that a change in mass also means a change in energy, and vice versa; the changes are linked by the formula $\Delta E = c^2 \Delta m$.

Energy is not an easy idea to understand, but it is usually simpler to think about and describe the processes by which work is done and by which energy is transferred, rather than to worry about what energy *is*, or what *form* it may take.

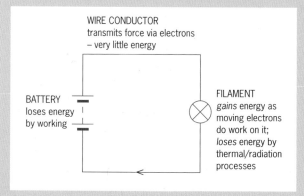

WIRE CONDUCTOR
transmits force via electrons
– very little energy

BATTERY
loses energy
by working

FILAMENT
gains energy as moving electrons do work on it; *loses* energy by thermal/radiation processes

Fig E.1 **Energy in a simple circuit**

4 Newton's laws at work

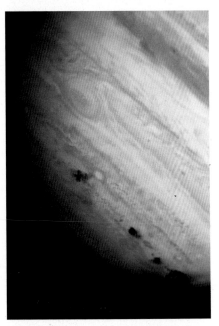

In July 1994, fragments of Comet Shoemaker-Levy hit Jupiter, a planet over 300 times the mass of the Earth. The fragments exploded on impact, and astronomers round the world recorded the fireballs which rose up through the dense atmosphere of cloud. The dark brown patches, lower left, mark the sites of impact of the fragments

IT IS VERY RARE for an asteroid or a comet to hit the Earth. But smaller pieces of matter arrive from space all the time. Each day, 400 tonnes is added to the Earth's mass, from all the pieces of metal, rock and ice travelling through 'empty' space which are trapped by the Earth's gravity and fall into its atmosphere.

Meteorites – chunks of matter in space – fall very fast, between 10 and 70 kilometres per second. Friction with the atmosphere heats them to extremely high temperatures, and that is why we can see them at night as 'shooting stars'. Even objects of 50 metres in diameter generate enough heat to melt and vaporise them completely in the upper atmosphere. This is just as well, as an object of that size has the kinetic energy of a 10-megaton nuclear bomb (4.2×10^{16} J).

Massive objects have collided with the Earth about once in every 100 million years. The Earth and the other inner planets of the Solar System would have experienced many more collisions but for the giant planets Jupiter, Saturn, Uranus and Neptune, whose gravitational pull tends to divert the path of large objects. In 1994, for example, fragments of Comet Shoemaker-Levy crashed into Jupiter. It is estimated that, without the giant planets, similar collisions with the Earth could rise to once every 100 000 years.

Energy on a massive scale

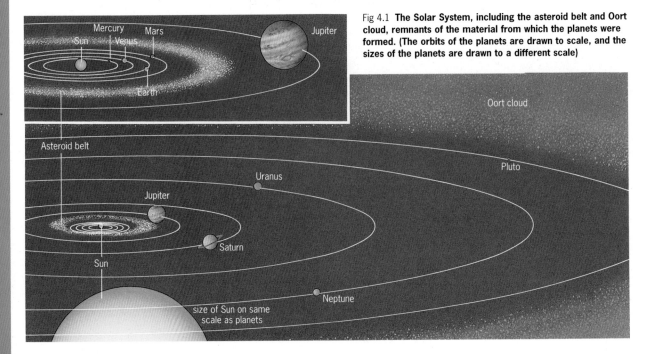

Fig 4.1 **The Solar System, including the asteroid belt and Oort cloud, remnants of the material from which the planets were formed. (The orbits of the planets are drawn to scale, and the sizes of the planets are drawn to a different scale)**

We can think of the Solar System as an arrangement of planets in orderly orbits round the Sun. In addition, countless objects smaller than the planets take other paths, and sometimes collide with the planets. They include, for example, asteroids, and meteorites which are smaller bodies. Asteroids are straying members of the asteroid belt, a mass of material that lies in a wide band between the orbits of Mars and Jupiter (shown in Fig 4.1).

Meteorites are mostly fragments of comets, consisting of metal and rock dust embedded in frozen water and gases, which come from the Oort cloud, on the outermost fringes of the Solar System. Many of these bodies do not have the regular orbits of the planets. There are over 150 asteroids which might one day collide with the Earth. The largest is estimated to be 8 km in diameter.

Fig 4.2 **Travelling at great speed, a meteorite heats up and vaporises because of friction with particles in the atmosphere. We call the streak of light we see in the sky a meteor**

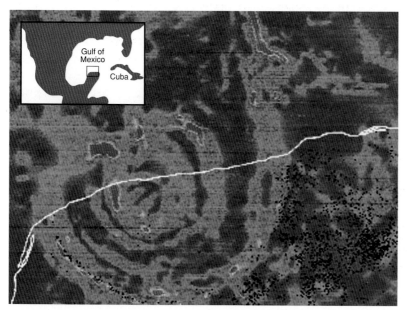

Fig 4.3 **The 180 km wide Chicxulub crater in Mexico, the largest known crater caused by an asteroid impact. The crater was gradually filled in by less dense deposits. Its size and shape are revealed by gravity measurements: the presence of less dense rock causes the gravity field to change (opposite to the effect of mountains, see page 46) and so allows us to produce an image of the crater**

Fig 4.4 **In 1991, Mr Pettifor from Peterborough just missed being hit by a 0.68 kg meteorite. He later donated it to the Natural History Museum**

An asteroid hitting the ground makes a crater far larger than itself, an indication of the colossal energy of its impact. An example is the 180 kilometre-wide crater at Chicxulub, Yucatan, in Mexico (Fig 4.3). The asteroid is thought have been just over 10^{12} tonnes in mass and 10 kilometres in diameter. It collided with the Earth 65 million years ago, at the end of the geological period known as the Cretaceous, and with an estimated kinetic energy of about 4×10^{23} J. This is about a thousand times greater than the total annual global energy used by human beings today (4×10^{20} J).

The asteroid impact at Chicxulub caused a major upheaval to life on Earth. Such a massive impact sends volumes of dust into the atmosphere, which circulates round the globe for years and cuts out sunlight. Plants fail to grow, and animals that depend on them die. The decline of the dinosaurs is the best known effect thought to have been caused by the Chicxulub impact, but it is thought that about half of all marine species died out at the same time. Similarly, 160 million years earlier, 95 per cent of sea life disappeared, probably as result of a larger impact by an asteroid whose remains have been found in the South Atlantic Ocean.

?

A How did the Chicxulub asteroid gain its vast energy? Would it be possible to change the path of a similar asteroid on target for the Earth if we knew about it soon enough?

The ideas in this chapter

The physics

To work out the energy of an asteroid, you need to understand the ideas of **kinetic energy, gravitational potential energy** and also the **principle of the conservation of energy**. You will go on to use the notion of a field of **gravitational force** obeying the inverse square law that was developed in Chapter 3. You will learn how these forces and fields determine the **orbital motion** of asteroids, satellites and planets. You will also meet the physical principles of how **rockets** can put satellites into orbit, and so gain a good understanding of **momentum** and its **conservation**.

The mathematics

On the whole, simple algebra is all you need. You will be handling large numbers, interpreting graphs and learning of the importance of the area under a graph. Though not essential, integral calculus or the use of a spreadsheet would be helpful as well.

1 ENERGY AND WORK

We have looked at the consequences of an asteroid, an object of massive energy, hitting the Earth. Now, we turn to the concept of energy on a very much smaller, everyday level. We base our idea of energy on the concept of **work**, since the energy of a system is both defined and measured in terms of the work a system can do. The word 'work' is defined by physicists to mean something more precise than its common use, because we say that:

Work is done only when a force moves something.

For example, work is done:

- when you use muscular force to lift an object upwards against gravitational force
- when you saw through a piece of wood, again using muscular force to tear the fibres of the wood apart
- when an electrical force moves electrons through a resistor
- when water under pressure turns a turbine in a hydroelectric power station
- when the petrol–air mixture in the cylinder of a car engine explodes and pushes the piston outwards.

Forces doing no work

You are not doing any work when you stand still with a heavy rucksack on your back. You get tired because your muscles are stressed. But if the rucksack stays still, no work will be done on it. In the same way, a table wouldn't be doing any work if you took a rest and put your rucksack on it.

The work formula

See question 1. ■ The quantity of work done by a force is measured in joules:

Work done = **force** × **distance moved in the**
(in joules) (in newtons) **direction of action of**
  the force (in metres)

$$W = Fd$$

B Think of another four examples of work done, and write them down.

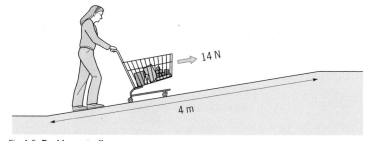

Fig 4.5 **Work is being done when a force is used to raise a load. The forces *F* in (a) and (b) depend on the load and the direction in which the force can be applied**

(a) **Raising a load vertically, using a pulley**

(b) **Pushing a load up a ramp**

As shown in Fig 4.5(a) and (b), the direction of the force *F*, and the movement of the point where the force is applied, have to be taken into account.

■ See questions 2 and 3.

EXAMPLE

Q A person pushes a shopping trolley up a ramp 4 m long, as in Fig 4.6. She applies a force of 14 N along the ramp. How much work has been done?

A Work done = force × distance moved
(J)   (N)   in direction of
                              motion (m)

= 14 × 4

= 56 N m (56 J)

Fig 4.6 **Pushing a trolley up a ramp**

---

2 THE CONSERVATION OF ENERGY IN A GRAVITATIONAL FIELD

Because we are able to measure work, we often describe the energy of a system in terms of how much work it does – or is capable of doing. Then we can define energy in this way:

A system has energy if it is capable of doing work.

We use this idea below, and in many other topics in this book.

Gravitational potential energy

When you lift something, you are doing work. Think of taking a bag of sugar from the floor and putting it on a shelf (Fig 4.7). To do this, you have to exert a force that is just greater than the force of gravity on the sugar bag (its weight, *mg*), and you have to raise the bag through a height, *h*.

The work done on the bag = force × distance

$$W = mgh \text{ (in joules)}$$

In a system of two objects, gravity provides a pair of forces of attraction that are equal in size but opposite in direction – Earth pulls on the bag and the bag pulls on the Earth. In the Example above, if the sugar bag fell off the shelf, both bag and Earth would move towards each other (though the Earth would move only a tiny distance!).

Work done on sugar bag = *Fh* = *mgh*

Fig 4.7 **The force to lift the sugar bag just exceeds the force of gravity**

The sugar bag could do a little useful work while it is falling – for example, we could make it briefly operate a very small generator and produce an electric current to light a small bulb! On a much larger scale, the energy of waterfalls is used to produce electricity in hydroelectric power stations.

So we can think of the Earth–sugar bag system as having 'stored' energy when the bag is resting on the shelf. We call this energy **gravitational potential energy**. 'Potential' reflects the fact that the sugar bag doesn't look very energetic when it is just sitting on the shelf. But we know that it is capable of doing work when it falls. We have noted that work is done when a force moves something. If we don't use the gravitational force on the bag to do useful work (such as turning a tiny generator), then the work done simply increases the **kinetic energy** – the 'energy of movement' – of the bag: as it falls, it moves faster.

3 THE PRINCIPLE OF THE CONSERVATION OF ENERGY

It is not *obvious* that the work done in lifting the sugar bag to the shelf is equal to the potential energy stored when it gets there, and that this potential energy is equal to the kinetic energy it would gain as it fell back to the floor.

In any case, you can't actually measure the potential energy directly. If you carried out a careful experiment to measure the work done and the final kinetic energy acquired, your results would probably show that the work done and the final energy were only approximately equal. It is, for example, hard to allow for work done against friction, or for experimental errors.

Nevertheless, one of the most fundamental principles of science states that:

Energy cannot be created or destroyed.

Belief in this principle grew stronger during the nineteenth century as a result of many increasingly accurate experiments, and scientists gradually came to accept it. It became known as the **principle of the conservation of energy**.

The kinetic energy formula

You have already used the formula for kinetic energy:

$$\text{Kinetic energy} = \frac{1}{2}mv^2$$

which describes the kinetic energy of an object in terms of its mass and speed. We can derive this formula by looking at the kinetic energy of an object on which work is done.

Fig 4.8 shows a car having work done on it, and gaining speed and so kinetic energy. A net accelerating force F acts on the car, mass m, and the car moves a distance x from rest, gaining speed, v.

time = 0
speed = 0

time t later
speed v

net accelerating
force F

m

Fig 4.8

distance x

According to the principle of conservation of energy, the work done, Fx, is transferred to the kinetic energy of the car (ignoring friction and other energy losses):

$$\text{Work done} = \text{kinetic energy gained} = Fx = max$$

$$(\text{since force } F = \text{mass} \times \text{acceleration})$$

We now relate ax to speed v using the kinematic equation of motion, $v^2 = 2ax$, for an object starting at zero speed (as Example, page 17).

Rearranging this gives: $$ax = \tfrac{1}{2}v^2$$

So we can write:

$$\text{kinetic energy gained} = m(ax)$$

$$= m \times \tfrac{1}{2}v^2$$

which we normally write as: $$\tfrac{1}{2}mv^2.$$

■ See questions 4, 5 and 6.

E You pick up a stone and throw it vertically in the air. It lands back at exactly the same place from which you picked it up.

(a) Did you transfer any energy in this situation?

(b) What has happened to it?

(c) What assumptions about energy have you had to make in giving your answers?

EXAMPLE

Q A rock of mass 25 kg falls from a cliff to a beach 30 metres below. What is its **a)** kinetic energy, **b)** its speed just before it hits the beach?

A Relative to the beach, the gravitational potential energy of the rock before falling can be expressed as:

$$mg\Delta h = 25 \times 9.8 \times 30 \text{ (in joules)} = 7350 \text{ J}$$

a) Energy is conserved so, ignoring friction with the air, the kinetic energy of the falling rock is also 7350 J.

b) Therefore: $$\tfrac{1}{2}mv^2 = \tfrac{1}{2} \times 25 \times v^2$$
$$= 7350 \text{ J}$$
$$v^2 = 7350 \times \tfrac{2}{25}$$

So: $$v^2 = \tfrac{2 \times 7350}{25}$$
$$v = 24 \text{ m s}^{-1}$$

Speed of rock just before it hits the beach = 24 m s^{-1}

JOULE'S WORK ON HEAT AND ENERGY

EXPERIENCE TELLS US that, when work is done on an object, energy can be transferred to make it hotter. Hitting metal with a hammer is an example. Conversely, we can use a substance which is hotter than its surroundings to do work, such as burning petrol in a car engine.

In the nineteenth century, the debate was over whether 'heat was a form of energy'. We now say that 'an energy input is required to make something hot, and an object hotter than its surroundings can be used to do work'.

James Prescott Joule (1818–1889) was an amateur scientist interested in heat and energy. Between 1843 and 1878, he carried out very careful experiments to establish the principle of the conservation of energy as it applied to heating and working. He set out to show that, when frictional forces do work on a substance, it gets hotter, and that there is a direct relationship between the work done and the rise in temperature of the substance.

In his experiments, Joule measured the work done in stirring some water and the rise in temperature it produced. A paddle-wheel, which was driven by a falling weight, rotated in the water – doing work via friction on the water and heating it at the same time Joule

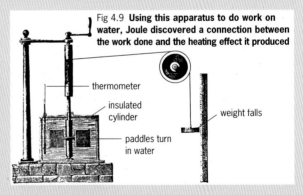

Fig 4.9 **Using this apparatus to do work on water, Joule discovered a connection between the work done and the heating effect it produced**

thermometer

insulated cylinder

weight falls

paddles turn in water

showed that the work done to lift the weight up to the starting position was always proportional to the 'heat energy' gained by the water as the weight fell.

At that time, 'heat' was measured, not in joules but in calories, a unit still used on food packaging today. We now remember the work of Joule by the unit for energy named after him.

The principle of the conservation of energy – that energy cannot be created or destroyed – is now thought of as one of the most fundamental laws of the material world. It has been extended by Albert Einstein to include the idea that mass is a form of energy.

4 ENERGY CONSERVATION IN GRAVITATIONAL FIELDS

Gravitational potential energy in a uniform field

As we have seen, gravitational potential energy is measured simply as the work done to move an object directly against the force of a gravitational field, that is, upwards (refer back to Fig 4.7).

Change in gravitational = **work done when moving**
potential energy   **something against a**
                    gravitational force

$$\Delta E = \textbf{gravitational force} \times \textbf{distance}$$

or $$\Delta E = mg\Delta h$$

This formula for a change in gravitational potential energy is correct only if the value of g, the gravitational field strength, is constant

---

EXAMPLE

Q A holiday park has a roller-coaster shown in the diagram of Fig 4.10. The ride follows a loop path that begins and ends at D. The car is hauled from D to point A, the top of the ride, and then moves from rest, under gravity, to arrive at point D again.

a) Ignoring frictional forces, what are the speeds of the car at:
point B;
point C;
point D?

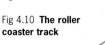
Fig 4.10 **The roller coaster track**

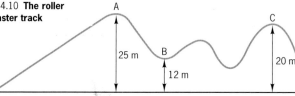

b) To stop at point D, the car has to be braked from a speed of 10 m s⁻¹. In practice, friction has reduced the car's speed during the ride. By point D, what proportion of the energy which the car had at point A has it lost because of friction?

A

a) The whole of this part is solved using the principle of the conservation of energy – we can just look at the changes in potential energy and kinetic energy at the various points. We do not need to know how the force varies at different points on the track. We don't even need to know the mass of the car, since it is the same throughout the ride, and cancels out, as we can see in the first solution.

Speed at point B
Gain in kinetic energy = loss in potential energy

$$\tfrac{1}{2}mv^2 = mg\Delta h \text{ (where } g = 9.8 \text{ m s}^{-2})$$

$$v^2 = 2g\Delta h \text{ (where } \Delta h = 25 - 12 = 13, \text{ in metres)}$$

$$v^2 = 2 \times 9.8 \times 13 = 254.8$$

giving: $v = 16 \text{ m s}^{-1}$

Speed at point C
The net loss in potential energy at point C is due to a drop of 5 m from point A.

(Check for yourself that the speed at point C is just under 10 m s⁻¹.)

Speed at point D
The drop is 25 m. So, using $v^2 = 2g\Delta h$,

$$v^2 = 2 \times 9.8 \times 25$$

$$v = 22 \text{ m s}^{-1}$$

b) The fraction of energy lost due to friction

Before braking, the car is travelling at 10 m s⁻¹.

So, before braking, the kinetic energy $\tfrac{1}{2}mv^2$ per unit mass (that is, $m = 1$) is:

$$0.5 \times 10^2 = 50 \text{ J kg}^{-1}$$

For a theoretical value for the car's kinetic energy we assume that the ride is frictionless, so the gain in kinetic energy per unit mass equals the potential energy, $mg\Delta h$, lost per unit mass:

$$9.8 \times 25 = 245 \text{ J kg}^{-1}$$

So the energy wastage due to friction per unit mass of car is:

$$245 - 50 = 195 \text{ J kg}^{-1}$$

The proportion of energy wasted during the ride

$$= 195 \div 245 = 0.80$$

over the whole distance Δh. The formula is accurate enough for objects moved close to the surface of the Earth, but will not be correct for distances comparable with the size of the Earth, let alone distances on the scale of the Solar System. This situation is considered more fully below.

So far, you have solved some quite complicated dynamic and kinematic problems using the ideas of force, mass, acceleration and the equations of motion. These techniques are easy to use with uniform forces on objects, and hence uniform accelerations of the objects. But in most everyday situations, things aren't so simple, and it is often more convenient to solve problems using the idea of the conservation of energy, as in the Example on the previous page.

5 GRAVITATIONAL POTENTIAL ENERGY IN THE EARTH'S FIELD

As Chapter 3 explains, the gravitational field strength of an object varies with the distance from that object, and the field strength obeys the inverse square law. Earth has a **radial gravitational field** shown in Fig 4.11. Over short distances close to the Earth's surface, the variation in gravitational field strength is too small to worry about, so here we assume that the field is uniform. Table 4.1 summarises the relevant formulae. G is the universal gravitational constant (see Chapter 3) and M is the mass of the Earth. Some of the formulae still need to be explained.

The universal gravitational constant, G, is 6.67×10^{-11} N m^2 kg^{-2}
The Earth's mass $M = 5.98 \times 10^{24}$ kg

Table 4.1

	Uniform field	Radial field
Force	$F = mg$	$F = -GMm/r^2$
Change in gravitational potential energy	$mg\Delta h$	$GMm(1/r_1 - 1/r_2)$
Gravitational potential	(not a useful idea in a uniform field!)	$-GM/r$

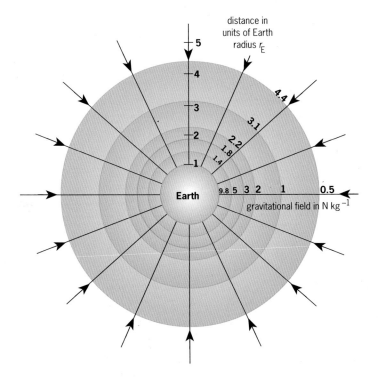

Fig 4.11 **The radial gravitational field of the Earth, showing how the value varies with distance from the Earth's surface. Radial lines are force lines, and circular lines are lines of equal gravitational field**

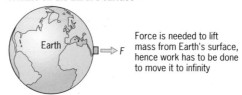

A mass on the Earth's surface

Earth

Force is needed to lift mass from Earth's surface, hence work has to be done to move it to infinity

A mass at infinity

The mass has zero force exerted on it, hence its gravitational potential energy is zero

Fig 4.12 **Gravitational potential energy of a mass at infinity and at the Earth's surface**

'Negative energy' – a very useful idea!

Before looking at the details of the radial field, we have to clear up a matter which often causes confusion – *the negative sign given to the value of gravitational potential energy.*

Imagine an object of mass m in empty space an infinite distance r_∞ from any other massive body. Force in a radial field is GMm/r_∞^2. But since r_∞ is at infinity, the object will have zero force acting on it. It cannot 'fall' towards anything, so it has no potential energy and it cannot gain kinetic energy – or do any work. So it must have *zero gravitational potential energy.*

Now imagine the same mass m sitting on the Earth. To move it to infinity (beyond the gravitational field of the Earth or any other body), you would have to *give* it energy – kinetic energy if you propelled it in a rocket. (Soon, we shall calculate just how much energy it needs.) But when the mass gets back to infinity again, it will of course have zero gravitational potential energy.

The only way to 'balance the books' (and so conserve energy) is to accept that when the mass rested on the Earth it had *negative* energy, and that it needed the same quantity of *positive* energy to reach infinity and stop there (with zero energy):

some positive + same amount of negative = zero energy
kinetic energy   potential energy

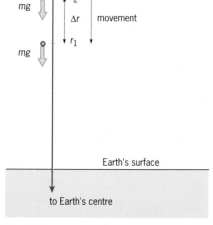

F (a) When objects fall towards the Earth, they lose potential energy and gain kinetic energy. Does this mean that, when an object moves in a gravitational field, the change in kinetic energy is always opposite in sign to the change in potential energy? (Look at the word equation in the text.)

(b) A mass falls from a point A to a point B. In doing so, it gains 50 kJ of kinetic energy. If the mass had 5000 kJ of gravitational potential energy (GPE) at point B, how much GPE did it have at A?

Finding a value for gravitational potential energy in a radial field

See question 7. ■

As in Fig 4.13(a), an object of mass m is at a point r_2 from the centre of the Earth. Then it is moved to a nearby point r_1, a small extra distance Δr closer to the Earth's centre.

For a small movement, the force does not vary within distance Δr. The object is, for all practical purposes, in a **uniform field**. So the change in gravitational potential energy would simply be $mg\Delta r$, as shown in the graph of Fig 4.13(b). The change in potential energy per unit mass is $g\Delta r$, and is represented by the area ABCD under the graph.

mg r_2
        Δr movement
mg r_1

Earth's surface

to Earth's centre

Fig 4.13(a) **An object near the Earth is moved a small distance, over which the gravitational field does not change**

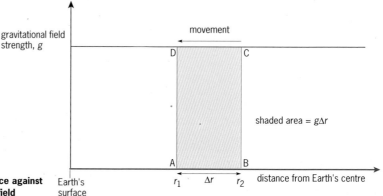

gravitational field strength, g

movement

D   C

shaded area = $g\Delta r$

A   B

Earth's surface

r_1 Δr r_2 distance from Earth's centre

Fig 4.13(b) **Graph of force against distance for a uniform field**

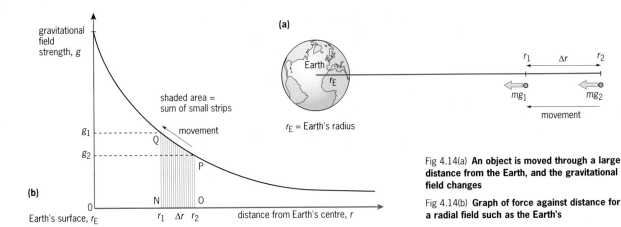

Fig 4.14(a) **An object is moved through a large distance from the Earth, and the gravitational field changes**

Fig 4.14(b) **Graph of force against distance for a radial field such as the Earth's**

Imagine a very much larger distance, on the scale of Fig 4.14(a). The curved graph of Fig 4.14(b) represents a field of varying strength, getting less as the distance from the Earth increases. So, as the object is moved from r_2 to r_1, it experiences an increasing gravitational force. The change in potential energy per unit mass is represented by the area NOPQ under the graph.

■ See question 8.

Calculating the change in gravitational potential energy

1. Graphically and numerically
We can find the value of the change in gravitational potential energy from the graph of Fig 4.14(b) by counting the squares in each small shaded strip. This is tedious! Instead, we can add them up numerically using a spreadsheet, which you can do in the assignment task set at the end of this chapter.

2. Algebraically, using calculus
We can find area PQRS in Fig 4.14(b) more easily using a simple formula for the result. We employ calculus to get the formula, recognising the fact that, for large distances, g depends upon r according to the inverse square law. The symbol E_p is used for gravitational potential energy.

The change in gravitational potential energy, ΔE_p, is given by using the definition of work (force × distance). The potential energy gained equals the work done in moving mass m a distance Δr between points at distances r_2 and r_1 from the centre of the Earth.

So:
$$\Delta E_p = -\frac{GMm}{r^2}\Delta r$$

As explained on page 64, the negative sign is there to balance the books and keep potential energy negative.

The equation may be integrated to give:

$$E_p = -GMm \int_{r_2}^{r_1} \frac{1}{r^2}\,\mathrm{d}r = GMm\left(\frac{1}{r_2} - \frac{1}{r_1}\right)$$

This gives:
$$E_p = \frac{GMm}{r_2} - \frac{GMm}{r_1}$$

This is the **potential energy difference formula**, where E_p is the difference in potential energy – the kinetic energy gained – by a mass m moving from a point at distance r_2 to a point at distance r_1 from the centre of the Earth.

6 GRAVITATIONAL POTENTIAL

The **gravitational potential** of an object at a point in space is a useful idea. It allows us to calculate the energy changes, or the work done, when a satellite moves to a different orbit. Gravitational potential can be compared to the concept of electrical potential, and is defined as follows:

Gravitational potential is the change in potential energy for a unit mass that moves from infinity to a point at less than infinity.

If we make $r_2 = $ infinity, $r_1 = r$ and $m = 1$ (kg), the potential energy difference formula above simplifies to:

$$-\frac{GM}{r}$$

For the field of the Earth, we can use this formula to calculate the potential at a point at a distance r from the centre of the Earth.

G The gravitational potential at the Earth's surface is found by inserting the following values in the formula $-GM/r$:

GM for Earth $= 4.0 \times 10^{14}$ N m² kg⁻¹

Radius of Earth, $r_E = 6.38 \times 10^6$ m

Check that this calculation results in a value of -6.3×10^7 J kg⁻¹.

EXAMPLE

The energy gained by an asteroid falling to Earth

Q Estimate a value for the kinetic energy of an asteroid colliding with the Earth.

A Assume that the asteroid has a diameter of 10 km and is made of rock of density 3000 kg per cubic metre. This gives it a mass of about 1.6×10^{15}kg.

Assume, too, that it falls from infinity, that is, from a place where the gravitational potential is zero. At the Earth's surface it is at a place with a gravitational potential of $-GM/r_E$. In other words, its potential energy per kilogram has been reduced by GM/r_E.

Loss in gravitational potential (energy per unit mass):

$$= -\frac{GM}{r_E} = -\frac{(6.7 \times 10^{-11}) \times (6.0 \times 10^{24})}{6.4 \times 10^6}$$

$$= -6.3 \times 10^7 \text{ J kg}^{-1}$$

This energy has of course been converted to kinetic energy. So the total kinetic energy of the asteroid is equal to the potential energy lost by its mass of 1.6×10^{15} kg:

Gain in KE = loss in GPE (gravitational potential × mass)

$$= 6.3 \times 10^7 \text{ (J kg}^{-1}) \times 1.6 \times 10^{15} \text{ (kg)}$$

$$= 1.0 \times 10^{23} \text{ J}$$

We can calculate its impact speed from the formula for kinetic energy:

$$E_K = \tfrac{1}{2}mv^2 = 1.0 \times 10^{23}$$

$$v^2 = \frac{2 \times E_K}{m}$$

$$v = \sqrt{\frac{2 \times 10^{23}}{1.6 \times 10^{15}}} = 11.2 \times 10^3 \text{ m s}^{-1}$$

So impact speed is 11.2 km s⁻¹.

A typical meteorite enters the Earth's atmosphere at about 20 km s⁻¹, almost double the speed calculated in the Example. This is because meteorites are already moving when they start falling towards the Earth, and the Earth is also moving. We have also ignored the gravitational effect of the Sun, which is very large.

7 ESCAPE SPEED

H Suggest some practical difficulties that might arise if a spacecraft is given the energy for its escape speed all at once as it leaves the ground.

Looking the other way round at the Example above, we can see that if an object is at rest on the Earth's surface, we would have to *give* it some kinetic energy before it could escape completely from the Earth. To reach infinity, where the Earth's gravitational field doesn't affect it, we would have to give it 6.3×10^7 J of kinetic energy for every kilogram of its mass. This would convert the negative gravitational potential it had on the Earth's surface (its GPE per unit mass), of value $-GM/r_E$, to zero.

You should now use the kinetic energy formula to check what is obvious, that each kilogram of an object has to be given the right amount of energy to reach a speed of 11.2 km s⁻¹ – in the right direction – for it to be able to escape from the Earth. This speed is called the **escape speed** (or escape velocity).

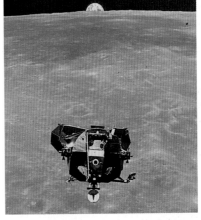

?

I A neutron star has a mass of 4×10^{30} kg and a radius of 10 km. Show that the escape speed from the star is about two-thirds the speed of light.

■ See question 9.

EXAMPLE

Q The Moon has a mass of 7.3×10^{22} kg and a radius of 1.7×10^6 m. What is the escape speed for an object on the Moon?

A The gravitational potential at the Moon's surface is:

$$-\frac{GM}{r_E} = -\frac{(6.7 \times 10^{-11}) \times (7.3 \times 10^{22})}{1.7 \times 10^6}$$

$$= -2.88 \times 10^6 \text{ kg}^{-1}$$

Any object has to be given an equivalent quantity of kinetic energy per unit mass to escape the Moon's gravitational pull.
So, kinetic energy per kilogram is:

$$\tfrac{1}{2}v^2 = 2.88 \times 10^6$$

$$v = \sqrt{2 \times 2.87 \times 10^6}$$

Escape speed $v = 2.4 \times 10^3$ m s⁻¹

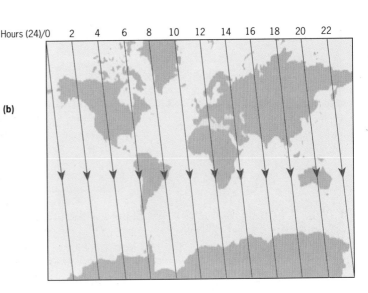

Fig 4.15 **The NASA lunar module leaving the Moon to dock with the command module**

---

8 MOVING IN AN ORBIT

The simplest orbit for an Earth satellite is circular, centred on the Earth's centre. A typical satellite used to monitor the Earth's surface has the **circumpolar orbit** shown in Fig 4.16(a). While the satellite orbits, the Earth spins underneath it. So, if a complete orbit takes 2 hours, the satellite will overfly the whole of the Earth in 12 orbits, as Fig 4.16(b) shows.

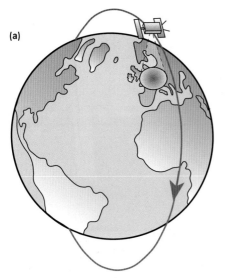

(a)

(b)

Hours (24)/0   2   4   6   8   10   12   14   16   18   20   22

Fig 4.16 **Satellites in circumpolar orbits can monitor the whole Earth as it spins beneath them**

EARTH SATELLITES PROVIDE REMOTE EYES AND EARS

THE FACT THAT EARTH satellites can be put into orbit fairly cheaply has revolutionised everyday communications and information systems. On our television screens, we can now see events unfold in the remotest parts of the world as they happen. All a news reporter needs is a small dish aerial and a portable power supply, and access to a **communications satellite.** These satellites are in a high, **geosynchronous** orbit above the equator.

Communications satellites also carry international telephone messages that are far clearer than those transmitted by the old radio or land-line systems. (Chapter 21, Communications, describes how information is sent as radio waves between satellites and ground stations, and explains the digital technology that allows a huge number of messages to be transmitted at the same time.)

Weather satellites provide the pictures of cloud systems we see on television weather reports. These satellites have polar orbits and are closer to the ground than geosynchronous satellites. The recording of land and sea temperatures, and the forecasting of climatic changes and trends in the greenhouse effect, have all become much more reliable, thanks to weather satellites. It was a British satellite collecting data over Antarctica that first provided evidence of a 'hole' developing in the ozone layer.

Links between satellite and ground computers allow us to know the positions of satellites very accurately. Similarly, satellites give us precise positions for objects on Earth. For many years, ships and aircraft have relied on **navigation satellites** to pinpoint their locations. On land, systems are being developed using navigation satellite data, which will tell the drivers of vehicles their exact locations, and the routes to take to avoid traffic jams.

Military satellites keep an eye on the movement of ships, troops and vehicles. They can detect when weapons are being fired, and can eavesdrop on conversations over a wide range of electronic communications systems.

The **infrared satellite sensors** which detect land and sea temperatures can also monitor the growth of crops – whether they are healthy or diseased, or lack water, for example. In the European Union, agricultural land is monitored to check that farmers who claim money for leaving land uncultivated ('set aside') are not in fact growing crops on it.

Satellites are sent into orbit either by rocket systems used once only, such as the European Ariane (see page 72), or by the re-usable Space Shuttle (USA). An increasing number of countries are putting satellites into orbit, for their own national uses such as weather monitoring, or on a commercial basis, hiring out the satellite facilities to research and communications organisations such as universities and television companies.

Fig 4.17(a) **Infrared satellite image of a Bedfordshire farm showing that different types of crops can be identified. This picture was taken from Landsat at an altitude of 705 km**

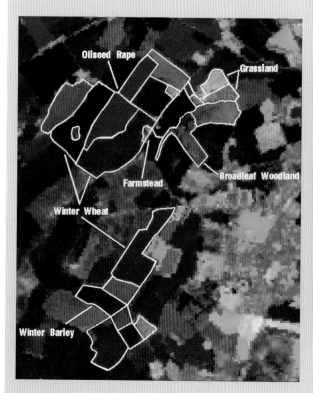

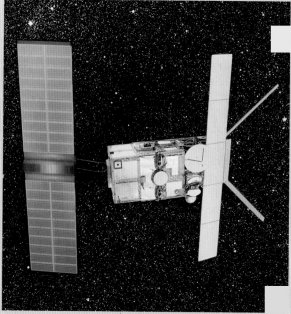

Fig 4.17(b) **European Remote Sensing satellite, ERS1, surveys the structure of the Earth's surface, watches shore lines and ocean currents (it can detect oil spills), monitors crop and vegetation growth and surveys and maps the polar ice caps**

Newton explained how, in theory, an object shot at great speed from the Earth's surface could become an orbiting satellite. He imagined a cannon on a mountain, firing cannonballs parallel to the Earth's surface at ever-increasing speeds. Each ball lands further from the cannon than the previous one, travelling in a curved path. Newton imagined that, eventually, the curved path of a ball would match the curvature of the Earth, and *then, the ball would be in orbit.*

> Remember, we say that an object is accelerating if its speed is changing, or its direction of motion is changing, or both.

Fig 4.18 **Isaac Newton's orbiting cannon ball.** In his book, *System of the World*, Newton wrote: 'The greater the velocity with which [a cannonball] is projected, the farther it goes before it falls to the Earth. We may therefore suppose the velocity to be so increased, that it would describe an arc of 1, 2, 5, 10, 100, 1000 miles before it arrived at the Earth, till at last, exceeding the limits of the Earth, it should pass into space without touching.'

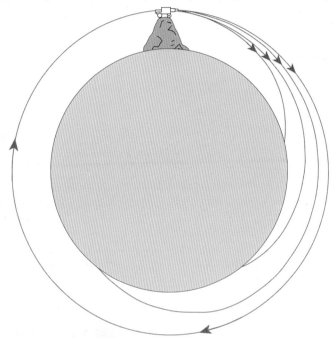

It is of course the force of gravity acting on a cannon ball that makes it fall back to Earth. A ball which remains in orbit also experiences this force. Its speed may be constant, but its velocity continuously changes – because the *direction* of the ball's motion is always changing. The force of gravity is responsible, making the orbiting ball accelerate towards the centre of the Earth – in free fall.

All orbiting satellites are in free fall. Their acceleration is equal to the value of *g* at that distance from the Earth, and this acceleration is directed towards the centre of the Earth.

Clearly, the speed of a satellite has to be just right to stay in a circular orbit. If it is too slow, the satellite falls to Earth like the 'unsuccessful' cannon balls in Fig 4.18. Too fast, and the satellite's orbital radius may increase, or its orbit may become an ellipse. Much too fast, and the satellite leaves Earth orbit completely.

> **J (a)** Suggest some practical difficulties that would arise in using Newton's cannon method for getting a satellite into orbit from the top of a high mountain. (Assume it does not then collide with anything.)
>
> **(b)** To what speed would the satellite have to be accelerated if fired horizontally?
> (Radius of Earth = 6.4×10^6 m)

Speed and the radius of the orbit

As shown in the box on the next page and in Fig 4.19, a satellite that moves in a *circular* orbit has an inward gravitational acceleration *g* and speed *v*, which are linked by the equation:

$$g = v^2/r$$

where *r* is the radius of the orbit.

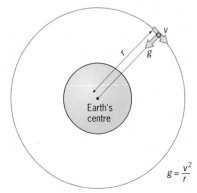

Fig 4.19 **Illustrating the quantities in the formula** $g = v^2/r$

Formula for the inward acceleration of an object moving in a circle

The formula developed here applies to any object moving in a circle – for example, the rim of a wheel, a ball whirling on the end of a string, a car going round a bend or a satellite in orbit.

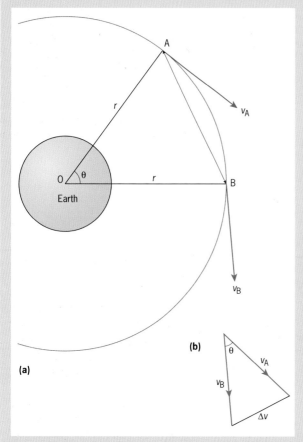

(a)

(b)

Fig 4.20 **(a) A satellite moving in a circle. (b) Triangle of vectors for the satellite**

Fig 4.20(a) shows two positions, A and B, of a satellite moving in a circle round the Earth at constant speed v. The radius line, r, moves through an angle, θ, measured in radians, in a time t, and the object moves a distance $r\theta$. The gravitational force always acts along a radius, inwards towards the centre of the Earth.

Fig 4.20(b) shows how the vector representing the *velocity* of the satellite has changed in time t. Since the speed of the satellite doesn't change, vectors v_A and v_B are of equal length.

Δv shows the change in velocity using a triangle of vectors such that:

$$v_B = v_A + \Delta v$$

In these two diagrams the value of t, and therefore of θ, are large. Now see Fig 4.21, in which θ is very small, so that Δv is also very small. This also means that the *time* involved is small.

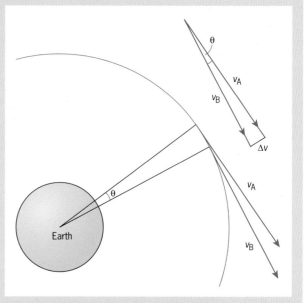

Fig 4.21 **Triangle of vectors with a smaller angle**

The acceleration of the body is given by

$$a = \Delta v/t \qquad [1]$$

Referring back to Fig 4.20(a), we can see that the object covers a distance $r\theta$ in time t at a speed v, so that

$$v = r\theta/t, \quad \text{hence } 1/t = v/r\theta \qquad [2]$$

Substituting for $1/t$ in equation 1 gives

$$a = v\Delta v/r\theta \qquad [3]$$

v is directed along a tangent to the circle, so is always at right angles to the radius line. (Note that for small θ, $\sin \theta = \tan \theta = \theta$ radians.)

When the radius turns through an angle θ so does v. So, with the usual small-angle approximation:

$$\Delta v = v\theta \text{ (See Fig 4.21)} \qquad [4]$$

Substituting for Δv in equation 3 gives:

$$a = v^2/r \qquad [5]$$

As angle θ gets smaller and smaller, the direction of Δv becomes closer and closer to being at right angles to v. This means it is in the direction of the radius and acting towards the centre of the circle. This is, of course, the direction of the force causing the acceleration.

We can therefore represent the force F required to keep a body of mass m moving in a circle, using the relation:

$$\text{force} = \text{mass} \times \text{acceleration}$$

as: $\qquad\qquad F = mv^2/r$

The speed of a satellite in a circular orbit

We saw in Fig 4.17 that a satellite in polar orbit sweeps over certain areas of the Earth. The areas depend on the number of its orbits in 24 hours, hence on its speed. The speed is set before launch, so that the orbital time matches the purpose of the satellite – whether it is a near-Earth weather monitoring satellite, or a distant communications satellite, for example.

The satellite's speed depends both on the radius of its orbit and on the strength of the gravitational field at that height. A typical height for a satellite in near-Earth orbit is 100 km. This gives an orbital radius of 6.47×10^6 m. The inward acceleration is the value of g at that height, which is 9.53 m s^{-2}.

Fig 4.22 **A meteorologist studies the image from a weather satellite, to provide reliable forecasts**

Thus, from equation 5 above, its speed $v = \sqrt{gr}$.

Hence, $v = 7.85$ km s^{-1}.

The circumference, $2\pi r$, of its orbit is 4.07×10^4 km. The satellite therefore completes an orbit in 86.3 minutes.

Now reread this section and work through the calculations for yourself.

Geostationary satellites

Communications satellites are put into **geostationary** or **geosynchronous** orbits, meaning that the satellites stay still relative to the Earth. As a result, ground aerials which send and receive data to and from a satellite need only point in one direction, rather than having to steer to follow the satellite.

To be geostationary, an orbit has to be directly above the equator, and the satellite must complete its orbit in exactly the time for the spot underneath to rotate through *its* orbit, namely 24 hours. In theory, a geostationary satellite 'keeps up with' the point on the rotating Earth. In practice, its position shifts a little, as the satellite's speed is affected by slight changes in the Earth's gravitational field. Small adjuster rockets are fired to regain the correct position.

Using equation 5 on the previous page, we can calculate the height at which a satellite must orbit in order to remain geostationary.

Let the radius of the orbit be r_s. The satellite must have an orbital speed v such that

$$2\pi r_s = vt = v \times 86\ 400 \quad (t \text{ is 24 hours} = 86\ 400 \text{ seconds})$$

So $\qquad\qquad\qquad v = 2\pi r_s/86\ 400 \qquad\qquad$ [6]

From equation 5, $\qquad\qquad v = \sqrt{gr_s} \qquad\qquad$ [7]

At distance r_s, $\qquad\qquad g = GM/r_s^2$.

Putting this value for g into equation 7,

$$v = \sqrt{GM/r_s}$$

Equating the expressions for v in equations 6 and 7, we have

$$2\pi r_s/86\ 400 = \sqrt{GM/r_s}$$

Inserting the values of constants G and M, we get a value for the radius, r_s, for a geostationary orbit of 4.23×10^7 m. This is over six times the radius of the Earth and, at that distance, the value of g is 0.223 N kg^{-1}.

K An Earth satellite has a circular polar orbit at a height of 400 km above the Earth's surface.

Calculate **(a)** its orbital speed, **(b)** the time it takes to complete an orbit, and **(c)** how many orbits it makes in a 24-hour period. (Use $g = 8.7$ m s^{-2} and Earth radius $= 6.4 \times 10^6$ m.)

L (a) Work out for yourself the values calculated for the geostationary satellite. Take $GM = 4.0 \times 10^{14}$ N m^2

(b) Use the data for the Moon given in the Example on page 67 to find the orbital radius of a lunostationary satellite.

Putting satellites into orbit

Fig 4.23(a) shows how a rocket such as the Ariane rocket lifts satellites in orbit. Fuel makes up most of the mass of the main rocket with its various stages, and much of the fuel is used to get the final stage into near-Earth orbit – about 100 km above the Earth's surface. The remaining fuel propels the comparatively small payload of one or more satellites into their final orbits.

Fig 4.23(a) **A satellite is launched into a geo-stationary orbit in several stages. (More than one satellite can be put into orbit in the same launch)**

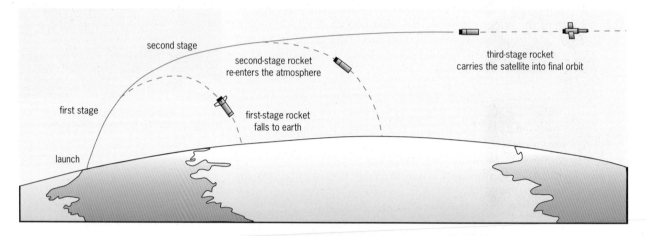

second stage

second-stage rocket re-enters the atmosphere

third-stage rocket carries the satellite into final orbit

first stage

first-stage rocket falls to earth

launch

Fig 4.23(b) **The Ariane rocket. It has either two or four booster rockets, depending on the mass of the satellites being put into orbit**

Fig 4.24 **The Olympus communications satellite, used for television and radio broadcasting**

Weightlessness

Weight is simply the force due to gravity acting on an object. We sense 'weight' when we lift something up, and we are aware of the weight of our own bodies as sensations in our muscles. Occasionally, we 'feel weightless', on a roller coaster or in a plane that suddenly loses height through air turbulence, because for a moment we are in free fall and our bodies do not need muscular support. Yet we still *have* weight, from the pull of Earth's gravity.

M An astronaut is in a spacelab orbiting Earth. She has the task of moving a metal box labelled '50 kg' from one side of the lab to the other.

(a) Suggest how she could check whether the box was empty or not.

(b) Assuming that the box and contents did in fact have a mass of 50 kg, outline and explain a procedure which would allow the box to be moved across the lab effectively and safely.

Fig 4.25 **Astronauts in Earth orbit feel 'weightless' because, like their spacecraft, they are in free fall. What indicates 'weightlessness' in the top picture?**

Astronauts in a spacecraft orbiting the Earth also experience 'weightlessness'. This is because, like the orbiting cannon ball on page 69, they are in continuous free fall towards the centre of the Earth, as explained above. Some people believe that this sensation is due to the absence of the force of gravity. This is clearly false. If there were no gravity force acting on the astronauts, they and their spacecraft would all fly off in a straight line – obeying Newton's laws of motion.

9 MOMENTUM – ROCKETS AND COLLISIONS

In any interaction involving forces, *momentum is always conserved* (see also pages 33–34 in Chapter 3 on momentum). A rocket uses this principle. It gains speed by ejecting material at high speed. Fig 4.26 shows what happens. The total momentum for the whole system during flight, rocket + fuel, is constant, and in fact zero.

Fig 4.26 **A rocket at rest, and 1 second after fuel ignition**

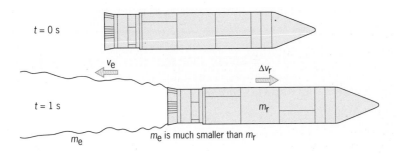

$t = 0$ s

v_e Δv_r

$t = 1$ s m_r

m_e m_e is much smaller than m_r

Initially, the rocket is at rest. Suppose that, in 1 second, the rocket ejects a mass m_e of hot gases at a velocity v_e. In this second, the ejected material carries momentum of $m_e v_e$ to the left. The rocket gains an equal quantity of momentum, moving to the right with a speed Δv_r. The system is isolated, meaning that no other forces or bodies are involved, so there is no net change in momentum. We can write:

$$m_e v_e + m_r \Delta v_r = 0$$

and the rocket gains in speed by $\Delta v_r = -(m_e/m_r)v_e$

EXAMPLE

Q A rocket starts firing and ejects 5.0 kg of gas per second at a speed v_e of 5.0×10^3 m s^{-1} relative to the rocket. The mass of the rocket is 4 tonnes.
a) What is the velocity of the rocket, v_r, after 1 second?
b) Will a spacecraft powered by a rocket engine move with a constant acceleration?

A

a) After 1 second, the rocket's mass has been reduced by 5 kg, but we will ignore this mass change (of about 0.1%) for the first second. Velocity is a vector, and we follow the convention that the direction of the ejected gas is *negative:*

$$\text{total momentum} = \frac{\text{momentum of}}{\text{rocket}} + \frac{\text{momentum of}}{\text{ejected gas}} = 0$$

$$(4 \times 10^3 \times v_r) - (5 \times 5 \times 10^3) = 0$$

So, after the first second, $v_r = 6.3$ m s^{-1}

b) No, the acceleration of a rocket-propelled spacecraft will not be constant. The rocket ejects gases at a steady rate, so that, each second, the gain in momentum of the rocket itself is constant over time. As its mass gets less, the gain in *velocity* per second has to increase in order to maintain the constant increase in momentum.

Suppose that in the short time Δt, the rocket ejects a mass of gas, Δm_e, at a velocity v_e relative to the rocket. At this time, the mass of the rocket is m_r. As a result, the rocket gains in velocity by Δv_r. We know that, in the short time Δt, by the momentum law,

$$\frac{\text{change in momentum}}{\text{of rocket}} = \frac{-\text{ change in momentum of}}{\text{gases ejected}}$$

$$\begin{array}{c}(\text{mass } m_r \text{ of} \\ \text{rocket})\end{array} \times \begin{array}{c}(\text{change in velocity} \\ \Delta v_r)\end{array} = \begin{array}{c}(\text{mass } \Delta m_e \text{of} \\ \text{ejected fuel})\end{array} \times \begin{array}{c}(\text{velocity } -v_e \text{ of} \\ \text{ejected fuel})\end{array}$$

$$m_r \Delta v_r = -\Delta m_e v_e$$

Thus, the *change* in velocity of the rocket in the short time Δt is:

$$\Delta v_r = -\Delta m_e v_e/m_r$$

This equation confirms that the change in velocity – the acceleration – will get bigger as time goes on, because m_r is decreasing.

■ This may be a good time to tackle the spreadsheet Assignment about rocket motion on page 79. This is a numerical, non-calculus method for investigating rocket propulsion.

It is clear from the Example that the mass of the rocket, and consequently its acceleration, is changing constantly and uniformly. The mathematics of rocket motion is given more fully in the box on the next page.

Rocket propulsion

In any situations of rocket propulsion, we have to consider the total system, that is, the rocket plus the ejected fuel gases. At any given time, the rocket is moving with a velocity v_r, as shown in Fig 4.27, where v_r is measured by a stationary observer so that we record what happens from a stationary frame of reference.

The rocket is about to eject a mass m_e of gas at a velocity v_e. It simplifies the mathematics to take the rocket's mass as $m_r + m_e$ before the gas is ejected. As gas is ejected, the velocity of the rocket increases by Δv_r. (As usual, we use the symbol Δ to signify a small fraction of the quantity involved.)

Relative to the rocket, the mass m_e of fuel is ejected at a speed $-v_e$. The rocket starts to move, and the observer sees the gas moving at a velocity $v_r - v_e$ relative to the stationary frame of reference. Note that v_e is likely to be much larger than v_r.

At any two points in time, and applying the conservation principle:

total momentum before = total momentum after

Therefore:

$$\begin{matrix}\text{momentum of} \\ \text{(rocket + some fuel)}\end{matrix} = \begin{matrix}\text{momentum of rocket +} \\ \text{momentum of ejected fuel}\end{matrix}$$

$$(m_r + m_e)v_r = m_r(v_r + \Delta v_r) + m_e(v_r - v_e)$$

This expression simplifies to:

$$m_r \Delta v_r - m_e v_e = 0$$
$$\Delta v_r = v_e \left(\frac{m_e}{m_r}\right)$$

But m_e is also the *mass loss of the rocket*, and can be replaced by the quantity $-\Delta m_r$ (taking care with the sign) to give the more general result:

$$\Delta v_r = -v_e \left(\frac{\Delta m_r}{m_r}\right) \qquad [1]$$

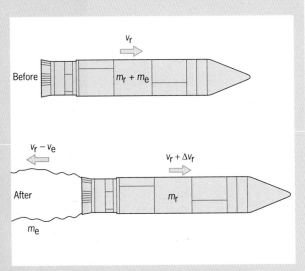

Fig 4.27 **A rocket at rest, and 1 second after fuel ignition**

This is the same formula as the one obtained by the simple treatment given in the Example on page 74.

Changes over time

Now take the longer view, in which the mass of the rocket changes from an initial value m_i to a final value of m_f, and changes its velocity from v_i to v_f as it does so. Integrating equation 1 above between these limits:

$$\int_{v_i}^{v_f} dv = v_e \int_{m_i}^{m_f} \frac{dm_r}{m_r}$$

This expression integrates to:

$$v_f - v_i = v_e \ln \frac{m_i}{m_f}$$

Where ln is the logarithm to base e.

Thus, given the initial mass of a rocket travelling at any speed, we can calculate its new velocity when it has changed its mass to a new value.

Rocket and jet engines

Rocket engines are designed to work in the vacuum of space. Their energy source is a fuel–oxygen mixture which is usually as a liquid.

▇ See questions 10, 11 and 12.

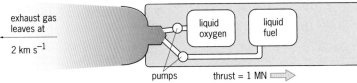

combustion chamber: fuel reaches it at the rate of 500 kg s⁻¹

exhaust gas leaves at

2 km s⁻¹

liquid oxygen

liquid fuel

pumps   thrust = 1 MN ▭▷

Fig 4.28 **A simplified rocket engine**

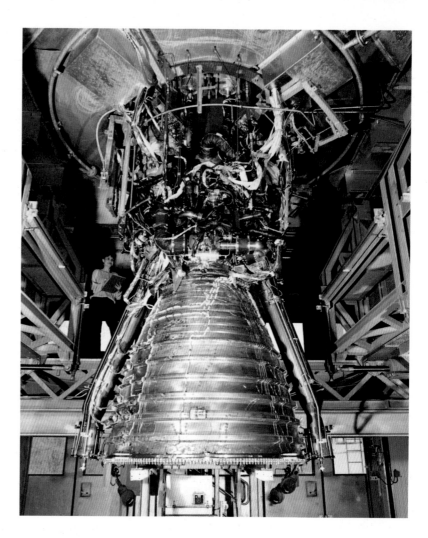

Fig 4.29 **The engine of the Ariane rocket**

For the US Apollo missions to the Moon (1969–1972), the Saturn rocket of the first launch stage was larger than the tower of Big Ben. It was powered by five huge engines, the most powerful ever made, and the second stage rocket by five smaller engines. Each of the first-stage engines was 5 metres long with a mass of 8.4 tonnes, generating a thrust of 6.7 meganewtons. After the first stage was completed, and all the fuel used up, five rocket engines took over for the second stage. The kerosene–liquid oxygen fuel mixture generated a thrust of 1.14 meganewtons, and its exhaust velocity in space was 2500 metres per second.

A third stage then put the spacecraft bound for the Moon into Earth orbit. The spacecraft included its three-man astronaut team, their life support systems, the Lunar Lander and the rockets and fuel to get the craft to the Moon and back.

Thrust

A rocket engine using hydrogen and oxygen as fuel can eject hot gases at a rate of 500 kg s^{-1} and a speed of 3 500 m s^{-1}. This is a change of momentum of 500×3500 kg m s^{-1} *per second*, giving a *rate of change of momentum* of 1.75×10^6 kg m s^{-2}. From Newton's second law, this is equivalent to a force of 1.75×10^6 newtons. In rocketry, this is the **thrust** produced by the engine.

A jet engine uses the same principle as a rocket engine, but can only work in the atmosphere because it uses fuel that needs oxygen in the air to burn.

A meganewton, MN, is 10^6 newtons.

See question 13. ■

SUMMARY

After studying this chapter, you should be able to do the following.

■ Explain energy changes using the idea of work, $W = Fd$, and use the definition of work to make calculations involving force, $F = mg$, and distance in gravitational examples.

■ Derive the formulae for a change in gravitational potential energy, $\Delta E = mg\Delta h$, and for kinetic energy, $\frac{1}{2}mv^2$, and use these ideas and formulae to explain, and make calculations for, bodies moving in uniform gravitational fields.

■ Use the ideas and formulae for a change in gravitational potential energy, $GMm(1/r_1 - 1/r_2)$, and kinetic energy to explain, and make calculations for, bodies moving in radial gravitational fields.

■ Have a deeper understanding of the principle of the conservation of energy as it applies to gravitational situations.

■ Draw and interpret force–distance graphs for both uniform and radial gravitational fields.

■ Use the idea of gravitational potential and its formula, $-GM/r$, and distinguish between gravitational potential and gravitational potential energy.

■ Derive the formula for escape speed and use it to make calculations.

■ Explain how gravitational force is linked to satellite motion in circular orbits and use the formula: $g = v^2/r$.

■ Explain the terms circumpolar orbit, geostationary orbit, 'weightlessness', free fall.

■ Use the concept of momentum to explain the action of a rocket and a jet engine, and calculate changes of speed and of thrust for a rocket in simple situations.

QUESTIONS

1 There are two ways of lifting the barrel in the diagram from the cellar to the pavement (Fig 4.Q1):
a) rolling it up the ramp, **b)** lifting it up directly, using a rope. Write a paragraph about which method is likely to involve doing more work. (You need to use the definition of work, considering the forces of friction and gravity and the distances moved.)

Fig 4.Q1

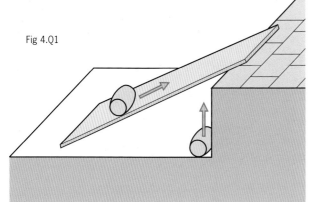

2 This question is about work and the conservation of energy. You push a loaded supermarket trolley with a total mass of 25 kg horizontally, in a straight line, for a distance of 50 metres with a steady force of 20 N.

a) How much work do you do on the trolley?

b) Doing work on a system gives it energy. What systems (objects) are likely to gain energy as you push the trolley?

c) Assuming that there are no losses of energy due to friction, estimate the speed of the trolley at the 50 m point.

d) Why is it a good idea for car parks at supermarkets to be made as level as possible?

3
a) You do some hard work using a hand saw to cut through a log of firewood. What has gained energy as the result of your work? Justify your answer.

b) What facts would you have to know to estimate whether or not the energy you transferred in sawing through the log is greater or less than the energy that would be released when the log was burned?

4 A tennis ball is dropped from a height of 5 m and rebounds to a height of 3.2 m.

a) At what speed did it hit the ground?

b) How much energy was lost in the bounce?

5 Can:

a) the gravitational potential energy,

b) the kinetic energy of an object, ever have negative values? Explain your answers.

6 A student is reading a physics textbook of mass 2.5 kg while travelling in a train at a speed of 125 km h^{-1}. What is the kinetic energy of the book relative to **a)** the student, **b)** a passenger in another train as it passes in the opposite direction at a speed of 125 km h^{-1}?

7 The force needed to stretch a rubber band varies as shown in the diagram. Estimate the energy stored in the rubber band when stretched as illustrated.

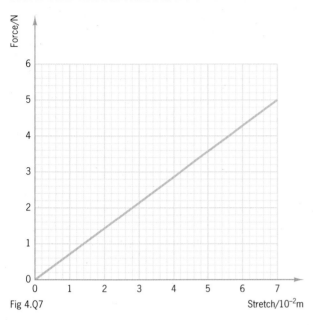

Fig 4.Q7

8 Use the data in the table for Mars and its gravitational field to answer the following questions.

Mass of Mars = 6.40×10^{23} kg
Radius = 3.39×10^6 m
Period of rotation = 8.862×10^4 s

a) Plot a graph of the data in the table to show how the gravitational field of Mars varies with distance.

Table 4.Q8 **g = gravitational field strength in N kg⁻¹, R = distance from the centre of Mars in metres × 10⁶**

g	3.82	2.74	1.76	1.22	0.90	0.69	0.54	0.43	0.20	0.10
R	3.39	4.00	5.00	6.00	7.00	8.00	9.00	10.00	15.00	20.00

b) Estimate from your graph the change in potential energy per unit mass for a Mars Lander spacecraft as it moves between 20×10^6 m and the surface of Mars.
   (i) Use your answer to estimate the gravitational potential at the surface of Mars.
   (ii) Check this answer by using the formula for gravitational potential

$$V = -GM/r.$$

c) Calculate the escape velocity for Mars.
   (i) What must be the period of a Martian communications satellite that remains above a Martian landing site in an 'areostationary' orbit? (Ares is Greek for Mars.)
   (ii) What would be the radius of the satellite's orbit?

9 A neutron star has a mass of 10^{30} kg and a radius of 10^4 m.

a) Assuming that Newton's laws still apply, and that G is still 6.7×10^{-11} N m–2 kg⁻², show that the escape velocity from the star is about 1.2×10^8 m s⁻¹.

b) A black hole is an object for which the escape velocity is at least equal to the speed of light, 3×10^8 m s⁻¹. To what maximum size would this neutron star have to shrink to make it a black hole?

10 A spacecraft is accelerated by a rocket engine that provides a steady thrust in a region of space where gravitational fields are negligible. Sketch the graph you would expect of speed against time for the spacecraft. Explain your graph.

11 A spacecraft has a mass of 400 tonnes and is powered by rockets which emit a mass of fuel at a speed of 5000 m s⁻¹ relative to the rocket.

a) What is the increase in speed of the spacecraft in a time in which it uses up **(i)** 100 tonnes of fuel, **(ii)** 300 tonnes of fuel?

b) The spacecraft as described, of mass 400 tonnes, leaves the outskirts of the Solar System with a speed of 12.0×10^3 m s⁻¹ in the direction of the star Alpha Centauri. It then uses 300 tonnes of fuel to gain more speed. How long will it take to reach the Alpha Centauri system? (Distance to Alpha Centauri from the Solar System 1.37×10^8 light-seconds.)

12 A rocket engine can accelerate a space station in the vacuum of space, where a jet engine would be useless. Explain these facts.

13 A rocket engine ejects 200 kg of hot gases at a speed of 3000 m s⁻¹. The gas is ejected at a rate of 5 kg s⁻¹. What thrust (force) would this engine exert on the rocket?

Assignment

MOVING IN A GRAVITATIONAL FIELD

Motion in a uniform gravitational field

This assignment simulates the motion of an object propelled vertically upwards from the ground, in a uniform gravitational field such as exists close to the surface of the Earth.

As for the Assignment in Chapter 2, the details of using spreadsheets vary slightly from one program to another. If you find difficulty with the instructions below, ask someone who is familiar with the program you are using.

1 Begin by setting up the spreadsheet, or re-using the one described for the Assignment in Chapter 2, page 28. The symbols used in this Assignment are those shown in Fig 4.A1. Note $D \equiv \Delta$. Use an initial speed V1 of 50 m s^{-1} and a value for acceleration of free fall g of -9.8 m s^{-2}.

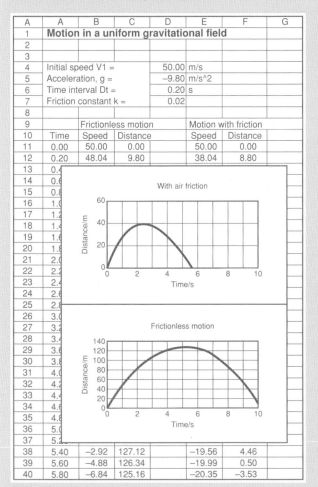

A	A	B	C	D	E	F	G
1	**Motion in a uniform gravitational field**						
2							
3							
4	Initial speed V1 =			50.00	m/s		
5	Acceleration, g =			−9.80	m/s^2		
6	Time interval Dt =			0.20	s		
7	Friction constant k =			0.02			
8							
9		Frictionless motion			Motion with friction		
10	Time	Speed	Distance		Speed	Distance	
11	0.00	50.00	0.00		50.00	0.00	
12	0.20	48.04	9.80		38.04	8.80	
13	0.4						
14	0.6						
15	0.8						
16	1.0						
17	1.2						
18	1.4						
19	1.6						
20	1.8						
21	2.0						
22	2.2						
23	2.4						
24	2.6						
25	2.8						
26	3.0						
27	3.2						
28	3.4						
29	3.6						
30	3.8						
31	4.0						
32	4.2						
33	4.4						
34	4.6						
35	4.8						
36	5.0						
37	5.2						
38	5.40	−2.92	127.12		−19.56	4.46	
39	5.60	−4.88	126.34		−19.99	0.50	
40	5.80	−6.84	125.16		−20.35	−3.53	

Fig 4.A1

a) Produce a graph of distance travelled plotted against time. Explain the shape of the graph.

b) Then find out **(i)** how long it takes for the object to return to the ground, **(ii)** how high the object went.

c) Investigate what happens when you change the initial speed.

Introducing air friction

2 Objects travelling through the air experience air resistance. In general, the air resistance depends on the shape of the object, but is usually modelled as a resistive force whose magnitude is proportional to the square of the speed of the object at any instant. In simulating the effect of air resistance, we have to remember that its force always opposes the object's motion, so the force changes sign when the object changes from upward to downward motion.

a) Add a line to the spreadsheet as shown in Fig 4.A1, labelled 'Friction constant k = '. Begin by inserting the value for k of 0.02 in the appropriate cell. Head two more columns 'Speed' and 'Distance' and, if you wish, tidy up the presentation to conform to Fig 4.A1.

The algorithm for distance is identical to that for frictionless motion.

b) To allow for the effect due to friction on speed, we must insert a deceleration effect which opposes the speed at any time, that is, it *adds to* gravitational deceleration going up but *subtracts from* acceleration going down. But as the resistive force is proportional to (speed)^2, and since a negative number squared is always positive, we cannot automatically rely on a change in the direction of speed to produce a change in the direction of friction. To allow for this, we use the trick that the resistive acceleration is given by

$$k \times \text{actual (signed) value of speed} \times \text{unsigned speed value}$$

The unsigned value is the absolute value of speed, which the program returns using the 'formula' @ABS( ). Thus, the algorithm for the 'Speed' column, with air friction, is

$$=E11+\$D\$6*(\$D\$5+\$D\$7*(-E11)*@ABS(E11))$$

Insert this formula into cell E12, remembering also to put the initial speed value (as $D$4) in cell E11.

c) Using the cell locations illustrated in Fig 4.A1, copy down the formulae by columns, and plot a graph of distance against time for motion with air friction. Typical graphs are shown in Fig 4.A1.

d) As an extra, plot the speeds against time as well.

e) Think about these points:
Do negative distance values have any meaning?
What values of friction constant k give realistic results (if at all!) for **(i)** a tennis ball, **(ii)** a table tennis ball, **(iii)** a badminton shuttle?

f) How might you use the simulation to help you measure k for any given object?

Rocket motion

The mathematics of rocket motion is quite complex. It generally involves a knowledge of integral calculus, though many real-life situations either cannot be solved by calculus, or require equations which are very difficult to solve.

By contrast, spreadsheets can produce solutions to quite complex analytical problems, to as high a degree of accuracy as needed. They can be based on simple mathematics and the underlying physics.

3 In rocket motion, the problem is to find what happens when an accelerating object is losing mass. The basic algorithms involved (and shown in Fig 4.2A) are as follows, where D is used to indicate a change in a value.

mass loss in time Dt: DM = flow rate per second × Dt

rocket mass M = M − DM

the momentum law gives M.DV = DM.Ve

hence DV = DM/M × Ve

speed V = V + DV

a) Begin by using the values of the constants shown in Fig 4.A2, for: emission speed (jet velocity), the time calculating interval Dt, the initial velocity of the rocket, the jet velocity Ve, the initial mass of the rocket, and the rate of mass emission per second (flow/s). For a realistic simulation, the total time of rocket firing should be about 600 s.
b) Plot graphs of speed against time and rocket mass against time.
c) Investigate the effect of increasing the flow rate, the initial mass of the rocket and the speed of ejection.

Gravitational potential energy

This assignment is to illustrate how gravitational potential varies with distance from the centre of a massive sphere, the Earth. It assumes Newton's inverse square law for gravity. The energy transferred to potential energy in moving an object a distance Dr along a line of force in a gravitational field can be calculated from the work done as:

force × distance.

For a unit mass, the force is g N. g starts at a value of 9.827 N kg^{-1}.

4

a) Use a spreadsheet to model the variation of g with distance:

Put Dr = 500 000 m

Use GM = 4×10^{14} J m kg^{-1}

Initial value of r = Earth radius = 6.38×10^6 m

Set up a column to calculate g from r = 6.38×10^6 m to about 5r (3.19×10^7 m) in steps of Dr = 5×10^5 m, using the inverse square law:

$$g = GM/r^2$$

Set up another column to calculate the energy transfer at each step:

DU = Dr × (mean value of g in the step interval)

In another column, calculate the increasing value of gravitational potential U (potential energy per unit mass) by successively summing the newly calculated values of DU.

b) To find how U varies with distance, plot U against r.
c) What happens when you plot U against 1/r?

Fig 4.A2

A	A	B	C	D	E	F	G	H
1	SIMULATION OF ROCKET PROPULSION							
2	Interval Dt =		1.5	second	flow/s =	60	kg/s	
3	Initial velocity =		0	m/s	DM =	90	kg	
4	Jet vel, Ve =		4000	m/s				
5	Rocket mass, M =		40000	kg				
6			Algorithms		V=V+DV	M=M-DM	DV=Ve x DM/M	
7								
8	t	M	DV	V				
9	0	40000	9.0000	0.0000				
10	1.5	39910	9.0203	9.0000				
11	3	39820	9.0407	18.0203				
12	4.5	39730	9.0612	27.0610				
13	6	39640	9.0817	36.1221				
14	7.5	39550	9.1024	45.2039				

FORCE AND MOTION IN A NEWTONIAN UNIVERSE

This Chapter Map brings together the central ideas from Chapters **2 Motion in space and time**, **3 Newton's universe** and **4 Newton's laws at work**. It shows the main concepts and equations which you have studied in these chapters, and how they are interlinked and depend on one another. The size of the type indicates the level of the concepts: those in the largest type are the main concepts, and these are subdivided into associated ideas.

The map is also a guide to the key information you need to know and understand for the syllabus you are following, and you can use it as a revision checklist.

Looking through the chart, you will be able to identify the concepts and the mathematics you feel confident about. You can also see which concepts you may need to study further, and the connected ideas which will support you in working on the areas you find difficult.

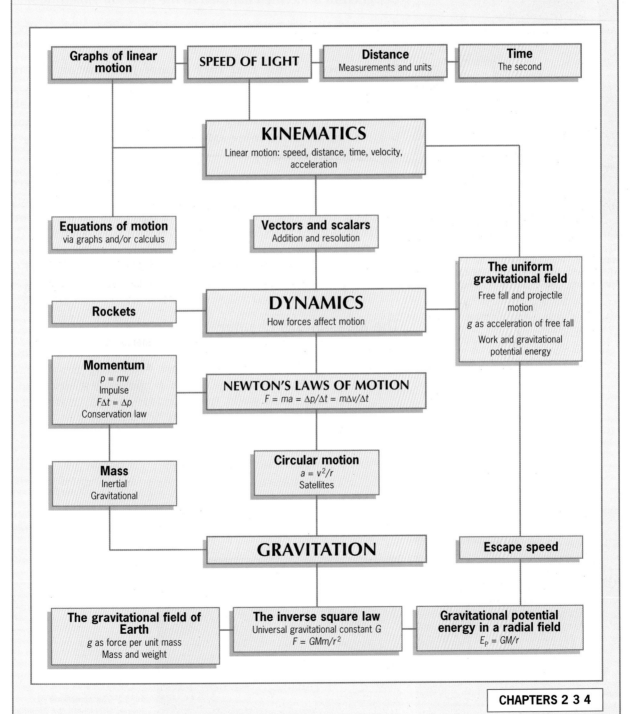

CHAPTERS 2 3 4

5 Materials and forces: Structures and microstructures

The Tokyo Bay Bridge. With a span of 570 metres, it is much shorter than the 1990 metres of the world's largest suspension bridge, between Honshu and Shikoku Islands of Japan

THE TACOMA NARROWS BRIDGE in Washington State was opened in 1940. After only four months it collapsed – not as a result of an earthquake, but simply because it was badly designed. It proved not strong enough for the loads it was designed to bear when exposed to the forces of violent winds.

The need to understand how materials behave when they are subjected to forces is literally a matter of life and death.

But engineers know that they cannot just rely on theory when they are building these massive structures. They have learnt from the mistakes of the past and now test their designs on models accurate to the smallest detail before they begin to build. Then more safety tests are made at all stages during construction.

These tests can be quite spectacular. When the Japanese were building their new suspension bridge in Tokyo Bay, they dropped 100-tonne hammers from a height of 60 metres on to the middle of the span. The impacts sent waves backwards and forwards along the bridge, and made it twist and turn like a corkscrew. Only because the structure survived this battering did they decide that the bridge was safe enough.

The ideas in this chapter

When we consider structures, we often think of huge engineering projects, such as the Channel Tunnel connecting Britain and France, or the Humber Bridge which is the longest suspension bridge anywhere in the world, spanning 1410 metres. Or we might think of the world's tallest building, the Sears Building in Chicago (520 metres high, including its TV antennae).

The principles behind the strength and stability of these impressive structures, and the forces affecting them, also apply on a much smaller scale to cups and saucers, fishing rods and tennis rackets, and even animal skeletons and plants.

In this chapter, we shall be looking at these forces, and at the composition and properties of the materials in the objects, noting

Fig 5.1 **The same kind of forces affect both the Channel Tunnel and a cup and saucer**

that all materials are built up from individual atoms. We shall see that the properties of different materials arise from the ways their atoms are arranged on the molecular scale, and how the atoms and molecules pack together to form a **microstructure**. We shall explain the differences by creating simple models to describe the materials and then compare these models with reality.

1 STATICS – OR HOW STRUCTURES STAY UP

Large structures, such as bridges and buildings, are stationary – or move only slightly in the wind. They are subject to very large forces due to their own mass, and also due to the ground and any other point of attachment.

The forces acting on these structures must be in equilibrium – that is, their *resolved components* in any direction must balance and cancel. If the forces on an object do not balance out, then, by Newton's second law, some acceleration takes place.

> **The study of forces in equilibrium – of forces acting on a stationary object – is called statics.**

Forces are also in equilibrium if a body is moving uniformly – that is, in a straight line without acceleration.

In order to see how forces act on an object, we must examine all the forces which are acting on it. You have seen that force is a **vector**: it has both magnitude and direction. To add two forces, we draw a parallelogram of forces as in Fig 5.2. The **resultant** force is represented by the diagonal of the parallelogram. The size and direction of the resultant force can be obtained either by scale drawing or by calculation.

Newton's second law:
force = mass × acceleration

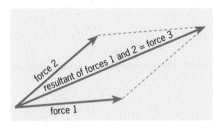

Fig 5.2 **A parallelogram of two forces acting on a body. One side of the parallelogram represents the magnitude and direction of one force. An adjacent side represents the magnitude and direction of the second force. The diagonal between them is the resultant**

Calculating the resultant of forces acting at right angles

It is much easier to calculate the resultant force when the two forces act on the object at right angles. The Example shows two ways of working out the resultant force.

EXAMPLE

1 Scale drawing

In Fig 5.3, the forces acting at a point on an object are drawn to scale (1 cm ≡ 2 N).

a) The two forces are drawn acting at a single point.
b) The parallelogram (rectangle in this case) is completed.
c) The diagonal of the parallelogram is drawn starting from the point at which the forces act. This diagonal represents the resultant force.

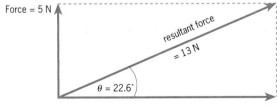

Force = 5 N

resultant force
= 13 N

θ = 22.6°

Force = 12 N

2 Calculation

Alternatively, the magnitude and direction of the resultant force can be calculated by using trigonometry. Using Pythagoras' theorem:

$$(\text{resultant force})^2 = (5)^2 + (12)^2 = 169$$
(values in newtons)

$$\text{resultant force} = \sqrt{169} = 13\ \text{N}$$

The angle θ between the resultant and the 12 N force has tangent equal to 5/12:

$$\theta = \tan^{-1}(5/12)\quad \text{(inverse tan on calculator)}$$
$$= 22.6°$$

Fig 5.3 **Obtaining the resultant of two forces acting on a body at right angles by scale drawing (here, $\times\frac{1}{2}$)**

Coplanar forces acting at a point

Suppose there are three forces acting at a point. We can demonstrate the effect using three spring gauges linked by strings to a common point and with the opposite ends of the gauges anchored to three fixed points (Fig 5.4). The force in gauge 3 must be equal and opposite to the resultant force of gauges 1 and 2, as represented by the diagonal of the parallelogram. Similar parallelograms can be drawn for the pair of gauges 1 and 3 and also for the pair of gauges 2 and 3. These resultants in their turn will be in equilibrium with the forces on gauges 2 and 1 respectively.

Fig 5.4 **Three forces in equilibrium**

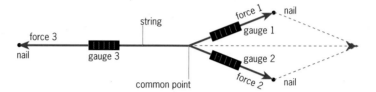

Suppose several forces all act in the same plane and we wish to find the overall resultant force. The forces are **coplanar** – they can all be represented by vectors drawn to scale on a flat sheet of paper. To find the resultant force, we can take the forces in pairs and draw parallelograms of forces for each pair. We then get a resultant for each pair. We can pair up these resultants and repeat the process until we are left with one overall resultant force. With a lot of forces to deal with, this method could be time-consuming and tedious.

A better method is to take all the forces acting on a body and resolve each force into components that are at right angles. To do this graphically, set up x and y axes on graph paper and draw in the forces to scale. In Fig 5.5:

- Draw to scale forces F_1, F_2 and F_3 acting in the correct directions: Fig 5.5(a).

- Draw resolved components F_{1x} and F_{1y} along the x and y axes. Similarly, draw the components for F_2 and F_3.

- Add forces as vectors along the x and y axes, taking into account plus and minus directions: Fig 5.5(b).

- Measure the lines to get the magnitude and direction of the overall force.

Fig 5.5 **Resolving forces by drawing to scale and adding forces**

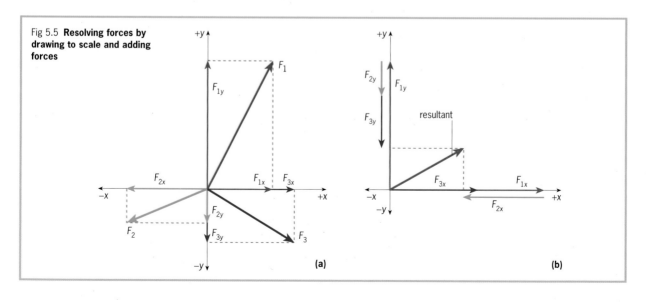

We could extend to forces in three dimensions, and resolve in the *x*, *y* and *z* directions. But even for three-dimensional bridges and buildings, we often consider two-dimensional cross-sections. So we won't complicate our analysis here with the third dimension.

Returning to forces acting in a single *x*–*y* plane, provided these forces act through a single point in a body and the body is not being accelerated, the resultant force must be zero. We can draw the forces acting on the body as vectors and we can join up the vectors, tail-to-head, one after another. *For a body experiencing zero resultant force*, the vectors will form the sides of a *closed* polygon (Fig 5.6).

If the vectors do not close up, then the line F_x that closes the polygon represents both the magnitude and direction of the extra force needed for equilibrium (Fig 5.7). Without this additional vector force, the resultant force *R* on the body will be equal and opposite in direction, and will produce an acceleration on the body.

We can say that:

Any point object acted on by a number of forces will only be in equilibrium (ie stationary or moving with uniform velocity) if the force vectors form a closed polygon.

We can now see how this works in some practical situations.

Radio mast

The mast is held in place by two cables (support wires) as shown in Fig 5.8. These give tension forces pulling down on the mast. If we split the tension force in each cable into vertical and horizontal components, we can see from the force diagram that the horizontal components of these forces will oppose each other. The vertical components are opposed by an upward force F_{ground} exerted by the ground up though the mast.

The parallelogram of forces is shown for the three forces (these are F_{cable}, F_{cable} and F_{ground}) and beneath is the closed polygon of forces (a triangle as there are three forces). In practice, a mast is held in place by two more cables acting at right angles to the plane of the paper. The force F_{ground} must oppose the vertical components of the forces exerted by all four cables.

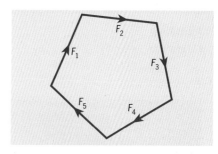

Fig 5.6 **Five forces forming a closed polygon. If these five forces act at a single point within a body, they will be in equilibrium. They will produce no acceleration of the body**

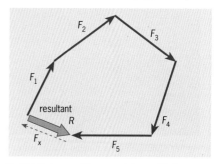

Fig 5.7 **Polygon of forces not in equilibrium. Resultant force *R* is in the opposite direction to the force needed to close the polygon (ie to achieve equilibrium)**

A **point object** is an object which is so small that we can consider that all the forces acting on it meet at a point. Often the forces acting on a larger body all meet at the centre of mass, so that we can consider the larger body to be equivalent to a single point.

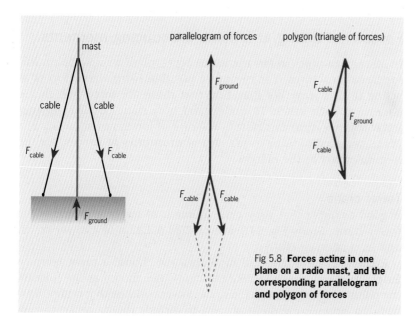

Fig 5.8 **Forces acting in one plane on a radio mast, and the corresponding parallelogram and polygon of forces**

EXAMPLE
Cable car

Fig 5.9 **Calculating the tension (force) within the suspension cable of a cable car**

See question 1. ■

?

C In practice, the cable in the Example on the right will probably make a smaller angle with the horizontal. Why is it not possible for the cable car to be slung from a cable which is exactly horizontal?

Q A cable car of total load 15 kN is suspended beneath a cable (Fig 5.9). The tension (force) T in the cable is uniform, and the cable makes equal angles of 20° to the horizontal in either direction. What is the magnitude of the tension T?

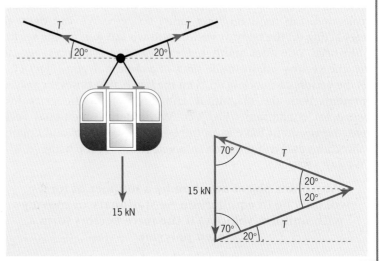

A We can find the value of T by the **scale drawing** in Fig 5.9. Draw a vertical line representing the downward force of 15 kN. From the centre of the line, draw a horizontal (perpendicular) line. Then draw lines from each end of the vertical line at 70° (90° − 20°) to meet the horizontal, to intersect and make an isosceles triangle. The length of each of these equal sides gives the magnitude of T.

It is fairly simple to find T by **calculation**. By sketching the triangle described above, we see that:

$$\sin 20° = (7.5 \text{ kN})/T$$

or: $\qquad T = (7.5 \text{ kN})/\sin 20° = 22 \text{ kN}$

Notice that the tension in the wire is greater than the downward force exerted by the cable car.

✔

Moments of forces:
When a body is in equilibrium, the sum of anticlockwise moments about the pivot (balance point) is equal to the sum of clockwise moments. For example:
Where M and m are in newtons and x and y are in metres:

$$M_1x_1 + M_2x_2 = m_1y_1 + m_2y_2 \text{ (Nm)}$$

If an object is in **static equilibrium**, no resultant force acts on it and it remains at rest or moving at constant velocity.

Forces not acting at a point: introducing moments

In all the examples so far, the forces act through a single point. But for large structures it may not be possible to treat the forces as acting in this way. We may find that there are forces which are parallel to each other. To understand how static equilibrium (net force acting = zero) can exist in the case of forces not all acting at a point, we need to bring in the idea of **moments of forces**, see the Reminder box in the margin.

Tower crane

A good example that shows the moment of forces in action is the tower crane. Study Fig 5.10, a simplified diagram of a tower crane, and its caption carefully.

Remembering our use of moments, we choose a pivot point or **fulcrum** about which to take moments. We choose as the fulcrum the point where the jib is fixed to the tower. (Although it is important from an engineering point of view to include the tie cables to take

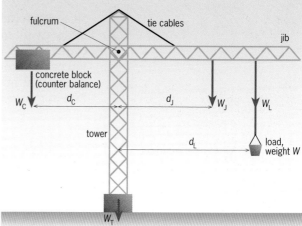

much of the strain on the jib, we will ignore these in the calculation of moments.) We can assume that they are arranged symmetrically, so that the tensions in each are equal. From the principle of moments, we obtain:

$$\underset{\substack{\text{concrete}\\\text{block}}}{W_C \times d_C} = \underset{\text{jib}}{(W_J \times d_J)} + \underset{\text{load}}{(W_L \times d_L)}$$

For the tower crane to remain in equilibrium and not topple, the turning moment of the block on the left of the equation must balance the sum of the turning moments of the jib and the load on the right.

Watch a tower crane in operation and you will notice that the load is moved into the required position not only by turning the crane but by moving the load either towards or away from the fulcrum. This moves the load away from the equilibrium position we have calculated.

Now we see the need for the tie cables. As the load moves away from the equilibrium position, the tie cables take up unequal tensions so that the turning moments remain balanced. The forces in the tie cables must be balanced by forces exerted by the crane tower. The concrete block counterbalance can move, but it cannot always be precisely positioned to balance the load. In addition, there will be some distortion (extension and bending) of the crane parts to 'take up' the forces. A full calculation is complicated. In practice, a tower crane is tested for stability with appropriate loads before being put to use.

Suspension bridge

An engineer designing a suspension bridge, like the one in Fig 5.11, takes account of forces acting at points within the structure and the turning moments of the forces. The engineer considers the **forces of tension** within the steel wires supporting the roadway, and the **forces of compression** within the reinforced concrete towers, and then chooses materials that will withstand the huge forces exerted on the bridge.

Fig 5.10 **A tower crane, showing the main forces involved. The effect of gravity on the mass of the tower will produce a downward force W_T which acts directly through the tower base.**

Next, the horizontal jib extends outwards from the tower on two opposite sides. But note that the jib extends further outwards on the side used for lifting the load. So the centre of gravity of the jib lies a distance d_J outwards on this side of the tower, and there is a force downwards of W_J due to the weight of the jib. Tie cables are provided to help support the jib.

A load W is being lifted at a distance d along the jib from the centre of the tower. A concrete block acting as a counterbalance is positioned at distance d_C on the opposite side, creating a downward force W_C

?

D A tower crane is used to lift a load of 1500 kg. The concrete counterbalance distance d_C of 12 m from the crane's fulcrum is of mass 2000 kg. The jib of the crane creates a downward force W_J of 4000 N at a distance d_J of 3 m from the fulcrum. At what distance from the fulcrum must the load be lifted for the tower crane to be in equilibrium?

E Write a (non-mathematical) description of the forces acting on different parts of the bridge in Fig 5.11. Identify the parts under tension and under compression. Consider how static equilibrium is achieved. For equilibrium, forces acting at a point must balance. Turning moments about any point on the bridge must also balance. (Consider only the forces acting in the plane of the paper.)

Fig 5.11 **Structure of a suspension bridge**

Steel can withstand large tension forces, whereas concrete is suitable only for compression forces. An engineer would not suspend a bridge with concrete rods, though inserting prestressed steel into concrete does enable the resulting composite material to withstand forces of tension as well as compression. It is clear, then, that to choose the best materials for a particular application, we have to find out more about the properties of the materials.

2 HOOKE'S LAW AND SPRINGS

Nearly all the components of a bridge that we looked at in the last section are of rigid substances – in other words, solids. They retain their shape when forces are applied to them. To understand why a solid retains its shape, we look at the interactions of its individual atoms or molecules. A stretching force, or tension, applied to the solid tends to pull the atoms or molecules apart. Usually, the stretching is very small because of the huge forces holding atoms close together. In the same way, compressing a solid pushes the atoms closer together, but then they start to repel each other with great force.

So when we apply a force to such a solid material, we find that it distorts in some way. If the force is one of tension, the material extends. This extension is proportional to the applied force. The proportionality was first found by Robert Hooke (1635–1703) for materials shaped in the form of a spring. It was because such springs had been developed that Hooke was able to discover the law. The same effect is found when the materials are shaped as wires or rods, although the extension is much smaller.

Let's start by considering springs. When a force is applied to a spring, the coils move apart and the spring is extended. Hooke stated his law for a spring as follows:

Extension is proportional to the applied force that causes it.

Fig 5.12 shows the relationship between force and length for a typical steel spring. To apply a force to the spring, you fix the spring at one end and suspend a sequence of masses from the other. Then take measurements of the length of the spring as you change the masses, altering the applied force. Usually, we plot the quantity that we are varying along the *x* axis and what we measure up the *y* axis. But not here. In this case, the quantity we vary, the force, is plotted on the *y* axis, and the length we measure is on the *x* axis, as in Fig. 5.12.

The important result is that the graph is a straight line, *provided we keep the applied force reasonably small*. The straight line shows that the spring extends by equal amounts for equal masses added. If the applied force is too great, the spring will not return to its original shape when the force is removed. We say that the spring has gone past its **elastic limit** and Hooke's law no longer applies.

We can replot the graph in the form of force against **extension** of the spring, meaning a *change* in its length (also called **deformation** which means change of shape). The new graph is shown in Fig 5.13. For the linear region of the graph, we have:

$$\text{gradient} = \frac{\text{force}}{\text{extension}}$$

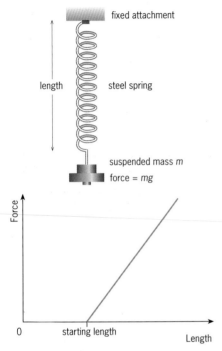

Fig 5.12 **Force–length graph for a steel spring obeying Hooke's law.**

Springs are made from materials other than steel. Each material usually has a straight-line region of the graph, but its *slope* will be different for different materials. For example, springs made from some materials need much larger forces to produce the equivalent extension, so the gradient of the graph for these materials will be steep

?

F (a) Suppose you obtain graphs of force against length for a number of *different* springs made from the *same* metal. The springs may have different cross-sectional areas and different starting lengths. Would you expect the gradients and intercepts to be equal for all the springs? Explain your answer.

(b) Read the caption to Fig 5.12, sketch the diagram and then add a line for a spring (of the same starting length) made of a material that needs smaller forces than the steel spring to produce the equivalent extension.

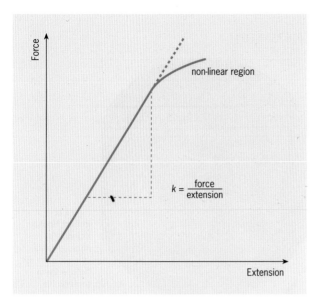

Fig 5.13 **Force–extension graph for a spring**

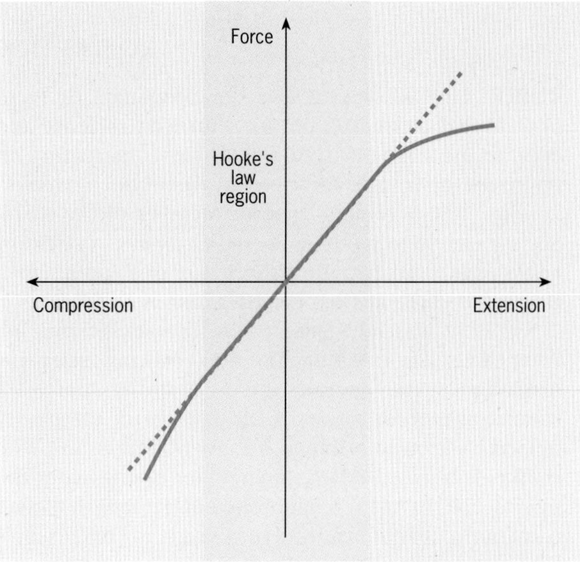

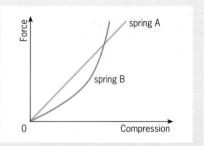

Fig 5.14 **Compression and extension of a spring. The Hooke's law region is a continuous straight line, but note that the positive and negative regions are not symmetrical – the extent of the linear region is not the same, nor are the curvatures of the non-linear regions the same**

We can relate force F and extension x by a constant k to give:

$$F = kx$$

where k is the gradient of the linear part of the graph. k is the **spring constant**, or **force constant**. It has the units of force divided by length, that is, $N\ m^{-1}$.

Springs can be compressed as well as extended. You can think of compression as a negative force causing a negative distortion, and show the *reduction* of length negatively along the x axis. The whole picture for both extension and compression of a spring is shown in Fig 5.14 and described in the caption. The slope of the Hooke's law region of the graph is the same for extension and compression.

'Non-linear' springs

Springs are usually designed to be as elastic as possible – that is, we want the relation between force and extension to be linear over as wide a range of extension and compression as possible. But some springs are designed to become harder and harder to compress by increasing forces. Such a spring can be used to stabilise vibrating machinery. In these springs, instead of compression being proportional to load (applied force) as for spring A in Fig 5.15, compression gets less with an increase in load, as shown by spring B.

Coil spring

The coil spring is shown in Fig 5.16. Coil springs consist of wire wound in a helix. When a coil spring is stretched, the spring extends as each coil twists.

Hooke's law applies to the linear deformation (the change of shape in one dimension) of a spring. But it also applies, as we shall see later, to linear deformation of objects that are rods or other shapes. In addition, the law applies to other types of deformation such as twisting – **torsional deformation** – as long as the applied load is small.

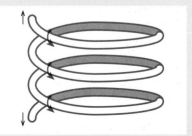

Fig 5.15 **Spring A obeys Hooke's Law. Spring B is non-linear – it gets harder to compress as the force increases**

Fig 5.16 **Stretching a coil spring**

ROBERT HOOKE (1635–1703)

ROBERT HOOKE was the son of a clergyman. He went to Oxford as a chorister but soon turned to science and became the assistant to Robert Boyle, constructing an improved air pump for him. In 1660 he moved to London, and became a founder member of the Royal Society (set up to promote the development of scientific knowledge) and became its 'curator of experiments'. He proposed the law named after him in 1678.

Not only was he ingenious in his experiments, he was clever at thinking up and making scientific instruments (he devised the first spring-controlled balance wheel for a watch). Hooke's work on gravity helped Newton to arrive at his law of gravitation, and Hooke is recognised in biology for work with the microscope. In 1665, he published *Micrographia*, a book containing many beautiful drawings of microscopic observations and also details of his work on optics.

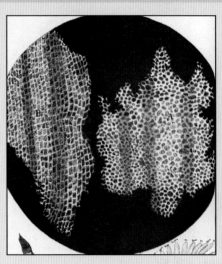

Fig 5.17
Robert Hooke's drawing in *Micrographia* of cork cells as he saw them through his microscope

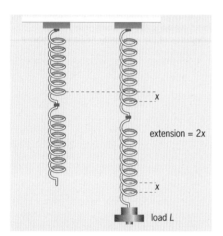

Fig 5.18 **Springs in series. Each spring has the same tension and extension. The total extension is 2x**

See question 2. ■

?

G A load is suspended from several identical springs in series. If another spring is added in the series and the load remains the same, what happens to the extension of each spring?

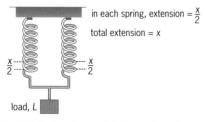

load, L

Fig 5.19 **Springs in parallel. The springs share the load, so the tension and extension in *each* spring are halved compared with the series case. The total extension is x**

Springs in series and parallel

Springs can be combined to carry a single load. The effective force constant k of the system of springs will depend on how these springs are arranged. In Fig 5.18, there are two identical springs attached end to end (in series):

● Each spring has its own force constant k, so force F equals kx.

● If a load L is suspended from the lower spring, a force F acts on both springs.

● Using $x = F/k$, we get an extension x in each spring.

● The total extension is $2x$.

So the force constant k_s of the system of **springs in series** is:

$$k_s = \frac{\text{force}}{\text{total extension}} = \frac{F}{2x} = \frac{kx}{2x} = \frac{k}{2}$$

Fig 5.19 shows the same two springs supporting the same load, but now the springs are in parallel:

● The force F exerted by the load L is shared between the springs so that each has a force of tension $F/2$.

● Extension for each spring is given by the force of tension divided by its spring constant, that is, $F/2$ divided by k. Hence the extension is $F/2k$.

● This extension $F/2k$ is the extension for the parallel system of two springs.

As we might expect, the extension is found to be half of the extension if the load was suspended from just one of the springs. The force constant k_p of the system of **springs in parallel** is:

$$k_p = \frac{\text{force}}{\text{extension}}$$

If we insert $F/2k$ for the extension which results from applying a force F, we get:

$$k_p = \frac{F}{F/2k} = 2k$$

Measuring with springs

Fig 5.20 **A spring balance (left) and a newton meter (right) in use**

Springs are used in many force-measuring instruments such as spring balances and newton meters. An object's weight is proportional to its mass, so the scale on a spring balance can be graduated in mass units. Newton meters are graduated in force units. Only the sizes of the graduations and the labelling of the scales differ in the two instruments, even though they are designed to measure different quantities. Both instruments are calibrated using standard forces – usually the gravitational force exerted on standard masses.

3 FORCES THAT DEFORM

As we have seen already, materials do not have to be made into springs to be deformed by forces. Most objects distort slightly when a force is applied to them. The applied forces may act in different directions and on objects of any shape. Hooke's law applies within certain limits to most everyday materials such as rubber, glass, wood and metals. But for some materials such as clay and plasticine the law does not apply at all because forces easily change their shape.

Fig 5.21 shows common effects of forces acting on a body. These effects are:

- **Tension** – forces act outwards in opposite directions and tend to lengthen the body.

- **Compression** – forces act inwards in opposite directions and tend to shorten the body.

- **Shear** – forces act in opposite directions along parallel faces, producing a tendency for parallel sections of the body to slide.

- **Torsion** – a type of shearing which twists the body lengthwise.

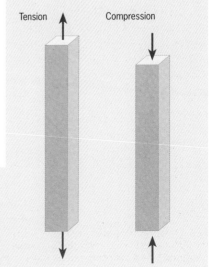

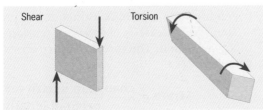

Fig 5.21 **Forces acting on bodies to produce tension, compression, shear and torsion**

Fig 5.22 **A concrete road bridge and an iron rail bridge over the Thames. Engineers take all the forces and their effects into account when designing and choosing materials for bridges**

In a loaded structure, there may be several effects at the same time. The response of an object to a particular force depends on its size, its shape and the material it is made from. A strong pillar may be either a thin column of a strong material or a thicker column of a weaker material. Differences in the properties of stone and iron account for the differences in design between the two bridges shown in Fig 5.22, for example.

It is not just the stiffness of the stone and iron that affects how they react to forces. Another important factor is the density of the material. Bridges have to support not only the traffic passing across them but also their own weight, and this may be many times greater than the weight of the traffic.

Stress and strain

Which is stronger – a bridge girder or a cup? They cannot be compared directly since they are made of different materials and also have very different sizes and shapes. Before we look at the properties of objects, we should compare the properties of the materials they are made of. To do this we use measurements called **stress** and **strain**.

Stress is a relationship between the applied force and the area over which it acts:

$$\text{Stress = applied force per unit area} = \frac{\text{force}}{\text{area}} = \frac{F}{A}$$

The units of stress are $N\ m^{-2}$. Notice that these are the same units as those for pressure.

In each example shown in Fig 5.23, a force is applied over a surface area. So in each case a stress is applied to the body. The shape of the body changes as a result of applying the stress. To measure this change, we compare the size or shape of the body after the forces are applied with the size or shape before. **Strain** is the measurement which does this. Precisely how we define strain depends on the type of deformation we produce in the body.

The simplest case is of forces of tension applied to opposite ends of a wire or rod. The forces produce an extension and the strain is given by:

$$\text{Strain} = \frac{\text{change in length (extension)}}{\text{original length}}$$

As this is a ratio (of two lengths), strain has no units – we say it is *dimensionless*. We can express strain as a percentage change in length by multiplying the strain ratio by 100.

The same rules of stress and strain apply when a body has forces of compression applied to it.

Now we can see what happens if we apply the same force to two wires of different cross-sections, as in Fig 5.23(a). Although the two forces are the same, there is a smaller *stress* applied to the wire of larger cross-section, so it extends by a smaller amount. To extend this thicker wire by the same amount as the thinner wire, we must apply a larger force, as Fig 5.23(b) shows: for equal extensions, the forces need to be in the same ratio as the cross-sectional areas, and then the stress is the same for each wire.

We see from this why engineers working with structural materials find the ideas of stress and strain more useful in comparing the effects of a force on different-sized pieces of a material than merely looking at extensions using different forces.

Strain may occur through shearing or twisting an object as well as extension or compression. These different measures of strain are shown in Fig 5.24.

EXAMPLE

Q The ropes of a child's swing are 3 m long. They extend by 30 mm when a child sits on the swing. What is the strain in the ropes?

A
$$\text{strain} = \frac{\text{extension}}{\text{original length}} = \frac{x}{L}$$

$$= \frac{30 \times 10^{-3}}{3} = 0.01$$

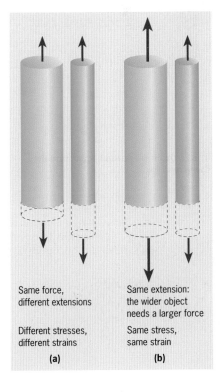

Fig 5.23 **Strain: the area of cross-section affects the relationship between force and extension**

Same force, different extensions

Different stresses, different strains

(a)

Same extension: the wider object needs a larger force

Same stress, same strain

(b)

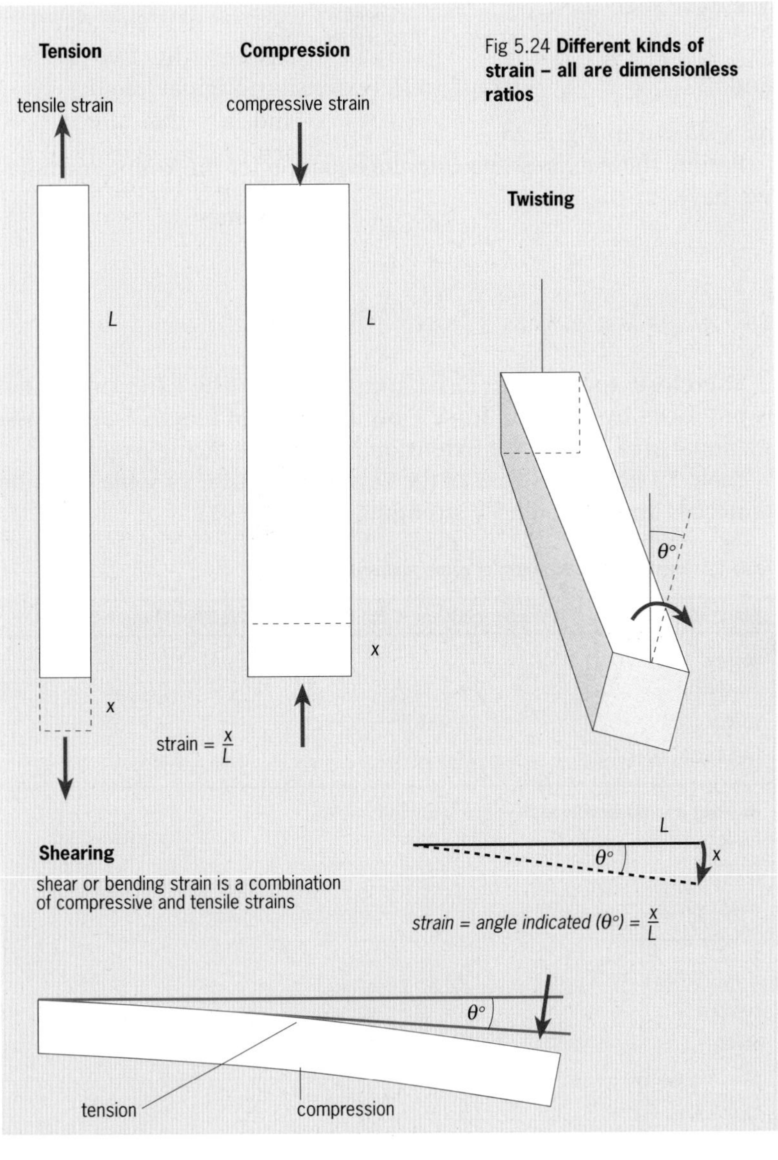

Tension

tensile strain

L

x

Compression

compressive strain

L

x

$\text{strain} = \frac{x}{L}$

Fig 5.24 **Different kinds of strain – all are dimensionless ratios**

Twisting

$\theta°$

Shearing

shear or bending strain is a combination of compressive and tensile strains

L

$\theta°$

x

$\text{strain} = \text{angle indicated } (\theta°) = \frac{x}{L}$

$\theta°$

tension

compression

The Young modulus is also referred to as Young's modulus.

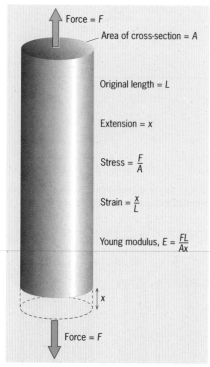

Fig 5.25 **Quantities needed to measure the Young modulus**

Force = F

Area of cross-section = A

Original length = L

Extension = x

Stress = $\frac{F}{A}$

Strain = $\frac{x}{L}$

Young modulus, $E = \frac{FL}{Ax}$

x

Force = F

See questions 3 and 4. ■

See question 7. ■

The Young modulus

We have seen that for springs subjected to small forces the ratio of force to extension is a constant:

$$F/x = k.$$

Similarly, we can say for many materials:

The ratio of stress to strain is a constant.

This is the same as saying that many solid materials as well as springs obey Hooke's law. So, for materials that obey the law, engineers can predict the effect of applying tensile forces, whatever the size or shape of the object made from the material. Tensile forces are longitudinal forces which tend to stretch the material and pull it apart. The ratio is called the **Young modulus** of the material:

$$\textbf{Young modulus, } E = \frac{\textbf{tensile stress}}{\textbf{tensile strain}}$$

Strain has no units, so the Young modulus has the same units as stress, namely N m^{-2}.

We have seen that:

$$\text{stress} = F/A \text{ (force/area)}$$

and:

$$\text{strain} = x/L \text{ (extension/original length)}$$

as is shown in Fig. 5.25.

Putting these expressions into the equation for the Young modulus, we have:

$$E = \frac{F}{A} \div \frac{x}{L}$$

$$= \frac{FL}{Ax} \text{ (in N m}^{-2}\text{)}$$

Therefore, to measure E, we must measure the extension x that is produced by a tensile force F on an object of length L and cross-sectional area A (see the Extension box on the facing page).

Table 5.1 gives the values of the Young modulus E for some common materials and their tensile strengths (see page 97).

Table 5.1 **Elasticity and strength of some materials**

Material	Young modulus, E/N m^{-2}	Tensile strength/N m^{-2}
Metals		
aluminium	7.0×10^{10}	7×10^{7}
copper	11×10^{10}	1.4×10^{8}
high tensile steel	21×10^{10}	1.5×10^{9}
Building and household materials (approximate values)		
rubber	7×10^{6}	3×10^{7}
brick	7×10^{9}	5×10^{6}
wood (spruce):		
along the grain	1.3×10^{10}	1×10^{8}
across the grain	3×10^{6}	
concrete	1.7×10^{10}	4×10^{8}
bone	2.1×10^{10}	1.4×10^{8}
glass (eg window)	7×10^{10}	$3–7 \times 10^{7}$
carbon fibre	7.5×10^{11}	2×10^{9}

?

H Suggest reasons why wood is stronger in one direction than another.

Measuring the Young modulus for a wire

The standard way to measure the Young modulus for a wire is shown in Fig 5.26(a). The sample wire is suspended from the ceiling and is stretched by adding masses to the lower end.

Because the amount of stretch is small, the extension is usually measured with a vernier scale. A reference wire of the same material as the sample wire holds the main scale of the vernier, usually calibrated in millimetres. The vernier attachment is fixed to the sample wire.

A mass is suspended from each wire to keep them taut. The position of the vernier on the main scale is noted. Then, extra mass added to the sample wire produces an increased downward force on the wire and the vernier scale moves downwards relative to the main scale. The new position of the vernier scale is noted and the extension of the wire calculated. Further masses are added to increase the downward force, and the new values of the extension are noted.

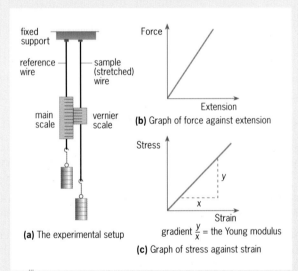

(a) The experimental setup

(b) Graph of force against extension

(c) Graph of stress against strain

gradient $\frac{y}{x}$ = the Young modulus

Fig 5.26 **Experiment to measure the Young modulus**

Graph 5.26(b), of force against extension, will be linear as long as the wire remains within its elastic region. Values of strain (extension/original length) can be calculated by measuring the length of the wire at the start of the experiment.

A micrometer screw gauge measures the diameter of the wire, so its area of cross-section is found. Graph 5.26(c), of stress against strain, can then be drawn. This graph is also linear. Its slope gives a direct value for the Young modulus for the wire.

Because the reference wire is made of the same material as the sample wire, any increase or decrease of the laboratory temperature will produce equal expansion or contraction in both wires.

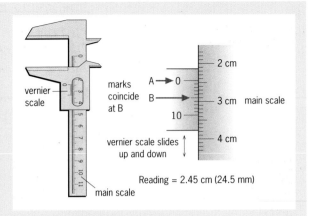

Reading = 2.45 cm (24.5 mm)

Fig 5.27 **Reading the vernier scale**

Using the vernier scale

The vernier scale, Fig 5.27, measures small distances. It has *two* scales: the zero position on the *vernier scale* gives the reading from the *main scale* to within 1 mm. The nearest lower-value millimetre reading is taken, so in Fig 5.27 it is 2.4 cm (24 mm).

The vernier scale now increases the precision of the value to ±0.1 mm. The correct value is shown by the mark *on the vernier scale* which lines up best with any mark on the main scale. In the diagram, the fifth mark (arrowed B) lines up with a mark on the main scale. This gives a correct overall reading of 2.45 cm (24.5 mm). If no one mark quite lines up and the setting lies between two marks, you can make an estimate to ±0.05 mm.

Using the micrometer screw gauge

This is used to measure thicknesses of materials accurately. The main scale gives a reading to 1 mm as with a vernier scale. The screw inside the gauge moves by 1 mm during one turn, that is, it has a pitch of 1 mm. The barrel of the screw is divided into 100 divisions, and you make a reading to 0.01 mm by noting the division which is in line with the scale line drawn on the main shaft. The reading in Fig 5.28 is 8.92 mm. Care is needed: some screw gauges have a pitch of 0.5 mm with the barrel divided into 50 divisions. In these gauges, the main scale is used first to read to 0.5 mm.

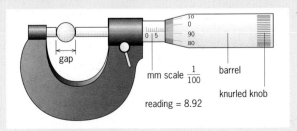

mm scale $\frac{1}{100}$

reading = 8.92

Fig 5.28 **Reading a micrometer screw gauge**

EXAMPLE

Q A lift is supported by a steel cable of diameter 2.5 cm. The maximum length of the lift cable when the lift is at the ground floor is 36 m. The Young modulus for steel is 2.1×10^{11} N m^{-2}. By how much does the cable stretch when six people of total mass 420 kg enter the lift at the ground floor?

A First, we use the equations for the Young modulus and for stress:

$$\text{Young modulus} = \frac{\text{stress}}{\text{strain}} \qquad \text{stress} = \frac{\text{force}}{\text{area}}$$

So:

$$\text{strain} = \frac{\text{force}}{\text{area} \times E} = \frac{F}{AE}$$

Since force = mg:

$$\text{strain} = mg/AE$$

where m is the total mass of the 6 people, and A is the area of cross-section of the cable.

Now use the equation for strain:

$$\text{increase in length} = \text{original length} \times \text{strain}$$
$$= \text{original length} \times mg/AE \qquad (A = \pi r^2)$$
$$= \frac{36 \times 420 \times 9.8}{\pi \times (1.25 \times 10^{-2})^2 \times 2.1 \times 10^{-2}}$$
$$= 1.6 \times 10^{-3} \text{ m}$$

The cable extends by 1.6 mm.

?

I A steel wire has a diameter of 0.57 mm and is 1.5 m long. What tension is required to produce an extension of 1.5 mm in the wire?

Here are some useful points to note about the Young modulus:

- Materials which are difficult to stretch have large values for the Young modulus. It takes a large stress to produce a small effect (strain).
- The values of E in Table 5.1 apply only if the *internal structure* of the object is not changed in any way (see below).
- The Young modulus for a material is only constant over the limited range of strain when the material is obeying Hooke's law – when extension (or distortion) is proportional to the applied force.

Stress–strain curves

Many materials obey Hooke's law only for small loads. Also, different materials have different stress–strain curves. Steels with a wide range of properties are made by varying the composition, particularly the amount of carbon which is added. Figs 5.29 and 5.30 show the stress–strain curves for two very different steels.

The steel wires of Figs 5.29 and 5.30 both show a linear region AB where stress is proportional to strain. If the load is removed from either specimen anywhere along AB, the wire will return along the path BA to its original length A.

Beyond B, the force needed to extend either steel is no longer proportional to the extension. In this region, each specimen increases in length by a larger amount for the same increase in force compared with the linear region. What is more, if the load is removed at, say, point C, the extension (and therefore the strain) does not return to

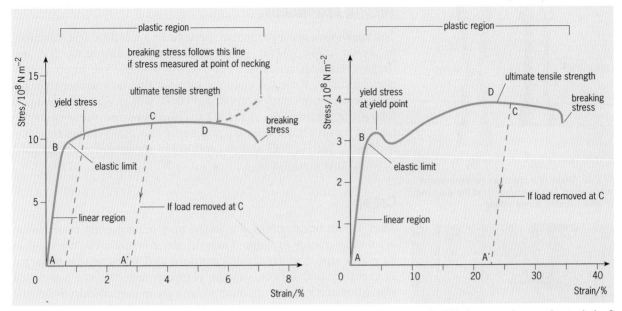

Fig 5.29 **Stress–strain curve for steel wire 1.**

Fig 5.30 **Stress–strain curve for steel wire 2**

zero but stays at A'. The steel wire has a permanent extension because of its inelasticity. In the curves, we can see what happens as either wire is unloaded. The stress–strain curve follows CA'. This line is parallel to the original elastic line BA. When a material behaves like this and fails to return to its original length when the load is removed, we say that **plastic strain** has occurred.

The yield point

Beyond the elastic limit, the stress–strain curves of Figs 5.29 and 5.30 for the two steels show different shapes. Curve 5.30 shows a peak at the point when the stress (the force required to go on extending the sample) falls very rapidly. This is the **yield point** and the stress at this point is called the **yield stress**.

Steel wire 1 also yields, but the exact yield point is not clear. In this case, we choose an arbitrary value of plastic strain: 0.5% is commonly used, meaning that AA' equals 0.5%. Follow the line parallel to AB upwards from the 0.5% point on graph 5.29 and you meet the point of the curve marked as the yield stress point.

Tensile strength

For both specimens there is a maximum stress, different for the two samples. This maximum stress is called the **ultimate tensile strength**. The force needed to produce further extension then actually gets less. If we just leave the load on the specimen it will go on extending until it breaks. Knowledge about such possible failure is important for large-scale structures.

The tensile strength of steel is especially important because steel is used so much in construction. But notice in Table 5.1 (page 94) the value for bone – it is a factor of 10 less than steel, yet higher than the value for many materials. The skeletons of humans and other creatures are a balance between lightness and capacity to withstand the stresses of body weight and movement. Notice also that the value for carbon fibre is especially large, which is why it is used in rackets, fishing rods and high performance bicycles.

?

J A single steel wire is used to lift a cradle of mass 50 kg containing **(a)** a woman and **(b)** an elephant. The type of steel used is that in Fig 5.29. Assume that, for safety, the stress in the steel wire should not exceed $0.1 \times$ the ultimate tensile strength of the steel. Calculate the minimum diameter of wire required in each case. (Take the mass of the woman as 55 kg and of the elephant as 4000 kg.)

Fig 5.31 **Chris Boardman, 1992 Olympic Games gold medal winner on the Team Lotus racing bicycle**

See questions 5 and 7.

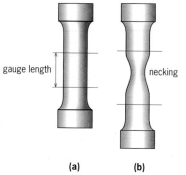

gauge length

necking

(a)   **(b)**

Fig 5.32 **Shape of a specimen for commercial stress–strain measurements** (a) **at the start and** (b) **after necking during testing**

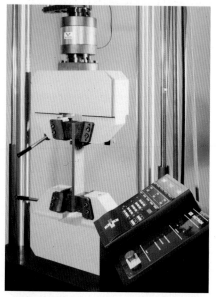

Fig 5.33 **A servo-hydraulic machine stress-testing a metal sample.**

Commercial machines for making standard test measurements stretch samples by applying a variable load. The apparatus instantly detects any increase or decrease of force due to change in length. In this way, the equipment is able to identify the parts of the curve in Fig 5.29 **where the required stress is getting less**

See questions 8 and 9. ■

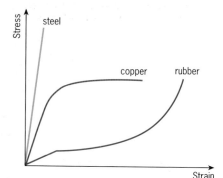

Fig 5.34 **Stress–strain graphs for steel, copper and rubber (not to scale)**

Testing specimens

Test specimens are a particular shape to fit a standard test machine, as shown in Fig 5.32(a). As the load increases during the test, the shape eventually alters, see Fig 5.32(b), and the specimen undergoes 'necking'. If we calculate the stress at the neck, rather than for the overall sample, the stress–strain curve will go on rising and will follow the dashed line in Fig 5.29. This is because the area in the necked portion has decreased, and hence the strain (force/area) has increased. See Fig 5.33 and its caption for the operation of standard stress-testing equipment.

Creep

After a time under stress, a material may permanently retain some or all of its change in shape when the stress is removed. We say that its plastic distortion is time dependent and describe this property as **creep**. When it occurs in the plastic region, the creep is called **viscoplasticity**. Some specimens show creep even within the elastic region and return with time to their original undeformed state. This type of creep is called **viscoelasticity**.

EXAMPLE

Q A cable is made from 100 strands of high-tensile steel wire, each of diameter 1 mm. The strain on the cable should never exceed one-fifth of the breaking strain. What is the maximum load that should be supported by the cable?
Breaking strain 0.1%. Young modulus 2.1×10^{11} N m^{-2}.

A   Total area of cable = 100 times area of one strand

$$= 100 \times \pi \times (0.5 \times 10^{-3})^2 \text{ m}^2$$

$$= 7.85 \times 10^{-5} \text{ m}^2$$

By definition: Stress = $E \times$ strain

Breaking strain = 10^{-3} (ie 0.1%)

Maximum allowed stress = $E \times (0.2 \times$ breaking strain)

$$= 2.1 \times 10^{11} \times 0.2 \times 10^{-3} \text{ N m}^{-2}$$

$$= 4.2 \times 10^7 \text{ N m}^{-2}$$

Maximum load = stress × area

$$= 4.2 \times 10^7 \times 7.85 \times 10^{-5} \text{ N}$$

$$= 3.3 \times 10^3 \text{ N}$$

Fig 5.34 shows stress–strain graphs for steel, copper and rubber. Although these are not to scale, they do compare relative strain. Copper can be extended more easily than steel and has passed through its plastic stage. The plastic region of steel has not been reached in the graph. Rubber stretches very easily at first, but then becomes harder to stretch. On unloading, it returns to its original length. There is no permanent strain, so rubber is also behaving elastically. Some materials do not show plastic behaviour – these are said to be **brittle**. Two brittle materials are glass and concrete.

Behaviour of rubber

What are the energy implications when materials are deformed? We will consider rubber, which deforms easily. When it is stretched and compressed repeatedly and rapidly, as happens in vehicle tyres, it demonstrates what is known as **hysteresis**, when the strain, which is the effect, lags behind the stress, which is the cause. A graph of stress against strain for rubber in a car tyre is shown in Fig 5.35. Read the caption carefully.

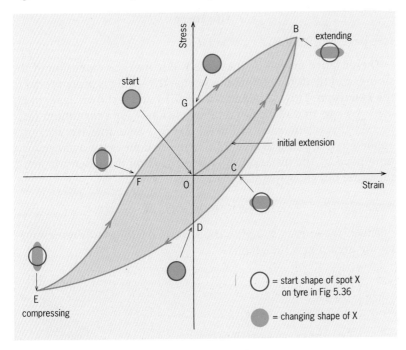

Fig 5.35 **Hysteresis in a rubber car tyre.**

After the initial stretching along AB, stretching ceases and the rubber contracts along curve BCD. When the stress has been reduced to zero, the rubber has contracted to C, with a strain remaining that corresponds to an extension.

The rubber returns to its original size only when it experiences a compressive stress OD.

Further compression takes the stress–strain relationship to E. Compression gradually reduces, and increasing extension then takes the rubber though EFGB, and so on.

If the repetitive cycling ceases, the rubber object quickly returns to its original form.

K Study the graph of Fig 5.35. In terms of this graph, describe the path traced out by spot X in Fig 5.36 as the car tyre makes half a revolution.

The curve shows clearly that the strain in the rubber lags behind the change in the stress in the process of hysteresis. The shaded area inside the stress–strain curves represents work done on the rubber. This work generates heat within the rubber.

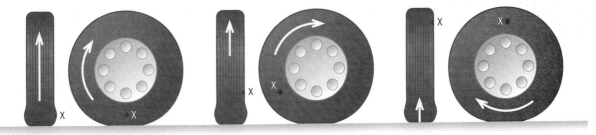

Fig 5.36 **Deformation of point X on the side of a car tyre during half a revolution. Notice the change in the shape of that part of the tyre**

Fig 5.37 **The continual deformation of this racing tyre generates a great deal of heat**

4 THE ENERGY STORED IN A STRETCHED ELASTIC MATERIAL

When an object is deformed by a force, work is done by the force. As a result, energy is transferred to the material. As long as deformation continues, we say that the energy is 'stored' in it. In the simple case of stretching a spring:

Work done = force × distance moved in the direction of the force

(see page 58). That is: $W = Fd$

The distance moved by the force is equal to the extension of the spring, and the force required to stretch the spring is proportional to the extension. This was shown in Fig 5.13 and is shown again in Fig 5.38. We see that there is a total extension X for a maximum force F_m. As the relationship is linear, we can take the average force applied to the spring to be $F_m/2$. The work done is the average force multiplied by the distance, since the force and the extension are in the same direction (parallel to each other):

Work done = $\frac{1}{2}F_m X$

This is the energy stored in the stretched spring. It is represented by the area under the graph.

Fig 5.38 **Calculating the energy stored in a stretched spring**

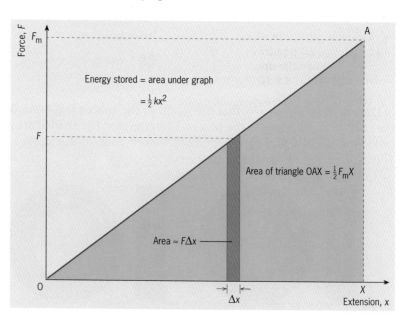

Another way of calculating the energy stored is to consider a small extension Δx of the spring, see Fig 5.38 again. By taking a very small extension, the force changes very little: we can take its average value as F. The work done in stretching the spring by this small distance Δx is $F\Delta x$. The work done is represented by the area of the dark strip. This strip is approximately a rectangle of area $F\Delta x$.

To find the work done in extending the spring, we can divide the whole of the area under the straight-line graph into strips from the origin where $F = 0$ to the point on the line where $F = F_m$. The total area of these strips must equal the triangular area under the graph (area of triangle OAX). This area gives us the total work done in stretching the spring, where X is the total extention at F_m.

See question 6.

$$\text{work done} = \tfrac{1}{2}F_m X$$

Its value is the same as that calculated above for the work done. The work stored in the spring is sometimes called **elastic potential energy**.

Now, $F_m = kX$ where k is the force constant of the spring. By substituting this expression for F_m in the formula above, and for any given extension x (not just maximum X corresponding to force F_m), we can express the energy stored in a stretched spring in general as:

Energy stored in a stretched spring $= \tfrac{1}{2}kx^2$

Springs may of course be compressed, and x can also refer to compression.

Similar arguments apply to solid objects which are stretched or compressed. The formulae for stretching or compressing a rod or wire are:

$$\textbf{Energy stored } = \tfrac{1}{2}\frac{EAx^2}{L}$$

and: **Energy stored per unit volume** $= \tfrac{1}{2}$ **stress** $\times$ **strain**

Question 9 at the end of the chapter leads you through the derivation of these relationships.

E is the Young modulus and A and L are the area and length of the rod which has an extension or compression x.

EXAMPLE

Q Concrete pillars are used to support the upper floor of a building. Each pillar is 200 mm in diameter and is 2.5 m tall. When the upper floor is loaded, each pillar supports a load of 8×10^4 N.

a) What is the compression of each pillar due to this load?

b) How much energy is stored in each pillar as a result of the compression?

E for high strength concrete $= 4 \times 10^{10}$ N m^{-2}.

A

a) Strain $= x/L = \text{stress}/E$

$$\text{compression } x = \frac{\text{stress} \times L}{E} = \frac{(\text{load/area}) \times L}{E}$$

$$= \frac{8.0 \times 10^4 \times 2.5}{4 \times 10^{10} \times \pi \times (0.1)^2}\,\text{m} = 1.6 \times 10^{-4}\,\text{m}$$

b) Energy stored $=$ average force $\times$ compression

$$= \tfrac{1}{2}(8 \times 10^4) \times (1.6 \times 10^{-4})\,\text{J} = 6.4\,\text{J}$$

This is easier here than using:

$$\text{energy stored} = \tfrac{1}{2}EAx^2/L$$

$$= \tfrac{1}{2}\frac{4 \times 10^{10} \times \pi \times (0.1)^2 \times (1.6 \times 10^{-4})^2}{2.5}\,\text{J}$$

$$= 6.4\,\text{J}.$$

---

5 MICROSTRUCTURES AND THE PROPERTIES AND BEHAVIOUR OF MATERIALS

Think of common objects around the house. In the kitchen, for example, we use copper saucepans and glass bowls. Their properties are very different, as shown if you drop them on the floor!

Everyday materials may be either simple pure chemicals or complex mixtures of chemicals. Copper is a single metallic element. Glass is several chemicals, the main one being the compound silica. *Natural* materials such as wood, leather and stone are more complex. *Artificial* materials or modified natural materials are often complex, such as brick, rubber, metal alloys, fabric mixtures such as polyester/cotton, and cardboard. *Composites* are very useful materials, combining the properties of two or more materials. Everyday examples are reinforced concrete, glass-reinforced plastic (GRP), carbon fibre and plywood.

What affects the way a given material behaves? To answer this, we have to consider many factors, including:

● the forces between atoms – chemical bonds,

● the arrangement of the atoms – whether they are in the form of molecules or form a complex extended structure,

● the nature of the molecules (if the material is molecular) – whether they are long and interlinked as in rubber and plastics, or small and weakly bonded to each other such as between atomic layers within graphite,

● its microstructure (its structure on a very small scale) – whether it is uniform or not on a small scale and whether it contains cracks and imperfections,

● its macrostructure (large-scale structure) – whether it is single crystal, polycrystalline, fibrous, composite.

Fig 5.39 **Scanning electron micrograph of a silicon chip connector in contact with microcircuitry, ×383 (false colour)**

When we talk about the microscopic structure of a material we mean the structure that we can identify using an optical microscope or an electron microscope (Fig 5.39). Macroscopic structure, on the other hand, we can see with the naked eye or with a hand lens.

The forces that hold solid materials together are electrical forces. They are due to the charges on the fundamental particles of matter which make up atoms, namely **electrons** and **protons**. In a pure chemical, the atoms are either all the same – the chemical is an *element*; or different types of atoms are combined to form the molecules of one kind of *compound*.

The different types of atoms in a material, and the way these atoms link together, determine mechanical properties such as density, elasticity and strength. As a general rule, the forces (called bonds) that hold the atoms together within the molecule of a compound are stronger than the forces that hold the molecules together to form the solid material. You will learn more about the bonding of atoms in Chapter 8.

Molecules and giant structures

Not all pure chemicals have their atoms arranged as molecules. Chemists use the word **molecule** to describe combinations of atoms that can exist and move about independently, particularly when the material is in liquid or gaseous form. For example, we can have molecules of water, oxygen, carbon dioxide and iodine. When these substances solidify, the forces between the molecules are weak and the solids are also weak. We find them as liquids or gases at room temperature because the forces between molecules are small and they can move apart easily.

In other compounds, the forces holding the molecules together are almost as strong as the forces holding the atoms together within the molecules. Silica is a good example. Its molecules of silicon dioxide (SiO_2) are connected by forces in a way that makes it impossible to tell where one molecule ends and another begins. This makes for a very strong structure which is called a **molecular giant structure**.

See questions 10 to 14.

Quartz

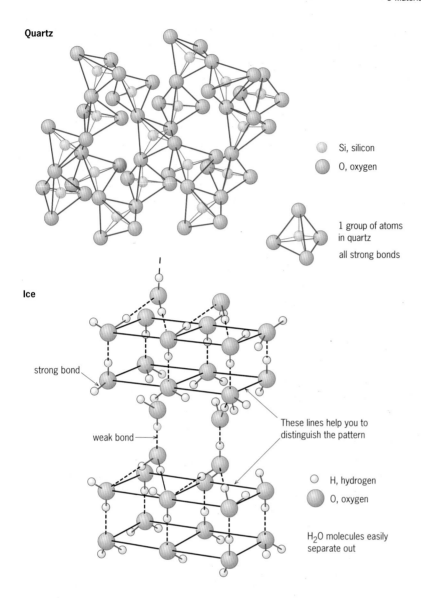

Si, silicon

O, oxygen

1 group of atoms
in quartz

all strong bonds

Fig 5.40(a) **The structure of quartz, a form of silica which has a molecular giant structure. Each silicon atom is in the centre of four oxygen atoms arranged tetrahedrally round it**

Ice

strong bond

weak bond

These lines help you to distinguish the pattern

H, hydrogen

O, oxygen

H_2O molecules easily separate out

Fig 5.40(b) **The structure of ice, which has weak intermolecular bonds**

■ See question 15.

Silica in its impure form, sand (an essential building material), comes from weathered rocks. Silica is hard and strong, so it remains after weathering has worn the other components of the rocks to fine particles. Fig 5.40 shows the structure for silica in the crystalline form of quartz, and compares it with ice which has very weak bonds between the hydrogen atoms. The hydrogen bonds can easily break, leaving separate water molecules (H_2O).

Sodium chloride – rock salt – is an example of an **ionic giant structure**. Salts in general are often compounds of metals and acid radicals (examples are metal chlorides, nitrates and carbonates). The building blocks are **ions** – that is, atoms or groups of atoms that have gained or lost **electrons**. Ions can exist separately in aqueous solution (dissolved in water). In solids, they combine in a lattice of particles held together by the electrical forces between oppositely charged ions. They make up familiar crystals such as copper sulphate, familiar in the chemistry laboratory.

Many materials with an ionic giant structure are weak and brittle. Others are strong and brittle, such as calcium carbonate. As limestone, it is a common building material. The problem with using ionic structures as building materials is that they tend to slowly dissolve in water and may be eroded by acids (Fig 5.41).

Fig 5.41 **Severely eroded limestone archway. (The area at top left has been restored)**

Fig 5.42 **The behaviour of rubber is determined by the arrangement of its molecules**

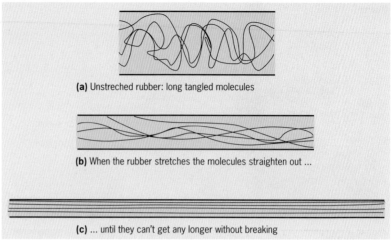

(a) Unstreched rubber: long tangled molecules

(b) When the rubber stretches the molecules straighten out ...

(c) ... until they can't get any longer without breaking

Rubber is a natural polymer made of long-chain molecules. In its normal state the molecules are bent back and forth, tangled together as in Fig 5.42(a) with weak forces acting between the molecules. As in (b), when the rubber in a rubber band is pulled, the molecules straighten out and allow the rubber to become longer. Eventually, as in (c) the molecules are fully stretched and almost parallel with each other. By that time the rubber may be up to 10 times its original length. Extra force will make the rubber break. On the other hand, if the rubber is not stretched to breaking, and the force is then removed, the molecules tend to curl back again because of the attraction and cross-links between adjacent molecules. Unless there been some reorganisation of the cross-links, the return is elastic.

Structures and properties of metals

Each atom in a metal has one or more outer electrons which can move freely between atoms. We can picture these electrons forming a 'sea' of negative charges that move amongst an ordered arrangement – a lattice – of positive metal ion spheres. It is because these electrons are so mobile that metals are good conductors of electricity.

The force of attraction between the ions and the free electrons gives rise to a 'metallic' bond. The most stable arrangement for the atoms is to occupy the minimum volume, forming a regular array of close-packed, hexagonally arranged layers. This arrangement gives a metal its regular crystalline structure.

There is no *directional* bonding between the ion 'spheres' to restrict the positions of neighbouring atoms. So, though the bonds (forces of attraction) between metal ions are strong, they can still be easily made to move to new positions. In general, then, metals are not only strong but also flexible. They can be hammered and flattened – they are **malleable** (from the Latin *malleus*, a hammer) – and can be drawn out into wires – they are **ductile**.

See question 15.

Usually, when a liquid metal solidifies, regions of crystal structure start to grow ('nucleate') in lots of different places. This means that many small crystals or **grains** (also called crystallites) form to make up the solid metal. These grains pack together with irregular **grain boundaries** in different directions. The material is said to be **poly-crystalline** (Figs 5.43 and 5.44). The sizes of the grains and the presence of their boundaries affect the mechanical properties of the metal, as we shall see later.

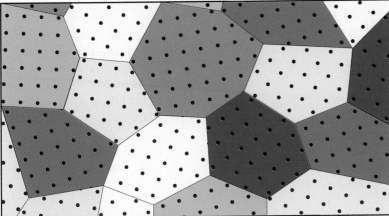

Fig 5.43 **Polycrystalline titanium alloy, showing grains (crystallites) and grain boundaries. Taken with polarised light, the micrograph records different orientation of atoms**

Fig 5.44 **A polycrystalline metal, showing that there is the same regular lattice of atoms within the boundary of each grain, but that the lattice is differently orientated in neighbouring grains**

What happens when a metal is stretched

When a load is applied to a metal, the atoms (strictly, metal ions) move very slightly apart. Imagine the bonds between the atoms as a network of springs. The extension of the springs – the distance the atoms move apart – is proportional to the load. When the load is removed, the atoms move back to their original positions. The metal obeys Hooke's law.

Just as with a spring, once the metal is stretched to the point where the ions are separated beyond the linear region of a force–extension curve, plastic behaviour takes over. The atoms must rearrange in some way. In fact, planes of atoms move across each other (Fig 5.45) – we say they **slip**. The force must be large enough to break the bonds that link all the atoms in one plane to the atoms in the adjacent plane.

Once a plane has slipped by a small amount, the bonds between neighbouring atoms can reform. This is what happens when a specimen necks, but does not break, during a stress–strain test. We might expect that slip requires very large forces. But we shall see later that a much smaller force than we might expect can produce slip in a large metal crystal.

In a polycrystalline metal sample, the atoms in one small grain are not in the same plane as atoms in neighbouring grains: the grain boundaries interrupt these planes. This prevents slip continuing through the metal sample in any one direction. In large single metal crystals, slip occurs very much more easily because the atom planes are continuous.

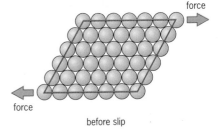

force

force

before slip

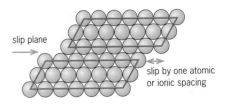

slip plane

slip by one atomic or ionic spacing

Fig 5.45 **Slip occurring in a perfect crystal**

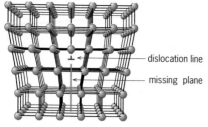

Fig 5.46 **Part of a metal crystal with a dislocation**

Let's look at this more closely. Fig 5.45 shows whole planes of atoms sliding across each other. But, however pure a metal sample might be, it still contains imperfections. One type of imperfection is called a **dislocation**, a plane of atoms (or ions) missing over part of the lattice, as shown in Fig 5.46. The bonding is already weaker in the region of the missing plane. A shearing force applied to the crystal planes (Fig 5.47) will move one plane of atoms by one lattice spacing. This requires less force than it takes to move a large number of planes of atoms.

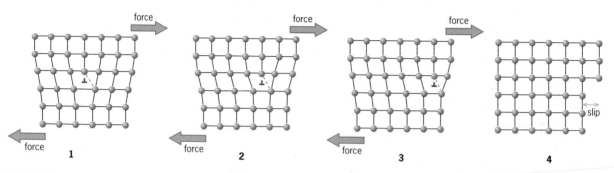

Fig 5.47 **A shear force can move a dislocation one lattice spacing at a time. In these two-dimensional drawings, the disolcation, indicated by the conventional symbol of an inverted T, moves from one group of atoms to another**

These dislocations help to give metals their properties of being ductile and malleable. We have seen that the metal atoms move and rebond easily (we can think of the moving layers flowing over each other). This allows a metal to be drawn out into a wire – it is ductile. In general, also, a metal does not crack easily when hammered (it is not brittle), therefore it is malleable.

Models for movement of dislocations

Moving a ruck in a carpet
Suppose a carpet is laid flat on the floor, and we try to pull it across the floor. It is very difficult: it takes a lot of strength to overcome all the forces keeping carpet and floor together. Instead, we can create a ruck across the carpet close to one side, as in Fig 5.48 (reducing the carpet–floor forces), and then move the ruck through the carpet and across the floor. Like the steps of dislocation in a crystal in Fig 5.47, this gradual movement requires much less force.

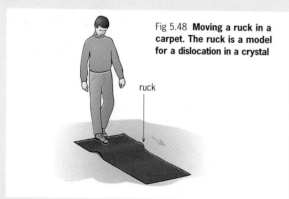

Fig 5.48 **Moving a ruck in a carpet. The ruck is a model for a dislocation in a crystal**

ruck

Ball-bearing crystal model
For a simple model of a crystalline structure in two dimensions, we can place small ball bearings between parallel glass plates, as in Fig 5.49. A single layer is held in place by the frame of slides, with enough ball-bearings to pack the enclosed space almost completely to form a lattice, yet allowing movement in two dimensions.

When the model is gently tapped, dislocations and crystal boundaries appear in the lattice. There will also be small gaps. These are like the gaps, called **vacancies,** for missing atoms which always occur in a three-dimensional crystal lattice.

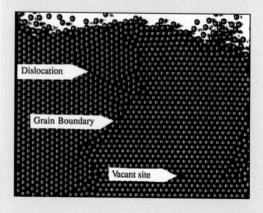

Dislocation

Grain Boundary

Vacant site

Fig 5.49 **A ball-bearing model of a crystalline material showing vacancies, crystal (grain) boundaries and dislocations**

6 MAKING MATERIALS STRONGER

In the previous section, we saw that a material with no impurities may not be strong, particularly when it contains only a few grain boundaries. If a material is stressed by applying forces that distort it, the number of boundaries can be increased considerably. The resulting tangle of boundaries can increase the strength.

Introducing atoms of other elements, such as atoms that are larger than those in the metal lattice, can stop dislocations from spreading and strengthen the material. We have seen that in a polycrystalline material, the boundaries between the small grains can also hold back dislocation movement. It is therefore more difficult to deform materials consisting of small grains than materials with large crystals.

■ See question 15.

Faults and cracks in microstructures

As a rule, a material is brittle if cracks easily form and spread in it. In manufacturing, materials are often made by chemical reactions or by using heat to mould material into shapes (think of the blacksmith at his forge). During these processes, exposure to air can weaken the surface of a material. Millions of tiny imperfections on the surface can add up to form cracks. Then, these tiny cracks may **propagate** (spread) through the object when a load is applied.

This was a problem with cast iron, used extensively during the Industrial Revolution in the nineteenth century. Early railways made great use of cast iron for rails and bridges. But girders developed cracks, and many accidents occurred when bridges collapsed. It was almost as dangerous when the iron rails themselves cracked – steam engines exerting enough traction to pull loaded carriages or wagons produced a stress within the rails close to the breaking stress.

More recently, in the early days of jet passenger planes, cracks in the body of aircraft caused catastrophic accidents. Cracks develop where the strain is greatest. In the case of aircraft, cracks started at the edges of the windows. It is essential also to avoid cracks in the blades and casings of turbines such as those used in hydroelectric power stations to generate electricity. Engineers test for potential problems by using Perspex models and applying suitably scaled-down loads. In testing, the model is viewed using polarised light. The resulting interference colours (see Chapter 16) show how the strain is distributed within the model. Fig 5.50 shows an example of this for the strain in an artificial femur.

Cracks are dangerous because they are unpredictable. They are likely to occur when a material is repeatedly loaded and unloaded. For instance, in an aircraft the pressure on the outside of the body changes every time the aircraft takes off and lands. The wings are continually flexing up and down, as you may notice if you sit in a window seat in a plane. When an object fails under these conditions it is called **fatigue failure**.

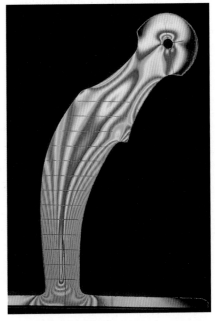

Fig 5.50 **Strain patterns in the plastic model of an artificial hip bone**

Hardening glass

Often, an engineer prefers to use a weaker material with larger cross-section rather than a stronger material which is more liable to crack. Wood is not as strong as glass but is less likely to break, so it has traditionally been more useful as a building material. Now, methods for treating the surface of glass make its behaviour more predictable, so it is more widely used in large buildings.

Fig 5.51 **A car windscreen shatters in small pieces, so that the danger to passengers is reduced**

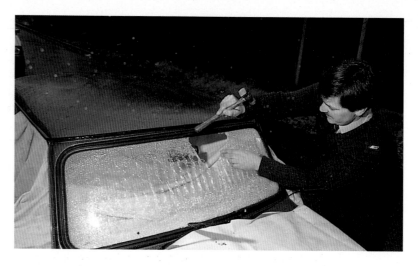

Different methods are used for making windscreens. In one method, some of the surface ions of glass are exchanged chemically with other ions of a larger diameter. These larger ions create compressive stress on the surrounding glass ions. In another method, the glass is heated until it goes soft, and cold air jets make the outer surface cooler than the interior. A compressive stress is set up in the surface, with a tensile stress in the interior.

With both methods, the surface of the glass is in compression. This makes crack propagation difficult, because whenever a crack starts, compression forces at the surface tend to close it up. But once a crack does penetrate beyond the surface, it spreads very rapidly indeed as the built-up stress is large and there is no resistance to spreading. Fig 5.51 shows what happens.

Glass is an example of an **amorphous** structure. This means that the molecules do not exist in a regular pattern. You will find out more about crystalline and amorphous materials in Chapter 8.

Composite materials

Cracks may stop when they meet a tougher material. For this reason, **composites** are made with more than one material. The continuous component is the **matrix**, and one or more other tougher materials are distributed in it. When a crack develops in the matrix, it extends until it meets the tougher component. Concrete is an example: a crack spreads in brittle cement but stops when it meets a particle of sand or gravel (Fig 5.52). (Two quite brittle materials can also be combined to make a tougher composite.)

Glass is widely used to reinforce composites. An example is glass-reinforced plastic (GRP), used for the hulls of dinghies. A crack forming in the resin matrix spreads until it meets the glass fibre. This fibre then separates slightly from the matrix, stress around the fibre is reduced and the crack stops. A sudden blow will create thousands of tiny cracks, but they do not develop into a single catastrophic crack. In fact, the material remains almost as strong as it was before the damage.

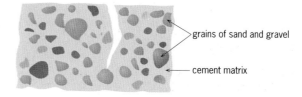

Fig 5.52 **Crack-stopping in concrete**

grains of sand and gravel

cement matrix

Describing the properties of materials

The technical words used to describe the properties of materials have precise meanings which may not match their everyday use.

Term	Meaning and example
brittle	breaks suddenly and catastrophically with rough edges; once started, cracks propagate easily; bricks and biscuits are brittle
tough	the opposite of brittle; the material resists crack propagation and so distorts rather than breaks; nylon, rope, bones, tendons and most textiles are tough
strong	the material needs a large stress to distort it; steel, titanium alloys, rubber, glass, wood (along the grain) and cotton are strong (the metals are about ten times stronger than the non-metals)
elastic	the material will return to its original length (or shape) when any load is removed; rubber, steel, glass and wood are usefully elastic – they withstand everyday forces without permanent distortion
plastic	the material distorts easily with quite small stresses but does not fracture; plasticine and wet clay are typical plastic materials; metals and ice show plastic behaviour if stresses act for a long time (metals creep, glaciers flow)
hard	hard materials are not easily cut; diamond is hard, graphite is soft (hardness is measured on a scale from 1 to 10 using the *Moh scale*)
soft	the opposite of hard
malleable	material changes shape but does not crack when subjected to sudden large forces (eg being hit with a hammer); many metals (eg copper) are malleable
ductile	material changes shape rather than cracking when subjected to a large but steadily applied force; describes the behaviour of many metals which can be drawn out into wires when forced through a small hole

Examples
Biscuit is weak and brittle, *glass* is strong and brittle. A piece of *string* is soft, tough and fairly strong. *Rubber* is soft and elastic, *steel* is hard and elastic. *Wood* is soft, tough, fairly elastic and strong.

K Describe as concisely and clearly as possible the mechanical properties (response to forces) of any three of the following materials: a cheese biscuit, a piece of newspaper, a piece of string, a plastic ruler, a piece of food-wrap film. Use the words listed in the box, plus any others that seem relevant.

SUMMARY

By studying this chapter you should have learnt and understood the following about structures and materials:

■ Forces act at a point, and moments of force are in equilibrium when a body is not accelerating.

■ Hooke's law states: extension x is proportional to applied force F: $F = kx$, where k is the spring constant.

■ Understand the behaviour of springs in series and in parallel.

■ The key concepts of **stress** include: force/area, F/A; and **strain** as extension/length, x/L.

■ The Young modulus is the ratio stress/strain for an elastic material, $E = FL/Ax$.

■ The types of stress that objects and structures are subject to include: tension, bending, shear.

■ A metal shows elastic and plastic deformation and has an elastic limit.

■ To find the energy stored in a stretched object (such as a spring), use $E = \frac{1}{2}kx^2$.

■ The terms used to describe the mechanical properties and types of materials include: brittle, tough, strong, hard, elastic, plastic, malleable, ductile, crystalline, polycrystalline, amorphous, composite.

■ The properties of a material can be explained by considering its microstructure at the level of forces between particles (atoms, molecules) and between larger features (crystals, grains); the significance of crystal defects.

■ Understand the importance of cracks in causing the failure of a material to withstand forces, specific methods used to stop cracks propagating.

■ Know ways in which materials are used in everyday structures: buildings, bridges, household objects.

QUESTIONS

1 Fig 5.Q1 shows a proposed design for a pylon to carry an overhead conductor for a new high-speed railway system. The joints A, B and C are all freely pivoted. When the line is not in use, the effective load placed on the system by the conductor is a vertical force of 0.80 kN. The rigid link XY exerts no force on the conductor or on the member CB in normal conditions.

(a) **(i)** Calculate the forces in the insulators P and Q under the conditions described.
    (ii) For each force, state whether it is tensile or compressive.

b) Suggest a reason for the inclusion of the link XY.

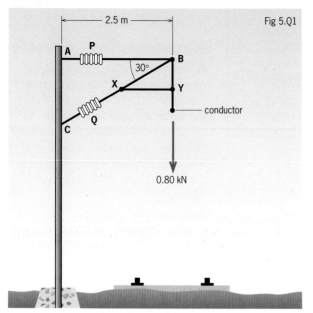

Fig 5.Q1

[AEB: 1994, A/AS Specimen Paper 1, Section A, q. 1]

2 A particular type of spring obeys Hooke's law and has a force constant of 2×10^2 N m^{-1}.

a) How much will it extend when a load of 100 N is attached to it?

b) Two such springs are connected in series. What will be the total extension for a load of 100 N?

c) A designer wishes to use springs of this type in an application in which a load of 200 N produces an extension of about 3 cm. How could the springs be arranged to get this result?

3 A nylon climbing rope is 10 mm in diameter and 30 m long. When it supports a climber of mass 80 kg, it gets 1.0 m longer.

a) Calculate the Young modulus for nylon.

b) Imagine a situation in which the climber falls the full length of the rope. Discuss what would happen if the rope were made of material with a Young modulus **(i)** one-tenth that of nylon, **(ii)** ten times that of nylon (other dimensions being the same).

4 A metal rail is 20 m long and has an area of cross-section of 10^{-2} m^2. When heated, each metre length of the metal rail expands by 10^{-5} m for every Celsius degree rise in temperature.

Take the Young modulus of the metal as 18×10^{10} N m^{-2}.

a) By how much does the rail expand when the temperature increases from 10 °C to 35 °C?

b) If the rail is fixed so that it is not allowed to expand, how much force will it exert on the fixing?

5 Fig 5.Q5 shows a simplified view of a suspension bridge.

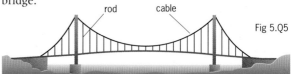

Fig 5.Q5

The road is supported by cylindrical rods attached to the suspension cables. There are two identical cables and sets of rods on each side of the roadway.
  Each pair of rods effectively supports 5.0 m of road which has a mass per unit length of 2.0×10^3 kg m^{-1}. The longest rods are 55 m long and have a cross-sectional area of 5.0×10^{-3} m^2.

The Young modulus for steel is 2.0×10^{11} Pa. Take acceleration of free fall, $g = 10$ m s^{-2}.

a) Assuming that the limit of proportionality is not exceeded, calculate:
    (i) the tension in each rod,
    (ii) the tensile stress in the longest rods,
    (iii) the tensile strain in the longest rods.

(b) **(i)** What changes will occur in a rod when a lorry goes over the bridge?
    (ii) Explain why the rods used in bridges may fail due to cracking.

6 In a test on a steel spring designed to be used as a bed-spring, the following results were obtained:

Force/ N	0.00	15.00	30.00	45.00	60.00	75.00	90.00	105.00	120.00
Length/ mm	100.00	90.00	80.00	70.00	60.00	52.00	46.00	42.00	40.00

a) Draw a graph which best displays the behaviour of the spring.

b) Suggest why the spring is designed to have non-linear behaviour.

c) An average person has a mass of 60 kg. How many springs are needed so that the body sinks 3 cm into the mattress?

d) How much energy is stored in each spring when it is depressed by 3 cm?

e) Calculate the force constant of the spring over its linear region.

7 Use the data given in the table of elasticity and strength (page 94) to test if there is a relationship between the Young modulus and tensile stress for the listed materials.

8 Sketch the stress–strain graph you would expect for each of the materials used for the following purposes: **(i)** a seat belt; **(ii)** the bolts on an aircraft ejection seat; **(iii)** the cone of a loudspeaker; **(iv)** the support strings for a baby bouncer. Justify each graph.

9 This question is on energy stored in a stretched material. See pages 100–1 for the background. The question leads you through the proof that the energy stored in unit volume of stretched material is $\frac{1}{2}$ **stress** × **strain**.

a) Show that the energy stored in a stretched wire of length L, area of cross-section A and Young modulus E, is given by the formula: $\frac{1}{2} Eax^2/L$, where x is the extension produced by the stretching force.

b) Which part of the formula may be replaced by strain? Complete the formula:

$$\text{Energy stored} = \tfrac{1}{2} \text{ strain} \times ?$$

c) Now eliminate E from the formula by putting $E = $ stress/strain.

d) Now reinsert strain by using the definition: strain = x/L.

e) Explain how the formula obtained in **d)** is equivalent to the statement that the energy stored in a stretched object per unit volume is: $\frac{1}{2}$ stress × strain.

10 Concrete beams used in buildings are usually prestressed. This means that steel wires under tension are inserted into the wet concrete and remain in tension when the concrete sets. In a typical beam 20 such steel wires each 6 m long and 2 mm in diameter are stretched to a strain of 0.15% while embedded in the concrete. The Young modulus of the steel used is 2×10^{11} N m^{-2}.

a) How much energy is stored in the stretched wire?

b) Suggest why special care has to be taken in demolishing buildings which contain prestressed concrete beams.

11 The density and Young modulus of some useful structural materials are listed in the table.

	Young modulus/10^{10} N m^{-2}	Density/10^3 kg m^{-3}
aluminium	70	2.7
steel	20	7.8
molybdenum	27	10.5
titanium	12	4.5
magnesium	42	1.7
wood (spruce)	1.3	0.5
glass (eg window)	7	2.5
carbon fibre	7.5	2.3

a) Use the data to plot a graph of Young modulus against density for the materials.

b) Comment on the main features of the graph.

c) Suggest why one of the materials is increasingly used in applications which require light, strong materials, and give three examples of such applications.

d) Suggest a relationship between density and elasticity for the listed materials. Test your suggested relationship by adding and plotting any data you can find about other materials. Is your 'law' generally valid?

12 Glass, ceramic tiles and plasterboard may be cut in a straight line by making a small cut along the desired line and then applying forces on both sides of the cut (Fig 5.Q12). Explain the principle underlying this process.

Fig 5.Q12

13 Explain why concrete is much stronger in compression than it is in tension.

14 Describe any composite material that you know about. Explain why it has been made a composite and relate its properties to its uses.

15 Why are the forces between atoms (the metallic bond) usually less important for predicting the strength of a metal than other aspects of the metal specimen's structure?

16 At a depth of 400 km below the Earth's surface the pressure is about 1.4×10^{10} Pa. If a cubic metre of solid rock (Young modulus 7×10^{10} N m^{-2}) were brought to the surface, what would be its new volume? Give two reasons to explain why the rock is likely to melt.

Assignment

THE YOUNG MODULUS AND TENSILE STRENGTH OF METALS

In this Assignment you will use experimental data to determine the Young modulus for metal wires. Two identical wires are suspended using a vernier arrangement as shown in Fig 5.A1.

Fig 5.A1 **Readings of the vernier scale as loads are added to the right wire**

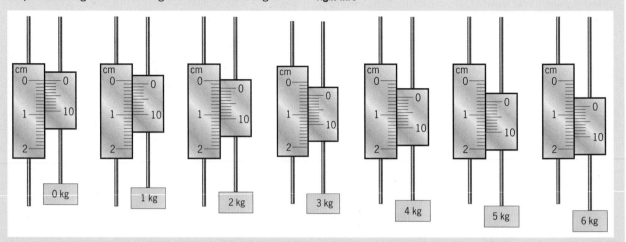

The left wire is held taut with a fixed mass. Different loads are applied to the right wire and the stretching of the wire is measured using the vernier. The test wire has a length of 2.10 m between support and vernier and a uniform diameter of 0.41 mm.

Table 5.A1 **Data for wire specimen 2**

Load/ kg	0.0	1.0	2.0	3.0	4.0	5.0	6.0	6.5	7.0	7.2
Extension/ cm	0.00	0.020	0.036	0.056	0.076	0.095	0.120	0.220	0.552	0.822

1

a) By reading the vernier settings, determine the extensions produced by the different loads. Tabulate these values and plot a graph of load against extension. (You may wish to use a spreadsheet to tabulate your data and plot your graph.)

b) Without replotting your graph, describe how you would recalculate your data so that stress is plotted on the y axis and strain on the x axis.

2

Calculate the Young modulus for the wire. Look up tabulated data for the Young modulus of metals and suggest the type of metal used. (It is likely to be a common metal.)

3

Table 5.A1 gives data for a particular specimen of steel wire 2 of the same length 2.10 m and diameter 0.80 mm. Plot the data and compare your graph with that for the wire 1. Describe any significant differences and suggest reasons for the differences. What do you expect to happen when wire 2 is unloaded?

4

Calculate the Young modulus for specimen 2.

5

The load on specimen 1 is increased from 0.0 kg to 6.0 kg. Calculate:

a) the change in elastic potential energy of the wire,
b) the change in potential energy of the mass,
c) the total change in potential energy of the mass and wire together.

Comment on your values.

6

A wire made of steel as used for wire 2 is used for raising a load of 1 tonne.

Assuming a safety factor of 5, estimate the required diameter of the wire. Do you consider the steel alloy to be suitable for this task?

MATERIALS AND FORCES

The main ideas that relate to the forces that act on materials, and to aspects of the structure of materials, are brought together in this Chapter Map, which shows how the ideas are linked. You can use the map to cross-match with the needs of your syllabus. The map should also help you to identify the areas you may need to study further.

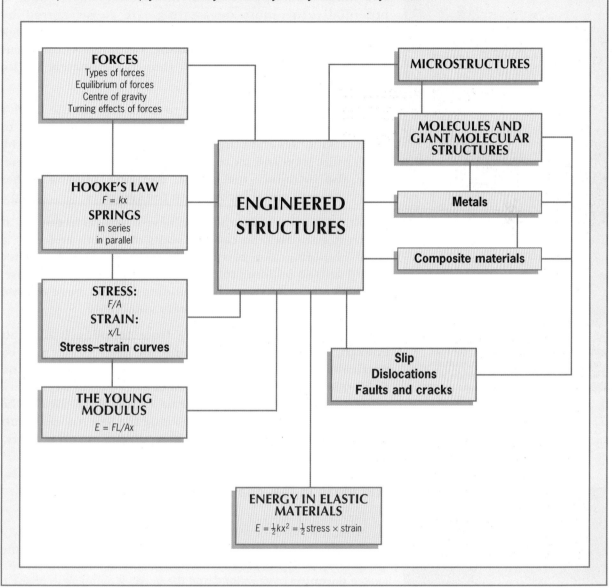

6 Oscillations and mechanical waves

The upper deck of a two-tier elevated roadway came crashing down on to the lower deck, crushing 80 people in their cars, when the freeway at Oakland on the San Andreas fault in California was hit by an earthquake in 1989

THE MOST POWERFUL natural mechanical waves, and so potentially the most devastating, are those caused by earthquakes. They are called seismic waves from the Greek word *seismos* which means a shaking. Earthquakes can occur almost anywhere in the world, though fortunately we do not have major ones in Britain.

Severe earthquakes occur close to fault lines in the Earth's crust. Here, architects and engineers go to great lengths to protect buildings and other structures. They add strength by using more steel and more concrete than is usual elsewhere. They also introduce mechanical damping, putting rubber inserts or even metal springs into the supports.

An object, a bell for instance, can have a natural frequency at which it resonates. So can the walls of a room and even a whole building. An earthquake at its natural frequency would cause a building to oscillate wildly and then collapse, so architects make sure that buildings do not have natural frequencies that would make them vulnerable.

Preventing a high-rise building from toppling is difficult enough, but stopping the collapse of flyovers has proved particularly tricky, as highway engineers in Japan and California have discovered after major earthquakes which have caused chaos and death. California in particular awaits the 'big one' – a massive earthquake along the San Andreas fault line which it is predicted could occur in the next few years.

The ideas in this chapter

As the chapter opener indicates, the study of oscillations and mechanical waves is of crucial importance to our everyday safety.

In this chapter we look in some depth at the theory behind oscillations and at **simple harmonic motion** in particular. We also look at **damping** of vibrations, and the need for damping in practical applications such as car suspension systems.

Stationary, or **standing**, waves have special importance in music. We look at the theory behind these waves and consider their application to such instruments as guitars and organ pipes.

1 SIMPLE HARMONIC MOTION

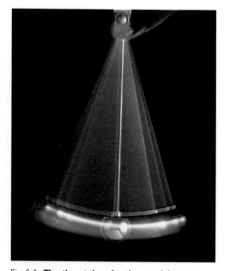

Fig 6.1 **The time taken for the pendulum to swing is constant, provided the angle of swing is small. Galileo noticed this as he watched a lamp swing on a chain in Pisa Cathedral**

Most of us at some time have seen a grandfather clock with a pendulum (Fig 6.1). The reason for using a pendulum is that the time taken for each swing through a small angle is constant. So, as the pendulum swings back and forth, it controls very precisely the movement of the clock's mechanism.

The pendulum is an example of an **oscillator** – it moves backwards and forwards in a regular (periodic) way, as in Fig 6.2(a). The **period** of the pendulum is the time it takes to go from one position in its swing through a complete oscillation back to its original position. The number of such oscillations in a unit of time is called the **frequency**. The distance the bob has swung away from its central equilibrium position is called its **displacement**, and the maximum displacement is the **amplitude** of swing.

Another example of an oscillator is a mass vibrating up and down on the end of a spring, see Fig 6.2(b). Here also, the mass moves up and down past a central position.

Displacement is movement in a particular direction and therefore is a vector. For the mass on the spring, displacement x (and also amplitude) is measured in metres. For the pendulum, displacement is measured as an angle, θ, usually in radians.

The period T is measured in seconds. Frequency f is the number of vibrations in a second, and so we have the inverse relationship:

$$f = \frac{1}{T}$$

We can see that the unit of frequency is second^{-1} (ie s^{-1}). This unit of frequency is given the special name hertz (Hz). This means that:

$$1 \text{ Hz} = 1 \text{ s}^{-1}$$

To find out more about oscillations, we can fix a pen to a pendulum and swing the pendulum over a moving strip of paper on, for example, a dynamics trolley moving at right angles to the swing of the pendulum. As shown in Fig 6.3, the pen plots a trace which represents the displacement of the pendulum with time. This is the sequence of events:

- At time $t = 0$ the pendulum is set swinging by letting it fall away from a position of maximum displacement. This displacement is the amplitude A of swing.

- The pendulum has zero displacement (it hangs vertically) at time $t = T/4$.

- At time $t = T/2$, the pendulum has maximum displacement on the side opposite its start side.

- The pendulum again has zero displacement at $t = 3T/4$ as it returns back through the vertical position.

- The pendulum takes time T to do a complete swing and return to its original position.

If you are familiar with a plot of cos θ against θ, then you will recognise this as the curve in Fig 6.3.

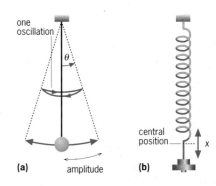

Fig 6.2 **Examples of oscillating systems:** (a) **a pendulum;** (b) **a mass suspended from a spring**

A radian is the angle subtended at the centre of a circle (in this case the pivot of the pendulum) by an arc equal in length to the radius. (Here, the radius is the length of the pendulum.)

A What is the frequency (in Hz) of an oscillator vibrating with a period of 1.25×10^{-3} s?

B (a) Plot cos θ against θ between 0° and 360°, by hand or by graphical calculator.

(b) State two ways in which the movement of the shadow in Fig 6.4 resembles the simple harmonic motion of a pendulum.

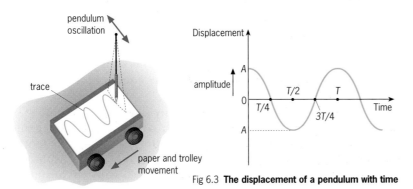

Fig 6.3 **The displacement of a pendulum with time**

The movement of such an oscillating system is called **simple harmonic motion** because many musical instruments create vibrations like this. We say that the path of the motion varies **sinusoidally** – that is, like a sine (or cosine) curve.

Mapping simple harmonic motion onto a circle

It helps us to understand simple harmonic motion if we link it to movement in a circle. As in Fig 6.4, we can place a sphere near the edge of a rotating turntable and project a shadow of the sphere onto a screen. As the turntable rotates, the shadow of the sphere moves across the screen with a motion similar to that of the bob of a pendulum moving back and forth.

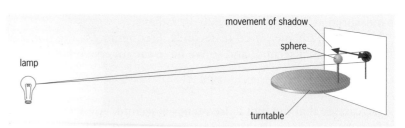

For another demonstration, we can use a vibrating dynamics trolley with a ticker-tape machine to calibrate the movement, as in Fig 6.5.

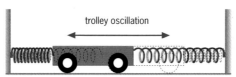

Fig 6.5 **An oscillating trolley attached to two identical anchored springs**

The trolley is attached to two fixed posts using two identical springs. At equilibrium, the trolley sits centrally between the posts with the springs pulling equally in opposite directions. When the trolley is pulled to one side, the two springs both act, with a force proportional to the displacement x, to return the trolley to the central equilibrium position.

This restoring force F is given by:

$$F = -kx$$

where k is the spring constant for the combination of the two springs. Note that when one spring is compressed and tending to lengthen, the other is stretched and tending to shorten. When the trolley returns to its central position, it has a velocity which makes it overshoot. It goes on to achieve maximum displacement on the other side.

We can trace the change of position of the trolley with time by connecting the trolley to a ticker-tape machine (Fig 6.6). We get a series of dots at equal time intervals as the trolley moves from maximum displacement on one side of the equilibrium position to maximum displacement on the other side.

We can see that the dots are not equally spaced along the strip of paper, and we can match them up with points which are equally spaced around a semicircle, as Fig 6.6 shows. It is as if a spot is moving with constant angular speed around the circle, just as the sphere in Fig 6.4 was moving uniformly around with the turntable.

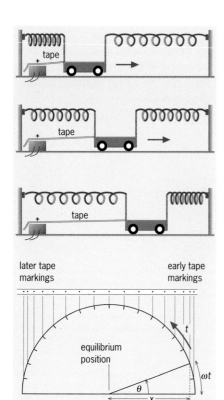

Fig 6.6 **Relating movement of the trolley to uniform motion in a circle**

The dynamics trolley in Fig 6.5 is demonstrating Hooke's law – see page 88.

Fig 6.4 **The movement of the shadow of an object rotating in a circle illustrates simple harmonic motion**

The spot would make one complete revolution, covering an angle of 2π radians, in period T. Its angular speed ω is obtained from angle turned through, divided by time. That is:

$$\text{Angular speed } \omega = \frac{2\pi}{T} = 2\pi f$$

From Fig 6.6, we see that the displacement of the oscillating trolley is given by:

$$x = A \cos \theta = A \cos \omega t$$

where A is the maximum amplitude of the oscillation.

C The dots along the paper in Fig 6.6 are unequally spaced. Explain this as fully as you can.

EXAMPLE

Q A trolley oscillates with amplitude 5.0 cm and period 2.5 s. What are its displacements relative to its equilibrium position 0.4 s and 1.0 s after being set off from a position of maximum amplitude?

A We first need to find ω in radians. (Your calculator needs to be in radian mode.)

We have seen that $\omega = 2\pi f$ and $f = 1/T$. So:

$$f = \frac{1}{2.5} = 0.4 \text{ Hz} \quad \text{and} \quad \omega = 2\pi \times 0.40 \text{ s}^{-1}$$

This gives us a displacement after 0.4 s of:

$$x = A \cos \omega t$$
$$= 5.0 \cos(2\pi \times 0.40 \times 0.4)$$
$$= 5.0 \cos(1.01)$$
$$= 2.7 \text{ cm}$$

That is, the trolley is 2.7 cm from the equilibrium position, on the same side as it started.

After 1.0 s:

$$x = 5.0 \cos(2\pi \times 0.40 \times 1.0)$$
$$= 5.0 \cos(2.51)$$
$$= -4.0 \text{ cm}$$

That is, the trolley is 4.0 cm from the equilibrium position, on the opposite side from where it started.

Displacement, velocity and acceleration

You have already encountered relationships between distance, velocity and acceleration (see Chapter 2). In simple harmonic motion, though, the displacement varies sinusoidally with time – and this is a lot more complicated than if displacement were linear with time.

Fig 6.7(a) shows the displacement for an oscillator behaving sinusoidally with time. It is plotted with maximum displacement at time $t = 0$ so that $x = A \cos \omega t = A$. The velocity of the oscillator at any time is given by the small change in displacement Δx which occurs during a small change in time Δt. That is:

$$\text{Velocity} = \frac{\Delta x}{\Delta t}$$

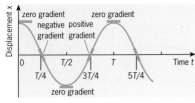

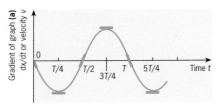

(a) Displacement–time curve

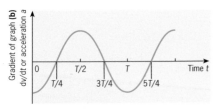

(b) Velocity–time curve

(c) Acceleration–time curve

Fig. 6.7 **Curves for simple harmonic motion**

This velocity is the gradient at any point on the displacement-time curve in Fig 6.7(a). We can see the following:

- At $t = 0$, the gradient is zero.
- At $t = T/4$, the gradient is negative and is the maximum possible negative value.
- At $t = T/2$, the gradient is again zero.
- At $t = 3T/4$, the gradient has the maximum possible positive value.
- At $t = T$, the oscillator has returned to its starting position and the gradient of the curve is zero.

We can plot the complete range of values of the gradient. This gives us the velocity–time curve shown in Fig 6.7(b). We see that the shape of the curve is similar to that of Fig 6.7(a) but is displaced along the time axis. The velocity is zero at time $t = 0$. It is in fact given by $\sin \omega t$; and since the velocity starts off in a negative direction, the dependence is $-\sin \omega t$. The velocity *lags* the displacement by $\pi/2$. We say that it is out of phase by $-\pi/2$; we discuss phase on page 129.

Now we take the velocity–time curve and consider the gradient $\Delta v/\Delta t$ in a similar way. This gives us the variation of acceleration with time, as shown in Fig 6.7(c). We see that acceleration is in opposite phase to the displacement – that is, they are out of phase by π, so that when one is at a maximum positive value the other is at a maximum negative value, and so on. Acceleration can be represented by $-\cos \omega t$.

The physics of the curves

Do the curves make sense? We can see that the velocity is maximum at zero displacement. Think of a pendulum. It swings fastest at the centre of its swing. What about the trolley connected to the two springs? It is also moving fastest at its central position. On the other hand, in both examples, velocity is zero at maximum displacement. The pendulum and the trolley are at rest for an instant as they reverse direction of movement. But the force is a maximum here and the acceleration must be a maximum. The force is returning the pendulum or the trolley back to its central equilibrium position and so the acceleration must be negative compared with displacement. In the central position, there is no horizontal force on the pendulum. In the case of the trolley, the horizontal forces exerted by the springs exactly balance out.

Proof by calculus

The relationships for velocity and acceleration can also be shown by calculus. Displacement is given by:

$$x = A \cos \omega t$$

To find the velocity, v, we differentiate:

$$v = \frac{dx}{dt} = \frac{d}{dt} (A \cos \omega t)$$

$$= -A\omega \sin \omega t$$

We differentiate again to find the acceleration a:

$$a = \frac{dv}{dt} = \frac{d}{dt} (-A\omega \sin \omega t)$$

$$= -a\omega^2 \cos \omega t$$

But we can replace $A \cos \omega t$ by x. So:

$$a = -\omega^2 x$$

So what *is* simple harmonic motion?

Simple harmonic motion (s.h.m.) is motion in which the acceleration is directly proportional to the displacement from a fixed point and is directed towards this point.

We can see both from the graphical treatment and from calculus that displacement varies sinusoidally with time and that acceleration is proportional to the displacement. Acceleration is oppositely directed, as shown by the negative sign.

To show that the motion of a pendulum is simple harmonic motion

We need to show that the acceleration of the bob is proportional to its displacement from the centre of motion, which is taken as the vertical position.

Suppose we have a bob of mass m. A gravitational force mg will act on the bob vertically downwards. There is also a tension T acting in the string. When the bob is displaced from the vertical by an angle θ, the tension acts on the bob at an angle θ to the vertical (Fig 6.8). We resolve forces along and perpendicular to the string. Along the string we have:

$$T = mg \cos \theta$$

The tension in the string must be equal and opposite to the gravitational force on the bob *as resolved in the direction of the string*.

Perpendicular to the string is an *unbalanced* force of $mg \sin \theta$. This force acting on the bob of mass m must give it an acceleration a in the direction of decreasing θ (that is, it acts in the opposite direction to the displacement). From this we can say that:

$$mg \sin \theta = -ma$$

or:

$$g \sin \theta = -a$$

If θ is very small and measured in radians, then $\sin \theta \approx \theta$, and we get:

$$g\theta = -a$$

So we see that acceleration is proportional to the angular displacement θ.

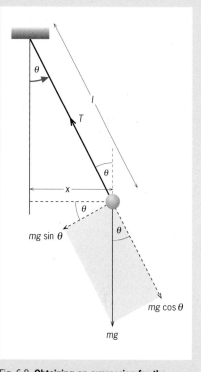

Fig. 6.8 **Obtaining an expression for the period of a simple pendulum**

The period of a simple pendulum

Let x be the displacement of the bob from the vertical in a horizontal direction and l be the length of the pendulum. Then for small θ:

$$\theta = \frac{x}{l}$$

We know from the previous section that $g\theta = -a$, so:

$$\frac{gx}{l} = -a$$

$$a = -\left(\frac{g}{l}\right)x \qquad [1]$$

The acceleration is proportional to angular displacement θ, in line with the definition of s.h.m. We compare [1] with our equation for s.h.m on page 118:

$$a = -\omega^2 x$$

so that:

$$\omega^2 = \frac{g}{l}$$

Since $\omega = 2\pi f$: $4\pi^2 f^2 = \frac{g}{l}$

giving:

$$f = \frac{1}{2\pi}\sqrt{\frac{g}{l}}$$

The period of the pendulum is $T = 1/f$, so:

$$T = 2\pi\sqrt{\frac{l}{g}}$$

Note that this expression is only true provided θ is small (less than approximately 10°, which is less than 0.2 radians).

The period of oscillation of a mass suspended from a spring

The force exerted by the spring is given by:

$$F = -kx$$

where x is the extension or compression of the spring from its equilibrium length, and k is the spring constant. When a mass m is suspended from the spring, it will already be stretched through an extension x_0 given by $kx_0 = mg$, see Fig 6.9(b).

After pulling the mass down a further distance x and letting go of the mass, as in (c), we have the following forces:

● a force downwards equal to mg due to the gravitational force on the mass,

● a force upwards equal to $k(x + x_0)$ due to the tension of the stretched spring.

These are not equal and so there is a net force acting on the suspended mass giving it an acceleration a. As force equals mass × acceleration, assuming that we give a downward force a plus sign and an upward force a minus sign, we have:

$$\underset{\substack{\text{force exerted} \\ \text{by spring}}}{-k(x + x_0)} + \underset{\substack{\text{force due} \\ \text{to gravity}}}{mg} = \underset{\substack{\text{resulting (net)} \\ \text{accelerating force}}}{ma}$$

As the upward force $k(x + x_0)$ of the spring is greater than the downward force on the mass mg, the acceleration is upwards and hence has a negative value. The equation continues to apply as the spring contracts through its equilibrium extension.

Now, by Hooke's law we can replace mg by kx_0 in the equation to give:

$$-k(x + x_0) + kx_0 = ma$$

So:

$$-kx = ma$$

$$a = -\frac{k}{m} x$$

Comparing this with the equation for s.h.m:

$$a = -\omega^2 x$$

$$\omega^2 = \frac{k}{m}$$

This gives the period of oscillation for the suspended mass on the spring as:

$$T = \frac{2\pi}{\omega} = 2\pi\sqrt{\frac{m}{k}}$$

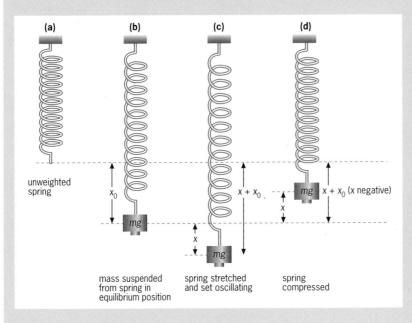

(a) unweighted spring

(b) x_0 · mg · mass suspended from spring in equilibrium position

(c) $x + x_0$ · x · mg · spring stretched and set oscillating

(d) mg · $x + x_0$ (x negative) · spring compressed

Fig. 6.9 **Obtaining an expression for the period of oscillation of a mass suspended from a spring**

2 ENERGY CHANGES IN AN OSCILLATING SYSTEM

In all the examples of s.h.m. which we have considered – that is the pendulum, the mass on a spring and the vibrating dynamics trolley – the velocity v of the oscillating mass m varies from a maximum when it passes through the equilibrium position to zero at maximum displacement when the direction of motion is instantaneously changing.

For each system, the kinetic energy at any time is $\frac{1}{2}mv^2$. We can now look at how this energy varies.

Energy–time curves

Since v changes sinusoidally with time as shown in Fig 6.7(b), the variation of kinetic energy with time can be shown by squaring the velocity (Fig 6.10). The total energy of the oscillating system must remain constant. So, as the kinetic energy rises and falls, it is being transferred back and forth into another form of energy. This form must be potential energy. We can see this energy as the difference (shaded areas) between the constant level of energy and the kinetic energy as in Fig 6.10(b). Alternatively, potential energy can be plotted positively in the same way as kinetic energy, as Fig 6.10(c), and is seen to be out of phase with kinetic energy by a quarter of a period.

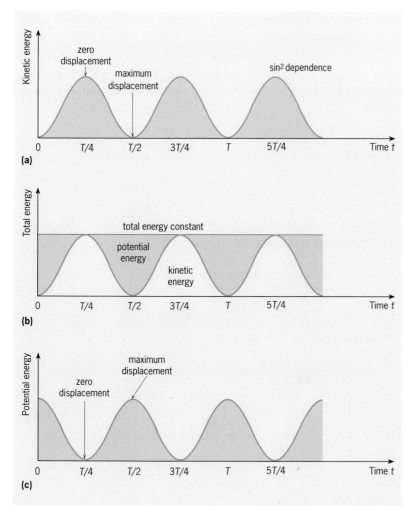

Fig. 6.10 **Variation of** (a) **kinetic energy,** (b) **total energy and** (c) **potential energy of an s.h.m. oscillator with time**

Energy–displacement curves

We can also show the variation of kinetic energy and potential energy with displacement x, rather than with time t, and this gives us curves of a different shape. When working out the form of these curves, it is easiest to start with the potential energy contribution to the energy.

As before, let's consider a mass connected horizontally between two springs. (This is so that we do not need to consider gravitational potential energy.) We will take the force constant as k for the two springs acting on the mass. To move the trolley from its equilibrium position at $x = 0$ to displaced position x requires an applied force $F_{applied}$ given by:

$$F_{applied} = kx$$

This applied force is equal and opposite to the restoring force exerted by the springs. The work done is force × distance and is equal to the triangular area under the applied force plot of Fig 6.11:

$$\text{work done} = \tfrac{1}{2}F_{applied}\, x_{max} = \tfrac{1}{2}kx_{max}^2$$

But x_{max} is the maximum displacement or amplitude of vibration. We called this A earlier. So:

$$\text{work done} = \tfrac{1}{2}kA^2$$

This is the potential energy now stored in the springs.

At maximum displacement, the kinetic energy is zero and this means that $\tfrac{1}{2}kA^2$ is the total energy of the vibrating system. The total energy of a vibrating system is therefore proportional to the square of the amplitude:

$$E_{total} \propto A^2$$

For a displacement $x < A$, the potential energy is $\tfrac{1}{2}kx^2$. Taking the difference between the total energy and the potential energy we obtain:

$$\text{kinetic energy } E_K = E_{total} - \tfrac{1}{2}kx^2 = \tfrac{1}{2}k(A^2 - x^2)$$

Fig 6.12 shows how the potential energy and kinetic energy vary as displacement x varies.

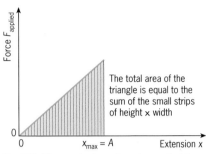

Fig. 6.11 **The graph shows how the applied force needs to increase in order to extend a spring system. Its maximum extension = x_{max} = A**

Fig 6.12 **The variation of potential energy and kinetic energy with displacement x for an oscillator**

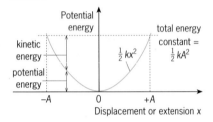

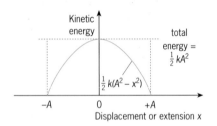

See questions 1–6. ■

Alternative expression for velocity

$$E_K = \tfrac{1}{2}k(A^2 - x^2)$$

$$= \tfrac{1}{2}mv^2$$

So $mv^2 = k(A^2 - x^2)$, and therefore:

$$v = \pm \sqrt{\tfrac{k}{m}(A^2 - x^2)}$$

But:

$$\sqrt{\tfrac{k}{m}} = \omega$$

since $\omega^2 = k/m$, see Extension box on page 120. This gives us velocity in the form:

$$v = \pm\, \omega \sqrt{A^2 - x^2}$$

3 DAMPING

So far, we have assumed that there is no friction in our oscillating systems. However, in reality this can never happen – there will always be some friction, causing energy to be lost from the system.

The total energy of a vibrating system depends on the square of the amplitude. If energy is removed from the system, the amplitude of vibration is reduced. How rapidly this happens depends on the amount of friction, which causes what is known as **damping**.

Damping, causing a reduction in the amplitude of vibration, is often an advantage (Fig 6.13). With *light damping*, the oscillations gradually reduce in amplitude but take a long time to disappear. With *very heavy damping*, the system does not oscillate but returns to its equilibrium position very slowly. With *critical damping*, the amount of friction is just right, so that the system returns as quickly as possible to equilibrium without overshooting. The time for this to happen is approximately a quarter of the period for free oscillation of the system.

?

F Give some other practical examples of damping.

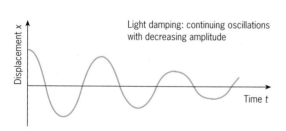

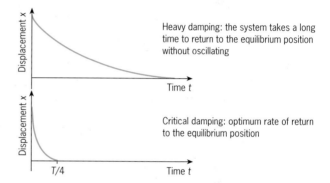

Fig. 6.13 **The effect of damping on an oscillating system**

SMOOTHING OUT A BUMPY RIDE

CAR SUSPENSIONS are fitted with springs and shock absorbers to damp out the vibrations of a moving car. The shock absorbers are pistons moving up and down in a viscous fluid (usually an oil). Without the shock absorbers, the car would bounce up and down repeatedly on the springs.

The amount of damping is set by the dimensions and materials for the spring and absorber. Car suspensions need to be slightly less than critically damped. If there is too much damping, then when the system has responded to a bump or rut in the road it does not return to its equilibrium position in time to respond to the next bump or rut. If the damping is too light, the passengers experience a lot of oscillations before the system recovers. When shock absorbers become worn, the ride of a car becomes bouncy and rather uncomfortable.

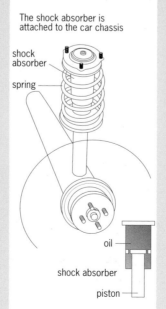

Fig. 6.14(a) **A car suspension**

Fig 6.14(b) **A rally car: the suspension allows the rough ride to be smoother than if there were no suspension in the vehicle**

Fig 6.15 **A mother pushing a child on a swing takes care not to reach a dangerous amplitude**

4 FORCED OSCILLATIONS AND RESONANCE

If the vibrating system is losing energy, it is possible to replace the energy or even increase the total energy by applying a force to the system periodically and in the correct direction. If the frequency of the applied force is the same as the natural frequency of the system, the amplitude of vibration builds up.

Think about pushing a child on a swing (Fig 6.15). You have to be careful not to push too much or too often as this would cause the the amplitude to get too large and the child could fall off.

It is often necessary to prevent vibrations building up in mechanical systems. When the applied force has the same periodic frequency as the natural frequency of the system, we say **resonance** occurs. When this happens, a large amount of energy is given by the force to the system. (You may have felt intense vibrations when in a car next to a large vehicle with a powerful engine running.)

The vibrations that were set up by strong winds in the Tacoma Narrows Bridge led to its collapse (Fig 6.17) in a spectacular case of resonance. The amplitude of vibration of the bridge became larger and larger until the bridge broke up.

Soldiers marching across a bridge can make it resonate. In 1850, 200 soldiers marching in step across a bridge in Angers, France lost their lives when the resonance between the frequency of their steps and the natural frequency of the bridge caused the bridge to collapse. To avoid this, soldiers break step across bridges.

In a car, you get uncomfortable vibrations due to resonance at particular speeds of either the engine or the vehicle itself. If resonance were set up in a turbine or a jet engine, it is possible that the engine would disintegrate. So we can see that it is important to avoid resonance in structures and vehicles.

Damping prevents the build-up of amplitude. Fig 6.16 shows how the amplitude of vibration of an oscillating system varies with the frequency of the driving force and also shows how damping affects the resonance curves. When the frequency of the driving force equals the natural frequency of the system being driven, the oscillations build up and can become totally uncontrollable.

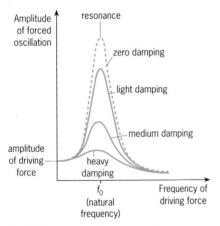

Fig 6.16 **Resonance curves for a driven system with natural frequency f_0**

Fig 6.17 **Tacoma Narrows Bridge, known during its short life as Galloping Gertie because it oscillated in the wind. It took a wind of only 42 mph to make it collapse. (The newspaper reporter who owned the car crawled to safety)**

Coupled pendulums

If two identical oscillators can interact then they will exchange energy, just as the soldiers marching in Angers exchanged energy with the bridge with disastrous results.

A pair of pendulums of the same length can be connected to a common support and set swinging, as shown in Fig 6.18(a). It helps to lightly link the pendulums, but often enough energy can pass through the support itself. When pendulum 1 is set swinging, its energy gradually leaks through to pendulum 2 until all energy is transferred and it is stationary. Pendulum 2 has now achieved maximum amplitude and starts to pass the energy back to pendulum 1; and so on.

As a variation of this demonstration, the so-called Barton's pendulums are suspended from one length of string, as in Fig 6.18(b). There is one heavy pendulum (the driver), and a series of pendulums with bobs of smaller equal mass. The lengths of the pendulums differ but one is of equal length to the driver pendulum. It is this one which starts to vibrate with the largest amplitude.

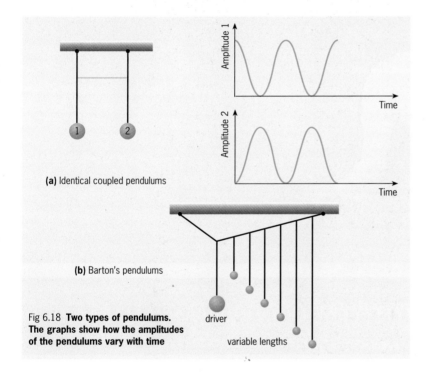

(a) Identical coupled pendulums

(b) Barton's pendulums

Fig 6.18 **Two types of pendulums. The graphs show how the amplitudes of the pendulums vary with time**

driver

variable lengths

G Why is the driver made with a heavy bob?

5 WAVES AS OSCILLATIONS

We have seen that oscillations are periodic. We know that waves are also periodic. Clearly there is a relationship between the two. In fact, an oscillating system can be used to set up waves in a mechanical system (Fig 6.19). Once the waves have been set off, they travel through the spring or other medium for long distances until the energy is dissipated.

There are different types of waves. Some need a substance to pass through – these are called **mechanical waves**. Such waves displace the material which they pass through. Examples of mechanical waves are seismic waves in the Earth, water waves on the surface of ponds and oceans, and sound waves through air. Seismic waves

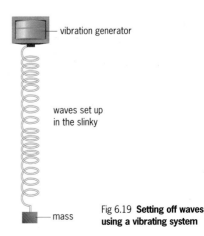

vibration generator

waves set up in the slinky

mass

Fig 6.19 **Setting off waves using a vibrating system**

are usually set up by earthquakes but can be caused by nuclear explosions. The speed with which they travel depends on the medium and the type of wave.

There are other waves, called **electromagnetic waves**, which require no medium to travel in. These waves are covered in Chapter 16.

Transverse and longitudinal waves

Mechanical waves can be classified as **transverse** or **longitudinal** according to how they travel (Fig 6.20). Both types of wave can be demonstrated using a slinky (a long steel spring). The transverse wave occurs when the coils move *at right angles* to the direction of motion of the wave, with the motion along the length of the slinky. To produce a transverse wave, the slinky is rested on a flat surface and one end is moved from side to side, setting up the oscillation and hence the travelling wave.

Fig. 6.20 **Transverse and longitudinal waves in a slinky**

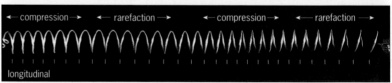

The end of the slinky can also be moved in and out along its axis. The coils undergo **compression**, followed by **rarefaction** when the coils open out. Displacement of the coils is now *along* the axis of the spring.

Progressive and stationary waves

When we start a wave in the slinky, either transverse or longitudinal, we can watch it travel from one end to the other. Because it progresses along the slinky it is called a **progressive** wave. However, if the far end of the slinky is fixed, waves are reflected back. These can combine with the next waves which are travelling forwards. At the right combination of frequency and speed, the waves travelling in opposite directions can produce a **stationary** or **standing** wave. In this chapter we shall consider both types.

6 SETTING UP PROGRESSIVE WAVES

In general, waves spread out once they are generated, like the ripples in a pond when a pebble is dropped into it. The spreading of waves, both transverse and longitudinal, are looked at in detail in Chapter 16. Here, we restrict ourselves to setting up linear waves such as along a slinky, in a string or in an organ pipe.

Transverse waves are easier to show, so let us look at transverse waves set up in a stretched string. With care, we can set up a single pulse in the string and watch it travel, as shown in Fig 6.21(a). Although the pulse maintains its shape, its amplitude is likely to decrease as it passes along the string. Such pulses can continue for long distances.

We can also set up a wave train consisting of a number of waves. It too will travel down the string and preserve its shape – see Fig 6.21(b). Alternatively, by continuing to vibrate one end of the string, we can establish a continuous wave travelling along the string, as shown in Fig 6.21(c).

Fig 6.21 **Transverse waves travelling along a stretched string**

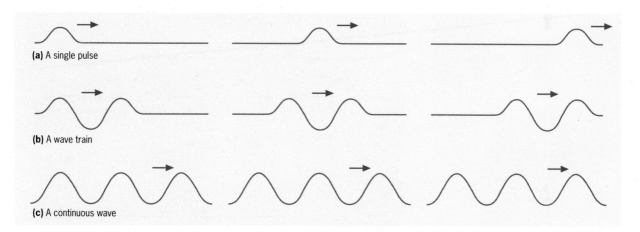

(a) A single pulse

(b) A wave train

(c) A continuous wave

BOAT SETS UP A SOLITON

A SINGLE PULSE has been given the special name of **soliton**. It is not easy to set one up in a string, but the transmission of solitons along optic fibres is important in optical signal communications.

In 1834, John Scott-Russell was walking by the Forth–Clyde canal when he saw a large-scale soliton and wrote this description of it.

'I was observing the motion of a boat which was rapidly drawn along a narrow channel by a pair of horses, when the boat suddenly stopped – not so the mass of water in the channel which it had put in motion; it accumulated round the prow of the vessel in a state of violent agitation, then suddenly leaving it behind, rolled forward with great velocity, assuming the form of a large solitary elevation, a rounded, smooth and well-defined heap of water, which continued its course along the channel apparently without change of form or diminution of speed. I followed it on horseback, and overtook it still rolling on at a rate of some eight or nine miles an hour, preserving its original figure some thirty feet long and a foot to a foot and a half in height. Its height gradually diminished, and after a chase of one or two miles I lost it in the windings of the channel. Such, in the month of August 1834, was my first chance interview with that singular and beautiful phenomenon.'

Fig 6.22

There are a number of features of these waves to notice. First, the speed of the pulse, wave train or continuous wave passing down the string is independent of the frequency with which the end is vibrated, and is also independent of the amplitude of the waves established. The speed depends on the nature of the string only. It is because the speed is constant that the shape of the pulse or wave remains the same.

Fig 6.23(a) shows a single wave as it progresses along a string. Suppose the wave moves by exactly its own length during a time T, as shown in Fig 6.23(b). The wave has moved by a single wavelength λ. The velocity of the wave is the distance travelled divided by time and so we have:

$$v = \frac{\lambda}{T}$$

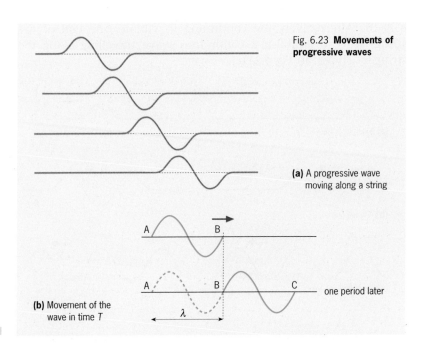

Fig. 6.23 **Movements of progressive waves**

(a) A progressive wave moving along a string

(b) Movement of the wave in time T

one period later

See question 8. ■

As the time T is the time of travel through one complete oscillation, the wave will have moved from AB to BC and a succeeding wave will have moved into position AB, as shown by the broken line in the drawing. T is the **period** of vibration. Also, the **frequency** of vibration f is given by:

$$f = \frac{1}{T}$$

so that we have:

$$\text{speed} = \frac{\text{distance}}{\text{time}} = \text{frequency} \times \text{wavelength}$$

This simple **wave formula** applies to all waves. The frequency of the wave is the frequency of vibration used to set up the waves from the end of the string. As the speed is a characteristic of the string, wavelength in a given string will depend on the frequency of the oscillating source.

Another important property of the wave is its amplitude A, which is the maximum displacement of any point on the wave, or any point on the string, from the zero displacement position. The properties are summarised in Fig 6.24.

?

H A wave moves along a string with velocity 0.60 m s^{-1}. Its wavelength is 20 cm. Calculate the frequency and period of the wave.

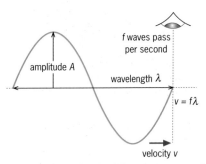

Fig. 6.24 **Properties that define a wave**

7 WAVE EQUATION

When a vibration generator causes a wave train to travel along a string, there is a sideways displacement to the string given by:

$$y = A \sin \omega t \quad \text{where} \quad \omega = 2\pi f$$

Note that y is varying with time t and so we can write it as $y(t)$ to mean that y depends on t. We can use x for distance along the string. The generator is at the end where $x = 0$. So we can write the displacement of the end of the string connected to the generator as:

$$y(0,t) = A \sin \omega t$$

On page 117 we said that the displacement y of an oscillating trolley could be represented by $A \cos \theta$ where θ is the angular displacement around a circle. However, this time we need to start from the zero displacement position so that $y = A \sin \theta$. We can write:

$$y(0,t) = A \sin \theta$$

Further along the string, the displacement will be out of step by an amount which depends on how far one goes along the string. It is as if we have moved by an additional angular displacement corresponding to a distance x along the string. With **phase angle** ϕ, we have:

$$y(x,t) = A \sin(\theta + \phi)$$

As $\theta = \omega t$, we can rewrite this as:

$$y(x,t) = A \sin(\omega t + \phi)$$

?

I At time $t = 0$, the maximum transverse displacement for a small segment of a vibrating string is 2 cm. What is the displacement for a segment of the string which is vibrating out of phase by 35°?

Showing the meaning of phase angle

Consider a shadow arrangement similar to that shown on page xx, but this time with *two* spheres mounted on the turntable. The spheres are both at the same distance from the centre, with an angular separation ϕ. In Fig 6.25 we use the example $\phi = 90°$.

Sphere A is shown in line with the lamp and the screen. This means that at time $t = 0$, the displacement x_A of the sphere is zero, that is $x_A = 0$ at $t = 0$.

We now consider a later time t when sphere B has come directly into line between the lamp and the screen. The displacement of the shadow of sphere A is now given by:

$$x_A = A \sin \omega t$$
$$= A \sin(\pi/2)$$

as the turntable has moved through 90°.

Meanwhile, the displacement of the shadow of B is given by:

$$x_B = A \sin 0$$

since the sphere is directly between the lamp and the screen. If we write the displacement of B in the form:

$$x_B = A \sin(\omega t + \phi)$$
$$= A \sin(\pi/2 + \phi)$$

then: $0 = \pi/2 + \phi$ or $\phi = -\pi/2$

The phase angle for B is the angle that it lags A around the turntable. Displacements for A and B are plotted in Fig 6.25 for ϕ equal to $-\pi/2$.

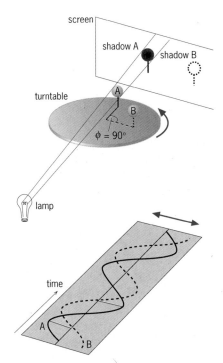

Displacement of the shadows on the screen
B lags A by an angle ϕ (=90° here)

Fig 6.25 **Demonstrating phase angle**

Fig. 6.26 (a) **A guitar being played.** (b) **This high-speed photograph shows the bass and two top strings vibrating. The tension is least in the bass string, hence the amplitude of the wave is greatest**

8 PRODUCING STATIONARY WAVES

Stationary waves are set up in stringed instruments such as a guitar (Fig 6.26). What we see is the string vibrating from side to side. At the moment that the string is plucked, a progressive transverse wave is set up travelling out from that point. It meets the fixed end of the string and is reflected back. The amplitudes of the two waves add together as they meet.

The string vibrates naturally at certain frequencies because it is fixed at both ends. When the outgoing and reflected waves are added together subject to this condition, a stationary wave is set up in the string. If the string is plucked centrally we get the **fundamental** mode (shape of wave). In this case, the string vibrates with maximum displacement at the central position (called the **antinode**) and the displacement falls away to zero at the two ends (called **nodes**).

We can investigate stationary waves in the laboratory using a stretched string or a long rubber band. The string is stretched from a fixed support over a pulley and is held taut by a suspended mass (Fig 6.27). A vibration generator is tied to the string near one end.

The frequency of vibration is altered starting from a low value, usually with little effect on the stretched string. As the frequency increases, the vibrator reaches a particular frequency at which the string suddenly starts vibrating strongly, see (a). **Resonance** is occurring. The natural frequency of the string is now equal to that of the vibrator. This resonance frequency is the **fundamental frequency** f_0 of the string, sometimes called the **first harmonic**.

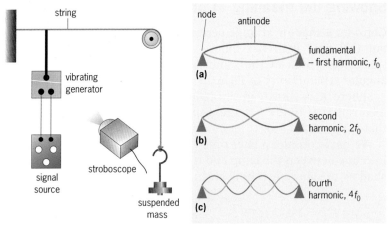

Fig. 6.27 **Stationary waves on a stretched string (or rubber band)**

As we increase the frequency of the vibrator further, oscillation of the string is rapidly reduced. We notice little movement until the vibrator frequency has doubled to $2f_0$. Now the envelope of the string (the shape the vibrating string encloses) has altered. There are three nodes, one at each end and one in the middle, see (b). This means that there are two antinodes positioned a quarter of the string length from each end. These two antinodes are vibrating out of phase and there is a phase difference of π. This frequency of vibration is referred to as the **second harmonic** or the **first overtone**.

If we increase the frequency further, we identify other resonant frequencies. The next is $3f_0$ with three antinodes on the string, and so on.

J A string vibrates with a fundamental frequency of 350 Hz. What is the frequency of **(a)** the first overtone and **(b)** the fourth harmonic?

Reflected waves

Let us look in more detail at how to set up standing waves. We set off a short wave on a slinky which has been firmly fixed at its far end. Assume that the wave consists of one and a half wavelengths, as in Fig 6.28(a). The wave travels along the slinky until it reaches the far end.

At this point, the wave can travel no further forwards and is reflected back, as in (b). This means that the velocity has changed sign. In addition, the phase of the wave has changed. If the displacement of the forward wave is upwards at the instant of time when it reaches the far end, then its displacement is downwards on reflection. This makes sense. At the fixed end, the displacement of the incoming and outgoing waves sum to zero. This must be so because there can be no displacement of the string at the fixed point. The reflected wave is out of phase by π. It passes back 'through' the forward wave (think how ripples can pass through each other on the surface of a pond). Where the two waves overlap, the displacement of the slinky is the sum of the two waves. But, eventually, we see the reflected wave emerge complete and pass back along the slinky, as in (c).

The frequency, velocity and wavelength of the wave all remain the same in reflection. If no energy is lost at the far end, the amplitude of the reflected wave equals that of the incoming one. The phase difference of π which we have identified and which is illustrated by Figs 6.28(a) and 6.28(c) is crucial to the setting up of standing waves.

When waves pass through each other, the displacement at any point is the sum of the individual displacements of the two waves passing in opposite directions. Fig 6.29(a) shows the relative positions of two waves travelling in opposite directions. Work out for yourself what pattern you will see. The answer is given in Fig 6.29(b).

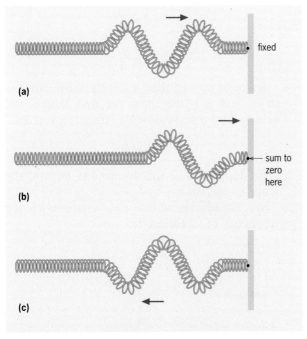

(a)

(b)

(c)

Fig 6.28 **Forward and reflected wave on a slinky**

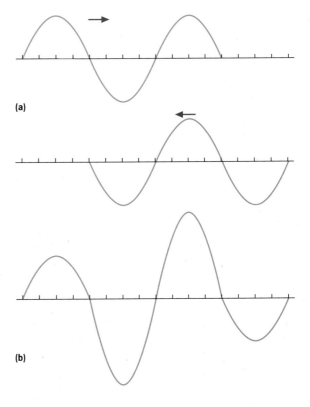

(a)

(b)

Fig 6.29 (a) **Relative positions of two waves travelling in opposite directions.** (b) **Sum of the displacements of the two waves in** (a)

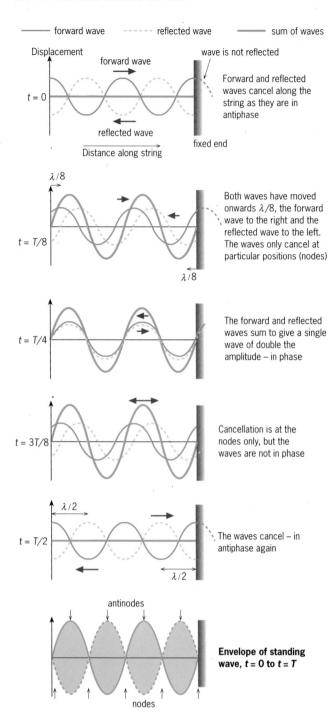

forward wave ---- reflected wave —— sum of waves

Displacement

wave is not reflected

$t = 0$

forward wave

reflected wave

Distance along string

fixed end

Forward and reflected waves cancel along the string as they are in antiphase

$\lambda/8$

$t = T/8$

$\lambda/8$

Both waves have moved onwards $\lambda/8$, the forward wave to the right and the reflected wave to the left. The waves only cancel at particular positions (nodes)

$t = T/4$

The forward and reflected waves sum to give a single wave of double the amplitude – in phase

$t = 3T/8$

Cancellation is at the nodes only, but the waves are not in phase

$\lambda/2$

$t = T/2$

$\lambda/2$

The waves cancel – in antiphase again

antinodes

Envelope of standing wave, $t = 0$ to $t = T$

nodes

Fig. 6.30 **How a standing wave is set up from the combination of forward and reflected waves**

Standing waves on a string

Let us consider a progressive transverse wave on a string with an exact number of complete wavelengths in its length.

It is not easy to see what happens at all times, so let us select certain special times. We start with $t = 0$ when the displacement of the *forward* wave is maximum at both the initial and far ends, though of course this displacement will be modified by the reflected wave. Fig 6.30 shows the situation with the forward, reflected and summed waves at this time $t = 0$, and at a series of times during the first half-period. We could go on to show the waves for the second half-period in a similar way in order to complete the overall pattern. The envelope for all possibilities is shown in the final diagram.

Frequency of standing waves on a string

We have seen that, as there are always nodes at the ends of the string, there must be a whole number, n, of half-wavelengths between each end. With a string length l, this means that:

$$n\frac{\lambda}{2} = l$$

or:

$$\lambda = \frac{2l}{n}$$

It can be proved that, if T is the tension in the string and μ is the mass per unit length, the velocity c of a wave down a string is given by:

$$c = \sqrt{\frac{T}{\mu}}$$

(The derivation of this formula is beyond the scope of this book.)

We already know that $c = f\lambda$, where f is the frequency of vibration. So:

$$f = \frac{c}{\lambda} = \frac{n}{2l}c = \frac{n}{2l}\sqrt{\frac{T}{\mu}}$$

See question 7.

?

K Show that the units of $\sqrt{\frac{T}{\mu}}$ are the same as those for c.

Standing waves using sound waves or microwaves

Sound waves are longitudinal waves in air (or another medium) similar to longitudinal waves in a slinky. They consist of compressions and rarefactions in the air. By sending sound waves from a loudspeaker and reflecting them from a *hard* surface, standing waves are established in the air. There will be regions where the air molecules are vibrating back and forth strongly (antinodes) and other regions where there is no movement of the air (nodes). By moving a microphone between the loudspeaker and the reflecting surface (Fig 6.31), the variation of displacement can be shown on an oscilloscope.

Instead of a loudspeaker, we can use a microwave transmitter. In this case, instead of a microphone and oscilloscope we then use a microwave receiver and ammeter. A metal plate acts as reflector. Although the experiment is similar, the nature of the waves themselves is very different. Microwaves are electromagnetic waves which you will meet in Chapter 16.

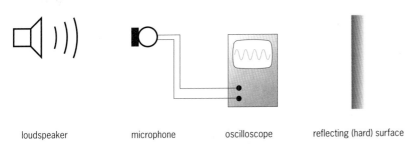

Fig. 6.31 **Demonstration of standing waves using sound waves**

loudspeaker   microphone   oscilloscope   reflecting (hard) surface

Standing waves in an organ pipe

These are sound waves and therefore longitudinal waves. They are set up by the compression and rarefaction of the air within the pipe. Nodes and antinodes occur just as they do for a vibrating string. But there is a difference. In the case of a string, both ends are effectively held fixed in order to establish the standing wave. In an organ pipe, vibrations can be set up either with both ends of the pipe open or with one end closed and the other open. At a closed end there must be a node present, and at an open end there must be an antinode (the air vibrates freely back and forth). The two situations are summarised in Fig 6.32.

See questions 9 and 10.

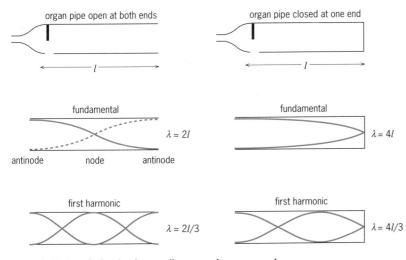

Fig. 6.32 **Modes of vibration for standing waves in an organ pipe**

L (a) An organ pipe is open at both ends. Calculate the length required such that the lowest frequency of vibration is 512 Hz.

(b) What would be the required length for an organ pipe closed at one end?

Take the velocity of sound in air as 330 m s^{-1}.

Beats

Suppose we tune two strings of a guitar to vibrate at almost, but not quite, the same frequency. Plucked simultaneously, the volume of the sound produced by them appears to rise and fall continuously. This rise and fall has a fixed frequency called the **beat frequency**. What is happening is that the sound waves produced by the two guitar strings interfere and our ears detect the variation of the resultant intensity. Maximum intensity is heard when the waves add together (interfere **constructively**) and minimum intensity is heard when the waves cancel each other out (interfere **destructively**).

We can see what is happening by adding together the two separate waves as shown in Fig 6.33(a). The resultant, obtained by the **principle of superposition**, is shown in Fig 6.33(b).

To work out the beat period and the beat frequency we need to know the number of cycles between each maximum of the beat pattern. We do this for each string.

The time between each maximum is T, and string 1 is tuned to produce a frequency f_1. During time T, string 1 emits f_1T cycles. Similarly, string 2 emits f_2T cycles at frequency f_2. During the period when the beat pattern goes from one maximum to the next maximum there must be one cycle difference between the waves from the two strings. The waves start in phase at the maximum, gradually go out of phase, and come back into phase, with a difference of one cycle (and only one cycle).

So:
$$f_2T - f_1T = 1$$

This gives:
$$T(f_2 - f_1) = 1$$

$$T = \frac{1}{f_2 - f_1}$$

Alternatively, we can find the beat frequency f_B where:

$$f_B = \frac{1}{T}$$

?

M A tuning fork vibrating at 256 Hz and a vibrating guitar string produce beats at a frequency of 2.5 Hz. Obtain two possible values for the frequency of vibration of the guitar string. Suggest how you might determine experimentally which value is correct.

Fig 6.33 **Setting up beats. Check that string 2 goes through one more cycle than string 1 during the beat period**

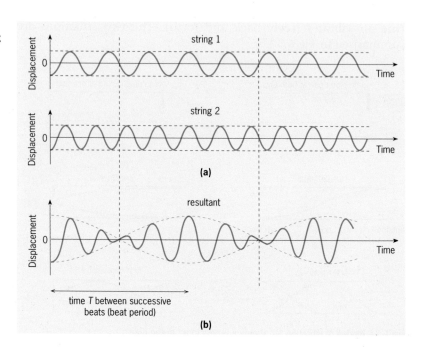

time T between successive beats (beat period)

(b)

SUMMARY

After studying this chapter you should understand the concepts and be able to use the equations:

■ Know the meaning of amplitude A, period T and frequency f of a wave and the relationship $T = 1/f = 2\pi/\omega$.

■ Know the meaning of speed v and wavelength l of a wave and the relationship $v = f\lambda$.

■ A mechanical wave needs a medium through which to pass. Waves may be longitudinal or transverse.

■ The period of a simple pendulum is given by $T = 2\pi\sqrt{l/g}$ and the period of a mass–spring system is given by $T = 2\pi\sqrt{m/k}$.

■ In simple harmonic motion (s.h.m.), acceleration is directly proportional to displacement from a fixed point and directed towards this point: $a = -\omega^2 x$.

■ The velocity of a mass m under s.h.m. is given by $v = \pm\omega\sqrt{A^2 - x^2}$.

■ Know the meaning of the terms: oscillator, displacement, node, antinode, harmonic, overtone, phase difference, damping.

■ There is a transfer of energy between kinetic and potential forms during s.h.m.

■ Understand the difference between stationary (standing) and progressive (travelling) waves.

■ A progressive wave can be represented by $y(x,t) = A\sin(\omega t + \phi)$ where ϕ is the phase angle.

■ Energy can be exchanged between a driving system and a driven vibrating system. Resonance occurs when their frequencies are equal.

■ The superposition of two waves of almost equal frequencies f_1 and f_2 where $f_2 > f_1$ produces beats of frequency $f_2 - f_1$.

QUESTIONS

1 Define simple harmonic motion. Explain why a clock can be designed around a system executing simple harmonic motion. Give one example of such an oscillating system.

A mass oscillates at the end of a spring. Describe the energy changes that occur when the mass goes from the lowest point of its motion, through the midpoint, to the highest point.

[ULEAC, A-level Physics, June 1994, Paper 2, Q10]

2 Fig 6.Q2 shows a simple pendulum, of length l, whose point of support is a height H above the floor and whose bob is a distance h above the same floor.

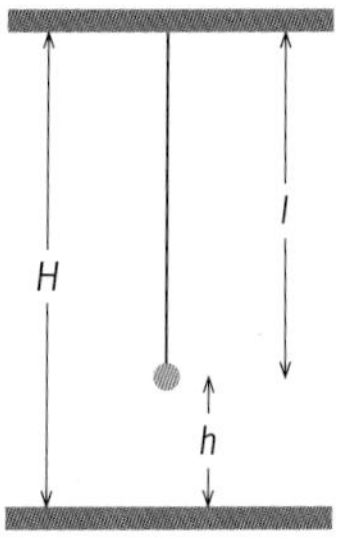

Starting from the relation:

$$T = 2\pi\sqrt{l/g}, \text{ show that:}$$

$$T^2 = \frac{4\pi^2}{g}H - \frac{4\pi^2}{g}h$$

Fig 6.Q2

What is represented by T?

Assume that you have found by experiment a series of corresponding values of T and H. Sketch and label the graph you would draw to determine g and H.

Explain briefly how you would determine the values of g and H from your graph.

[ULEAC, A-level Physics, Jan 1993 Paper 2, Q3]

3 The period of oscillation of a mass m suspended from a spring depends on the mass m, whereas in the case of a vibrating pendulum the mass of the bob does not affect the period. Why is this so? (Consider the exchange of kinetic and potential energies in each case.)

4 A spring for which the force constant is $k = 6$ N m^{-1} hangs vertically from a vertically oscillating support. A mass of 0.15 kg is attached to the lower end of the spring. The frequency of the support increases steadily from 0.1 Hz to 10 Hz. Describe the subsequent motion of the suspended mass, including a sketch of how the amplitude of vibration of the supported mass changes with frequency.

5 An atom in a molecule vibrates with simple harmonic motion of amplitude 8.0×10^{-11} m and period 4.0×10^{-13} s. Find its maximum speed and its maximum acceleration.

[WJEC, A-level Physics, 1989, paper A1, Q8]

6 A light helical spring of force constant 12 N m^{-1} hangs vertically and carries at its lower end a scale pan of mass 20 g and, resting on it, a mass of 100 g. When the system is set in vertical motion, calculate the maximum amplitude the motion can have if the 100 g mass is to remain in the pan throughout the motion. Clearly state your reasoning. Assume that the pan is rigidly attached to the spring.

Identify and comment upon the stage of the motion at which contact is lost.

[WJEC, A-level Physics, 1994, paper A2, Q9 (part)]

7 This question is about standing waves on a stretched string.

The string is fixed at one end; the other end passes over a pulley and the string is held in tension by a weight. Standing waves are set up in the string (length 0.75 m) by a vibration generator. The three lowest frequencies at which standing waves are produced are 20 Hz, 40 Hz and 60 Hz.

a) Draw the appearance of the 60 Hz standing wave.

b) What is the wavelength of the wave on the string at 60 Hz?

c) The standing wave is produced by progressive waves travelling in opposite directions.
    (i) Calculate the speed of the progressive wave along the string.
    (ii) Will this speed be changed when the frequency of the vibration generator is doubled? Give your reasoning.

d) 60 Hz is in the audible range yet this apparatus would not produce much sound. Give a reason why the same frequency played by a stringed instrument like a double bass would be much louder.

[O & C Nuffield, A-level Physics, 1994, paper 2, Q3 (part)]

8 A progressive wave moves along a stretched spring. Fig 6.Q8 shows the variation of displacement with distance along the spring at one instant.

Fig 6.Q8

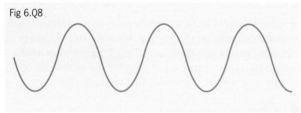

a) Copy Fig 6.Q8 and indicate on it: **i)** the amplitude of the wave; **ii)** the wavelength.

b)   **(i)** Why does a stretched string which has both ends fixed produce stationary waves when it is plucked?
    (ii) Sketch a diagram to illustrate the third harmonic vibrations of a stretched string for a fixed length and tension.
    (iii) State and explain the relation between the frequencies of the first and second harmonics.

[AEB, A-level Physics, Summer 1994, paper 2, Q10 (part)]

9

a) An organ pipe of length 0.280 m, closed at one end, is made to sound its fundamental note. Calculate the frequency emitted if the speed of sound in air is 340 m s^{-1} and the end correction is 0.015 m.
(Note that an end correction is an effective additional length at the open end that should be added to the pipe length.)

b) A steel piano wire of length 0.400 m and diameter 0.30 mm vibrates transversely in its fundamental mode. If it emits a note of the same frequency as the pipe in **a)**, calculate the tension in the wire. [Density of steel = 8.00×10^3 kg m^{-3}.]

c) Due to a fall in temperature the speed of air drops to 334 m s^{-1}. When the pipe and the wire sound their fundamental notes together, beats are heard because the pipe's frequency has fallen while the frequency of the wire has remained unchanged.
    (i) Calculate the beat frequency.
    (ii) State what property of the sound varies at this beat frequency.

[NEAB, A-level Physics, June 1993, Syllabus A, paper I, section 2, Q2]

10 A church organ contains many pipes, which are not all made in the same way.

a) The basic design of one type of pipe, used in organs, is shown in Fig 6.Q10.

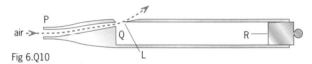

Fig 6.Q10

Air is blown in through the end P and passes a sharp lip L. The effect is to create a disturbance Q, near the end of the trapped column of air in the pipe. The other end of the pipe is closed by an adjustable plunger at R.
    (i) Indicate why stationary waves are set up in the pipe. Assume that the pipe behaves as if the end Q is open.
    (ii) Draw two labelled sketch-graphs for the stationary wave at the fundamental frequency of the pipe, the first showing the variation in amplitude of the motion of air particles between Q and R, and the second showing the variation of air pressure between Q and R.

b) In another type of pipe used in an organ, air enters in a similar way but there is no plunger closing the pipe at R.
    (i) Describe and explain the difference in the fundamental notes produced by two pipes of equal length, one being the closed pipe of the type described in **a)** and the other the open pipe in **b)**.
    (ii) Frequencies other than the fundamental can be produced when a pipe is blown. State what series of frequencies are possible, using both closed and open pipes, and explain briefly how they arise in each case.

c) The limited space available for a particular organ means that the longest pipe which can be fitted has an effective length of 3.0 m. Calculate the lowest frequency which can be played on this organ. [Velocity of sound in air = 330 m s^{-1}.]

[O & C, A-level Physics, 1993, paper 3/4, Q6]

Assignment

COMBINING WAVES: THE PRINCIPLE OF SUPERPOSITION

By the principle of superposition, two waves can pass through each other, and the forward and reflected waves add together. Also, two waves of slightly differing frequencies add together to produce beats. In this assignment we shall see how sinusoidal waves can be added together to produce other wave shapes.

In the laboratory
Experimentally, we would sum the waves coming from two or more loudspeakers using a microphone and look at the resultant on the screen of an oscilloscope (Fig 6.A1). However, instead of using the microphone to sum our sound signals, we shall use a spreadsheet to sum the amplitudes of our waves numerically, and instead of an oscilloscope screen we shall use the graphics capability of the spreadsheet.

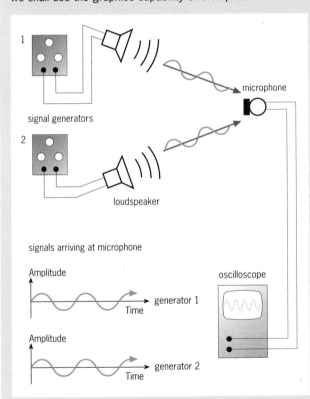

Fig 6.A1 **Superimposing waves on an oscilloscope screen via a microphone**

1 Suppose both signal generators are producing waves of 250 Hz and that the oscilloscope is calibrated so that 1 cm along the x axis corresponds to 1 ms.
a) What is the period (in s) of the sound waves?
b) What is the period (in cm) of the waves on the screen?

Using the spreadsheet
Open a new spreadsheet and set up cells by entering:
A1: Time/s   B1: Resultant/cm   C1: Y2/cm
D1: Time for Y2/s   E1: Y2/cm
A2 to A18: Enter values of time 0 to 4 in 0.25 intervals.
D2 to D18: Copy in the same entries as for A2 to A18.
C2 and E2: =sin(PI()*A2/2)   or enter: 3.142 for PI().

Check that the entries as quoted for C2 and E2 will give you the period you calculated in question 1.

Return to B2 and enter: =C2+E2.
Select C2 to C18 and use **Fill down** to complete the cells.
Do the same for E2 to E18 and B2 to B18.

Using the chart facility
Select cells A1 to B18 (ie select two columns of 18 cells).

Select **New** in the **File** menu followed by **Chart** and **X-values for X–Y chart**.

(The instructions might differ slightly for different spreadsheets but you need to choose x–y plotting facilities.)

You should now see the amplitude of the resultant wave plotted versus time just as you would see on the screen of the oscilloscope. It is the resultant for two in-phase waves of equal amplitude.

Varying the relative phase and the relative amplitude
A repeat listing of time can be inserted in column D so that you can vary the phase of Y2 compared with Y1. For instance, in D2 to D18, insert values 2 to 4, followed by 0.25 to 2 in 0.25 intervals. This will obtain Y2 π out of phase with Y1. To change relative amplitudes, for instance to change to an amplitude ratio of 3:1, fill the cell with: sin((PI()*A2/2))/3.

Square wave and saw-tooth wave
Open a new spreadsheet and set up cells:
A1: Time/s   B1: Resultant/cm   C1: A1/cm
D1: A2/cm   E1: A3/cm   F1: A4/cm
A2 to A18: As before, enter times 0 to 4 in 0.25 intervals (copy from the previous spreadsheet).
C2: =sin(PI()*A2/2)   D2: =(sin(PI()*3*A2/2))/3
E2: =(sin(PI()*5*A2/2))/5   F2: =(sin(PI()*7*A2/2))/7

Describe how the sine waves represented by the entries in columns C2, D2, E2 and F2 differ.

Set up cell B2:   =C2+D2+E2+F2.
Fill down for cells B1 to F18.
Select A1 to B18 and plot an x–y graph as before.

2 You should obtain a plot which is beginning to look like a square wave.
a) How would you suggest that an improved square wave plot can be obtained?
b) What feature of the square wave is likely to be particularly difficult to reproduce accurately?

For obtaining a saw-tooth wave you need:

C2: =sin(PI()*A2/2)   D2: =–(sin(PI()*2*A2/2))/2

E2: =(sin(PI()*3*A2/2))/3   F2: =(sin(PI()*4*A2/2))/4

(Note that the start of the saw-tooth does not come at the origin of the graph.)

Think how other wave patterns can be obtained. Any signal can be reproduced by using enough sinusoidal terms.

Sinusoidal variations and response of the human eye

Breaking up waveforms into their sinusoidal components is very important in advanced physics. The technique is named of **Fourier analysis**. It is used particularly in optics and electronics. Detectors, including the human eye, respond to patterns according to the components making up the incoming signal. Many detectors respond poorly to the high frequencies, that is, they respond badly to the extra terms which you did not include in the spreadsheet but which you deduced were necessary to obtain a good waveform.

We now look at an experiment using patterns like the one in Fig 6.A2(a) – a sinusiodal variation across the page in cycles of light intensity as reflected off the stripes. Each pattern has a different width of stripes.

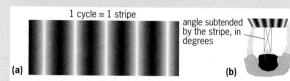

(a)   **(b)**

Fig 6.A2 (a) **The pattern with sinusoidal variation of light intensity.** (b) **Viewpoint of the subject, from above**

The subject views each pattern in turn at reading distance, and the experimenter records the ease with which the subject distinguishes the lines. (We do not need to go into the details of the procedure.) This gives a measure of the eye's sensitivity, as shown in the graph of Fig 6.A3. Note that both the x and the y scales are logarithmic.

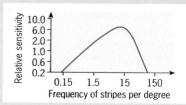

Fig 6.A3 **Graph showing how the eye responds to sinusoidal stripes of different spatial frequency**

At the left of the graph, at about 7° per stripe, the eye cannot detect intensity differences. On the right, it detects no light intensity variation at about 100 stripes per degree. The

eye, then, is poor at detecting variations which happen over large angles, and also poor at detecting very rapid variations when the stripes are very close together.

3 Estimate how many times more sensitive the eye is when lines repeat 15 times per degree, than when the lines repeat only once every 6.5 degrees (approximately 0.15 times per degree).

Staircase intensity pattern

Using a curve like Fig 6.A3, we can work out how the eye responds to certain intensity patterns and see how the eye can be fooled, as in **optical illusions**. Fig 6.A4 shows an example of such an illusion, the staircase intensity pattern.

Fig 6.A4
A staircase intensity pattern; an example of an optical illusion

4 Describe what you actually see with your eyes.

The staircase has a 'width of tread' which, again, we can measure as the angle subtended by the eye when the page is placed at reading distance (taken to be 0.3 m).

5 Estimate this angle from the pattern in Fig 6.A4 and hence the number of treads per degree of angle.

The pattern of Fig 6.A5(a) can be broken into a single saw-tooth and a series of small saw-teeth as in (b); check that this is so when the two plots are added. When the single saw tooth and the small teeth of (b) are broken into spatial sine waves, as described in **Square wave and saw-tooth wave**, these add together as in (c) to give a variation of brightness that the viewer perceives.

6 Does this brightness distribution correspond to your description in question 4?

It is important to realise that the eye is an exceptionally good measuring device but that it can distort the input signal it receives, just as manufactured electronic and optical detectors may also distort their input signals.

Fig 6.A5 **The staircase intensity pattern** (a) **can be broken into saw-tooth components** (b) **and appears to the eye as in** (c).

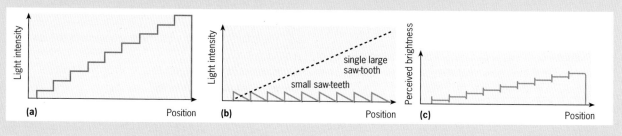

(a)   **(b)**   **(c)**

OSCILLATIONS AND MECHANICAL WAVES

The Chapter Map below brings together the main ideas from this chapter, and shows you how they are inter-connected. In your studies, cross-match the entries in the map with the syllabus you are following, and ensure that you understand the ideas and the way that they are interlinked.

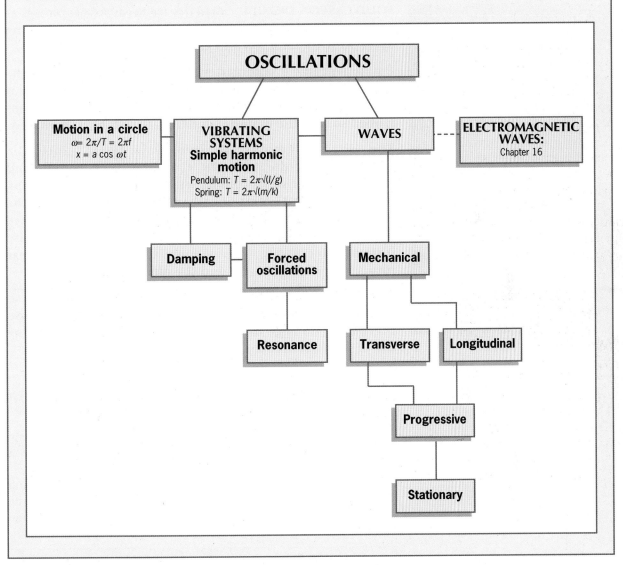

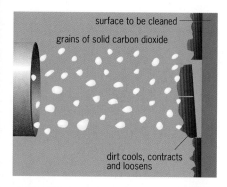

surface to be cleaned

grains of solid carbon dioxide

dirt cools, contracts and loosens

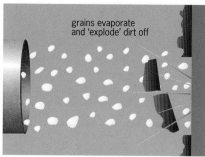

grains evaporate and 'explode' dirt off

Cleaning machinery with dry ice

SOLID CARBON DIOXIDE, called dry ice because it vaporises at −78 °C without first melting, is now being used as a cleaning agent to strip paint, oil and grime from machinery parts.

Equipment like a sandblaster sprays grains of dry ice at the parts. If paint is being removed, the number of layers that come away depends on the strength of the jet. The jet is controlled so that the material underneath will be undamaged.

The dirt is removed in two ways. First, since the temperature of the ice is −78 °C, the dirt cools rapidly and contracts, breaking its contact with the underlying material. Second, the hard grains of carbon dioxide chip away at the grime. Grains manage to pass through the dirt to the surface being cleaned. The impact heats the grains and they vaporise instantly, increasing their volume by 800 times. The gas forms little explosions behind the dirt and so strips it off.

The dry ice evaporates completely in the cleaning process, so it generates no waste, apart from the material it removes. So there is no damage to the environment, in contrast to the use of solvents in more conventional cleaning methods.

1 INTRODUCTION TO THE STATES OF MATTER

Every day, we see water moving in a river or pouring from a tap. We say it is **liquid**. Although water keeps the same volume as it moves, it does not keep its shape.

When water is cooled below 0 °C, it loses its ability to flow and turns to ice. Ice retains its shape as well as its volume. Before the days of refrigerators, blocks of ice were cut in winter (Fig 7.1) and

Fig 7.1 **Cutting blocks of ice on the St Lawrence River in the ninetenth century**

stored in special ice-houses for keeping food cool in summer. The properties of ice are so different from those of water that we say that water changes its 'state of matter' (or 'phase') when it transforms to ice. It has become a **solid**.

When we heat water in a kettle to 100 °C, we get yet another state of matter – it becomes a **gas**. Water is composed of molecules that are constantly moving. These molecules begin to move around so rapidly during the heating that they leave the kettle and join the molecules of air in the room. We say that the water is **evaporating**. To some extent this would happen to the water even if it wasn't heated – heating simply speeds up the process.

A gas does not keep to a fixed volume – the molecules disperse throughout the room.

Note that the 'steam' you see coming from a boiling kettle is not a gas. In fact, this 'steam' is made up of tiny droplets of water, produced when water vapour **condenses** in the cooler air. These droplets are then carried along in the gas flow.

We can see that the main difference between the different states of matter is how the individual atoms or molecules are held (bound) together.

Plasma – a fourth state of matter

If we heat gas molecules to a very high temperature, we obtain a fourth state of matter, a **plasma**. It has been known for hundreds of years that chemical reactions are not sufficient to produce the energies emitted by stars, including the Sun. In the Sun, temperatures are so high that the outer electrons of the atoms are separated from the protons and neutrons of the nuclei. The resulting positively and negatively charged particles move around separately and rapidly as a plasma.

This is how plasmas form. Because stars consist of large amounts of matter, the gravitational force tends to compress the particles into a smaller volume. The pressure increases and the electrons and protons move faster and faster. The temperature rises further and energy is radiated as electromagnetic rays (photons). Energy can only continue to be emitted if a nuclear reaction takes place. With fusion of the nuclei of elements at the low end of the Periodic Table, a minute proportion of their mass is converted to very large quantities of energy. Thus the high temperature is sustained.

Plasmas do not normally occur on Earth, outside the laboratory. A lot of research is carried out on them in the hope of producing large quantities of energy through nuclear fusion for the twenty-first century. It is difficult to produce sufficiently high temperatures to sustain high-energy plasmas (although a plasma is produced by argon-arc welding equipment). In an attempt to create fusion conditions, small capsules of hydrogen or

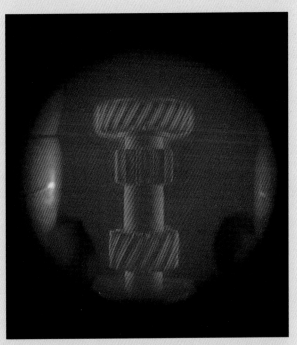

Fig 7.2 **A plasma chamber in which a vehicle transmission gear is being hardened: ions from the plasma 'dissolve' into the steel surface (which is given a negative charge to attract them), and a modified, harder surface is formed**

helium are bombarded by laser beams from several directions to create high temperatures that will produce a plasma, and magnetic fields hold the charged particles of the plasma at high density. But sustained nuclear fusion is yet to be achieved.

There is more about nuclear fusion and the processes in stars in Chapter 27.

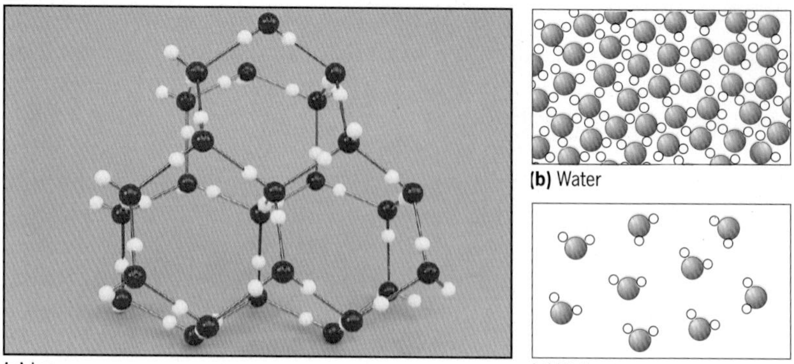

Fig 7.3 **Pattern of atoms in ice, water and water vapour (gas)**

(a) Ice

(b) Water

(c) Water vapour

In a solid they are held so tightly that little movement is possible. The atoms or molecules can vibrate back and forth – the higher the temperature, the more they vibrate. But it is very difficult for them to push past each other. They take up a specific pattern in ice, for example, as shown in Fig 7.3(a).

In a liquid, the atoms or molecules are attracted but cannot move past each other easily. Water consists of molecules composed of two atoms of hydrogen and one of oxygen, although a minute proportion of the molecules (1 in 10^7) may separate into oppositely charged particles called **ions**.

When the water becomes a gas, the gas molecules travel freely at high speeds, as we shall see later. The molecules are spaced wide apart – a gas is nearly one thousand times less dense than a solid.

So, the different states of matter have very different properties. One important property is density, usually labelled ρ (Greek 'rho'). Density is the mass per unit volume. That is, if M is the mass and V is the volume:

$$\text{Density} = \frac{\text{mass}}{\text{volume}}$$

$$\rho = \frac{M}{V}$$

Table 7.1 **Densities of common materials at room temperature (20 °C)**

Substance	Density (kg m^{-3})
Solids	
Aluminium	2 700
Copper	8 930
Gold	19 300
Ice (0 °C)	917
Plastic (PVC)	1 300
Platinum	21 500
Uranium	19 00
Liquids	
Glycerol	1 260
Mercury	13 500
Olive oil	9 200
Water	998
Gases	
Air	1.29
Carbon dioxide	1.98
Helium	0.177
Hydrogen	0.0899
Oxygen	1.43

If we look at the densities of common materials at room temperature we should be able to identify their states. Table 7.1 gives this information for a selection of materials. The table also gives us some surprises.

There are large differences in density between gases and liquids and between gases and solids. The difference in density between a solid and a liquid is much less clear cut and there are some anomalies (peculiarities). For instance, the density of mercury at room temperature is higher than the density of copper – but mercury is a liquid because it can flow and copper is a solid because it is hard and can be cut. As another instance, glycerol as a liquid has a similar density to a typical plastic.

Most substances are denser as a solid than as a liquid. Yet compare water and ice: at 0 °C ice must be less dense than water, since ice forms on the surface of ponds and icebergs float on the surface of the sea. If the ice in the pond did not form on the surface but sank to the bottom, the entire pond would eventually freeze. We go to considerable lengths to insulate water pipes in winter because water expands as it freezes. When the water becomes ice its volume increases, so it breaks open the pipes. Because the ice cannot flow, a householder knows there is a problem when water doesn't flow – or when the ice turns back to water!

?

A What would happen to pond life if ice were less dense than water?

2 CHANGING THE STATE OF A SUBSTANCE

Let us continue to consider water. In order to turn it into ice we must reduce its temperature to 0 °C (zero Celsius) and then take away more energy thermally so that freezing begins. The usual method of extracting this heat is to cool the surroundings to below 0 °C so that the water loses energy to these surroundings.

The energy that water at 0 °C must lose so that it turns to ice at 0 °C is called the **latent heat of fusion**. It is energy that must be taken away from the molecules to reduce their ability to move around; its removal does not actually go towards reducing the temperature. The temperature at which this takes place is called the **temperature of fusion**. For water, the temperature of fusion is 0 °C. If we go in the other direction, that is, if we warm ice to turn it to water, this critical temperature is called the **temperature of melting**. Temperatures of fusion and melting are commonly called melting points.

To turn water into its gaseous state, we first heat it to 100 °C, the boiling temperature. We continue heating to provide the **latent heat of vaporisation**. This is the energy we must give to the water molecules to increase their separation and provide them with the kinetic energy they need to achieve in the gaseous form. This kinetic energy is large, and the latent heat of vaporisation is greater than the latent heat of melting.

Fig 7.4 shows the change in temperature of a substance over time as it is heated at a constant rate through its different phases. During the periods of melting and vaporisation the temperature remains constant. The energy input provides the required latent heat.

Table 7.2 shows the latent heats required to convert 1 kg of some materials from one state to another. This called *specific* latent heat. As latent heat is measured in joules, specific latent heat is measured in joules per kilogram (J kg^{-1}).

For a substance of total mass m and specific latent heat L, we shall need an amount of energy E given by:

Energy = mass × latent heat for 1 kg

$$E = mL$$

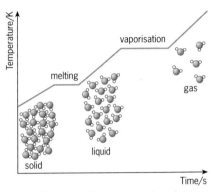

Fig 7.4 **Variation of temperature of a substance as it is heated at constant rate through its states of matter**

Table 7.2 **Latent heats and melting and boiling points of common substances**

Substance	Melting point (°C)	Latent heat of fusion (MJ kg^{-1})	Boiling point (°C)	Latent heat of vaporisation (MJ kg^{-1})
Aluminium	661	0.38	2520	10.8
Copper	1085	0.21	2590	4.7
Gold	1065	0.07	2850	1.7
Mercury	−39	0.016	357	0.29
Oxygen (O$_2$)	−219	0.014	−183	0.22
Water	0	0.333	100	2.257

Dependence of melting and boiling temperatures on pressure

So far, we have assumed that the water is at standard pressure. In fact the temperatures of melting and boiling alter if there is a change in pressure of the surroundings. The change in pressure needs to be large for us to notice the effect, but we do commonly meet it.

If we increase the pressure on a block of ice, the freezing temperature of the ice decreases. This is what happens when we ice-skate. The

B For water, find the ratio of the latent heats of fusion and vaporisation. Compare this ratio with the ratio for other substances. Comment.

C An electric kettle contains three litres (3 × 10^{-3} m^3) of water at 100 °C.

(a) Calculate the amount of energy required to boil the kettle dry assuming that all the energy is used in the boiling process and none is lost to the kettle or its surroundings.

(b) Given that electrical energy costs 7.5p per kilowatt-hour, find the cost of boiling all the water.
(Take the density of water to be 10^3 kg m^{-3}.)

Fig 7.5 **Skating on ice. The pressure under the blade causes the ice to melt**

sharp blades of the skates apply a large force to the ice over a small area (Fig 7.5). There is a very large pressure under the blades which helps the ice melt under the blades. The water effectively lubricates the skating surface.

The temperature of the ice and the width of the blades both need to be carefully controlled. If the skates were designed with very broad blades, they would not apply enough pressure. Even using narrow blades, it is not possible to skate on ice that is at too low a temperature.

The temperature of vaporisation of water is lowered by a *decrease* in pressure, so that water boils at a lower temperature on a mountain top, where the atmospheric pressure is lower, than at sea level. A mountain is not the place to make a good cup of tea: though the boiling temperature is not lowered much, the drop is enough to reduce the amount of flavour extracted from the tea leaves!

Sublimation

Some substances can change directly from the solid state to the vapour state. This process is called **sublimation**. The temperature and pressure may need adjusting for this to happen. For example, solid carbon dioxide (dry ice) remains solid up to −78 °C, then it changes directly into the gas. Dry ice is used in research laboratories for keeping equipment and samples cold. It is also used in the theatre for creating an artificial fog – the evaporating carbon dioxide molecules cool the air and water in the air condenses out as droplets.

3 ATOMS AND MOLECULES

In Chapter 5, we saw that solids can form many different structures at the microscopic level. In all cases, the solids consist of atoms held closely together, though the arrangements of these atoms can differ considerably. Here we are going to look at some of these patterns, especially those that are regular. For this, we use a very simple model in which the atoms are considered to be identical hard spheres. Bringing the atoms close together is similar to packing marbles or polystyrene spheres next to each other.

If we push the marbles or polystyrene spheres together, they resist compression once their outer surfaces touch. Since marbles are hard, they resist compression more than the polystyrene spheres. In both cases, there is a force opposing the movement.

In a similar way, there will be a force of repulsion when two atoms are brought together. Here it is due to the outer electrons repelling each other as the atoms become close. The force of repulsion, shown by the red curve in Fig 7.6, gets much larger as the separation between atoms is reduced.

We have no difficulty in separating the marbles or polystyrene spheres – there are no attractive forces to overcome. However, atoms do resist separation, because there is an attractive force. It has a negative value. But this attractive force must get rapidly smaller as the separation of the atoms increases – the blue curve in Fig 7.6 shows approximately how the force of attraction varies with distance. It will be zero at large separations. Similarly, the force of repulsion falls off as distance increases. Note that this fall-off is more rapid than for the attractive force, so atoms still attract each other after repulsion has approached zero.

We must add together the attractive and repulsive forces to find the net force at any separation. This is the same as adding the two curves together. The resulting force–separation curve will depend on the relative sizes of the repulsive and attractive forces. In Fig 7.6, the final force–separation curve, the purple curve, is a good representation of the force between a pair of adjacent atoms.

What can we tell from this curve?

- At a separation d_0 the net force between the two atoms is zero. At this separation, the repulsive force exactly balances the attractive force. This will be the *equilibrium* position of the two atoms. At this separation, they remain stable.

- At very large separations the force between the atoms approaches zero. But the force of attraction does start to have an effect when approaching particles are relatively far apart.

- There is a large force of attraction for separations slightly greater than d_0. This attractive force must be overcome if we wish to separate the atoms.

- If the two atoms are brought closer together than d_0, the force of repulsion becomes very large.

Pulling atoms apart or pressing them closer together is similar to extending or compressing a spring. This is why we can apply Hooke's law (page 88) both to extending springs and to deforming solids.

If we pull the atoms apart against the attractive force trying to hold them together, we must do work. This is like doing work to raise a mass upwards in a gravitational field. Just as the mass being raised in the gravitational field gains potential energy, so these two atoms **gain potential energy** as they separate.

We can plot the potential energy gained by these atoms as their separation changes. As is usual (see page 64 in Chapter 4), we set the potential energy to be zero for very large separations. When the atoms are moved closer together, the potential energy drops, just as the potential energy of a mass decreases as it falls to earth. Since it decreases from a zero value, the potential energy must become more negative (see Fig 7.7). The equilibrium position d_0, where the net force on the atoms is zero, corresponds to the minimum in the potential energy curve. For a very small change of separation from d_0 the potential energy does not change.

Take care! The potential energy minimum in Fig 7.7 does not correspond to the minimum on the resultant purple force curve in Fig 7.6. The minimum on the force curve corresponds to the separation where there is the biggest net attractive force, not zero net force.

To decrease the separation further, the atoms must now be moved against the force of repulsion. The potential energy starts to rise again. Eventually it reaches a very large positive value.

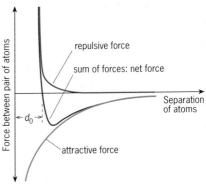

Fig 7.6 **Forces of repulsion (red) and attraction (blue) between two atoms and also their sum (purple)**

D Redraw the three curves shown in Fig 7.6 for a pair of atoms in which the force of repulsion is weaker and the force of attraction is stronger.

Energy change equals force multiplied by distance moved. Since the net force is zero, there can be no change in energy.

See questions 1, 2 and 3.

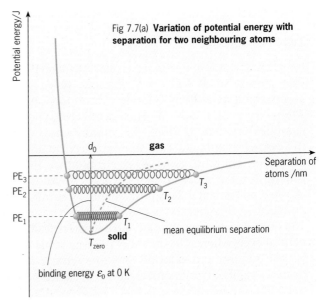

Fig 7.7(a) **Variation of potential energy with separation for two neighbouring atoms**

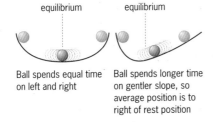

Fig 7.7(b) **Model for the potential energy of a pair of atoms**

equilibrium — Ball spends equal time on left and right

equilibrium — Ball spends longer time on gentler slope, so average position is to right of rest position

The potential energy curve and properties of materials

We have seen that at equilibrium the separation of two atoms corresponds to the position of minimum potential energy. The atoms will remain at separation d_0 provided the temperature is at absolute zero. (Later in your studies you will discover that reaching absolute zero is not *quite* possible because of the 'uncertainty principle'.)

If we give the atoms some energy by heating them to a finite temperature T_1, the potential energy (PE) increases to a value PE_1, see Fig 7.7(a). The atoms are bound together by a potential energy $\in_0$ equal to the maximum depth of the curve. The energy that heating gives to the atoms makes them vibrate towards and away from each other, in response to the repulsive and attractive forces. We have pictured the distance between the atoms as a spring between two points on the curve at T_1. These two points represent the minimum and maximum separations of the atoms at that temperature.

Alternatively, we can think of a marble oscillating backwards and forwards in a bowl within a gravitational field, as in Fig 7.7(b). The more energy we give the marble, the bigger the oscillations as it rises further up either side of the bowl. It is quite helpful to think in terms of an oscillating marble in a bowl that has the same shape as the potential energy curve for our two atoms.

Increasing the temperature of the atoms, say to T_2 and then T_3, giving total potential energies PE_2 and PE_3 respectively, makes them vibrate relative to each other by larger amounts. This is shown on the diagram as stretched springs at T_2 and T_3. The higher the energy of the atom pair, the wider their range of separations.

The mean separation of the atoms is the mid-point of these lines of vibration. We immediately see that, because the potential energy curve is not symmetrical about the minimum, the mean separation increases as the temperature increases. Look at the marble oscillating in the similarly shaped bowl. When this happens for many atoms, the average separation of all the atoms has increased. Our solid has *expanded* with increase in temperature.

If the temperature of the atoms is raised until their potential energy is zero, the atoms can fly apart – the gaseous state has been reached. Somewhere between the gaseous state and the solid state (and we cannot be sure where this is on our figure), there is sufficient movement for the atoms to be in the liquid state. The depth (ε_0) of the potential energy curve will be related to the sum of the latent heats of the substance.

The type of force–separation curve we have looked at is very helpful for describing the interaction between atoms. Atoms in solids bond together in different ways, and the precise form of the interaction will vary according to the type of bonding involved. A summary of the main types of bonding found in solids is given in the next Extension box.

Types of bonding

Ionic bonding is found in many crystalline materials that consist of at least two different types of atoms, such as sodium chloride – common salt. Atoms try to fill their outer shell of electrons. For example, one type of atom, the sodium atoms in the case of sodium chloride, loses an electron and the other, chlorine, gains an electron. Once the sodium has lost an electron it becomes positively charged and is said to be a positive ion; by gaining an electron the chlorine becomes a negative ion. The crystal of sodium chloride must be electrically neutral everywhere. The only way this is possible is for the sodium and chlorine ions to alternate in nearest neighbour directions, as shown in Fig 7.8(a).

Electrical forces act over a relatively long distance, so bonding is not directional but it is strong. It is difficult for ions of opposite charge on different planes to slip past each other, so ionic crystals are not plastic (see page 98). They usually fracture before reaching their elastic limit.

Ionic crystals are poor conductors of electricity as neither the ions themselves nor individual electrons are free to move. When put into water, ionic crystals dissolve: the water reduces the electrical forces between the ions and so the ions can separate. They now move around and can conduct electricity (they move, carrying charge).

In **covalent bonding** atoms share, rather than

exchange, electrons in the region between the two atoms. Unlike the ionic bond, the covalent bond is highly directional. Molecules produced by such bonding retain a definite shape. Oxygen, O_2, is a covalently bonded molecule. Both oxygen atoms would like two more electrons in their outer shells and, as shown in Fig 7.8(b), they achieve this by each providing two electrons for sharing (making a total of four shared electrons).

Covalent bonding can lead to highly extended 'molecules'. Diamond is such an example, as shown in Fig 7.8(c). Carbon has four electrons in its outer shell and by sharing a further four electrons, one from each of four other carbon atoms, it forms a continuous structure. The four bonds are symmetrically arranged in space and are called tetrahedral bonds.

Silicon and germanium also have four electrons in their outer shells (carbon, silicon and germanium are all from Group IV of the Periodic Table, see Table 18.2), and so they bond in the same way. The arrangement of the outer electrons of silicon and germanium gives rise to their semiconducting property. This property has been the basis of the development of semiconductor devices and the advance of much of the electronics industry. Unlike ionic compounds, covalent materials do not conduct in aqueous solution.

In metals, each atom has one or more outer **delocalised** electrons that are free to move between atoms. The metal can be pictured as a giant structure of positively charged ions in a sea of electrons, as in Fig 7.8(d). This is **metallic bonding**.

We might expect these ions to repel each other and fly apart. Not so. The interaction between the electrons and the ions sets up a counteracting attractive force. Why this should happen was not properly understood until physicists came up with a new theory called quantum physics (see Chapter 17).

The fact that the electrons can move freely around accounts for the good electrical conductivity of metals. The interaction between the large number of ions and the 'sea' of electrons is not directional. The ions, though, stay fixed, often close packed together. If a tensile or shearing force is applied to the metal, it is easy for the planes of ions to move past each other. So metals exhibit plasticity as we saw in Chapter 5 (page 97).

Even neutral atoms attract each other weakly. The atoms are composed both of positively charged protons in their nuclei and moving electrons in a surrounding cloud that constantly changes shape. In a pair of like atoms, the positive and negative charges from one atom and the positive and negative charges from the other produce a small instantaneous force, as in Fig 7.8(e). Although the force changes as the electrons move, it averages over time to produce an attraction that is called **van der Waals bonding** and occurs between all atoms and molecules.

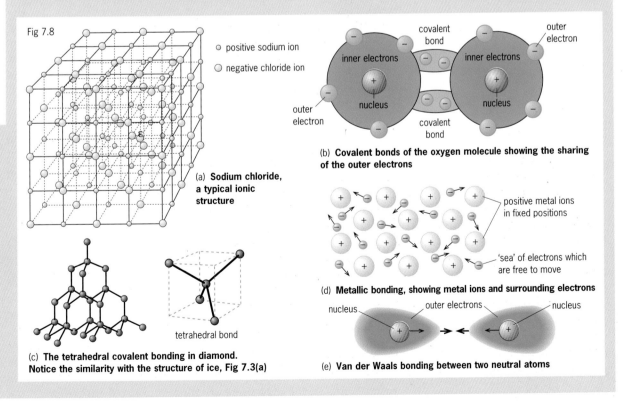

Fig 7.8

○ positive sodium ion

○ negative chloride ion

(a) **Sodium chloride, a typical ionic structure**

(b) **Covalent bonds of the oxygen molecule showing the sharing of the outer electrons**

covalent bond

outer electron

inner electrons

inner electrons

nucleus

nucleus

outer electron

covalent bond

tetrahedral bond

(c) **The tetrahedral covalent bonding in diamond. Notice the similarity with the structure of ice, Fig 7.3(a)**

positive metal ions in fixed positions

'sea' of electrons which are free to move

(d) **Metallic bonding, showing metal ions and surrounding electrons**

nucleus

outer electrons

nucleus

(e) **Van der Waals bonding between two neutral atoms**

4 CRYSTAL STRUCTURES

Adding together layers of spheres is similar to adding together layers of atoms. The final structure will depend on how we build up the layers and whether these layers themselves are made up of identically sized spheres or spheres of different sizes. Substances that are built from an ordered stacking of atoms are called crystals. These often possess flat faces that match regular planes of atoms. Most crystals can be **cleaved**. This means that a hard blow with a sharp edge will split the crystal along a plane to create a pair of flat surfaces. Because they can be cleaved, diamonds and other gemstones have attractive, shiny surfaces and are valuable for jewellery.

E In the simple cubic structure, what is the total number of atoms per cubic cell?

Simple cubic structure

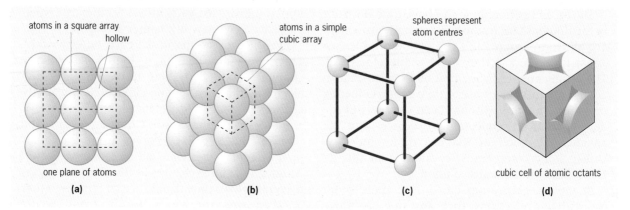

Fig 7.9 **Different ways of representing a simple cubic structure**

Let us continue to think of atoms as hard spheres. They can be arranged so that they touch in a single layer as a square array – see Fig 7.9(a). Note that there are four atoms round a hollow. Next we can arrange another layer of spheres vertically above the first, and so on – see Fig 7.9(b). Such a three-dimensional arrangement is called **simple cubic**.

Taking eight atoms from the arrangement in Fig 7.9(b), and reducing their size, their centres are shown joined by lines in Fig 7.9(c). Note that each atom will also be at the corners of eight cubes, the one in (b) and a further seven cubes. Alternatively, look at Fig 7.9(d) with the atoms (spheres) as in (b). It is clear that the part of an atom at each corner is an eighth of an atom.

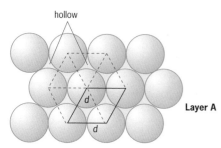

Fig 7.10 **Close packing spheres in a single layer**

F By considering spheres of diameter d, compare the densities of packing within a single plane for the simple cubic arrangement and for the close packing arrangement.

Hint: find the area of a square joining the centres of four atoms in Fig 7.9(a) and the area of a parallelogram joining four atoms in Fig 7.10.

Close packing structures

The spheres in Fig 7.9(a) are not packed together as closely as possible. Fig 7.10 shows a different arrangement of spheres in a single layer. Here, a second row of touching atoms is displaced relative to the first one. The third row is a repeat of the first row. Comparing with Fig 7.9(a), note that there are only three atoms at a hollow but that the closest separation of centres of the atoms remains the same, equal to the diameter of the atom. The result is that in Fig 7.10 there are more atoms per unit area.

Building close packed structures: hcp and fcc

We can build a regular close packed structure on the Fig 7.10 layer in two different ways, as follows:

Method 1

- We call our first layer of close packed spheres (atoms) layer A. Notice that each sphere touches six others (nearest neighbours) and is surrounded by six hollows (Fig 7.11).

- Now we start to build up a three-dimensional structure. We place the next plane of spheres into hollows. Notice that, even if we make the next plane of spheres close packed, we can only centre spheres in half the hollows. At this stage it doesn't matter which half-set of hollows we use. We label this second layer B.

- Next we add the third layer of spheres and again we can only fill half the hollows. This time the choice makes a difference.

- By filling one set of hollows, we can place all the spheres in this third layer vertically above spheres in the first layer. So we label this layer as another A layer. Note the hollows in layer B of Fig 7.11 in which the third-layer spheres are placed.

We can picture a hexagonal cell made up within these three layers, as shown in Fig 7.12(a). Fig 7.12(b) shows the conventional picture of the hexagonal cell with the spheres (atoms) shrunk to a small size. It is easy to see that we can continue with this ABABAB... arrangement of the layers. We call this pattern a **hexagonal close packed (hcp) structure**.

Method 2

- Let us return to packing our third layer and look again at Fig 7.11. We already have layers A and B. We now choose the alternative half-set of hollows to the set we chose previously. The spheres of the third layer no longer lie vertically above the spheres of layer A, nor do they lie above the spheres of layer B. So we label this third layer C. We go on to consider a fourth layer. Again there is a choice of hollows, but we choose to place the spheres in hollows such that they lie vertically above the atoms in layer A. This produces another A-type layer. Packing on upwards produces an ABCABCABC... pattern. Fig 7.13 shows such layers packing

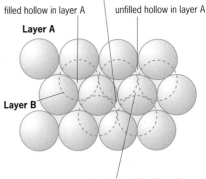

1 The second layer hollow is vertically above first layer hollow - filling it creates a new layer C

filled hollow in layer A   unfilled hollow in layer A

Layer A

Layer B

2 The second layer hollow is vertically above the atom in the first layer - filling it creates a further A layer

Fig 7.11 **Two different ways of adding the third layer for close packing of spheres**

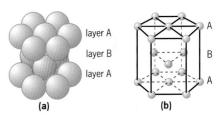

layer A

layer B

layer A

A

B

A

(a)   (b)

Fig 7.12 **Hexagonal close packing (hcp)**

■ See question 4.

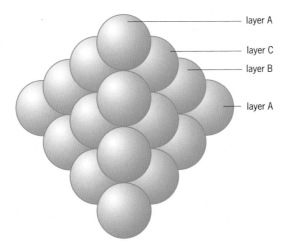

layer A

layer C

layer B

layer A

Fig 7.13 **Face-centred cubic packing (fcc) within a pyramid with a triangular base**

G By considering sharing of atoms between adjacent cubic cells, work out the number of atoms per cubic cell for the face-centred cube.

in a pyramidal form starting from a triangular base. The structure is still close packed and the arrangement is called **face-centred cubic (fcc) close packing**.

As the name suggests, a face-centred cube has atoms at the corners of a cube and atoms at the centres of each face, as in Fig 7.14(c). The close packing structure that we have described can be related to a series of face-centred cells by tilting the close packed planes. You should be able to see the relationship in Figs 7.14(a) and (b). The close packing planes link up three diagonally related corners of a cube. Try to identify how an atom in such a plane will be surrounded by six nearest neighbours within the same plane, and by three atoms in each of the two adjacent planes. All the outer surfaces of the pyramid shown in Fig 7.14(b) are of close packed planes of atoms.

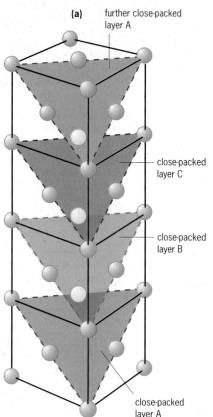

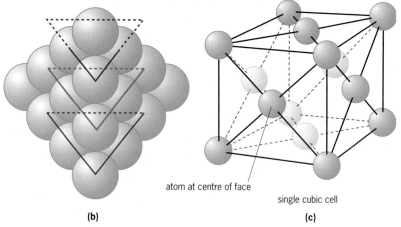

(b)

(c)

Fig 7.14 **Face-centred cubic (fcc) packing structure showing the close packed layers**

Body-centred cubic structure

Another common crystal structure is called **body-centred cubic (bcc)** because there is an atom at the centre of the cube (that is, at the centre of its 'body') in addition to the atoms at the corners (Fig 7.15). Iron has this structure up to 800 °C, at which temperature it changes to fcc. This is an example of a solid changing its phase but remaining a solid. Heating iron to above 800 °C and quenching it in water prevents the iron from returning to a bcc structure. The face-centred form of iron has different properties from the body-centred form – in particular it is very brittle. Iron containing carbon as an impurity gives steel. Such variations in the properties of iron are immensely important for the science of metallurgy and for the iron and steel industry.

Determining the structures of crystals

We find out about the structure of crystals by probing them with radiation. The spacing of the layers of atoms is about 0.5 nm, so we need radiation with a wavelength of similar size. X-rays are electromagnetic waves with a wavelength of about 1 nm. When the X-rays hit the crystal, they are diffracted by the planes of atoms, just as light is diffracted by a narrow slit or by a diffraction grating.

Max von Laue in 1912 was the first to observe this effect. Photographic film is used to detect the diffraction pattern (Fig 7.16), seen as a series of dots on the film. We can see a similar pattern

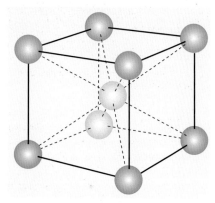

Fig 7.15 **Body-centred cubic (bcc) structure**

when we pass a beam of light from a laser through two diffraction gratings crossed at right angles. In a crystal, it is the many parallel planes of atoms that produce the diffraction effect, just like the large number of rulings of the gratings. By measuring the distances between spots on the film, the spacing of the planes can be calculated.

Nowadays, we use solid state detectors, not film, to detect the X-rays. Measured intensities are digitised to store as a computer record, and computer software packages calculate the crystal structures and the positions of the atoms in the planes. Even the molecular structures of complex molecules such as DNA have been worked out from X-ray analysis.

We can use electrons instead of X-rays for studying crystal structures. Electron microscopes give pictures similar to those from an optical microscope. Alternatively, electron microscopes can be set to produce diffraction patterns, because electrons can behave as waves. (You will learn more about particle–wave duality in Chapter 17.)

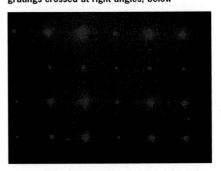

Fig 7.16 **A Laue diffraction pattern, above. Compare it with the diffraction pattern for two gratings crossed at right angles, below**

The structures of other solids

Although many materials have a regular crystalline structure, others do not. Such materials are described as **amorphous**, which means 'without form'. Alternative descriptions used are **glassy** and **solid liquid**. So, not surprisingly, ordinary glass is an amorphous material. It is made from silica (SiO_2) with added oxides.

A diffraction pattern for glass does not show spots. Instead, it shows a couple of very diffuse rings (Fig 7.17) that arise because the separations of nearest and next nearest atomic neighbours have different average values. But there is no order at longer range.

Glass is hard and brittle and surface cracks weaken it considerably (see also page 107). Glass cutters use this property extensively. They scratch the surface and break the glass along this scratch. As a demonstration, a glass rod is supported at its ends and loaded centrally (Fig 7.18); the load is increased until the rod breaks. The procedure is repeated, but using a glass rod with a small scratch on the under-surface. The rod now breaks, but with a much smaller load.

Glass softens when heated and then it is possible to shape it, as glass blowers do. It can be drawn out into the very fine fibres that carry light waves in fibre optic communications. Although not a liquid in the true sense, glass flows very slowly. Very old glass windows are thicker at the bottom than at the top as a result of slow flow in the Earth's gravitational field.

Polymers are made up of chains of atoms forming long molecules. The backbone of the molecule is usually a series of carbon atoms covalently bonded together with other types of atoms attached. There are many examples of polymers in the home – polythene (made from polyethene molecules), nylon, PVC and Perspex are a few examples. The molecular chains can bend and become tangled. Some regions of the polymer may be entirely amorphous, while other regions may be composed of molecules arranged in a regular (crystalline) form. Rubber is such a polymer – Chapter 5 (page 99) describes how the molecules of rubber are more tangled in its original shape than when stretched.

There are different types of polymer. **Thermosetting polymers** can be moulded: they change to the required shape when heated and then retain this shape when cooled. If reheated, they decompose rather than soften and distort. Others, called **thermoplastics**, soften on reheating, become flexible and then melt.

Fig 7.17 **X-ray diffraction photograph for glass (an amorphous material)**

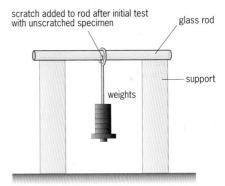

scratch added to rod after initial test with unscratched specimen

glass rod

support

weights

Fig 7.18 **Testing the strength of a glass rod**

6 LIQUIDS

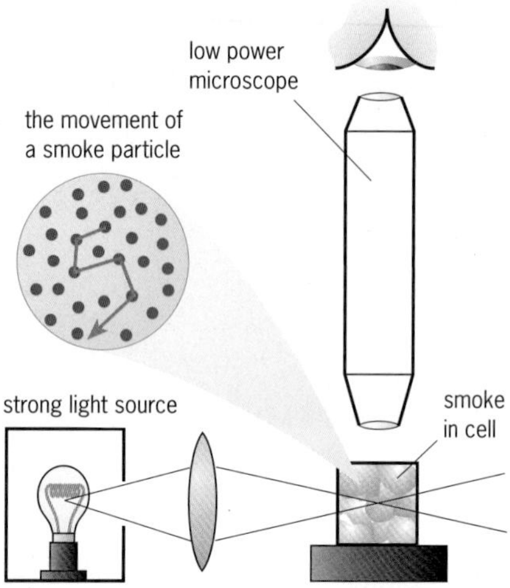

We have seen that there is no long-range order between molecules within a liquid. An X-ray picture of a liquid is similar to that for an amorphous solid such as glass. Molecules move around relative to each other and do not keep the same molecules as neighbours. This ability of the molecules to move easily past each other allows liquids to flow. A liquid does not maintain its shape if shear forces are exerted on it, although a liquid can withstand compressive (hydrostatic) forces, because it maintains its volume.

We can see that liquid molecules move easily by observing **Brownian motion**. The botanist Robert Brown first noticed the effect in 1827 when looking at pollen grains suspended in water. The pollen grains have sufficiently small mass to be buffeted around by the unseen water molecules. A similar arrangement is shown in Fig 7.19. Smoke particles can be seen moving in air, demonstrating that gas molecules also collide with their neighbours.

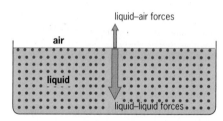

Fig 7.19 **Demonstration of Brownian motion**

Surface tension

Liquids behave as if there is a very delicate skin over their surface. For example, a pin can be balanced on the 'skin' of a water surface, and some insects can walk on the surface of ponds. This is possible because of **surface tension**: surface molecules tend to be further apart than underlying molecules, and attractive forces between neighbouring liquid surface molecules are stronger than forces between underlying molecules, as shown in Fig 7.20.

We shall return to liquids later to study other properties. For instance, we shall find out about the importance of buoyancy and fluid flow to forms of transport (Chapter 8) and more about the passage of thermal energy through a liquid (Chapter 14).

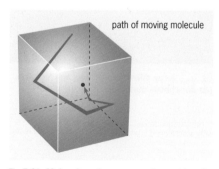

Fig 7.20 **The surface of a liquid acts like a skin**

7 GASES

We have seen that the atoms in a solid vibrate but do not move around. In a liquid, atoms or molecules are able to drift around, but they only alter their mean positions slowly. In contrast, the atoms or molecules in a gas move around rapidly over large distances. There is no pattern to the movement.

As they speed around, the molecules (or atoms) of a gas keep colliding with the walls of their container (Fig 7.21). As the molecules rebound off the walls, there is a change in their momentum.

Each molecule exerts a small force on the walls during a very small interval of time (Fig 7.22). All these small forces add up so that a large number of such collisions produce a total average force on the walls that is measurable. Assuming that the molecules are moving *equally in all directions*, the force per unit area of wall will be the same over all the walls of the container. This force per unit area is the **pressure** exerted by the gas and it is equal in all directions.

Fig 7.21 **Molecules of a gas speed around and collide with the walls of their container**

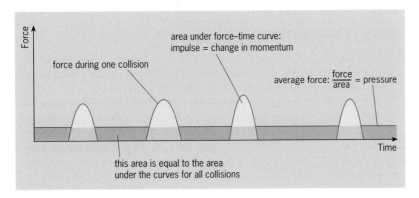

The ideal gas equation

In the middle of the seventeenth century, Robert Boyle showed experimentally that:

The volume of a fixed mass of gas is inversely proportional to the pressure applied to it if the temperature is kept constant.

That is, at constant temperature and mass:

$$\text{pressure} \propto \frac{1}{\text{volume}}$$

If p is the pressure and V the volume of the gas, for constant temperature T and constant mass we have:

$$p \propto \frac{1}{V} \quad \text{or} \quad pV = \textbf{constant} \qquad \text{[Law 1]}$$

The law is known as **Boyle's law** and applies provided the pressure of the gas is low. In addition to Boyle's law, there are two other gas laws.

The **pressure law** states that:

The pressure of a fixed mass of gas at constant volume is proportional to temperature as measured on the Kelvin scale.

That is, for constant volume and mass:

$$\text{pressure} \propto \text{temperature}$$

$$p \propto T \quad \text{or} \quad \frac{p}{T} = \textbf{constant} \qquad \text{[Law 2]}$$

■ See questions 5 and 6.

This is a fundamental relationship for an ideal gas (real gases need to be at a low pressure). Note that we have not so far said precisely what we mean by temperature. The Extension box on page 155 describes how a temperature scale is obtained.

The other law is called **Charles' law**:

The volume of a fixed mass of gas at constant pressure is proportional to temperature as measured on the Kelvin scale.

That is, for constant pressure and mass:

$$\text{volume} \propto \text{temperature}$$

$$V \propto T \quad \text{or} \quad \frac{V}{T} = \textbf{constant} \qquad \text{[Law 3]}$$

BOYLE AND CHARLES – ORIGINATORS OF THE GAS LAWS

ROBERT BOYLE (1627–1691) was the youngest child of the Earl of Cork. He was privately taught and then went to Eton College before going on a Grand Tour of Europe. His main work was as a chemist, and he is known as the scientist who established chemistry as a separate science.

While working at Oxford, he improved an air pump made by Hooke and showed that objects in a vacuum fall at the same velocity under gravity. Besides publishing his gas law, he proposed the idea of chemical elements. He incorrectly believed that base metals could be transformed into gold and his attempts to convert metals led to the repeal of an English law against producing gold and silver from other substances.

Jacques Charles (1746–1823) was originally a clerk in the civil service. The French physicist became interested in gases from his experience of ballooning. In 1783 he made the first ascent in a hydrogen balloon. A few years later he formulated his famous law, and sent it to Gay-Lussac who made more accurate measurements to support it. Hence, in France, the law is called Gay-Lussac's law.

Demonstrating Boyle's law

As shown in Fig 7.23, a small amount of air is trapped at the closed end of a glass tube by an oil column. The tube is calibrated for volume. The other end of the tube connects with a chamber containing air. The air pressure in the chamber, and hence the pressure applied to the oil column, can be varied using a foot-pump. Whenever an adjustment is made to the pressure, time must be allowed for the gas to achieve thermal equilibrium with its surroundings at constant room temperature. To show Boyle's law, it is necessary to plot p against $1/V$, or pV against V.

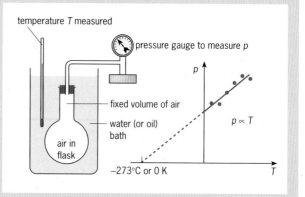

Fig 7.24 **Demonstrating the pressure law**

for higher temperatures) and the variation of pressure with temperature is measured.

Demonstrating Charles' law

A small amount of gas (air is suitable) is trapped in a calibrated capillary tube, usually by a bead of concentrated sulphuric acid to ensure that the gas remains dry (Fig 7.25). The capillary tube is inserted in a water bath. The length of the gas column (proportional to the volume of the gas) is measured as the temperature of the water bath is varied.

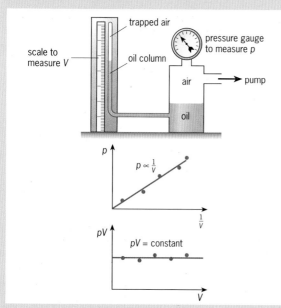

Fig 7.23 **Demonstrating Boyle's law**

Demonstrating the pressure law

As shown in Fig 7.24, a volume of gas is trapped in a flask that is connected to a pressure gauge. The flask is inserted in a water bath (or oil bath

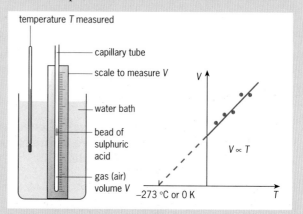

Fig. 7.25 **Demonstrating Charles' law**

Putting the three laws together we obtain the combined **gas law**. For a constant mass of ideal gas:

$$\frac{pV}{T} = \text{constant}$$

or:

$$\frac{p_1 V_1}{T_1} = \frac{p_2 V_2}{T_2}$$

Subscript 1 refers to the initial conditions for the ideal gas and subscript 2 refers to a later set of conditions.

If a real gas obeys the combined gas law equation, then it is behaving ideally. The equation assumes that there is no attraction, no repulsion and no collisions between the atoms or molecules of the gas. Real gases usually cool if allowed to expand – this property is exploited, for example, to liquefy helium from the gaseous phase at very low temperatures. But an ideal gas will not cool down if it is allowed to expand freely (that is, if it is allowed to expand into a vacuum) and an ideal gas cannot be turned into a liquid.

H Suggest why the pressure of a gas needs to be low for it to obey the gas laws.

Ideal gas temperature scale – Kelvin temperature scale

We have seen that we can put $p_1 V_1/T_1 = p_2 V_2/T_2$, where subscripts 1 and 2 refer to two sets of conditions for our ideal gas. Alternatively, $pV \propto T$. The following explains how the accurate calibration of temperature can be made.

We measure pV as we vary T (using a thermometer) over a range in which the gas behaves ideally. We get a straight line, showing this proportionality, as in Fig 7.26. The plot is that of a linear equation:

$$y = mx + c$$

where m is the gradient and c is a constant.

To establish the calibration line we use straight proportionality to extend the line through to the origin $pV = 0$, $T = 0$. This sets $c = 0$ and defines one fixed point.

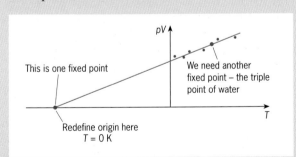

Fig 7.26 **pV is proportional to T for an ideal gas**

We now need only fix one other point to define the calibration line. This second fixed point must be reproducible by any scientist anywhere on Earth. The fixed point is chosen as the triple point of water, a unique temperature at which water vapour, liquid water and ice coexist in equilibrium, see Fig 7.27(c). To obtain this point experimentally, a triple point cell, shown in Fig 7.27(c), is used. Air has been removed, leaving a pressure due to water vapour alone. This pressure is 0.61 kPa, which is, of course, very much lower than atmospheric pressure (101.4 kPa). For historical reasons related to earlier definitions of the Celsius and centigrade scales, this triple point is defined as 273.16 on the Kelvin scale, temperature unit **kelvin** (K). With the Celsius values of –273.16 °C for absolute 'zero' and 0 °C for the triple point, this makes the interval between kelvins the same as between degrees Celsius.

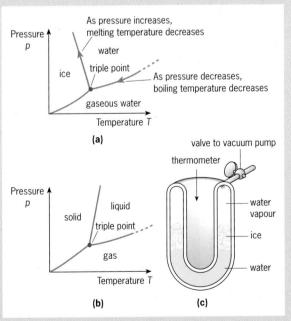

Fig 7.27 **Phase diagrams for** (a) **water**, (b) **a more typical substance.** (c) **A triple point cell: temperature = 273.16 K, pressure = water vapour pressure (0.16 kPa)**

We now have, for a particular mass of ideal gas:

$$\frac{(pV)_{\text{at unknown temperature}}}{\text{unknown temperature}} = \frac{(pV)_{\text{at triple point temperature}}}{273.16}$$

If we call the unknown temperature T and rearrange the equation we get:

$$T = \frac{(pV)_T}{(pV)_{\text{tr}}} \times 273.16 \text{ K}$$

where the subscript 'tr' refers to the triple point. This gives us a primary scale of temperature.

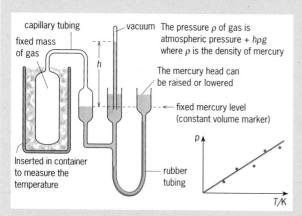

Fig 7.28 **A constant-volume gas thermometer**

We could go on to show that this scale is identical to a thermodynamic scale defined by the use of a reversible heat engine (see Chapter 15). But, while the thermodynamic scale is a theoretical concept, here we have a practical method of

measuring temperature. We use a gas thermometer containing a small mass of a gas and calibrate the product pV to obtain temperature. To simplify, we keep the gas volume constant and change only the pressure: see Fig 7.28.

Once we have obtained our primary scale, we can cross-check it using other physical properties to measure temperature, such as the expansion of mercury with temperature, the variation of the resistance of a metal or the electromotive force (e.m.f.) set up between the junctions of two different metals at different temperatures (as in a thermocouple). We return to practical aspects of temperature measurement in Chapter 14.

The Avogadro constant and molar mass

The **mole** is the unit used to measure the amount of a substance. One mole of any substance contains 6.02×10^{23} 'elementary entities', which may be atoms or molecules. This number is the Avogadro constant, symbol N_A. It was derived by finding experimentally the number of atoms in 0.0120 kg (12 g) of carbon–12. The mass of any other element containing 6.02×10^{23} atoms or molecules will then be the **molar mass**, symbol M, of that element. For example, the molar mass of iron is 0.0585 kg and it contains 6.02×10^{23} atoms.

Alternatively, gaseous oxygen has two atoms per molecule, and here the molar mass, which is 0.0320 kg, contains 6.02×10^{23} molecules, but twice as many atoms.

Finally we need to know the constant of proportionality in the gas equation. From experiment we find that the constant depends on the mass of gas involved. If we double the mass of gas (this means doubling the number of moles of gas) at constant pressure and temperature, then the volume of gas doubles. So pV/T is proportional to the amount of gas enclosed, that is, proportional to the number of moles n:

$$\frac{pV}{T} = nR \quad \text{or} \quad pV = nRT$$

where R is the constant of proportionality and is called the **molar gas constant**. It is a fundamental quantity that can be measured to high accuracy and it has units. The constant can be found from the fact that 1 mole of an ideal gas at standard temperature (273.16 K) and pressure (1.014×10^5 Pa) occupies 0.0224 m³:

$$R = \frac{1.014 \times 10^5 \times 0.0224}{1 \times 273.16} \frac{\text{Pa m}^3}{\text{mol K}} = 8.31 \text{ J K}^{-1} \text{ mol}^{-1}$$

(Note that Pa m³ ≡ N m⁻² m³ ≡ N m ≡ J.)

This equation, $pV = nRT$ for n moles of gas, or the alternative form, $pV_m = RT$ for one mole of gas, where V_m refers to molar volume, is called the **universal gas law equation**.

EXAMPLE

Q A constant-volume gas thermometer shows a difference in height in its mercury column of 8.00 cm at the triple point of water.

a) What difference in height does it read at
(i) the boiling point of water and
(ii) the melting point of lead (327 °C)?

b) With the density of mercury 1.35×10^3 kg m^{-3}, convert the readings to values of pressure in kPa.

A
a) We have:
$$\frac{p}{T} = \frac{p_{tr}}{T_{tr}}$$

where p and T are the pressure and temperature read by the thermometer when in thermal contact with a substance, and p_{tr} and T_{tr} are the pressure and temperature when it is in contact with water at its triple point of 373 K. (Remember to convert temperatures to kelvin.) As the difference in height of the mercury column is proportional to pressure, we can obtain height

difference for **(i)** and **(ii)** by direct proportionality:

i) p (of boiling water) $= \dfrac{8.00 \times 373}{273} = 10.9$ cm Hg.

ii) p (of melting lead) $= \dfrac{8.00 \times 600}{273} = 17.6$ cm Hg.

b) A 1 cm column of mercury creates a pressure given by:

mass of column per unit area $\times g$
$= $ density $\times g \times$ height
$= (1.35 \times 10^3) \times 9.8 \times (1 \times 10^{-2}) \times 1 = 132$ Pa
$= 0.132$ kPa

(The 10^{-2} above is a value which converts to cm.)

Hence:

	Pressure (cm Hg)	Pressure (kPa)
Triple point water	8.0	1.06
Boiling point water	10.9	1.44
Melting point lead	17.6	2.32

8 THE KINETIC THEORY OF GASES

The behaviour of gases is related to the microscopic movement of the atoms or molecules, and we use **kinetic theory** to give a model of this behaviour.

We are going to obtain an expression for the pressure of an ideal gas, and shall go on to find out how the speeds of the gas molecules vary. We shall use the model of a large number of molecules moving around in a closed box, making many collisions with the walls. There are a number of underlying assumptions.

Simplifying assumptions used in the kinetic theory

- The gas consists of a very large number N of molecules.
- The molecules are moving rapidly and randomly.
- The motion of the molecules can be described by Newtonian mechanics.
- Collisions between the molecules themselves and between the molecules and the walls are perfectly elastic.
- There are no attractive intermolecular forces.
- The only intermolecular forces that act are those during collisions and these are effectively instantaneous (that is, of small duration compared with the time interval between collisions).
- Molecules have negligible volume compared with the volume of the container (the molecules are, in effect, points in space).

If the collisions were not elastic, the kinetic energy of the molecules would be converted to other forms of energy and the gas pressure would decrease with time. Attractive intermolecular forces would, in particular, affect collisions at the walls by pulling molecules away from the walls. To assume no attractive intermolecular forces, we must have a gas at a low density.

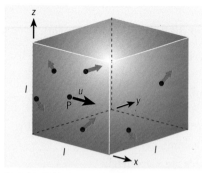

Fig 7.29 **Cubic box containing molecules of an ideal gas. Molecule P is moving with speed *u* in the +*x* direction**

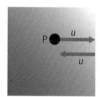

Fig 7.30 **Change of momentum when molecule P collides elastically with the wall of the container**

change of momentum
= *mu* − (−*mu*) = 2*mu*

Pressure of an ideal gas

Let us consider a cubic box with sides of length *l* containing molecules of mass *m* (Fig 7.29). We start by working out the pressure that the molecules exert on the right-hand wall (points **1–6**), then the total pressure of all the molecules (points **7–9**).

1 First consider molecule P moving in the +*x* direction with speed *u*. The momentum when it approaches the right-hand wall is *mu*. The momentum after an *elastic* collision is −*mu*. So, the total change of momentum = *mu* − (−*mu*) = 2*mu* (Fig 7.30).

2 Now consider the interval of time until the molecule makes another collision with the *same* wall. This is the time for the molecule to travel across the box, bounce off the opposite wall, and return. The distance travelled by the molecule is 2*l* at speed *u*. The time between collisions at the same wall is therefore 2*l*/*u*. So the molecule makes one collision with the right-hand wall in each interval of time 2*l*/*u*. The number of such collisions per unit time (i.e. per second) is *u*/2*l* (which is the inverse of the time interval).

3 We are now able to find the *rate of change* of momentum for the molecule by multiplying the change of momentum for one collision by the number of collisions in unit time (ie the number in 1 s). The rate of change of momentum is:

$$2mu \times \frac{u}{2l} = \frac{mu^2}{l}$$

4 By Newton's second law, the rate of change of momentum of the molecule equals the force which the wall exerts on the molecule to cause it to rebound. This is equal and opposite to the force the molecule exerts on the wall. The force on the wall is mu^2/l. This is of course an averaged force, as Fig 7.22 showed.

5 We now find the overall force on the right-hand wall by adding forces from the impacts of all *N* molecules. The total force on the wall is:

$$\frac{mu_1^2}{l} + \frac{mu_2^2}{l} + \frac{mu_3^2}{l} + \dots$$

where u_1 is the velocity in the +*x* direction of molecule 1, u_2 of molecule 2, and so on.

Hence the total force is:

$$\frac{m}{l}(u_1^2 + u_2^2 + u_3^2 + \dots) = \frac{mN\overline{u^2}}{l}$$

where: $$\overline{u^2} = \frac{u_1^2 + u_2^2 + u_3^2 + \dots}{N}$$

is the average over the square of all the velocities in the +*x* direction. (It is called the *mean square velocity* for molecules moving in the *x* direction.)

6 The pressure exerted on the walls by these molecules is given by:

$$\text{pressure} = \frac{\text{force}}{\text{area}} = \frac{mN\overline{u^2}/l}{l^2} = \frac{mN\overline{u^2}}{l \times l^2} = \frac{mN\overline{u^2}}{V}$$

where *V* is the volume of the box.

The total mass of the gas is *mN* and the density *ρ* of the gas is *mN*/*V*. This enables us to express the pressure as $p = \rho\overline{u^2}$.

7 The molecules are moving with components of velocity in the y and z directions in addition to a component in the x direction. We find the overall velocity c_1 for molecule 1 using Pythagoras' theorem:

$$c_1^2 = u_1^2 + v_1^2 + w_1^2$$

where u_1 is the velocity component of molecule 1 in the $+x$ direction as previously, and v_1 and w_1 are the velocity components of molecule 1 in the $+y$ and $+z$ directions (Fig 7.31).

8 Just as we obtained an average value of mean square velocity $\overline{u^2}$ for the velocity components of the molecules moving in the x direction, we can also obtain mean square velocities for the components in the y and z directions:

$$\overline{v^2} = \frac{v_1^2 + v_2^2 + v_3^2 + \dots}{N} \qquad \overline{w^2} = \frac{w_1^2 + w_2^2 + w_3^2 + \dots}{N}$$

This gives us an overall mean square velocity of:

$$\overline{c^2} = \overline{u^2} + \overline{v^2} + \overline{w^2}$$

9 We now obtain the final expression for pressure. The molecules on average are moving uniformly in all directions. So we can write:

$$\overline{u^2} = \overline{v^2} = \overline{w^2}$$

This gives:

$$\overline{c^2} = \overline{u^2} + \overline{v^2} + \overline{w^2}$$

$$= 3\overline{u^2}$$

or:

$$\overline{u^2} = \frac{\overline{c^2}}{3}$$

Hence:

$$p = \rho\overline{u^2} = \tfrac{1}{3}\rho\overline{c^2}$$

No account has been taken of collisions between molecules. However, provided these collisions are elastic, both total momentum and total kinetic energy are conserved. Exchange of momentum and kinetic energy between molecules has no overall effect on the pressure.

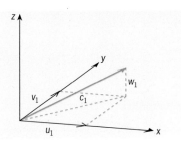

Fig 7.31 **Resolving velocity c_1 into x, y and z components u_1, v_1 and w_1**

■ See questions 7 and 8.

EXAMPLE

Q The density of air at 20 °C is 1.20 kg m^{-3}. Assuming atmospheric pressure of 1.01×10^5 Pa, calculate the root-mean-square speed of molecules in air.

A
$$p = \tfrac{1}{3}\rho\overline{c^2}$$

Hence:
$$\overline{c^2} = \frac{2p}{\rho} = \frac{3 \times 1.01 \times 10^5}{1.20} = 2.53 \times 10^5 \text{ m}^2 \text{ s}^{-2}$$

$$\sqrt{\overline{c^2}} = 502 \text{ m s}^{-1}$$

?

I The passage of sound through air relies on movement of the air molecules. Look up the speed of sound in air and comment on the value compared with the value calculated for $\sqrt{\overline{c^2}}$.

We can compare our expression for pressure from kinetic theory with that for pressure in the ideal gas equation. The first expression for pressure is based on the microscopic movement of the molecules and the second is based on the macroscopic quantities volume and temperature:

$$p = \tfrac{1}{3}\frac{Nm}{V}\overline{c^2} \quad \text{and} \quad p = \frac{n}{V}RT$$

Remember:
$$\rho = \frac{Nm}{V}$$

Equating the two expressions for pressure gives:

$$\tfrac{1}{3}\frac{Nm}{V}\overline{c^2} = \frac{n}{V}RT$$

That is:

$$\tfrac{1}{3}Nm\overline{c^2} = nRT$$

This equation creates a bridge between the small-scale and the large-scale models.

We know that $\tfrac{1}{2}m\overline{c^2}$ is the kinetic energy of one molecule and $\tfrac{1}{2}Nm\overline{c^2}$ is the total kinetic energy E_k of all the molecules. By using the bridging equation we see that the kinetic energy of the molecules is given by:

$$E_K = \tfrac{3}{2}nRT$$

This shows us that the kinetic energy of all the molecules is directly proportional to temperature. The energy must be the internal energy U of the gas. So:

$$U = \tfrac{3}{2}nRT$$

Remember that the above equation refers to n moles of gas. To obtain the kinetic energy of the molecules in 1 mole we must divide by n. One mole of gas contains N_A molecules, where N_A is the Avogadro constant. We divide by N_A to obtain the energy for one molecule. These two steps give us:

$$U_{\text{one molecule}} = \frac{3}{2}\frac{nRT}{nN_A} = \frac{3}{2}\frac{RT}{N_A}$$

R/N_A is the **Boltzmann constant**, k, the gas constant for a single molecule. Its value is 1.380×10^{-23} J K^{-1}, and it is a fundamental constant that is very important in physics.

We now see that the average kinetic energy per molecule:

$$= \frac{3kT}{2}$$

and that we can associate kinetic energy $\tfrac{1}{2}kT$ with each degree of freedom – that is, with the freedom to move in each of the x, y and z directions.

See question 9. ■

Avogadro's hypothesis

This states:

Equal volumes of a gas under the same conditions of temperature and pressure have an equal number of molecules.

We consider two gases A and B and use subscripts A and B to identify them.

For gas A we have: $p_A V_A = \tfrac{1}{3}N_A m_A \overline{c_A^2}$

For gas B we have: $p_B V_B = \tfrac{1}{3}N_B m_B \overline{c_B^2}$

For the gases to be at the same pressure and volume, these expressions must be equal:

$$N_A m_A \overline{c_A^2} = N_B m_B \overline{c_B^2} \qquad [1]$$

Also, as they are at the same temperature:

$$\tfrac{1}{2}m_A\,\overline{c_A^2} = \tfrac{3}{2}kT \ \text{ and } \ \tfrac{1}{2}m_B\,\overline{c_B^2} = \tfrac{3}{2}kT$$

giving: $\tfrac{1}{2}m_A\,\overline{c_A^2} = \tfrac{1}{2}m_B\,\overline{c_B^2}$

From equation [1], we can see that we must have:

$$N_A = N_B$$

which is the justification for Avogadro's hypothesis.

What has been described here concerns ideal gases with negligible forces between molecules that have negligible size. In real gases, the molecules have a finite size and small forces act between them. But under suitable conditions, gases do match the behaviour of ideal gases very closely.

We have emphasised that the gas laws apply best at low pressures. It is important that our real gas is not close to the conditions for liquefaction. So its temperature should be well above the critical temperature and its pressure well below the critical pressure at which liquefaction takes place. The real test of kinetic theory is that it gives satisfactory results for gases in most practical circumstances.

9 DISTRIBUTION OF MOLECULAR SPEED IN AN IDEAL GAS

In the calculation of the pressure of an ideal gas we used $\bar{c^2}$, the *mean* of the squared velocities. However, there will be wide variation in speed among the molecules. The situation is analogous to cars travelling along a motorway. A few will be travelling slowly. There will be many travelling close to the speed limit, others faster and some at an excessive speed. In a gas, too, there will be fast and slow molecules with a large proportion moving somewhere near the mean speed. This speed is determined by the temperature of the gas. Change the speed limit on a motorway and the distribution of the speed of the traffic changes. Alter the temperature of the gas and the speed (velocity) distribution alters.

Fig 7.32 shows a typical speed distribution for the molecules of a gas. Fig 7.32(a) shows the distribution as a histogram and Fig 7.32(b) shows it as a continuous plot. In both figures, the x axis shows the spread of speeds of the molecules. But what is plotted on the y axis needs more explanation.

The histogram shows the number of molecules moving within specific ranges of speed. Histogram intervals are set at 100 m s^{-1} so that the histogram represents the number of molecules with speeds 0 to 99 m s^{-1}, 100 to 199 m s^{-1} and so on. We can draw a curve to represent the outline of the histogram but it will not be very accurate.

If we take much smaller intervals we obtain more columns to the histogram and can draw a more precise curve. If we make the intervals of width 1 m s^{-1}, we obtain a distribution where the y axis is the number of molecules per unit interval of speed (that is, per m s^{-1}) and we obtain the continuous curve of Fig 7.32(b). The calibration of the y axis depends on the quantity of gas involved. This quantity is arbitrary, but a million molecules is assumed in Fig 7.32, so that the area under the graph is one million.

The graph shows distribution of *speed*. However, in calculating pressure, it is *speed squared* that is important and this means that the molecules at the upper end (right-hand side) of the distribution have the largest influence. Similarly, on a motorway, the seriousness of an accident usually depends on the kinetic energy of the vehicles as well as their mass. Kinetic energy is proportional to speed squared and accidents become proportionately more serious as speed rises. This is an important reason why speed restrictions are imposed.

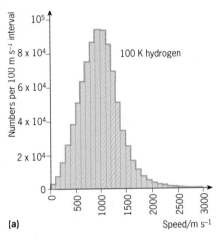

(a)

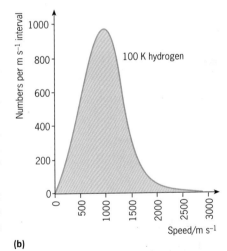

(b)

Fig 7.32 **Typical speed distribution for 1 million molecules of hydrogen at 100 K shown** (a) **as a histogram at 100 m s^{-1} intervals and** (b) **as a continuous curve**

J Using Fig 7.32(b):

(a) State the most probable speed of the gas molecules.

(b) Estimate what fraction of the molecules travel between 500 m s^{-1} and 1000 m s^{-1}.

(c) Estimate what fraction of the molecules of the gas travel at twice the most probable speed or more.

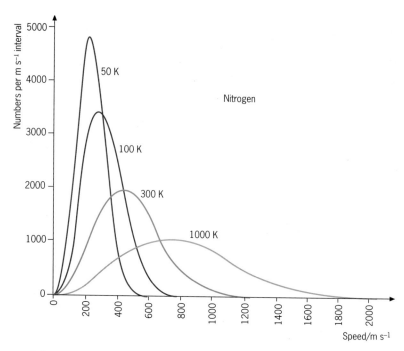

Fig 7.33 **Variation of speed distribution for 1 million molecules of nitrogen as the temperature of the gas changes**

If the gas is heated, the curve showing the distribution of speeds spreads out horizontally (Fig 7.33), but the area under each curve remains constant. The total area under the curve has to remain the same as the area represents the total number of molecules. Because the curve spreads out horizontally, it is clear that the number of molecules at the peak speed must decrease with increase in temperature.

SUMMARY

After studying this chapter you should be able to:

■ Describe qualitatively the differences between states of matter.

■ Describe different types of bonding between atoms.

■ Explain the term latent heat and use values of latent heat in calculations.

■ Draw and interpret force against separation and potential energy against separation curves for adjacent atoms in a solid or liquid.

■ Describe simple cubic crystal structures in terms of packing of spherical atoms and carry out simple calculations of atomic spacing.

■ Describe Brownian motion.

■ State the gas laws:

$$pV = \text{constant} \quad \frac{p}{T} = \text{constant} \quad \frac{V}{T} = \text{constant}$$

and outline how they are proved experimentally.

■ Be aware of the relation between the Kelvin temperature scale and the ideal gas equation and describe the use of the constant-volume gas thermometer to measure Kelvin temperature.

■ Use the ideal gas equation:

$$\frac{pV}{T} = \text{constant} \quad \text{or} \quad \frac{p_1 V_1}{T_1} = \frac{p_2 V_2}{T_2}$$

to make simple calculations of pressure, volume and temperature of a gas.

■ Understand that in an ideal gas the equation of state is given by $pV = nRT$, where R is the molar gas constant.

■ State the main assumptions used in the kinetic theory of an ideal gas and carry out simple calculations using the theory.

■ Understand the relation between the Avogadro number N_A and the mole.

■ Know that the average kinetic energy per molecule is $3kT/2$, where k is the Boltzmann constant R/N_A.

QUESTIONS

1 Fig 7.Q1 shows how the force F between a pair of molecules varies with their separation r.

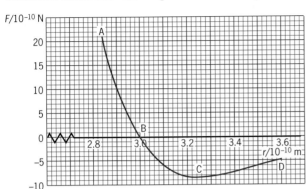

Fig 7.Q1

a) Describe how F varies with r in the regions A to B, B to C and C to D. Explain how a solid rod made of molecules which behave in this way will behave for small applied longitudinal forces.

b) Using the information on the graph, write down the equilibrium separation of the molecules, and calculate the strain beyond which the intermolecular bond would break.

c) Using the graph, estimate the energy required to decrease the separation of the molecules from 3.0×10^{-10} m to 2.9×10^{-10} m.

[ULEAC: 1993, Paper 3, Topic B, Q8 (part) and ULEAC: 1996, Specimen Papers, PH4, Topic 4A, Solid Materials, Q6]

2
a) Sketch the form of the **(i)** potential energy vs. separation, and **(ii)** force vs. separation graphs for a pair of atoms. Explain the shapes of the graphs. Identify the positions of equilibrium on the two graphs.

b) This model may be applied to the case of a metal solid.
  (i) State how each of the four quantities, molar volume, molar latent heat of fusion, linear expansivity and compressibility, is related to the shapes of the potential energy vs. separation and force vs. separation graphs.
  (ii) The molar volume of aluminium is 1.0×10^{-5} $m^3\ mol^{-1}$. Estimate the mean separation of the atoms in aluminium metal. The molar latent heat of fusion of aluminium is 11 kJ mol^{-1}. Assuming that melting of the metal results in 1 of the 12 attraction bonds between aluminium and its neighbours being broken, estimate the depth of the potential energy minimum.

[O&CSEB: 1994, Physics 3/4, Q11]

3
a) Fig 7.Q3 shows the variation of potential energy of a pair of molecules with their distance apart r. What is the significance of ε, σ and r_0?

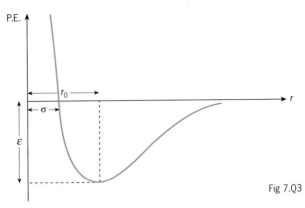

Fig 7.Q3

b) (i) The molar latent heat of vaporisation of water is 4.07×10^4 J mol^{-1}. If the number of near neighbours in water is 10, estimate a value for ε. Prove any formula that you use.
  (ii) When water is boiled under standard atmospheric pressure of 1.0×10^5 Pa, one mole has its volume increased from 1.90×10^{-5} m^3 (as water at 100 °C). Calculate the energy required to push back the atmosphere when boiling (evaporating) 1 mole of water at 100 °C.
  (iii) In the light of your answer to **b)(ii)**, revise your estimate of ε in **b)(i)**. Explain your reasoning.

c) If the surface energy of water at 100 °C is 0.058 J m^{-2}, estimate the number of molecules per unit area in the surface of water.

d) Assuming that the molecules of the water are packed in a simple cubic manner, estimate r_0.

[WJEC: 1994, Physics A2, Q10]

4 The mechanical properties of a sample of copper are generally very different from the theoretical values derived by considering a perfect single crystal (fcc) model of the material, where the atoms form cubic close packing.

a) Sketch the arrangement of atoms in such a perfect copper crystal.

b) How does the tensile strength of an actual copper specimen compare with that predicted for an ideal single crystal of copper? Explain the difference in terms of microstructure.
[Refer also to Chapter 5.]

[NEAB: June 1993, Syllabus A, Paper II, Option H, Q1 (part)]

5 The equation p/T = constant applies to an ideal gas. State the meaning of the symbols p and T. Under what conditions will this equation hold?

Describe an experiment to check whether or not the equation applies to a gas such as air under the conditions you have stated.

[ULEAC: June 1994, Paper 2, Q8 (part)]

6 The tyres of a car are pumped up to a pressure of 26 lb in^{-2} above atmospheric pressure. The air in the tyres is at a temperature of 18 °C. What is the pressure in the tyres after a journey such that the air in the tyres has risen to a temperature of 25 °C? (Assume the volume within each tyre remains constant.)

7

a)   **(i)** State four assumptions of the kinetic theory of an ideal gas.

      (ii) A gas molecule in a cubical box travels with speed c at right angles to one wall of the box. Show that the average force the molecule exerts on the wall is proportional to c^2.

b) The table gives measured values of pressure and density for a fixed mass of gas at constant temperature of 27 °C.

Pressure (10^5 Pa)	0.60	0.80	1.00	1.20	1.40
Density (kg m^{-3})	0.68	0.91	1.14	1.37	1.60

     (i) Plot a graph of pressure against density. Does your graph indicate that the gas behaves as an ideal gas under these conditions? Justify your answer.

    (ii) Use your graph to calculate the root-mean-square speed of the molecules of the gas.

   (iii) The temperature of the gas is raised to 57 °C. Calculate the pressure when the density is 1.00 kg m^{-3}, and hence draw the corresponding graph of pressure against density at 57 °C, using the same axes as before.

c) A container holds a mixture of helium and argon, relative masses 4 and 40 respectively.

     (i) For the gases in the container, calculate the ratio:

$$\frac{\text{root-mean-square speed of helium atoms}}{\text{root-mean-square speed of argon atoms}}$$

    (ii) There are approximately equal numbers of helium and argon atoms in the container, when gas starts to leak slowly out of a small hole in the side. After a short time, will the number of argon atoms remaining in the container be greater or less than the number of helium atoms? Give a reason for your answer.

[NEAB: June 1993, Syllabus A, Paper II, Q12]

8 A pressurised tank at 23.0 °C contains 3.00 mole of argon gas (relative atomic mass of argon = 40.0). Calculate **a)** the root-mean-square speed of the argon atoms and **b)** the total kinetic energy of the gas.

9

a)   **(i)** Explain how molecular movement causes a pressure to be exerted by a gas.

      (ii) Use a simple model to derive the expression $p = \frac{1}{3}\rho c^2$ for the pressure exerted by an ideal gas.

b) By using the ideal gas law show that $\overline{c^2} = 3RT/M$, and calculate the value of the root-mean-square velocity for the molecules of oxygen at 20 °C. (Relative molecular mass of oxygen $M = 32$.)

c) Specialised processes used in the fabrication of microelectronic devices require conditions where the number of molecules colliding with the solid surface per unit area per second, N, is less than the critical value, in order to keep the surface free from contamination. By using the simple model used in **a)**, show that $N = nc_x/2$ where c_x is the average molecular velocity along the x direction and n is the number of molecules per unit volume.

d) Use your answer to **b)** as an approximate value for c_x, and therefore estimate the pressure such that N is less than 1×10^{13} collisions m^{-2} s^{-1}. (Note that $n = N_A p/RT$.)

e) Estimate the time for the surface to become essentially covered with oxygen. (Note that the spacing between the atoms of a solid surface is of the order of 3×10^{-9} m.)

[WJEC: 1994, Physics S, Q12]

Assignment

MAXWELL-BOLTZMANN SPEED DISTRIBUTION AND MOLECULAR BEAMS

Here, we use a spreadsheet to demonstrate graphically the distribution of speeds of gas molecules in a chamber, for different temperatures and different species of molecule. Such speed distributions are not easily measured directly, but it is possible to measure the distribution of speeds of the molecules as they escape through a tiny hole.

The Maxwell Boltzmann speed distribution

We consider N identical molecules in a closed chamber. It can be shown that the number of molecules dN having a speed within the v to $v + dv$ is given by:

$$dN = N\left(\frac{2}{\pi}\right)^{\frac{1}{2}}\left(\frac{M}{k}\right)^{\frac{3}{2}}\frac{1}{T^{3/2}}\,v^2\exp\left(-\frac{Mv^2}{2kT}\right)dv \qquad [1]$$

where N is the total number of molecules in the chamber,
M is the mass of one molecule
k is Boltzmann's constant
and T is the kelvin temperature.

We can write M as mu where m is relative mass of the atom or molecule and u is one atomic mass unit. We rewrite the equation in the form:

$$\frac{dN}{N} = \text{constant } 1 \times \left(\frac{m}{T}\right)^{\frac{3}{2}} v^2 \exp\left(-\frac{\text{constant } 2 \times mv^2}{T}\right)dv \qquad [2]$$

where constant $1 = \left(\frac{2}{\pi}\right)^{\frac{1}{2}}\left(\frac{u}{k}\right)^{\frac{3}{2}} = 1.053 \times 10^{-6}\ \text{m}^{-3}\ \text{s}^2\ \text{K}^{\frac{3}{2}}$

and constant $2 = \frac{u}{2k} = 6.014\ \text{m}^{-2}\ \text{s}^2\ \text{K}$.

1 Show that the units are given correctly by looking up the units of u and k, and using the formulae for constant 1 and constant 2.

	A	B	C	D	E	F
1	INPUT DATA					
2	Speed increment	Temperature 1	Temperature 2	Temperature 3	Temperature 4	
3	100	300	500	1000	2000	
4						
5	Relative mass	Constant 1	Constant 2			
6	2	1.0526E-06	0.00006014			
7						
8	CALCULATION of MAXWELL-BOLTZMANN SPEED DISTRIBUTION					
9		Temperature 1	Temperature 2	Temperature 3	Temperature 4	
10	Speed v	Distribution dN/N	Distribution dN/N	Distribution dN/N	Distribution dN/N	
11	0	0.0000	0.0000	0.0000	0.0000	
12	100	0.0006	0.0003	0.0001	0.0000	
13	200	0.0023				
14	300	0.0050				
15	400	0.0086				
16	500	0.0130				
17	600	0.0179				
18	700	0.0231				
19	800	0.0284				
20	900	0.0335				
21	1000	0.0384				
22	1100	0.0427				
23	1200	0.0463				
24	1300	0.0492				
25	1400	0.0512				
26	1500	0.0523				
27	1600	0.0526				
28	1700	0.0520				
29	1800	0.0506	0.0396	0.0207	0.0089	
30	1900	0.0486	0.0403	0.0220	0.0097	
31	2000	0.0461	0.0407	0.0233	0.0105	

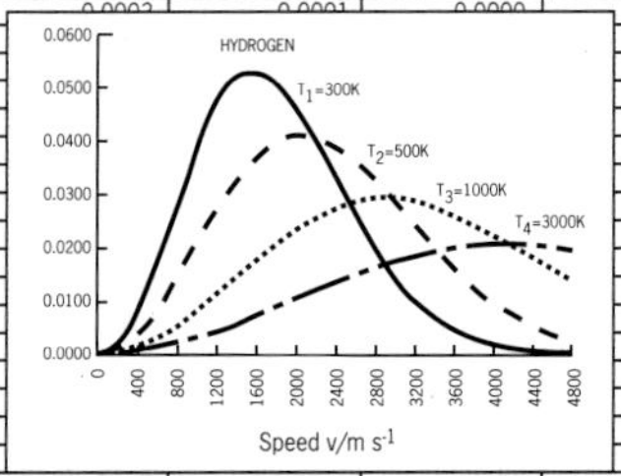

Fig 7.A1 Spreadsheet and graphs for Maxwell–Boltzmann speed distribution

2 Set up a **spreadsheet** as in Fig 7.A1 to compare distributions for four different temperatures. Type into cells as follows (variables are not italic on spreadsheet).

A1: Heading – Input Data
A2: Label – Speed increment
B2 to E2: Headings – Temperature 1 to 4
A3: A suitable speed increment (eg 100)
B3 to E3: Appropriate temperatures
A5 to C5: Labels – Relative mass, Constant 1, Constant 2
A6: A value for relative mass (eg 2 for hydrogen, H_2)
B6 and C6: Values for the two constants
A8: Heading – Calculation of Maxwell–Boltzmann speed distribution (the spreadsheet allows the heading to run along the row)
B9 to E9: Headings – Temperatures 1 to 4
A10: Column heading – Speed v
B10 to E10: Column headings – Distribution dN/N

3 Now enter **data** and **formulae** as follows.

A11: Speed 0
A12: = A11 + $A$3 (this increments the speed to the next value)

Now click and highlight cell A12 plus as many cells as you need for the range of speed values; eg highlight down to A58. Open Edit and click Fill Down.
Into cell B11 insert the formula:

= $B$6*($A$6/$B$3)^1.5*A11^2*EXP(-$C$6*$A$6*A11^2/$B$3)*$A$3

a) Compare this expression with equation 2 above and check that it enters data for the relative number of molecules in a speed interval dv.

Highlight cell B11 and use the Fill Down facility to produce entries down to B58.
Do the same for the C, D and E columns.
Copy the formula of cell B11 across to C11, D11 and E11 and change the two $B$3 parts of the formula.

b) What do you change them to?

Fill down as before.
Now make a check on your table. Each cell gives the fractional number of molecules within the speed interval, so each column of cells sums to unity unless there are not enough cells to include nearly all the molecules.

c) Suggest why we say *nearly all* rather than *all* molecules.

To sum column B (Temperature 1), enter =SUM(B11:B58) and Fill Right for columns C, D and E.

d) Comment on the values obtained.

4

a) Now plot line graphs for the speed distributions using your chart facility. Describe how the distributions change with change of temperature. By writing out a table, compare the speeds at which peak numbers of molecules occur for each temperature.

b) The average (root mean square) speed of the molecules is obtained from $\frac{1}{2}Mv^2 = \frac{3}{2}kT$. Add these values to your table and comment on why these values are different from those in **4a)**.

c) Use your spreadsheet to compare the speed distributions of other molecules (see Table 7.A1) with those for H_2. Note that you may need to alter the value of the speed increment as well as the relative mass m. Describe how the distributions change.

Molecule	Relative mass
Helium, He	4
Argon, Ar	18
Oxygen, O_2	32
Chlorine, Cl_2	70

Table 7.A1 **Relative masses for gas molecules**

Effusion and molecular beams

To confirm the Maxwell–Boltzmann speed distribution within the chamber, the molecules are allowed to escape through a small hole: the process is called **effusion**. The speeds of the escaping molecules are measured. but the speed distribution for the escaping molecules is not the same as in the chamber. The expression must be multiplied by an extra v.

5

a) Why does a larger proportion of fast molecules escape than of slow ones? (Consider the time a molecule takes to be travel back and forth across the chamber until it finds the pin-hole.)

b) Set up the spreadsheet for effusion.

For this, alter the title in cell A8. The formula for the speed distribution in the beam turns out to be:

$$\frac{dN}{N} = \text{constant } 1 \times \left(\frac{m}{T}\right)^2 v^3 \exp\left(-\frac{\text{constant } 2 \times mv^2}{T}\right)dv \quad [3]$$

where constant $1 = \left(\frac{1}{2}\right)^{\frac{1}{2}} \left(\frac{u}{k}\right)^2 = 7.235 \times 10^{-7} \text{ m}^{-4} \text{ s}^4 \text{ K}^2$

and constant 2 is unchanged. You may wonder why constant 1 changes, and also the dependence on m and T. This happens because we again make the area under our curves equal to unity. The entry for cell B11 becomes:

=$B$6*($A$6/$B$3)^2*A11^3*EXP(-$C$6*$A$6*A11^2/$B$3)*$A$3

Check that this expression agrees with equation 3. Alter cells C11, D11 and E11 appropriately. Check that entries in

cells B11 to B58 etc sum to nearly unity. If they do not, you have not included all the molecules, and you must enlarge the speed increment.

c) Plot the distributions and compare them with the Maxwell-Boltzmann distributions. You may wish to do this for different molecules. Have the coordinates of the peaks of the curves changed?

Checking the distribution experimentally

This is carried out using a speed selector, shown in Fig 7.A2. If the two discs of the selector are in line along the stationary axle, molecules pass directly through the slits to the detector. When the discs are rotated, molecules only reach the detector when their time of flight equals the time for the next slit of the disc to rotate into position.

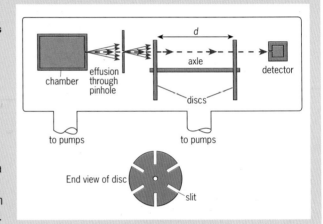

Fig 7.A2 **Speed selector for molecules in a beam**

6

a) Assuming the distance between the two discs is d, the frequency of rotation is f, and there are six slits in each disc, obtain a formula for the speed v of those molecules passing through adjacent slits of the two discs.

b) With a disc separation of 10 cm and molecules of a chosen gas emitted from the chamber at 300 K, calculate the disc frequency f for the detector to give a peak signal.

As the speed distribution depends on the mass of the molecules, effusion through small holes in a membrane can be used to separate isotopes. This is how uranium isotopes for nuclear devices were originally separated, using the gas UF_6.

c) Suggest why a very large number of membranes and stages were required.

Effusion cells are widely used to provide fine beams of molecules for constructing crystal layers by 'molecular beam epitaxy' (MBE), where epitaxy means building up layers on top of a crystal surface.

ORDINARY MATTER

The Chapter Map below includes the main ideas that relate to the nature and behaviour of matter in its solid, liquid and gas states. You can use the map to cross-match ideas with the needs of your syllabus. The map should also help you to identify the areas you may need to study further.

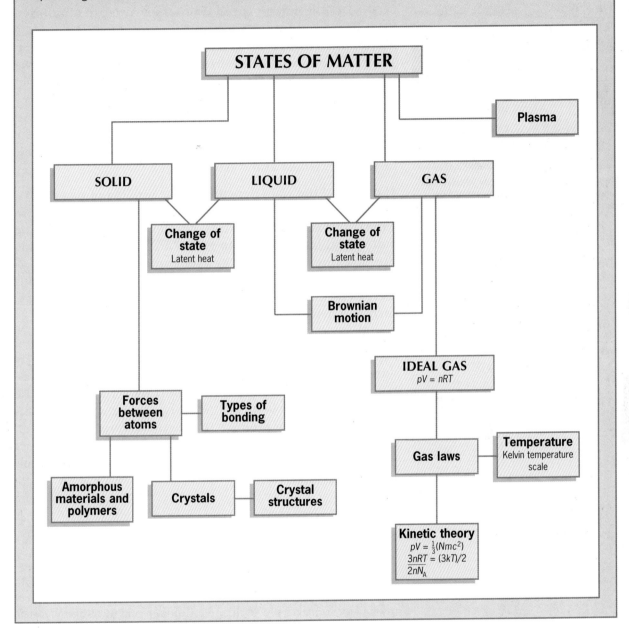

STATES OF MATTER

Plasma

SOLID

LIQUID

GAS

Change of state
Latent heat

Change of state
Latent heat

Brownian motion

IDEAL GAS
$pV = nRT$

Forces between atoms

Types of bonding

Gas laws

Temperature
Kelvin temperature scale

Amorphous materials and polymers

Crystals

Crystal structures

Kinetic theory
$pV = \frac{1}{3}(Nmc^2)$
$3nRT = (3kT)/2$
$2nN_A$

8 Transport

The French TGV (Train à Grande Vitesse), the world's fastest passenger train, with a front designed to reduce air friction

IN THE DESIGN of high speed trains, fuel costs are minimised by reducing friction: friction with the air is kept low, while friction with track is set at a sufficiently high level for safety. The photograph is of the front of a high speed train. The train is shaped like this to reduce the friction force (drag) caused by moving through a fluid (air). The resulting low 'drag coefficient' reduces the energy loss at high speed and so increases the train's maximum speed. The rear of the train, too, needs to be the right shape, to minimise air turbulence, which also wastes energy.

The mass of the train depends on the number of passengers, so the applied forces needed to accelerate and brake have to adjust to this. In emergencies, deceleration may have to be quite rapid but still safe, so great care is taken in designing the braking systems. The geography of the track is important: a longer route may save energy if it avoids inefficient changes between kinetic and potential energy, such as when a loaded train goes up and down hills.

A modern vehicle with low drag travelling at high speed experiences the same kind of force that allows aircraft to fly – aerodynamic lift. Trains and cars are unlikely to take off – but the effect is to reduce the 'supporting' force between the ground and the vehicle's wheels. There must be this supporting force to produce the frictional forces that a train needs in order to accelerate or decelerate. If friction is reduced too much, traction is lost and the driving wheels will simply skid.

The ideas in this chapter

Newton's laws of motion are particularly important for an understanding of the topic of **transport**. The modern world relies on an ability to move materials and people quickly, efficiently and safely from place to place. We shall apply fundamental ideas about **mass, force, velocity, acceleration, momentum, impulse** and **energy** not only to the motion of vehicles. We shall also look at the behaviour of **fluids** (liquids and gases) to understand how ships float, aeroplanes fly and fluids are transported through pipes. Transport systems have to take account of the fact that vehicles move through a fluid (usually air) and that the fluid is an important restraint on motion.

But a moving fluid carries kinetic energy and so can be used to do work. Moving water is used to drive turbines in hydroelectric power stations. The energy of moving air is used in wind turbines, a cheap source of energy but with an impact on the environment which can be controversial.

1 CONSERVATION OF MOMENTUM AND ENERGY IN TRANSPORT SYSTEMS

Chapters 3 and 4 dealt with the key ideas of momentum and energy in relation to moving objects. Both total energy and momentum are **conserved** when engines are used to move vehicles such as cars, trains and rockets – but *kinetic* energy is not conserved.

Think about a rocket: the exhaust gases and the rocket itself gain kinetic energy from burning fuel, but a significant fraction of the energy in the fuel mixture is transferred to the random motion of gas particles and does not appear as useful kinetic energy of the rocket vehicle itself. The same applies to road and rail transport vehicles, which gain their energy from burning fuels. Even electric traction engines ultimately depend on the fuels used in power stations to generate electricity.

Not all the energy transferred by a transport system appears as kinetic energy. Some energy is usually transferred to or gained from gravitational potential energy, for example when aircraft climb and land, and ground vehicles go up and down hills. But what goes up must come down and there is no net change in gravitational potential energy for a vehicle that always returns to its starting level. There are usually some friction losses: for example, energy is lost to the surroundings when vehicles brake (the brake discs get hotter).

Nevertheless, in all these changes both *total* energy and momentum are conserved – although it is more difficult to keep track of the energy transfers than the momentum changes.

■ See question 1.

Traction and braking

The control of a car – in accelerating, turning and braking – relies on its contact with the road surface. The key factor is **friction** between the tyres and the road surface. The area of contact is quite small, as is shown by the Example.

EXAMPLE

Q The mass of a given car is 1000 kg. Its tyre pressure is rated as 2.4×10^5 Pa above normal atmospheric pressure: that is, about 3.4×10^5 Pa. What area of the tyre is in contact with the ground?

A Using the definition, pressure = force/area, we get:

$$\text{area of contact} = \frac{\text{weight of car}}{\text{tyre pressure}} = \frac{1000 \times 9.8}{3.4 \times 10^5} = 0.029 \text{ m}^2$$

This gives a total area of 290 cm². Thus the area of contact *per tyre* is 72 cm² – roughly the area of the palm of your hand.

?

A Estimate the pressure you exert on the ground when standing barefoot on two feet.

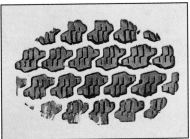

Fig 8.1 (a) **The patterned surface of tyres;** (b) **The print of a tyre showing the area in contact with the ground**

Fig 8.2 **Forces acting on an accelerating car (the force arrows are not drawn to scale)**

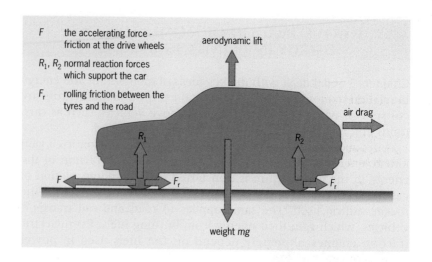

F the accelerating force - friction at the drive wheels

R_1, R_2 normal reaction forces which support the car

F_r rolling friction between the tyres and the road

aerodynamic lift

air drag

R_1 R_2

F F_r F_r

weight mg

As the car wheel turns, the part of the tyre in contact with the ground is instantaneously at rest with reference to the ground. If the car is accelerating or braking, the force that ultimately acts on the car as a whole must be due to the friction between the tyres and the ground. If there were zero friction the wheels would spin freely and the car would skid out of control – as tends to happen on icy roads. Fig 8.2 shows the forces acting on an accelerating car.

The maximum value of the frictional force F is decided by the nature of the two surfaces in contact and the **normal** force N between them. Here the word 'normal' has the special mathematical meaning of 'at right angles to the surfaces at the point of contact'. This force is often called the normal *reaction*. The relationship is simple:

$$F = \mu N$$

where μ is a dimensionless number called the coefficent of friction. Its value depends on the nature of the surfaces: μ is usually lower for surfaces sliding against each other than it is for the same surfaces when they are at rest relative to each other. We therefore distinguish between the coefficient of *static friction* and the coefficient of *sliding friction*. For rubber against a paved road, the static coefficient varies between 0.7 and 0.9. The coefficient for sliding friction is between 0.5 and 0.8; on a wet road this reduces to between 0.25 and 0.7. Note that the force F does not depend on the area of the surfaces in contact. Figs 8.3 to 8.6 illustrate some aspects of the coefficient of friction. Table 8.1 gives examples of coefficients of friction for various surfaces.

See question 2. ■

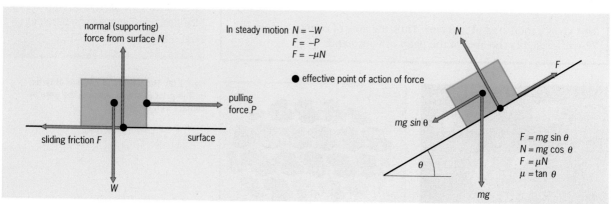

normal (supporting) force from surface N

In steady motion $N = -W$
$F = -P$
$F = -\mu N$

● effective point of action of force

pulling force P

sliding friction F surface

W

N

F

$mg \sin \theta$

θ

$F = mg \sin \theta$
$N = mg \cos \theta$
$F = \mu N$
$\mu = \tan \theta$

mg

Fig 8.3 **Friction on a flat surface**

Fig 8.4 **For friction on a slope, $\mu = \tan \theta$**

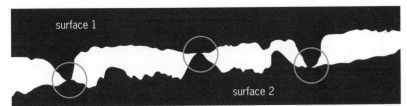

surface 1

surface 2

Fig 8.5 **Frictional force is independent of area. No real surface is perfectly flat – it has many bumps and valleys. Just three bumps have to be in contact to support one object on another. If one of these is rubbed away by relative movement, another one takes its place. The actual area of contact is therefore very small**

Table 8.1 **Coefficients of friction; these vary with temperature and humidity, so the values are approximate**

Surfaces	Static friction	Sliding friction
Steel on steel	0.74	0.57
Rubber on concrete	0.9	0.7
Wood on wood	0.25–0.50	0.2
Ice on ice	0.1	0.03
Waxed wood on dry snow	–	0.04
Teflon on Teflon	0.04	0.04
Synovial joints in humans	0.01	0

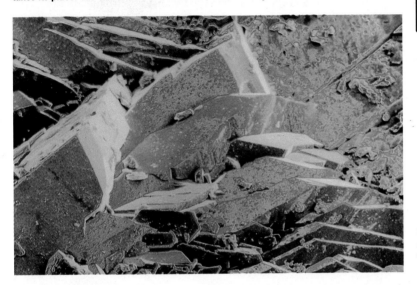

Fig 8.6 **Photomicrograph showing the roughness of a coated metal surface which is smooth to the naked eye. Used in engine components, such a surface is designed to retain lubricants**

Sliding down a slope

Fig 8.4 shows the forces acting on an object that is sliding steadily down a slope. It is not accelerating, so the component of the object's weight *down* the slope must equal the frictional force acting *up* the slope. The frictional force is μN where N is the supporting force between the object and surface of the slope. Thus:

$$mg \sin \theta = \mu N$$

But N is also due to the weight of the object. In fact, $N = mg \cos \theta$, so we can write:

$$mg \sin \theta = \mu mg \cos \theta$$

which simplifies to:

$$\mu = \sin \theta / \cos \theta = \tan \theta$$

This relationship can be used as a simple way of measuring the coefficient of friction between two surfaces.

Friction and acceleration

Over-keen drivers 'burn rubber' at traffic lights when they try to accelerate with forces greater than can be provided by static friction between tyres and road. The tyre moves relative to the road surface and work done against sliding friction produces local heating. Similarly you may see skid marks when vehicles brake so strongly that wheels 'lock', so that again there is relative movement between road and tyre as the tyres slide along the ground.

B A bicycle has a mass of about 15 kg, and a typical adult has a mass of about 60 kg. Assuming a coefficient of friction between tyre and road of 0.8, estimate the maximum deceleration of a bicycle. Explain why both front and rear brakes should be used when stopping a bicycle as safely as possible in an emergency.

C The estimate of F is more accurate for the greatest *deceleration* that could be produced without the car skidding. Why is this?

EXAMPLE

Q A typical value for the coefficient of static friction between a particular tyre and a road is 0.8. Estimate the maximum acceleration that a car of mass 1000 kg can have.

A Maximum frictional force $F = \mu N = 0.8 \times 1000 \times 9.8 = 7.8$ kN. The acceleration produced by this force is:

$$a = F/m = (7.8 \times 10^3)/1000 = 7.8 \text{ m s}^{-2}$$

But this is a very rough estimate – the two drive wheels share the weight of the car with the other pair of wheels, so the normal reaction N is less than the weight of the car. This means that the maximum possible acceleration is less than that calculated. Motor engineers still find the simple formula $F = \mu W$ useful, however, where W is now the load on the driving axle, which may be measured (with difficulty at speed) or calculated from theory.

Fig 8.7 **These drag car race vehicles are designed to accelerate as much as possible. The acceleration is helped by large back wheels (which have a gearing effect) and the fact that much of the weight of the car is placed directly above the back wheels**

Fig 8.8 **The forces acting on a car as it goes round a bend.** (a) **Simple case;** (b) **when air resistance (drag) is taken into account**

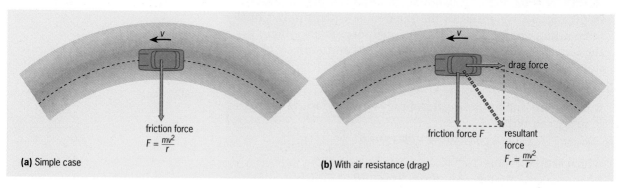

friction force
$$F = \frac{mv^2}{r}$$

(a) Simple case

friction force F

drag force

resultant force
$$F_r = \frac{mv^2}{r}$$

(b) With air resistance (drag)

See questions 3 and 4. ■

A car may skid if it is made to turn too sharply, that is, in too small a circle. A central force is needed to make a car move in a circle, and again this must be provided by the force of friction between car and ground. Fig 8.8(a) shows a simple version of the situation, and (b) shows the more realistic case where air friction is taken into account.

Wheels and rotational inertia

A spinning wheel has both kinetic energy and momentum due to its rotation – even when the wheel is not moving forwards. A turning force (or **torque**) is needed to get a wheel to rotate, and

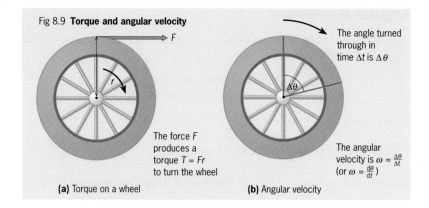

Fig 8.9 **Torque and angular velocity**

> F

r

The force F produces a torque T = Fr to turn the wheel

(a) Torque on a wheel

The angle turned through in time Δt is Δθ

Δθ

The angular velocity is $\omega = \frac{\Delta\theta}{\Delta t}$ (or $\omega = \frac{d\theta}{dt}$)

(b) Angular velocity

the force does work that appears as kinetic energy, as shown in Fig 8.9(a). Torque is measured, like the moment of a force, in newton metres (N m). Tests show that it is harder to spin a wheel of large diameter than a wheel of smaller diameter – even when both wheels have the same mass. Just as we call the resistance of a mass to being accelerated its *inertia* (*m*), so we can also define the resistance of a wheel to being rotated by a turning force as **rotational inertia** *I* (also called **moment of inertia**). The units of rotational inertia are of mass × (distance)^2, that is kg m^2.

Rate of spin is measured by the angle through which a radius line turns per second, that is, its **angular velocity** ω, as in Fig 8.9(b). Angular velocity is measured in radians per second. Just as velocity is a vector, so is angular velocity. With α as **angular acceleration**, comparing linear and rotational motion:

$$\text{Force} = \text{mass} \times \text{acceleration}$$
$$F = ma$$

Torque = rotational inertia × angular acceleration
$$T = I\alpha$$

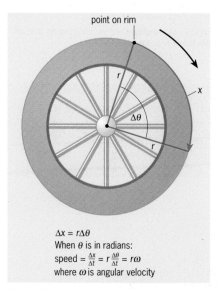

point on rim

r

x

Δθ

r

$\Delta x = r\Delta\theta$
When θ is in radians:
speed $= \frac{\Delta x}{\Delta t} = r\frac{\Delta\theta}{\Delta t} = r\omega$
where ω is angular velocity

Fig 8.10 **The distance moved and the angular velocity of a point on the rim of a wheel**

EXAMPLE

Q A torque of 20 N m acts on a wheel that is initially still and has a rotational inertia of 600 kg m^2.

a) What is the angular acceleration produced?

b) At what angular velocity is the wheel spinning after 2 minutes?

c) The wheel is 30 cm in radius. How fast (in m s^{-1}) is a point on the rim moving at this time?

A

a) Angular acceleration = torque/rotational inertia:

$$\alpha = 20/600 = 0.033 \text{ rad s}^{-2}$$

b) Angular velocity = angular acceleration × time

$$= 0.033 \times 120 = 4 \text{ rad s}^{-1}.$$

c) See Fig 8.10. By the definition of a radian:

distance covered per second by a point on the rim
= angle turned through per second × radius

So the speed of a point on the rim is 4 × 0.30 = 1.2 m s^{-1}.

?

D A spinning ride in an activity park is 15 m in radius and has a maximum rotation in which it spins once in 2 s.

(a) What is its angular velocity?

(b) How fast in metres per second will a chair on the rim move?

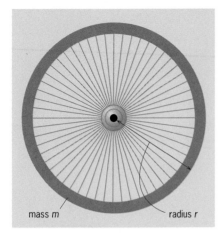

mass m radius r

Fig 8.11 **A simplified wheel: we assume that all the mass is in the rim**

Angular momentum

Just as linear momentum is defined as mass × velocity, so we define angular momentum as:

**Angular momentum
= rotational inertia × angular velocity = $I\omega$**

As with linear motion, Newton's second law applies:

linear force = rate of change of momentum $F = \dfrac{\Delta(mv)}{\Delta t} = \dfrac{\Delta P}{\Delta t}$

Rotational torque = rate of change of angular momentum

$$T = \dfrac{\Delta(I\omega)}{\Delta t}$$

Rotational inertia of a simple wheel

The structure of a wheel may be simplified as shown in Fig 8.11. The spokes are assumed to have negligible mass compared with the rim.

As all the mass (m) is at the same distance r from the centre of rotation, the rotational inertia is simply mr^2.

Rotational inertia of everyday objects

In many cases the mass of a spinning wheel is not evenly distributed on the rim. The rotational inertia can be calculated if the distribution is fairly simple, but may in practice have to be measured experimentally. The value of the rotational inertia of a car wheel is quite difficult to calculate mathematically, for example. Table 8.2 gives the rotational inertias of some simple shapes. Note that the value also depends on how the object is spun round an axis.

?

E A torque of 25 N m is applied to an object. After 30 s it has gained an angular velocity of 2 rad s^{-1}. What is its rotational inertia?

See question 5. ■

Table 8.2 **Rotational inertia for simple objects**

Object	Axis of rotation	Rotational inertia (m = mass of object)
Thin ring (simple wheel) r m	through centre, perpendicular to plane	mr^2
Thin ring m r	through a diameter	$\frac{1}{2}mr^2$
Disc and cylinder (solid flywheel) m r	through centre, perpendicular to plane	$\frac{1}{2}mr^2$
Thin rod, length d m d	through centre, perpendicular to rod	$\frac{1}{12}md^2$

Rotational energy

The kinetic energy of a rotating body is $\frac{1}{2}I\omega^2$. Compare the kinetic energy formula for a rotating body with the linear formula $\frac{1}{2}mv^2$.

If the wheel is also rolling forwards with speed v it will also have *translational* kinetic energy $\frac{1}{2}mv^2$.

The total kinetic energy E_K of a rolling wheel is therefore:

$$E_K = \frac{1}{2}I\omega^2 + \frac{1}{2}mv^2$$

Large vehicles (like some electric locomotives) can reduce the waste of energy in braking by storing linear kinetic energy as rotational energy in massive rotating wheels (flywheels). The track wheels are linked through a gearing system to the flywheel instead of to the usual friction brakes. As the track wheels slow down, the flywheel speeds up, and vice versa.

Conservation of angular momentum

Hold the axis of a bicycle wheel in your hands and get someone to spin it in the vertical plane. Then try to tip the wheel through a small angle away from the vertical. You will experience a mysterious force that seems to oppose the change in direction of the axis. In fact, this is a consequence of the conservation of angular momentum – and Newton's third law of motion. Changing the direction of the axis of spin means altering the angular momentum, not in size but in direction. This requires a force, and as forces occur in pairs there is an equal and opposite force exerted on you as the wheel changes direction. This 'resistance' of a spinning wheel to change its direction of action is what makes a bicycle so stable – as long as the wheels are spinning! The importance of the conservation of angular momentum to the origin and nature of the Solar System is described in Chapter 27.

Newton's third law states that, when two bodies interact, the force exerted by one body (here, the holder tipping the wheel) brings about an equal and opposite force exerted by the other body (the wheel).

2 COLLISIONS

Most traffic accidents involve **collisions** – between vehicles, between vehicles and fixed structures or between vehicles and pedestrians. We describe collisions in which the structures are permanently distorted as inelastic (see page 97 for more about **elasticity**). The original kinetic energy of the colliding bodies is usually dissipated in heating the materials and the surroundings. Kinetic energy is conserved only in collisions between perfectly elastic bodies. No real objects we come across in everyday life are perfectly elastic, but rubber balls, 'superballs', steel balls and snooker balls approach elastic behaviour.

See question 6.

Collisions between *particles* (molecules, atoms, electrons, etc.) underpin a great deal of what happens in physics and these are often perfectly elastic, provided, that is, they collide at low enough speeds. If speeds are too high, we get chemical changes or, in specially built particle *colliders*, atoms are smashed and other dramatic events occur. But in developing our model of the kinetic theory of gases (see page 157) we assume that the collisions between the atoms or molecules of a gas are perfectly elastic.

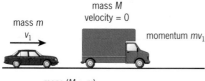

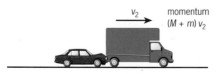

Fig 8.12 **An inelastic collision: a small car colliding with a lorry**

?

F Much expensive damage is caused in traffic accidents as inelastic materials are bent and distorted. This damage would not occur if the materials used were perfectly elastic. Would this be a good idea?

See question 7. ■

Fig 8.13 **Measuring skid marks at the site of an accident enables the speed of vehicles before impact to be calculated**

Fig 8.14 **The crumple zone slows down deceleration in a crashing car, and so reduces injury to passengers**

Collisions in one dimension

1 Inelastic collisions

The simplest example of an inelastic collision is when a moving object (such as a car) collides with a stationary object: momentarily they crunch together and then they move away from each other at the same time. This is a typical situation in vehicle collisions (see Fig 8.12).

The system is isolated so the total momentum is unchanged by the collision:

momentum before collision = momentum after collision

With the symbols in the diagram:

$$mv_1 = (M + m)v_2$$

Suppose the car has a mass of 1000 kg and hits the lorry at a speed of 30 m s^{-1} (this is just less than the maximum motorway speed of 70 mph). The lorry has a mass of 9000 kg. Then:

$$1000 \times 30 = 10\,000 \times v_2$$

giving $v_2 = 3$ m s^{-1}.

The combined vehicles move off at a speed of 3 m s^{-1} but the lorry would quickly stop if it was braked. Traffic accident experts can estimate this initial speed of the combined vehicles from measurements made at the site of the accident and hence calculate the speed of the incoming vehicle to see if it was exceeding the speed limit (Fig 8.13).

Energy changes in an inelastic collision

The incoming vehicle had a kinetic energy ($\frac{1}{2}mv^2$) of 450 kJ. The combined vehicles move off with a kinetic energy of just 45 kJ. Thus only 10 per cent of the input energy remains as kinetic energy in the motion of the combined vehicles. (Check these values yourself.) The lost kinetic energy has gone to make a lot of noise, but mainly to deform (do work on) and eventually heat the material of the vehicles. Vehicles have design features that minimise both the *quantity* of this energy delivered to human bodies in the vehicles and the *rate* at which it is delivered. These features include protective shells, crumple zones, seat belts, collapsible steering wheels, laminated windscreens and rapid-inflation airbags.

Momentum, force and energy in a traffic collision

The key relationships involved here are between:

- kinetic energy and work done in deforming structures,
- impulse and change of momentum (defining the force produced).

Imagine a car crashing into a very massive structure such as a concrete wall. Fig 8.14 shows what happens to a car in this situation.

The forces acting on the car deform it, doing work. The work done is given by the formula $W = Fd$, where we simplify a complex situation by considering F to be the average force acting until the car is stopped, and d the distance moved by the centre of mass of the car after the front of the car hits the wall. Some of the car's kinetic energy is transferred to sound and to small pieces that fly off, but these are tiny compared with the energy transferred to produce the main structural deformation. Thus we can say:

$$\frac{1}{2}mv^2 = Fd$$

For a given car travelling at a given speed, the force F can be reduced if d can be made larger. This is the role of the crushable 'crumple zone' at the front of the car. As the engine compartment crumples it not only absorbs some energy but also increases the distance in which the main protective metal 'shell' behind it is brought to a stop. This metal shell is strong enough to with-stand the forces developed in a crash without breaking or extreme deformation. This means that the passengers are not injured by pieces of the car body. However, this feature would not help if the passengers were free to move and collide with the hard metal of the rapidly slowing car.

See questions 8, 9 and 10.

The significance of impulse

Somehow or other, the driver and passengers have to be brought to a stop, and safety features in a car reduce the decelerating forces to as harmless a size as possible. The fascia (front panel) of the car is padded by soft material. The steering wheel is designed to collapse rather than crush the driver's chest or head. Most new cars now have airbags that inflate very quickly when the car rapidly decelerates. The bag surrounds the steering wheel and decelerates the driver more slowly than contact with the steering wheel would. The purpose of both seat belts and airbags is simply to allow the body to decelerate more slowly. We can best understand the physics of these features by using the concepts of momentum and impulse:

Impulse = change of momentum

$$F\Delta t = \Delta(mv) = m\Delta v \ (m \text{ is constant})$$

Therefore:

$$F = m\Delta v/\Delta t$$

F is the force required to change the momentum. In a car collision the human body of mass m is brought to a stop from the collision speed v. The force needed to do this is applied to the body mainly by the seat belt, with some small help from the feet pressing against the floor. We start by considering the forces involved when a car is braked to a stop.

Fig 8.15 **Stopping distances at various speeds (source: *Highway Code*)**

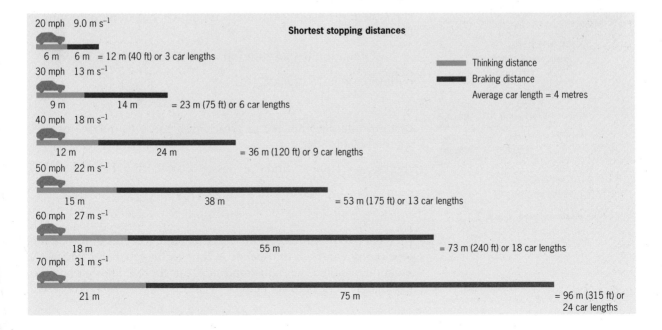

20 mph   9.0 m s^{-1}

Shortest stopping distances

━━ Thinking distance
━━ Braking distance
Average car length = 4 metres

6 m   6 m = 12 m (40 ft) or 3 car lengths
30 mph   13 m s^{-1}

9 m   14 m   = 23 m (75 ft) or 6 car lengths
40 mph   18 m s^{-1}

12 m   24 m   = 36 m (120 ft) or 9 car lengths
50 mph   22 m s^{-1}

15 m   38 m   = 53 m (175 ft) or 13 car lengths
60 mph   27 m s^{-1}

18 m   55 m   = 73 m (240 ft) or 18 car lengths
70 mph   31 m s^{-1}

21 m   75 m   = 96 m (315 ft) or 24 car lengths

EXAMPLE

Q Estimate the size of the force needed to stop the driver of a car when the car is braked to a stop from a speed of 13.6 m s⁻¹ (ie 30 mph). The *Highway Code* assumes that a car travelling at 13.6 m s⁻¹ is stopped in a distance of 14 m by normal braking, as in Fig 8.15. (A typical driver has a mass of 65 kg.)

A The time taken to stop is t, which we can estimate from the relationship:

$$\text{Distance covered} = \text{average speed} \times \text{time}$$

$$14 = \frac{13.6 + 0}{2}\, t$$

giving: $t = 2.06$ s.

Using the impulse = change of momentum relation $F\Delta t = m\Delta v$, we have:

$$F \times 2.06 = 65 \times 13.6$$

or: $F = 429$ N.

This force is less than the weight (say about 600 N) of the driver, which the feet are quite accustomed to coping with.

In a collision, both the car and the driver's body may be brought to a stop in 0.1 s.

The time has decreased from 2.06 s to 0.1 s, a factor of about 20, so the force is increased by the same factor to 8600 N – over 13 times the body weight.

In practice, things would be more complicated, with some of the force being applied to other parts of the body. For example, unless restrained by a seat belt, the body would pivot round the feet and would hit the steering wheel and/or the windscreen.

Seat belts

A perfectly unstretchable seat belt would be not be a good idea. It would stop the body in the same time as the car stopped, and a large force (as calculated above) would act on the body. The material of the seat belt is chosen so that it stretches just enough without breaking to allow the body to keep moving forward after the car has stopped. The driver takes a longer time than the car to stop, and it is obvious from the relation $F = m\Delta v/\Delta t$ that the applied force F decreases.

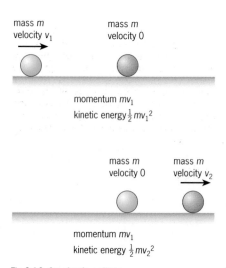

mass m velocity v_1

mass m velocity 0

momentum mv_1
kinetic energy $\frac{1}{2}mv_1^2$

mass m velocity 0

mass m velocity v_2

momentum mv_1
kinetic energy $\frac{1}{2}mv_2^2$

Fig 8.16 **An elastic collision: $v_1 = v_2$**

2 Elastic collisions

Imagine two elastic balls of equal mass colliding. Fig 8.16 shows the situation just before and just after the collision.

The system is isolated, so there is no net change in momentum. That is:

$$\text{momentum before collision} = \text{momentum after collision}$$

With quantities as in Fig 8.16:

$$mv = mv_1 + mv_2$$

This formula does not tell us which way the balls move – or their relative speeds.

Suppose $v = 10$ m s⁻¹ and the masses m are both 1.0 kg.

Then: $1.0 \times 10 = 1.0 \times v_1 + 1.0 \times v_2$

which simplifies to: $v_1 = 10 - v_2$ [1]

We have two unknowns in the equation, and it seems that we could choose any values of the speeds so long as the left-hand side of the equation equalled the right-hand side. But there is another law that applies here – the balls are perfectly elastic, so *kinetic energy is conserved*:

$$\tfrac{1}{2}mv^2 = \tfrac{1}{2}mv_1^2 + \tfrac{1}{2}mv_2^2$$

See question 11. ■

We can solve these simultaneous equations by inserting the value of v_1 (in terms of v_2) from equation 1 and simplifying:

$$(10)^2 = (10 - v_2)^2 + v_2^2 \qquad [2]$$

This results in the quadratic equation:

$$v_2^2 - 10v_2 = 0 \qquad [3]$$

This has two solutions: the speed of the second ball v_2 is either 0 or 10 m s^{-1}. The solution $v_2 = 0$ is physically impossible: the first ball would have to move through the second at 10 m s^{-1}. But the second ball is free to move, and the second solution tells us that this ball will move off with the original speed of the first. This effect is illustrated by the toy known as Newton's cradle (Fig 8.17), but flicking one coin at another on a flat table gives you some idea of the theory. There is no momentum left for the first ball, so it will stop.

Simple experiments with steel balls or coins suggest that, when a small mass hits a larger mass at rest, the small mass is likely to bounce back and the large mass to move on. When a large mass collides with a small mass at rest, both move on with the smaller mass having the greater speed.

Two-dimensional collisions

Fig 8.18 shows two examples of typical collisions between two particles in a cloud chamber. This device is explained in Chapter 18. It shows the paths of ionised particles, such as hydrogen and helium nuclei, as thin clouds of water. These form along the paths that the particles take through a damp gas.

In each photograph in Fig 8.18, a particle entered from below and collided with the nucleus of an atom of the gas in the chamber. The atom was ionised and its nucleus shot off, leaving a trail. The original particle was deflected in the collision and forms the other arm of the forked trail. Fig 8.19 illustrates a similar event occurring on a snooker table. Note that after the collision the snooker balls move off on paths that are at right angles to each other. This is because they are equal in mass. It is a consequence of the conservation laws of momentum and energy that the angle must be a right angle. Fig 8.18(b) shows a similar 90° fork, suggesting that the two particles were also of equal mass. In fact, the incoming particle was an alpha

Fig 8.17 **Newton's cradle. The far right ball has just hit its neighbour and the far left ball has moved out**

?

G A large steel ball and a small steel ball travelling in opposite directions collide and bounce apart elastically. Sketch graphs of force against time showing the forces acting on each ball during the collision.

Fig 8.18 **Cloud chamber pictures, showing the paths of moving particles. Left, alpha particles (green) stream up through hydrogen gas in the chamber and one alpha particle (yellow) collides with a hydrogen nucleus (red). Right, an alpha particle collides with a helium nucleus and both move off at 90°, because an alpha particle *is* a helium nucleus and both have the same mass**

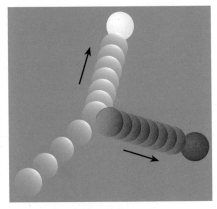

Fig 8.19 **Two snooker balls after a collision moving apart at a 90° angle**

particle emitted from a radioactive material. The gas in the cloud chamber was mostly helium. This is direct evidence that alpha particles are identical to helium nuclei. Results like this were used in the early days of nuclear physics to measure the relative masses of particles and so identify them.

Fig 8.18(a) shows that the incoming alpha particle was not deflected as much as the one in Fig 8.18(b). This suggests that it probably collided with a less massive nucleus. The nucleus was in fact hydrogen. These are carefully chosen pictures, taken from just the correct angle – the collisions occur in three-dimensional space so that you will only see the 90° angle by chance. In practice, investigators take stereoscopic photographs, which allow them to work out the angles correctly whatever the position in space.

See question 12.

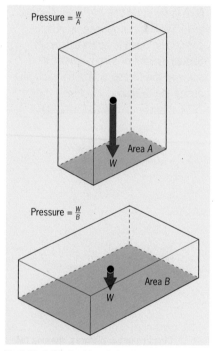

Fig 8.20 **A block with weight W can exert different pressures on a surface**

3 FLUIDS, FLOATING AND FLYING

The movement of air over a curved surface produces unexpected forces that allow objects heavier than air to fly. These forces can be explained using Newton's laws, and can also be related to the way water flows, in pipes and rivers for example.

Newton's laws of motion are best expressed for fluids in a statement of the Swiss mathematician Daniel Bernoulli (1700–1782) and known as **Bernoulli's principle**. Other important ideas that we use in this section of the chapter are **fluid pressure**, **thrust** and **viscosity** (the name for the *internal friction* of a fluid). We consider **hydrostatics** and **hydraulics**, which deal with the behaviour and uses of liquids under pressure. We also deal with the behaviour of fluids in response to forces. This behaviour is best explained using the **kinetic theory of matter**, which is dealt with in Chapter 7.

Pressure

Pressure is defined as the *force exerted on a surface per unit area* (see Fig 8.20):

$$\text{Pressure} = \text{force/area}$$

$$P = F/A$$

Pressure is a scalar quantity with units of newtons per square metre. The unit of pressure has a special name, the pascal (Pa): 1 pascal = 1 N m^{-2}. Fig 8.21 shows a simple example of the same force producing different pressures, due to the weight of a skier with and without skis. The large area of the ski 'spreads the load', and the weight of the body is more easily supported by the soft snow.

Fig 8.21 **A ski reduces the pressure on snow**

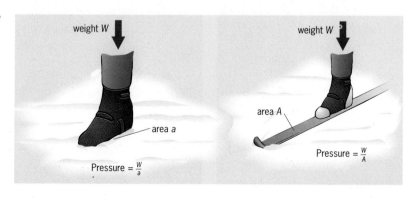

Pressure in liquids

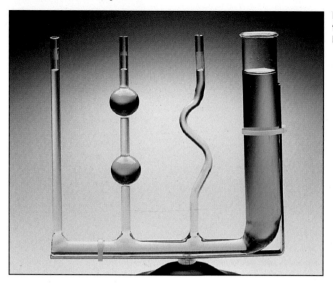

Fig 8.22 **Liquid finds its own level**

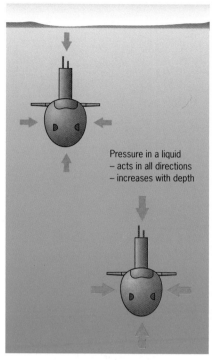

Fig 8.23 **Two submarines at different depths have different pressures acting on them**

Pressure in a liquid has special features that are shown in Fig 8.23. Theory and experiment show that in a liquid:

● Pressure increases with depth.

● At a given point in a liquid the pressure acts equally in all directions.

● A liquid can *transmit* a pressure exerted on it at one place so that it acts at some other place.

Look ahead to Fig 8.31, which shows a hydraulic press that illustrates the third point.

Hydrostatic pressure

The pressure at the bottom of a cylindrical jar of liquid is caused by the weight of the liquid and of the atmosphere above the liquid surface (Fig 8.24).

We now consider the pressure P *due to the liquid alone.*

The liquid has density ρ. The mass of water in the cylinder is thus:

$$m = \text{volume} \times \text{density} = Ah\rho$$

The weight of this mass is $mg = Ah\rho g$ and is a force acting over the area A. Thus the pressure P (force per unit area) caused by the liquid on the base of the cylinder is:

$$P = mg/A = h\rho g$$

Any object placed at this depth will be acted upon by this pressure. If you made a small hole in the side of the cylinder, water would be forced out sideways. It is perhaps surprising that this pressure acts *in all directions*, not just downwards as happens with, say, skis in the example opposite.

The force due to **fluid pressure** is usually called **thrust**:

$$\textbf{Thrust} = \textbf{pressure} \times \textbf{area}$$

$$\boldsymbol{F = PA}$$

At a level at some height above the base we might expect that the pressure is less. For example, at a depth y the pressure should be $y\rho g$.

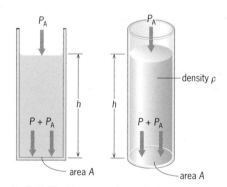

Fig 8.24 **Liquid pressure in a cylinder**

■ See questions 14 and 15.

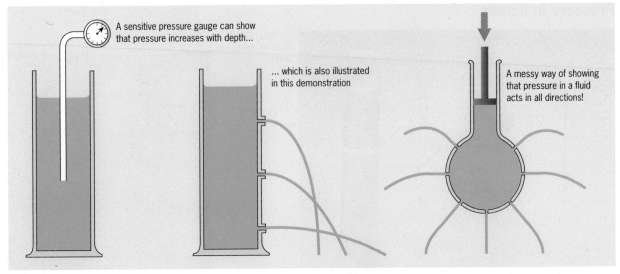

Fig 8.25 **Pressure increases with depth and also acts in all directions**

Simple experiments as shown in Fig 8.25 confirm that pressure increases with depth in a fluid and also that pressure acts in all directions.

Consider a very thin layer of liquid at a depth y. The downward force on the layer due to the liquid is the weight of the volume above it, that is $Ay\rho g$. The layer is in equilibrium (it is stationary), therefore there must be an upward force on it equal to the downward force. This can only come from the thrust due to the pressure P in the liquid at depth y. This means that:

$$PA = Ay\rho g$$

That is:

$$P = y\rho g$$

where y is the distance of the layer below the surface. This is the basic relation in the science of **hydrostatics**, and is a simple formula that works for liquids, which have a constant density. (Gases under pressure are compressed so that the density increases. This complication will be dealt with later.)

Measuring pressure in fluids

The simple **manometer** is being used to measure the pressure of a gas supply in Fig 8.26. The excess pressure of the gas supply compared with the pressure of the atmosphere is balancing the pressure exerted by the column of water.

The pressure exerted by the liquid column, $h\rho g$, equals the excess pressure of the gas supply.

A **barometer** is used to measure atmospheric pressure. The original barometers used a column of mercury as shown in Fig 8.27. Air pressure acts on the surface of the mercury in the dish. This pressure is transmitted through the liquid and produces a thrust, acting upwards at the base of the mercury column, which exactly balances the weight of the column. The space above the column is (almost) a vacuum, created as the mercury column fell to its balancing level of about 76 cm above the surface of the mercury in the dish.

The atmospheric pressure is given by $h\rho g$ where ρ is the density of mercury. (Remember that this is independent of the area of the dish.)

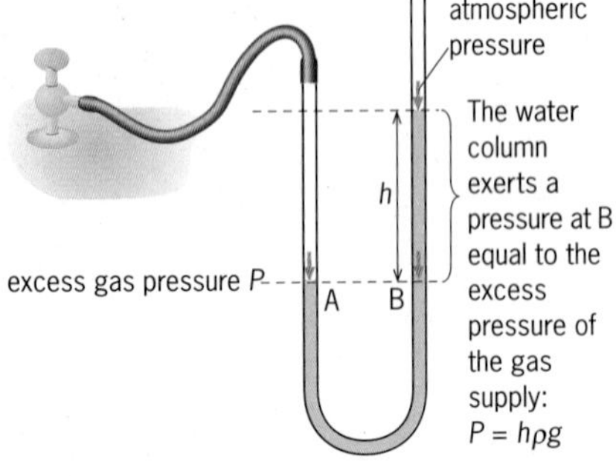

?

H Estimate the extra pressure exerted on your body when you have dived into a swimming pool and are at a depth of 3 m.

(The density of water is 1000 kg m^{-3}.)

atmospheric pressure

The water column exerts a pressure at B equal to the excess pressure of the gas supply:

$P = h\rho g$

excess gas pressure P

h

A   B

Fig 8.26 **Simple U-tube manometer**

?

I A simple water manometer connected to the laboratory gas supply shows a difference of 120 mm in the water levels in the two arms. What is the excess pressure of the gas supply?

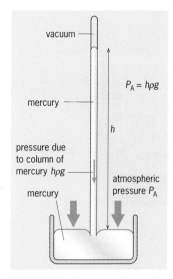

Fig 8.28 **An aneroid barometer**

Fig 8.29 **An altimeter**

Fig 8.27 **A simple mercury barometer**

For accurate measurements using a mercury barometer, a correction has to be made for the liquid's variation in density with temperature, and *h* has to be measured with great care. Barometers play an important part in weather forecasting: blocks of warm (usually damp) air and cold dry air have different densities, causing different atmospheric pressures at ground level. These pressure changes appear some distance ahead of the associated weather systems.

A cheaper instrument for measuring atmospheric pressure is an **aneroid** (liquid-free) barometer. It has a partially evacuated metal box with a corrugated shape that is kept from collapsing under the pressure of the atmosphere by a strong spring (Fig 8.28).

Changes in atmospheric pressure cause small movements in the surface of the box. These movements are amplified mechanically or electronically for display. Aneroid barometers have to be calibrated from time to time against the more direct measurements made by a mercury barometer. The aneroid principle is also used in an **altimeter** (Fig 8.29), since air pressure varies with height.

See questions 16 and 17.

?

J Before taking off, the pilot of an aircraft should check and possibly reset the aneroid altimeter. Give two reasons for doing this.

Hydraulics

Modern industrial machinery relies heavily on the techniques of **hydraulics**. The moving surfaces of aircraft wings, car brakes, and many types of large machines used for manufacturing and construction all rely on liquids to transmit pressure. The key idea is that although the *pressure* is the same throughout a hydraulic fluid, the *force* (ie thrust) it produces at any surface depends on the area of the surface. This means that a hydraulic system can act as a *force-multiplying machine* (Fig 8.30).

Fig 8.30 **Left: an earth-moving machine; the cylindrical rods are part of the hydraulic system. Right: hydraulic pipes in an aircraft**

Fig 8.31 **The principle of a hydraulic press**

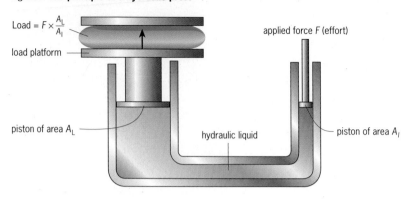

The main features of a hydraulic press are shown in Fig 8.31. A hydraulic press produces a large crushing force for a much smaller input force.

The input force F acts on the area A_I and creates a pressure P in the hydraulic fluid (usually a special oil) of size:

$$P = F/A_I$$

The pressure is transmitted through the fluid and acts on the surface of area A_L, which is much larger than the area A_I. Thus the thrust (force) exerted on the hydraulic press surface is:

$$\text{force} = \text{pressure} \times \text{area}$$
$$= PA_L$$
$$= FA_L/A_I$$

If area A_L is 100 times area A_I, then the input force is increased by the same factor, so is 100 times larger.

Are liquids incompressible?

It is often said that hydraulics rely on the fact that liquids are 'practically incompressible'. It is certainly hard to squash them, compared with gases. But solids are usually even harder to squash. However, solids have a tendency to shear, bend or shatter when large forces are exerted on them.

These problems do not apply to liquids, hence the popularity of hydraulic machinery. However, liquids do have their disadvantages – they evaporate, and they also have to be contained in solid containers and pipes. Large pressures may cause the solids to fail and the liquid to leak away.

Why ships float: Archimedes' principle

Like many people, the Greek scientist and mathematician Archimedes of Syracuse (about 287 to 212 BC), had one of his best ideas in the bath – or so the fable tells us. But wherever he got the idea, Archimedes was the first scientist to realise the significance of the everyday event of the water level rising when you get into a bath. He linked this effect to a problem that he had been given to solve.

It was suspected that a royal crown was made of a silver–gold alloy, and not the pure gold supplied to the jeweller who made it. The crown's density would give Archimedes the answer. He could weigh the crown, but how could he measure its volume? Then he realised that the volume of the crown would equal the volume of water it displaced.

Archimedes also considered an object that floated either because it was hollow or because it was less dense than water. He stated that the displaced water would exert an upward force on the object. These ideas are brought together in **Archimedes' principle**, which is now stated thus:

> **When a body is wholly or partly immersed in a fluid it experiences an upward force (upthrust) equal to the weight of the fluid displaced.**

The effect is due to fluid pressure.

Consider a block of material (assumed regular for the sake of simplicity) floating in a liquid (Fig 8.32).

The pressure on its upper surface, P_U, is $h\rho g$, where ρ is the density of the fluid. This pressure acts vertically on the surface and presses down with a force (thrust) of $Ah\rho g$.

The pressure on the lower surface is $P_L = (h + \Delta h)\rho g$ and acts upwards with a thrust $A(h + \Delta h)\rho g$. The thrusts on the vertical sides cancel out.

The net thrust on the block is thus upwards and is $A\Delta h\rho g$. $A\Delta h$ is the volume (V) of the block and the expression for the **upthrust** simplifies to $V\rho g$, which is of course the weight of the fluid displaced.

Whether the block sinks or floats depends on its own weight W.

If $W > V\rho g$ it will sink. If the block is of density ρ_B, its weight is $V\rho_B g$, so the condition for sinking is simply $\rho_B > \rho$.

If ρ_B is less than ρ, the block has a net upward force and will rise until it displaces just enough of the fluid for the upthrust to equal its weight.

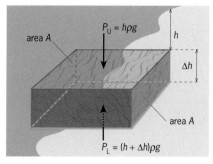

Fig 8.32 **Forces acting on a floating object**

?

K The Example below gives the proportion of wood under water as 0.57. The ratio of the densities of wood and water is also 0.57. Is this a coincidence? Explain.

L Ships are usually made of steel, of density 7700 kg m^{-3} – much denser than water. Why do steel ships float?

See question 13.

EXAMPLE

Q A block of cedar wood has a mass of 200 kg and a density of 570 kg m^{-3}. What fraction of the wood will be under the surface when the block is floating in water (density 1000 kg m^{-3})?

A The wood will float when it displaces a weight of water equal to its own weight, which is equivalent to saying that the wood and the *displaced* water have the same mass.

Mass of water displaced = 200 kg.

Volume of water displaced = mass of water/density of water = 200/1000 = 0.20 m^3.

The volume of the block = mass of block/density of block = 200/570 = 0.351 m^3.

Thus the proportion of the wood under the water surface is 0.20/0.351 = 0.57.

Fig 8.33

Fluid flow

Fig 8.34(a) shows a light ball suspended in the flow of air from the exit end of a vacuum cleaner. The weight of the ball is balanced by an upward force caused by the air flow hitting the bottom of the ball and being deflected. The momentum of the air mass changes,

Fig 8.34 **Polystyrene ball: (a) held in a vertical air stream from a vacuum cleaner nozzle; (b) in an angled air stream**

resulting in the force. This force acts in the direction of the air flow. It may be a puzzle to you how the ball seems to stay neatly inside the air flow without falling away sideways. Now look at Fig 8.34(b): it shows the same ball in an air flow tilted at an angle. The simple 'change of momentum' force would not support the ball at an angle, and it is obvious that there must be another force (or set of forces) stopping the ball from falling.

This is an easy demonstration of unexpected effects that happen when a fluid moves relative to a solid surface. These effects are important in a wide range of applications, from the flight of aircraft to hair sprays, via the design of windmills and the ability of base-balls, tennis balls and cricket balls to 'swing' in the air. The principles of fluid flow were first explained over two hundred years ago by Daniel Bernoulli, although it is doubtful whether he saw *all* the consequences of his theories.

Bernoulli's principle

Imagine a liquid flowing through a tube shaped as in Fig 8.35. The flow is continuous – just as much liquid enters at X as leaves at Y. For this to be true, the liquid must flow *faster* in the narrow part of the pipe than it does in the wider parts.

For a liquid flowing through a tube of cross-sectional area A at a speed v, the volume passing any point each second is Av. As A varies, so will v – provided the liquid is **incompressible**. At the pressures normally found in practical situations, liquids are incompressible. Thus, for a liquid flowing in a pipe of varying cross-sectional area, the quantity Av stays constant. This is the **principle of continuity** for a liquid.

Gases are more easily compressed, however, and if the pressure changes as the pipe widens or narrows the gas *density* will change. What must be constant is the mass of gas that enters and the mass that leaves any section of pipe; it will help to think of this in terms of the number of molecules that enter and leave the section. At any given values of area A and speed v, the mass of gas of density ρ in the section v metres long is $\rho A v$. In this case, we express the principle of continuity in its more general form as:

$$\rho A v = \textbf{constant}$$

Pressure in a moving fluid

The liquid moves because somewhere back along the pipe there is a pump (say) producing a pressure, which is the same everywhere within the liquid. Now think about what must be happening at Z in Fig 8.35. Here the liquid is being accelerated to increase its speed. Acceleration requires a net force acting in the direction of the acceleration. There must therefore be a greater pressure to the left of Z than there is to the right.

Fig 8.36 shows a demonstration of the pressure differences in a model of the arrangement. The vertical tubes act as manometers, with the heights of the liquid columns measuring pressure. The columns show that the pressure in the narrow tube is less than that in the wide tube in front of it, as theory predicts. The pressure increases again in the second wide tube – because the liquid is being slowed down and a force is required to decelerate it.

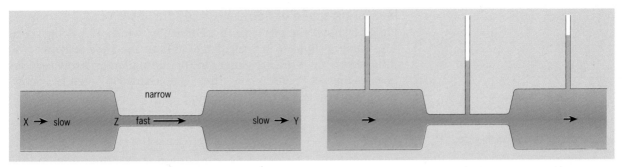

Fig 8.35 **The volume of liquid flowing per second is the same in both the wide and narrow sections of the tube. This means that the liquid must flow faster in the narrow section**

Fig 8.36 **Pressure variation in a liquid flowing through a tube of varying cross-section**

Energy changes in a flowing fluid

Bernoulli developed a mathematical treatment of the energy changes that occur when a liquid flows in a pipe. This is shown in the Extension box on page 192. He proved the **Bernoulli equation**, linking the pressure P, the velocity v and the density ρ of a fluid at a point h above a given reference point:

$$P + \tfrac{1}{2}\rho v^2 + \rho gh = \text{constant}$$

In words:

pressure + kinetic energy per unit mass + gravitational potential energy per unit mass

is constant throughout a moving fluid. **Bernoulli's principle** follows from this. That is, in a moving fluid:

The greater the speed, the lower the pressure.

Applications of Bernoulli's principle

Fragrance spray

Fig 8.37 shows a hand-operated fragrance spray. When you press the bulb, air is forced through the tube, which has a constriction in it – that is, part of the tube is narrower than the rest. The air travels faster in the narrow part and so its pressure decreases. Liquid is drawn up the vertical tube, breaks into droplets as it meets the air jet and is carried out of the nozzle in a fine spray.

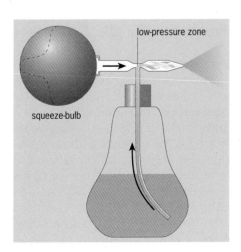

Fig 8.37 **A fragrance spray**

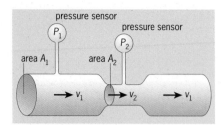

Fig 8.38 **The principle of the Pitot tube**

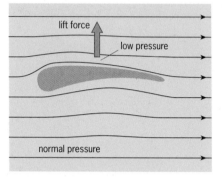

Fig 8.39 **Aerofoil with lines of air flow**

?

M Mrs Frisbee sold her pies in dish-shaped plastic containers. Young customers discovered that a spinning pie dish (now called a *Frisbee*) had aerodynamic properties. Explain: **(a)** why Frisbees fly; **(b)** why they fly in a more controlled manner when they are spinning.

Fig 8.40 **A car in a wind tunnel, with streamlines shown by streamers**

Measuring speed of flow

Fig 8.38 shows the principle of a Pitot tube, which allows the speed of a moving fluid to be calculated. Where the tube narrows, the fluid speeds up. Pressure gauges in the normal and narrow sections give readings of P_1 and P_2, say. There is no change in potential energy, so we can use Bernoulli's equation:

$$P_1 + \tfrac{1}{2}\rho v_1{}^2 = P_2 + \tfrac{1}{2}\rho v_2{}^2 = \text{constant}$$

so that:
$$P_1 - P_2 = \tfrac{1}{2}\rho(v_2{}^2 - v_1{}^2) \qquad [1]$$

The principle of continuity also applies, so that:

$$A_1 v_1 = A_2 v_2 \qquad [2]$$

The areas are known, so if we substitute:

$$v_2 = \frac{A_1}{A_2} v_1$$

in equation 1, then the speed v_1 of the fluid in the main pipe can be calculated from the pressure and area measurements.

Aerofoils: the principle of flight

An **aerofoil** is the shaped wing of an aircraft (Fig 8.39). The top surface of an aerofoil is convex, which means that while the aircraft is in flight, air travels faster over the top surface than over the bottom. This is because the air stream has to travel a greater distance over the top surface in the same time (as with a liquid, the air mass has to be continuous). The result is a smaller pressure on the upper surface than on the lower, and the aerofoil experiences a net upward thrust – called **lift**. The wing is usually angled to the direction of motion so that it has to push against the air. This produces an additional force (reaction) with an upward component to add to the lift. In general, the angle the aerofoil makes with the forward direction of motion (angle of attack) is large enough for this reaction force to be more effective in maintaining height than the aerofoil's Bernoulli effect.

Streamlines and turbulent flow

The simple theory of fluid flow described above assumes that the flow is steady, so that particles of the fluid flow in smooth paths as shown by the continuous streamlines made visible by streamers in Fig 8.40. But above a certain speed the flow becomes chaotic and we have **turbulent** flow. Energy is dissipated in sound and heating; and extra pressures are produced which can affect the straight-foward motion of a vehicle.

There is often turbulent flow when two air streams that have diverged over a moving object join together again. This effect can be seen in the air behind a car being tested in a wind tunnel. The disturbed turbulent air produces a low-pressure region, which tends to pull the car backwards, contributing to the drag forces that oppose the motion of any object through a fluid. The turbulent drag force is reduced by vehicle design. The best shape is like an aerofoil, of course, but drivers would worry if, at a certain speed, the car started flying! The rear spoiler fitted to some cars, as shown in Fig 8.41(a), is a way of reducing the drag force by making the air flow less turbulent at the back of the vehicle. As in (b), racing cyclists and downhill skiers wear specially shaped helmets to reduce the effect of turbulence.

Other drag forces act on a surface moving through a fluid. For example, air has to be pushed away as a car drives through it, and this produces a reaction force that acts on the car. Next we look at frictional forces between the vehicle's surface and the fluid.

Friction and viscosity

Layers of fluids move over each other very easily; for this reason a fluid cannot resist a shearing stress (see Chapter 5). But fluids do have a kind of internal friction called **viscosity**. This varies: moving a knife blade through water is much easier than moving it through honey. Similarly, viscous forces arise when adjacent layers of liquid move against each other.

In a simple model of what happens, we assume that when a fluid moves through a pipe the fluid layer next to the wall of the pipe is at rest, and that the fastest stream is at the centre of the pipe, as in Fig 8.42. There is a constant **velocity gradient** in the fluid, $\Delta v/\Delta x$, where x is the distance measured radially in the pipe. A liquid that behaves like this is called a *Newtonian liquid*. Friction between adjacent layers in the fluid determine how fast the fluid can flow – compare pouring water and treacle. A fluid has a **coefficient of viscosity** η (the Greek letter eta) which determines the size of the viscous force. The resistive viscous force is greater for a wider pipe. These factors are combined in **Newton's law of viscosity**. When F is the force due to viscosity acting on a fluid stream of cross-sectional area A, against the direction of flow, then:

$$F = \eta A \frac{\Delta v}{\Delta x}$$

This force is also the drag force on the sides of the container.

Viscosity has units of N S m^{-2} and values of the viscosity of some fluids are given in Table 8.3 – blood *is* thicker than water! Viscosity varies with temperature: as a general rule the coefficient decreases as temperature increases, so that liquids are more 'runny' when hot.

Fig 8.41 **Above: A car with rear spoiler. Below: Special helmet worn by most cyclists in time trials**

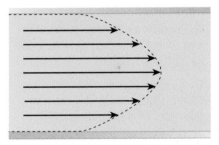

Fig 8.42 **Streamline flow with viscosity. The speed at the wall is zero, and increases to a maximum at the centre of the pipe**

Table 8.3 **Viscosities of some common fluids**

Fluid	Viscosity (N s m^{-2})
Water at 20 °C	1.0×10^{-3}
Water at 100 °C	0.3×10^{-3}
Blood at 37.5 °C	2.7×10^{-3}
Typical motor oil at 20 °C	830×10^{-3}
Glycerine at 20 °C	850×10^{-3}
Air at 23 °C	0.018×10^{-3}

See question 18.

Stokes' law

Viscous forces also affect the speed of streamline motion of an object in the fluid. For a sphere in streamline motion at a speed v through a fluid, the Irish physicist George Stokes (1819–1903) proved that it would experience a drag force:

$$F = 6\pi\eta r v$$

where η is the coefficient of viscosity of the fluid and r is the radius of the sphere (Fig 8.43).

This idea may be linked to the *terminal speed* of a falling sphere (see page 39). In this case the viscous drag on the sphere equals its weight:

$$mg = 6\pi\eta r v$$

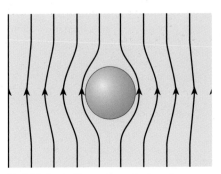

Fig 8.43 **Streamline movement of a sphere through a fluid. The arrows show the velocity of the fluid relative to the sphere**

EXAMPLE

Q Calculate the terminal speed of **a)** a lead sphere, **b)** a hailstone, both having a radius of 2 mm.

(The density of lead is 1.14×10^4 kg m^{-3}; the density of ice 9.2×10^2 kg m^{-3}.)

A

a) As above, the condition for terminal speed is that $mg = 6\pi\mu rv$, or terminal speed v is:

$$v = mg/6\pi\eta r$$

The mass of the lead sphere is:

$$m = \text{volume} \times \text{density} = \tfrac{4}{3}\pi r^3 \times \rho$$

so the equation can be simplified to:

$$v = \tfrac{2}{9}r^2\rho g/\mu$$

$$= (\tfrac{2}{9} \times 4 \times 10^{-6} \times 1.14 \times 10^4 \times 9.8)/(1.83 \times 10^{-5})$$

$$= 5.5 \times 10^3 \text{ m s}^{-1}$$

b) The only difference in the conditions of the problem is the density of the materials. Thus the ratio of terminal speeds is the same as the ratio of densities. Ice is less dense than lead, so it falls more slowly.

The terminal speed of the hailstone is:

$$v = (5.5 \times 10^3 \times 9.2)/114 = 443 \text{ m s}^{-1}$$

Note that in practice both spheres would fall more slowly, since turbulence would set in at much lower speeds.

?

N Calculate the terminal speed of the lead sphere in the example when it falls through water at 20°C.

O A parachutist has a mass of 60 kg. Assuming that Stokes' law applies in this case, what radius of circular parachute would be required to ensure a terminal speed of 4 m s^{-1}?

The spinning ball

The streamlines of Fig 8.43 will be different if the ball is spinning. In some ball games such as cricket, football and tennis, a player will put a spin on the ball to make it *swing*. There is a viscous drag at the surface of the ball. Fig 8.44 shows what might happen. There is a difference in the relative speeds of the fluid at each side of the ball, and the fluid causes a greater pressure where its speed is lower. The ball tends to move sideways, causing confusion to batsmen, goalkeepers and the receiver of a tennis service.

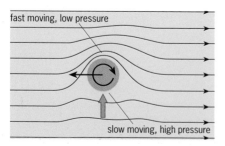

Fig 8.44 **Streamlines for a spinning ball**

See question 19. ■

Drag effects for a vehicle

Energy can be 'saved' in transportation by reducing a variety of losses. Nowadays the **drag factor** is taken into account in car design. This factor indicates how much energy a moving vehicle 'loses' as a result of air friction. This kind of friction loss is due to the vehicle working against a frictional *drag force* F_d. When v is the vehicle's speed, A its cross-sectional area and ρ the density of air, then:

$$F_d = \tfrac{1}{2}CA\rho v^2$$

where C is its drag factor, a constant which depends upon the shape of the vehicle.

The car is also subject to *rolling friction*, due to the contact of the tyres with the road surface. This creates a force given by μN where N is the normal reaction between car and road. The coefficient of rolling friction for a car is about 0.02. The normal reaction N gets slightly less at higher speeds as the car behaves like an aerofoil (luckily not very successfully at usual road speeds!).

Rolling friction is much more important than drag for a car travelling at low speeds. However, drag increases as the square of the speed. Drag and rolling friction become equal at a speed of about 18 m s^{-1} for a typical car, and eventually drag is the main source of frictional loss.

Extracting energy from moving fluids

Wind power has gained a great deal of attention as a source of renewable energy. Originally windmills transferred energy from moving air directly to the working machinery – usually for milling grain to make flour. They are now used in large groups on **wind farms** to drive electric generators.

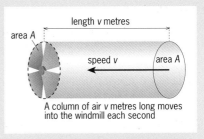

A column of air v metres long moves into the windmill each second

Fig 8.45 **Air column hitting rotating blades**

The energy available from a moving mass of fluid depends on its speed and density – that is, on the kinetic energy it carries. Consider a windmill of circular cross-section. It will remove kinetic energy from a cylinder of moving air reaching it. We assume that its blades cover the whole circular area, as do the blades of the turbine shown in Fig 8.45.

Each unit volume of air has a mass ρ where ρ is the density of air (1.3 kg m^{-3}). Its kinetic energy per unit volume is thus $\frac{1}{2}\rho v^2$, where v is the velocity of the air. The volume of air flowing through the windmill per second is Av where A is the area swept by the arms of the windmill. Thus the maximum energy *per second* (joules per second = watts) that is available to the windmill is $Av \times \frac{1}{2}\rho v^2$. So the maximum power available is $P_{max} = 0.5 A\rho v^3$.

The windmill cannot extract all the energy available. This would require all the kinetic energy of the air column to be transferred, and even in theory the best that can be done is to extract about 60 per cent of it. In practice there are other losses due to blade design, friction in gearing and generator losses that reduce the practical output to about 15 per cent of P_{max}.

Notice that the power available varies as the *cube* of the wind speed. This means that above-average wind speeds contribute far more than half the energy extracted. The overall effectiveness of wind farms depends on the distribution of wind speed over the year.

The two main types of wind generator are the horizontal axis and the vertical axis generators (Fig 8.46).

Fig 8.46 **Types of wind generator: above, vertical axis; below, horizontal axis**

EXAMPLE

Q A wind farm has 100 wind generators each of blade diameter 50 m. Estimate the power that could be generated by the farm at a time when the wind speed is 12 m s^{-1}. Assume that the windmills are 15 per cent efficient.

(Air diameter = 1.3 kg m^{-3}.)

A The acceptance area of a wind generator is $\pi r^2 = \pi \times 2500 = 7.85 \times 10^3$ m^2.

The total working area of the 100 generators in the wind farm = 7.85×10^5 m^2.

Thus:
$$P_{max} = (0.5 \times 7.85 \times 10^5) \times 1.3 \times (12)^3 = 880 \text{ MW}$$

The likely output is 15 per cent of this: 130 MW.

The main energy losses from a car engine are not due to these effects at all. As explained in Chapter 15, heat engines are inefficient, and dissipate almost 70 per cent of the energy that comes from the burning fuel. There are other friction losses in the transmission, so that the energy available for moving the car is only about 14 per cent of the energy supplied by the fuel.

EXAMPLE

Q How much power needs to be delivered to the wheels of a car travelling at a typical motorway speed of 31 m s⁻¹ (70 mph)?

(Data: At this speed, rolling friction provides a force of 210 N; the car has a cross-sectional area of 1.8 m²; the density of air is 1.3 kg m⁻³; the drag coefficient of the car is 0.4.)

A The drag force on the car is calculated from:
$$F_d = \tfrac{1}{2}CA\rho v^2$$
$$= 0.5 \times 0.4 \times 1.8 \times 1.3 \times (31)^2$$
$$\text{Drag force} = 450 \text{ N}$$

Thus the total frictional force on the car is 210 N + 450 N = 660 N.

This force does work that requires energy to be supplied by the car's engine.

The work done per second = force × distance moved per second (speed), that is:

$$\frac{\text{power}}{\text{needed}} = \frac{\text{work done}}{\text{per second}} = \frac{\text{frictional}}{\text{force}} \times \text{speed}$$

$$= 660 \text{ N} \times 31 \text{ m s}^{-1}$$
$$= 19.8 \text{ kW}$$

Bernoulli's equation

Consider an incompressible fluid moving along a pipe. The diameter of the pipe is not uniform – its dimensions are as in Fig 8.47. The fluid moves because of the pressure exerted by an outside system such (eg pump). Assume the fluid is 'perfect', so no energy is lost through viscous forces.

There is a force F_1 acting on A_1, a cylindrical surface of fluid at position 1. This force is due to the pressure in the fluid to the left of A_1. A short time later, the surface has moved to position 2, so an element of fluid with a particular volume has passed from 1 to 2. In another part of the tube an *equal* volume of fluid has also moved, since the fluid is incompressible. This is shown at positions 3 and 4. Here the tube is wider and at a higher level. Note that this second element of fluid is moving against a pressure force F_2, due to the fluid to its right. Now we consider the work done by the forces and the energy changes in the moving fluid.

The **work done by the forces** is given by:

$$W = F_1 \Delta x_1 - F_2 \Delta x_2$$

The negative sign is there because F_2 acts in the opposite direction to F_1. The forces are thrusts, that is $F_1 = P_1 A_1$ and $F_2 = P_2 A_2$, where P_1 and P_2 are the pressures at the surfaces A_1 and A_2 respectively. So we have:

$$W = P_1 A_1 \Delta x_1 - P_2 A_2 \Delta x_2$$

or, if ΔV is the total volume of both the cylindrical elements of fluid:

$$W = (P_1 - P_2)\Delta V$$

For the energy changes, what happens over time? The lower cylinder at position 1–2 eventually gets to the upper position 3–4, its kinetic energy changing from $\tfrac{1}{2}\Delta m v_1^2$ to $\tfrac{1}{2}\Delta m v_2^2$. It has changed its gravitational potential energy by $\Delta mg(h_2 - h_1)$. Δm is the mass of the volume element ΔV.

Now comes a tricky idea. We can ignore what happens to the fluid in between our two selected positions: any bits of work done cancel out – and so do the energy changes. Only the 'loose ends' count so that the **total net work done** is simply as we derived above, and it is this that produces the energy changes. So the net work done is:

$$(P_1 - P_2)\Delta V = \tfrac{1}{2}\Delta m(v_2^2 - v_1^2) + \Delta mg(h_2 - h_1)$$

But $\Delta m = r\Delta V$ where r is the density of the fluid. So the above equation simplifies to:

$$P_1 - P_2 = \tfrac{1}{2}r(v_2^2 - v_1^2) + rg(h_2 - h_1)$$

which can be neatly rearranged as:

$$P_1 + \tfrac{1}{2}rv_1^2 + rgh_1 = P_2 + \tfrac{1}{2}rv_2^2 + rgh_2$$

This is the Bernoulli equation. In words, we can say that at points along a streamline in a moving, non-viscous incompressible fluid:

Pressure + kinetic energy per unit volume + gravitational potential energy per unit volume = a constant at all points

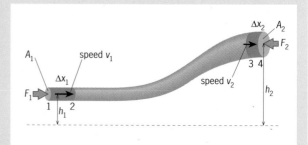

Fig 8.47 **An incompressible fluid moving through a non-uniform pipe**

P Put $v_1 = v_2 = 0$ in the Bernoulli equation. Comment on the result.

SUMMARY

After studying this chapter you should:

■ Have improved your understanding of Newton's laws of motion as applied to transport and moving fluids.

■ Understand the role of friction in transportation, and be able to apply the laws of static and sliding friction and coefficient of friction.

■ Understand the ideas of viscosity, streamline flow and drag factor, relating these to Stokes' law and terminal speed.

■ Understand the physics of rotating bodies and be able to use the ideas of angular velocity, angular acceleration, torque, rotational inertia and rotational kinetic energy.

■ Be able to apply the laws of conservation of momentum and conservation of energy as appropriate to elastic and inelastic collisions.

■ Be able to use the relationship: impulse = change of momentum ($F\Delta t = \Delta P$).

■ Understand and be able to apply the concept of pressure in fluids and relate this to practical situations involving hydraulics and flotation.

■ Understand and be able to use Archimedes' principle.

■ Understand the Bernoulli principle and how fluid pressure, potential energy and kinetic energy are related in the Bernoulli equation.

QUESTIONS

1 Two vehicles on the same road have equal kinetic energies. Does this mean that they must have equal momenta? Explain your answer.

2 Ice hockey and ice dancing both rely on a very low friction between skates and the ice. Would these activities be possible if the contact between participants and ice were *completely* frictionless? Explain your answer.

3 A car of mass 2200 kg is at rest. The coefficient of friction between its tyres and the ground is 0.7. The driver intends to accelerate the car from rest as quickly as possible.

a) What is the maximum force that could be exerted on the car by friction?

b) Draw a sketch showing the two main forces *acting on the car* just after it begins to accelerate.

c) Calculate the resulting acceleration of the car.

d) Use your answer to part **c)** to estimate how long the car takes to reach a speed of 30 m s^{-1}.

e) In practice your answer to part **d)** is bound to be optimistic: the car takes longer to reach this speed. Why?

4 According to the *Highway Code* a car travelling at 13.6 m s^{-1} should be stopped in a distance of 14 m by normal braking.

a) What value of the coefficient of friction between tyres and ground is required to produce this result?

b) How far would the car travel if the coefficient of friction was reduced by icy conditions to 0.2?

5 Another name for *angular momentum* is *moment of momentum*. By considering a particle of mass m, moving with speed v in a circle of radius r, suggest why this is a suitable alternative title for angular momentum.

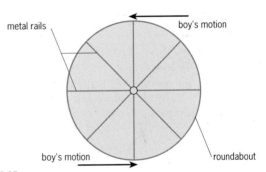

Fig 8.Q5

Two boys, each of mass 50 kg, play recklessly on a playground roundabout, of mass 100 kg and diameter 4.0 m. They run at 6.0 m s^{-1} in opposite directions tangentially towards opposite ends of a diameter of the roundabout, jumping on at the same instant. See Fig 8.Q5. The roundabout is initially at rest. Treat the roundabout as a disc and each boy as a point mass.

a) Find the initial angular velocity of the roundabout when the boys jump on.

b) The boys move radially to the centre. What is the new angular velocity of the roundabout?

c) By what factor does the kinetic energy of the roundabout increase when the boys move to the centre? Explain, in as much detail as you can, where this extra energy comes from.

[O&C: Physics 3/4, 1993]

6 Explain the difference between an *elastic* and an *inelastic* collision, with examples of each type.

7 One way of measuring the speed of a bullet fired from a gun under test is to use a ballistic pendulum. This is simply a large block of wood suspended from long strings: Fig 8.Q7.

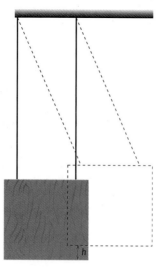

a) A bullet of mass 8.0 g is fired into the block, mass 6.0 kg. Bullet + block move off with a speed of 0.50 m s⁻¹. What was the speed of the incoming bullet?

direction of bullet

Fig 8.Q7

b) In practice, the testers would not measure the speed of the block of wood directly, but measure the height h the block rose before stopping. How could they use this measurement to find the initial speed of the block?

8 A moving car collides with a container lorry that is stopped, but not braked, at the side of the road. Explain why the combined vehicles move off with less kinetic energy than originally possessed by the car. What has happened to the missing kinetic energy?

9 A superball is dropped on a hard floor and rebounds with a speed 90 per cent of that with which it hit the floor. Choose the correct value of the percentage change in momentum produced by the bounce and explain why you chose it: a) 10 per cent, b) 90 per cent, c) 190 per cent.

10 Irrespective of your answer to Question 5, you should be able to show that the change in kinetic energy of the ball is about 20 per cent. Do so, showing any working.

11 A collision between an electron and a helium atom is illustrated by Fig 8.Q11. The collision is perfectly elastic.

A helium atom is about 7000 times as massive as an electron. Check and comment on the following correct statements about the result of the collision:

a) The helium atom would gain twice as much momentum as was carried by the incoming electron

b) The helium atom would take away about 0.1 per cent of the electron's kinetic energy.

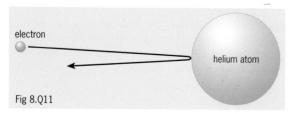

electron

helium atom

Fig 8.Q11

12 Fig 8.Q12 shows a generalised two-dimensional collision between particles of equal mass.

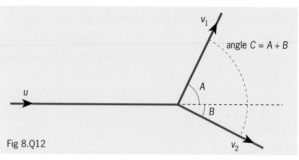

Fig 8.Q12

Both momentum and energy are conserved. The aim of the analysis is to prove that when the particles have equal mass, and whatever the size of the angles A and B, the angle C is always a right angle. To do this we generate three equations – one for kinetic energy and two for momentum – then use the techniques for solving simultaneous equations to eliminate the velocities. The resulting trigonometrical relationship between the angles A, B and C should show that C must be a right angle.

a) Write down the equation relating the kinetic energies before and after the collision – equation 1.

b) Momentum is a vector quantity. Write down the relationships between the masses and velocities in (i) the direction of travel of the incoming particle equation 2 – and (ii) a direction at right angles to this (where the incoming momentum is zero) – equation 3.

c) Step 1: Eliminate u from the system by squaring equation 2 and equating the right-hand side to the right-hand side of equation 1. Call the result equation 4.

d) Step 2: Equation 3 can be written as $v_1 = \ldots$ Eliminate v_1 from equation 4 by inserting its equivalent in terms of v_2. The result is a long collection of sines and cosines, but miraculously v_2^2 appears everywhere and can be cancelled out! Call the result equation 5.

e) Simplify equation 5. Try techniques like multiplying through by a denominator and collecting like terms together. Apply some trigonometrical insight. (Hint: $\cos^2 A + \sin^2 A = 1$.)

f) You should now have something like:
$\sin A \sin B - \cos A \cos B = 0$
Check that this must mean that $A + B = 90°$.

13 The size of a merchant ship is usually defined by its *tonnage*. A certain merchant ship is said to have a tonnage of 15 000 tonnes when loaded to its certificated capacity. This means that the fully laden ship displaces that mass of water when it floats. The density of fresh water is 1000 kg m⁻³, while sea water in the North Atlantic has a density of 1024 kg m⁻³.

a) Does the ship displace a different mass of water in the North Atlantic than in the River Thames?

b) In which water will the ship sink deeper?

14 A submarine is at a depth of 800 metres in sea water of density 10 030 kg m⁻³. Assuming standard atmospheric pressure in the submarine, what is the pressure difference between the inside and the outside of the hull?

15 A fuel tank has dimensions 2 m long by 1.5 m wide by 1.8 m deep. It is filled with oil of density 850 kg m⁻³.

a) What is the pressure at the base of the tank?

b) What force acts on the base of the tank?

c) What force acts on the largest side of the tank?

16 Fuel tanks often have gauges at the side which show how full the tank is (Fig 8.Q16).

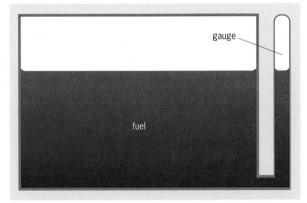

Fig 8.Q16

Explain why the pressure at the bottom of the gauge is the same as the pressure on the bottom of the tank.

17 A manometer can be used in a simple way of comparing the densities of two liquids that do not mix, such as oil and water. Paraffin and water are poured into a U-tube and reach equilibrium as shown in Fig 8.Q17. The U-tube has uniform cross-section.

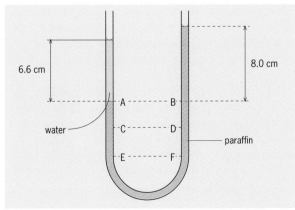

Fig 8.Q17

a) What must be true about the pressures at points A and B ? (And must also be true at points C, D, E and F.)

b) The density of water is 1000 kg m⁻³. Use the values in the diagram to calculate the density of paraffin.

18

a) (i) Write down Newton's law of viscosity and define the symbols used.

(ii) Show how the units of the coefficient of viscosity may be deduced from Newton's law.

b) A smooth steel plate of diameter 10 cm rests with one of its flat faces on a smooth horizontal surface. With the surfaces in contact and clean and dry, a horizontal force of 5.0 N is required to move the disc across the surface. However, when the horizontal surface is covered with a layer of oil, a constant horizontal force of 0.50 N enables the disc to achieve a terminal velocity of 7.5 cm s⁻¹ across the surface. The thickness of oil between the disc and the horizontal surface is observed to be 0.1 mm.

(i) How does the oil film reduce the force needed to move the plate?

(ii) Explain why the plate achieves a terminal velocity.

(iii) Determine the work done per second in moving the plate at the terminal velocity across the oil-covered surface. What happens to the energy transformed in doing the work?

(iv) Determine the coefficient of viscosity of the oil, stating any assumptions made.

c) Discuss how the effectiveness of the oil as a lubricant might change if:

(i) the viscosity increases

(ii) the temperature increases.

[NEAB 1993]

19

a) The drag force on a moving vehicle is given by the equation $F = 0.5C_dA\rho v^2$.

(i) Explain the meaning of each of the terms in this equation.

(ii) Describe an experiment that will verify the relationship between F and v.

b) For a particular car, $A = 3.0$ m² and $C_d = 0.35$.

(i) Calculate the drag force on the car when it is travelling at 144 km h⁻¹ through air of density 1.2 kg m⁻³.

(ii) Calculate the power required to overcome the drag force at the speed of 144 km h⁻¹.

(iii) The output of the car's engine, as specified by the manufacturer, is 55 kW. Suggest why this value differs from your answer to b) (ii).

[UCLES: Modular Sciences Physics, 1994]

ENERGY AND CARS

The power unit in a car is the engine, and its output is given in kilowatts. It is measured in 'horse-power', a relic of the days when manufacturers of the new-fangled steam engines wanted to impress potential customers by comparing their engines with the alternative and obsolescent horse technology of the day!

The engine of a typical small car is rated at about 70 horse power (hp). This is equivalent to about 53 kW, since 1 hp = 0.746 kW. This of course represents the maximum power that the engine can produce, say, when the car is travelling at its maximum speed on a level road on a windless day. This leads to the following question.

What decides the maximum speed of a car?

Table 8.A1 **Forces acting and power requirement for a car at different speeds**

Speed /m s^{-1}	Rolling friction /N	Air resistance /N	Total frictional force/N	Power requirement /kW
9.0	.228	52	280	2.5
18	223	206	449	8.1
27	219	470	689	18.6
36	212	834	1 046	37.7
45	204	1 300	1 504	67.7

The density of air is 1.29 kg m^{-3}.

Table 8.A1 shows the power needs of a typical car at various constant speeds on a flat road, as explained on page 190. The engine produces forces, acting at the wheels, which have to do work against frictional forces. The two main frictional forces are **air resistance** and **road** or **rolling friction**.

Air resistance, *R*.
The front of the car has to push air aside and also 'slide' through it. This means that the air gains kinetic energy, and also exerts a frictional force on the surfaces of the car. In addition, there is a serious energy loss caused by the fact that when the air stream breaks away from the car at the rear it becomes **turbulent** (see page 188).

Rolling friction, *F*.
Energy losses due to the effect of the tyres rolling over the road also have a variety of causes. Rolling friction is caused by unevenness in the road surface, and some of this is due to the surface being distorted by the weight of the car. The tyre itself flexes and there is 'internal friction' which makes the tyre hot.

The rate at which work is done against resistive forces must equal the effective power output of the car. The fifth column in the table has been calculated with the definition of:

$$\text{work} = \text{force} \times \text{distance}$$

and from the relationship:

$$\text{power} = \text{rate of working (in N m s}^{-1}\text{, or J s}^{-1}\text{)}$$

ie: $$P = W/t = Fd/t = Fv$$

This means that the maximum speed depends upon the maximum power available and the frictional force *F* at that speed.

The resistive forces
The most obvious point about the table is that, as the speed increases, air friction becomes the dominant energy loss. Rolling friction is almost a constant (actually getting slightly less as speed increases). The effect of air friction depends not only on speed but also on the shape and size of the car – in particular the cross-sectional area. This is usually expressed in a formula:

$$R = \tfrac{1}{2}CA\rho v^2$$

where *R* is the air resistance (or 'drag') for a vehicle travelling at speed *v*, *A* is the cross-sectional area, ρ is the density of the air, and *C* is a 'shape constant' called the **drag coefficient**. A streamlined sports car has a drag coefficient of about 0.3, an estate car about 0.45.

This means that the total force resisting motion for a car, F_r, can be written as:

$$F_r = F + R = \text{constant} + \tfrac{1}{2}CA\rho v^2$$

Overall efficiency
This calculation ignores other energy uses in the car, such as the radiator fan, use of lights, windscreen heaters, radio, air conditioning etc. The useful energy is produced by burning fuel in the engine, and the energy transfer from this provides the major system loss.

Not only does the combustion heat the engine parts and so the surroundings at a rate that needs a radiator (really a forced convector) and fan to keep them at a suitably low working temperature, but the laws of thermodynamics require an even larger energy loss (see Chapter 15). Less than 15 per cent of the energy available from burning the fuel can be used to power the car.

1 A car magazine quotes the power of the engine in a BMW 530i V8 as 218 hp. What is this power in kW?

2

a) What is the difference between streamline and turbulent motion for a fluid?

b) Why should turbulence be a source of energy loss?

3 Running a car with the windows open increases the fuel used per kilometre by about 3 per cent.

a) Explain why this happens.

b) Using the air conditioner in a car reduces the distance covered per litre by about 12 per cent. Suggest the advantages and disadvantages of using open windows instead of air conditioning on a hot day.

4

a) Use the data in Table 8.A1 to plot graphs of:
(i) air resistance against speed,
(ii) power required against speed.

Comment on any similarities and differences between the two graphs.

b) Using the graphs, or otherwise, estimate the engine power that would be needed at speeds of:
(i) 20 m s^{-1},
(ii) 50 m s^{-1}.

5

a) Give two reasons why a container lorry needs a much more powerful engine than a medium sized car.

b) A small car travels about 14 km per litre of fuel, while a large estate type car manages only about 6 km. What is the main reason for this difference?

6 A sports car has a cross-sectional area of 2.0 m^2 and a drag coefficient of 0.32. Estimate its maximum speed for an engine power of 90 kW. Estimate the maximum speed of a Land Rover type vehicle which uses the same engine but has a front area of 3 m^2 and a drag coefficient of 0.48.

7 Use any of the data from the table to find the value of CA for the car involved. Is it likely to be a large car or a small car? Give reasons for your answer.

TRANSPORT

The main ideas connected with the movement of solid objects and of fluids (liquids and gases) appear in this Chapter Map. Notice that Newton's laws of motion occupy a central place. You can use the map to cross-match items with the requirements of your syllabus. The map should also help you to identify the areas you may need to study further.

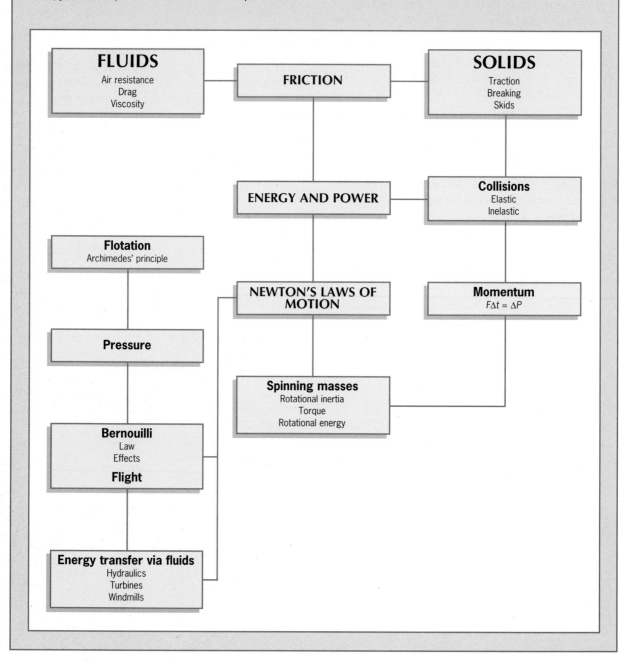

CHARGE: CURRENT AND FIELDS

IN TODAY'S WORLD we rely increasingly on electricity, for communications as well as to operate the equipment we use. Here, we look at the fundamental properties of electric charge and the motion of charged particles through matter. There are two kinds of charge, positive and negative, and charges interact with each other by means of two main forces: the electric force and the magnetic force.

Electric currents allow us to transfer energy over great distances and deliver the energy to where it is needed. For this we use devices, such as generators, transformers and motors, that operate by electromagnetic interactions. You will learn how the flow of electric charge is controlled by the design of circuits that use resistors, diodes, capacitors and other components.

Most of our use of electricity relies on electromagnetic devices, so the principles of electromagnetism are covered, the way that these principles are applied at home and in industry, and the importance of alternating currents in these applications.

Ordinary matter is held together by the electric charge on electrons and protons, and the movement of charge can produce the moving, interconnected fields we call electromagnetic radiation (this is covered in further detail in the fourth section of the book).

9 Charge and current

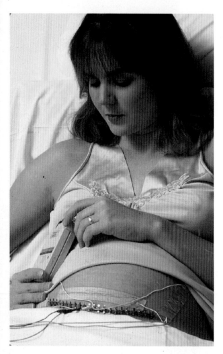

WHEN YOU VISIT the dentist or have treatment in hospital for an accident, the chances are that you will be given drugs to combat the pain. Modern drugs are very effective for treating short-term pain. But some people suffer from conditions that give them continuous pain, and the prolonged use of pain-killing drugs can cause unwanted side effects.

One of the more effective methods of dealing with continuous pain is called transcutaneous electric nerve stimulation, or TENS for short. (Transcutaneous means 'across the skin'.) Users of TENS carry a box the size of a personal stereo on a belt. It runs from a 9 V battery and has two electrodes which the user can tape to the area of the body that has pain. The TENS unit supplies short pulses of current at low voltage through the electrodes to the painful area and the user can adjust the strength of the current and the length of pulse time. The pulses of current stimulate nerve endings and this inhibits the sensation of pain.

Many pain sufferers find TENS to be very effective in treating acute pain, such as the pain in the stump of a limb after it has been amputated.

Pulses of current from a TENS unit can help relieve continuous pain and avoid the side effects of pain-killing drugs

Introduction

Electricity is the development in physics during the twentieth century that has perhaps had the greatest effect on the largest number of people. Electric current to the homes of many millions of people has given them energy at the flick of a switch. In the United Kingdom, electricity is distributed on the national grid – a complex circuit connected to every home, factory, school and hospital.

Countless time- and labour-saving devices are run on electricity: examples range from our vast modern communications systems to the electric sensors that allow doctors to monitor the vital functions of the human body. Hundreds of amperes of current are needed to run our railway systems; it takes a current of only a microampere to keep a wristwatch working.

The basic ideas described in this chapter are relevant to these and many other circuits. To begin with, there are some basic facts about circuits that can be established easily by experiment.

A battery consists of a stack of several cells. Each cell has two electrodes, one positive and one negative. The word 'battery' is used to describe a single cell as well as a collection of cells.

1 CURRENTS AND CIRCUITS

The simplest circuit is a closed path starting from a battery (or cell) and returning to the battery, as in Fig 9.1(a). The fact that this complete loop is required suggests that something is moving round the circuit. We call what moves an **electric charge**, and the rate of flow of the charge an **electric current**. The size of a current is measured by an **ammeter** in **amperes**, symbol A. (Later in the chapter, we shall look in more detail at the nature of current, and of resistance.)

It is easy to show by experiment how currents behave in circuits. The simplest circuit is a **series** circuit. The charge has only one path to follow, and it passes through each component placed in the circuit, such as the bulb in Fig 9.1(a).

For a simple series circuit, the current is the same at every point in the circuit. A component has a **resistance** which lowers the current: when a component is added to the circuit, the current decreases as the resistance of the circuit increases. Each component, such as each of the bulbs in Fig 9.1(b), then receives the same lower current, as recorded by the ammeter.

Instead of wiring components in series, we can connect them in **parallel**, as shown in Fig 9.2. Each bulb offers a different path for the charge to flow through. As each bulb is switched on, the total current drawn from the power source increases. The total current I is equal to the sum of the currents in the parallel arms, that is:

$$I = i_1 + i_2 + i_3$$

2 VOLTAGES IN CIRCUITS

A voltmeter measures a quantity called a **voltage**. (We shall see later in this chapter exactly what a voltage is.) In a series circuit, the voltage across the battery is equal to the sum of voltages across all parts of the circuit, as shown in Fig 9.3(a). When components are connected in parallel, the voltage across one component is the same as the voltage across any other component.

Parallel circuits are very important in everyday life. In the home, there is probably one circuit for lighting, and one with sockets in the wall for electrical equipment (known as the power circuit or ring main). The components – lights or equipment – are connected in parallel to their circuit.

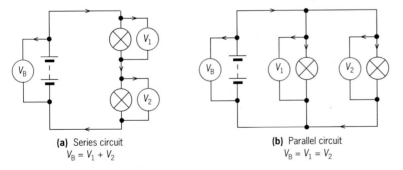

(a) Series circuit
$V_B = V_1 + V_2$

(b) Parallel circuit
$V_B = V_1 = V_2$

Fig 9.3 **Measuring voltage in series and in parallel circuits. The voltmeters are connected across the battery and across each bulb**

What is an electric current?

An electric current is a flow of electric charge; any moving charge is a current. Charge is carried by particles such as **electrons** or **ions**. When these particles are electrons, they usually travel in a wire. When they are ions, they are moving usually in a water solution. If charged particles are moving, a current exists. If the charged particles do not move, there is no current.

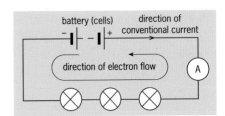

Fig 9.1 **A simple circuit with a battery and three bulbs connected in series. The convention is that the direction of the current is from the plus to the minus side of the battery**

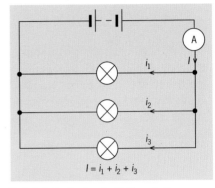

$$I = i_1 + i_2 + i_3$$

Fig 9.2 **A simple circuit with three bulbs connected in parallel**

?

A (a) Sketch a circuit for a hair-dryer and TV plugged into the power supply.

(b) One of the light bulbs in the lighting circuit of a house fails. What effect has this on the rest of the light circuit, and why?

?

B State whether or not a current exists in the following situations.

(a) Your hair stands on end after you comb it on a dry day.

(b) You are eating chocolate with a small piece of metal foil attached to it. The foil touches a filling and you feel pain.

(c) A bolt of lightning flashes through the sky.

(d) A gardener has cut through a cable of a lawnmower while mowing the lawn. The ends of the cable are lying apart on the ground while the gardener goes to switch off at the mains plug.

Electrons moving in a circuit are negatively charged. They move from the negative terminal to the positive one. In spite of this, the *direction of the current* in a circuit is taken as being from *positive to negative* (look back at Fig 9.1). This is the conventional way to represent current, agreed by early experimenters working on current electricity. As we know, they were wrong about current in a wire, but it doesn't matter, as we shall see later.

The unit of electrical charge is the **coulomb**, symbol C, and it relates to amperes in this way:

> **A coulomb is the amount of charge that flows when there is a current of one ampere for one second.**

So, for example, if an ammeter in a circuit measures a current of 1 A, a charge of 1 C will pass through it during each second. Since the measured current I is the rate of flow of charge Q the amount of charge that flows in a circuit can be calculated.

Since: $I = Q \div t$

then: $Q = I \times t$

where t is in seconds and Q is in coulombs.

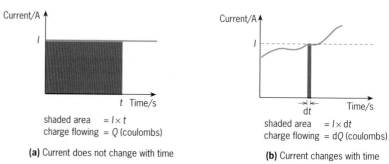

shaded area $= I \times t$
charge flowing $= Q$ (coulombs)

(a) Current does not change with time

shaded area $= I \times dt$
charge flowing $= dQ$ (coulombs)

(b) Current changes with time

Fig 9.4 **Graphs of current against time**

Graph 9.4(a) is for a constant current. The area under the graph is equal to the charge flowing in the time shown. Graph 9.4(b) shows a changing current. The shaded area indicates the charge flowing in a small period of time, dt. Only a small amount of charge flows; we call it dQ. Even though the current is changing, we can consider it to be steady over the short time dt. Therefore we can write:

$$dQ = I \times dt$$

$$I = \frac{dQ}{dt}$$

We can now give a general definition of current:

> **Current is equal to the rate of flow of charge.**

We shall use this definition later in the chapter when we examine the behaviour of capacitors.

Common units used for electric currents include:

$$1 \text{ A} = 1000 \text{ mA} = 1\ 000\ 000\ \mu\text{A}$$
(microampere)
$$1 \text{ mA} = 10^{-3} \text{ A}$$
$$1\ \mu\text{A} = 10^{-6} \text{ A} = 10^{-3} \text{ mA}$$

C The charge carried by one electron is 1.6×10^{-19} C. If the current in a wire is 1 A, calculate the number of electrons that pass any point each second.

D Calculate the current in each of the following:

(a) 2 C flows through a bulb in 10 s.

(b) 2 μC flows through a light-emitting diode in 1 ms.

(c) 20 nC flows into an integrated circuit in 500 ms.
(1 nC $= 10^{-9}$ C)

E How much charge flows in the following examples?

(a) 5 A for an hour.

(b) 50 mA for a day.

(c) 5 μA for 20 s.

The total charge can be found by taking the area under the graph in small strips (one strip is seen in Fig 9.4) and adding the areas of all the strips together. This is done using calculus and integrating:

$$dQ = I\ dt$$

$$Q = \int I\ dt$$

3 CURRENT AND FREE ELECTRONS

The current in a material (such as a wire) is due to the movement of **free electrons**. Copper, used for wires connecting components, is a good **conductor** of electrons. (So are other metals.) A copper wire consists of millions of identical copper atoms. Each atom has electrons. Most of them are tightly bound to the atomic nucleus by the electrical attractive force between the positive nucleus and the negative electrons.

One or two electrons on the outside of the atom are less tightly bound; being furthest from the nucleus, the attractive force on them is lowest. In a solid, the atoms are closely packed and each atom may have up to twelve other atoms next to it. An electron on the outside of an atom may be the same distance from a neighbouring nucleus as from its own nucleus. So which atom does it belong to? Such electrons can move between atoms and are known as free electrons. They behave like an electron 'cloud' and move inside the metal in random directions.

When the wire is in a circuit and a voltage is placed across it, the free electrons are attracted towards the positive side of the power supply and there is a net drift of the electron 'cloud' in that direction. This is a **current**. Each electron carries an electric charge of 1.6×10^{-19} C.

Materials with many free electrons are good conductors; those with few are poor conductors. The number of free electrons per cubic metre – the number of *charge carriers per unit volume* – indicates how good a conductor a material is.

The current also depends on the speed at which the electron 'cloud' travels in a particular material – their **drift speed**. The cross-sectional area of the conductor also has an effect on the current: a thick wire offers an easier path for the electrons than a thin one. The way that the current depends on these factors is described by the **transport equation**, as explained in the box.

?

F Two wires X and Y are made from the same material. X has twice the cross-sectional area of Y and the current in X is twice that through Y. Work out the ratio of the average drift speed of electrons in X to the average drift speed of electrons in Y.

Derivation of the transport equation

Fig 9.5 shows a section of wire carrying a current of I amperes. As you read on, write down an expression for each step, and note the units.

Each electron carries a charge of e coulombs and travels with an average drift speed of v metres per second. There are n free electrons per cubic metre of wire, and its cross-sectional area is A square metres.

Suppose it takes t seconds for an electron to pass from X to Y. The distance from X to Y is vt metres and the volume of wire between X and Y is Avt cubic metres.

The number of electrons between X and Y is equal to the volume between X and Y multiplied by n, that is, $nAvt$. If each electron carries a charge of e coulombs, then the total charge that passes across point Y in t seconds is $nAevt$.

Now, the current is equal to the charge flowing per second, that is, $nAevt/t$. Therefore $I = nAev$ in coulombs per second.

The expression applies to a single electron of charge e. We can use Q to represent any charge, when the flow of charge $I = nAQv$.

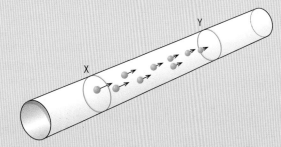

Fig 9.5 **Current in a wire: the flow of free electrons carrying negative charge. (Conventionally, the current is in the opposite direction)**

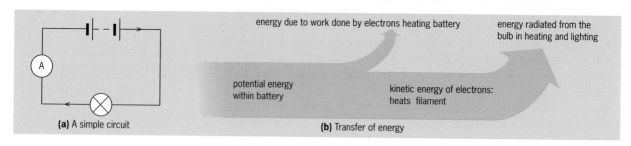

4 POTENTIAL DIFFERENCE AND ENERGY IN CIRCUITS

Fig 9.6 (a) **A circuit with a light bulb.** (b) **The transfer of energy in circuit** (a)

All circuits deliver energy from their source (a battery or power supply) to the components in the circuit. The amount of energy transferred depends on the current. In a circuit containing a bulb, as in Fig 9.6, electrons pass through the bulb filament. The electrons collide with filament atoms, work is done on these atoms, energy is transferred to them and they heat up and give out light. Electrons can flow easily through the copper connecting wire – there are fewer collisions – and so it does not get hot.)

We measure the energy delivered to the filament (the work done in it) in terms of joules per coulomb of charge passing through it. This is the **potential** (energy) **difference** (**p.d.**) across the filament. Now we can give this definition of a volt:

1 joule per coulomb = 1 volt

If the p.d. across a bulb is 6 V, this means that, for each coulomb that passes through the bulb, 6 J is the amount of energy transferred from the current to do work heating and lighting the bulb.

5 RESISTANCE

Why is energy transferred when charge flows through a bulb or through a motor? Why does the current get smaller when bulbs are connected in series? Why is the current from the supply greater when bulbs are connected in parallel?

The answers to all these questions are best understood if we use the idea of **resistance**:

Resistance is the electrical property of a material that makes the moving charges dissipate energy.

Resistance restricts the flow of charge, so a resistance makes the current smaller. Connection wires have very low resistance; the wires used in lamp filaments have higher resistances. But they are all conductors. Insulators like the plastic round the wires in a cable have very much higher resistance – ideally an infinite resistance.

Some very useful materials are both poor conductors and bad insulators. They are **semiconductors**. There is more about semiconductors on page 216 and in Chapter 22.

Resistance is measured in **ohms**. The symbol is Ω (the Greek letter omega) and the ohm is named after Georg Simon Ohm who found the relationship between current and p.d. He measured the current through wires as he changed the voltage across the wires.

G **(a)** Explain why conductors are designed to have very low resistances.

(b) Give an example of an insulator and its purpose in an electrical circuit in the home.

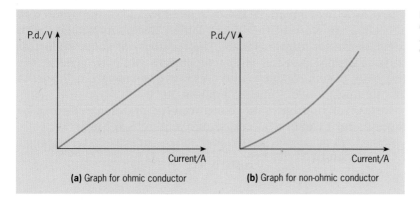

Fig 9.7 **Graphs for p.d. against current for** (a) **a conductor material that obeys Ohm's law and** (b) **a conductor material that does not obey Ohm's law**

He discovered a very important relationship which he published in 1826. At a steady temperature:

p.d. across a = current through conductor × a constant conductor

$$V = I \times R$$

The constant is the resistance R of the conductor and the equation is known as **Ohm's law**. We can also say that the p.d. across the conductor is directly proportional to the current flowing through it at constant temperature. The equation can be rearranged as:

$$R = \frac{V}{I}$$

where V is the p.d. across the conductor and I is the current through it. From this equation it is clear that:

$$1\,\Omega = \frac{1\,V}{1\,A}$$

The graphs in Fig 9.7 show the behaviour of two types of conductor. Not all obey Ohm's law. Those that do are called **ohmic conductors**. At a constant temperature their resistance is constant over a wide range of currents, so their p.d.–current graphs are straight lines.

But many non-ohmic conductors are useful. Their resistance depends on p.d., even at a constant temperature, so their p.d.–current graphs are not a straight line and often show sharp changes in gradient. The semiconductor diode is a good example of a non-ohmic component, which we shall look at in Chapter 20.

> **H** Calculate the resistance of a 6 V bulb that takes a current of 0.06 A.

See questions 1 and 2.

Resistivity

Ohm also investigated how the resistance of a conductor depends on its dimensions and the material that it is made of. For the same p.d. across a wire, its resistance increases with length l (so $R \propto l$), and decreases with cross-sectional area A (so $R \propto l/A$). If you increase the length of a wire in a circuit, the current will decrease. If you increase the thickness of the wire, the current will increase.

The resistance of a wire at constant temperature depends on its length, its cross-sectional area and its **resistivity**, symbol ρ (the Greek letter rho). The resistivity is a constant for the particular material of a wire and has the units ohm metres (Ω m). The three factors give the following equation for resistance:

$$R = \rho \frac{l}{A}$$

> **I** Two wires, A and B, are made of the same material. A is twice as long as B but has twice its diameter. Which wire has the greater resistance?

Table 9.1 The resistivity of materials at 20 °C

Table 9.1 **The resistivity of materials at 20 °C**

Material	Resistivity ρ/Ω m
Conductors:	
silver	1.6×10^{-8}
copper	1.7×10^{-8}
aluminium	2.8×10^{-8}
iron	8.9×10^{-8}
mild steel	14×10^{-8}
constantan	49×10^{-8}
graphite	3000×10^{-8}
Semiconductors:	
germanium	0.6
silicon	2300
Insulators:	
porcelain	10^{11}
glass	10^{12}
PVC	10^{12}
PTFE	10^{16}

The resistivity of a material can be altered by adding impurities that affect the structure of the crystal lattice. Resistivity changes with temperature for most materials. The resistivity of good conductors such as metals increases with temperature while, in general, the resistivity of semiconductors decreases with temperature. Table 9.1 lists the resistivity of some materials at 20 °C.

Instead of resistivity, the **conductivity** of a material may be more useful, symbol σ (Greek: sigma) and units Ω^{-1} m^{-1}:

$$\text{conductivity} = \frac{1}{\text{resistivity}}$$

$$R = \frac{l}{\text{conductivity} \times A} = \frac{l}{\sigma A}$$

6 ENERGY AND POWER IN CIRCUITS

The energy transferred to a component in a circuit, such as a bulb, depends on the current in it and the p.d. across it. The current in amperes is the number of coulombs passing through the bulb per second, and the p.d. in volts is the amount of energy in joules transferred by each coulomb as it passes through the bulb. If we multiply the two quantities, p.d. and current, we get an interesting result:

joules per coulomb × coulombs per second = joules per second

Joules per second is the power, or rate of energy transfer, measured in watts.

We now have a very useful equation for electrical circuits:

power (watts) = p.d. (volts) × current (amperes)

$$W = V \times I$$

Energy is transferred when a current passes through something with electrical resistance. We can think of the energy transferred as the **work** done to overcome the resistance.

?

J A 40 W, 240 V mains bulb has a filament that is the same length as the filament in a 40 W, 12 V car headlamp bulb. The filaments are both made from tungsten. Compare their thicknesses.

K If two identical lengths of copper and iron wire of equal diameter are connected in series to a power supply, the iron wire reaches a higher temperature. When they are connected in parallel, the copper wire reaches the higher temperature. Compare the resistances of the two wires and explain these observations.

It is often convenient to use other forms of the **power equation**, which we can derive from $V = IR$ or $I = V/R$.

If the *current* and resistance are known, then:

$$W = V \times I = IR \times I = I^2R$$

So: $\quad\quad W = I^2R$

Engineers refer to wasteful energy transfers (eg energy wasted in a power cable or in the windings of the rotor in a dynamo) as I^2R losses. If the *voltage* and resistance are known, then:

$$W = V \times I = V \times \frac{V}{R} = \frac{V^2}{R}$$

So: $\quad\quad W = \frac{V^2}{R}$

Energy transferred in a particular time

If we know the power (the rate of energy transferred by motor, for example, then it is easy to calculate the amount of energy that is transferred in a particular time:

energy transferred = power × time

$$= IV \times t \text{ or } I^2Rt \text{ or } \frac{V^2}{R}t$$

?

L Calculate the energy radiated by a 6 V, 0.3 A torch bulb in half an hour.

7 INTERNAL RESISTANCE AND E.M.F.

In the circuit (a) of Fig 9.8, the switch is open and there is no current in the bulb, but the reading on the voltmeter is 12.1 V. In circuit (b), the switch is closed and there is a current through the bulb. The voltmeter shows the p.d. across the bulb *and* across the battery. The reading on the voltmeter is now slightly less, at 11.8 V.

In the first case, the only current is a few microamperes through the voltmeter. This current is negligible compared with the current in the bulb in the second circuit. Why does the voltage decrease? The difference in voltage can be explained by considering the **internal resistance** of the battery.

When the circuit is complete and charges move, there are energy transfers within the battery. As chemical changes occur, potential energy is transferred to the charged particles: they move and a current is generated. Energy is also transferred to overcome resistances inside the battery – some 'joules per coulomb' are lost across the internal resistance, indicated by r in Fig 9.9. This resistance is due to collisions between charged particles and other atoms in the battery. The fact that a battery heats up is evidence for this.

A good battery has a low internal resistance, typically about 1 Ω or even less. When charge flows through the bulb, the current is quite high. For a 24 W bulb it would be about 2 A. Using $V = IR$, the voltage across an internal resistance of 0.5 Ω is 1 V. We have to consider the *whole* circuit, with the internal resistance in series with the load resistance (the total resistance outside the battery).

Electromotive force, e.m.f.

The rate at which energy is transferred within the battery is measured in joules per coulomb, that is, volts. The battery voltage is responsible for forcing the current round the whole circuit, including through the battery, and is called the **electromotive force** or **e.m.f.** ($\in$).

Since energy must be conserved, the total energy transferred in the circuit must be equal to the energy transferred in the battery, so:

energy transferred = energy transferred in the circuit
per second within   and in overcoming the internal
the battery   resistance every second

So:
$$\in I = I^2R + I^2r$$

Dividing both sides by the current I:
$$\in = IR + Ir$$

or:
$$IR = \in - Ir$$

Putting this in words:

p.d. across load = battery e.m.f. – voltage across
internal resistance

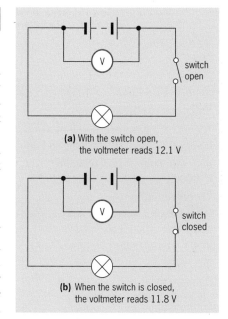

(a) With the switch open, the voltmeter reads 12.1 V

(b) When the switch is closed, the voltmeter reads 11.8 V

Fig 9.8 **Reading voltage with the switch open and closed**

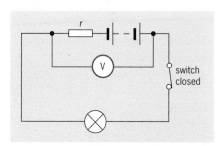

Fig 9.9 **The circuit of Fig 9.8 with the internal resistance of the battery represented by** *r*

✓ The word equation refers to *voltage* across internal resistance, rather than p.d. This is to distinguish p.d. as a voltage that can be measured with a voltmeter, from the p.d. across the internal resistance which cannot be measured directly.

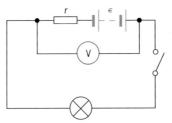
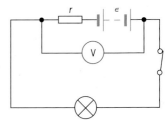

(a) Switch open: voltmeter reading = 12.1 V   **(b)** Switch closed: voltmeter reading = 11.8 V

Volts: p.d. and e.m.f.

The e.m.f. measures the rate at which energy is transferred per coulomb from an energy source such as a battery or a dynamo, in the kinetic energy of the moving charge.

P.d. measures the rate at which energy is transferred per coulomb from the kinetic energy of the charge to do work in a circuit.

Both are measured in joules per coulomb, that is, in volts.

See questions 3 and 4.

M An extra high tension (e.h.t.) power supply for use in a laboratory is designed to provide very high voltages and is set to give 6000 V. A voltmeter capable of measuring this voltage is connected to measure the p.d. across the output of the supply. When a milliammeter is connected across the output it shows a current of 6 mA and the reading on the voltmeter drops to almost zero. What can you deduce about the internal resistance of the power supply?

N Battery A has an e.m.f. of 2.0 V and an internal resistance of 1 Ω. For battery B, the values are 1.0 V and 2 Ω. A and B are connected to a 2 Ω resistor as shown. Using Kirchhoff's laws, calculate the current through the 2 Ω resistor.

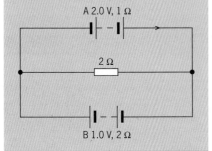

A 2.0 V, 1 Ω

2 Ω

B 1.0 V, 2 Ω

In Fig 9.10(a), with the switch open, the voltmeter measures IR. I is small because the voltmeter has a large resistance. Also, r is very small compared with R. Therefore, Ir is small and the voltmeter reading IR is close to the e.m.f. ϵ.

In Fig 9.10(b), with the switch closed, resistance R is small, so the current I is high. (For a bulb of 12 V and 24 W, R would be 6 Ω.) Ir is small, but significant, compared with IR.

If you think about it, we cannot measure the battery e.m.f. accurately. The voltmeter itself will always require some current, however tiny, and some voltage will be lost across the internal resistance. As we use voltmeters of higher and higher resistances, the reading gets closer and closer to the actual e.m.f., but it never quite gets there.

Kirchhoff's laws

The behaviour of simple circuits is summarised in Kirchhoff's two laws. The first law states:

Current in a circuit is conserved.

This means that, in a parallel circuit, all the current entering a junction leaves the junction, as shown in Fig 9.11(a). Kirchhoff's first law also means that charge does not collect at the junction.

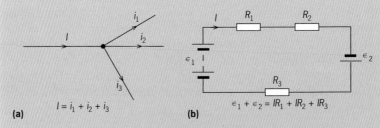

(a) $I = i_1 + i_2 + i_3$ **(b)** $\epsilon_1 + \epsilon_2 = IR_1 + IR_2 + IR_3$

Fig 9.11 (a) **The circuit junction, illustrating Kirchhoff's first law.** (b) **E.m.f.s and resistances in a circuit, illustrating Kirchhoff's second law**

The second law is the law of conservation of energy applied to circuits and is shown in Fig 9.11(b). The law states that:

The sum of the e.m.f.s is equal to the sum of the IR products.

The IR products are the p.d.s across each load in the circuit, including any internal resistances in the battery. The sum of the p.d.s is the total energy transferred as each coulomb does work around the circuit. The sum of the e.m.f.s is the total energy transferred to each coulomb within the current sources.

Accepting that e.m.f.s are positive and p.d.s are negative, then an alternative statement is:

The sum of the voltages around a circuit is zero.

Fig 9.10 **Looking at the e.m.f. ϵ of a battery**

PROFESSOR GEORG SIMON OHM (1789–1854)

TODAY WE CAN confirm Ohm's law by experiment with modern apparatus. We have reliable ammeters and voltmeters and stable sources of electrical energy. Ohm had none of these.

In the 1820s when he did his experiments, batteries were unreliable – their e.m.f.s changed unpredictably. To overcome this problem, he used an effect discovered in 1822 by another scientist called Thomas Johann Seebeck.

Seebeck had demonstrated that an e.m.f. is set up between the two junctions where two unlike metals join. This is known as a **thermocouple**. So, as in Fig 9.12(a), a voltage is produced between the two junctions in the circuit made from copper and bismuth wires. When one copper/bismuth junction is in freezing water and the other is in steam, a voltage is produced that is proportional to the difference in temperature between the two: the greater the temperature difference, the larger the voltage. Ohm was therefore able to use a thermocouple as a source of e.m.f. that could be accurately controlled.

Thermocouples are used extensively in industry to measure temperature. Their great advantage is that the active junctions are very small and little energy is needed to produce the thermo-e.m.f. This is ideal when the temperature of very small bodies is being measured. In the home, a gas central heating boiler has a thermocouple to detect water temperature. It regulates the system for turning off the gas when the water is hot enough.

In his investigations, Ohm passed a current through copper wires of different lengths, and also used another recent discovery to measure the current. In 1819, the Danish scientist Hans Oersted had discovered that a small magnet suspended near a wire was deflected when a current was passed through the wire. Ohm used this deflection to measure the current accurately. With these apparently crude techniques, he discovered the law named after him, one of the principal experimental laws in physics.

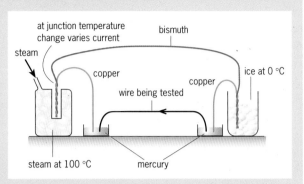

Fig 9.12(a) **The thermocouple apparatus which Seebeck devised and which Ohm used in his investigations. Arrows indicate the direction in which charge flows**

Fig 9.12(b) **Making a thermocouple: two wires, one of platinum and the other of rhodium–platinum alloy, are being welded together to form one junction of the thermocouple. When the two junctions are at different temperatures, there is a current between them. When joined to a suitable meter, the thermocouple can be used as a thermometer or a thermostat**

8 SIMPLE CIRCUITS WITH RESISTANCES

A resistor is a component designed to have a stated resistance. Resistors are manufactured to preferred values, from ohms to megohms, and to a specified accuracy or **tolerance**. They are made from a variety of materials, such as carbon, metal film on a ceramic base, or wound wire. Modern electronic microcircuits have resistors that are painted on to a circuit board using resistive paint. Some of the uses of resistors are now described.

Resistors used to limit current

Some devices require only a small current. A good example is the **light-emitting diode** (LED), used as an on/off indicator on many electronic devices such as televisions, video players and computers.

Fig 9.13 **Modern electronics circuits use miniature 'chip' components. This section of circuit board shows components including chip surface-mounted resistors (Z310 etc) and capacitors (C203 and C204)**

An LED shines brightly with a voltage of only 2.0 V across it and the maximum current through it should be 20 mA, so as not to overload it.

EXAMPLE

Q We want to use an LED with a 9 V supply and so we use a resistor in series with the LED to limit the current. What is the value of the resistor?

A The circuit is shown in Fig 9.14. The voltage across the resistor is:

9.0 V – 2.0 V = 7.0 V

and the current in the circuit needs to be 20 mA. Using $V/I = R$, the resistance of the limiting resistor should be:

7.0 V/20 mA = 350 Ω

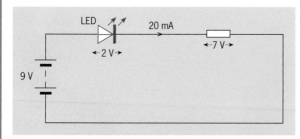

Fig 9.14(a) **A circuit showing an LED connected to a 9 V battery**

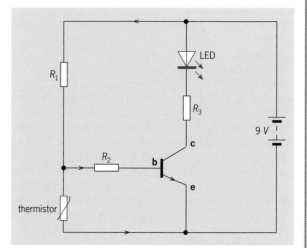

Fig 9.14(b) **The circuit for a simple temperature-sensitive switch. When the temperature of the thermistor changes, the transistor is 'switched on'. Then, the p.d. across the transistor (contacts c and e) drops to nearly zero. the p.d. across R_3 and the LED is then nearly 9 V, just as in** Fig 9.14(a). **The LED comes on when the temperature increases**

Resistors are available in preferred values, the nearest values being 330 Ω or 390 Ω. The use of 330 Ω in the Example above would mean the current would exceed 20 mA and therefore damage the LED. The 390 Ω should therefore be used. This means that the LED may be a little dim but can be used without danger of being damaged.

Combinations of resistors

Resistors in series
To find the overall resistance of several resistors in series, called the **equivalent resistance**, we *add* the individual resistances together. The total resistance R_T of three resistors with resistances R_1, R_2 and R_3, connected in series is given by:

$$R_T = R_1 + R_2 + R_3$$

Resistors in parallel
The total resistance R_T of three resistors with resistances R_1, R_2 and R_3 connected in parallel is given by:

$$\frac{1}{R_T} = \frac{1}{R_1} + \frac{1}{R_2} + \frac{1}{R_3}$$

If you are unsure of this, substitute some values into the equations and work out R_T. It is always smaller than the resistance of any of the separate resistors used in parallel.

?

O Three resistors of 10 Ω, 100 Ω and 1000 Ω are connected in parallel. Calculate the equivalent resistance of this arrangement.

Deriving equations for resistors

Resistors in series

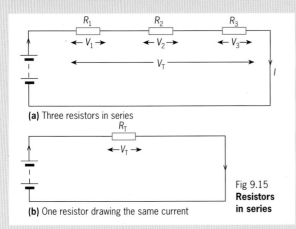

(a) Three resistors in series

(b) One resistor drawing the same current

Fig 9.15
Resistors in series

Resistors in parallel

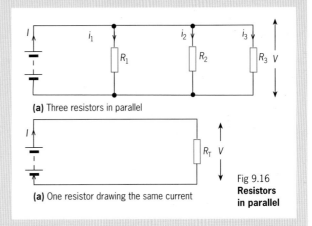

(a) Three resistors in parallel

(a) One resistor drawing the same current

Fig 9.16
Resistors in parallel

Fig 9.15(a) shows three resistors in series with a battery. The total voltage across the three resistors is V_T and the current in the circuit is I amperes. Now, if the p.d.s across the resistors are V_1, V_2 and V_3 respectively, we can write:

$$V_T = V_1 + V_2 + V_3$$

We also know that the p.d. across one resistor is given by IR (current same everywhere in the circuit).

Therefore: $\qquad V_T = IR_1 + IR_2 + IR_3$

We now wish to replace the three resistors in this circuit with one that draws the same current as in Fig 9.15(a). Call its value R_T. It follows that:

$$V_T = IR_T$$

or: $\qquad IR_T = IR_1 + IR_2 + IR_3$

The current I can be eliminated to give:

$$R_T = R_1 + R_2 + R_3$$

Fig 9.16 shows three resistors R_1, R_2 and R_3 in parallel with a battery that is assumed to have zero internal resistance. This time, the p.d. across each resistor is the same: call it V. The current I from the battery splits so that:

$$I = i_1 + i_2 + i_3$$

where i_1, i_2 and i_3 are the currents in R_1, R_2 and R_3 respectively. Suppose we replace the three resistors by one, with resistance R_T, that draws the same current from the battery and will have the same p.d. V across it. Then:

$$I = \frac{V}{R_T} = \frac{V}{R_1} + \frac{V}{R_2} + \frac{V}{R_3}$$

This time we can eliminate p.d. V to give:

$$\frac{1}{R_T} = \frac{1}{R_1} + \frac{1}{R_2} + \frac{1}{R_3}$$

The potential divider

The circuit in Fig 9.17 on the next page shows two resistors connected in series to a battery. With this little circuit, we can obtain any voltage between zero and the supply voltage at the point between the two resistors (the p.d. across R_2). This voltage turns out to depend on the size of the two resistors and the supply voltage.

The voltage, or e.m.f., across the battery is V_B. The circuit is a simple series circuit, so that:

$$V_B = V_1 + V_2 = IR_1 + IR_2 = I(R_1 + R_2)$$

The p.d. across R_1 is given by:

$$V_1 = IR_1$$

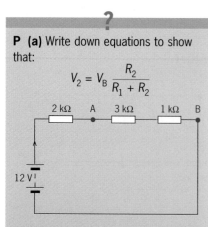

P (a) Write down equations to show that:

$$V_2 = V_B \frac{R_2}{R_1 + R_2}$$

(b) Show that the p.d. across the 3 kΩ resistor in the circuit is 6 V. How would you rearrange the resistors so that the p.d. between points A and B is 6 V?

If we take the ratio of the p.d.s across R_1 and the battery, then:

$$\frac{V_1}{V_B} = \frac{IR_1}{I(R_1 + R_2)} = \frac{R_1}{R_1 + R_2}$$

or:

$$V_1 = V_B \frac{R_1}{R_1 + R_2}$$

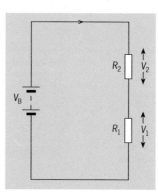

This arrangement is called a **potential divider** since it can divide the supply voltage V_B in any ratio that is required. In the box on the comparator, a potential divider is used to 'set' the voltage at one of the two inputs into the operational amplifier (op-amp). This voltage is set by choosing appropriate values of the two resistors.

Fig 9.17 **Two resistors are used in a potential divider to divide the voltage**

The comparator

A comparator is an electronic switch. It compares two voltages. The circuit in Fig 9.18 turns the light-emitting diode on when a changing voltage V_{in} is equal to or greater than a reference voltage. The comparator has two inputs called the 'inverting' (−) and 'non-inverting' (+) inputs. The reference voltage is fixed at the non-inverting input by using a potential divider. The circuit in the diagram uses a 0 to 12 V supply, so the voltage at the non-inverting input will be given by:

$$V_{ref} = 12 \times \frac{R_2}{R_1 + R_2}$$

So, suppose we want the reference voltage to be 4 V. (This would mean that the LED would light when the input voltage is equal to or greater than 4 V.) We now have:

$$4 = 12 \times \frac{R_2}{R_1 + R_2} \quad \text{or} \quad \frac{R_1}{R_1 + R_2} = \frac{1}{3}$$

Fig 9.18 **A comparator circuit**

Comparators require very small currents, typically less than a microampere. This means that the current in the resistors will be very small, so their values should be high. In this circuit, values of R_1 = 200 kΩ and R = 100 kΩ would be suitable.

The light-dependent resistor

The potential divider is widely used in electronics to fix the voltage at a point in a circuit. It can be used when one of two resistors is a special resistor such as a light-dependent resistor (LDR).

The resistance of an LDR depends on the intensity of the light falling on it. In the dark, its resistance is very high, typically tens of kilohms. In bright light, its resistance drops to about 300 Ω. So, as the light changes, the p.d. across the fixed resistor with a value of 1 kΩ will vary. But whatever the resistance of the LDR, the p.d.s across it and the fixed resistor must add up to the supply voltage.

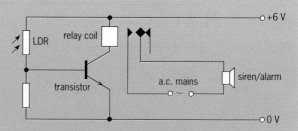

Fig 9.19 **A potential divider circuit serving as a burglar alarm. It uses a light-dependent resistor and fixed-value resistors**

In Fig 9.19, the supply voltage is 6 V. When the resistance of the LDR is high (in the dark), the p.d. across it will be higher, probably about 5.5 V, with only about 0.5 V across the other resistor. In bright light, the situation is reversed, with only 1 V across the LDR and so about 5 V across the fixed resistor.

When light falls on the LDR, the increase in voltage across the fixed resistor switches the transistor on, which in turn energises the relay. A mains-operated alarm is switched on by the relay.

Variable resistors – rheostats and potentiometers

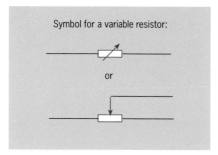

Fig 9.20 **Variable resistors**

Variable resistors differ from fixed resistors. Variable resistors usually have three connections, one at each end of the resistance element, the third as a slider (Figs 9.20 and 9.21). Variable resistors can be used in two ways: to control current or to control voltage.

In the circuit of Fig 9.20, the simple variable resistor is used as a **rheostat** to control current. The sliding (adjustable) contact is used, with one of the end connectors; the third terminal is not used. As the sliding contact moves, the current in the circuit varies and is measured on the ammeter.

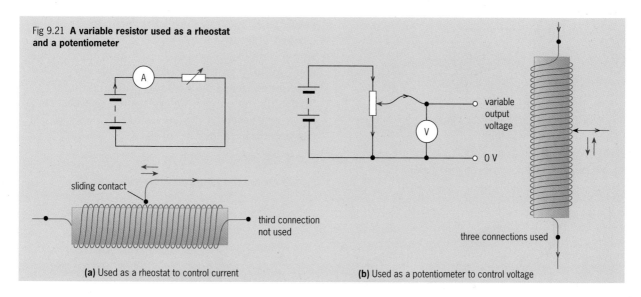

Fig 9.21 **A variable resistor used as a rheostat and a potentiometer**

(a) Used as a rheostat to control current

(b) Used as a potentiometer to control voltage

The variable resistor in Fig 9.21(b) is used to control voltage. It is a **potential divider** (or **potentiometer**), and uses all three connections. The sliding contact effectively divides the variable resistor into two parts. The output voltage from a potentiometer can be any value from zero to the supply voltage as the variable contact is moved, and so the resistance is varied from its maximum to zero. Again, this is a very useful circuit to use in experimental work when you need a variable voltage supply. In electronics, such a circuit is used as a simple **volume control** in audio amplifier circuits. When you adjust the volume control, you are moving the slider.

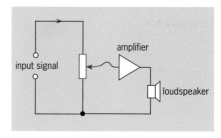

Fig 9.22 **A potentiometer used as a volume control**

More on potential dividers

Circuits with potential dividers are very useful – but, be careful. So far we have made the assumption that the potential divider and the potentiometer do not provide any current. In Fig 9.17, we assumed that the current is the same in both resistors. In Fig 9.21(b), we assume that there is a very small current in the wire from the slider of the variable resistor.

In practice, of course, some current passes from the potential divider to whatever is connected to it. Suppose you want to light a 3 V, 0.2 A bulb and you only have a 6 V battery. The circuit shown in Fig 9.23(a) might be thought a solution, but it is not.

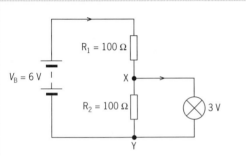

(a) A potential divider used in a circuit with a 6 V supply and a 3 V bulb. Will the bulb light up?

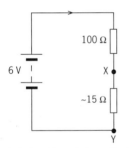

(b) The equivalent resistance between X and Y is about 15 Ω.

See questions 5 and 6.

In the circuit of Fig 9.23(a), the voltage across one of the two resistors is given by:

$$6 \text{ V} \times \frac{100}{100 + 100} = 6 \text{ V} \times \tfrac{1}{2} = 3 \text{ V}$$

When the bulb is connected across one of the resistors, we might expect it to light. But the bulb itself also has a resistance that affects the current. When it lights normally with 3 V across it, the current through it will be 0.2 A. Therefore the resistance of the bulb is 3 V/0.2 A = 15 Ω.

When the bulb is connected it changes the circuit. The effective resistance between X and Y is reduced since the bulb is in parallel with one of the 100 Ω resistors. Without doing an accurate calculation, we do know that the resulting resistance between X and Y will now be *less* than 15 Ω. The p.d. *V* across XY will therefore be much less than 3 V.

Using:

$$V_2 = V_B \frac{R_2}{R_1 + R_2}$$

it is approximately: $6 \times \dfrac{15}{100 + 15} = 0.8 \text{ V}$

With 0.8 V across it, there will not be enough current to light the bulb. In this application, the potential divider does not work. Also, the available current is limited by R_1 to be less than 0.06A! A simple rheostat would be better, although not the ideal solution to this problem.

What we learn is that for a potential divider to be useful, the current drawn from it must be very small compared to the current in the resistors of the potential divider.

Fig 9.23 **Circuits containing a potential divider**

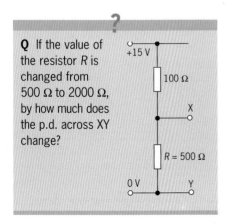

9 USING AMMETERS AND VOLTMETERS

It is an important principle that, when we take measurements, the act of measuring should not change what is being measured. But we sometimes have to compromise. In electrical measurements, for example, it is usually impossible to measure current without changing it a little. Provided the change is small enough to be negligible, we accept the change. The question to ask in any experiment is: how small is negligible? As a rule of thumb, anything of 1 per cent or less is acceptable. We may have to put up with larger errors in some circumstances, so it is important that we are aware of the error.

Ammeters

Ammeters measure current. The ideal ammeter should have zero resistance so that, when connected correctly in series in a circuit, it will not affect the current but will just measure it. However, real ammeters always do have a resistance, which should be as small as possible – small compared with the resistance in the circuit as a whole.

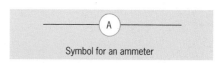

Symbol for an ammeter

Fig 9.24 **An ammeter**

EXAMPLES

Q The circuit in Fig 9.25(a) shows a 12 V battery connected to a powerful motor. With no load, this motor draws a current of about 10 A. Estimate the resistance of the motor.

Fig 9.25

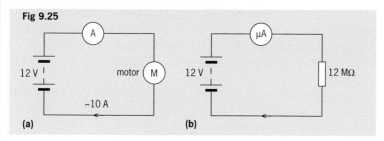

(a)   (b)

A Assuming the internal resistance of the battery is negligible:

$$\text{resistance} = \text{voltage/current} = 12\,/10 = 1.2\ \Omega$$

Q An ammeter in the circuit should have a resistance of about one hundredth of this for the error to be acceptable. Calculate the resistance of a suitable ammeter to measure the current in the circuit.

A Resistance of ammeter = 1.2/100 = 0.012 Ω.

Q The motor is replaced with a 12 MΩ resistor, see Fig 9.25(b). What is the current now?

A 'New' current = $12/(12 \times 10^6) = 1.0 \times 10$ A = 1 μA.

Q The resistance of an ammeter to measure the current in this circuit can be larger. How large?

A The resistance of the 'new' ammeter = $(12 \times 10^6)/100$
$$= 120\,000\ \Omega = 120\ \text{k}\Omega.$$

Voltmeters

A voltmeter is connected to a circuit in parallel to measure the p.d. across a part of the circuit. The ideal voltmeter would have such a high resistance that it would draw no current at all. The very high resistance electronic digital voltmeters approach this, using a negligible current. In practice, however, a voltmeter must draw some current; otherwise it wouldn't work.

The current drawn by the voltmeter in Fig 9.27 must be very small compared with the current in the resistor. In practice, we ensure that the resistance of the voltmeter is much bigger than that across the part of the circuit for which the p.d. is being measured. A typical moving coil voltmeter has a resistance of 100 kΩ, while a digital voltmeter's resistance is typically 10 MΩ – a hundred times greater.

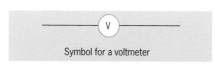

Symbol for a voltmeter

Fig 9.26 **The voltmeter**

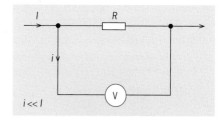

Fig 9.27 **A voltmeter across a resistor**

See the Assignment at the end of this chapter for more on electrical measurement.

10 MORE ABOUT CONDUCTORS

The ideas about free electrons discussed at the start of this chapter give us a simple model that we can use to explain resistance. These ideas also explain why the resistance of good conductors tends to increase with temperature.

When there is a current in a wire, some electrons move 'freely'. A force acts on an electron due to the electric field created by the applied voltage. This accelerates it and it gains kinetic energy. It is not likely to travel very far before it collides with an atom, which itself has vibrational kinetic energy. In the collision, the electron loses some of its kinetic energy to the atom. After the collision, it will again be accelerated until it collides once more, losing energy and so on. The collisions with atoms provide the resistance that prevents an electron from accelerating along a clear path. With the energy transferred from the electrons, the atoms vibrate more (the wire gets hotter) and collisions are more likely. This increases the resistance.

Most metals behave in this way – their resistance decreases as the temperature decreases. Titanium behaves in a similar way, but when the temperature becomes very low (0.39 K, −272.8 °C) the resistance becomes zero. At this temperature, titanium becomes what is called a **superconductor**.

In recent years, materials have been discovered that become superconductors at much higher, more attainable temperatures than 0.39 K. Materials based on oxides of bismuth become superconductors at about 77 K, −196 °C, a temperature that can be obtained quite easily using liquid nitrogen as a coolant. These are called high temperature superconductors.

Semiconductors

Semiconductor materials include the best known, silicon. Others are germanium, lead sulphide, selenium and gallium arsenide. At ordinary temperatures, pure silicon is not a good insulator but not a good conductor, because at these temperatures silicon has few free electrons in it. But when silicon is heated, the energy gained frees more electrons.

Semiconductor materials, then, are not much use at ordinary temperatures in their pure state. To make a semiconductor that is useful, the pure material is 'doped' by adding tiny amounts of impurity atoms. Semiconductors are dealt with in greater detail in Chapter 22.

Symbol for a capacitor

11 CHARGE AND CAPACITORS

The flash on a camera takes time to store enough energy for the high intensity bulb to give a flash. To store this energy, the flash unit uses a **capacitor**, as shown in Fig 9.28, a component designed to store a particular amount of charge and release it when required. A capacitor consists simply of two conducting plates separated by an insulating layer, sometimes called a dielectric.

Fig 9.28 **Capacitors**

Charging and discharging

A capacitor in a simple d.c. circuit is shown in Fig 9.29 and its behaviour during charging and discharging is shown in Fig 9.30(a) and (b). In each, centre zero microammeters record any current. Read the captions to these carefully, and then read on.

The current is a flow of charge. In each case, the flow of charge is high to begin with, and quickly gets less and less. In Fig 9.30(a), the battery drives a current round the circuit. One plate becomes positively charged and the other becomes negatively charged. When the free wire is moved (Fig 9.30(b)) to point A, the capacitor is 'discharged'. The charge on the plates returns to where it came from!

Looking again at Fig 9.30(a), the battery redistributes the charge around the circuit. Electrons move away from one plate of the capacitor and on to the other plate, but not by crossing between the plates: the electrons reaching the negative plate in Fig 9.30(a) come from the adjacent wire; those that leave the positive plate move into the wire connected to it.

The changing current is a consequence of the forces between charges. When the capacitor is charged, electrons flow to the plate that becomes negative. At first this is easy, so the current is large. The electrons spread out and the repulsive forces between them is small. As more electrons arrive, these forces increase, making it harder for more electrons to get on to the plate (the current decreases), until eventually there are so many electrons that repulsive forces prevent more from arriving. The capacitor is now fully charged.

(At the same time, there is a similar process on the positive plate, except that here electrons leave the plate and the forces are attractive.)

The battery does work moving the charge, and energy is stored in the capacitor. When the capacitor is discharged, the stored energy 'does work' moving the charge back.

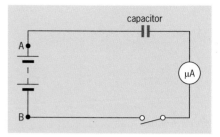

Fig 9.29 **A capacitor in a circuit**

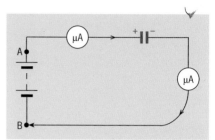

Fig 9.30(a) **Charging: A current is shown on both meters when the free wire is touched to side B of the battery to complete the circuit. The needle on both meters moves sharply to one side and quickly returns to zero. After this, no further current flows. The capacitor has been 'charged'**

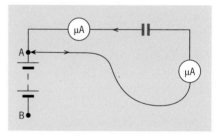

Fig 9.30(b) **Discharging: When the wire is moved from point B to point A, both meters show a current, but this time in the opposite direction. Again, the readings on the meters fall quickly to zero. The capacitor has now been 'discharged'**

Capacitance

When fully charged, the potential difference across the capacitor plates is the same as that across the battery.

So the maximum amount of charge that a capacitor can store is proportional to the battery voltage:

$$Q \propto V$$

where Q is the magnitude of charge on the plates measured in coulombs, and V is the voltage across the capacitor.

Therefore the ratio of Q/V is constant and is a measure of the amount of charge that is stored on the capacitor per volt. This is called **capacitance**, symbol C, and is a constant value for any particular capacitor. That is:

$$C = \frac{Q}{V}$$

Capacitance C is measured in **farads**, and one farad is one coulomb per volt:

$$1 \text{ F} = 1 \text{ C V}^{-1}$$

Be careful to distinguish between the italic C for the variable value of capacitance, and the upright C, the symbol for the unit the coulomb.

One farad (1 F) is a very large amount of capacitance, and most practical capacitors have much smaller value measured in microfarads (μF), nanofarads (nF) and picofarads (pF):

$$1 \text{ F} = 1\,000\,000 \text{ μF} = 10^6 \text{ μF}$$
so: $\quad 1 \text{ μF} = 10^{-6} \text{ F}$
$$1 \text{ μF} = 1000 \text{ nF} = 10^3 \text{ nF}$$
Also: $1 \text{ μF} = 1\,000\,000 \text{ pF or } 10^6 \text{ pF}$
so: $\quad 1 \text{ pF} = 10^{-12} \text{ F}$

Working voltage

If the voltage across a capacitor is too high, the insulator between the plates fails to insulate, and charge passes between the capacitor plates. This is why capacitors are marked with a working voltage, the maximum voltage that should be placed across the capacitor. A good rule of thumb is to make sure that the potential difference across a capacitor is never bigger than about two-thirds of the working voltage, especially in alternating current circuits. (See Chapter 13 for a more detailed description of peak voltages and r.m.s. voltages to understand why this is so.)

Capacitors in parallel and series

For capacitors connected in parallel, the total capacitance can be found by adding the individual capacitances:

$$C_T = C_1 + C_2 + C_3$$

So, as the number of capacitors connected in parallel increases, so does the total capacitance.

When capacitors are connected in series, the equivalent capacitance can be found from the equation:

$$\frac{1}{C_T} = \frac{1}{C_1} + \frac{1}{C_2} + \frac{1}{C_3}$$

This means that when capacitors are connected in series, the total capacitance is reduced. (Look at the diference between these equations and those for resistors given earlier in the chapter.)

R What is the total capacitance between the points A and B in this circuit?

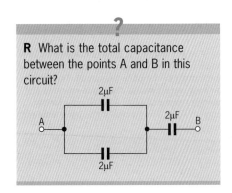

Capacitors in parallel

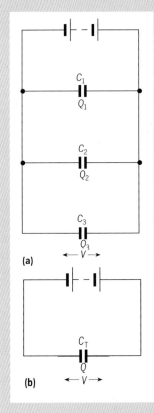

(a)

(b)

Fig 9.31 shows three capacitors with capacitances C_1, C_2 and C_3 connected in parallel with a battery. The p.d. across each capacitor is the same: call it V. The capacitors will have charges of Q_1, Q_2 and Q_3 respectively. The total charge on the three is simply:

$$Q = Q_1 + Q_2 + Q_3.$$

In Fig 9.31(b), the three capacitors are replaced with one of value C_T, which stores the same charge, Q, when there is V volts across it. Therefore, since $Q = CV$:

$$C_T V = C_1 V + C_2 V + C_3 V$$

The p.d. across all the capacitors is the same and can be eliminated:

$$C_T = C_1 + C_2 + C_3$$

Fig 9.31 **Measuring voltage across capacitors in parallel**

Capacitors in series

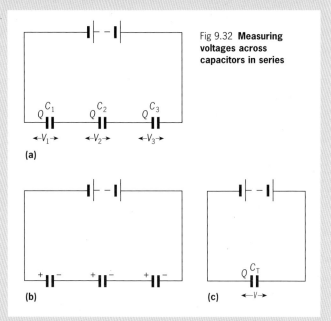

Fig 9.32 **Measuring voltages across capacitors in series**

S Some capacitors are marked '5.0 µF, 12 V'. What does this mean? Describe how some of these capacitors could be used to make the following:

(a) A capacitance of 2.5 µF with a working voltage of 24 V.

(b) A capacitance of 5.0 µF with a working voltage of 24 V.

T The circuit shows four arrangements, A to D, of three 1 µF capacitors.

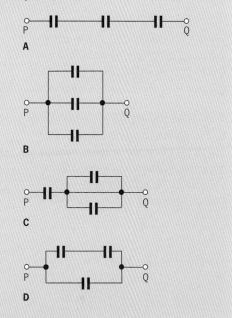

Which arrangement has:

(a) the smallest capacitance?

(b) the largest capacitance?

Fig 9.32 shows three capacitors of capacitances C_1, C_2 and C_3 in series with a battery. At any time while they are charging, the current in the circuit will always be the same at all points in the circuit, and it will take the same time for each capacitor to charge. This means that the charge on each capacitor will be the same – call it Q_C.

Since charge does not ordinarily cross between plates in a capacitor, you may be wondering how the middle capacitor becomes charged. Electrons leave the left plate of capacitor 1. This 'induces' the movement of electrons from the left plate of capacitor 2 on to the right plate of capacitor 1. This in turn induces electron flow from the left plate of capacitor 3 on to the right plate of capacitor 2. Electrons move on to the right plate of capacitor 3 to complete the process. The end result is that each left plate becomes positively charged and each right plate is negatively charged.

The p.d. across each capacitor will be V_1, V_2 and V_3 respectively. Suppose we replace the three capacitors with one of size C_T that will carry the same charge, Q, and have a p.d. of V across it. Now this p.d. V will equal the sum of the p.d.s across the three capacitors:

$$V = V_1 + V_2 + V_3$$

Also:

$$V = \frac{Q}{C}$$

Therefore:

$$\frac{Q}{C_T} = \frac{Q}{C_1} + \frac{Q}{C_2} + \frac{Q}{C_3}$$

The charge Q can be eliminated to give:

$$\frac{1}{C_T} = \frac{1}{C_1} + \frac{1}{C_2} + \frac{1}{C_3}$$

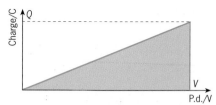

Fig 9.33 **Graph of charge on a capacitor against voltage**

See question 7. ▨

✓

Compare the equation for energy stored in a capacitor with that for the energy stored in a spring that obeys Hooke's law:

$$E = \tfrac{1}{2}Fx = \tfrac{1}{2}kx^2$$

where x = extension of spring in metres produced by force F newtons, and k = force constant of the spring.

?

U The batteries in the circuits shown are all the same and all the capacitors are identical. Which arrangement stores the greatest amount of energy when the capacitors are all fully charged?

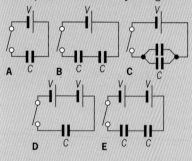

?

V Some portable computers use capacitors as a short-term power backup. The circuits are designed so that the capacitor can discharge at a constant current. A 4 F capacitor charged to 5 V is required to supply 50 μA to a memory board. For how long would it be able to do this?

See questions 8 to 12. ▨

Energy stored in a capacitor

Getting more charge on to an already charged capacitor requires you to do work against the repulsive forces between the charges. If the capacitor is at a p.d. of V volts – a measure of joules per coulomb – then adding a charge Q (measured in coulombs, C) requires an amount of energy in joules given by:

$$\text{energy} = \text{joules/coulomb} \times \text{coulombs}$$
$$E = V \times Q$$

Charge on a capacitor is proportional to the voltage across it, so a graph of voltage against charge is a straight line through the origin, as shown in Fig 9.33. This means that adding the same amount of charge to a capacitor requires more work to be done. The total work done in charging the capacitor from 0 V is given by the shaded area under the graph. This is also the energy (in joules) stored in the capacitor:

$$\text{area} = \tfrac{1}{2}QV$$

But from the capacitance equation: $\quad Q = CV$

So: $\qquad\qquad$ **Energy stored** $= \tfrac{1}{2}QV = \tfrac{1}{2}CV^2$

Capacitors and resistors

When a resistor is placed in series with a capacitor, as in Fig 9.34(a), the time taken for the capacitor to charge and discharge is increased because the current in the circuit is decreased. In the circuit in Fig 9.34(b), the oscilloscope (CRO) shows how the p.d. across the resistor changes as the capacitor discharges. The p.d. across the resistor is proportional to the current through it, so the trace on the oscilloscope shows how the current in the circuit changes.

The curve is an *exponential* decay curve. It has an interesting property: the current in the circuit decays by the same ratio in successive equal intervals of time. This kind of curve appears frequently in nature. You may be familiar with it in the radioactive decay curve. The constant ratio we measure in that case is called the **half-life**, that is, the time taken for the activity of the radioactive source to halve. In the case in Fig 9.34(c), the sizes of both the resistor and the capacitor determine the rate of discharge. This is shown from the units of resistance and capacitance:

$$R \times C = \frac{\text{volts}}{\text{amps}} \times \frac{\text{coulombs}}{\text{volts}}$$

$$= \frac{\text{coulombs}}{\text{coulombs/second}} = \text{seconds}$$

The product RC is called the **time constant** and is used to measure the rate of decay of current in a capacitor.

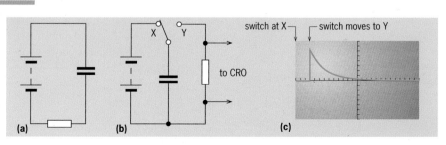

Fig 9.34 (a) **A resistor and capacitor in series.** (b) **With the switch at X the capacitor charges. With the switch at Y, it discharges and produces the CRO trace shown in** (c)

The mathematics of the exponential decay of a capacitor

Imagine a capacitor of capacitance C farads that is initially charged to a p.d. of V volts and then discharged through a resistor of R ohms. Although we know that the current decays, we shall assume that the average current over a small interval of time, Δt, is I amperes. The amount of charge, ΔQ, that flows in this time is given by:

$$\Delta Q = -I\Delta t$$

The current can also be given in terms of the resistance and the p.d. across it from Ohm's law ($V = IR$) so:

$$\Delta Q = -\frac{V}{R}\Delta t$$

The p.d. is also related to the charge on the capacitor from $CV = Q$, so:

$$\Delta Q = -\frac{Q}{RC}\Delta t$$

Another way of writing the equation is as:

$$\frac{\Delta Q}{\Delta t} = -\frac{Q}{RC}$$

If Δt is very small, it is written as dt and the equation becomes:

$$\frac{dQ}{dt} = -\frac{Q}{RC}$$

or:

$$\frac{1}{Q}dQ = -\frac{1}{RC}dt$$

This is a first-order differential equation. To solve it we have to integrate both sides:

$$\int_{Q_0}^{Q}\frac{1}{Q}dQ = -\int_0^t\frac{1}{RC}dt$$

or:

$$\ln\left(\frac{Q}{Q_0}\right) = -\frac{t}{RC}$$

$$\frac{Q}{Q_0} = \exp\left(-\frac{t}{RC}\right)$$

$$Q = Q_0\exp\left(-\frac{t}{RC}\right)$$

A graph of charge Q against time t is an exponential decay curve. The rate of decay, or the steepness of the curve, depends on the initial charge Q_0 and on the values of R and C. This is as shown in Fig 9.35.

If we substitute the time t in the equation with RC, then Q will be the charge after RC seconds:

$$Q = Q_0\exp\left(-\frac{RC}{RC}\right) = Q_0\exp(-1)$$

So:

$$\frac{Q}{Q_0} = \frac{1}{e} = \frac{1}{2.718} = 0.37$$

That is, in RC seconds the charge on a capacitor decays to 37 per cent of the initial charge.

Testing for exponential change – method 1

If a quantity decreases in an exponential manner, then over equal intervals of time there is a *constant ratio* between the starting and finishing values. For example, if you record a quantity x at equal intervals, then consecutive readings (x_1, x_2, x_3, x_4 . . .) should be in the same ratio to one another ($x_1/x_2 = x_2/x_3 = x_3/x_4 = \ldots$).

Testing for exponential change – method 2

The following equation was derived above:

$$\left(\frac{Q}{Q_0}\right) = \exp\left(-\frac{t}{RC}\right)$$

Taking the log of both sides, we get:

$$\ln\left(\frac{Q}{Q_0}\right) = -\frac{t}{RC}$$

or:

$$\ln Q - \ln Q_0 = -\frac{t}{RC}$$

A graph of $\ln Q$ against time t will therefore be a straight line – a distinctive mark of an exponential change over time. The slope of the graph will be $-1/RC$.

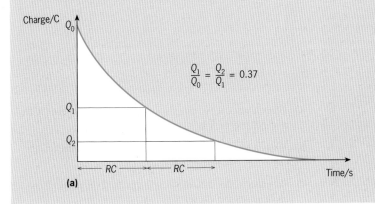

$$\frac{Q_1}{Q_0} = \frac{Q_2}{Q_1} = 0.37$$

(a)

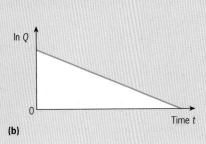

Fig 9.35 (a) **Exponential decay curve for a discharging capacitor.** (b) **Plotted as ln Q against time**

(b)

SUMMARY

In studying this chapter, you should have learned and understood the following:

■ An electric current is the flow of electric charge and can be described by the equations:

$$I = dQ/dt \text{ and } I = nAQv.$$

■ Energy is transferred when a charge flows through a circuit. The potential difference is the energy transferred per coulomb as charge passes through a component or device. This can be described as 1 volt = 1 joule/coulomb.

■ The size of the current in a circuit depends on the resistance in the circuit.

■ Resistance $R = \rho l/A$, where ρ is the resistivity, which depends on the material.

■ Ohm's law states that resistance = p.d./current.

■ The power in a circuit is given by $P = IV = I^2R = V^2/R$.

■ Electromotive force (e.m.f.) is the energy transferred per coulomb in a source of current. Some of the energy gained by each coulomb is transferred in doing work against the internal resistance of the power supply.

■ Potential dividers and potentiometers can be used to control voltages in circuits.

■ The availability of free electrons decides whether materials are conductors, semiconductors or insulators. At low temperatures, some materials become superconductors.

■ Capacitors store charge Q when there is a p.d. V across them. The capacitance C of a capacitor by is given by $Q = CV$.

■ Energy can be stored in capacitors. The energy stored E is given by $E = \frac{1}{2}QV = \frac{1}{2}CV^2$.

■ The rate at which a capacitor charges or discharges depends on the time constant RC.

■ The decay of charge on capacitors is exponential.

QUESTIONS

1 The diagrams in Fig 9.Q1 show the characteristics of some useful electronic devices.

a) Describe how the resistance of each one changes.

b) The graphs illustrate the behaviour of the following components: diode, light-dependent resistor, thermistor, copper wire. Match each graph to one of these components.

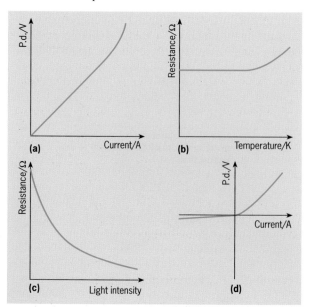

Fig 9.Q1

2 A four-terminal box is tested using the circuit in Fig 9.Q2. Below the circuit there are four different boxes, A to D. The resistors used in each are identical.

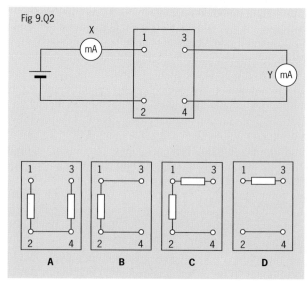

a) Which circuit would show a reading on meter X but not on meter Y?

b) Which circuit would result in both milliammeters showing the same reading?

c) Which circuit would show a reading on meter X twice as large as on meter Y?

3 The battery in the circuit in Fig 9.Q3 has an e.m.f. of 5.4 V. The current in the bulb is 0.30 A. The voltmeter reads 4.8 V.

a) Explain why the voltmeter reading is smaller than the e.m.f.

b) Use the information provided to calculate the internal resistance of the battery.

c) Calculate the energy transferred per second in the bulb.

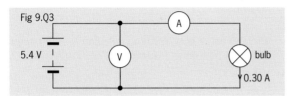

Fig 9.Q3

4 A 12 V battery of negligible internal resistance is connected across AD in Fig 9.Q4.

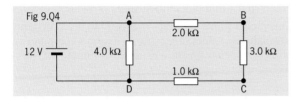

Fig 9.Q4

a) Calculate the value of the potential difference across the terminals BC.

b) **(i)** A calibrated oscilloscope used as a very high resistance voltmeter is connected across BC. Why would you expect it to record the same potential difference as your answer to **a)**?
   (ii) The oscilloscope is replaced by a moving coil voltmeter of resistance 6.0 kΩ. Calculate what this voltmeter would read.

c) In order to check that the moving coil voltmeter is reading accurately, the oscilloscope is now added in parallel with it. Assuming that they are both accurately calibrated, state the reading on the oscilloscope and the voltmeter.

5 A student wanted to light a lamp labelled 3V, 0.2 A, but had only a 12 V battery of negligible internal resistance. In order to reduce the battery voltage, she connected up the circuit shown in Fig 9.Q5(a). She included the voltmeter so that she could check the voltage before connecting the lamp. The maximum value of the resistance of the rheostat CD was 1000 Ω.

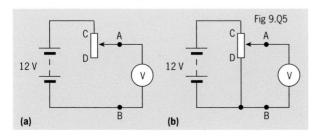

Fig 9.Q5

a) She found that, when the sliding contact of the rheostat was moved down from C to D, the voltmeter reading dropped from 12 V to 11 V. What was the resistance of the voltmeter?

b) She modified his circuit as shown in Fig 9.Q5(b) using the rheostat as a potentiometer, and was now able to adjust the rheostat to give a meter reading of 3 V. What current would now flow through the voltmeter?

c) Assuming that this current is negligible, compared with the current through the rheostat, how far down from C would the sliding contact have been moved?

d) The student then removed the voltmeter and connected the lamp in its place, but it did not light. How would you explain this? (The lamp itself was not defective.)

6 Fig 9.Q6(a) and (b) show meters that have been incorrectly connected in circuits with a battery (e.m.f. 3 V, internal resistance 1 Ω) and a bulb (of 3 V, 0.2 A). The voltmeter has a range of 0–3 V and a resistance of 30 kΩ. The ammeter has a range of 0–3 A and a resistance of 0.1 Ω.

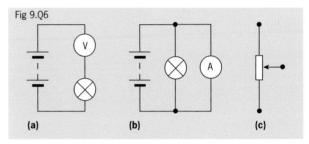

Fig 9.Q6

Using the information provided above, answer the following questions. Give your reasoning in each case. (Precise calculations of meter readings are *not* required.)

a) **(i)** Would the bulb in Fig 9.Q6(a) be lit?
   (ii) Would the reading on the voltmeter be near zero, about half full scale or close to full scale?

b) **(i)** Would the bulb in Fig 9.Q12(b) be lit?
   (ii) Would the reading on the ammeter be near zero, about half full scale or close to full scale?

In order to plot a graph of current against potential difference for a bulb over the range from zero to about 2.5 V, all the original components are provided, together with a 10 Ω, three-terminal variable resistor, shown in Fig 9.Q7.

Complete the drawing of a circuit which would enable the required readings to be taken using the components in Fig 9.Q6(a) to (c).

7 The flash unit of a camera uses a 5000 μF capacitor which is charged to 20 V. The capacitor is discharged across the flash bulb in 1 ms (1 × 10⁻³ s).

a) How much energy is stored in the capacitor?

b) What is the power rating of the flash bulb?

c) Estimate the average current that flows when the capacitor discharges.

8 The equation $\Delta Q = (Q/RC)\Delta T$ can be used to plot a theoretical graph of the discharge of the capacitor in Fig 9.Q8. Assume that the capacitor has a value of 500 µF, the resistor 100 kΩ, and that there is a charging p.d. of 10 V.

a) How much charge would be stored on the capacitor when it is fully charged?

b) Taking ΔT as 5 s, calculate $\Delta T/RC$ and the amount of charge, ΔQ, that flows from the capacitor in this time.

c) How much charge will be left on the capacitor after the first 5 s?

d) How much charge leaves the capacitor in the next 5 s? How much charge will be left after this time?

e) Repeat part **d)** for 5 s intervals for $2 \times RC$ seconds.

f) Use your results to plot a graph of charge against time. Check the graph for a constant ratio property.

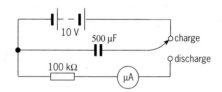

Fig 9.Q8

9 Question 8 took you through a method for calculating the charge left on the capacitor at intervals. The process is repetitive and probably rather boring after you've made a few dozen calculations, but it is just the kind of job that computers are very good at.

A simple computer model of capacitor discharge could be written to perform the following operations in a loop:

1) $Q = CV$
Calculate Q from values of V and C. The computer would need to know an initial value for V and the size of C.

2) $I = V/R$
Use the value of Q and R to calculate I. The computer needs to be told the value of R.

3) $\Delta Q = -I\Delta T$
This works out ΔQ. The computer needs to be told ΔT.

4) $Q = Q + \Delta Q$
The new value of Q = old value of Q plus ΔQ.

5) $T = T + DT$
Start by telling the computer that: $T = 0$. Now go back to step 1 and repeat each step.

This model shows the steps that need to be carried out to calculate the values of Q as time passes. A computer could also construct a spreadsheet, as shown in Fig 9.Q9.

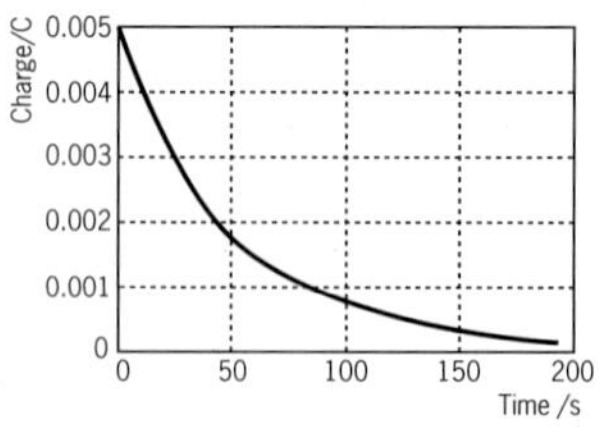

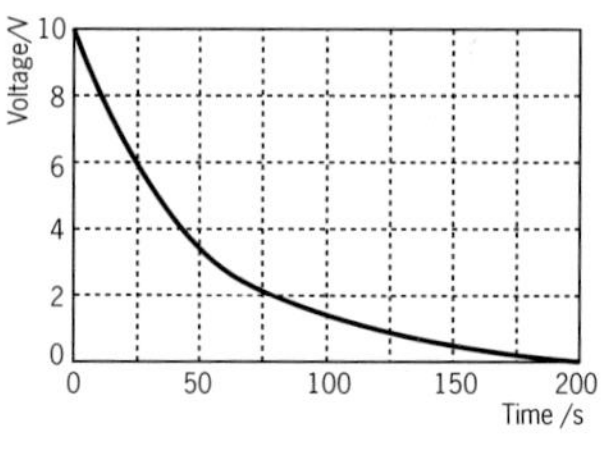

Fig 9.Q9

Use the spreadsheet to obtain enough data to plot graphs of charge against time for about $3 \times RC$.

By changing the initial values of R and C, the precise nature of the decay curve can be examined.

Discharging a capacitor

	Data		You can change values in this section	
		Series resistance, R =	100000	ohm
		Capacitance, C =	0.0005	F
		Supply voltage, V =	10	volt
		Time interval, dt =	0.5	s
		Charge on capacitor, Q=CV	0.005	coulomb
	Formulae	Current I = V/R		
		Change in charge dQ = -Idt		
		New charge = old charge + change	Q=Q+dQ	
		Voltage across C=Q/C		

Time, t	Charge Q	Change in charge, dQ	Current in R, I	Voltage across C
0	0.005000	-0.00005000	0.000100	10.00
0.5	0.004950	-0.00004950	0.000099	9.90
1	0.004901	-0.00004901	0.000098	9.80
1.5	0.004851	-0.00004851	0.000097	9.70
2	0.004803	-0.00004803	0.000096	9.61
2.5	0.004755	-0.00004755	0.000095	9.51
3	0.0047			9.41
3.5	0.0046			9.32
4	0.0046			9.23
4.5	0.0045			9.14
5	0.0045			9.04
5.5	0.0044			8.95
6	0.0044			8.86
6.5	0.0043			8.78
7	0.0043			8.69
7.5	0.004300	-0.00004300	0.000086	8.60
8	0.0042			8.51
8.5	0.0042			8.43
9	0.0041			8.35
9.5	0.0041			8.26
10	0.0040			8.18
10.5	0.0040			8.10
11	0.0040			8.02
11.5	0.0039			7.94
12	0.003			7.86
12.5	0.003			7.78
13	0.003			7.70

10 In the circuit in Fig 9.Q10, both capacitors are initially uncharged. Assume that $R_2 = 2R_1$ and $C_2 = 2C_1$.

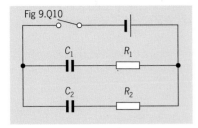

Fig 9.Q10

When the switch is closed, which of the following statements are true?

a) The initial current in R_1 is twice as large as that in R_2.

b) The final p.d. across both capacitors will be the same.

c) When fully charged, both capacitors will store the same amount of charge.

11 A capacitor is charged from a 6 V direct current supply, and is then discharged through a 10 kΩ resistor using the circuit shown in Fig 9.Q11. the diagram also shows the appearance of the resulting cathode ray oscilloscope trace. The sensitivity of the CRO was 1 V cm^{-1} and its time base was set to 0.1 s cm^{-1}.

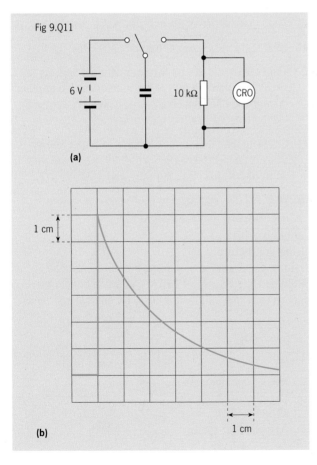

Fig 9.Q11

(a)

1 cm

(b)

1 cm

a) Showing your working, calculate the initial value of the discharge current.

b) **(i)** Use the drawing of the CRO trace to estimate a value for the charge which flowed through the resistor during one discharge.
(ii) From this, calculate a value for the capacitance.

c) This capacitor is now replaced by one of half its capacitance. Draw and label the trace as it would now appear.

12 In an experiment to measure the speed of a revolver bullet, the circuit in Fig 9.Q12 is set up so that when the revolver is fired the bullet first breaks conductor A and then conductor B, 0.02 m behind A, before being stopped in the block behind B.

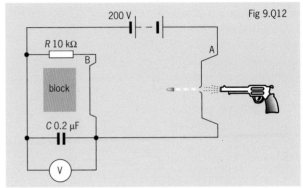

Fig 9.Q12

Before the revolver is fired, the reading on the voltmeter, whose resistance is much higher than that of R, is 200 V. After firing, the voltmeter reading is found to have dropped to 150 V.

a) What is the current through R the instant the bullet has broken the first conductor A? Show how you arrive at your answer.

b) Describe what happens to the current through R as the bullet travels from A to B and then breaks the conductor B. Give a reason for your answer.

c) How much charge is lost by the capacitor as the bullet travels from A to B?

d) Calculate an approximate value for the time it takes the bullet to pass from A to B. Show how you arrive at your answer.

Assignment

MAKING ELECTRIC MEASUREMENTS

We may need to use measuring instruments within a circuit to record values such as the current or the p.d. Electrical sensors, too, are useful for measuring a wide variety of environmental factors, monitoring the functions of the body and helping engineers to measure physical changes.

Measuring current

The typical laboratory ammeter is in fact a moving coil microammeter that is capable of measuring currents up to 100 µA. The maximum current that can be measured is referred to as the full-scale deflection or FSD of the meter. An ammeter can be used to measure larger currents by inserting a **shunt resistor**. The shunt resistor is connected in parallel with the meter so that most of the current is shunted past the meter.

Fig 9.A1

1 The ammeter in Fig 9.A1 has a resistance of 1000 Ω.

a) What is the p.d. across the microammeter when there is a current of 100 µA in it?

b) What is the p.d. across the shunt resistor?
The meter in Fig 9.A2 is required to measure a maximum current of 1 mA. That is, when the meter is placed in a circuit where the current is 1 mA, the meter should show a full-scale deflection.

Fig 9.A2

c) What will the actual current in the microammeter be?

d) What will be the current in the shunt resistor?

e) Work out the p.d. across the microammeter and the resistor.

f) Show that the value of the shunt resistor required is 111.11 Ω.

g) The meter is now placed in a different circuit and shows a current of 5.6 mA. What is the current in the microammeter?

h) Calculate the sizes of the shunt resistors required to measure the following maximum currents:
(i) 100 mA, **(ii)** 1 A, **(iii)** 5 A, **(iv)** 10 A.

Voltmeters

The microammeter in question 1 can be used as a voltmeter. Remember that voltmeters are always connected in parallel.

2 What is the maximum voltage that the microammeter can measure if it is used as a voltmeter?

To measure higher voltages a resistor, called a **multiplier resistor**, is connected in series with the microammeter. A voltmeter is needed to measure voltages up to 1 V, as shown in Fig 9.A3. To show a full-scale deflection the current in the microammeter must again be 100 µA.

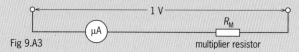

Fig 9.A3   multiplier resistor

3

a) Work out the current in the multiplier resistor.

b) What will the p.d. across the microammeter be?

c) What will the p.d. across the multiplier resistor be?

d) Show that the size of the multiplier resistor will be 9.0 kΩ.

e) What is the resistance of the voltmeter?

f) The voltmeter is placed across a different part of the circuit and shows a p.d. of 0.4 V. How much current is there in the microammeter? What is the p.d. across the microammeter?

g) Calculate the sizes of the multiplier resistors needed to convert the microammeter into voltmeters measuring up to **(i)** 5 V, **(ii)** 10 V, **(iii)** 50 V, **(iv)** 1000 V.

h) What will the resistance of each voltmeter in the previous part of the question be?

Using voltmeters

The two circuits in Fig 9.A4 show the same voltmeter, which has a resistance of 100 kΩ, in two different circuits. In each case, the battery has negligible internal resistance.

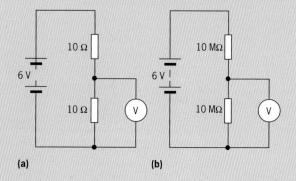

(a)   (b)

Fig 9.A4

a) What is the p.d. across each 10 Ω resistor in Fig 9.A4(a)?

b) What is the p.d. across each 1 M Ω resistor in Fig 9.A4(b)?

c) When the voltmeter is placed across one of the 10 Ω resistors in Fig 9.A4(a), it shows a p.d. of 3 V. When it is placed across one of the 1 MΩ resistors in Fig 9.A4(b), it shows a p.d. of about 0.5 V. Explain why the readings are different.

Measuring e.m.f.

It is not possible to measure e.m.f. without some error by simply placing a voltmeter across the power supply. This is because a voltmeter draws some current, even though it may be very small, and there will always be some volts lost across the internal resistance of the power supply. If we consider the equation:

$$\epsilon = V - IR$$

the measured voltage is only equal to the e.m.f. when the current is zero.

A potentiometer can be calibrated to measure in volts. If the variable resistor used is linear, then the angle swept through will be proportional to the resistance, as is shown in Fig 9.A5. A high resistance voltmeter connected to it will show that the voltage is also proportional to the angle.

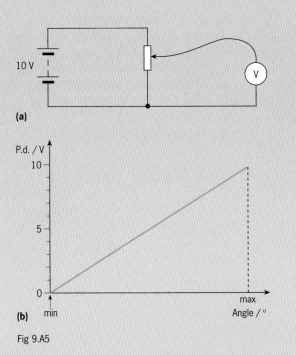

(a)

(b)

Fig 9.A5

The scale of the variable resistor could also be calibrated in volts.

The potentiometer in Fig 9.A6 is connected to a second battery, which has an e.m.f. that is known to be about 6 V but

is not known accurately. A sensitive centre zero microammeter is also included in the circuit.

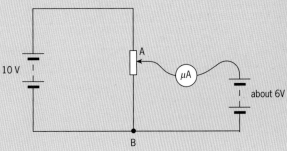

Fig 9.A6

 5

a) The potentiometer is set so that the voltage across AB is about 8 V. What will the direction of the current in the microammeter be?

b) The potentiometer is now set so that the p.d. across AB is about 4 V. What will the direction of the current in the microammeter be now?

c) The potentiometer is carefully adjusted until the reading on the microammeter is zero. The reading on the potentiometer shows that the voltage across AB is 5.6 V. What is the e.m.f. of the second battery? Explain why the current will be zero under these conditions.

 6

a) Draw a circuit showing how the resistance could be measured using an ammeter and a voltmeter.

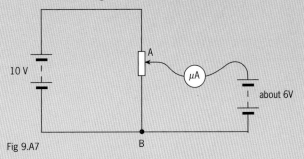

Fig 9.A7

The circuit in Fig 9.A7 shows two potential dividers. Each battery has the same e.m.f. and internal resistance. To begin with, all the resistances are equal.

b) What will be the reading on the meter? Why?

The circuit is now changed so that an unknown resistance R_x replaces one resistor, and a variable resistor replaces another, as in Fig 9.A8.

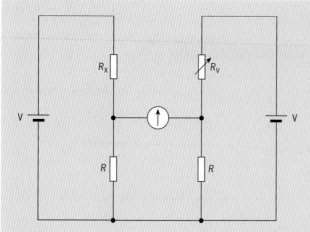

Fig 9.A8

c) Under what conditions will the reading on the meter now be zero?

As both batteries are the same, we can make do with just one of them. The circuit now becomes that in Fig 9.A9. This arrangement is known as a Wheatstone bridge.

Fig 9.A9

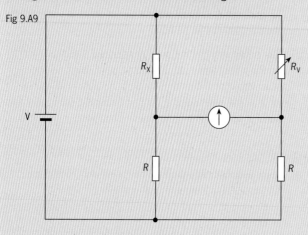

This arrangement is very useful for detecting small changes in resistance. Strain gauges can be used to detect forces and movement. A strain gauge is a special resistor: its resistance changes under strain.

Generally, two identical gauges are used in the bridge; one is under strain and the other is not. Changes in resistance of the gauge under strain are compared to the other, causing the bridge to become unbalanced. The small change in current is used to measure the strain of the force causing it.

Often, the change in current is very small. In Fig 9.A10, this will lead to a small difference in the voltages across each resistor (value R), and therefore at the two inputs to the operational amplifier.

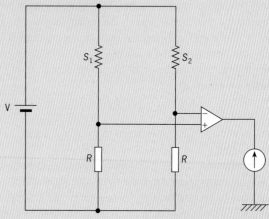

Fig 9.A10

Amplifiers like this can multiply the 'difference' voltage at the inputs by as much as a million times. So, for example, if the change in resistance of the strain gauge causes the voltages at the two inputs of the amplifier to change by 1 μV (just 1 microvolt), the voltage at the output will change by 1 V. The combination of the bridge circuit and the amplifier produces a very sensitive device.

For more on operational amplifiers, see Chapter 22.

CHARGE AND CURRENT

This chapter contains a large number of concepts that are closely interrelated. The Chapter Map below brings together these ideas and shows how they connect.

Use the map to check that you understand all the ideas and the go back through the chapter and study the text again.

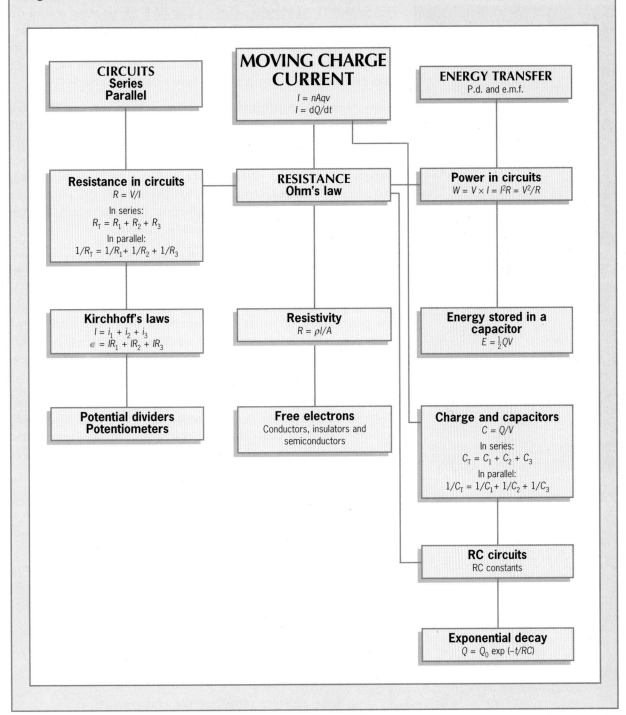

CIRCUITS
Series
Parallel

MOVING CHARGE CURRENT
$I = nAqv$
$I = dQ/dt$

ENERGY TRANSFER
P.d. and e.m.f.

Resistance in circuits
$R = V/I$
In series:
$R_T = R_1 + R_2 + R_3$
In parallel:
$1/R_T = 1/R_1 + 1/R_2 + 1/R_3$

RESISTANCE
Ohm's law

Power in circuits
$W = V \times I = I^2R = V^2/R$

Kirchhoff's laws
$I = i_1 + i_2 + i_3$
$\epsilon = IR_1 + IR_2 + IR_3$

Resistivity
$R = \rho l/A$

Energy stored in a capacitor
$E = \frac{1}{2}QV$

Potential dividers Potentiometers

Free electrons
Conductors, insulators and semiconductors

Charge and capacitors
$C = Q/V$
In series:
$C_T = C_1 + C_2 + C_3$
In parallel:
$1/C_T = 1/C_1 + 1/C_2 + 1/C_3$

RC circuits
RC constants

Exponential decay
$Q = Q_0 \exp(-t/RC)$

10 Charge and field

Sensitive microelectronic devices need to be protected from everyday electrostatic charges

THE SPARK THAT jumps across a gap of 5 cm or so in a Van de Graaff generator is the dramatic effect of static electricity. Less conspicuous, but the cause of many problems in microelectronic circuits, are the tiny quantities of charge that build up and jump only a few micrometers.

Many items of equipment that we rely on every day are controlled by microelectronic devices – just think of telephones, radio and television, traffic lights, domestic appliances such as washing machines and freezers, control systems in all forms of transport systems, railway signalling and aircraft control. On top of this there is all the computer equipment in use. It is very important that these devices are not made unreliable by electrostatic effects.

The amount of charge that can damage circuits is very small – fractions of a microcoulomb are enough. The relatively small voltages that do damage are easily reached with everyday activities. Shifting position on a foam-filled chair can generate 1500 volts in a humid atmosphere and up to 18 000 volts if it is very dry. Just walking on some carpets can generate up to 35 000 volts: that's why you may feel a shock when you touch a metal door handle!

Voltages of this size would certainly destroy sensitive microelectronic circuits. So, built into the design of equipment are means to protect their delicate microelectronics from static electricity. It is no wonder that microelectronics engineers earth themselves before going near a circuit board!

The ideas in this chapter

Most objects in the world around us consist of matter in which charge is balanced, and to the surroundings such matter is electrically neutral. However, the forces between the charges within materials are responsible for the strength of the materials and all their mechanical and electrical properties.

Charged bodies exert forces on each other: some are pulled towards each other and some are pushed away from each other. This is because there are two types of charge – we call them positive and negative. Their action can be summarised as:

Like charges repel, unlike charges attract.

In this chapter we consider these charges in more detail. Of particular importance are the **electric fields** that charge produces, and the movement of charge from one place to another. We then go on to consider **capacitors** and finally we show the similarity between the equations for an electric field and the equations for a gravitational field.

1 ELECTROSTATIC FORCES

Atoms are made from tiny negatively charged particles called electrons, and larger positively charged particles called protons. Each carries the same (yet opposite) quantity of charge. The charges are responsible for the electrostatic behaviour of materials. Atoms (except hydrogen) also contain neutral particles called neutrons. When two materials rub together, a few atoms of one material lose electrons, becoming positively charged. These electrons 'stick' to atoms on the other material, giving it a net negative charge. (In a television, a beam of electrons strikes the screen. You are aware that negative charge has built up if you touch the screen and hear a crack.)

A very important property of electrostatic forces is that they act at a distance. Hair or fur can be raised if a charged object (such as a comb) is brought near. Any charged body in the space around another charged body is acted on by an **electric field**, and will 'feel' a force. (Compare with gravitational fields and magnetic fields – see Chapters 4 and 11.) The direction of the force depends on whether the charges are alike or unlike. The size of the force depends on the size of the charges and their distance apart.

The behaviour of electric fields is very similar to that of gravitational fields. It is easiest to begin with **uniform** fields.

2 UNIFORM ELECTRIC FIELD

The field between two parallel charged plates is uniform. When a charged body such as charged foil on an insulated handle (Fig 10.1) is held in the space between such plates, the foil is deflected, since a force acts on the foil to make it move. The amount of the foil's deflection is a measure of the field strength, and is the same everywhere, because the force is the same at any point between the plates: the field is uniform.

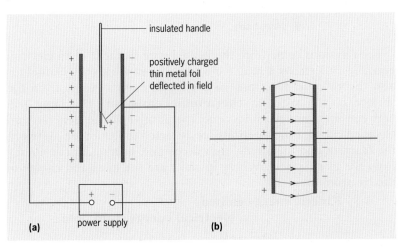

(a)

power supply

(b)

Fig 10.1 (a) **Investigating the field between two parallel plates.** (b) **The electric field pattern between the plates. In the centre the even spacing of the field lines indicates a uniform field. Near the edges of the plate the field is no longer uniform**

The strength of the field is defined as the force on each coulomb of charge:

Electric field strength = force in newtons on a charge of one coulomb.

Electric field strength is measured in newtons per coulomb (N C^{-1}). ▓ See questions 1 and 2.

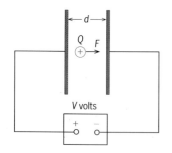

Fig 10.2 **Calculating the electric field strength between two parallel plates**

EXAMPLE

Q Estimate the electric field between the 'points' of a spark plug.

A First, estimate the distance between the points. This is usually very small, say 0.5 mm. The voltage across the points will be about 3000 V. So, electric field strength:

$$= \frac{V}{d} = \frac{3000 \text{ volts}}{0.5 \text{ mm}}$$

$$= \frac{3000}{5 \times 10^{-4}}$$

$$= 6 \times 10^6 \text{ V m}^{-1}$$

See questions 3, 4 and 5. ■

?

A The beginning of this chapter described the problems static electricity can cause in microelectronic circuits. Within such a circuit, the insulating property of the material (usually a layer of glass) between the layers of conducting track breaks down in an electric field of 10^6 V m^{-1}. If the insulation is 10 μm thick, what size voltage will cause this field?

B A volt is 1 joule per coulomb. Show that V m^{-1} is equivalent to N C^{-1}.

C Calculate the energy gained by an electron in (a) joules and (b) eV, if it has been accelerated through a p.d. of (i) 1 V, (ii) 1000 V, (iii) 1 000 000 V. Repeat the calculations for an alpha particle, which has a charge of +2e.

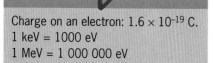

Charge on an electron: 1.6×10^{-19} C.
1 keV = 1000 eV
1 MeV = 1 000 000 eV

If a charged body is free to move, then it will move in the field – the field does work on the body and the charged body gains energy as it accelerates between the plates.

When the separation of the plates is d metres, then the work done by a force of F newtons in moving the charge from one plate to the other will be:

$$F \times d \text{ (joules)}$$

The work done by the field is equal to the energy gained by the charge. If the potential difference (p.d.) across the plates is V volts, then the energy gained by Q coulombs will be:

$$Q \text{ coulombs} \times V \text{ joules per coulomb} = QV \text{ joules}$$

Therefore: $$Fd = QV$$

Rearranging this gives: $$\frac{f}{Q} = \frac{V}{d}$$

We now have an alternative and very useful way of measuring the strength E of an electric field:

E is measured in newtons per coulomb = volts per metre

The electron gun and electronvolts

Television tubes, cathode-ray oscilloscope tubes, X-ray tubes and electron microscopes are all devices that use beams of electrons.

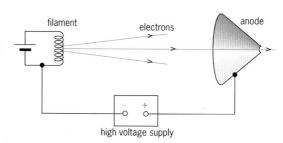

Fig 10.3 **An electron gun. Electrons are emitted from the heated filament and attracted to the positive anode**

The electrons come from the heated wire filament of an electron gun. Fig 10.3 shows a simple electron gun in which the filament is also the cathode. Electrons with enough energy escape from the surface of the wire, in **thermionic emission**. The large p.d. between the filament and a cone, the positive anode, generates an electric field, and the electrons are attracted to the anode. Between the filament and the anode, they accelerate, gaining kinetic energy from the electric field. Where m and e are the mass and charge of the electron, then:

Kinetic energy gained
= electrical energy from field.

$$\tfrac{1}{2}mv^2 = eV$$

where v is the speed of the electron after it has been accelerated through a p.d. of V volts. The energy gained by the electron is eV joules or V electronvolts. One electronvolt is the energy gained by a particle carrying a charge of 1.6×10^{-19} coulombs after it has been accelerated through a p.d. of one volt. The electronvolt is a more convenient unit of energy than the joule when we are dealing with the energies of particles on an atomic scale.

EXAMPLE

Q Calculate the velocity of a 5 MeV alpha particle, which has a mass of 6.646×10^{-27} kg.

A The energy of the alpha particle is all kinetic energy. Therefore:

$$\tfrac{1}{2}mv^2 = 5\,\text{MeV}$$

To carry out this calculation the energy must be in joules:

$$5\,\text{MeV} = (1.6 \times 10^{-19}) \times (5 \times 10^6)$$

$$= 8.0 \times 10^{-13}\,\text{J}$$

We can now write:

$$0.5 \times (6.646 \times 10^{-27}) \times v^2 = 8.0 \times 10^{-13}$$

$$v^2 = \frac{8.0 \times 10^{-13}}{0.5 \times (6.646 \times 10^{-27})}$$

$$= \frac{8.0}{0.5 \times 6.646} \times 10^{14} = 2.4 \times 10^{14}$$

Therefore: $\qquad v = \sqrt{(2.4 \times 10^{14})} = 1.6 \times 10^7\,\text{m s}^{-1}$

See questions 6, 7 and 8.

The charge on the electron – Millikan's experiment

In 1917, Robert Millikan, an American physicist, devised a clever way of measuring the charge on the electron. He sprayed tiny oil droplets between two parallel charged plates and watched their motion carefully through a microscope (Fig 10.4).

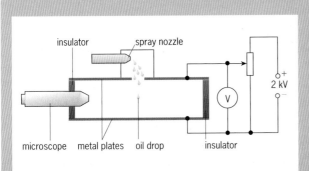

Fig 10.4 **Apparatus for Millikan's oil-drop experiment**

We find with Millikan's apparatus that many of the oil droplets become charged as they are forced through the tiny hole of the spray. Most of the charged droplets are negative and these experience a force in the electric field between the plates. All the oil droplets are acted upon by gravity and the buoyancy force of the air.

By carefully adjusting the voltage across the plates, Millikan was able to make some of the charged droplets stand still. Then, he knew that the resultant force on such a droplet was zero. The weight of the droplet and the resistance due to air molecules was balanced by the electric force on the droplet.

According to Stokes' law (see page 189), the force F acting on a spherical oil droplet of radius r which is moving with velocity v through a medium of viscosity η is given by:

$$F = 6\pi r \eta v$$

For oil density ρ, the mass of the drop will be:

$$\text{mass} = \tfrac{4}{3}\pi r^3 \rho$$

and its weight: $\qquad \tfrac{4}{3}\pi r^3 \rho g$

There is an upthrust acting on the droplet equal to the weight of the air displaced:

$$\text{upthrust} = \tfrac{4}{3}\pi r^3 \rho_A g$$

where ρ_A is the density of air.

The apparent weight W of the oil droplet is therefore given by:

$$W = \tfrac{4}{3}\pi r^3 g(\rho - \rho_A)$$

Now, if the droplet is moving at a terminal velocity of v_t, then apparent weight is:

$$W = \tfrac{4}{3}\pi r^3 g(\rho - \rho_A) = 6\pi r \eta v_t$$

Electric field strength E between the plates d metres apart is given by:

$$E = \frac{V}{d}$$

If an electric force F_E is applied so that the droplet with charge Q moves at constant velocity, then:

$$F_E = QE = W = \tfrac{4}{3}\pi r^3 g(\rho - \rho_A)$$

Millikan observed many droplets, carefully measuring the voltage required to give a constant velocity for each droplet. From his measurements he was able to calculate the smallest charge on a droplet. This was 1.6×10^{-19} C. He also discovered that other charged droplets carried either the same amount of charge or a whole-number multiple of this charge. He concluded that 1.6×10^{-19} C was the smallest possible charge and that it must be the charge on one electron.

Fig 10.5 **Examples of some electric fields**

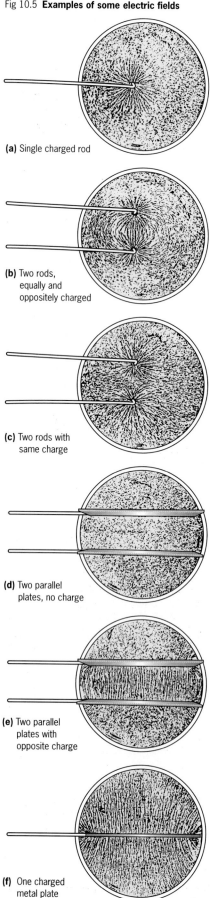

(a) Single charged rod

(b) Two rods, equally and oppositely charged

(c) Two rods with same charge

(d) Two parallel plates, no charge

(e) Two parallel plates with opposite charge

(f) One charged metal plate

Direction of the electric field

A charged foil between two parallel plates is deflected in the same direction wherever it is placed between the plates – that is, the electric field has a direction. If the p.d. across the plates were reversed, the direction of the field would also be reversed and the foil would be deflected in the opposite direction.

The direction of the electric field between the plates is the direction of the force on a positive charge, that is, from positive to negative. The direction of any electric field is the same; the idea can even be used to describe the field around an isolated charge.

We use lines to describe the field, as indicated by small particles in Fig 10.5 and drawn as in Fig 10.6. A field line represents the orientation of the force and the arrows shows the direction of the force (positive to negative). The strength of the field is shown by the spacing of the lines – the closer they are, the stronger the field. Between the parallel plates they are evenly spaced, indicating a field of uniform strength.

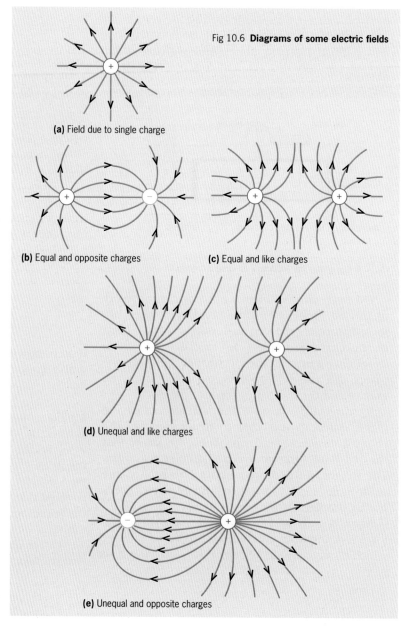

Fig 10.6 **Diagrams of some electric fields**

(a) Field due to single charge

(b) Equal and opposite charges

(c) Equal and like charges

(d) Unequal and like charges

(e) Unequal and opposite charges

3 ELECTRICAL POTENTIAL AND EQUIPOTENTIALS

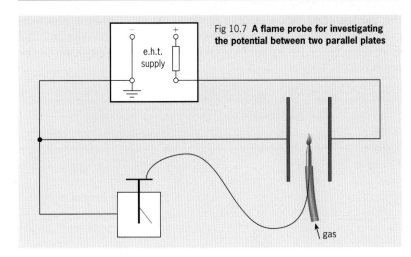

Fig 10.7 **A flame probe for investigating the potential between two parallel plates**

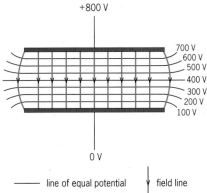

Fig 10.8 **Equipotentials and field lines between two parallel plates. The field lines always cut the equipotentials at right angles**

A voltmeter measures the p.d. across the parallel plates. A flame probe is used to measure voltage at points in the space between the plates (Fig 10.7): the voltage increases as the probe is moved from the left plate, which is at 0 V, to the right plate. More precisely, the probe measures a voltage *in the field between the plates*. This voltage is called the **electric potential**. The potential can be defined as the p.d. between the plate at 0 V (or zero potential) and the probe.

When the probe is moved parallel to the plates the potential remains the same: the probe is moving along an **equipotential**, that is, a plane of equal voltage or potential. The equipotentials are always at right angles to the field lines (Fig 10.8). Notice that at the edges of the plates where the field is no longer uniform, the equipotentials are still perpendicular to the field lines.

The potential changes gradually in a uniform field. Fig 10.9 is a graph of potential against distance between the plates. The plot shows that the potential gradient is constant and negative. Thus, where *r* is the distance, the field strength (the force on unit charge) is given by the equation:

Field strength = −(potential gradient)

$$E = -\frac{\mathrm{d}V}{\mathrm{d}r}$$

Although this relationship has been derived for a uniform field, we shall see later that it is true for all fields.

▦ See question 9.

?

D Copy the four diagrams of the electrodes and sketch the electric fields between them. Show the direction of the fields. Add equipotentials to your diagrams.

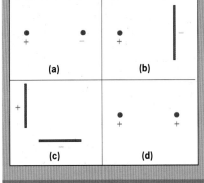

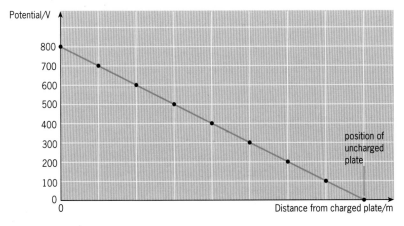

Fig 10.9 **Potential between two charged parallel plates plotted against distance from positive plate**

Non-uniform fields

In non-uniform fields the potential gradient varies. Fig 10.10(a) shows the electric field below a thundercloud. Notice how the potential gradient increases around the pointed church spire. The pointed spire distorts the field pattern, making the field stronger around the point.

Fig 10.10(a) **Equipotentials below a typical thunder cloud. Notice how the equipotentials are closer around the pointed spire of the church**

ground

Fig 10.10(b) **Rapid discharge of a thundercloud**

?

E On a dry day, over flat country, measurements taken above the ground show that the electric potential can increase by about 100 volts per metre.

(a) What is the p.d. between the ground and a distance 1.8 m above it?

(b) A person is 1.8 m tall. Why isn't there such a voltage between that person's nose and feet?

(c) Sketch the equipotentials around the person standing in the open.

▓ See the Assignment for more on the charge distribution in a thundercloud, and the propagation of lightning.

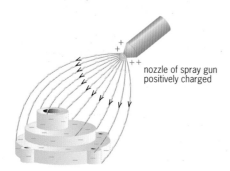

nozzle of spray gun positively charged

Fig 10.11 **Electrostatic paint spraying: the charged droplets follow the field lines to the object to be painted, which is earthed**

This property is used in a lightning conductor, which is basically a pointed metal rod. One end is earthed and the pointed end extends above the top of the building to which it is attached. A thundercloud often has a concentration of negative charge at its base. This generates an electric field between the base and the ground. Below the thundercloud, the field at the point of the lightning conductor is strong enough to ionise air molecules around the point.

The resulting charged molecules – ions – move in the field: positive ions go towards the cloud base and electrons go to earth through the lightning conductor. This mechanism allows the thundercloud to discharge slowly before enough charge builds up to cause lighting.

Electrostatic paint spraying

Particles of paint are given a positive charge as they leave the nozzle of a spray gun (Fig 10.11). The object to be painted is earthed so that there is an electric field between the nozzle and the object. The charged paint droplets follow the field lines and are deposited evenly over the surface of the object.

4 PARALLEL PLATE CAPACITOR

So far, we have been looking at the electric field between two parallel plates. This arrangement can in fact *store electric charge*, which makes it a **capacitor**. We have seen that the field between the capacitor plates depends upon the p.d. across them and the distance between them. Let us now look at other factors that affect the strength of a field.

There is a p.d. across the plates when they are charged (see page 217). In fact, the charge on the plates is proportional to the p.d. ($Q = CV$). The quantity of charge on the plates also depends on the area of the plates. If the area of the plates is doubled, the quantity of charge it can store is doubled. So:

$$Q \propto A$$

The ratio of charge to area (Q/A) is called the charge density. Since the charge density is proportional to p.d., it is proportional to the field strength:

$$\text{charge density} \propto \text{field strength}$$

$$\frac{Q}{A} \propto \frac{V}{d}$$

There is one other factor that affects the field strength. This is the medium between the plates, the **dielectric**. The dielectric is an insulator, the simplest being a vacuum. Often, the insulating material is air, oil or paper. For a vacuum between the plates:

$$\frac{Q}{A} = \varepsilon_0 \frac{V}{d}$$

where ε_0 is a constant called the **permittivity of free space**, which links charge density and field strength. Its units are farads per metre and it tells us how good free space (a vacuum) is at allowing an electric field to be established.

The value of ε_0 is 8.85×10^{-12} F m^{-1}, or alternatively 8.85×10^{-12} C^2 N^{-1} m^{-2}.

Another way of arranging the equation gives expressions for **capacitance**, symbol C:

$$C = \frac{Q}{V}$$

$$= \varepsilon_0 \frac{A}{d}$$

The second line of the equation gives the capacitance in terms of the physical dimensions of the dielectric medium and the capacitor.

In practice, different dielectric materials are used for capacitors, depending on their size (see Table 10.2 on the next page).

The equation for the capacitance is slightly modified to account for the dielectric:

$$C = \varepsilon_0 \varepsilon_r \frac{A}{d}$$

where ε_r is called the **relative permittivity** of the dielectric used. Table 10.1 gives values of relative permittivity for some common materials.

The plates of a capacitor are covered with a 'blanket' of charge. The charge spreads out evenly over most of the plates – only at the edges is the distribution more uneven with more charge concentrated at any 'sharp' edges and corners.

✔

The farad, symbol F, is the unit of capacitance: a 1 farad capacitor charged by a p.d. of 1 volt carries a charge of 1 coulomb. In practice, capacitors have much smaller values, ranging between microfarads and picofarads.

See questions 10 and 11.

?

F (a) An air-spaced capacitor of 1 µF is needed. Estimate the size of the plates if they are 1 mm apart.

(b) Mica has a relative permittivity of 6. What would be the size of the plates with mica as the dielectric?

Table 10.1 **Relative permittivity for some common materials**

Substance	Relative permittivity
air	1
paper	between 2 and 3
water	about 80

PRACTICAL CAPACITORS

THERE ARE MANY different types of capacitor available, made from different materials and suitable for different applications. They come in a range of shapes and sizes, in values from picofarads up to farads. Capacitors are used to store charge (and energy), to 'smooth' a fluctuating voltage (see page 288) or to allow high frequency signals to 'bypass' parts of a circuit. They are essential in circuits designed to respond to certain frequencies.

Examples are the 'tuned circuit' (see page 293) in radio receivers used to select a signal of a particular frequency, and 'filters' which use the fact that the behaviour of a capacitor in a.c. circuits depends on frequency. With accurate capacitors, filters will respond to very narrow bands of signal frequencies which can be either filtered out or allowed to pass.

Sometimes very accurate capacitors are needed, usually of low value, typically pF or nF. Capacitors of very high value are used, such as in 'smoothing', where accuracy is not important.

All capacitors lose charge over time, that is, they 'leak'! Most capacitors have a fixed value, but others vary over a range of values, as in tuned circuits. Table 10.2 shows some of the common types of capacitor that are used.

Fig 10.12

Upper: small value ceramic capacitors

Centre: aluminium electrolytic capacitors

Lower: variable capacitor

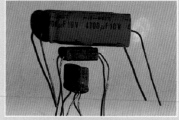

Table 10.2

Type	Capacitance range	Maximum voltage	Accuracy	Leakage	Comments
mica	1 pF–10nF	100–600	good	good	very useful at radio frequencies
ceramic	10 pF–1 μF	50–30 000	poor	fair	cheap, small
polystyrene	10 pF–2.7 μF	100–600	excellent	excellent	high quality, used in accurate filters
polycarbonate	100 pF–30 μF	50–800	excellent	good	high quality, small
tantalum	100 nF–500 μF	6–100	poor	poor	high capacitance, polarised
electrolytic (aluminium)	100 nF–2 F	3–600	horrible	awful	smoothing power supply filters, polarised

Effect of the dielectric

Placing an insulating material between the plates of a capacitor increases the capacitance. That means that more charge can be stored for the same voltage across the plates, and this effect is greater when materials with greater values of ε_r are used. For more about dielectrics, see Chapter 20.

?

G Two spheres of equal size and mass carry identical charges. Explain what would happen to the repulsive force between them if:

(a) the charge on one of them is doubled,

(b) the charge on both of them is doubled,

(c) the charge on both of them is doubled and the distance between them is doubled.

H In a hydrogen atom the electron and proton are about 10^{-10} m apart. Calculate the attractive force between them.

5 NON-UNIFORM ELECTRIC FIELD – COULOMB'S LAW

In 1785, Coulomb measured the forces between small charged bodies and summarised his results in the form of an equation that is now known as Coulomb's law. He discovered that force F depends on the size of the charges Q_1 and Q_2 on the bodies, and on the distance r between them. As the distance increases the force decreases – in fact, if the distance doubles the force decreases by a factor of four. That is, the force obeys an inverse square law:

$$F = k\frac{Q_1Q_2}{r^2}$$

where k is a constant.

This equation is almost identical in form to Newton's law of gravitation:

$$F = -G\frac{m_1 m_2}{r^2}$$

Unlike gravity, which is always attractive, the electric force can be either attractive or repulsive. See Fig 10.13: when the charges are opposite, the force is attractive and defined as negative, like gravity; when the charges are of the same sign, the force is repulsive and defined as positive.

Fig 10.14 shows the field around an isolated charge. Fig 10.14(a) is a two-dimensional representation, but the field around the charge is three-dimensional. Fig 10.14(b) attempts to show this.

See questions 12 and 13.

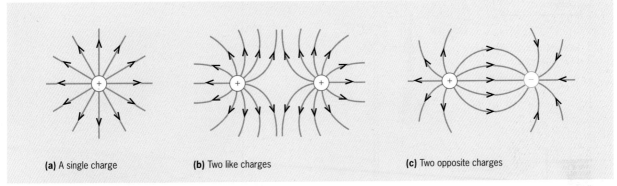

| (a) A single charge | (b) Two like charges | (c) Two opposite charges |

Fig 10.13 **The electric fields around point charges**

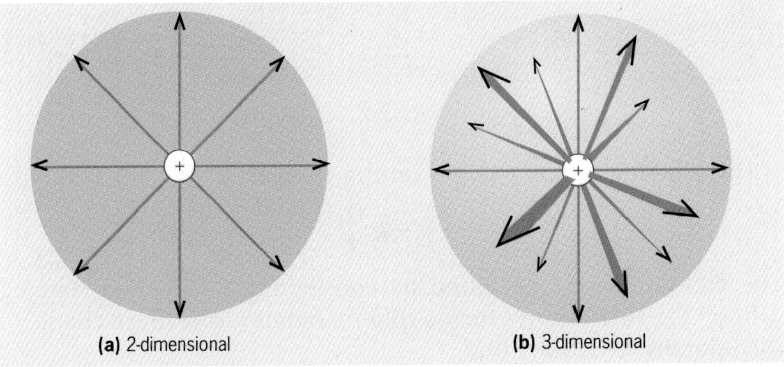

| (a) 2-dimensional | (b) 3-dimensional |

Fig 10.14 **The field around an isolated charge**

The field lines spread out radially from the centre. Yet again, this 'map' of field lines helps us visualise the field, but it is limited. It is very difficult to show on such a 'map' the precise way that the magnitude of the field changes. The equations below help solve this problem.

Field strength

The strength E of the electric field can be measured by considering the force on a second 'test' charge, Q_2, at a distance r from a charge Q_1:

field strength = force per unit charge

$$= \left(\frac{kQ_1 Q_2}{r^2}\right) Q_2$$

$$E = \frac{kQ_1}{r^2}$$

That is, the field strength in a non-uniform electric field also obeys the inverse square law.

EXAMPLE

Q The electric field strength 50 cm from the centre of a small charged sphere is 18 N C^{-1}. Calculate the charge on the sphere. ($k = 9 \times 10^9$.)

A The field strength is:

$$E = k\frac{Q}{r_2} = 18\,\text{N}\,\text{C}^{-1}$$

Rearranging:

$$Q = \frac{Er^2}{k}$$

$$= \frac{18 \times (5 \times 10^{-1})^2}{9 \times 10^9}$$

$$= \frac{18 \times 25 \times 10^{-2}}{9 \times 10^9} = 50 \times 10^{-11}$$

$$= 5 \times 10^{-10}\,\text{C}$$

Potential in a radial field

The flame probe used to investigate the potential between parallel plates was described earlier (see Fig 10.7). As shown in Fig 10.15, it can also be used to investigate the way electrical potential varies around an isolated charged body, in this case a sphere. The potential decreases, but it follows a $1/r$ law.

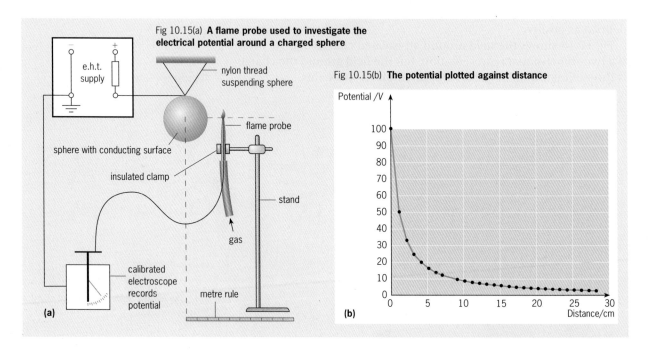

Fig 10.15(a) **A flame probe used to investigate the electrical potential around a charged sphere**

Fig 10.15(b) **The potential plotted against distance**

The potential again depends on the charge:

$$V = k \frac{Q_1}{r}$$

The constant k can be found by considering an isolated charged sphere. Fig 10.16 shows such a sphere, radius r, carrying a charge of Q coulombs. We know that:

$$\text{charge density} = \varepsilon_0 \times \text{field strength}$$

$$\frac{Q}{A} = \varepsilon_0 \frac{kQ}{r^2}$$

where $A = 4\pi r^2$, the area of the sphere. Therefore we have:

$$\frac{Q}{4\pi r^2} = \varepsilon_0 \frac{kQ}{r^2}$$

The charge and radius cancel, leaving us with:

$$k = \frac{1}{4\pi\varepsilon_0}$$

Q coulombs

Fig 10.16 **An isolated charged sphere**

See questions 14 and 15. ■

I Fig 10.16(a) shows how potential varies around a charged sphere. Sketch a graph to show how potential varies **(a)** close to a flat surface, and **(b)** far away from that surface.

Relationship between field and potential

Earlier in this chapter on page 235, we found that the field strength E was equal to minus the potential gradient:

$$E = -\frac{dV}{dr}$$

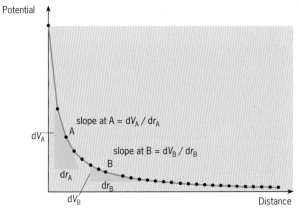

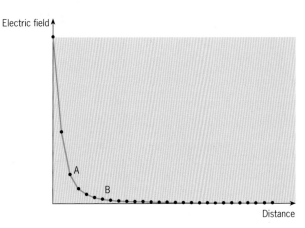

(a) Potential plotted against distance

(b) Electric field plotted against distance

The same is still true with fields that vary by an inverse square law. We can show this mathematically:

$$V = \frac{1}{4\pi\varepsilon_0} \frac{Q}{r}$$

$$E = -\frac{dV}{dr}$$

$$= -\frac{Q}{4\pi\varepsilon_0} \frac{d(1/r)}{dr}$$

Therefore:
$$E = \frac{1}{4\pi\varepsilon_0} \frac{Q}{r}$$

This means that the slope of the potential–distance graph of Fig 10.17(a) gives the value of the field strength at that point.

The equations for inverse square electric fields are summarised in Table 10.3, together with the equivalent gravitational field equations.

Fig 10.17 **The relationship between potential and electric field. The gradient of the potential–distance curve gives the field strength at any point**

Table 10.3 **Comparison of electric field and gravitational field equations**

Gravitational field	Electric field
$F_G = -G\dfrac{m_1 m_2}{r^2}$	$F_E = \dfrac{1}{4\pi\varepsilon_0}\dfrac{Q_1 Q_2}{r^2}$
$E = -G\dfrac{m}{r^2}$	$E = \dfrac{1}{4\pi\varepsilon_0}\dfrac{Q}{r^2}$
$V_G = -G\dfrac{m}{r}$	$V_E = \dfrac{1}{4\pi\varepsilon_0}\dfrac{Q}{r}$

SUMMARY

By the end of this chapter you should know and understand the following:

■ The field between charged plates is uniform.

■ Electric field strength is the force per unit charge measured in newtons per coulomb.

■ The equations for uniform fields are $E = F/Q = V/d$.

■ The work done in moving a charge of Q coulombs through a p.d. of V volts is QV joules.

■ The kinetic energy of an electron emerging from an electron gun is given by $\frac{1}{2}mv^2 = eV$.

■ The electronvolt is the energy gained by a particle carrying a charge of 1.6×10^{-19} coulombs when it has been accelerated through a p.d. of 1 V.

■ The potential at a point in an electric field is the work done per coulomb in moving a unit charge from zero or earth potential to that point.

■ Equipotentials can be thought of as surfaces in space of equal potential which cut electric field lines at right angles.

■ Field strength E is equal to the potential gradient $-dV/dr$.

■ Charge density = permittivity × field strength of free space, and capacitance $C = \varepsilon_0 A/d$.

■ The force between two charges is given by Coulomb's law, $F = kQ_1 Q_2/r^2$, and field strength $E = kQ/r^2$, where $k = 1/4\pi\varepsilon_0$.

■ The potential around a charge Q is given by $V = kQ/r$.

QUESTIONS

1 Calculate the field strength in the following cases:

a) charge of 2 C, force of 500 N;

b) charge of 2 mC, force of 5 N;

c) charge of 5 μC, force of 20 mN.

2 Calculate the force on the following charges in a field of strength 50 N C^{-1}. (1 nC = 10^{-9} C; 1 pC = 10^{-12} C.)

a) 0.2 C; **b)** 2 mC; **c)** 5 μC; **d)** 2 nC; **e)** 5 pC.

3 Calculate the field strength for the following plate arrangements:

a) two plates 2 m apart with a p.d. of 10 000 V across them;

b) two plates 2 cm apart with a p.d. of 100 V across them;

c) two plates 0.2 mm apart with a p.d. of 1 V across them.

4 Two parallel plates are 2 cm apart and have a voltage of 1000 V across them. What is the force on a small, light body with a charge of 5 nC which is placed between the plates?

5 A charged polystyrene sphere of mass 2 g is suspended by a fine nylon thread between two plates 5 cm apart (Fig 10.Q5). When there is a p.d. of 2000 V across the plates, the thread holding the sphere makes an angle of 30° with the vertical. Calculate the charge on the sphere. (*g* = 9.8 N kg^{-1}.)

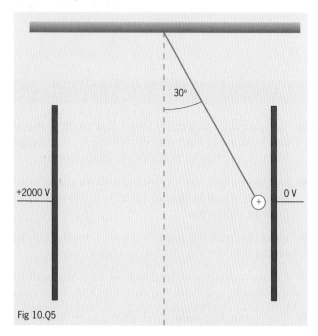

Fig 10.Q5

6 Calculate the velocity of a 5 MeV alpha particle. (Mass of an alpha particle is 6.646 × 10^{-27} kg; charge on an alpha particle is 3.2 × 10^{-19} C.)

7 A beam of electrons produced by an electron gun is displaced a vertical distance *s* metres on passing through the electric field between two parallel charged plates of length *l* metres in a vacuum, as shown in Fig 10.Q7.

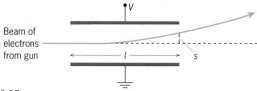

Fig 10.Q7

The potential difference *V* between the deflecting plates is doubled. State, giving your reasoning, precisely what effect, if any, this has on:

a) the force on an electron between the plates;

b) the time an electron takes to travel the distance *l*;

c) the vertical displacement *s*.

[NSP 209: Short Answer Paper 1988]

8

a) A student is given six sealed opaque bags containing different numbers of identical ball bearings. She measures the mass of each bag on a top-pan balance. The readings are as follows:

30.8 g, 92.4 g, 61.6 g, 38.5 g, 77.0 g and 53.9 g

Calculate the mass of a ball bearing and the number of balls in each bag.

b) In a version of Millikan's experiment, charged polystyrene spheres were balanced between two horizontal parallel plates. All the spheres had the same mass but had different charges due to different numbers of extra electrons on each sphere. The following voltages were recorded for six different spheres:

400 V, 560 V, 240 V, 880 V, 720 V and 640 V

What was the smallest possible number of extra electrons on each sphere?

9 Plot a graph of potential against distance using the information in Fig 10.Q9.

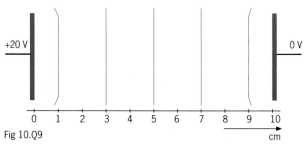

Fig 10.Q9

a) From your graph, calculate the potential gradient.

b) What is the field strength? How does this compare with the potential gradient?

10

a) Calculate the area of a pair of plates required to make a 1 nF air-spaced capacitor if the plates are 0.1 mm apart. If the plates are made from 2 cm wide aluminium foil, how long will the strips be? (Permittivity of free space $\varepsilon_0 = 8.85 \times 10^{-12}$ C^2 N^{-1} m^{-2}.)

b) The capacitor is charged from a 10 V supply. Calculate the charge on the capacitor.

c) This charge is measured using a coulomb-meter which has an internal capacitance of 1 μF. Comment on the accuracy with which this can measure the charge on the 1 nF capacitor.

d) The gap between the plates is filled with paper which has a relative permittivity of 2. How does this affect the length of the strips?

e) In practice, the long strips are rolled into a cylinder, and a second layer of paper is placed on top of the top plate to prevent the plates from touching. It is found when this is done that the measured capacitance is higher than the calculated value. Explain how the capacitance is increased.

11
A student hopes to measure the force between two capacitor plates. He places one plate on the pan of a top-pan balance and clamps the other plate above it using an insulating support, as shown in Fig 10.Q11.

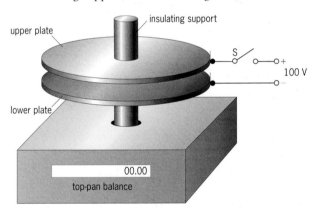

Fig 10.Q11

The plates, each of area 3.0×10^{-2} m^2, are initially 10 mm apart. The lower plate is earthed. A 100 V d.c. source is connected for a few seconds across the plates, and is then disconnected.

a) Calculate:
   (i) the capacitance of the parallel plate capacitor, and hence
   (ii) the energy stored in the capacitor.

b) It can be shown that if the plate separation were changed by 0.10 mm, the electrical energy stored would change by 1.4 nJ.
   (i) Hence calculate the force of attraction between the plates.
   (ii) Would such a force be detectable on a balance designed to resolve mass to 10 mg? Justify your answer.

[Nuffield: Short Answer 1987, ref. NSP200]

12
Two small spheres each of mass 50 mg are suspended from a common point by nylon threads of length 0.8 m. When the two balls are given equal charge they move so that the thread holding each ball makes an angle of 5° with the vertical. Start by drawing a diagram showing all the forces acting and then calculate the charge on each ball. ($k = 9 \times 10^9$ N m^2 C^{-2}; $g = 9.8$ N kg^{-1}.)

13
Calculate the repulsive force between two protons 1 mm apart. Compare this force with the gravitational force of attraction between them. (Mass of a proton, 1.67×10^{-27} kg; charge on a proton, 1.6×10^{-19} C; $G = 6.67 \times 10^{-11}$ N m^2 kg^{-2}.)

14
The electric field strength near a very small charged body is 150 N C^{-1}. The potential at the same point is 300 V. How far from the body were these measurements made, and what is the charge on the body?

15
In an alpha particle scattering experiment, a 5 MeV alpha particle has a head-on collision with a gold nucleus. The alpha particle 'rebounds' back along the same path. Calculate the closest distance of approach of the alpha particle to the gold nucleus. (Atomic number of gold, 79; atomic number of alpha particle, 2; charge on an electron, 1.6×10^{-19} C; $k = 9 \times 10^9$ N m^2 C^{-2}.)

Assignment

LIGHTNING

The thundercloud
Study the typical thundercloud (Fig 10.A1).

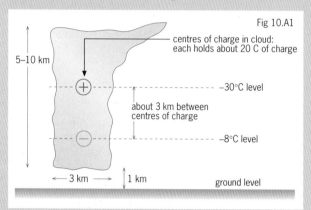

Fig 10.A1

- centres of charge in cloud: each holds about 20 C of charge
- 5–10 km
- $+$ —30°C level
- about 3 km between centres of charge
- $-$ —8°C level
- 3 km
- 1 km
- ground level

The cloud remains like this for about 30 minutes, in which it may produce a lightning flash as often as every 20 seconds. About two-thirds of all flashes are between charges in the cloud, not between the cloud and earth. As each flash will destroy most of the 20 C of charge, the cloud must be a continual generator of static electricity.

a) Why would you expect there to be *two* regions of charge within a cloud, one positive and one negative?

b) Observations suggest that there are strong vertical winds within a thundercloud. Suggest what causes these winds.

c) Show that the average current delivered by each lightning stroke to earth is about a third of an ampere.

The lightning flash
An electrical discharge will occur in dry air at atmospheric pressure in electric fields of above 3×10^6 V m^{-1}. The average field under a thundercloud is only about 3×10^4 V m^{-1}. This suggests that other factors play a part in developing a lightning stroke. The electric field may be higher in small regions in the cloud. Water droplets in it become polarised in the field; polarisation deforms their shape and they become elongated. Larger droplets will deform more easily – if droplets of several millimetres in diameter are present, electrical breakdown can happen with fields of 5×10^5 V m^{-1}. Once started, the lightning will continue through the cloud, even though the fields below may be weaker.

a) A region of charge of 20 C is spread over the base of the cloud of diameter 3 km. Estimate the electric field between the cloud and the ground.

b) With the cloud base 1 km above the ground, estimate **(i)** the p.d. between the cloud and ground, and **(ii)** the energy dissipated in a single stroke.

c) Why will a polarised water droplet allow a lightning stroke to start when the field round it is less than 3×10^6 V m^{-1}?

The lightning stroke is quite complex: study Fig 10.A2. The discharge starts with a **stepped leader** in which negative charge moves in steps of 10 to 200 metres.

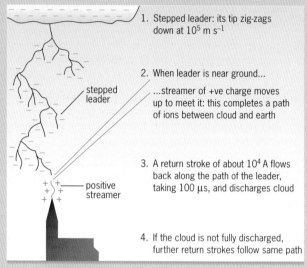

1. Stepped leader: its tip zig-zags down at 10^5 m s^{-1}

2. When leader is near ground...
   ...streamer of +ve charge moves up to meet it: this completes a path of ions between cloud and earth

stepped leader

3. A return stroke of about 10^4 A flows back along the path of the leader, taking 100 µs, and discharges cloud

positive streamer

4. If the cloud is not fully discharged, further return strokes follow same path

Fig 10.A2 **A lightning stroke**

Point discharge currents
Since Benjamin Franklin invented the lightning conductor, we have known that a well designed lightning conductor is very effective, but we have only recently begun to understand how it works.

If the electric field is strong enough, then an exposed earthed metal rod will pass a current to the atmosphere. This is called the **point discharge current** and provides one way for a conductor to discharge a cloud. Point discharge occurs when the lines of force near a point are sufficiently concentrated to accelerate electrons to a speed at which they ionise neutral air molecules by collision.

a) A nitrogen atom needs about 15 eV of energy to ionise it. How fast does an electron need to be moving to provide this much energy in a collision?

b) Estimate the acceleration of an electron in a field of 3×10^6 V m^{-1}. Assuming the electron does not collide with another particle, how far would it travel from rest before it gains 15 eV of energy?

c) Why will such ionisation occur at the point of a conductor when the average field below a cloud is far less than 3×10^6 V m^{-1}?

(Permittivity of free space, $\varepsilon_0 = 9 \times 10^{12}$ C^2 N^{-1} m^{-2}; charge on an electron, 1.6×10^{-19} C; mass of an electron, 9.1×10^{-31} kg.)

CHARGE AND FIELD

This Chapter Map brings together the principles underlying charge and the electric field it generates, indicating how the ideas are linked. You can use the map to cross-match with the needs of your syllabus. The map should also help you to identify what you are confident about and the areas you may need to study further.

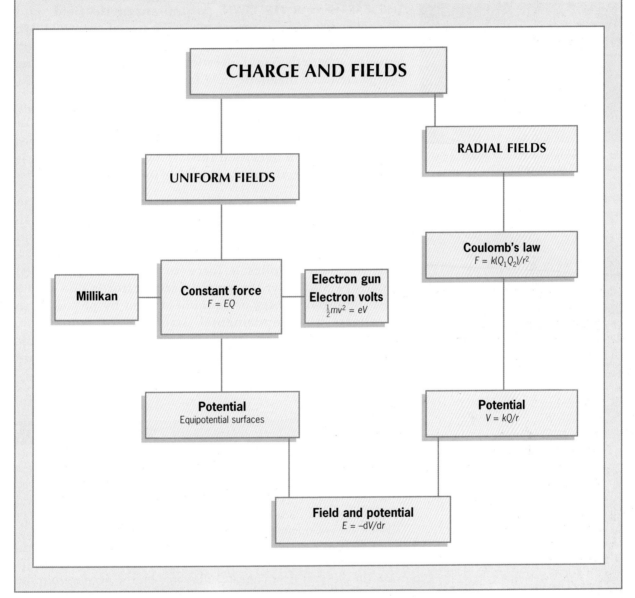

CHARGE AND FIELDS

UNIFORM FIELDS

RADIAL FIELDS

Millikan

Constant force
$F = EQ$

Electron gun
Electron volts
$\frac{1}{2}mv^2 = eV$

Coulomb's law
$F = k(Q_1 Q_2)/r^2$

Potential
Equipotential surfaces

Potential
$V = kQ/r$

Field and potential
$E = -dV/dr$

11 Electromagnetism

The Fermi particle accelerator, Illinois, a ring tunnel of 6.3 km circumference.

Particles travel at very high energies between magnets contained in the red and blue casings

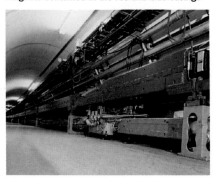

THE EARTH HAS ITS OWN natural magnetic field, and whenever we use a compass we make use of this field. We can also create artificial magnetic fields, by using the link between magnetism and electricity, which is that a moving charge creates a magnetic field (and vice versa).

We use artificial magnetic fields in a wide range of everyday applications – in the starter motor on a car, for example, or for controlling the direction of the electron beam that gives us the picture in a television tube. In fact, wherever there is an electric motor, there is a magnetic field.

In the world of science, artificial magnetic fields play an essential role. A mass spectrometer, for example, is used to analyse chemical compounds by ionising the molecules and sorting the charged bits according to mass. To sort them, a magnetic field applies a force to the moving charged fragments. Similarly, the particle accelerator is one of the most important pieces of equipment used by physicists to investigate the fundamental structure of sub-atomic particles. In it, atoms are smashed and the sub-atomic particles are given immense speeds by forces that come from magnetic fields.

1 FIELDS AROUND CURRENTS

If we pass an electrical current along a wire and then bring a compass near to it, the compass needle (which is a strong magnet) will be deflected (Fig 11.1). This is because the current has a magnetic field around it, as first demonstrated by Hans Christian Oersted in 1820.

What do we mean by a magnetic field? The compass needle is deflected because it experiences a force. With a constant current, the size of the force depends only on how far the needle is from the wire: the closer it is, the greater the force on the needle – but the needle does not have to touch the wire. The effect of the magnetic field, then, acts at a distance. So we describe the space around the wire as containing a magnetic field. (See Chapter 3, page 17, for more about the idea of a field.)

A magnetic field can be represented by 'field lines' that show the shape of the field (see also Chapter 10). We draw the lines close together if we want to show a strong field, and further apart for a weaker field.

The field has a **direction of action**. We define the direction of the field at a point as the direction of the force that would act on an isolated north pole placed there. (So far no-one has ever found such a thing as an isolated north pole – magnetic poles always occur in pairs – but it is a useful idea to help define field directions.) Magnetic field is a vector quantity.

Fig 11.1 **The magnetic field around a wire carrying a current**

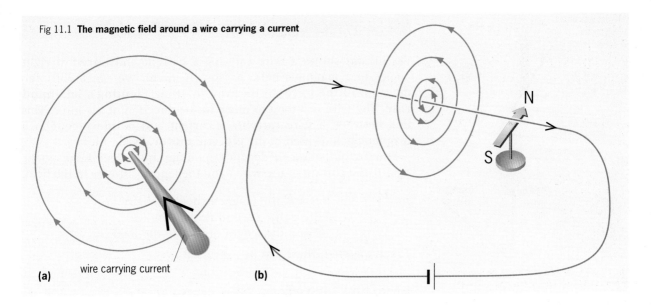

(a) wire carrying current

(b)

Field lines are always in complete loops. The direction of the magnetic field is from north to south, as shown in Fig 11.2(a). The Earth has its own magnetic field, which has been used for hundreds of years in navigation – see Fig 11.2(b).

See question 1.

Fig 11.2 (a) **The magnetic field around a bar magnet.** (b) **The Earth's magnetic field behaves as if there were a large bar magnet inside the Earth. This 'magnet' has its south pole towards the Earth's magnetic north: when a compass points north, its north pole is attracted towards the Earth's 'magnetic north' which is in fact a south pole! To check this, note the direction of arrows on the magnetic field lines in** (a) **and** (b)

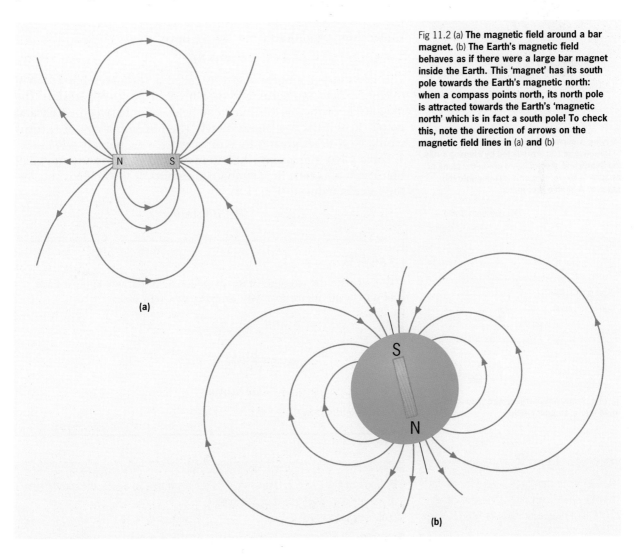

(a)

(b)

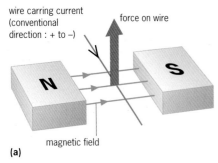

wire carring current
(conventional
direction : + to –)

force on wire

N

S

magnetic field

(a)

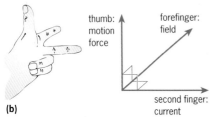

thumb:
motion
force

forefinger:
field

second finger:
current

(b)

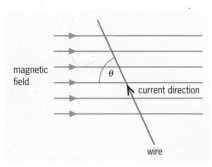

wire coil   rotation

N   b   c   **S**

a   d

brush

–   +

brush   commutator

(c)

Fig 11.3 **Fleming's left-hand rule. (a) A wire
carrying a current in a magnetic field.
(b) Direction of force predicted by Fleming's rule.
The thumb and second figure are in the plane of
the page – the forefinger is pointing into the
page. (c) A simple d.c. motor**

See questions 2 and 3. ▨

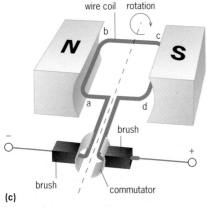

magnetic
field

θ

current direction

wire

Fig 11.4 **A current-carrying wire placed at an
angle θ to a magnetic field**

2 FORCES ON WIRES CARRYING CURRENTS

Fig 11.3(a) shows a wire carrying a current and placed at right angles to a magnetic field. As shown in (b), we can predict the direction of the force on the wire by using **Fleming's left-hand rule**. The wire will tend to move in the direction of the force. This link between a wire carrying a current and its movement in a magnetic field is used in the **electric motor**, as Fig 11.3(c).

The size of the force F depends upon the strength of the magnetic field B, the current in the wire I and the length l of wire in the field. That is:

$$F \propto B, \text{ the magnetic field strength}$$
$$F \propto I, \text{ the current in the wire}$$
$$F \propto l, \text{ the length of wire in the field}$$

Taken together, these can be expressed as:

$$F = kBIl$$

where k is a constant.

We choose the units of B to make $k = 1$, as follows. Suppose the magnetic field strength is such that a wire 1 metre long carrying a current of 1 ampere feels a force of 1 newton. We shall define this field strength B as 1 newton per ampere metre (N A^{-1} m^{-1}). Therefore we have:

$$1 \text{ N} = k \times (\text{N A}^{-1} \text{ m}^{-1}) \times 1 \text{ A} \times 1 \text{ m}$$

Under these conditions $k = 1$. So we have:

$$\boldsymbol{F = BIl}$$

The newton per ampere metre is called a **tesla**, symbol T. Small magnetic field strengths are measured in μT (micro-tesla). The magnitude of the Earth's magnetic field varies between 24 μT and 66 μT. In the UK it is about 49 μT. The force on a coil of N turns would be N times greater than on a single wire, namely $NBIl$.

Force F is a maximum when the field and current are perpendicular (with magnetic field strength constant). If the angle is smaller, the force is reduced (Fig 11.4).

$$F = BIl \sin \theta$$

EXAMPLE

Q Calculate the strength of the magnetic field that will apply a force of 0.01 N per metre to a wire carrying a current of 5 A.

A
$$F = BIl \quad \text{or} \quad B = \frac{F}{Il}$$

$$B = \frac{0.01 \text{ N}}{5 \text{ A} \times 1 \text{ m}}$$

$$= 0.002 \text{ N (A m)}^{-1}$$

Field strength = 2 mT

?

A Calculate the force on a 100 m power cable carrying a current of 10 A in a magnetic field of 50 μT.

?

B Look at Fig 11.3(c). Use Fleming's left-hand rule to check the movement of sections a–b and c–d of the wire.
Hint: see also drawing (a).

The current balance

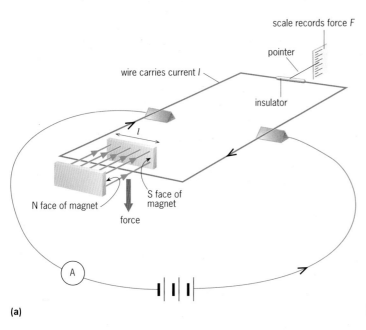

scale records force F

pointer

wire carries current I

insulator

l

N face of magnet

S face of magnet

force

A

(a)

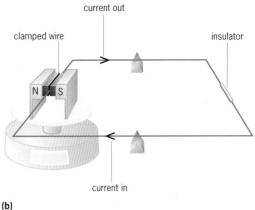

Fig 11.5(a) **A current balance for measuring the field between two magnets**

Fig 11.5(b) **A top-pan balance for measuring the field between two magnets. The force on the wire is measured as the weight shown on the balance**

current out

clamped wire

insulator

N S

current in

(b)

The current balance, shown in Fig 11.5(a), is used to measure magnetic field strength. The force on a known length of wire carrying a known current in an unknown magnetic field is measured and is used to determine B, the field strength, using the equation $F = BIl$ in the form:

$$B = \frac{F}{Il}$$

The force can be measured using a counterweight and scale as in (a), or directly using a sensitive top-pan balance as in (b).
(See the worked Example on page 248.)

■ See question 4.

?

C Examine the magnetic fields in Fig 11.6 carefully and use Fleming's left-hand rule to show that the directions of the forces on the wires are correct.

The ampere

Two wires carrying currents also exert forces on each other, as shown in Fig 11.6.

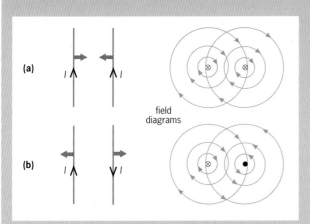

(a)

I

I

field diagrams

⊗ ⊗

(b)

I

I

⊗ ●

Fig 11.6 **Two wires carrying current exert forces on each other: (a) like currents attract, (b) unlike currents repel (direction of current in lower right wire: out of page)**

The unit of electric current, the ampere, is defined in terms of its magnetic effect, that is, the force between two wires each carrying a current of 1 A (Fig 11.7). When the wires are parallel and 1 m apart, the force between them should be 2×10^{-7} N (in theory, if they are infinitely long). This is the force used to define the ampere.

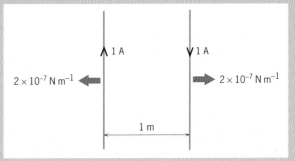

1 A

1 A

2×10^{-7} N m^{-1}

2×10^{-7} N m^{-1}

1 m

Fig 11.7 **The definition of the ampere**

3 THE MOVING COIL GALVANOMETER

The current in a coil of wire can be measured using a moving coil galvanometer (Fig 11.8). To be exact, the force on the coil of wire is measured and the current is determined from this measurement.

Fig 11.8 **A moving coil galvanometer**

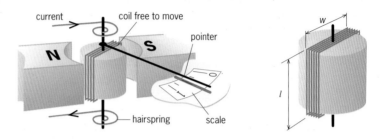

The lightweight coil is suspended so that it is free to move without friction in a uniform magnetic field. When a current flows, the coil turns against a hairspring. The spring tightens until the elastic force of the spring is equal to the magnetic force on the coil. A pointer attached to the axis moves against a scale.

The magnet has concave ends: the field is always radial to the cylinder on which the coil is mounted, so that the moving coil is always perpendicular to the magnetic field. Now consider a coil carrying a current of I amperes in a magnetic field of B tesla. If the length of the side of the coil is l metres, then the force on each side will be BIl. These two forces produce a turning force or torque on each turn of the coil.

Where its width is w metres, the torque is given by:

$$\text{torque} = BIl \times w \quad \text{(in newton metres)}$$

Since $l \times w$ is equal to the area of cross-section of the coil, A:

$$\text{torque} = BIA$$

This is for one turn of wire. The coil will have N turns. Therefore:

$$\text{total torque} = NBIA$$

The springs at either end of the coil, that carry the current to it, provide an opposing torque that prevents the coil from spinning. The torque of a spring is proportional to the angle it turns through. The coil will turn until the two torques are equal. That is:

$$\text{magnetic torque} = \text{spring torque}$$
$$NBIA = k\alpha$$

where k is a constant and α is the angle turned through by the coil. From this we can see that the angle α is proportional to the current I. This means that the scale will be linear because equal changes in current will produce equal changes in deflection on the scale.

Moving coil versus digital meters

The moving coil meter has been the workhorse of laboratories for many years. It is used in 'moving coil' multimeters. However, for many purposes it has now been replaced by digital meters that are based on integrated circuit amplifiers. When used as voltmeters, these have resistances far higher than their moving coil equivalent. Even so, moving coil meters are still better for some purposes such as monitoring the charging or discharging of a capacitor.

4 FORCES ON CHARGED PARTICLES IN BEAMS

A television picture is produced from an electron beam controlled by magnetic fields. The beam scans across the television screen at high speed. The cathode ray oscilloscope also uses an electron beam. In fact, J.J. Thomson investigated 'cathode rays' in 1897 and observed how electric and magnetic fields affected them. Thomson showed that the cathode rays were tiny negatively charged particles. It was several years before they were called 'electrons' but Thomson is credited with their discovery.

Fig 11.9 shows a beam of identical charged particles, each with a charge Q coulombs and moving with an average velocity of v metres per second. The beam is moving at right angles to a uniform magnetic field of strength B tesla. We can work out the force exerted by the field on each particle in the beam. When the length of beam in the field is l and it takes t seconds for a particle to move this distance, then:

$$l = vt$$

If the length l of the beam contains N particles, then the number of particles passing a point in the field is also N, and therefore the total charge passing through any point in the field in t seconds is NQ coulombs. Therefore the current is given by:

$$I = \frac{NQ}{t}$$

If we now treat the beam like a wire, then using $F = BIl$ we get:

$$F = BIl$$
$$= B\frac{NQ}{t}vt$$
$$= BNQv$$

This is the force on a beam of N particles. Therefore the force on a single particle is given by:

$$F = \frac{BNQv}{N}$$

So, force on one particle: $\mathbf{F = BQv}$

> An electric current in a wire is a flow of charged particles confined in the wire. A beam of charged particles is also a current – for example, a beam of electrons is a current.

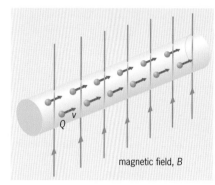

Fig 11.9 **A beam of charged particles in a magnetic field**

■ See questions 5 and 6.

EXAMPLE

Q Television tubes use magnetic fields to deflect the beam of electrons. Electrons are emitted from the electron gun in the tube at a speed of 2×10^7 m s^{-1}. They travel a horizontal distance of 20 cm to the screen and are deflected sideways by the magnetic field a distance of 10 cm.

Calculate an approximate value for the strength of the magnetic field required to do this. (Mass of an electron = 9.1×10^{-31} kg.)

A

Time taken to reach screen $= \dfrac{0.2 \text{ m}}{2 \times 10^7 \text{ m s}^{-1}}$

$= 1 \times 10^{-8}$ s

In this time the electrons are accelerated sideways a distance s of 10 cm with an acceleration a, given by:

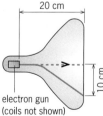

electron gun
(coils not shown)

$s = \frac{1}{2}at^2$

$a = \dfrac{2 \times 0.1 \text{ m}}{(1 \times 10^{-8} \text{ s})^2}$

$= 0.2 \times 10^{16}$ m s^{-2}

The force required is: $F = ma$
$= (9.1 \times 10^{-31}) \times (0.2 \times 10^{16})$
$= 1.8 \times 10^{-15}$ N

From $F = Bev$, the field strength is given by $B = F/ev$

Therefore: $B = \dfrac{1.8 \times 10^{-15}}{(1.6 \times 10^{-19}) \times (2 \times 10^7)}$

$= 5.6 \times 10^{-4}$ (tesla)

Approximate magnetic field strength = 0.56 mT

Measurement of *e/m* for electrons

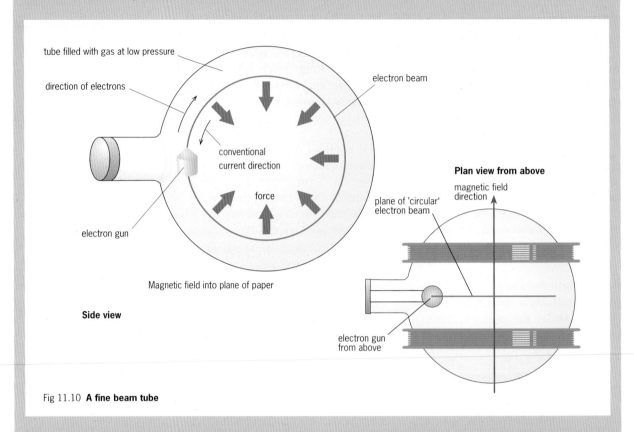

Fig 11.10 **A fine beam tube**

The apparatus in Fig 11.10 is called a fine beam tube, though it has a bulb rather than a tube. It includes an electron gun that fires a beam of electrons vertically into the spherical cavity of the bulb. As the drawing shows, the bulb is mounted between two coils (called **Helmholtz coils** – see Chapter 12). The coils produce a uniform magnetic field that is horizontal and at right angles to the path of the electron beam.

When the electrons emerge from the electron gun, the magnetic force is at right angles to both their direction of travel and the direction of the magnetic field. With the directions shown in Fig 11.10, the force is towards the centre of the tube. As the electron beam turns, its direction of travel is always perpendicular to the field and the force is always directed towards the centre of the tube. This results in the electrons moving in a circular path, with the magnetic force providing the centripetal force. So, using the equation $F = BQv$, derived on page 251 (in this case $Q = e$, the charge on the electron):

$$Bev = \frac{mv^2}{r}$$

where r is the radius of the circular path.

The electron gun provides us with more information about the energy of the electrons (see Chapter 10, page 232):

$$eV = \tfrac{1}{2}mv^2$$

Apart from e, m and v, all the other quantities in these two equations, that is B, R and V, can be measured by experiment.

The two equations can be rearranged to give the velocity v and the ratio e/m:

$$v = \frac{2V}{Br}$$

and:

$$\frac{e}{m} = \frac{2V}{(Br)^2}$$

Question 7 leads you through the derivation of these equations.

e/m for electrons is 1.76×10^{11} C kg^{-1}. With accelerating voltages of about 100 to 200 V, the velocity of the electrons is about 10^7 m s^{-1}. The apparatus described above gives results similar to these, using very ordinary laboratory equipment to make measurements. An ammeter and top-pan balance are needed to measure B, a voltmeter to measure the accelerating voltage V and a ruler to measure radius r of the beam.

▨ See question 7.

5 THE HALL EFFECT

An easy way to measure magnetic fields is to use a Hall probe. The probe is based on a small integrated circuit that produces an output voltage proportional to the magnetic field strength. This 'chip' depends on the effect discovered in 1879 by E.H. Hall. The Hall effect allows us to gain some important information about conduction in solids, including the speed of charge carriers and the number of charge carriers per unit volume.

If charged particles are moving in a solid placed in a magnetic field, then the magnetic force will affect their motion. Fig 11.11 shows a slice of conducting material carrying a current. A magnetic field is applied at right angles to the direction of the current. The moving electrons are pushed by a force towards the back edge of the block. They are unable to leave the block and therefore gather at the edge.

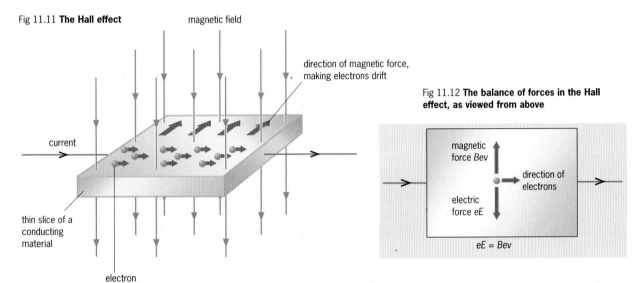

Fig 11.11 **The Hall effect**

magnetic field

direction of magnetic force, making electrons drift

current

thin slice of a conducting material

electron

Fig 11.12 **The balance of forces in the Hall effect, as viewed from above**

magnetic force Bev

direction of electrons

electric force eE

$eE = Bev$

So, in Fig 11.11 the magnetic force causes electrons to drift from the front edge towards the back edge. This makes the back edge negative and the front edge positive with a gradient in between. Like charges repel, so as the electrons, each of charge e, build up at the back, an electric field of strength E develops that opposes this sideways movement of electrons. Eventually the force from the electric field (Ee) on each electron exactly balances the magnetic force (Bev), as shown in Fig 11.12.

As the charges gather at the back edge, a voltage develops across the slice. This voltage is called the **Hall voltage**. The Hall voltage is largest for semiconductor materials like silicon and germanium. Good conductors have much smaller Hall voltages (the next Extension box explains the reason for this).

So far, we have considered the charge carriers to be electrons. In n-type semiconductors, current is certainly due to the movement of (**negative**) electrons. However, in p-type semiconductors, current is due to the movement of **positive** 'holes' (see Chapter 20). Electrons move into holes in one direction, making the holes arise in the opposite direction. Therefore, the description above is correct for an n-type semiconductor, but for a p-type semiconductor the polarity of the Hall voltage is reversed.

?

D Apply Fleming's left-hand rule to show that the movement of the electrons in Fig 11.11 is correct. If the current (in the same direction) were due to positive holes, which way would they move in the field? Why is the Hall voltage reversed?

Factors affecting the Hall voltage

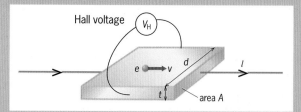

Fig 11.13 **The variables affecting the Hall voltage**

An equilibrium is quickly reached when the magnetic force on the electrons is exactly balanced by the electric force:

$$eE = Bev$$

But the electric field strength E can be expressed as the Hall voltage V_H divided by the width d of the block of material. Therefore:

$$e\frac{V_H}{d} = Bev$$

This gives: $\qquad V_H = Bvd$

The velocity of the electrons can be derived from the transport equation (see Chapter 9):

$$v = \frac{I}{nAe}$$

where n is the number of charge carriers per cubic metre.

The equation for the Hall voltage now becomes:

$$V_H = \frac{BId}{nAe}$$

This can be simplified further because $A/d = t$, the thickness of the block. So the final expression is:

$$V_H = \frac{BI}{nte}$$

We can now see why semiconductor materials are used for Hall probes. The number of carriers n in a semiconductor is many magnitudes smaller than for a good conductor, but is still large enough to allow for a significant current I. A small value of n produces a large Hall voltage V_H. Insulators have a very low value of n, but allow very little current to flow:

For doped semiconductors, $n \approx 10^{22}$ m^{-3}.

For a typical metal, $n \approx 10^{28}$ m^{-3}.

See question 8. ■

?

E A sample of aluminium 10 mm wide and 0.01 mm thick has 2×10^{29} free electrons per cubic metre. A current of 5 A flows through the chip and a magnetic field of strength 2 T acts at right angles to the surface of the sample. (The resistivity of aluminium is 2.7×10^{-8} Ω m.)
Calculate:

(a) the drift velocity of the electrons in the sample,

(b) the Hall voltage between the edges of the sample.

Why is it important that the Hall voltage is measured between two points which are exactly opposite each other? What would be the effect if the two contact points were misaligned by 0.1 mm?

SUMMARY

By the end of this chapter you should be able to:

■ Describe a magnetic field in terms of its shape and direction.

■ Describe the magnetic field around a wire carrying a current.

■ Measure magnetic field strength in tesla (N A^{-1} m^{-1}).

■ Use the equation $F = BIl$ to calculate the force on a wire carrying a current.

■ Describe the use of a current balance to measure magnetic field strength.

■ Describe the application of magnetic forces in devices such as the moving coil galvanometer and the simple d.c. motor.

■ Use the equation $F = BQv$ for the force on a moving charged particle in a uniform magnetic field.

■ Describe the Hall effect and use the Hall voltage to measure magnetic field strength.

QUESTIONS

1 Fig 11.Q1 shows the variation of magnetic field strength B with distance from a long wire carrying a constant current. To the right of the graph the small circle with a cross represents the wire perpendicular to the plane of the paper. Use the information from the graph to draw the magnetic field pattern around the wire. Your diagram should show at least four field lines. Label each line with the relevant size of the magnetic field.

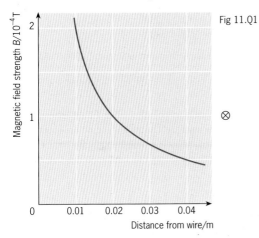

Fig 11.Q1

2 Fig 11.Q2 shows a coil placed between the poles of a magnet.

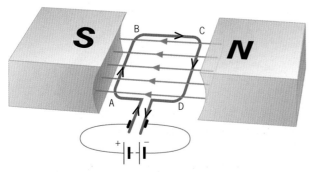

Fig 11.Q2

a) In which direction will the force act on the sides **(i)** AB, **(ii)** BC and **(iii)** CD? Describe the motion of the coil.

b) What will be the effect of **(i)** reversing the current, **(ii)** reversing the poles of the magnet?

c) The coil has ten turns and is 10 cm square. The magnetic field strength is 5 mT. When a current of 2 A flows through the coil, calculate **(i)** the force of the side AB, **(ii)** the torque on the coil.

3 This question is about the loudspeaker shown in Fig 11.Q3 which is to be used in a radio. These are its specifications:

Diameter of coil	25 mm
Number of turns in coil	240
B-field strength	0.50 T
Mass of cone and coil	30 g
Spring constant k of cone's suspension	2.0×10^4 N m^{-1}

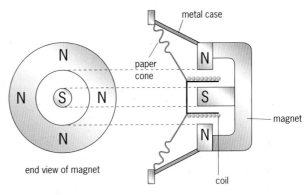

Fig 11.Q3 **A loudspeaker**

a) Calculate the length of wire in the loudspeaker coil.

b) This coil carries a steady current of 60 mA.
   (i) Show that the force on it is about 0.6 N.
   (ii) Calculate the displacement produced by this force, stating any assumption(s) that you make.

[O&C: 1991 Nuffield Science, Physics, Q5 (part)]

4 Fig 11.Q4 shows three long coplanar, parallel, equally spaced wires. The current in each wire is the same and their directions are shown in the diagram.

Fig 11.Q4

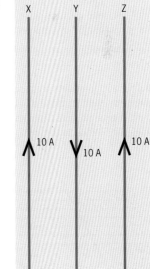

a) Compare the directions and sizes of the forces on each wire.

b) Describe how these would change if the same currents were all in the same direction.

5
a) A hydrogen ion of mass m and charge q travels with speed v in a circle of radius r in a magnetic field of flux density B.
   (i) Write down an equation, in terms of these quantities only, relating the magnetic forces on the ion to the required centripetal force.
   (ii) Hence show that the time T for one revolution of the ion is given by the expression:

$$T = \frac{2\pi m}{Bq}$$

Fig 11.Q5 shows plan and three-dimensional views of a simple form of particle accelerator known as a **cyclotron**.

It consists of two semicircular boxes called 'dees', after their D-shape. Hydrogen ions are injected near to the centre. The alternating potential difference is to accelerate the ions across the gap between the 'dees'. The ions are made to move in semicircular paths when inside the 'dees' by a magnetic field perpendicular to the plane of Fig 11.Q5(a). A charged plate P finally extracts the high-speed ions from the cyclotron.

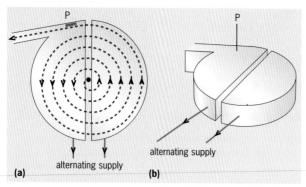

(a)   (b)

Fig 11.Q5

b) For a particular cyclotron accelerating hydrogen ions, the B field is 0.60 T.
   (i) Calculate the time period for one revolution of the hydrogen ion.
   (ii) Why does the fact that the time period is independent of the speed of travel of the ion and the radius of the orbit simplify the operation of the cyclotron?

[OCSEB: 1984, Nuffield Advanced Physics, Q7 (part)]

6 Fig 11.Q6 shows a charged particle entering a uniform magnetic field in the z direction. The field direction is in the x direction and the particle moves upwards in the y direction. What is the sign on the particle?

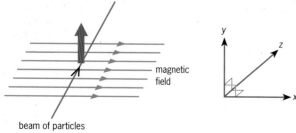

beam of particles

Fig 11.Q6

7 In a fine beam tube experiment to measure the charge-to-mass ratio (e/m) for electrons, an electron beam is produced from an electron gun that has a p.d. of V volts between the anode and cathode.

a) Write down an equation for the kinetic energy of the electrons emerging from the electron gun in terms of the p.d. V.

The electrons emerge into a uniform magnetic field of strength B tesla, perpendicular to the field direction. They move in a circular path of radius r metres.

b) Write down an equation relating the magnetic force on the electrons to the centripetal force required for the circular path.

c) By dividing the equation from part **b)** by that from part **a)**, show that the velocity of the electrons is given by:
$$v = 2V/BR$$

d) By substituting for v in either of the equations from parts **a)** or **b)**, now show that the charge-to-mass ratio is given by:
$$\frac{e}{m} = \frac{2V}{B^2r^2}$$

e) Calculate values for v and e/m using the following data:
$$V = 200 \text{ V}, r = 4 \text{ cm and } B = 1.2 \text{ mT}$$

8 Blood contains ions in solution. Fig 11.Q8 shows a model used to demonstrate the principle of an electromagnetic flow-meter, which is used to measure the rate of flow of blood through an artery by detecting the movement of ions.

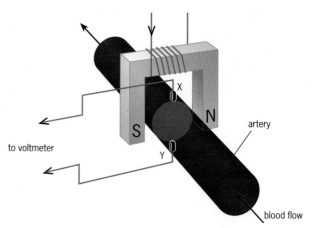

Fig 11.Q8 **An electromagnetic flow-meter**

When a magnetic field of 2.0 T is produced by the electromagnet, a p.d. of 600 μV is developed between the two electrodes X and Y. The cross-sectional area of the artery is 1.5×10^{-6} m². The separation of the electrodes is 1.4×10^{-3} m.

a) Write down an expression for the force on an ion in the blood which is moving at right angles to the magnetic field. Define the symbols you use.

b) An ion has a charge of 1.6×10^{-19} C. Show that the force on the ion due to the electric field between X and Y is 6.9×10^{-20} N.

c) A p.d. of 600 μV is developed when the electric and magnetic forces on an ion are equal and opposite. Hence calculate:
   (i) the speed of the blood through the artery,
   (ii) the volume of blood flowing each second through the artery.

Assignment

THE EARTH'S MAGNETIC FIELD

As well as giving off visible light, the Sun emits electro-magnetic radiation over a wide range of frequencies including ultraviolet and X-rays. The ultraviolet radiation is the main cause of the ionosphere, a region of ionised atmospheric molecules extending from about 40 km to about 300 km above the Earth, and reducing the ultraviolet that reaches its surface.

The Sun also emits charged particles, mostly positrons and electrons, in a stream called the 'solar wind': the magnetic field round the Earth greatly helps to protect us from it. The size and direction of the field at the surface of the Earth can be measured easily, but it is only since we have had remote sensing satellites that we have been able to measure the field around the Earth far out into space.

In 1984, the University of Surrey launched UOSAT 2 (also called OSCAR 11). It travels in a polar orbit around the Earth with a period of 98 minutes at an average height of 687 km, which is above the ionosphere.

UOSAT 2's sensors measure different properties of the space around the Earth, including temperature, light intensity, radiation intensity and magnetic field strength.

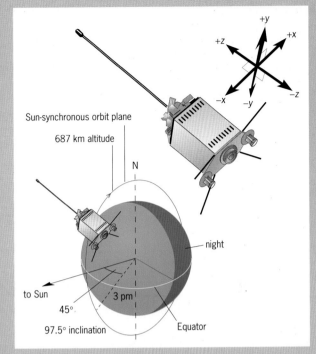

Fig 11.A1 **UOSAT 2 and its orbit geometry**

The satellite carries three magnetic field sensors, one in each plane of the satellite (Fig 11.A1). It also collects and transmits data in two ways. Data from sensors is transmitted 'live' so that receiving stations on Earth can monitor the readings in real time. Data is also collected and stored on board the satellite. This is transmitted to Earth period-ically as 'whole orbit data' (WOD). Fig 11.A2 shows examples of live and whole orbit data. The horizontal axis is time in all graphs, but minutes in WOD graphs and seconds in live data graphs.

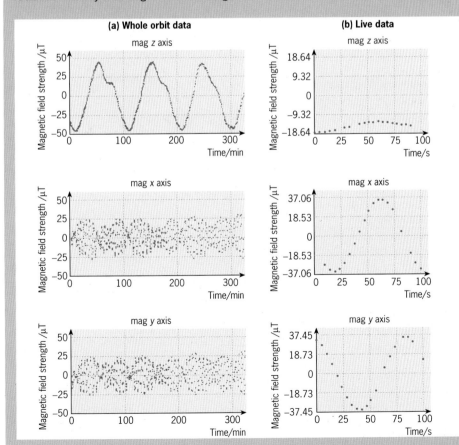

Fig 11.A2 **UOSAT data**

Question 1 is about the WOD data; question 2 is about both the WOD data and the live data.

1 Fig 11.A3 **The solar wind distorts the Earth's magnetic field**

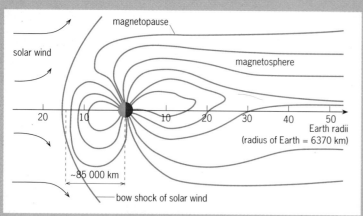

Fig 11.A4 **The aurora borealis – the 'northern lights' – produced when a fast stream of charged particles from the Sun interacts with the Earth's atmosphere**

a) How does the WOD data graph 'mag Z axis' confirm the period of the satellite?

b) Why does the polarity of the magnetic field in the z axis change during the satellite's orbit?

c) Make a sketch of Fig 11.A3, a diagram of the solar wind, and add arrows to show the direction of the Earth's magnetic field.

d) Mark on your diagram where in the orbit you think the magnitude of the 'mag Z axis' is a maximum.

e) At what times does the satellite cross the magnetic equator?

 2

a) From the WOD and live data graphs, estimate the period of rotation of the satellite about its z axis.

b) Why are the WOD plots for the x and y axes more complicated than the z axis plot?

The solar wind is a flux of particles from the Sun that reach Earth's atmosphere at speeds of about 400 km s⁻¹. This mixture of charged particles, which is also known as a **plasma**, collides with the outer layers of the ionosphere where the temperature can reach several thousand kelvin.

A property of the solar wind is that it carries some of the magnetic field at the Sun's surface away with it, through interplanetary space and towards the Earth. When the solar wind reaches the Earth, these magnetic fields interact and the Earth's field is distorted: as in Fig 11.A3, it is compressed on the day-side, with a very elongated tail on the night-side (away from the Sun).

The Earth's field is confined within a region called the **magnetosphere**. The boundary of this region where it meets the solar wind is called the **magnetopause**.

The magnetic field of the Earth at high latitudes (near the poles) allows a path for some of the particles in the solar wind to enter the Earth's atmosphere. Electrons accelerated to as much as 10 keV travel down the field lines. They interact with the neutral gas molecules in the atmosphere at an altitude of about 400–500 km, producing the spectacular aurora borealis ('northern lights') and aurora australis ('southern lights'), as shown in Fig 11.A4.

Electromagnetic radiation from the Sun, including ultraviolet rays and X-rays, penetrates the magnetosphere and ionises atoms in the upper parts of the atmosphere, creating a layer of free electrons and ions in the **ionosphere**.

3

a) Why are X-rays and ultraviolet radiations able to penetrate the magnetosphere?

b) Make a sketch copy of Fig 11.A3 and draw on it the path that such particles would take to enter the Earth's atmosphere in the polar regions.

c) When charged particles in the solar wind pass into the Earth's field, they often arrive at an angle to the field of less than 90°. Draw a diagram showing how these particles will move in the field. (Assume that the field in a small volume is uniform and represent the field lines as evenly spaced parallel lines.)

d) Suggest a mechanism which will explain the light emitted as aurora.

e) The ionosphere behaves like a mirror to some electromagnetic waves. Describe one application of this.

ELECTROMAGNETISM

The Chapter Map brings together the important ideas covered in this chapter, linking key aspects that relate to magnetic fields created by moving charged particles.

Use the map to plan your learning: consult your syllabus to see how it cross-matches with the key ideas you need to understand.

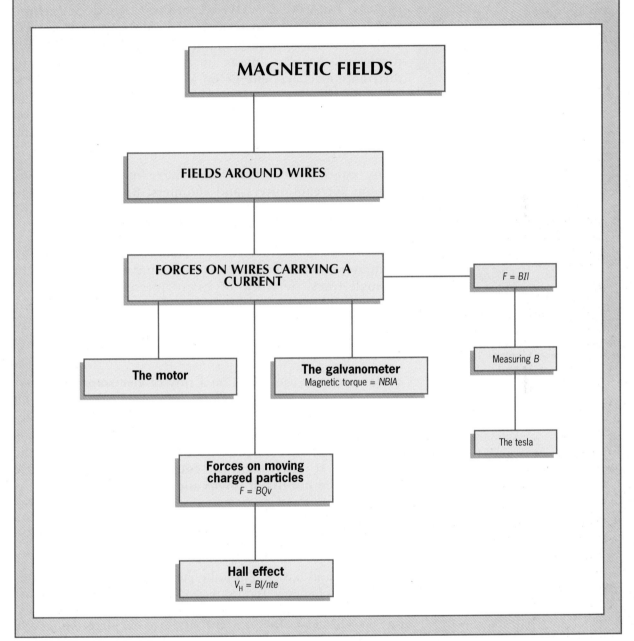

12 Electromagnetic induction

The Gatwick Airport monorail transports passengers between terminals

IN 1831, FARADAY DISCOVERED electromagnetic induction, which enabled him to transfer electrical power from one circuit to another by varying the magnetic linkage. With this discovery, Faraday made it possible to produce and distribute electrical energy on a large scale, and so society and industry were transformed as electricity was supplied to homes and factories far from its source. Over 150 years later, new applications of electromagnetic induction are still being developed.

Many cities in the world now have monorail transport systems for commuters and shoppers. But in the Chicago suburb of Rosemont this idea is being taken further, based on the latest improvements in linear induction motors.

This is the plan. Small, driverless cars controlled by computer will move on wheels within narrow 'guideways' in a single-track network of many interconnecting loops. People will arrive at a station, key in the station they want to travel to, and within 2 to 3 minutes a car will arrive to take them there. Nowhere in the suburb will be more than 400 metres from a station.

Under each car there will be a row of electromagnets that induce currents which drive the car at speeds of up to 50 kilometres per hour. The car can be braked simply by reversing the direction of the magnetic field. The project is designed to attract people out of their cars and onto public transport. If successful, it could greatly ease traffic congestion in our major towns and cities.

Introduction

Many electrical devices, especially portable ones, use batteries as their source of energy. But for most of our electrical needs we use electricity generated in power stations. To distribute the electricity, the power stations rely on the principle of **electromagnetic induction**.

Modern power stations have the capacity to produce many megawatts of electrical power – Britain's production of electrical power now exceeds 50 gigawatts (5×10^{10} watts). Electromagnetic induction is not just used in power stations. Car speedometers also use the principle, while induction heating is used extensively in industry. Some types of cookers induce a heating current within the base of cooking pans. Most motors in use in the world are induction motors, ranging from the small shaded pole motor in record decks to large powerful motors used in pumps that are the workhorses of the chemicals industry.

1 WHAT IS ELECTROMAGNETIC INDUCTION?

Michael Faraday's discoveries early in the nineteenth century have made the large-scale production of electrical energy possible. The date 29 August 1831 is acknowledged as the 'birth of the electrical industry' – the day on which Faraday discovered electromagnetic induction. He made the link between electricity and magnetism when he showed that an electric current was produced when a magnet was moved near to a conductor (a metal wire).

Simple experiments provide the evidence. A magnet is moved in and out of a coil connected to a galvanometer, as shown in Fig 12.1(a), and a current is observed whenever the magnet moves (or, if the magnet is kept stationary, whenever the coil moves). A wire moved through a magnetic field produces the same result, as in Fig 12.1(b).

Faraday's experiments also led to the first simple **transformer**. He wound two coils on an iron ring (Fig 12.2). One, the primary coil, was connected to a battery and the other, the secondary coil, to a galvanometer. The galvanometer showed a current in the secondary coil only when the current from the battery was switched on or off; that is, only when the current from the battery, and hence the magnetic field, was *changing*.

So, a current is induced whenever a wire in a closed circuit moves through a magnetic field *or* when the field changes through the wire.

The following helps to explain the important ideas in more detail (refer to Fig 12.3):

● The rigid straight wire XY moves with a velocity *v* perpendicular to a uniform magnetic field of strength *B* tesla. The wire is *l* m long.

● There is a force on the electrons in the moving wire (*Bev*, where *e* is the charge on a charge carrier, see page 252), and free electrons will move towards end X of the wire. As they build up at X, so does an electric field that opposes the movement of electrons (just as in the Hall effect, page 253).

● Eventually the electric force is equal to the magnetic force:

$$eE = Bev$$

where *E* is the electric field strength (see page 232). We can also write:

$$E = V/l$$

where *V* is the final voltage induced across the ends of the wire.

● This 'voltage' is an induced e.m.f., since it is the movement of the wire through the field that produces charge separation.

● Putting this together we have:

$$e \frac{V}{l} = Bev$$

or: $$V = Blv$$

● Thus the induced e.m.f. *V* is proportional to the magnetic field strength *B*, the length of the conductor and the speed at which it moves. (Note that if the conductor is at an angle *θ* to the field the induced voltage is $V = Blv \sin \theta$.)

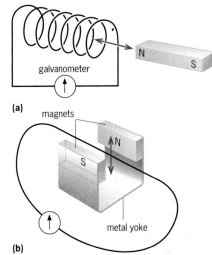

(a)

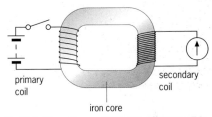

(b)

Fig 12.1 (a) **A current is recorded when a magnet is moved in and out of a coil.** (b) **The same result is recorded when a wire is moved through a magnetic field**

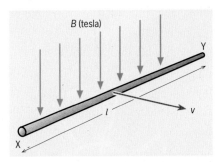

Fig 12.2 **When the current in the primary coil is switched on or off, the galvanometer shows a current in the secondary coil**

Fig 12.3 **A rigid straight wire XY moving with velocity v in a magnetic field of strength B tesla**

?

A A wire of length 10 cm moves at 5 m s^{-1} through a magnetic field of 1 T, cutting the field at right angles. Calculate the e.m.f. induced across the wire.

See questions 1 and 2.

The induced current

Once the e.m.f. is established, no more charge flows because the electric force is equal to the magnetic force. What happens when the wire is part of a complete circuit?

To answer that, suppose the wire XY shown in Fig 12.4 runs along two rails that are part of a simple circuit with a total resistance of R ohms and a uniform magnetic field B. The induced e.m.f. will drive a charge around the circuit. The direction of the 'conventional' current is shown.

Fig 12.4 **When the wire of Fig 12.3 forms a complete circuit, the induced e.m.f. drives a current round the circuit**

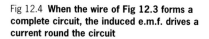

The direction of the conventional current is opposite to the direction in which the electrons move.

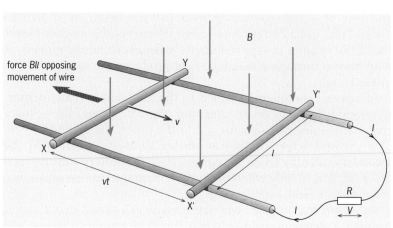

Now that there is a current (charge flowing), a force (BIl) affects the wire because it is in a magnetic field. This force acts in the opposite direction to the way the wire is moving. To keep the wire moving at a constant velocity v, a force equal to this must be applied in the direction of the movement. In t seconds the wire moves to X'Y' a distance of vt metres.

The work done in moving the wire is given by:

$$\text{work done} = \text{force} \times \text{distance moved}$$
$$= BIl \times vt$$

Since energy must be conserved, the work done is equal to the electrical energy transferred in the circuit:

$$\text{electrical energy supplied} = \text{power} \times \text{time}$$
$$= IV \times t$$

where V is now the p.d. across the load R. Therefore, we can write:

$$IVt = BIlvt$$

or:
$$V = Blv$$

This is the same as the formula obtained on page 261. What this tells us is that V, the induced e.m.f., is the same, whether a current is drawn or not.

Towards a more general law

With the help of Fig 12.4, we can look at this result in a slightly different way. When the wire XY moves it sweeps through an area of ($vt \times l$), so vl is the area swept through per second:

$$\frac{vt \times l}{t} = vl = \frac{\mathrm{d}A}{\mathrm{d}t}$$

where $\mathrm{d}A/\mathrm{d}t$ is the rate of change of area.

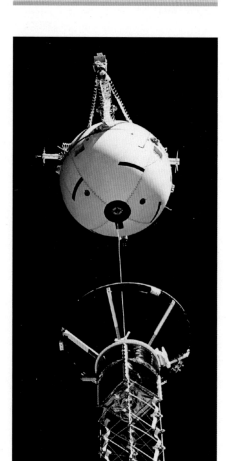

The Tethered Satellite System was launched by NASA in 1996 to see how much electricity could be generated by placing a wire in the Earth's magnetic field. The satellite contained a 20 km copper wire tether to cut through the field.

When the satellite was in orbit, 19.7 km of wire was reeled out. At that point the tether snapped. But by then a current of 1 A had been generated

We can now write the equation for the induced e.m.f. as:

$$V = B \frac{dA}{dt}$$

That is, the induced voltage is equal to the field strength multiplied by the rate at which the area swept through changes.

Faraday reached the same result, but he used the idea of **flux**, as explained in the next section.

EXAMPLE

Q A 10 cm wire is moved perpendicular to a steady magnetic field of strength 5 mT at a speed of 0.5 m s^{-1}. Calculate **a)** the rate of change of the area that the wire sweeps through, and **b)** the induced e.m.f.

A a) Speed of wire $\times$ length $= \dfrac{dA}{dt} = vl$

$\quad = 0.5$ m s$^{-1} \times 0.1$ m

So rate of change of area $= 0.05$ m^2 s^{-1}

b) $V = B \dfrac{dA}{dt}$

$\quad = 5 \times 10^{-3} \times 0.05 = 0.25 \times 10^{-3}$ V

So induced e.m.f. $= 0.25$ mV

See questions 3 and 4.

2 MAGNETIC FIELD AND FLUX

One of Faraday's important contributions to physics was the idea of **field lines**. The patterns we draw around magnets and coils to describe the shape and variation in strength of fields were first used by him. He imagined the field lines passing through a surface as shown in Fig 12.5, and he called this 'flux'.

The flux ϕ is defined as:

Flux = field strength $\times$ the area it passes through

$$\phi = BA$$

If the field is not perpendicular to the area, then the equation becomes:

$$\phi = BA \sin \theta$$

Flux is measured in T m^2, called webers (Wb).

The magnetic field can also be defined in terms of flux and is sometimes called the **flux density**:

$$\text{Flux density, } B = \frac{\phi}{A}$$

Fig 12.5 **In (a) and (b) the magnetic field is perpendicular to the area it is passing through, where field B' is less than B. In (c) the field is not perpendicular to the area and must be resolved into components – only the component perpendicular to the area is used in calculating the flux**

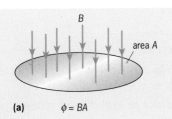

(a) $\phi = BA$

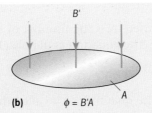

(b) $\phi = B'A$

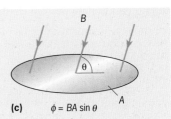

(c) $\phi = BA \sin \theta$

3 FARADAY'S LAWS OF ELECTROMAGNETIC INDUCTION

After careful experiment, Faraday set out his **law of electromagnetic induction**. For a wire moving through a field he found that the size of the induced e.m.f. is proportional to:

● the strength of the magnetic field,

● the speed at which the wire cuts the field,

● the number of turns of wire (as part of a coil, for example).

Flux cutting and flux linking

The action of a wire being passed through a field is often referred to as **flux cutting**.

Faraday's 'transformer' experiments, however, do not involve moving wires. Look back at Fig 12.2: a current is induced in a secondary coil when the current in the primary coil changes. The current in the primary produces a magnetic flux which links with the secondary coil. When this flux changes, a current is induced in the secondary coil. This is called **flux linking**. The induced current is proportional to:

● the strength of the field,

● the rate of change in the flux produced by the primary coil,

● the number of turns, N, on the secondary coil.

To induce a continuous current in the secondary, then, the current in the primary must continually change. For instance, if it is an alternating current (a.c.), then the induced current will also alternate.

The induced current will also depend on the a.c. frequency. An increase in frequency means that the flux is changing over a shorter time; that is, the rate of change of flux increases and the induced current is larger.

Conversely, if the frequency of the a.c. in the primary is reduced, the flux changes over a longer time, so this reduces the rate of change of flux and the induced current:

> **The size of the induced current in the secondary coil is proportional to the frequency of the alternating current in the primary coil.**

Faraday's laws can be summarised in one statement:

> **The induced e.m.f. is proportional to the rate of change of flux.**

Or, if ϵ is the induced e.m.f:

$$\epsilon = N\frac{d\phi}{dt}$$

where N is the number of turns of wire.

The equation could also be written as:

$$\epsilon = \frac{d(N\phi)}{dt}$$

The quantity $N\phi$ is called the **flux linkage** and is used when a particular coil with N turns is considered.

?

B To induce a current in a secondary coil, does the current in the primary have to change? Explain your answer.

See question 5. ■

4 LENZ'S LAW

When the north pole of a magnet is pushed into a coil, the direction of the induced current creates another 'north pole' in the coil because the current 'opposes' the approaching north pole. When the magnet is removed, the direction of the induced current is reversed, changing the end of the coil to a south pole. This again opposes the north pole, which is now leaving the coil (Fig 12.6).

Lenz's law states:

The direction of the induced current opposes the change that causes it.

Lenz's law is the 'electromagnetic' version of the law of conservation of energy. If Lenz's law were not true, energy would not be conserved.

To show that the effects are opposed, we introduce a negative sign. The equation for the induced e.m.f. ϵ should read:

$$\epsilon = -N\frac{d\phi}{dt}$$

C Explain why energy would not be conserved if Lenz's law were not true.

D When a magnet is pushed into a coil as in Fig 12.1, the galvanometer needle moves to one side and back to zero. When the magnet is removed, the needle moves the other way and back to zero. Use Lenz's law to explain this.

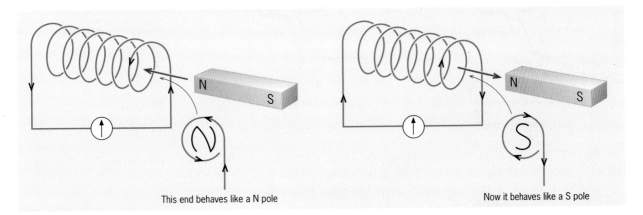

This end behaves like a N pole

Now it behaves like a S pole

Fig 12.6 **Demonstrating Lenz's law**

Mathematical treatment of Faraday's laws

The induced e.m.f. ϵ is given by:

$$\epsilon = -N\frac{d\phi}{dt}$$

$$= -N\frac{d(BA)}{dt}$$

$$= -N\left(B\frac{dA}{dt} + A\frac{dB}{dt}\right)$$

$$= -\left(NB\frac{dA}{dt} + NA\frac{dB}{dt}\right)$$

　　　　↑　　　　　↑
flux cutting term　　**flux linking term**

The **flux cutting** term describes the induced e.m.f. produced by a wire *moving* through a *constant* magnetic field such as in a dynamo. The rate at which the wire sweeps out an area as it passes through the field determines the e.m.f.

The **flux linking** term describes the induced e.m.f. produced by a *changing* flux linking with a *fixed* wire as in the transformer. The flux produced by one coil 'links' with a second coil. The rate at which the field changes, which often depends on the current in the primary coil, determines the induced e.m.f.

The dynamo

We have seen that the large-scale production of electrical energy has developed from Faraday's experiments. Power stations all over the world, whether burning fossil fuels or using wind power, produce induced currents. In the UK, electricity is generated as an **alternating current**. The equation for an alternating current is:

$$I = I_0 \sin 2\pi ft$$

where I_0 is the peak current. The simple bicycle dynamo and the generators used in power stations are basically the same.

Then:

$$\alpha = \omega t$$

and the velocity v is related to ω and the width of the coil x by:

$$v = \omega \frac{x}{2}$$

So the total e.m.f. due to both sides is given by:

$$\epsilon_{\text{tot}} = 2By\omega \frac{x}{2} \cos \omega t$$

$$= Bxy\omega \cos \omega t$$

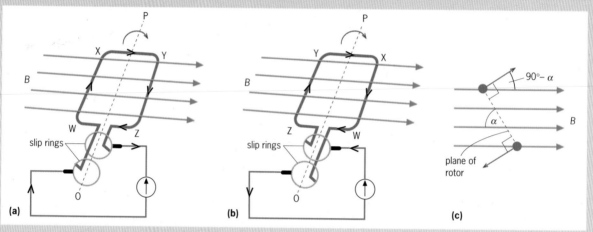

Fig 12.7 **How a dynamo works, and the definition of α**

Fig 12.7(a) shows a simple rectangular loop that is rotating about an axis OP in a uniform magnetic field of strength B tesla. Because the sides of the loop WX and YZ are moving at right angles to the field but in opposite directions, the e.m.fs induced in each side will drive a current in the same direction around the loop.

After half a cycle, WX and YZ will have changed places: Fig 12.7(b). The current passes out of the loop through slip rings. Each end of the circuit is always in contact with one slip ring, allowing for a continuous current. The current in the galvanometer will now be reversed. After another half-cycle the loop is back where it started and the current will be in the same direction as in the first half-cycle.

During each half-cycle the size of the current rises and falls. In Fig 12.7(c), suppose the plane of the loop makes an angle α with the magnetic field and that the sides WX and YZ have a length y and are moving at a speed v. The induced e.m.f. in each side will be given by:

$$\epsilon = Byv \sin(90 - \alpha) = Byv \cos \alpha$$

Now suppose the angular velocity of the loop is ω and it takes t seconds to move through an angle α.

But $xy = A$, the area of the loop, and $\omega = 2\pi f$, where f is the frequency of rotation. Therefore, for a coil of N turns:

$$\epsilon_{\text{tot}} = 2\pi fNBA \cos 2\pi ft$$

This is the equation of an alternating voltage of peak value:

$$\epsilon_0 = 2\pi fNBA$$

so that: $\epsilon_{\text{tot}} = \epsilon_0 \cos 2\pi ft$

The output of the dynamo is shown in the graph of Fig 12.8.

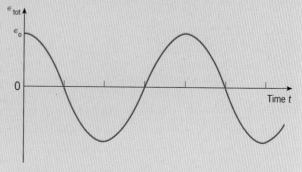

Fig 12.8 **The output of the dynamo**

5 MAGNETIC CIRCUITS

In the examples shown in Fig 12.9, the flux is set up by a current flowing through a coil, or **solenoid**. The flux, represented by the field lines, is in complete loops. This is always the case. In some cases, as in Fig 12.9(a), the loops are set up in the space around and through the coil, while in others, as in (b), they are contained within a solid core such as 'soft iron'. We can think of these complete loops as 'magnetic circuits'.

> ✔ 'Soft iron' refers to its magnetic properties – not its hardness! See page 270.

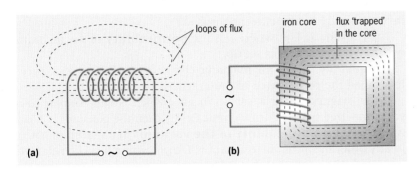

Fig 12.9 **The flux** (a) **around a solenoid and** (b) **in a transformer core**

We saw in Chapter 9, page 205, that the physical dimensions of the wire (cross-sectional area and length) in a simple electrical circuit affect the current. Similarly, we shall see here that the dimensions of a magnetic circuit affect the flux that can be set up.

The flux also depends on the size of the current and the number of turns on the coil. Without current and turns of the coil there would be no flux. We say the flux is 'generated' by the **current-turns**. If N is the number of turns, I is the current and ϕ is the flux, then:

$$NI \propto \phi$$

> **?** **E** Describe clearly how the physical dimensions of the core affect the flux.

To understand how the dimensions of the magnetic circuit affect the flux it is probably easier to think about a transformer. The flux is generated by a primary coil. The iron core has the interesting and useful property of confining nearly all the flux within the core. The physical dimensions of the core, its cross-sectional area and the length, affect the flux, as shown in Fig 12.10.

▦ See question 6.

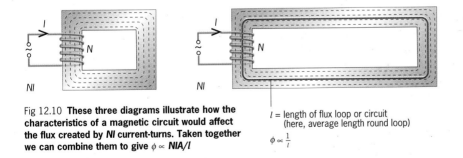
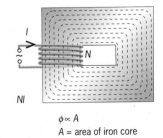

Fig 12.10 **These three diagrams illustrate how the characteristics of a magnetic circuit would affect the flux created by NI current-turns. Taken together we can combine them to give** $\phi \propto NIA/l$

l = length of flux loop or circuit (here, average length round loop)
$\phi \propto \frac{1}{l}$

$\phi \propto A$
A = area of iron core

Magnetic circuits compared with electrical circuits

At this point it is worth comparing magnetic circuits with simple electric circuits.

A voltage V causes a current I in a circuit that has electrical resistance. The resistance depends upon the physical dimensions of the wire and the material of the wire (see page 205).

When A is the cross-sectional area, l is the length of the wire and ρ is the resistivity of the material of the wire, then:

$$V = IR$$

$$= I\frac{\rho l}{A}$$

Looking at Fig 12.10, we can write a similar equation for a magnetic circuit:

$$NI = \phi \times \text{constant} \times \frac{l}{A}$$

Just as the current is 'set up' by a voltage in the electric circuit, the flux ϕ is set up by the current-turns, NI.

The constant is $1/\mu$, where μ is the **permeability** of the medium within which the flux is set up.

The quantity $l/\mu A$ is the magnetic circuit equivalent to resistance. It is called the **reluctance** of the magnetic circuit (R_{mag}).

If the magnetic field is set up in air (which is very similar to a vacuum – free space – for magnetic fields) then the constant is $1/\mu_0$, where μ_0 is the **permeability of the vacuum** (free space), in NA^{-2}. In any other medium, the constant is $1/\mu_0\mu_r$, where μ_r is the **relative permeability** of the medium. That is:

$$NI = \phi \frac{l}{\mu_0\mu_r A}$$

Electrical resistivity ρ is related to electrical conductivity σ by the equation:

$$\rho = \frac{1}{\sigma}$$

By comparison, we can see that 'electrical conductivity' is analogous to permeability. We think of *permeability* as a kind of 'magnetic conductivity' – it tells us how good a medium is at allowing a magnetic field to be established in that medium. The *relative permeability* tells us how much better that medium is compared to a vacuum.

F There is one important difference between electric and magnetic circuits. We know that an electric current is a flow of charge – coulombs per second. What 'flows' in a magnetic circuit?

See question 7. ▥

Summary – electrical circuits and magnetic circuits

Comparison of electrical circuits and magnetic circuits:

$$V = IR \qquad\qquad NI = \phi R_{mag}$$

$$V = I\frac{l}{\sigma A} \qquad\qquad NI = \phi \frac{l}{\mu_0\mu_r A}$$

In Table 12.1 we compare electrical and magnetic quantities.

Table 12.1

Electrical quantity	Magnetic quantity
Voltage V	Current turns NI
Resistance R	Reluctance R_{mag}
Conductivity σ	Permeability $\mu_0\mu_r$
Current I	Flux ϕ

EXAMPLE

Q Calculate the magnetic field in a long air-cored solenoid with 10 turns per centimetre and carrying a current of 1 A. (The permeability of free space $\mu_0 = 4\pi \times 10^{-7}$ N A^{-2}.)

A Use the equation for current-turns:

$$NI = \phi\frac{l}{\mu_0 A} = \frac{\phi}{A}\frac{l}{\mu_0}$$

$$= B\frac{l}{\mu_0} \quad \left(\text{since } B = \frac{\phi}{A}\right)$$

Rearranging:

$$B = \frac{N}{l}I\mu_0 \quad (\text{10 turns cm}^{-1} = 1000 \text{ turns m}^{-1})$$

$$= 1000 \times 1 \times 4\pi \times 10^{-7}$$

$$= 12.6 \times 10^{-4} \text{ T}$$

Magnetic field in the solenoid = 1.26 mT

6 USES OF MAGNETIC CIRCUITS

Field due to a long solenoid

Imagine a solenoid of length l metres with N turns, forming a circle as in Fig 12.11. Therefore, when there is a current the flux forms an enclosed loop inside the coil, also of length l. The cross-sectional area of the solenoid, and so of the tube of flux, is A m². We shall assume that the medium in the solenoid is air.

Starting with the equation for the flux:

$$NI = \phi \frac{l}{\mu_0 A}$$

Now, since $\phi = BA$ (flux density × area), we can write:

$$NI = \frac{BAl}{\mu_0 A} = \frac{Bl}{\mu_0}$$

We can rearrange this to obtain an equation for the field strength B in a solenoid:

$$B = \mu_0 \frac{N}{l} I = \mu_0 n I$$

where $n = N/l$, the number of turns per metre.

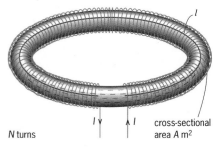

N turns

cross-sectional area A m²

Fig 12.11 **In a long solenoid, the flux forms an enclosed loop inside the coil**

Field due to a long straight wire

As in Fig 12.12, imagine a ring of flux of cross-sectional area A, around the wire at a distance r from the wire. The wire can be considered as part of a very large single-turn coil, that is, $N = 1$. Therefore current:

$$I = \phi \frac{l}{\mu_0 A}$$

The length l of the ring of flux is $2\pi r$, and since $\phi = BA$:

$$I = \frac{BA\, 2\pi r}{\mu_0 A}$$

Rearranging, we get:

$$B = \frac{\mu_0 I}{2\pi r}$$

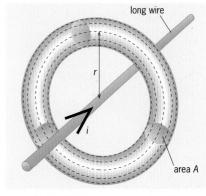

long wire

area A

Fig 12.12 **The field due to a long straight wire**

EXAMPLE

Q Calculate the the magnetic field strength around a straight wire carrying a current of 10 A at a distance of **a)** 10 cm, **b)** 20 cm, **c)** 100 cm. (The permeability of free space $\mu_0 = 4\pi \times 10^{-7}$ N A⁻².)

A
a) We use the equation for magnetic field strength:

$$B = \frac{\mu_0\, I}{2\pi r}$$

$$= \frac{4\pi \times 10^{-7} \times 10}{2\pi \times 0.1}$$

Magnetic field strength at 10 cm = 2×10^{-5} T

b) Since r is inversely proportional to B, and since r doubles here, B will halve to 1×10^{-5} T.

c) Now show that with a value for r of 100 cm, the field strength is 2×10^{-6} T.

Effect of iron

Iron has a very large relative permeability, which is why it is useful for making electromagnets, for example. When a coil is wound on a soft iron core, the magnetic field produced is about a thousand times stronger than it would be if the iron were not there. Iron also has the property of losing most of the magnetism as soon as the current stops flowing. This is why it is described as 'soft'. A magnetically 'hard' metal such as steel would retain much of the magnetism if it were used in the same coil. The ability of soft iron to have a strong induced magnetic field is given by its **relative permeability, μ_r**. The value of μ_r for iron is about 1000.

The magnetic field in a solenoid with an iron core is:

$$B = \mu_r \mu_0 N l$$

where N = number of turns and l = length of flux loop.

We can say that iron is a 'good conductor' of magnetism. Many other magnetic materials have high relative permeabilities but are less dense than iron. These find uses where light weight is an advantage, such as in miniature earphones and the small motors used inside portable audio tape players.

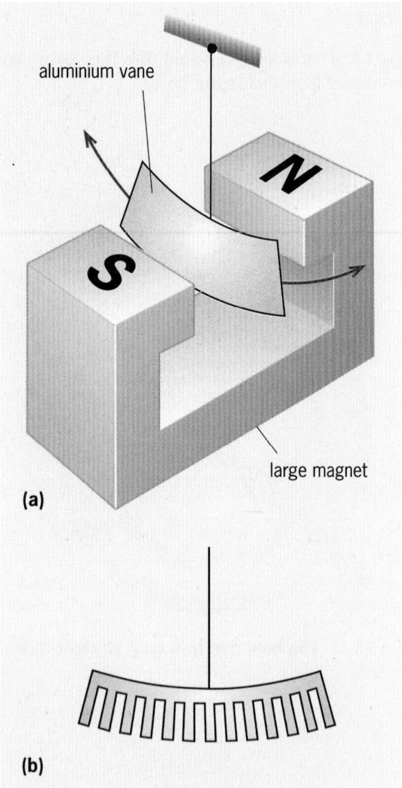

7 EDDY CURRENTS

Eddy currents are induced in blocks of conductor when a block is moving in a magnetic field or when the block is in a changing field. They are called 'eddy currents' because the charge moves in a swirling pattern rather like the eddy currents we see in river water.

Fig 12.13 shows a simple demonstration of eddy currents. The pendulum has an aluminium vane. When it swings between the poles of a magnet its motion is very heavily damped – it slows down and stops after a few swings.

This effect, sometimes called **magnetic braking**, is due to the currents induced in the aluminium as it cuts through the magnetic field. The currents flow in loops within the aluminium and, like all induced currents, they obey Lenz's law. That is, they flow in the direction that opposes the change that causes them.

As the vane enters the field (and B is getting stronger), an eddy current is induced in the vane. The moving electrons in the eddy current experience a force in the field of the magnet that opposes the motion of the vane, so slowing the vane. The force on the moving electrons is the same as the 'motor effect' (page 251).

When the vane moves out of the field (and B is now getting weaker) the direction of the induced current reverses, and so does the 'motor effect' force. This slows it down more. After a few swings the vane comes to rest.

If the solid vane is replaced by one with slots cut into it, the oscillating vane is affected far less and continues to swing for longer. The slots restrict the possible paths for eddy currents and effectively reduce the size of any currents induced.

Magnetic braking can be useful where damping is required. But eddy currents are a nuisance in many practical situations, because they dissipate energy due to electrical resistance of the material. This dissipation of energy can be useful, for example in induction heating. But in motors and transformers (see the page 276), the heating is wasteful and potentially dangerous.

Fig 12.13 (a) **The oscillations of an aluminium vane in a magnetic field illustrate the effect of eddy currents.** (b) **When a slotted vane replaces the solid vane, it oscillates for longer**

?

G A magnet allowed to fall freely down a plastic pipe accelerates to the ground. When the same magnet falls down a copper pipe of the same dimensions it takes longer to reach the ground. The magnet does not touch the sides of either pipe. Why does it take longer to fall down the copper pipe?

A transformer is an induction device. Primary and secondary coils are wound round a soft iron core, and an alternating e.m.f. in the primary coil induces an e.m.f. in the secondary coil. In such cases, iron cores are designed to reduce eddy currents. The soft iron core of a transformer is laminated – that is, it is cut into very thin layers perpendicular to the magnetic flux. These layers are lacquered, coated with an insulating material and then reassembled into a block with the layers very close together but not in electrical contact. This preserves much of the magnetic property of the metal but reduces the eddy currents.

Modern ferrite materials are made from a paste of small iron oxide particles baked into a solid shape. The particles are close enough to produce good magnetic properties (high μ_r) but eddy currents are reduced to a minimum.

H Give some practical examples of the use of magnetic braking.

8 BEHAVIOUR OF THE SIMPLE D.C. MOTOR

The current drawn by a simple motor (as described on page 248) varies in an interesting way. When a motor with no load is switched on, the initial current drawn by the rotor coil from its power supply is very high, but it quickly drops to a much lower value. If the motor is then loaded, the current drawn by the motor gradually increases. Another important fact is that when the motor does work its speed decreases.

The motor rotates because of the force on a wire carrying a current, BIl. Current I is driven by an applied e.m.f. ϵ_a (from the supply).

As the coil rotates, an e.m.f. ϵ_i is induced in the coil. This e.m.f. tends to drive a current in the opposite direction to the current from the power supply (Lenz's law). The size of this induced e.m.f. depends on the speed of rotation of the coil. The faster the coil rotates, the larger the induced e.m.f. The resultant current as measured by an ammeter arises from the sum of the two e.m.fs. Fig 12.14(b) shows a circuit diagram for a simple motor.

When the motor starts from rest there is no induced current. Therefore the initial current measured on the ammeter is high. As the motor's speed increases, so does the induced e.m.f., and the total current measured on the ammeter drops. If the motor does work and it slows down, the induced e.m.f. decreases and the measured current increases, delivering more energy to the motor.

If we consider the resistance in the rotor circuit (this includes any internal resistance of the power supply), we can say:

$$\text{total resistance} = \frac{\text{e.m.f. of supply} - \text{induced e.m.f.}}{\text{current}}$$

or:

$$R = \frac{\epsilon_a - \epsilon_i}{I}$$

We can rearrange this to give the current:

$$I = \frac{\epsilon_a - \epsilon_i}{R}$$

I is the current that is measured by the ammeter.

Assuming the resistance stays constant, which is a reasonable assumption if the coil does not get hot, then we can see that the current depends on the difference between the applied and induced e.m.f.s. The induced e.m.f. ϵ_i is often referred to as the **back e.m.f.**

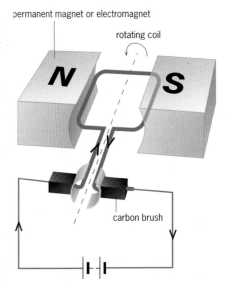

permanent magnet or electromagnet

Fig 12.14(a) **A simple d.c. motor (no load attached).**

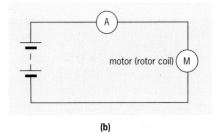

(b)

Fig 12.14(b) **Circuit diagram for motor**

I What will be the effect on **(a)** the resistance and **(b)** the current if the coil heats up considerably?

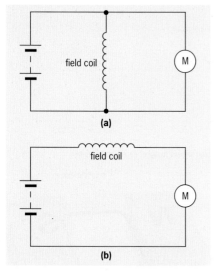

Fig 12.15 **Circuit diagrams for practical motors:** (a) **shunt wound**; (b) **series connection**

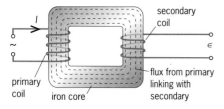

Fig 12.16 **Mutual inductance**

?

J Two coils are wound on the same core. The current in the first coil rises from zero to 20 mA in 0.5 ms. The e.m.f. induced across the secondary is 2 V. Calculate the mutual inductance.

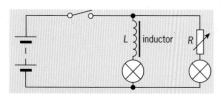

Fig 12.17 **A simple circuit to demonstrate self-inductance. The value R for the resistor is adjusted so that both bulbs are equally bright when the current is steady. When the circuit is switched on, the bulb in series with the resistor lights before the bulb in series with the inductor**

The following equation also helps us understand the energy balance of the motor:

$$IR = \epsilon_a - \epsilon_i$$

or:

$$\epsilon_a = \epsilon_i + IR$$

Multiplying both sides by the current, I, we get:

$$\epsilon_a I = \epsilon_i I + I^2 R$$

Put in words, this says:

> **Power from supply = useful power + power lost in heating rotor coil.**

This is a rather simplified picture of a real motor. (Power lost in heating the rotor coil is known as copper losses.)

In order to make the magnetic field as strong as possible, the coil in most motors is wound on a soft iron former. This introduces the possibility of some eddy current losses, even though the material will be designed to minimise them. Another possible source of energy loss is **flux loss**: the inevitable gap between the rotor and the poles of the magnet (or electromagnet) reduces the flux linking them as some of the flux spreads out or 'escapes' at the gap.

Instead of permanent magnets, most practical motors use electromagnets, usually called 'field coils': several coils are wound on a high permeability former to provide a strong field with the desired shape. Such a motor then has field coils and the rotor coil. These can be connected in parallel (shunt wound) or in series (Fig 12.15).

9 INDUCTANCE

Mutual inductance

In Fig 12.16, an e.m.f. ϵ_i is induced in the secondary coil whenever the flux produced by the primary coil changes. The flux from the primary will only vary with the current in the primary. Therefore:

$$\epsilon_i = -N_s k \frac{dI}{dt} = -M \frac{dI}{dt}$$

where N_s is the number of turns in the secondary coil and $M = N_s k$. The constant k depends on factors such as the number of turns on the primary coil and the relative permeability of the former for the rotor coil – all of which should be constant. M is called the **mutual inductance**. It is measured in V s A⁻¹, or henrys.

Self-inductance

Fig 12.17 shows a simple circuit. The two bulbs are identical and the resistance of the coil is the same as that of the resistor. The coil should have a soft iron core for the effect to be noticeable.

When the switch is closed, the bulbs both light up but do not come on together. There is a noticeable lag between the time when the bulb in series with the resistor lights up and when the other bulb lights up. When both are lit they have the same brightness. The effect of slowing down the growth of the current is described as the **inductive effect** of the coil. Hence the coils are also called **inductors**.

When the circuit in Fig 12.17 is first switched on, charge begins to flow in the inductor. This current creates a magnetic field in the coil where none existed before – that is, there is a changing

INDUCTION COILS

THE MOTIVE FORCE of a petrol engine is supplied when a mixture of petrol and air explodes. This explosive mixture is ignited by the spark from a spark plug, and the spark itself is produced when about 2 to 3 kV are applied across the spark gap.

Using induction, this high voltage comes from the 12 V car battery. The rotating cam in the distributor opens the 'points', and so cuts off the current in the primary coil. The sudden drop in the primary current means that the strong magnetic field due to that current also suddenly collapses. The rapid change then induces a very high voltage in the secondary coil. This voltage is applied to one of the spark plugs through the rotor arm. In one cycle of the cam, a high voltage is induced four times and is applied to each of the four plugs in turn.

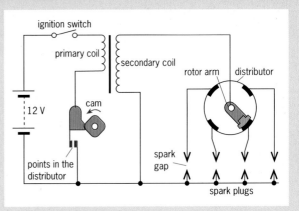

Fig 12.18 **The ignition system for a car**

magnetic field. The changing magnetic field induces a voltage that opposes the change (Lenz's law). It does this by creating an e.m.f. in the opposite direction to the e.m.f. from the applied voltage. As a result, the current in the coil builds up more slowly than in the simple resistor. However, eventually the current in the coil reaches a maximum. This maximum current is determined by the applied voltage and the total resistance in the circuit (Fig 12.19).

The final steady current depends on the supply and the resistance of the inductor:

$$I = \frac{\text{applied e.m.f.}}{\text{total resistance}}$$

Although the build-up of current may take only a small fraction of a second, we can see from the initial gradient of the curve in Fig 12.19 that the rate of change of current (dI/dt) at the start is constant.

This rate of change of current depends on two factors: the applied voltage and the inductive effect of the coil. Experiment shows that the initial rate of increase of current is directly proportional to the applied voltage:

$$V \propto \frac{dI}{dt}$$

The constant of proportionality is a property of the coil called the **self-inductance** of the coil and is given the symbol L:

$$V = L\frac{dI}{dt}$$

The unit of self-inductance is the henry, the same as the unit for mutual inductance.

▒ See question 8.

Fig 12.19 **Graph of current against time for the inductor circuit of Fig 12.17**

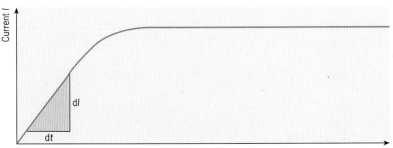

When the switch is opened, the inductor current collapses much more quickly. This reducing current induces a large e.m.f. in the opposite sense to the applied e.m.f. We can demonstrate the size of this e.m.f. by connecting a neon bulb across the inductor as shown in Fig 12.20. A neon bulb requires a voltage of at least 80 V across it for it to light. When the switch in this circuit is opened the bulb will light!

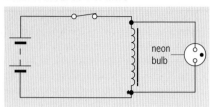

Fig 12.20 **When the circuit is opened, the neon bulb lights up, illustrating the high induced e.m.f. of about 80 V or more**

EXAMPLE

Q A 50 mH inductor is connected to a 3 V d.c. supply. By how much does the current increase in the first 10 ms?

A We can use the equation: $V = L\dfrac{\mathrm{d}I}{\mathrm{d}t}$

$$3 = 50 \times 10^{-3} \ (\text{henrys}) \times \frac{\mathrm{d}I}{0.01 \ (\text{s})}$$

The current increase in the first 10 ms = 0.6 A

Current variation in circuits containing inductance

We can treat the inductor as being similar to the motor we described on page 272. Like the motor, the e.m.f. of the supply is opposed by the e.m.f. induced in the inductor. The current in the circuit is given by:

$$\text{current} = \frac{\text{e.m.f. of supply} - \text{e.m.f. of inductor}}{\text{total resistance of circuit}}$$

or, in symbols: $\quad I = \dfrac{V - L\dfrac{\mathrm{d}I}{\mathrm{d}t}}{R}$

which can be rearranged as: $\dfrac{\mathrm{d}I}{\mathrm{d}t} = \dfrac{V - IR}{L}$

This is a first-order differential equation which can be solved using calculus. The solution to this equation is another equation, which describes the exponential growth of the current shown in the graph in Fig 12.21.

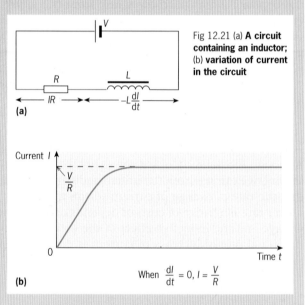

Fig 12.21 (a) **A circuit containing an inductor;** (b) **variation of current in the circuit**

(a)

(b) When $\dfrac{\mathrm{d}I}{\mathrm{d}t} = 0$, $I = \dfrac{V}{R}$

Inductance of a solenoid

By defining the e.m.f. induced in a solenoid in two ways – using Faraday's law and in terms of inductance – we can derive a formula for the inductance of a solenoid. This is very useful in making coils of a specified inductance. The e.m.f. induced in a solenoid is given by:

$$\epsilon_i = L\frac{\mathrm{d}I}{\mathrm{d}t} = N\frac{\mathrm{d}\phi}{\mathrm{d}t}$$

For a solenoid: $\qquad \phi = \mu_0\mu_r\dfrac{NA}{l}I$

Therefore: $\qquad \dfrac{\mathrm{d}\phi}{\mathrm{d}t} = \mu_0\mu_r\dfrac{NA}{l}\dfrac{\mathrm{d}I}{\mathrm{d}t}$

and: $\qquad N\dfrac{\mathrm{d}\phi}{\mathrm{d}t} = \mu_0\mu_r\dfrac{N^2A}{l}\dfrac{\mathrm{d}I}{\mathrm{d}t}$

By comparison with the first equation: $\quad L = \mu_0\mu_r\dfrac{N^2A}{l}$

EXAMPLE

Q A coil available in an electronic components catalogue has an inductance of 100 µH. It has 10 turns of wire wound over a length of 1 cm on a former of cross-sectional area 0.25 cm². Show that the material of the core has a relative permeability of about 320.

A Use the equation for the self-inductance of a coil:

$$L = \mu_0 \mu_r \frac{N^2 A}{l}$$

Rearranging gives: $\mu_r = \dfrac{Ll}{\mu_0 N^2 A}$

$$= \frac{1 \times 10^{-4} \times 0.01}{4\pi \times 10^{-7} \times 10^2 \times 0.25 \times 10^{-4}}$$

$$= 10^3/\pi$$

Relative permeability $= 318$

10 ENERGY STORED IN AN INDUCTOR

As the current through the inductor increases, energy is stored in the magnetic field around the inductor. Once the current reaches a steady value, no more energy is stored. When the current is switched off, the magnetic field collapses and the energy is released via the high induced e.m.f. which lights the neon bulb in Fig 12.20.

Calculation of energy stored in the magnetic field of an inductor

When an inductor is connected to a power supply, initially the power supply does work overcoming the e.m.f. induced in the inductor in order to set up a magnetic field. Energy is stored in the magnetic field as this happens. (When the current in the inductor is steady, the power supply continues to do work against the electrical resistance – that is, in heating.)

Therefore, to calculate the energy stored in an inductor we must consider what happens as the current rises to its maximum value.

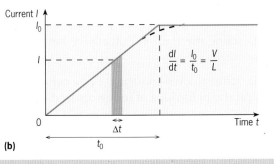

(a)

(b)

Fig 12.22 **Calculating the energy stored in an inductor**

The energy delivered per second, or the power in watts, is given by:

$$W = VI$$

where I is the current at that time.

The power is also equal to the small amount of energy ΔE that is stored during time Δt. Therefore:

$$VI = \frac{\Delta E}{\Delta t}$$

or:

$$\Delta E = VI\, \Delta t$$

$I\Delta t$ is equal to the shaded area shown on the *simplified* graph of current growth in the inductor in Fig 12.22(b).

The total energy is the sum of all the strips that make up the area under the curve as the current rises to its final value I_0:

$$\text{total area} = \tfrac{1}{2} I_0 t_0$$

where t_0 is the time it takes for the current to reach the peak current I_0.

This means that the total energy can be written as:

$$E = \tfrac{1}{2} V I_0 t_0$$

But the gradient, dI/dt, of this section of the curve is given by:

$$\frac{dI}{dt} = \frac{I_0}{t_0} = \frac{V}{L}$$

and so:

$$Vt_0 = I_0 L$$

We can now write the total energy as:

$$E = \tfrac{1}{2} I_0 I_0 L = \tfrac{1}{2} I_0^2 L$$

That is:

$$E = \tfrac{1}{2} I_0^2 L$$

The transformer

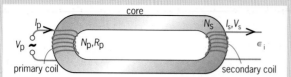

Fig 12.23 **A transformer**

Faraday's original experiments demonstrated the action of the transformer. If the current in the primary coil is continually changing, then the flux it produces will also change continuously. This means that a continuous but varying current will be induced in the secondary coil.

Transformers are used widely to change alternating voltages: **step-up** and **step-down** transformers increase or decrease voltages:

$$V_p/V_s = N_p/N_s$$

Most mains-powered electronic devices have a step-down transformer to provide the lower voltage that an electronic circuit needs to operate. Chapter 13 discusses both types of transformer in their use for transmitting electrical energy.

The e.m.f. induced in the secondary is given by:

$$\epsilon_i = -N_s \frac{d\phi}{dt}$$

The primary has N_p turns and resistance R_p ohms. So the applied voltage at any instant is:

$$V_p = I_p R_p + N_p \frac{d\phi}{dt}$$

where I_p is the corresponding value of the current in the primary. If the core is short and fat then a small current will produce a large flux, as $I_p R_p$ is small. In this case:

$$V_p \approx N_p \frac{d\phi}{dt}$$

If we divide the equation for ϵ_i by that for V_p, we get:

$$\frac{\epsilon_i}{V_p} \approx \frac{-N_s(d\phi/dt)}{N_p(d\phi/dt)} \approx -\frac{N_s}{N_p}$$

N_s/N_p is called the **turns ratio**.

Well-made transformers are very efficient, so we can replace the 'approximate' sign by 'equals':

$$\frac{\epsilon_i}{V_p} = -\frac{N_s}{N_p}$$

For a very efficient transformer, we can write:

Energy into primary = energy out of secondary.

The energy per second is the power (*IV*), therefore:

$$I_p V_p = I_s \epsilon_i$$

or:

$$\frac{I_s}{I_p} = \frac{V_p}{\epsilon_i} = \frac{N_p}{N_s}$$

In other words, if the transformer steps the voltage up then the current will be stepped down, and vice versa.

■ See question 9.

SUMMARY

By the end of this chapter you should be able to:

■ Describe Faraday's experiments on electromagnetic induction.

■ Understand Faraday's law of electromagnetic induction: $\epsilon = N(d\phi/dt)$

■ Use the equation $V = Blv$ to calculate the e.m.f. induced in a moving wire.

■ Define magnetic flux as the product of the vertical component of field strength and the area it is passing through.

■ Use Lenz's law to predict the direction of an induced current.

■ Describe the behaviour of a simple moving coil dynamo.

■ Compare magnetic circuits with electric circuits.

■ Use the equation $NI = (\phi l)/(\mu_0 \mu_r A)$ to calculate the magnetic field strength in a solenoid and around a long wire.

■ Describe how eddy currents are induced and give examples of their effects.

■ Describe the behaviour of a simple d.c. motor under variable loads.

■ Describe the mutual inductance between a pair of coils.

■ Describe the self-inductance of a coil, $V = L\, dI/dt$, and the behaviour of inductors in d.c. circuits.

■ Calculate the energy stored in the magnetic field of an inductor.

■ Describe the behaviour of a simple transformer.

QUESTIONS

1 An aircraft flying due North cuts the Earth's magnetic field at right angles. The wing span is 30 metres and the plane flies at a velocity of 200 m s^{-1}. The Earth's magnetic field strength is 4.3×10^{-5} T. Calculate the e.m.f. induced across the tips of the wings.

2 A magnet is dropped so that it falls vertically through a coil, as shown in Fig 12.Q2. The trace produced on the oscilloscope is shown.

Fig 12.Q2

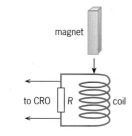

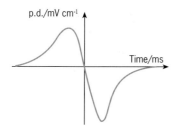

a) Explain the shape of the trace.

b) Why is the negative peak larger than the positive peak?

c) What is the significance of the areas enclosed under each peak?

3 A physicist decided to make an estimate of the current induced in her gold wedding ring when she quickly moved her hand through 90°. The calculation started as follows but is unfinished.

Earth's B field = 6×10^{-5} T [1]

cross-sectional area = 2 cm^2 = 2×10^{-4} m^2 [2]

Therefore the maximum possible change of flux of the B field when the hand is moved is:

$(6 \times 10^{-5}$ T$)(2 \times 10^{-4}$ m$^2) = 12 \times 10^{-9}$ T m^2 [3]

If the time taken for the movement is 0.2 s, the average induced e.m.f. is:

$$\frac{12 \times 10^{-9} \text{ T m}^2}{0.2 \text{ s}} = 6 \times 10^{-8} \text{ V}$$ [4]

a) Fig 12.Q3 shows a sectional diagram of the ring. Copy it and show on it the area referred to in the calculation as 2 cm^2.

Fig 12.Q3

b) What else must be estimated in order to complete the calculation of the induced current?

c) Modern gold wedding rings are sometimes thicker than that worn by this physicist. Explain how such a thicker ring would affect:
  (i) the induced e.m.f.;
  (ii) the induced current.

d) What assumption is being made in step 3 about the orientation of the plane of the ring in the Earth's B field
  (i) at the beginning,
  (ii) at the end of the hand movement?

[UCLES: 1976, Nuffield Advanced Physics, Paper 2 (short answers), Q3]

4 Fig 12.Q4 shows a copper disc of radius 120 mm situated in a uniform magnetic field B of 1.5×10^{-2} T. The plane of the disc is perpendicular to B, which points into the plane of the diagram. The disc is rotated about an axis through its centre O at 2500 revolutions per minute.

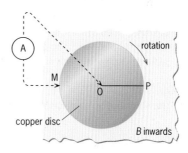

Fig 12.Q4

a) Calculate the magnetic flux cut by a radius OP in 1 s.

b) A stationary connection is made from a sliding contact touching the rim at M through an ammeter A to a similar contact at the centre of the disc. This is indicated by the dotted lines on the diagram. The resistance of this connection MAO is 0.10 Ω and the resistance between the rim and the centre of the disc is negligible. Calculate the magnitude of the current through A.

c) Explain whether a conduction electron in the copper disc will experience a force of magnetic origin towards the centre of the disc or towards the rim as a result of its motion in the magnetic field. Deduce the direction of the current through A.

[OCSEB: June 1994, Physics 3/4, Q3]

5 A plane rectangular coil with sides of length a and b is rotated at a frequency f about the axis AA' which is perpendicular to a uniform magnetic field B. The coil has n turns. From A to E (next page), choose in which circumstance the peak value of the alternating e.m.f. in the coil will be doubled.

A Both *a* and *b* are doubled.
B *f* is increased by a factor √2.
C *B* is halved.
D *n* is doubled.
E *f* is halved.

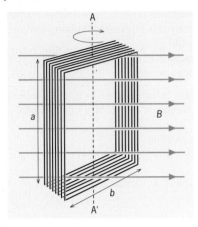

Fig 12.Q5

[OCSEB: June 1994, 9660/1, Q17]

6 Fig 12.Q6 shows a long 'slinky' solenoid with a constant alternating current in it. The small coil known as a search coil has a large number of turns and fixed area of cross-section. The search coil is placed to measure the magnetic field strength inside the solenoid. Describe how the following changes would affect the flux in the solenoid and the e.m.f. induced in the search coil:

a) Doubling the size of the current in the solenoid.

b) Stretching the solenoid so that the coils are twice as far apart.

c) Doubling the current *and* doubling the spacing between the coils.

d) Replacing the search coil with one with the same number of turns but twice the cross-sectional area.

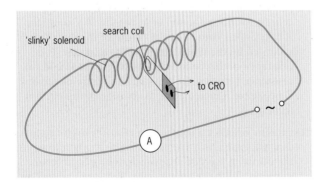

Fig 12.Q6

7 In Fig 12.Q7(a) a potential difference *V* across a long iron rod produces a current *I* in the rod, the resistance of the rod being *V/I*. In Fig 12.Q7(b), the current-turns *NI* around the rod produce a flux ϕ in the rod, the reluctance of the rod being *NI/ϕ*.
    If the area *A* of cross-section were doubled (the length staying the same and being large compared to the diameter), by which of the factors A to E below: **a)** is the

resistance of the rod multiplied, **b)** is the **reluctance** of the rod multiplied?

A $\frac{1}{4}$
B $\frac{1}{2}$
C 1
D 2
E 4

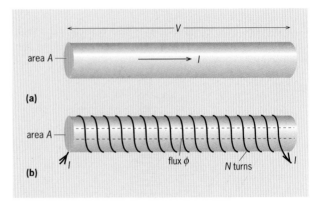

Fig 12.Q7

[UCLES: 1981, Nuffield Advanced Physics coded answer paper 1, Qs 7&8]

8 When the switch in the circuit in Fig 12.Q8 is closed, the current rises steadily to 0.25 A in the first 0.01 seconds. *R* is 0.1 Ω and *L* has negligible resistance.

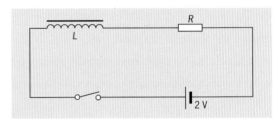

Fig 12.Q8

a) Show that the inductance of *L* is about 80 mH.

b) If *R* is increased to 0.2 Ω what will happen to the initial rise in current?

c) *R* is now increased to 2 Ω. What is the initial rate of current increase now?

d) *L* is replaced by a coil of inductance 1.0 H and resistance 1 Ω with *R* remaining at 2 Ω. Calculate the initial rise in current and the final current in the circuit.

9 A step-down transformer reduces 240 V to 12 V to light a 12 V, 36 W bulb to full brightness.

a) What is the turns ratio of the transformer?

b) How much current flows in the primary coil?

c) Show that the apparent resistance of the primary coil is given by:

(turns ratio)^2 × resistance in secondary

Assume that the primary and secondary coils themselves have negligible resistance.

Assignment

THE MOTORS THAT MAKE THE WORLD MOVE FAST

In April 1996, *New Scientist* described recent ideas in the competition to produce trains that can travel at speeds of up to 500 km per hour. The present record is held by Japan's Maglev train which has reached 517 km per hour. A similar design called the Seraphim is being developed in Mexico. Both Maglev and Seraphim use electromagnetic forces to propel them.

The original idea for all magnetically levitated trains is based on the **linear induction motor**. The idea behind it is this: use a magnet to apply a changing magnetic field to a conductor, and you induce a current in the conductor. With this current there is a field in the conductor that opposes the magnet's field, and the resulting force between magnet and conductor can be used to produce movement.

About 95 per cent of the world's motors are now induction motors, all based on the fact that a moving magnetic field can be used to cause motion.

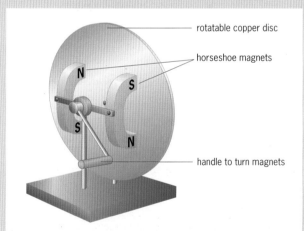

Fig 12.A1 **Setup to show that a moving field induces motion**

The magnets in Fig 12.A1 are rotated, and the moving field induces currents in the metal disc that give rise to an opposing field.

1 Suggest some possible current paths in the metal disc and use Lenz's law to explain why the metal disc should follow the rotating magnets.

To produce a practical motor based on this principle, we have to create a moving magnetic field in a different way from just physically moving magnets. We apply an alternating e.m.f. This produces an alternating magnetic field, that is, a changing field. Fig 12.A2 shows a possible arrangement.

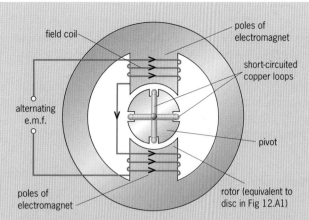

Fig 12.A2 **A single phase motor, which turns out to be a very poor motor because it doesn't turn!**

Although there are two coils, the magnetic field in each coil is in phase.

2 The arrows on the diagram show the direction of the current at an instant in time. What effect will that have on the two poles of the electromagnet 'field coils'?

The rotor of this motor consists of two short copper loops of very low resistance.

3 The changing field in the field coils will induce a current in only one of the two loops shown in Fig 12.A2. Which one, and why?

The induced current causes its own magnetic field, but there is no turning force (torque), so the rotor will not move by itself. If it is pushed it may continue to rotate. There are two ways in which this 'unsuccessful' motor can be made to work; that is, the rotor made to turn by itself!

The two-phase motor

The first method uses a capacitor to change the phase of the current in a second field coil which is placed at right angles to the first, as in Fig 12.A3.

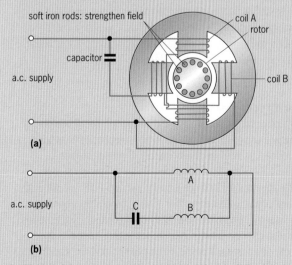

Fig 12.A3(a) **A two-phase machine and** (b) **the electrical circuit of the two coils**

The capacitor in series with the second coil makes the current in it out of phase with the current in the first coil (see page 289). The result is that the currents induced in the loops in the rotor are also out of phase, and a turning force is produced. The capacitor turns the motor into a two-phase machine. In practice, the phase difference is never as large as 90°. As the magnetic fields change out of phase, it is as if they are rotating about the rotor.

Motor with a 'shaded pole'

The second method of making a working motor is by using a 'shaded pole'.

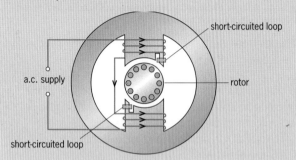

Fig 12.A4 **The shaded pole motor**

The shaded pole motor has two short-circuited loops on a slot in each pole of the electromagnet. The effect is to 'shade' the slotted section of the pole. This shorted loop is 'magnetically' linked to the field coil on each pole because they share the same iron core.

a) What will be the direction of the current induced in the shorting loop relative to that in the field coil on each pole?

b) How will the magnetic field over the shaded section compare to that over the unshaded part of each pole?

The shading ring converts the motor into a two-phase machine. Such motors are used widely where a motor is needed to run at a steady speed but not to do a lot of work. A record deck is a good example.

The three-phase motor

For applications requiring greater energy, the three-phase motor is used. An advantage is that electricity is actually produced as a three-phase supply in power stations.

Fig 12.A5 shows a simplified three-phase generator. The currents induced in three equally spaced coils will be 120∞ out of phase with each other. The coils on the generator are usually connected in a 'star' configuration. This means that the energy drawn from the generator requires four wires, one for each of the three phases and the fourth 'neutral' .

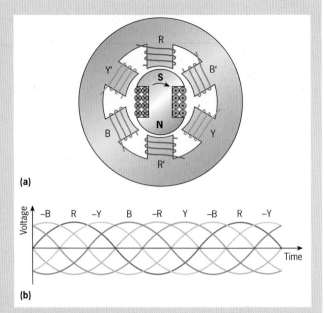

(a)

(b)

Fig 12.A5 **A three-phase generator. As the rotating magnet (here, an electromagnet) spins, it induces a current in each set of coils, RR', BB', YY' in turn: the three phases are red, blue and yellow**

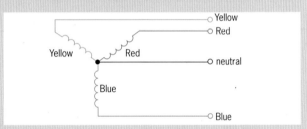

Fig 12.A6 **Star connection and wiring of a three-phase generator**

A three-phase induction motor has three sets of coils, each one connected to a different phase from the supply. This time, the phase between the magnetic field produced by each coil is exactly 120°. This produces a very good rotating magnetic field and a motor capable of providing very high turning forces.

Imagine the coils of this motor 'flattened' out, as in Fig 12.A7. The rotor is replaced by a flat aluminium plate which is placed above the coils.

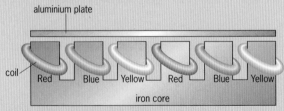

Fig 12.A7 **The linear motor: a 'flattened' induction motor showing the arrangement of coils on the iron core**

5 What will happen to the plate when the three-phase supply is switched on? Imagine the coils mounted on a train which sits on an aluminium rail. When power is supplied, the train will 'float' above the rail and move along it. This is the principle behind Maglev and Seraphim.

ELECTROMAGNETIC INDUCTION

The main ideas and equations that you have studied in this chapter are brought together in this Chapter Map. The arrangement of boxed items show how the ideas interlink and rely on each other. Use the map to plan your revision according to the syllabus you are following, and to identify any areas which you need to study further.

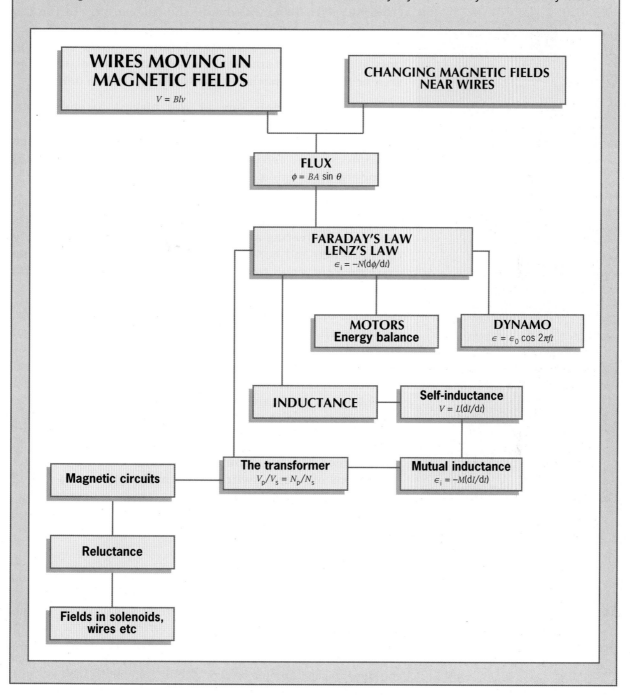

Continual monitoring and maintenance of the National Grid.
Above: an operations room
Below: repairing transmission lines

FEBRUARY 5TH, 1996, was one of the coldest days in the UK for many years. On that day the generators and distributors of electrical power were on standby to deal with the possibility that the amount of energy demanded would be greater than the capacity of the National Grid to deliver.

The National Grid is one of the biggest assemblies of equipment in the country. It is a highly flexible, integrated system with about 7000 kilometres of overhead high voltage transmission line carried by 20 000 pylons and over 500 km of underground cable.

By late afternoon on February 5th, the daily peak had passed and the operators could breathe a sigh of relief. The grid was capable of delivering a total of 57 gigawatts and fortunately the peak demand had reached only 48 750 megawatts. The extreme cold conditions and high electrical demand are fairly rare, and the grid had stood the test.

But there are more frequent, and usually more predictable, demands that the grid operators must cope with. Royal weddings and international football matches cause surges – demand increased by 10 per cent just after England's defeat by West Germany in the 1990 World Cup – making a consoling hot drink and other activities that use electricity.

Daytime demands are twice as high as at night, and the maximum demand in winter is four times the minimum in the summer. The surge immediately after Coronation Street as millions of people switch on their electric kettles is well known to the planners!

Introduction

Alternating currents play a very important role in electronic and physical systems. A microphone converts sound, caused by vibrations in the air, into an alternating current signal. When a loudspeaker is connected to an alternating current supply, it converts the signal back into sound. The electromagnetic waves radiated from a transmitting aerial are created by alternating currents and the signal picked up by a receiving aerial is an alternating current. The speed at which a computer carries out instructions is determined by the frequency of alternating current created by the vibration of a tiny quartz crystal.

On an average day the output of power stations in the UK is over 60 gigawatts and this energy is carried by alternating currents to the consumer. The nature of alternating current circuits is important to our understanding of both the National Grid and the design of a radio.

The UK mains supply is now nominally 230 V, but the distributors of electricity are allowed to vary it between 253 V (+10%) and 216 V (–6%).

1 ALTERNATING VOLTAGES AND CURRENTS

The 'voltage' from a battery is direct – that is, it drives a steady or **direct current** (d.c.) around a circuit. This means that the charge flows in one direction. Fig 13.1 shows that the current (or voltage) does not change with time. This is an idealised situation, since in reality the current from a real battery would gradually get smaller as the battery 'runs down'.

The electrical energy we get from the 'mains' is an alternating voltage. It drives an **alternating current** (a.c.) around a circuit.

An alternating current is continually changing direction: see Fig 13.2, which shows a typical current flowing in a mains circuit in the UK.

Alternating current shows a distinctive **waveform** that is continually repeated: the graph shows three complete cycles of the pattern. The pattern of the a.c. mains is repeated 50 times every second – it has a frequency of 50 Hz. One cycle takes one-fiftieth of a second or 0.02 s. This time is the **period** of the waveform. Another important characteristic of the waveform is that each cycle has a positive half-cycle and a negative half-cycle. During the positive half-cycle the current is in one direction; during the negative half-cycle the current direction is reversed.

The alternating current produced by a dynamo or generator varies in a very precise way – the current varies as a sine wave because of the way it is made by a spinning coil (see page 266). That is, where I_0 is the maximum current and f is the frequency of the current:

$$I = I_0 \sin 2\pi ft$$

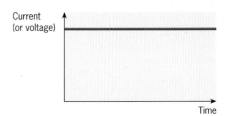

Fig 13.1 **A direct current does not change with time**

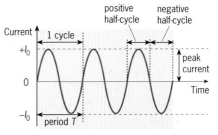

Fig 13.2 **An alternating current. This current varies sinusoidally, with alternate positive and negative half-cycles**

See question 1.

2 POWER IN A.C. CIRCUITS

Fig 13.3 shows two simple circuits in which a bulb is lit from an a.c. supply and from a d.c. supply. In each case the supply delivers energy to the bulb, the bulb gets hot and emits light.

In the d.c. circuit the energy transferred in the bulb each second (the power) can be found easily by using the equation:

Power = p.d. × current

Energy is also transferred in the a.c. circuit. However, there is a problem if we try to use the same equation. If the current and voltage are continually changing, which value of current and voltage do we use? The peak values are reached only twice each cycle. What about the average value? What is the average value of current or voltage over a complete cycle? Look at Fig 13.4 and decide.

The answer is zero! The positive half-cycle is cancelled by the negative one.

There is a solution: instead of current, we consider power. In a simple circuit like this, the current and voltage vary together, increasing and decreasing in time with each other. We say they are **in phase** with each other. If we multiply the value of the current at any instant by the corresponding voltage, we get the power at that same instant in time. The result is as shown in Fig 13.5.

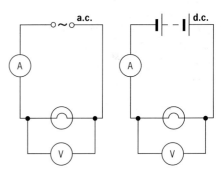

Fig 13.3 **Simple a.c. and d.c. circuits. In both cases the bulb lights: there is an energy transfer from the power supply to the bulb**

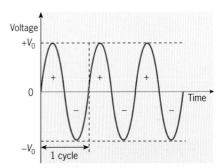

Fig 13.4 **Three complete cycles of alternating voltage. The average over a complete number of cycles is zero**

Fig 13.5 **Change in current, voltage and power over three cycles for a resistor. The power is always positive**

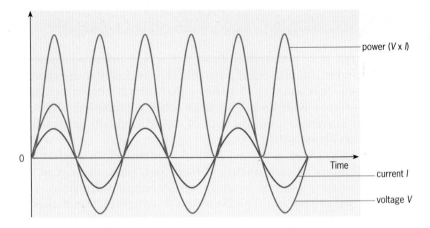

Since the current and voltage shown are sine waves, the corresponding power is a sine-squared wave. Notice that when the current and voltage are both negative the power is positive – the power is always positive, even though it is still varying. You can see from the graph that the average value for power is not zero. In fact, it works out that the average value of sine-squared is half of the maximum or peak value:

$$\text{average power} = \tfrac{1}{2} \times \text{peak power}$$
$$= \tfrac{1}{2} \times (\text{peak current} \times \text{peak voltage})$$
$$= (\tfrac{1}{\sqrt{2}} \times \text{peak current}) \times (\tfrac{1}{\sqrt{2}} \times \text{peak voltage})$$
$$= \text{r.m.s. current} \times \text{r.m.s. voltage}$$

where: $\quad$ r.m.s. current $= \tfrac{1}{\sqrt{2}} \times \text{peak current}$

$\qquad\qquad$ r.m.s. voltage $= \tfrac{1}{\sqrt{2}} \times \text{peak voltage}$

> ✔
> Note that the mathematical idea used here is:
> $$\tfrac{1}{2} = \tfrac{1}{\sqrt{2}} \times \tfrac{1}{\sqrt{2}}$$

The 'r.m.s.' alternating current and voltage are actually equal to the steady, direct current and voltage which transfer the same amount of energy in the bulb. The abbreviation r.m.s. stands for **root mean square**. It refers to the square root of the average of the squares of the current or voltage at every instant.

> ?
> **A** In the UK the mains voltage is 230 V. This is in fact the r.m.s. value of the voltage. What is the peak value?

Also: $\quad$ Peak current or voltage $= \sqrt{2} \times$ r.m.s. value

$\qquad\qquad\qquad\qquad\qquad = 1.4 \times$ r.m.s. value

See questions 2, 3 and 4. ▥

3 R.M.S., CAPACITORS AND WORKING VOLTAGES

The voltage stated for power supplies is usually the r.m.s. voltage. When using a capacitor we have to be sure not to exceed the working voltage of the capacitor (see page 218). In d.c. circuits it is wise not to use voltages over about two-thirds of the rated working voltage of the capacitor. With a.c. supply we have to be even more careful.

For example, suppose a capacitor has a working voltage of 16 V. If such a capacitor were used in a circuit operated from a 9 V d.c. supply we could be confident that it would behave well as a capacitor. However, if we have a 9 V *alternating* voltage across the same capacitor, the peak voltage across it would be (9 × 1.4) or 12.6 V. This is still below 16 V but is about 2 V over two-thirds of 16 V, which is 10.7 V. It would be better to use a capacitor with a higher rated working voltage.

> ?
> **B** What r.m.s. alternating voltage has a peak voltage which just exceeds 16 V?

4 DISTRIBUTION OF ELECTRICAL ENERGY
– THE NATIONAL GRID

The National Grid is the system used for distributing electrical energy around the country. It is a large and very complex circuit through which the power stations feed energy into every home, factory, hospital and school.

The size of the circuit creates problems. In more remote parts of the British Isles, consumers may be up to a hundred miles from the nearest power station. Cables of that length have substantial resistance. A current in these cables heats them – there is **resistive heating**. This heat represents energy lost, wasted energy that does not reach the consumer.

These energy losses are often called I^2R losses since, in a cable of resistance R carrying a current I, the resistive power loss is I^2R (see Chapter 9, page 206). The fact that the power loss is proportional to I^2 is significant: it means that halving the current reduces the power lost by a factor of four. This fact plays an important part in the design of the National Grid (Fig 13.6) – electricity is transferred at high voltages so that currents can be kept small for a given power load. **Transformers** are used to increase and decrease voltages, and can also interchange a.c. and d.c.

The largest modern generators produce electricity at 25 000 V. Heavy industries use electricity at 33 000 V while lighter industries use it at 11 000 V. Smaller users such as our homes use electricity at 230 V. Electricity is transmitted around the country at much higher voltages – 132 kV, 250 kV and 400 kV.

One of the advantages of generating electricity as an alternating current is that transformers change a.c. voltages easily and efficiently (large transformers are typically about 99 per cent efficient). Step-up and step-down transformers are used throughout the system to change voltages to obtain the required voltage. (Note that 'step-up' and 'step-down' refer to voltage and not current.)

Fig 13.6 **The National Grid, a large complex circuit which delivers energy from the power stations to the consumer. Different consumers require energy to be delivered at different voltages**

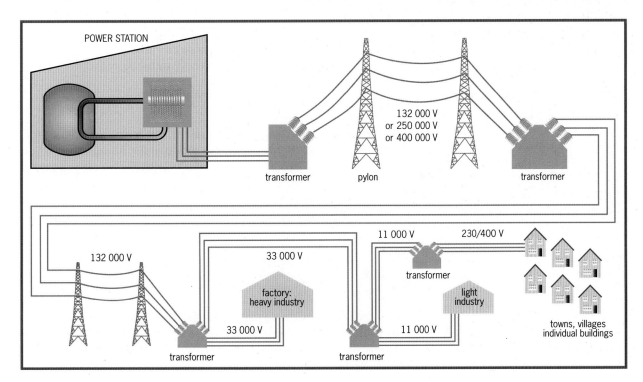

POWER STATION

132 000 V
or 250 000 V
or 400 000 V

transformer   pylon   transformer

132 000 V   33 000 V   11 000 V   230/400 V

factory:
heavy industry

light
industry

transformer

33 000 V   11 000 V

towns, villages
individual buildings

transformer   transformer

EXAMPLES

Q A power station generates a current of 100 A at 25 kV. The electricity is transferred along 100 km of power line to a chemicals factory. The power line has a total resistance of 50 Ω. Calculate the power lost in the power line and the percentage efficiency.

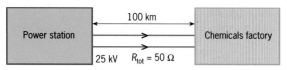

Fig 13.7 **The system described in the question**

A Power output of power station = $I \times V$

$\qquad$ = 100 A × 25 000 V

$\qquad$ = 2 500 000 W

$\qquad$ = 2.5 MW

Power lost from heating power line = I^2R

$\qquad$ = $100^2 \times 50$

$\qquad$ = 500 000 W

$\qquad$ = 0.5 MW

That means that only 2.0 MW is available at the factory. The system is only 80 per cent efficient.

Q The system is improved by adding two transformers, one to step up the voltage to 125 kV and a second to step it back down at the factory. The first transformer has a turns ratio of 1:5; the second has a turns ratio of 5:1. By how much does this reduce the power losses? What is the percentage efficiency? (Assume both transformers are 100 per cent efficient.)

Fig 13.8 **Two transformers added to improve the system shown in Fig 13.7**

A The voltage is stepped up by a factor of 5, therefore:

$\qquad$ current along power line = 100/5 = 20 A

$\qquad$ Power lost heating power line = I^2R

$\qquad$ = $20^2 \times 50$

$\qquad$ 20 000 W

$\qquad$ = 0.02 MW

A loss of 0.02 MW is smaller than 0.5 MW by a factor of 25 or 5^2. This means that 2.48 MW are now available at the factory. The system is now 99.2 per cent efficient.

Electricity in industry

Many industries that depend on electricity require vast amounts of energy. Take, for example, the large-scale production of aluminium, which is produced electrolytically from molten aluminium oxide obtained from the ore bauxite. The process requires a d.c. supply at nearly 1000 V and currents of 70 000 A! The d.c. voltage is derived from the secondary coil of a step-down transformer, which we will assume gives a secondary voltage of 1000 V. We can make some estimates of the turns ratio and primary current in the primary coil of the transformer.

Let us assume 100 per cent efficiency. The factory is connected to the National Grid via a local substation that gives an output voltage of 33 000 V. The required turns ratio would be:

$$\frac{1000}{33\,000} = 1 : 33 \quad \text{– a step-down transformer}$$

While the voltage is being stepped down, the current will be stepped up in the same ratio. That is:

$$\frac{\text{secondary current}}{\text{primary current}} = \frac{33}{1}$$

This gives: $\qquad$ primary current = 70 000/33

$\qquad\qquad\qquad\qquad$ = 2121.2 A

This is still a large current, but far smaller than the 70 000 A needed to produce the aluminium.

This example is quite typical of many heavy industries that need electricity supplied at such large voltages.

?

C (a) What is the turns ratio required to step up the 25 000 V from a generator to 250 000 V?

(b) By how much will the current be changed in this transformer? (Assume the transformer is 100 per cent efficient.) How will this affect the power losses in the cable carrying current from the transformer?

5 RECTIFICATION – CHANGING A.C. TO D.C.

Many electronic devices require a d.c. power supply. This means that, despite the advantages of transmitting electrical energy as an alternating current, the first thing that most devices need to do is to convert the a.c. to d.c. This process is called **rectification**. One of the simplest ways of rectifying current is to use a **diode**.

Diodes allow current to pass easily in one direction only (Fig 13.9). The two leads of the diode are called the **anode** and the **cathode**.

Examine the circuit in Fig 13.9. The diode will conduct only when the anode is positive – that is, during positive half-cycles. Fig 13.10(b) shows the alternating voltage from the supply and the voltage (p.d.) across the resistance R. The voltage can be shown on an oscilloscope connected across the resistor.

The voltage across the load and the current in it are not steady, but are direct. That is, they are always positive and the charge flows in one direction – it does not alternate. This simple circuit satisfies the aim of turning a.c. into d.c., but we do lose half the energy. For obvious reasons, this process is called **half-wave rectification**.

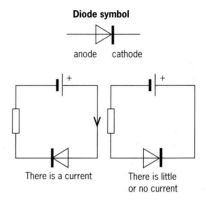

Fig 13.9 **The two circuits illustrate the action of a diode. The diode conducts only when the anode is positive**

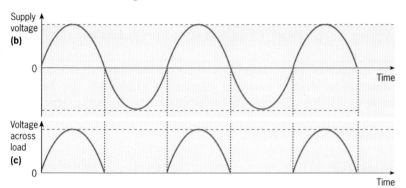

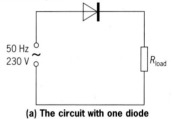

Fig 13.10 **Half-wave rectification using a single diode: as (c) shows, the diode conducts and allows a current through the load only during positive half-cycles**

(a) The circuit with one diode

The circuit in Fig 13.11(a) goes one step further. The arrangement of four diodes is called a **bridge rectifier** and it achieves full-wave rectification. The diodes are arranged so that a charge goes through the load in the same direction on each half-cycle. During the positive half-cycle the diodes D1 and D4 conduct; during the negative half-cycle the diodes D2 and D3 conduct.

The current is now direct. It is not steady but the current is in the same direction through the load during the whole cycle of applied a.c. As Fig 13.11(c) shows, the voltage has a 'ripple', with the ripple frequency being twice the frequency of the original alternating voltage shown in (b).

Fig 13.11 **Full-wave rectification using a bridge rectifier. This time, the current through the load is the same for positive and negative half-cycles of the supply**

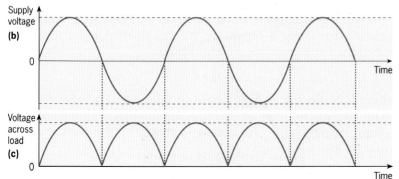

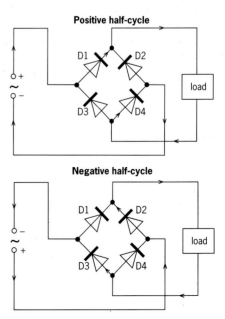

Smoothing out the current from a full-wave rectifier

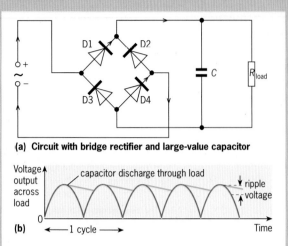

(a) Circuit with bridge rectifier and large-value capacitor

(b)

The varying d.c. can be smoothed out by placing a large-value capacitor across the load, as in Fig 13.12(a).

Fig 13.12(b) shows that during the first quarter of the cycle the capacitor charges. During the second quarter the voltage from the bridge rectifier drops and the capacitor starts to discharge. The rate of discharge depends on the capacitor and the resistance of the load – the *RC* constant (see Chapter 9, page 220).

If *RC* is large compared with the period of the ripple (1/ripple frequency), then the ripple will disappear and the voltage will be smooth.

Fig 13.12 Full-wave rectification with smoothing. The capacitor smoothes out the bumps in the full-wave rectified voltage

See question 5. ▨

6 REACTANCE AND IMPEDANCE

The behaviour of components such as resistors, capacitors and inductors in circuits carrying alternating current is the basis of many useful devices. The size of the current turns out to depend on frequency as well as the size of the component, which can be used to make circuits like filters.

Fig 13.13 The graphs show how the current in the resistor and the voltage across it vary with time. They are in phase

Resistors

Ohm's law is obeyed for a.c. just as it is for d.c. The alternating current in a resistor is in phase with the alternating potential difference across it (that is, they vary together as shown in Fig 13.13). The resistance in ohms is the same for d.c. and a.c.

Capacitors

In d.c. circuits we say that a capacitor 'blocks d.c.'. In other words, there is no current after the initial brief charging period. The space between the plates contains an insulator – charge does not flow from one plate to the other.

Similarly, charge does not flow between the plates when an alternating voltage is applied. However, charge flowing to and from one plate does induce a flow of charge on the other. That means that there are currents in the rest of the circuit.

So what does happen when we apply an alternating voltage to a circuit containing a capacitor? We can work this out from first principles. For a capacitor we know that:

$$Q = CV$$

where V is the p.d. at any instant in time across a capacitor of C farads and when there is a charge of Q coulombs on the plates.

Now, current is the rate of flow of charge, so the current at this instant is given by:

$$I = \frac{\mathrm{d}Q}{\mathrm{d}t} = \frac{\mathrm{d}(CV)}{\mathrm{d}t} = C\frac{\mathrm{d}V}{\mathrm{d}t}$$

This says that the current at any time is proportional to the rate of change of voltage.

If the p.d. across the capacitor changes sinusoidally, that is, if it is a standard alternating voltage, then:

$$V = V_0 \sin 2\pi ft$$

so:

$$I = C\frac{dV}{dt} = C\frac{d}{dt}(V_0 \sin 2\pi ft) = 2\pi fCV_0 \cos 2\pi ft$$

That is, if the p.d. is a sine wave then the current will be a cosine wave, as shown in Fig 13.14.

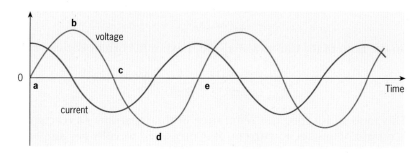

Fig 13.14 **The graphs show how the current in the capacitor and the voltage across it vary with time. Notice that the current 'leads' the voltage by 90°**

The first thing to notice is that the voltage and current are not in phase – they do not vary up and down together. Let us look carefully at what is going on as the applied voltage changes, referring to the letters in Fig 13.14.

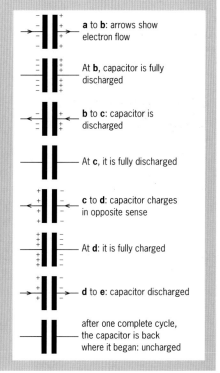

- *From a to b* At the start, when time $t = 0$, the voltage is zero, but at the same instant the current is a maximum value. As the voltage across the capacitor increases, the current decreases until it reaches zero when the voltage is a maximum. At this point the capacitor is fully charged.

  a to b: arrows show electron flow

  At b, capacitor is fully discharged

- *From b to c* The voltage now decreases and the current increases in the opposite direction – that means that charge is flowing in the opposite direction, discharging the capacitor. By the time the voltage reaches zero the current is a 'negative' maximum. The capacitor is now fully discharged.

  b to c: capacitor is discharged

  At c, it is fully discharged

- *From c to d* Charge continues to flow in the same direction as the direction of the voltage changes and gradually increases to a negative maximum. At this point the current is again zero and the capacitor is again fully charged but the charge on the plates is now reversed.

  c to d: capacitor charges in opposite sense

  At d: it is fully charged

- *From d to e* As the voltage returns to zero the capacitor discharges as charge now flows in the opposite direction again until at e the capacitor is uncharged.

  d to e: capacitor discharged

  after one complete cycle, the capacitor is back where it began: uncharged

Look back to Fig 13.14: the current is always a quarter of a cycle ahead of the voltage. (Remember: events to the left of the graph happen before events to the right.) A quarter of a cycle is equivalent to 90°, so we say that the current **leads** the voltage by 90°. The current is a maximum when the p.d. across it is zero. As the p.d. increases the current drops.

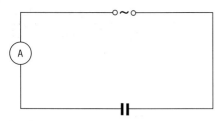

Fig 13.15(a) **The circuit shows a variable frequency supply connected to a capacitor**

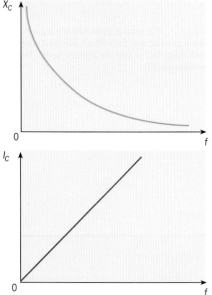

Fig 13.15(b) **The graphs show how the reactance of the capacitor and the current in the circuit vary with frequency**

?

D Calculate the reactance of a 1 μF capacitor at **(a)** 1 kHz, **(b)** 100 kHz and **(c)** 1 MHz.

In simple circuits the ratio of voltage to current tells us something important – which is the resistance of the conductor. Although it is a bit trickier, a ratio of voltage to current is also useful in circuits with capacitors, also known as **capacitive circuits**.

A simple ratio of voltage to current will not make much sense – when the voltage is zero the current is a maximum, giving a ratio of zero, and when the voltage is a maximum the current is zero, giving a ratio of infinity.

However, the ratio of maximum voltage to maximum current is useful. Let the maximum value of voltage be V_0. The maximum value of the current will be when the cosine term is 1. That is:

$$I_{max} = 2\pi f V_0 C$$

So the ratio of maximum voltage to maximum current is given by:

$$\frac{V_{max}}{I_{max}} = \frac{V_0}{2\pi f V_0 C} = \frac{1}{2\pi f C} = X_c$$

This ratio X_c is called the **reactance** of the capacitor. As reactance is a ratio of voltage to current, it is measured in volts per ampere, V A^{-1}, or ohms, Ω.

The equation and Fig 13.15(a) show that the reactance decreases as frequency increases. Or, put another way, as frequency increases, the current in the circuit will increase, as seen in Fig 13.15(b).

It appears as though the capacitor 'conducts' better at high frequencies. Remember, though, that charge does not move directly from one plate to the other.

Inductors

Charge does flow through an inductor, but a continually changing current will create a continually changing magnetic field. This changing magnetic field induces an e.m.f. that opposes the current. The equation that defines **inductance** (see page 273) gives us a relationship between the current and the voltage:

$$V = L\frac{dI}{dt}$$

where V is the applied voltage across the capacitor and I is the current in the inductor.

If the current is sinusoidal, that is:

$$I = I_0 \sin 2\pi f t$$

then:
$$V = L\frac{d}{dt}(I_0 \sin 2\pi f t) = 2\pi f L I_0 \cos 2\pi f t$$

As with the capacitor, the current and voltage are not in phase. Again, voltage and current are out of phase by a quarter of a cycle or 90°. This time, as Fig 13.16 shows, the voltage leads the current by 90°.

We can define an **inductive reactance** X_L in the same way as for a capacitor:

$$X_L = \frac{V_{max}}{I_{max}} = \frac{2\pi f L I_0}{I_0} = \mathbf{2\pi f L}$$

Inductive reactance is proportional to frequency. As frequency increases, the current in an inductor will get smaller, see Fig 13.17.

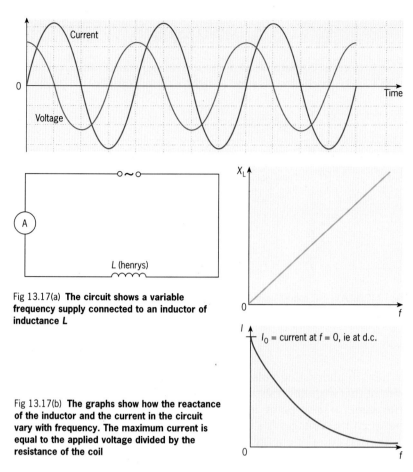

Fig 13.16 **The graphs show how the current in an inductor and the voltage across it vary with time. Notice this time that the voltage leads the current by 90°**

Fig 13.17(a) **The circuit shows a variable frequency supply connected to an inductor of inductance L**

Fig 13.17(b) **The graphs show how the reactance of the inductor and the current in the circuit vary with frequency. The maximum current is equal to the applied voltage divided by the resistance of the coil**

I_0 = current at $f = 0$, ie at d.c.

Phase differences of exactly 90° are found only in purely capacitive circuits and purely inductive circuits. In reality, it is almost impossible to get a 'purely inductive' circuit as the wire in the coil always has some resistance.

FILTER CIRCUITS

AUDIO SYSTEMS ARE DESIGNED to reproduce sounds signals in the range of 20 Hz to 20 kHz. Basic systems use two loudspeakers – one called a **woofer**, designed to reproduce low frequency or bass sounds, and a second called a **tweeter**, which reproduces the higher frequency or treble sounds. The output signal from the audio amplifier contains the whole range of frequencies. Filters are used to direct the 'correct' frequencies to the appropriate loudspeaker. Fig 13.18 shows a simple filter circuit called a **crossover** using capacitors and inductors.

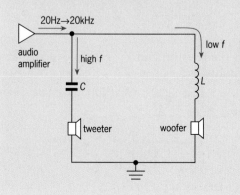

Low frequency signals find a lower reactance path through the inductor to the woofer and the higher frequency signals find it easier to pass through the capacitor.

Fig 13.18 **Crossover circuit**

?

E Calculate the reactance of a 1 μH inductor at **(a)** 1 kHz, **(b)** 100 kHz and **(c)** 1 MHz.

See questions 6 and 7.

7 IMPEDANCE

Many circuits contain a combination of resistance and reactance (Fig 13.19). Such circuits pose a slightly more complicated problem.

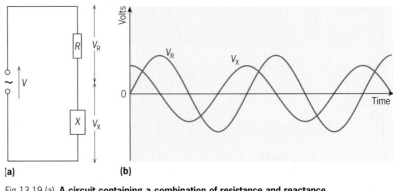

(a)   (b)

Fig 13.19 (a) **A circuit containing a combination of resistance and reactance.**
(b) **The graph shows that the peak voltages across each component are not in phase**

When an alternating voltage of fixed frequency is applied across a circuit that includes resistance and reactance in series, the current is the same at any point in the circuit. However, the relationship between the voltages is a little more complex.

Conservation of energy leads us to say, correctly, that the supply voltage at any instant is equal to the sum of the voltages across the resistance and the reactance at that instant. However, the peak voltages across each component are not in phase, so their sum cannot be equal to the peak voltage of the supply.

The combination of resistance and reactance is called the **impedance**, measured in ohms. Impedance can be calculated by treating the voltages across the different components in a similar way to vector quantities. We use vector diagrams to add together two forces that act in different directions, as shown in Fig 13.20(a). The voltages in Fig 13.19 may have different magnitudes but they also differ in phase. (The reactive voltage phasor is always 90° out of phase with the resistive voltage phasor.) Fig 13.20(b) is called a phasor diagram. The voltage phasors can be added in a similar way to vectors.

The impedance Z (in ohms) is defined by the equation:

$$Z = \frac{V}{I}$$

Resultant voltage: $V = \sqrt{V_R^2 + V_X^2}$

where V_R is the p.d. across the resistor, and V_X is the p.d. across any reactance, capacitance or inductive.

$$V = \frac{\sqrt{V_R^2 + V_X^2}}{I}$$

$$= \sqrt{\left(\frac{V_R}{I}\right)^2 + \left(\frac{V_X}{I}\right)^2}$$

$$Z = \sqrt{R^2 + X^2}$$

(a) Capacitor phasor diagram

(b) Inductor phasor diagram

(c) Phasor diagram for a series RC circuit

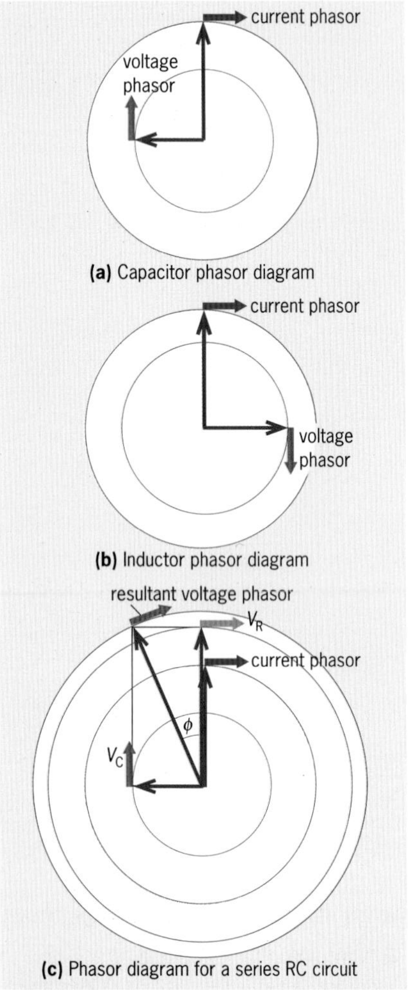

Fig 13.20 **Phasors are rotating vectors. They are very useful when analysing a.c. circuits containing capacitors and inductors.**
The phasor in the diagram is rotating clockwise. It rotates at the frequency of the supply. Note that for a capacitor, the current phasor is 90° ahead of the voltage phasor. For an inductor, the voltage leads the current by 90°.
In (c), the resultant voltage lags the current by an angle ϕ

?

F **(a)** Calculate the impedance of a 100 ohm resistor in series with a 10 μF capacitor at **(i)** 1 kHz, **(ii)** 1 MHz.

(b) Repeat the calculations replacing the capacitor with a 1 mH inductor.

8 LC CIRCUITS

LC circuits include combinations of inductance (L) and capacitance (C).

The circuit in Fig 13.21 is of great importance in electronics in general, and in radio in particular. It is called a **tuned circuit**. One of its uses is to select radio stations in a radio receiver (see Chapter 21).

As the frequency of the supply in Fig 13.21 is increased, the brightness of bulb L_1 gradually decreases until it reaches a minimum level of brightness. It then increases in brightness as the frequency is increased further.

The appearance of the two other bulbs also changes. Bulb L_2, which is in series with the capacitor, is unlit to start with, and gradually gets brighter as the frequency increases. L_3, which is in series with the inductor, starts off bright and gradually gets dimmer. This is shown graphically in Fig. 13.22.

The capacitor and inductor are in parallel so that the voltage across each component is always the same. The voltage across the capacitor lags the current by 90° but it leads the current in the inductor by the same amount. This means that the current in the inductor is 180° out of phase with the current in the capacitor arm.

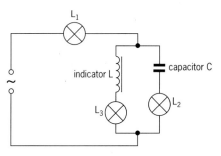

Fig 13.21 **An inductor and capacitor in parallel connected to a variable frequency supply**

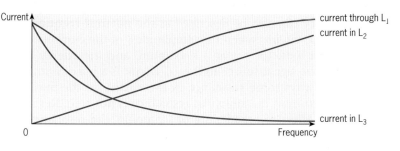

Fig 13.22 **Graphs showing how the currents change as frequency increases in the circuit in Fig 13.21**

As the frequency of the supply changes, there will be a frequency when the reactance of the capacitor and the inductor are the same:

$$X_L = X_C$$

That is:

$$2\pi f L = \frac{1}{2\pi f C}$$

This can be rearranged as:

$$f^2 = \frac{1}{4\pi^2 LC}$$

and so:

$$f = \frac{1}{2\pi \sqrt{LC}}$$

At this frequency the current drawn from the supply is a minimum, yet the currents in the capacitor and inductor are quite high. The currents in each arm are equal but they are also in antiphase (180° out of phase). They cancel out so that the actual current drawn from the supply is a minimum – see the curves of Fig 13.24 – and also explains why the bulb L_1 in Fig 13.21 reaches a minimum brightness.

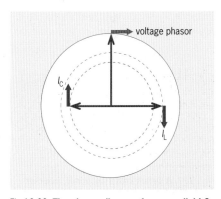

Fig 13.23 **The phasor diagram for a parallel LC circuit. The voltage is the same across both L and C. The currents in L and C are 180° out of phase. What happens when the currents are the same size?**

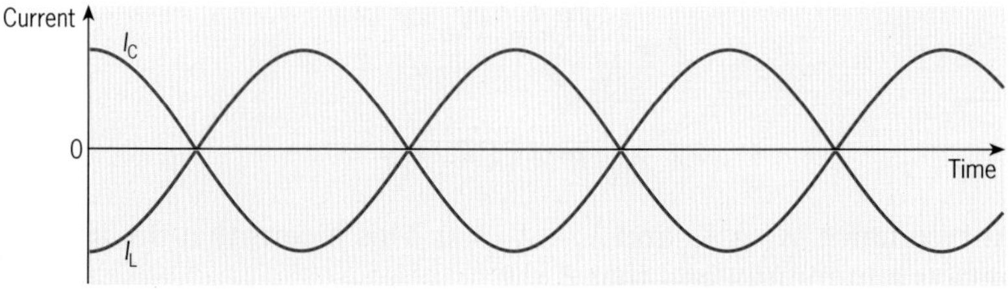

Fig 13.24 **Graphs of the current I_C in the capacitor and the current I_L in the inductor when the reactances X_C and X_L are the same. The currents are the same size but are 180° out of phase**

▨ See question 8.

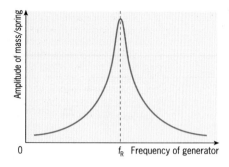

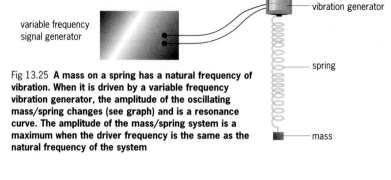

Fig 13.25 **A mass on a spring has a natural frequency of vibration. When it is driven by a variable frequency vibration generator, the amplitude of the oscillating mass/spring changes (see graph) and is a resonance curve. The amplitude of the mass/spring system is a maximum when the driver frequency is the same as the natural frequency of the system**

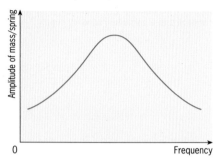

Fig 13.26 **If the mass/spring oscillator in Fig 13.25 is damped, the resonance curve is broadened. Adding resistance in a parallel LC circuit has the same effect on its resonance curve**

?

G Calculate the resonant frequency of a parallel LC circuit made from a 1 mH inductor and a 500 pF capacitor. (1pF = 10^{-12} F.)

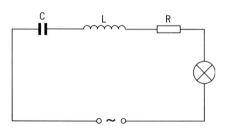

Fig 13.27 **A circuit with a capacitor, inductor and resistor in series to a variable frequency supply**

The behaviour of this circuit can be compared to the behaviour of mechanical oscillating systems. Fig 13.25 shows a mass and spring which is made to oscillate by means of a vibration generator. The graph shows how the amplitude of the oscillating mass changes as the frequency of the vibration generator is varied.

The characteristic peak on the graph shows resonance. When the frequency of the vibration generator is the same as the natural frequency of the mass–spring system, the energy transferred from the generator to the mass–spring is a maximum (see Chapter 6).

The frequency at which the capacitive reactance is equal to the inductive reactance is called the **resonant frequency** of the circuit. This frequency can also be described as the **natural frequency** of the parallel LC circuit, when the 'amplitudes' of the currents in the capacitor and inductor are at a maximum, and the energy transferred from the signal source to the LC circuit is a maximum.

There is always some resistance in the circuit, so the minimum current drawn from the supply at resonance is not zero. Resistance also has the effect of broadening or 'damping' the resonance (Fig 13.26).

9 IMPEDANCE OF A SERIES LCR CIRCUIT

In the circuit shown in Fig 13.27, as the frequency is increased the brightness of the bulb increases to a maximum and then decreases again. To analyse this circuit we need to know its impedance.

As with all series circuits, the current is always the same at any point in the circuit. The voltage across the inductor will lead the current by 90° and the voltage across the capacitor will lag the current by 90°. The voltage across the resistor will be in phase with the current. The phasor diagrams in Fig 13.28 show these voltages and how we can find their resultant.

The resultant voltage is given by:

$$V = \sqrt{(V_R)^2 + (V_L - V_C)^2}$$

The impedance is given by V/I, which gives:

$$Z = \sqrt{\frac{V_R^2}{I^2} + \frac{(V_L - V_C)^2}{I^2}}$$

$$Z = \sqrt{R^2 + (X_L - X_C)}$$

At resonance, $X_L = X_C$ and $X_L - X_C = 0$. This means that the impedance is a minimum and that $Z = R$.

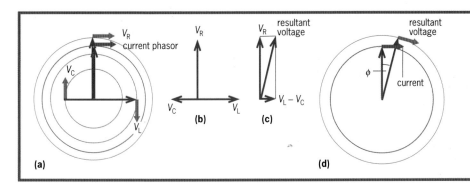

Fig 13.28 **Phasor diagrams for the series RCL circuit. (a) All the voltage phasors relative to each other. (b) The voltage phasors in isolation. (c) The resultant of the voltage phasors. (d) The resultant voltage in relation to the current. The current lags the voltage by angle ϕ**

The resonance curve for this circuit is shown in Fig 13.29. At the resonant frequency the current is a maximum.

As $X_L = X_C$ at resonance, the resonant frequency is given by the same equation we derived for the parallel LC circuit:

$$f = \frac{1}{2\pi\sqrt{LC}}$$

Fig 13.29 **The resonance curve for the series RCL circuit of Fig 13.27. In the series circuit the current from the supply is a maximum at resonance**

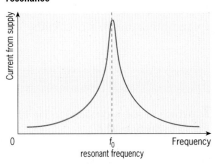

10 POWER IN REACTIVE CIRCUITS

The energy transfer in purely reactive circuits may seem curious. In fact when a *pure* capacitance or inductance is connected to an a.c. supply, there is no net energy transfer over a whole number of cycles.

Examine Fig 13.30 carefully. The current and voltage graphs are shown for a capacitor, that is, with the current leading the voltage by 90°. The power graph is also shown. This is derived from the other two by using the equation:

Power = current × voltage

for corresponding values of I and V throughout a whole cycle or so. As we can see from Fig 13.30, over a whole number of cycles there is no net energy transfer. What is delivered to the capacitors during the first quarter-cycle is returned during the next.

In practice this does not happen. It is virtually impossible to have a circuit without some resistance and the current working against the resistance will cause heating. The small resistance present in the leads of a capacitor causes some energy losses. The wire in the coils of an inductor will also have resistance. Real loads have impedance, and energy is transferred from the supply to the load.

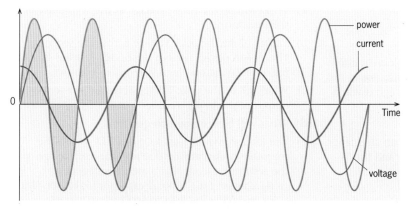

Fig 13.30 **Energy dissipation in a pure reactance – in this case a capacitor. The voltage and current are 90° out of phase. The resulting power goes through a complete cycle twice for each cycle of the applied current. The positive power half-cycles correspond to energy being delivered from the supply to the capacitor. During negative power half-cycles the energy stored in the capacitor is returned to the supply. Over a whole number of cycles the net transfer is zero!**

See question 9.

SUMMARY

By the end of this chapter you should be able to:

■ Describe the characteristics of alternating voltages and currents: peak voltage, frequency and period.

■ Use the relationships for r.m.s. and peak currents and voltages.

■ Describe the use of transformers in the National Grid and the advantages of transmitting electrical energy at high voltages.

■ Describe the use of diodes for rectifying alternating currents.

■ Describe the behaviour of resistors, capacitors and inductors in a.c. circuits.

■ Use the equations $X_C = 1/2\pi fC$ and $X_L = 2\pi fL$ for the reactance of capacitors and inductors in calculations.

■ Describe impedance as the combination of resistance and reactance, and calculate the impedance of LR, CR and LCR series circuits.

■ Calculate the resonant frequency of a parallel LC circuit.

■ Explain why pure reactances do not dissipate energy.

QUESTIONS

1 Which of the waveforms in Fig 13.Q1 (a) to (f) are a.c. and which are d.c.?

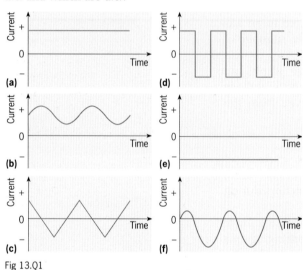

Fig 13.Q1

2 The output of an alternating supply is given by the expression: $v = 20 \sin 628t$. What is:
a) the maximum voltage of the supply?
b) the r.m.s. value of the voltage?
c) the frequency of the supply?

3 An electric fire has a resistive heating element and is rated at 230 V and 1 kW. Calculate the resistance of the element and the peak current flowing through it when it is connected to the mains.

4 Fig 13.Q4 shows an oscilloscope trace of an alternating voltage. The timebase is set to 0.1 ms cm⁻¹ and the Y-sensitivity to 2 V cm⁻¹. From the trace, calculate:
a) the peak voltage,
b) the frequency.

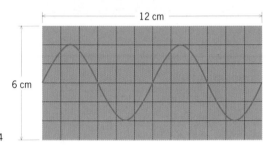

Fig 13.Q4

The trace is actually the p.d. across a 100 Ω resistor.

c) Calculate the r.m.s. current through the resistor.

5 Fig 13.Q5 shows the circuit of a simple power supply. The turns ratio of the transformer is 16:1.

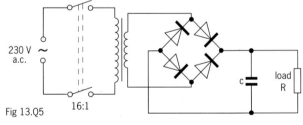

Fig 13.Q5   16:1

a) What will be the r.m.s. voltage at the secondary of the transformer?

b) What will be the peak voltage across the secondary? The silicon diodes used in the bridge rectifier have a p.d. of 0.7 V across them when they conduct.

c) What will be the peak voltage of the full-wave rectified output of the bridge rectifier?

d) What will be the ripple frequency?

e) The load shown in Fig 13.Q5 is 1 kΩ. When the smoothing capacitor is 1000 mF, the voltage across the resistor is steady. Estimate the effect of:
  (i) changing the 1000 μF capacitor for one of 10 μF,
  (ii) reducing the load to 10 Ω (C = 1000 μF).

6 Fig 13.Q6 shows how the reactance of a capacitor varies with frequency.

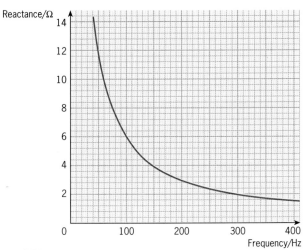

Fig 13.Q6

a) Use the information on the graph to calculate the value of the capacitor.

b) Copy the graph and, using the same axes, draw a graph of reactance against frequency for an inductor which has the same reactance as the capacitor at 150 Hz.

c) What is the inductance of the inductor in part **(b)**?

d) If this inductor were connected in series to a resistor of 10 Ω, what would be the impedance of the combination at: **(i)** 150 Hz, **(ii)** 300 Hz?

7 Fig 13.Q7 shows the circuit of a 'crossover' network used in a loudspeaker system. There are two loudspeakers in the system: a low frequency speaker or 'woofer' and a high frequency speaker or 'tweeter'. The crossover is designed so that signals below 800 Hz are directed to the woofer and those above 800 Hz to the tweeter.

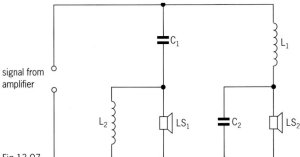

Fig 13.Q7

a) Which of the two speakers, LS_1 and LS_2, is the woofer and which is the tweeter? Explain your answer.

b) The inductor L_1 and the capacitor C_1 should both have a reactance of about 100 Ω at 800 Hz. Calculate suitable values for L_1 and C_1.

c) Calculate the reactance of L_1 and C_1 at **(i)** 5 kHz, **(ii)** 10 kHz.

d) The inductor L_2 and capacitor C_2 improve the performance of the system. Suggest suitable values for their reactance at 800 Hz, and evaluate suitable values.

8 A capacitor and an inductor are connected to a battery, as shown in Fig 13.Q8(a).

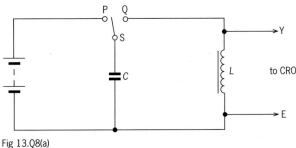

Fig 13.Q8(a)

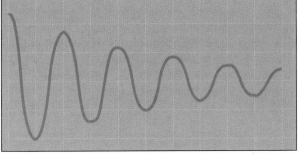

Fig 13.Q8(b)

When the switch S is moved from P to Q, oscillations are observed on the oscilloscope screen, as shown in Fig 13.Q8(b).

a) The timebase of the oscilloscope is set to 200 ms per division. Calculate the frequency of the oscillations.

b) The value of C is 470 µF. Calculate the value of the inductance.

[OCSEB: 1989, Nuffield Advanced Physics, Short answer paper]

9 Fig 13.Q9 shows how the alternating potential difference V applied to the ends of a coil made from thick copper wire, and the current I through the coil, vary with time t.

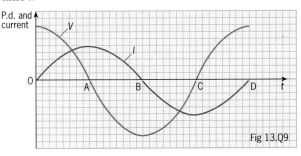

Fig 13.Q9

a) Why is the applied potential difference zero when the current I has a maximum value?

b) During which periods of time OA, AB, BC and CD would the source of power be **(i)** supplying energy, **(ii)** receiving energy?
How did you decide on these answers?
When the source of power *receives* energy, where does the energy come from?

[OCSEB: 1978, Nuffield Advanced Physics, Short answer paper]

Assignment

THE OSCILLOSCOPE

An oscilloscope is a high resistance a.c. and d.c. voltmeter, a timing device capable of measuring times in fractions of a microsecond. It can give a visual display of electrical wave-forms that allows the characteristics of those waves to be studied.

The cathode ray oscilloscope (CRO) is built around the cathode ray tube. In the tube a beam of electrons produced by an electron gun is accelerated towards a screen. The screen is coated with phosphor. When the electrons collide with the atoms in the phosphor their kinetic energy is transferred to the phosphor atoms and they radiate energy as light.

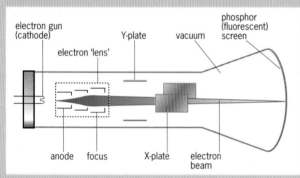

Fig 13.A1 **A cathode ray tube**

As the electrons travel as a beam from the electron gun, their acceleration can be varied. In this way the brightness of the beam on the screen can be changed. The beam can also be focused when it passes through an electrode.

Before striking the screen, the beam passes between two sets of parallel plates . The plates are at right angles and are called the X-plates and the Y-plates.

Fig 13.A2
X-plates and Y-plates

An accurate timing circuit called the **timebase** is connected to the X-plates. The signal from the timebase controls the movement of the electron beam from side to side, that is, in the X-direction. The timebase signal pulls the electron beam towards the right-hand X-plate so that the beam moves at a constant speed across the screen. The timebase control allows the speed at which the beam crosses the screen to be varied.

1 A typical oscilloscope screen is 10 cm wide and the time-base can be varied from 0.1 s cm^{-1} to 0.5 μs cm^{-1}. How long does it take the beam to cross the screen in each case?

The signal to be investigated is connected to the Y-plates. Any signal connected to the Y-plates via an amplifier makes the electron beam move vertically. The amplifier allows a wide range of voltages to be measured. The vertical deflection is always proportional to the voltage of the signal. The sensitivity of the amplifier is controlled by the Y-gain switch and would be measured in V cm^{-1} or mV cm^{-1}.

The CRO can measure direct voltages or alternating voltages. The a.c./d.c. switch is set accordingly. A direct voltage is connected directly to the Y-amplifier. An alternating signal reaches the amplifier via a capacitor.

Most CROs also have a third position on the switch to ground the input. This is very useful for finding out where the original trace is.

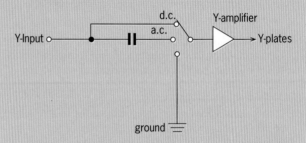

Fig 13.A3 **A.c./d.c. circuitry**

A direct voltage makes the trace (a dot of the timebase is switched off) move up or down, depending on its polarity.

2 With the Y-gain set at 2 V m^{-1}, what direct voltage would produce the following movements of the trace:

(a) up 2 cm,
(b) down 3 cm,
(c) down 0.5 cm,
(d) up 3.5 cm?

When an alternating voltage is applied to the input, the trace shown on the screen shows enough information to allow the voltage and frequency of the signal to be measured. It is often easier and more accurate to measure the peak-to-peak voltage, that is, the voltage from the top to the bottom of the signal. The peak voltage is then half of the peak-to-peak value.

Frequency can be found using the CRO by measuring the period of the signal and then 'calculating' the frequency using the equation:

$$\text{frequency (Hz)} = \frac{1}{\text{period (s)}}$$

The CRO should have a trigger control. This may be preset, but it helps you get a steady trace on the screen. The trigger synchronises the timebase with the Y-input signal.

3 The Y-gain of an oscilloscope is set at 2 V cm^{-1} and the timebase at 0.5 ms cm^{-1}. Work out the peak-to-peak voltage, the r.m.s. voltage and the frequency of the signals shown in Fig 13.A4.

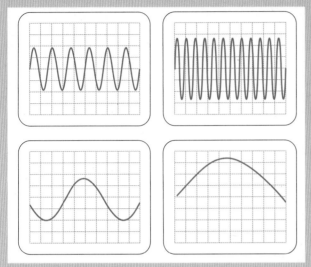

Fig 13.A4 **CRO traces**

A double-beam oscilloscope allows two signals to be compared. Two signals of the same frequency may be out of phase. By seeing the two traces on the screen together, the phase difference can be measured as the horizontal difference between equivalent points on the two traces.

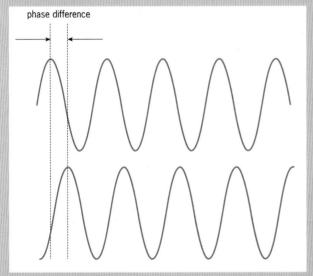

Fig 13.A5 **Phase difference**

4

a) Draw two a.c. signals which are **(i)** 90° out of phase, **(ii)** 180° out of phase.

b) With the timebase is set at 1 ms cm^{-1}, find the phase difference of the two signals shown in Fig 13.A6 in **(i)** ms and **(ii)** degrees.

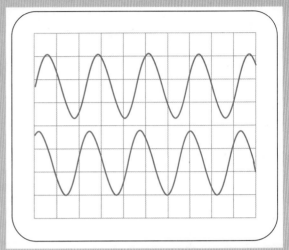

Fig 13.A6

ALTERNATING CURRENTS AND ELECTRICAL POWER

In this chapter, you have covered aspects of alternating current, as provided by the National Grid system: how electrical power is transmitted and used as a.c. (or converted to d.c.) in a range of circuits. The Chapter Map below brings together the key ideas you have studied and shows you how they interlink. Use the map alongside the syllabus you are following to check that you have covered alternating current thoroughly, and to identify areas where you may need to do further study.

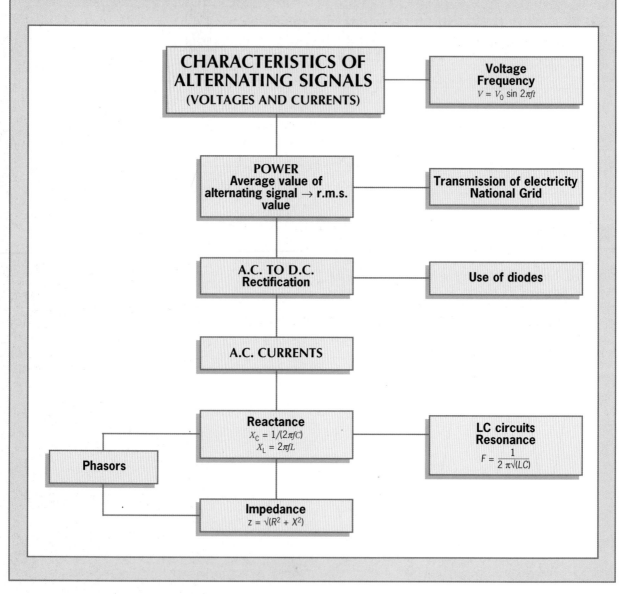

CHARACTERISTICS OF ALTERNATING SIGNALS (VOLTAGES AND CURRENTS)

Voltage Frequency
$V = V_0 \sin 2\pi ft$

POWER
Average value of alternating signal → r.m.s. value

Transmission of electricity National Grid

A.C. TO D.C. Rectification

Use of diodes

A.C. CURRENTS

Reactance
$X_C = 1/(2\pi fC)$
$X_L = 2\pi fL$

Phasors

LC circuits Resonance
$F = \dfrac{1}{2\,\pi\sqrt{(LC)}}$

Impedance
$z = \sqrt{(R^2 + X^2)}$

THERMAL PHYSICS

LIFE IS BASED on chemical reactions that can occur only within a limited range of temperatures – a range experienced by most places on Earth. However, these temperatures often don't suit humans: think of all the energy we use to heat (or cool) our homes and workplaces to a level that we find comfortable.

This section deals with thermal changes, meaning what happens to materials when they are heated, change state or change in temperature. We also see how thermal properties of different materials influence which we choose for different practical uses.

Thermal effects are important in devices that do work, such as steam turbines, petrol engines and diesel engines. The relation between heating and working is brought together in thermodynamics. In studying this area of physics, you will learn about entropy and the inevitable laws of chance which determine that, ultimately, all the energy we use ends up warming the surroundings.

14 Energy and temperature

LIKE MOST BOOKS, this book was written only after a lot of late nights and numerous cups of coffee – usually 'instant' coffee.

The instant coffee industry has developed into a multi-million pound market since it first began in the late 1930s. And the latest development is 'freeze dried', which gives the coffee a much better flavour than all the earlier versions.

Once the coffee beans have been roasted and ground, the soluble solids and flavour compounds are extracted. This mixture contains a lot of water, which would make granules stick together. The problem is to remove the water without removing the flavour.

In freeze drying, the mixture is cooled to just below 0 °C (it is still sticky) and air is added to make a 'foam' that is of the right density. This foam is then deep frozen and broken into the familiar airy granules we see in the coffee jar. The granules are then heated very rapidly in a drying cabinet at very low pressure so that the ice converts immediately to vapour without passing through the liquid stage, leaving behind almost all of the aroma and flavour of the original coffee.

The largest modern freeze-drying plants can produce up to 7 million cups of coffee each day – more than enough for several more physics books!

1 ENERGY AND WORK

When we switch on an electric fire in the home, a current passes through the element – usually a coiled wire. The electrical energy for this is provided at the power station by the chemical change when gas, oil or coal is burned. This energy is then transferred via the resistive coil of the fire to air molecules in the room. The air molecules in the room start to move around faster in every direction. This increased movement constitutes the so-called heating of the air.

We can transfer energy quite efficiently from the fire to the air molecules. But we cannot easily reuse the energy and change back this increased movement of the air molecules to energy that can be returned along our electrical circuit. There are certain restrictions on the transfer of energy. **Thermodynamics** is the science that describes how systems alter when their energy content changes. It describes and explains the transfer processes.

We can define heating as a process in which energy transfers from one body at a higher temperature to another body at a lower temperature. We saw how to obtain a temperature scale in Chapter 7 (page 155) and we shall look at practical methods of measuring temperature later in this chapter.

When thermal energy transfers *from* a body, there is a decrease in the amount of movement of the atoms and molecules within that body. The transfer of thermal energy *into* the second body means that an increase in movement of the atoms and molecules takes place there. There will also be a small change of potential energy because the distances, and hence the forces, between the molecules will change slightly.

Historically, the nature of heating greatly puzzled scientists. This is because they did not know then what we now understand about the structure of matter – that it is made up of atoms. Count Rumford in the late eighteenth century became interested in the way thick metal rods became hot when bored out to make cannons. But a great step forward came when Joule showed that mechanical work could be transferred to thermal energy, which resulted in an increase in temperature of water in a cylinder (see page 61).

We can demonstrate the transfer of mechanical work to thermal energy (as internal energy) using lead shot in a closed cylindrical tube. The tube is repeatedly inverted so that the lead shot falls from end to end each time. When the lead shot weighs 150 g (0.15 kg) and each fall is 0.75 m, we can calculate the rise in temperature of the lead shot after 100 falls.

● As the masses fall, work is done as a result of the gravitational force. This work is given by:

$$\text{work done} = \text{mass of lead} \times \text{acceleration due to gravity} \times \text{length of fall} \times \text{number of falls}$$

$$= 0.15 \times 9.8 \times 0.75 \times 100 \text{ J} = 110 \text{ J}$$

● The work is transferred to kinetic energy as the lead shot is falling.

● But the lead shot is brought to a sudden halt at the bottom of the tube. Now the energy transfers to internal energy within the shot. As a result, the amount of movement and the potential energy of the lead atoms is increased. Overall, the work done has been used to raise the temperature of the lead by ΔT.

For more about internal energy, see pages 54 and 319.

JOULE'S WORK ON THERMODYNAMICS

JAMES PRESCOTT JOULE came from a family of wealthy brewers in Salford and conducted his early experiments in a laboratory at the brewery. He was wealthy enough to fund his own experiments. On his honeymoon, he spent time devising an accurate thermometer to measure the temperature difference between the top and bottom of a scenic waterfall.

Joule's first thermometers measured to within ±0.02 °C and later ones to ±0.005 °C. It was a great achievement to measure temperatures so accurately, and necessary for Joule's study of the conversion of mechanical work to energy.

Joule had been educated privately at home, with no formal education in mathematics, so that he could not keep up with the new science of thermodynamics. However, he won the support of William Thomson (Lord Kelvin) and together they made great advances in thermodynamics, Joule providing the practical ability and Thomson the theory.

Fig 14.1
James Prescott Joule (1818–1889)

?

A Calculate the work done and the amount of energy produced after 100 falls if 250 g of lead shot are used in the same tube. Show that the temperature rise is again 5.7 K.

● The specific heat capacity (see page 309) for lead is 129 J kg⁻¹ K⁻¹. The amount of energy given to the lead shot is given by:

energy (in J) = mass of lead × specific heat capacity of lead
× temperature rise

$$= 0.15 \times 129 \times \Delta T = 19.4 \times \Delta T$$

● Equating work and energy (in J): $110 = 19.4 \times \Delta T$
So the temperature rise is: $\Delta T = 110/19.4 = 5.7$ K

This temperature rise will be detectable even when there are some energy losses to the surroundings. Note that the mass of the lead shot is used twice, once for the work done and once for the energy transferred to the lead shot. This means that the temperature rise does not depend on the mass of the shot used and need not have been given in the question. Increasing the mass of shot increases the amount of required work but also increases the mass to be heated. In the calculation we could use a symbol for mass and then cancel mass on either side of the equation.

2 THERMAL TRANSFER OF ENERGY

There are three types of thermal transfer of energy: **conduction**, **convection** and **radiation** (Fig 14.2).

Conduction

Energy is transferred within solids by the process called **thermal conduction**. The atoms in a solid vibrate, and as the temperature increases, the amplitude of vibration increases. The energy of the atoms increases and the internal energy of the solid is increased. The vibration of the atoms is rather like that of a set of masses connected to a lattice of springs.

Energy travels through a solid by way of vibrations through the system. Neighbouring atoms are set into increased vibration one after another. In metals, 'free' electrons travelling among the atoms also gain extra energy and travel a little faster as the temperature increases. Importantly, they help to transport the energy through the metal and pass on energy to more distant atoms.

Convection

There is movement of internal energy in a fluid (a liquid or a gas) by the process of thermal **convection** (conduction also occurs). The fluid particles, usually molecules, are free to move around independently. Their kinetic energy increases as more thermal energy is added to the fluid. Heating the fluid causes it to expand and become less dense. Because of the effect of gravity, the less dense (higher temperature) fluid rises up through the more dense fluid, which then falls to take its place. This movement is a very effective way of transferring those molecules that have been given extra kinetic energy. Currents of molecules are set up called convection currents.

Convection currents cause thermal energy to circulate in kitchen ovens. Cooks know that less dense hotter air rises to the top of the oven. When cooking more than one item of food, they choose the appropriate shelf for each item. Many modern ovens include an electric fan, which both increases the circulation rate and produces a more uniform temperature distribution. This is called **forced convection**, whereas if the thermal energy is left to distribute itself, the process is called **natural convection**.

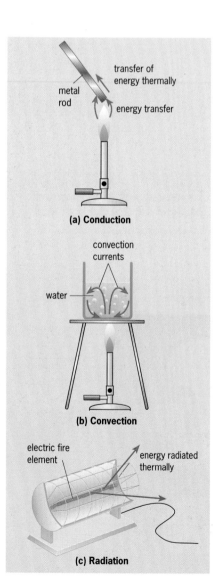

Fig 14.2 **Thermal conduction, convection and radiation**

Radiation

Energy can be transferred in the form of electromagnetic **radiation** (see Chapter 16). All objects emit and reflect electromagnetic radiation. We are aware of the radiation in the form of light waves. However, as the temperature of an object increases, both the range of frequencies and the quantity of radiation given out increase.

An electric fire radiates much of its energy in the visible region – we see the element glowing orange or red. Stars are much hotter and radiate much of their energy in the ultraviolet region.

On the other hand, thermal imaging cameras can pick up infrared radiation at longer wavelengths and smaller frequencies than the visible range. This means that hot bodies can be detected against a cooler background. Imaging cameras are used by the fire service to detect people overcome by smoke in buildings on fire (Fig 14.3) and, when slung from police helicopters, these cameras can detect suspected criminals on the run at night.

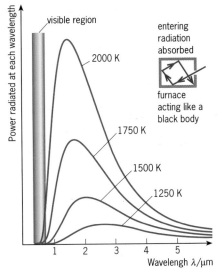

Fig 14.3 **In dense smoke or darkness, fire-fighters locate victims using a thermal imaging camera**

Black body

An object which completely absorbs thermal radiation at all wavelengths is called a **black body**. It reflects none of the radiation falling on it, yet it can emit radiation at all wavelengths. Fig 14.4 shows the radiation curves for a black body.

A furnace with a very small opening and thick, insulating walls approximates to a black body. Radiation directed into the furnace through the opening is absorbed by the inside walls. However, the furnace walls will be very hot (they will be at the temperature of the furnace) and so will be emitting radiation at all wavelengths, a small amount of which will escape through the opening.

Fig 14.4 shows that, as the temperature increases, the amount of radiation from the black body also increases (corresponding to the area under the curve). The wavelength at which peak emission occurs falls as the temperature increases. This is why when a piece of metal such as steel is heated, it first appears to be a dull red but, as it gets hotter and the peak in the emitted light moves to smaller wavelengths, its appearance becomes more orange.

Fig 14.4 **Black body radiation and diagram of a black body**

3 MEASURING TEMPERATURE

In Chapter 7 (page 155) we saw how to use the properties of an ideal gas to define a temperature scale that is reproducible anywhere. This scale is the Kelvin scale. We also saw that a constant-volume gas thermometer measures temperature. But such a thermometer is not easy to carry around and to use.

Temperature can be measured using *any physical property that varies with temperature in a reproducible way*. But we must be able to calibrate the variation against our standard Kelvin scale as obtained with the gas thermometer. We do not have to stay with the Kelvin scale itself, but the only other scales now in common use are the centigrade and Celsius scales (discussed shortly).

Examples of physical properties that we might use are:

- **length**, such as the length of a column of alcohol in a liquid-in-glass thermometer,
- **voltage**, such as that produced with a thermocouple,
- **electrical resistance** of a wire (often platinum) or semiconductor (thermistor),
- **pressure**, as in a gas thermometer,
- **thermal radiation** (using a pyrometer).

Temperature scales

The thermodynamic scale (or absolute scale or Kelvin scale)

This is the scale already discussed in Chapter 7. A single fixed point is taken at the triple point of water and defined as 273.16 K. The unit of temperature, the kelvin, is 1/273.16 of this triple point temperature. The size of the unit then determines at what stage we get down to the zero of the scale. There is no negative temperature and we can never quite reach zero in the laboratory.

Because scientists need to make measurements over a very wide range of temperatures, they have defined some additional fixed points for calibrating thermometers. However, only the triple point is fundamental to all scales.

The Celsius and centigrade scales

The **Celsius scale** is derived from the thermodynamic (Kelvin) scale using the same single fixed point as the thermodynamic scale. But, in Celsius, the ice point is 0.01 °C and the steam point is 100.00 °C.

The relationship is:

$$t = T - 273.15$$

where t is the temperature on the Celsius scale and T is the temperature on the thermodynamic scale.

In addition, a practical scale called the **centigrade scale** was devised for inexpensive thermometers. It is calibrated by using two fixed points (Fig 14.5): ice, defined as 0°, and steam (at standard pressure), defined as 100° on the centigrade scale.

It is common to ignore the small difference of 0.01 °C for defining the two scales at the melting temperature of ice, so that the scale with the temperature of ice as 0° is referred to as the Celsius scale, with the term centigrade becoming obsolete. The total interval is divided into a hundred sections, each section corresponding to one degree.

?

B Describe what you expect to happen at zero temperature.

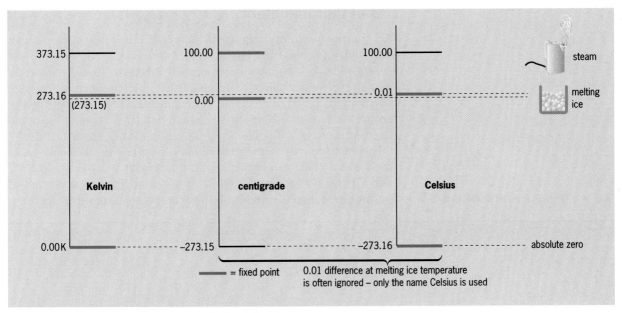

Fig 14.5 **The absolute (Kelvin), centigrade and Celsius scales of temperature**

Physical properties to calibrate temperature

Calibration is easy if we are measuring a change of length. More generally, though, we need to know the value of the property (here, length) at the fixed points and calibrate using a proportionality.

Let us see this in action for a mercury-in-glass **thermometer**:

- l_{100} is the length of mercury thread at 100°.
- l_0 is the length of mercury thread at 0°.
- l_θ is the length of mercury thread at θ°.
- Each degree interval will have a length $(l_{100} - l_0)/100$.
- The change of length due to a temperature rise from 0° to θ° is $(l_\theta - l_0)$.
- The temperature θ in degrees is given by change of length divided by length per degree interval:

$$\theta = \left[\frac{l_\theta - l_0}{(l_{100} - l_0)/100} \right] (\text{degrees}) = \frac{l_\theta - l_0}{l_{100} - l_0} \times 100°$$

We could also use a **platinum resistance thermometer** to measure the resistances R_0 at 0°, R_{100} at 100° and R_θ at an unknown temperature θ:

$$\theta = \frac{R_\theta - R_0}{R_{100} - R_0} \times 100°$$

All mercury-in-glass thermometers should agree precisely if they are calibrated correctly. So should platinum resistance thermometers. Also, all thermometers of either type should also agree with each other. However, we cannot assume that, for every degree change in temperature, the length of the mercury thread or the resistance of the platinum changes by exactly the same value. For example, the change in length of the mercury column between 9 °C and 10 °C and between 40 °C and 41 °C may differ. Similarly, resistance at different degree intervals may vary. So the two types of thermometer may not give precisely the same temperature readings between fixed points. Hence our need for the absolute scale.

> ✔ Mercury-in-glass thermometers are not often used in modern laboratories in case of breakage. Mercury has a significant vapour pressure and is slightly toxic.

> **?**
> **C** Readings using a resistance thermometer are 40.0 Ω at the ice point, 86.3 Ω at the steam point and 77.8 Ω at an unknown temperature θ. Calculate this unknown temperature as measured on the Celsius scale. What is the temperature on the Kelvin scale?

Use of thermometers

The choice of a thermometer will depend on the temperature range and where it is used. Mercury thermometers are no use below –39 °C, which is the freezing point of mercury. But they can be used over a wider range of temperature than the alcohol-in-glass thermometer. A large thermometer is not suitable for measuring the temperature of small amounts of material. This is because, while the thermometer achieves thermal equilibrium, energy could be transferred to or from the material and so temperature being measured could change. A resistance thermometer is a likely choice if high accuracy is required. Table 14.1 summarises the main types of thermometer, together with their advantages and disadvantages.

Table 14.1 **Different types of thermometer**

Thermometer	Thermometric property	Main advantages	Main disadvantages	Temperature range
Liquid-in-glass	Volume change (ie changing length of thread of mercury or alcohol)	Simple to use, cheap, portable	Fragile, limited range, not suitable for small objects	Mercury: 234–723 K; Ethanol: 173–323 K
Constant volume gas thermometer	Pressure of fixed mass of gas at constant volume	Absolute scale given, accurate, wide range	Bulky and inconvenient, slow response, not suitable for small objects directly	3–500 K
Resistance	Resistance of platinum wire	Accurate, wide range, useful for small temperature differences	Slow to respond, not suitable for small objects	15–900 K
Thermocouple	E.m.f. across junction of two dissimilar metals	Can measure small differences, fast response, wide range, can be read remotely	Small voltages, so need electronic amplification	25–1400 K (depending on metals)
Thermistor	Changing resistance of semiconductor	Provides electrical signal suitable for computer circuits	Calibration necessary, not very accurate	200 K–700 K
Optical pyrometer	Adjustment of current through lamp filament to match colour of object	No contact with hot object, simple to use, portable	Calibration necessary, not very accurate	Above 1250 K

4 HEATING AN OBJECT

We know from everyday experience that some things need more energy than others to raise their temperatures by the same amount. For instance, it takes more energy to heat water by one kelvin than the same mass of copper requires. The water molecules require more energy in order to move around fast enough to record a one kelvin rise than do the copper atoms, which mainly vibrate. The **heat capacity**, C, of an object is the amount of energy required to raise its temperature by one kelvin. The unit is kJ K^{-1}.

Specific heat capacity, molar heat capacity and latent heat

Heat capacity refers to a particular object of any mass. A more general quantity would be more useful. This is **specific heat capacity**, *c*, which refers to *unit mass* of a single substance. Each pure substance has its own particular value of specific heat capacity at a given temperature (though this value may change as the temperature of the substance varies).

We can compare expressions for heat capacity *C* and specific heat capacity *c*. Let's say we heat an object of mass *m* (kilograms) of a single substance by giving it energy ΔQ (kilojoules) so that the temperature rises by ΔT (degrees Celsius), then:

$$C = \frac{\Delta Q}{\Delta T} \quad \text{and} \quad c = \frac{\Delta Q}{m\Delta T}$$

EXAMPLE

Q A 3.0 kW electric kettle is used to bring 1.50 kg of water to the boil from a starting temperature of 18.0 °C. Assuming all the energy goes into heating the water, calculate the amount of energy and the time required to boil the water. Specific heat capacity of water = 4.20 kJ kg^{-1} K^{-1}.

A Energy required = specific heat × mass
× temperature rise
= 4.20 × 1.5 × 82 kJ
= 517 kJ

$$\text{Time required} = \frac{\text{energy required}}{\text{rate of heating}}$$

$$= \frac{517 \text{ kJ}}{3.0 \text{ kJ s}^{-1}}$$

= 172 s, or 2 min 52 s

We can use another thermal quantity, the **molar heat capacity**, particularly for gases. As the name suggests, this is the energy required to heat one mole of the substance through one kelvin.

We have already seen on page 143 that when a substance changes from one state to another, for example from solid to liquid, energy called **latent heat** is required to produce the change. The latent heat does not cause any change in temperature. Also, most substances have different specific heat capacities depending on whether they are solid, liquid or gas. If we add energy to the substance at a constant rate, we can see how these differences in heat capacity affect the rate of heating, and we can identify the temperatures where changes of state occur (Fig 14.6). We assume no energy loss to the surroundings.

?

D The kettle in the Example will transfer energy to the water with less than 100 per cent efficiency. List ways in which energy from the kettle is lost when increasing the temperature of the water.

See question 1.

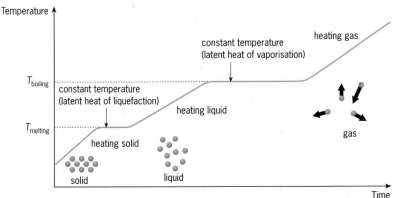

Fig 14.6 **Heating a substance uniformly to show effects of thermal capacities and latent heats**

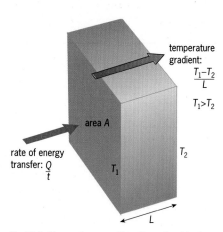

Fig 14.7 **Thermal energy flow through a block of material**

Thermal conductivity

When one side of a body is heated, it takes time for the internal energy to spread (by thermal conduction) to the other side. That is, there is a time lag between the heated side of the body changing its temperature by a significant amount and a corresponding change occurring on the far side.

Once thermal equilibrium has been reached, a temperature difference will be maintained through the object, assuming that heating continues and that energy can escape from the far side. The ease with which energy can transfer through the object depends on a property called **thermal conductivity**. However, as well as thermal conductivity, other factors determine how much energy passes through the object. We now consider all these factors.

Take the flow of energy through a block of material with a thickness of L metres (Fig 14.7). For energy to flow, there has to be a temperature difference between opposite faces of the block.

- Assuming the block is of exactly the same material throughout, the temperature gradient at any point between the opposite faces will be the same. We have:

$$\text{temperature gradient} = \frac{\text{temperature difference}}{\text{thickness of block}}$$

$$= \frac{T_1 - T_2}{L}$$

The larger the temperature gradient, the greater the rate of flow of energy.

- The flow will also depend on the area of cross-section A (m²) of the block. This is very similar to water passing through a pipe – the quantity of water flowing depends on the area of cross-section of the pipe and the pressure difference between the two ends. Transfer of thermal energy is also similar to flow of charge in an electrical conductor – electrical current is proportional to the cross-sectional area of the conductor and to the electric field (potential gradient) along the conductor (Fig 14.8).

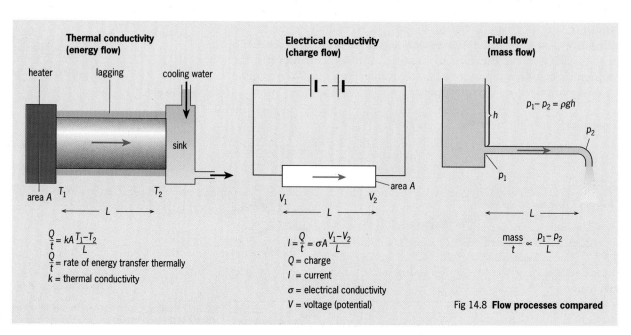

Fig 14.8 **Flow processes compared**

- The amount of energy that flows will depend on how well the material conducts it. Copper is a good conductor, while glass is a poor conductor. The rate of energy flow is proportional to the **thermal conductivity**, **k**, of the material. This dependence is the same as the way flow of electric charge depends on the electrical conductivity of a conductor.

The rate of flow of energy is the quantity of thermal energy Q (kilojoules) transmitted, divided by time t (seconds). We have:

$$\frac{Q}{t} \propto \frac{T_1 - T_2}{L}$$

$$\frac{Q}{t} \propto A$$

$$\frac{Q}{t} \propto k$$

So far, k has not been fixed, so we can make it the constant of proportionality for the combined expression:

$$\frac{Q}{t} = kA\frac{T_1 - T_2}{L}$$

Note that in this equation, T_1 is greater than T_2 and that energy goes *down* a temperature gradient from the side of higher temperature to the side of lower temperature.

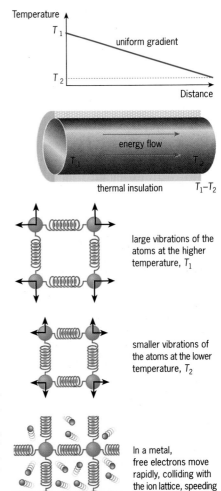 sidebar:

Units of k

k will have units which we work out from the equation.

$\frac{Q}{t}$ has units of energy/time, that is J s^{-1} or W.

$A\frac{T_1 - T_2}{L}$ has units of length^2 × temperature/length, that is m^2 K m^{-1}, equal to m K. (Note the space between m and K as otherwise the symbols would represent one thousandth of kelvin.)

W = (units of k) × m K

So the units of k must be W m^{-1} K^{-1}.

5 THE THERMAL FLOW OF ENERGY

Flow along a rod

Suppose we heat one end of a long rod that is thermally lagged (ie surrounded with insulation). We assume that energy flow is the same at all points along the whole length. We can represent this flow by a series of parallel lines (Fig 14.9), just as we can illustrate the uniform flow of water through a pipe.

What is happening on the atomic scale? We have already said that we can think of the atoms as solid spheres connected in all directions by springs (see pages 146 and 304). When we heat one end of the rod, the atoms at that end vibrate with increasing amplitude. Energy from the vibrations of these atoms is passed on to neighbouring atoms. Eventually the amplitude of vibration of all the atoms in the solid increases.

After a time, the rod reaches a state of equilibrium, with just as much energy leaving the far end as enters the heated end. Energy is transferred thermally through the rod, which means that, as you would expect, there is a temperature gradient in the rod. The far end is cooler than the heated end. We can think of the difference in temperature as driving the energy through the rod, just as a difference in voltage drives an electric current through a wire.

As already seen (pages 203 and 304), some of the energy in a metal is carried by free electrons. They move around faster at the hot end than at the cold end. By colliding with the lattice of ions (the spheres in our ball-and-spring model) they help rapid transfer of energy through the rod. We have already noted that copper, a typical metal, has a large thermal conductivity. This is because of its free electrons.

Fig 14.9 **Thermal energy flow in a lagged conductor**

In Chapter 17 you will learn about wave-particle duality. The vibrations of the atoms in the metal are transmitted as waves along the rod. However, just as sometimes light is described as the passage of particles called photons with no rest mass, so the passage of these vibrations can be also considered as the movement of particles with no rest mass, this time called *phonons*.

See question 2. ▨

?

E Why is the energy flow the same in each conductor in the Example?

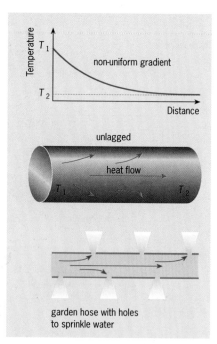

garden hose with holes to sprinkle water

Fig 14.11 Thermal energy flow in an unlagged conductor – similar to water flow in a garden hose with holes, though the conductor loses energy uniformly at all points on its surface

Table 14.2 **Thermal conductivities of common building materials**

Material	Thermal conductivity/W m⁻¹ K⁻¹
Brick	1.0
Concrete	1.5
Glass	0.8
Wood	0.1–0.4

EXAMPLE

Q An iron bar 0.50 m long and a copper bar 1.2 m long are joined end to end. One end of the iron bar is kept at 80 °C while the far end of the copper bar is maintained at 0 °C by a mixture of ice and water. The outer surfaces of the bars are lagged so that there are no thermal energy losses. Both bars are of circular cross-section, diameter 0.16 m. At thermal equilibrium the temperature at the junction of the metals is T_j. Calculate T_j and the rate of energy flow. Thermal conductivity of iron = 75 W m⁻¹ K⁻¹; thermal conductivity of copper = 390 W m⁻¹ K⁻¹.

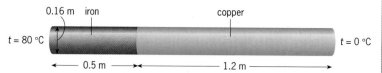

Fig 14.10

A The rate of energy flow through each conductor is the same.

Rate of energy flow through the iron bar is:

$$\frac{Q}{t} = k_{iron} \, A \, \frac{T_{80} - T_j}{0.50} \text{ (watts)}$$

$$= \frac{75 \times \pi \times 0.08^2}{0.50} \times (T_{80} - T_j)$$

$$= (3.02)(80 - T_j) \text{ W}$$

Rate of energy flow through the copper bar is:

$$\frac{Q}{t} = k_{copper} \, A \, \frac{T_j - T_0}{1.20} \text{ (watts)}$$

$$= \frac{390 \times \pi \times 0.08^2}{1.20} \times (T_j - 0)$$

$$= (6.53)(T_j - 0) \text{ W}$$

Equating gives:

$$(3.02)(80 - T_j) = 6.53(T_j - 0)$$

$$242 = 9.55 T_j$$

$$T_j = 25.3°C$$

Suppose that the rod in the Example above is not lagged. In this case, energy is lost by thermal processes along the sides of the rod. This is similar to water escaping from holes in the side of a hosepipe (as often used for watering gardens or orchards). The flow lines will no longer be parallel (Fig 14.11). More importantly, we can no longer assume a constant temperature gradient.

Flow through a glass window

The thermal energy flow through glass windows is of great practical importance. Glass has a low thermal conductivity (Table 14.2), comparable to that of brick. But because the thickness of glass in a window is so much less than the thickness of typical house brick, it is important to reduce the loss of energy through windows. The most common way is to use double glazing. Now work through the next Example.

EXAMPLE

Q A room has a single glass window of length 2.2 m, height 1.2 m, and thickness 5 mm. Assuming that the temperature in the room at the surface of the glass is 22 °C and that outside it is 3 °C, calculate the loss of energy from the room.
  Thermal conductivity of glass is 0.8 W m^{-1} K^{-1}.

A We use the equation:

$$\text{rate of energy flow} = \frac{Q}{t} = kA\,\frac{T_1 - T_2}{L} \text{ (watts)}$$

$$= 0.8 \times (2.2 \times 1.2) \times \frac{(22 - 3)}{0.005}$$

$$= 8026 \text{ W}$$

(Note that it is not necessary to convert °C to K because the equation involves a temperature difference and one degree Celsius has the same magnitude as one kelvin.)

At this point we should question the model. We are losing over 8 kW of thermal energy from the room, and would need a very large heater to maintain the temperature at 22 °C. Something must be wrong. Yet the numbers we have put in seem reasonable.

The error is that we have not allowed for layers of still air on either side of the window. The thermal conductivity of air is very low (approximately 0.025 W m^{-1} K^{-1}). Part of the temperature drop will be just inside the window and part will be just outside. Therefore the temperature difference between the opposite sides of the glass and the temperature gradient through the pane of glass will be rather less than we used in the calculation. The overall temperature profile is not as we have assumed in Fig 14.12(a) but is as shown in Fig 14.12(b).

Note that there has to be some temperature gradient at the two glass surfaces; otherwise, energy would be unable to enter or leave. We would need to take measurements to find the actual profile close to the window pane. Away from the windows of the room, energy is transferred mainly by convection.

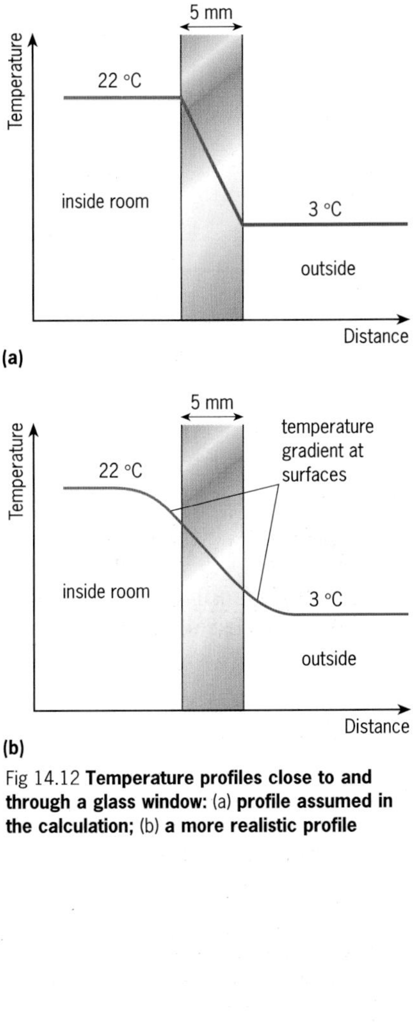

Double glazing and thermal resistance

Most new windows in Britain are now double glazed. The glass is commonly 4 mm thick, though the distance between the panes is less standard, often 5 mm for windows and 10 mm for patio doors. Using the analogy of flow of charge through a conductor, we can see that our double-glazed unit can be modelled as three resistances in series: the resistances of glass, air and glass. So the thermal resistance R of double glazing will be the sum of the thermal resistances of the three layers. The thermal resistance R for one layer is defined as t/kA where t is the layer thickness, A its area and k the thermal conductivity. With the three layers present, the summation is represented by:

$$\Sigma\,\frac{t}{kA}$$

where, the symbol Σ (capital Greek letter sigma), meaning 'sum of', refers to the three layers – glass, air and glass.

Now $k_{air} = 0.025$ W m^{-1} K^{-1} and $k_{glass} = 0.8$ W m^{-1} K^{-1}. Therefore with windows of thickness 4 mm, separation 10 mm and area 1 m^2 we obtain:

$$\text{thermal resistance} = \frac{4 \times 10^{-3}}{0.8} + \frac{10^{-2}}{0.025} + \frac{4 \times 10^{-3}}{0.8} \text{ K W}^{-1}$$

$$= 0.41 \text{ K W}^{-1}$$

(a)

(b)

Fig 14.12 **Temperature profiles close to and through a glass window:** (a) **profile assumed in the calculation;** (b) **a more realistic profile**

F Use the formula for thermal resistance to check that the units K W^{-1} are correct.

See questions 3 and 4. ■

For a single pane of glass, the thermal resistance will be 0.005 K W^{-1}, but remember that there will be some convection in the air gap also. The end-of-chapter Assignment enables you to investigate the use of glass in modern windows.

U-values

U-values are quantities used by architects and heating engineers for working out thermal energy flows within buildings – in particular, energy losses through the windows and walls. The *U*-value of a particular thickness of material is its **thermal transmittance**, that is, the thermal energy flow through the material per unit area for a temperature difference of one degree. Hence it is given by:

$$U\text{-value} = \frac{\text{rate of energy flow}}{\text{area} \times \text{temperature difference}}$$

Its units are W m^{-2} K^{-1}. *U*-values are available for different types of walls, windows, floors and roofs. The values have been obtained by *direct measurement* rather than theoretical calculation. Thus thermal transmittance values are more realistic than thermal conductivity values as they take account of the actual composition of the building

See question 5. ■

material and allow for any convection within hollow components.

The thermal transmittance from a room or building can be calculated by multiplying the *U*-value for each component of the surface structure by the corresponding area.

SUMMARY

After studying this chapter you should be able to:

■ Discuss the equivalence of work and energy and, in particular, the equivalence of work and the thermal energy required to heat a substance.

■ Describe the thermal energy transfer processes of conduction, convection and radiation.

■ State the advantages and disadvantages of different methods of measuring temperature.

■ Describe the difference between three scales of temperature: thermodynamic (absolute or Kelvin), centigrade and Celsius.

■ Explain the terms: thermal capacity, specific heat capacity, thermal conductivity and *U*-values.

■ Carry out calculations on the change of temperature of objects using thermal capacity or specific heat capacity.

■ Calculate thermal resistance using t/kA for a single layer and $\Sigma(t/kA)$ for a number of layers.

■ Use thermal conductivity (or *U*-values) for calculating thermal energy flow through materials.

QUESTIONS

1 A block of metal, of mass 103 g, is heated to 100 °C and then transferred to a polystyrene beaker containing 200 g of water at 19.8 °C. When thermal equilibrium has been reached, the water is at 21.6 °C.

a) Calculate **(i)** the energy gained by the water during this process and **(ii)** the specific heat capacity of the metal.

b) The experiment described is not a particularly good one for measuring the specific heat capacity of the metal.
   (i) Name two important sources of error.
   (ii) State a way of reducing *one* of them.

The specific heat capacity of water is 4.20 kJ kg^{-1} K^{-1}.

[UCLES, 1994 Paper 2, Q5]

2 Fig 14.Q2(a) shows an unlagged brass bar, length 20 cm and area of cross-section 0.32 cm². The two ends of the bar are maintained at the steady temperatures of 100 °C and 30 °C.

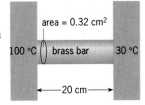

Fig 14.Q2(a)

The thermal conductivity of brass = 109 W m⁻¹ K⁻¹.

Fig 14.Q2(b) shows the temperature of the bar at different distances, x, from the hot end after the temperature has become steady.

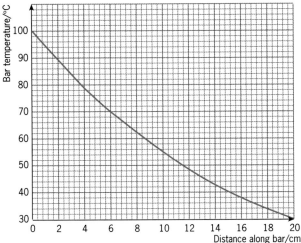

Fig 14.Q2(b)

a) Find from the graph the temperature gradient at the hot end of the bar. Calculate the rate of heat flow into the hot end of the bar.

b) Find from the graph the temperature gradient at the cool end of the bar. Calculate the rate of heat flow from the cold end of the bar.

c) What is the rate of heat loss from the sides of the bar?

d) State, and give one reason in each case, whether each of these five quantities would be higher or lower if the bar were perfectly lagged.

[UCLES: Jan 1995, Physics Paper 2, Q8 (part)]

3 A heating engineer leaves some incomplete calculations on the energy losses by conduction through unit area of single-glazed and double-glazed windows of pane thickness 4.0 mm. You are asked to complete them by answering the questions below. The engineer has already sketched the temperature gradients which occur in the air close to the inner and outer surfaces of the panes.

a) The single-glazed window, Fig 14.Q3(a), has a power transfer of 100 W m⁻². Show that the thermal conductivity of glass is 1.0 W m⁻¹ K⁻¹.

b) (i) Calculate the temperatures at the inner surfaces of the air cavity of the double-glazed window, Fig 14.Q3(b). The power transfer for the double glazed window is 60 W m⁻².

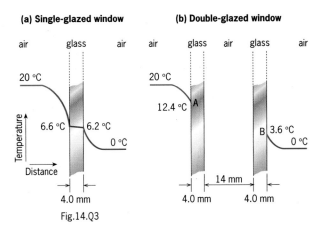

Fig.14.Q3

(ii) Complete Fig 14.Q3(b) by drawing a suitable line from A to B, marking the temperatures you have just calculated to show the gradient between these points.

c) Hence show that the 14 mm wide air cavity in the double-glazed window behaves as a conductor of 0.10 W m⁻¹ K⁻¹.

d) The thermal conductivity of air is, in fact, 0.024 W m⁻¹ K⁻¹. Suggest why the value for the air cavity is so much higher.

[OCSEB: 1993 Physics Paper 2, Q8]

4 The ground floor of a house has an area of 50 m². The floor is fitted with a carpet 15 mm thick which completely covers the floor. The carpet rests on a layer of concrete 200 mm thick. The top surface of the carpet has a temperature of 15 °C and the lower surface of the concrete has a temperature of 10 °C.

Thermal conductivity of concrete = 0.75 W m⁻¹ K⁻¹; thermal conductivity of the carpet material = 0.06 W m⁻¹ K⁻¹.

a) Calculate the rate at which energy transfer would occur without the carpet, assuming the temperatures of the top and bottom surfaces of the concrete are 15 °C and 10 °C respectively.

b) Calculate the temperature at the carpet/concrete boundary.

c) Calculate the energy transfer rate with the carpet in place.

[AEB: Winter 1994, paper 2, Q7]

5 Using data in Table 14.2, calculate U-values for the following:

a) Brick of face size 17.8 cm × 11.4 cm and thickness 7.6 cm.

b) Concrete beam of face size 3.00 m × 0.15 m and thickness 0.12 m.

c) Wood panel of area 5.0 m × 3.5 m and thickness 8 mm (use an average value for thermal conductivity).

See page 336 for the CHAPTER MAP that covers Chapters 14 and 15

Assignment 1

HEAT LOSS AND GAIN IN THE HOME

Heating bills are a big expense for most households, and people often wonder if installing double glazing is worth the cost. In this exercise, we calculate the thermal energy losses and gains of a house. Fig 14.A1 is the floor plan of a small bungalow, and Table 14.A1 gives the dimensions of its walls, windows and doors. Take other dimensions from the plan, marked out in 1-metre squares.

a) Calculate the total area of glass (windows plus patio door), the two external doors, the cavity walls, the floor and the ceiling.

b) Hence, obtain the total rate of energy loss in watts to the outside at 5 °C, when all the rooms are kept at 20 °C, by calculating the separate losses through glass, external doors, walls, floor and ceiling. Which is the largest loss?

c) Say whether the bungalow could be heated with:
(i) a typical small central heating boiler plus radiators supplying 40 000 British thermal units 1 BTU = 0.293 W (though some energy is lost directly through the flue);
(ii) electric heaters, each supplying 2 kW.

Fig 14.A1 **Floor plan of bungalow and conservatory elevation (grid ≡ 1 metre squares)**

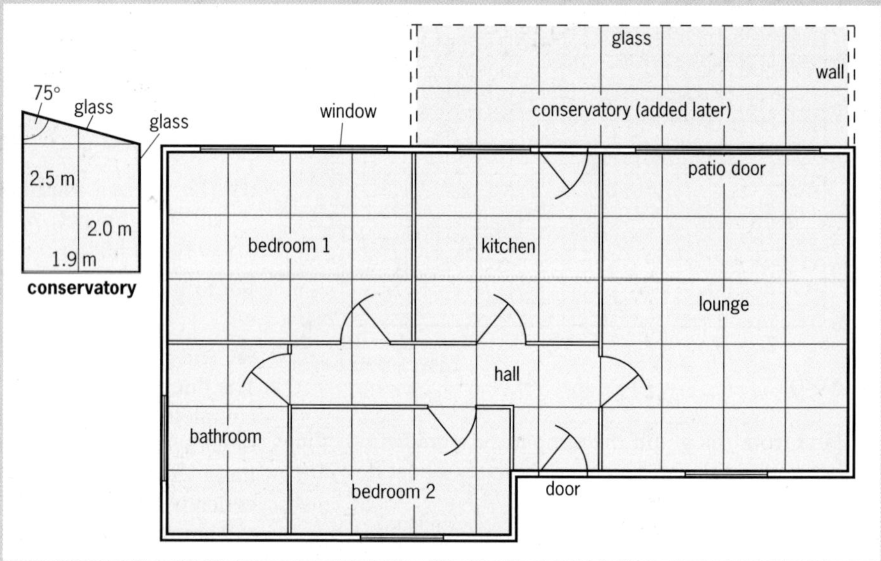

Table 14.A1 **Dimensions in metres**

	Height	Width
All walls	2.50	(as plan)
Windows	1.25	1.20
Doors	2.00	0.75
Glass patio door	2.00	3.00

U-*values*
Table 14.A2 gives the *U*-values for glass and other materials used in the bungalow.

Table 14.A2 **U-values of materials in W m⁻² K⁻¹ for materials used**

Glass	Single glazing 5.4 Double glazing 2.8 High performance double glazing: Suncool K 1.9
Other materials	Insulated external cavity walls 0.51 Internal walls 3.50 Doors 5.20 Floor 0.60 Insulated ceiling 0.35

U-values measure the rate of energy flow through a square metre of material for a temperature difference of 1 K (1 °C) between surfaces. For a given material, the energy flow in watts is given by:

U-value × area × temperature difference across thickness

Note that *U*-values are quoted for particular thicknesses of materials.

Heat losses from the bungalow without a conservatory
The bungalow has single glazing in the windows and patio door. At this stage, it has no conservatory.

d) Recalculate the energy losses for both ordinary and high performance double glazing (only losses through windows and patio door will change).

e) Calculate the cost per hour to heat the bungalow;
(i) with single glazing, and
(ii) with high performance double glazing. The energy cost is 8p per kilowatt-hour.

The bungalow is heated 15 hours a day for 90 days in the year under the same average conditions. Compare the yearly costs for the two types of glazing.

f) Double glazing is more expensive than single glazing. For a saving of about £100 a year on the heating bill, say whether it is worth:
(i) installing double glazing when the bungalow is built;
(ii) replacing relatively new single-glazed windows with double glazing.
(Estimate or research the cost of windows and their replacement.)

Solar energy gain of the conservatory

The owner adds a conservatory at the back, as shown on the plan, with a roof and long wall of glass. Next, we calculate the energy gain through the roof. (A similar calculation would apply to the glass wall.)

Engineers use **contour plots** and **overlays** to estimate solar heating. The solar energy a building absorbs depends on its geographical location (including latitude), the direction it faces, its dimensions and materials. Contour plots and overlays take account of these factors. They also allow for direct solar radiation, diffuse radiation from the sky and radiation reflected off the ground. A lot of calculation goes into producing curves for each set of factors.

In Fig 14.A2, the blue contour lines are for energy gain through the conservatory roof, and the red overlay lines are for different times of day and year at latitude 52°N (roughly the latitude of Milton Keynes). The contours of energy gain are for the roof slanted 75° to the vertical.

Because the conservatory faces south, the overlay is aligned with the Sun's maximum (noon) height, due south. The contours are drawn for totally transparent glass with a shading coefficient of 1.0.

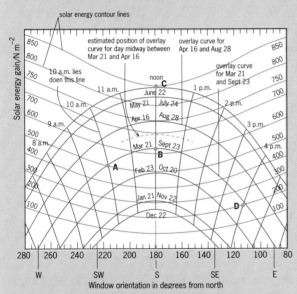

Fig 14.A2 **Contour plot (blue) and overlay (red) for the roof slanted at 75° on the south facing conservatory at latitude 52 °N**

You can read off energy gain from the contour plot. For example, at **A** at 10 a.m. on 21 March, the solar energy gain for single glazing is 500 W m⁻².

a) What is the solar energy gain per square metre at **B** (noon on the same day)?

Values of solar energy gain can be adjusted for the **shading coefficient** of any type of glass or glass-plus-blind. Table 14.A3 lists shading coefficients for some possible choices.

For instance, for single glazing with a shading coefficient of 0.95, the heating at 10 a.m. on 21 March becomes:

$$500 \times 0.95 \text{ W m}^{-2} = 475 \text{ W m}^{-2}.$$

Table 14.A3 **Shading coefficients, without and with blind**

Single glazing (6 mm glass)	No blind: 0.95	With blind: 0.55
Double glazing	0.53	0.50
Double glazing: Suncool K	0.30	0.29

b) Compare the maximum solar energy gain per square metre through the conservatory roof at noon on 22 June, at C for: (i) single glazing, (ii) Suncool K double glazing with a blind.

Energy gains are proportional to the shading coefficient, so values for the other combinations will lie between.

c) What are the corresponding energy gains at 4 p.m. at the end of the last week in September: **D**? Compare your values with those for 22 June in **b)**.

d) Calculate thermal loss per square metre through the glass roof when the temperature difference between inside and outside is 20 °C.

By comparing your answer with the values for heating in **b)**, describe what you would expect to happen to the June midday temperature in the conservatory. (Note that the vertical glass wall also contributes to the energy gains and losses.)

There is an identical bungalow at a turn in the road. It faces south-east and has the same conservatory. Fig 14.A3 shows the overlay moved to allow for the changed direction. The owner can afford only single glazing and no blinds.

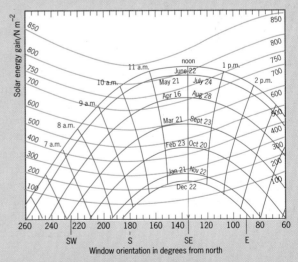

Fig 14.A3 **Contour map and overlay for the roof slanted at 75° on a south-east facing conservatory at latitude 52 °N**

e) Calculate the solar energy gain through this roof on 22 June. Compare it with the values from **b)** for the first conservatory.

(Acknowledgement: This Assignment is adapted from information supplied by Pilkington Glass Ltd, St Helens.)

15 The laws of thermodynamics

Professor George Pickett with the cryostat he uses to reach ultra-low temperatures. In 1993, he and his team cooled a piece of copper to just seven millionths of a degree above 0 K

BY THIS STAGE of your physics course, you will have spent a lot of time learning about properties of materials – conductors have resistance, fluids are viscous, and so on.

But at very low temperatures we cannot take these properties for granted. At 4 K, for example, mercury loses all its electrical resistance and so can carry a current for ever without any reduction in current. At very low temperatures, helium becomes a fluid that loses its viscosity and can flow freely. Oxygen even becomes magnetic.

Scientists are working to achieve lower and lower temperatures and to try to discover even more unusual properties in a wider range of materials.

The absolute zero temperature has been defined as 0 K (about –273.16 °C), when the only remaining energy is 'zero point' energy that arises from the so-called uncertainty principle and can never be removed. We also know that we will never be able to reach absolute zero. But that doesn't stop scientists from trying to get as close to it as possible.

The coldest temperature reached by 1995 was only 200 billionths of a degree above absolute zero – even outer space has a temperature of 'only' about –270 °C. So near, yet so far!

Introduction

Ever since physicists knew about absolute zero temperature, they have been trying to reach it. This is a problem of thermodynamics, that is, the study of energy transfer processes, including work and heating, and how these two processes are connected.

But thermodynamics is a lot more than trying to reach absolute zero. Thermodynamics explains why we can heat and expand a gas to drive internal combustion engines: most vehicles rely on some form of thermodynamic (heat) engine. It explains the limits to transferring energy, and shows, for example, that we can never build a totally efficient power station to supply electricity. In the home, thermodynamics explains the operation of refrigerators and other equipment that we use to do work for us.

1 THE ZEROTH LAW

When physicists first began to study thermodynamics, they identified three laws. Logically enough, they called these laws the first, second and third laws of thermodynamics. However, in the 1930s they realised that there was a much more fundamental law than these, so, rather than change the names of the others, they called the new one the **zeroth law of thermodynamics**. This states:

If two bodies are in thermal equilibrium with a third body, then they must be in thermal equilibrium with each other.

If two or more bodies are in this kind of equilibrium, as in Fig 15.1(a), they must have the same temperature because thermal energy is not transferred between bodies that are at the same temperature. As we saw in the previous chapter, for energy to be transferred thermally, a *temperature gradient* is required. Without the zeroth law a practical thermometer would not be possible.

We use the zeroth law in calibrating a thermometer. A practical thermometer (such as a mercury thermometer) and a gas thermometer are both placed in a liquid or similar reservoir of thermal energy, as in Fig 15.1(b). The gas thermometer is an 'ideal' thermometer used to define temperature (see page 156). Both thermometers reach thermal equilibrium with the liquid. Then the two thermometers are also in equilibrium with each other and so they are at the same temperature. This means that the practical thermometer can be calibrated against the gas thermometer for a fixed temperature. Further points can be established by altering the temperature of the liquid as measured using the gas thermometer.

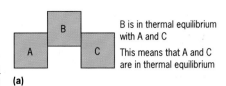

(a)

B is in thermal equilibrium with A and C

This means that A and C are in thermal equilibrium

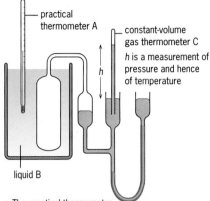

practical thermometer A

constant-volume gas thermometer C

h is a measurement of pressure and hence of temperature

liquid B

- The practical thermometer is in thermal equilibrium with the liquid
- The gas thermometer is in thermal equilibrium with the liquid
- Therefore the practical thermometer and the gas thermometer are in thermal equilibrium
- The thermometers will read the same temperature

Fig 15.1 **Calibrating a practical thermometer is possible as a consequence of the zeroth law**

2 THE FIRST LAW

We saw (pages 61 and 303) that Joule found he could increase the temperature of water by doing work on it. He kept the water thermally insulated from the surroundings so that no energy was thermally transmitted to the water from outside. But the internal energy of the water changed.

We use the symbol U for **internal energy** (measured in joules) and represent a small change by ΔU. We assume that a small amount of **work** ΔW is done to produce this change of internal energy. So the change of internal energy in joules is:

$$\Delta U = \Delta W$$

Alternatively, the internal energy of the water could be increased by allowing energy to enter from the outside, usually by thermal processes (such as conduction). We often call this energy 'heat' or 'thermal energy'. Assume that this amount of energy (in joules) entering the water is small, represented by ΔQ. The change of internal energy is given by:

$$\Delta U = \Delta Q$$

This assumes no work has been done.

But most changes in the internal energy of a body arise from a combination of both thermal transfer and mechanical transfer of energy or work. So more generally we write:

$$\Delta U = \Delta Q + \Delta W$$

We can use any combination of amounts of thermal energy or mechanical energy (work) to produce a given change of internal energy ΔU. The equation is an algebraic statement of the **first law of thermodynamics**. We can state the law in words:

The increase in internal energy of a system is the sum of the work done on the system and the energy supplied thermally to the system.

Beware! We are considering work done *on* the water so we use a *plus* sign in the equation. Some books refer to work done *by* the water so the relationship is written with a *minus* sign:

$\Delta U = \Delta Q - \Delta W$, as in Fig 15.2.

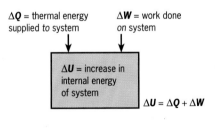

ΔQ = thermal energy supplied *to* system

ΔW = work done *on* system

ΔU = increase in internal energy of system

$\Delta U = \Delta Q + \Delta W$

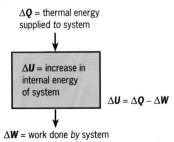

ΔQ = thermal energy supplied *to* system

ΔU = increase in internal energy of system

$\Delta U = \Delta Q - \Delta W$

ΔW = work done *by* system

Fig 15.2 **We write the expression for increase in internal energy differently depending on whether we are considering** (a) **work done *on* the system or** (b) **work done *by* the system**

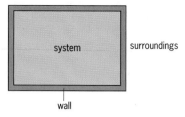

system + surroundings = universe

Fig 15.3 **The system, the wall and the surroundings**

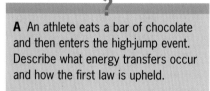

A An athlete eats a bar of chocolate and then enters the high-jump event. Describe what energy transfers occur and how the first law is upheld.

See question 1. ■

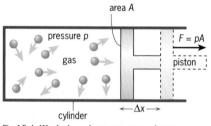

Fig 15.4 **Work done _by_ a gas _on_ a piston**

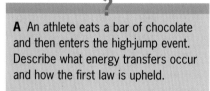

B Explain clearly why we must make each of the three assumptions.

It is probably easier for us to remember the law as the equation $\Delta U = \Delta Q + \Delta W$. Notice that in words, the law refers to a **system**. In our example, the system is the mass of water whose temperature has been raised by work or heating.

Sometimes we need to consider what happens to more than one object: we may have to consider the behaviour of a number of components. We refer to them all together as our system – we can think of the system as surrounded by an imaginary wall. All things outside this wall are the **surroundings**. The system plus surroundings taken together make up the **universe** (Fig 15.3).

Although energy can be transformed in different ways, it is not possible to create or destroy energy. (Later, in Chapter 23, you will read about how mass can be turned into energy and how energy can be transformed into mass. This is possible only with nuclear interactions; it does not happen in everyday thermodynamics.)

Applying the first law to a gas

Most real-life engines contain a gas which is heated and expands, and the gas (which does the work) follows a cycle of changes before it ends up under the same set of conditions as when it started the cycle. We shall look at the expansion of a gas, and other changes, restricting ourselves to an _ideal_ gas.

Because there are no forces of attraction or repulsion between the molecules of an ideal gas, the only way that the internal energy and the temperature of the gas can alter is by a change in the kinetic energy of the molecules. This makes an ideal gas system easier to consider than many other systems.

We look at what happens when work is done _by_ the gas. Suppose the gas is inside a cylinder with one end closed. At the other end is a piston (Fig 15.4). The molecules move around inside the cylinder, colliding with the walls, rebounding elastically off the walls and piston. This results in a pressure p on the walls and piston. The pressure acts on the piston of area A, producing an overall force:

$$F = pA$$

Suppose that, as a result of this force, the piston moves outwards slightly so that the volume of the gas increases. We then make the following assumptions:

● There is negligible friction between the piston and the walls of the cylinder.

● The force F exerted by the gas is balanced almost exactly by an external force F.

● But the external force is reduced by a very small amount.

This very slight imbalance of forces allows the piston to move. We assume it travels a small distance Δx so that there is negligible change of pressure of the gas. The volume of the gas changes by a small amount ΔV. The gas does work ΔW in expanding against the external force. We therefore have:

$$\text{work} = \text{force} \times \text{distance}$$
$$\Delta W = F\Delta x$$
$$= pA\Delta x$$

The product $A\Delta x$ is the increase in volume ΔV of the gas. So the work done _by_ the gas is:

$$\Delta W = p\Delta V$$

Note that if we consider the work done *on* the gas, the equation must have a minus sign:

$$\Delta W = -p\Delta V$$

EXAMPLE

Q A circular cylinder and piston of diameter 52 mm contains a gas at 2 atmospheres. Calculate the work done by the gas when the piston moves outwards a small distance (compared with the length of the cylinder) of 3 mm.

1 atmosphere = 1.01×10^5 Pa.

A Work done = pressure × change of volume = $pA\Delta x$

$$= 2 \times 1.01 \times 10^5 \times [\pi \times (26 \times 10^{-3})^2 \times 3 \times 10^{-3}] \text{ J}$$

$$= 1.3 \text{ J}$$

Isothermal changes in a gas

When the volume of a sample of gas changes, there is often a change of pressure. A common situation is for a gas to change its volume at constant temperature, that is, under *isothermal* conditions. Here we can use the equation of state for an ideal gas (see page 156):

$$pV = nRT$$

where R is the gas constant and n is the number of moles in the sample.

Experimentally, we can achieve an isothermal change by changing the volume slowly so that the gas remains in thermal equilibrium (that is at the same temperature) with the surroundings. With temperature constant, the pressure is related to volume by:

$$p = \frac{\text{constant}}{V}$$

This inverse dependence is shown on the pV diagram in Fig 15.5.

To find the work done by the gas when it expands, we plot the curve for actual values of p and V and measure the area under the curve. Remember: work done = $p\Delta V$. We measure the area by adding up the areas of the large number of strips (each a $p\Delta V$) making up the total area under the curve. Alternatively, we can obtain a mathematical expression for the work done. This does not require values of p to be calculated (see the Extension box below).

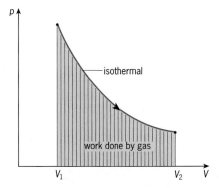

Fig 15.5 **The variation of *p* with *V* of an ideal gas during an isothermal expansion. The area under the curve equals the work done by the gas in joules**

■ See question 2.

Calculation of the work done *by* a gas during an isothermal expansion

For a small change in volume, the work done is given by:

$$\Delta W = p\Delta V \qquad [1]$$

Also we use the ideal gas equation:

$$p = \frac{nRT}{V}$$

Substituting for p from equation 1 we get:

$$\Delta W = \frac{nRT}{V}\Delta V$$

Instead of adding strips of width ΔV, we use calculus to integrate between the initial volume V_1 and the final volume V_2:

$$\Delta W = \int_{V_1}^{V_2} \frac{nRT}{V}\,dV = nRT \int_{V_1}^{V_2} \frac{dV}{V}$$

As the integral of dV/V is $\ln V$, we obtain:

$$\Delta W = nRT(\ln V)_{V_1}^{V_2}$$

$$= nRT(\ln V_2 - \ln V_1)$$

$$= nRT \ln \frac{V_2}{V_1}$$

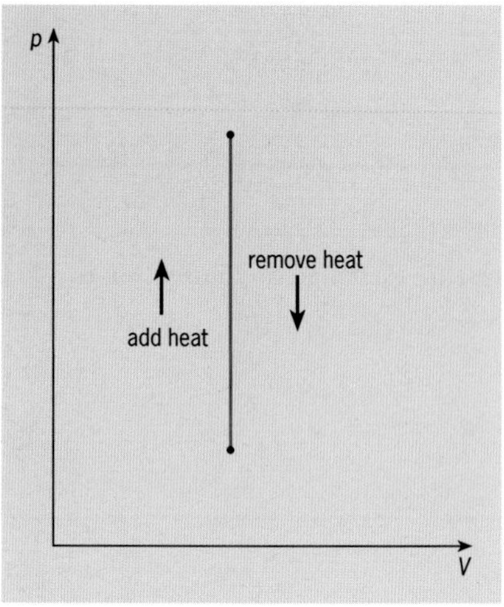

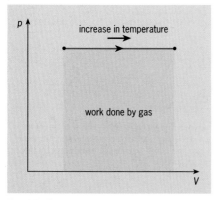

Fig 15.6 **Change of pressure at constant volume of an ideal gas. No work is done, but energy must be removed thermally for a reduction in pressure and added for an increase in pressure**

Change of pressure of a gas at constant volume

Rather than changing the volume of a gas at constant pressure, we might choose to alter the pressure at constant volume.

Suppose the gas is in a container with fixed walls and we change the temperature by either heating or cooling. Because the volume is fixed, the gas equation tells us we can expect the pressure to rise with a temperature increase: the kinetic energy of the gas molecules increases, their momentum increases, they make greater impacts as they collide with the walls, and so *pressure* on the walls increases.

Although we change the internal energy, no work is done. We can confirm this on the *pV* diagram of Fig 15.6 where the change is represented by a vertical line: below this line there is no area.

How much energy is required to raise the temperature of the gas by 1 kelvin? The amount will depend on the quantity of gas involved. We usually take one mole of gas and quote the **molar heat capacity**, which is the amount of energy in joules required to heat 1 mole by 1 kelvin. The symbol for this is C_V.

The amount of energy needed to increase the temperature of the gas by a small amount ΔT is $C_V\Delta T$. The small V tells us that this is the **molar heat capacity at constant volume**. No work is done on the gas (it does not change volume). All the thermal energy Q_V supplied goes to changing the internal energy of the gas. For 1 mole of gas:

$$\Delta U = \Delta Q_V = C_V\Delta T$$

Change of volume of a gas at constant pressure

We have seen that the work done by an ideal gas in expanding by a small amount ΔV is $p\Delta V$. If the expansion is large, the only way in which the pressure can remain constant (from the gas equation) is for the temperature to increase. The *pV* diagram of Fig 15.7 tells us that work done is still pressure × volume change, as this is the area under the horizontal line representing the change.

Again, we need to be able to work out by how much internal energy is increased. This time, we use the **molar heat capacity at constant pressure**, C_p. It is defined as the amount of energy in joules required to heat 1 mole of gas by 1 kelvin at constant pressure. For gases, the molar heat capacities at constant pressure and constant volume are not the same. When an ideal gas is heated at constant pressure its volume increases. It must do work against its surroundings as well as increase its internal energy (kinetic energy of the molecules). As a result, we expect the molar heat capacity at constant pressure to be larger than that for constant volume. The extension box on page 323 shows how much larger it will be for an ideal gas. For liquids and solids, the difference is very small because they expand very little on heating at constant pressure.

Fig 15.7 **Change of volume at constant pressure. The work done by the gas is the area under the line for p = constant. The temperature of the gas must be increased**

?

C Use the equation of state for an ideal gas to show that the temperature of a gas must increase for the pressure to remain constant as its volume increases. Suggest why this is so.

Summary of molar heat capacities for an ideal gas

Assume that Q_V is the thermal energy added to the ideal gas at constant volume, Q_p is the thermal energy added at constant pressure, and ΔT is the temperature change in each case. Then, for n moles of the gas:

$$\Delta Q_V = nC_V\Delta T \text{ and } \Delta Q_p = nC_p\Delta T$$

That is:
$$C_V = \frac{\Delta Q_V}{n\Delta T} \text{ and } C_p = \frac{\Delta Q_p}{n\Delta T}$$

EXAMPLE

A two-step process: isobaric plus isothermal change

Q A vessel contains 0.5 mol of an ideal gas which is taken through an isobaric (constant pressure) change AB, followed by an isothermal (constant temperature) change BC, as shown in Fig 15.8. At A, the gas has a pressure of 2.0×10^5 Pa (2 atmospheres) and volume 5.0×10^{-3} m³ (5 litres). The gas volume changes to 1.0×10^{-2} m³ at B, and to 1.5×10^{-2} m³ at C.

Calculate the work performed by the gas.

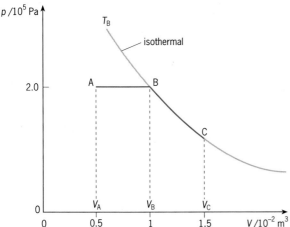

Fig 15.8 **pV diagram for a two-step process (isobaric plus isothermal change)**

A Stage AB

The work W_{AB} done by the gas will be the area under the path AB:

$$W_{AB} = \text{pressure (constant)} \times \text{change of volume}$$
$$= 2.0 \times 10^5 \times (1.0 - 0.5) \times 10^{-2} \text{ J} = 1000 \text{ J}$$

Temperature T at the isothermal

We need the temperature T_A at A to find the temperature T_B at B.

$$T_A = p_A V_A / nR = 2.0 \times 10^5 \times 5.0 \times 10^{-3}/(0.5 \times 8.31) = 240 \text{ K}$$

At B the volume is twice as large, and as the pressure remains constant the temperature doubles to $2T_A$, that is, 480 K.

Stage BC

To calculate the area under BC we need either to plot a graph and work out the area under the curve or we use calculus. Here we use calculus:

$$W_{BC} = \int_{V_B}^{V_C} p \, dV = nRT \int_{V_B}^{V_C} \frac{dV}{V} = nRT \ln \left(\frac{V_C}{V_B}\right)$$

$$= 0.50 \times 8.31 \times 480 \times \ln \left(\frac{1.5 \times 10^{-2}}{1.0 \times 10^{-2}}\right) \text{ J}$$

$$= 810 \text{ J}$$

The total work done by the gas in expanding from A to C is 1810 J.

The relationship between C_V and C_p for an ideal gas

We can apply the first law of thermodynamics to the change of volume of an ideal gas:

$$\Delta U = \Delta Q + \Delta W = \Delta Q - p\Delta V$$

where $\Delta W = -p\Delta V$ is work done, positive for work done *on* the gas; ΔV is negative since volume decreases.

We obtain the change of thermal energy of a mass of gas at constant pressure:

$$\Delta Q_p = \Delta U + p\Delta V$$

But we know (page 322) that $\Delta U = \Delta Q_V$.

So:
$$\Delta Q_p = \Delta Q_V + p\Delta V$$

We now express the changes of thermal energy at constant pressure and constant volume in terms of the corresponding specific heats and the change of temperature. We do this for *n* moles, giving:

$$nC_p\Delta T = nC_V\Delta T + p\Delta V$$

Using the state equation for an ideal gas, $pV = nRT$, and the condition of constant pressure, we substitute $nR\Delta T$ for $p\Delta V$:

$$nC_p\Delta T = nC_V\Delta T + nR\Delta T$$

We divide through by $n\Delta T$ to obtain $C_p = C_V + R$, which can be rearranged in the form:

$$C_p - C_V = R$$

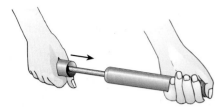

Fig 15.9 **The rapid compression of air in a bicycle pump is an example of adiabatic change**

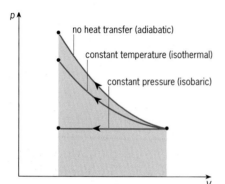

Fig 15.10 **Decrease in volume of an ideal gas compared on a *pV* diagram for adiabatic, isothermal and constant pressure changes**

Change of volume of a gas with no thermal energy transfer – an adiabatic change

When a gas is compressed *very rapidly*, there is no time for energy to leave the gas by thermal processes and go to the surroundings. But the temperature changes because work is done – temperature increases for a compression. The change is caused by an external force, so there will be an increase in pressure. This rapid change of volume is called an **adiabatic** change (no thermal transfer of energy). It is the opposite case to compressing very slowly where the gas remains at a constant temperature.

A common example of a process that is almost adiabatic is the compression of air in a bicycle pump (Fig 15.9). The mechanical work done in compression increases the internal energy of the air and so its temperature rises. If this is done very quickly, the energy does not have time to escape thermally through the walls of the pump.

There is also an adiabatic change when the gas is thermally isolated from its surroundings by an insulated wall preventing all thermal transfer. This contrasts with the isothermal case where the gas is contained by a wall that is a good thermal conductor of energy.

Zero thermal transfer is a very precise state, so what happens to an ideal gas in an adiabatic change is also precise. It is that:

$$pV^{\gamma} = \text{a constant}$$

where γ is the ratio of the principal specific heats of the ideal gas, C_p/C_V. γ is a constant.

This equation provides an important relationship between p and V during the change. In addition, it can be used with the ideal gas equation $pV = nRT$, so that it also provides relationships between T and p and between T and V during the change. It is an example of an exceptionally useful result in physics used by practising physicists who take for granted the derivation. But if you are interested in the mathematics, the next Extension box shows you how to obtain the equation.

We have now looked at different changes of volume that can occur to an ideal gas: change of volume at constant pressure, change of volume at constant temperature and change of volume with no thermal transfer. These are compared on a pV diagram in Fig 15.10 for decrease in volume from the same starting volume. Adiabatic changes are steeper on the pV diagrams than isothermal changes.

3 THE SECOND LAW

We saw from the first law that the change of internal energy of a system is related to the sum of the energy transferred to it thermally and to the work done on the system. We have also seen how work can increase internal energy.

So far, we have avoided discussing how to transfer internal energy to work. We find that we can transfer *some* of the energy to work, *but not all of it*. This gives rise to the **second law of thermodynamics**. One of several alternative statements for the law is:

Version 1 of the second law
It is impossible for thermal energy to transfer from a high temperature source to do an amount of work equivalent to that amount of thermal energy.

Proving pV^{γ} = constant for an adiabatic change

In an adiabatic change, pressure, volume and temperature all vary. We start from the equation of state for an ideal gas: $pV = nRT$.

Next we allow small changes of pressure p to $(p + \Delta p)$, volume V to $(V + \Delta V)$, and temperature T to $(T + \Delta T)$. The new values must also obey the gas equation:

$$(p + \Delta p)(V + \Delta V) = nR(T + \Delta T)$$

If we multiply out, we get:

$$pV + V\Delta p + p\Delta V + \Delta p\Delta V = nRT + nR\Delta T$$

The pV and nRT terms on opposite sides of the equation must be equal and can be cancelled. The $\Delta p\Delta V$ term is very small, so we can ignore it. This gives us:

$$V\Delta p + p\Delta V = nR\Delta T \qquad [1]$$

We have not yet used the adiabatic condition $\Delta Q = 0$. We do this in the equation for the first law $\Delta Q = \Delta U + p\Delta V$, giving:

$$\Delta U = -p\Delta V$$

But $\Delta U = nC_V\Delta T$, so we can write:

$$nC_V\Delta T = -p\Delta V \qquad [2]$$

From the Extension box on page 323 we have $nC_p\Delta T = nC_V\Delta T + nR\Delta T$, which becomes:

$$nC_p\Delta T = -p\Delta V + nR\Delta T \qquad [3]$$

Using equation 1 we replace the right-hand side of equation 3 to give:

$$nC_p\Delta T = V\Delta p \qquad [4]$$

Now we use equations 2 and 4 to find the ratio γ:

$$\gamma = \frac{C_p}{C_V} = -\frac{V\Delta p}{p\Delta V}$$

Rearranging, we get:

$$\frac{\Delta p}{p} = -\gamma\frac{\Delta V}{V}$$

Integrating either side we get:

$$\ln p = -\gamma\ln V + \text{constant of integration}$$

or: $\ln p + \gamma\ln V = \text{constant of integration}$

and removing the natural logarithm terms gives:

$$pV^{\gamma} = \text{constant}$$

where this constant is different from the original constant of integration. However, we are interested only in the fact that it is a constant, not in its value.

The industrial importance of this version of the second law is immediately obvious. Coal, gas or oil is burnt in a boiler in a power station to provide large quantities of thermal energy. This energy is transferred to energy within turbines and finally to energy used in the home. It is not possible for all the energy from the fuel to be transferred completely into work via the turbines.

It is a consequence of the second law that some energy is transferred to a reservoir of energy at a lower temperature than the temperature of the boiler. This reservoir may be the surrounding atmosphere or the water in a cooling system such as a river. There is no way of avoiding this transfer of lost thermal energy. Engineers have put a lot of effort into designing more and more efficient engines to do work but, as we shall see, there is an efficiency limit that arises from the second law.

There is an important alternative statement of the law:

Version 2 of the second law
It is not possible to have a thermal transfer of energy from a colder to a hotter body without doing work.

If the law did not apply, it would be very easy to take thermal energy from our garden and use it to heat the house in winter, and to do this without limit. We *can* transfer energy from a colder to a hotter body (or reservoir) but it is necessary to do some work. Both refrigerators and heat pumps do this.

See questions 3 to 6.

Entropy – a way of describing disorder

The second law is fundamental to what is happening to the Universe and to the direction of time, and is related to a measure of disorder called **entropy**.

When you wind a video to a new part of a film, you know your winding direction without checking whether you pressed the forward or rewind button. This is because some events cannot go backwards in real life. They are events in which disorder sets in. For instance, a waiter drops a stack of plates and they shatter: you know that shattered plates cannot rise up and reassemble themselves in the waiter's hands.

Fig 15.11 **Some events happen in everyday life; others do not**

As the Universe develops, the ordering of its constituents gets less and less. The second law may also be expressed as:

Version 3 of the second law

It is not possible to have a process in which there is an overall decrease in the entropy of the Universe.

From our everyday experience we know that some changes can occur and others cannot (Fig 15.11). We can understand why some events are so unlikely as to be impossible, by considering arrangements of molecules in a container. First, let us imagine a few molecules of a gas held in one half of the container by a partition. We then see what happens when we remove the partition. The molecules move around rapidly and very soon disperse throughout the container (Fig 15.12). In the laboratory we can demonstrate this effect with a coloured gas such as bromine.

As they move around, the molecules will on average spend half their time on one side of the container, half the time on the other. What is obvious is the extreme unlikelihood of the molecules all simultaneously returning to the original side. Let us use some numbers to try and see why this is so:

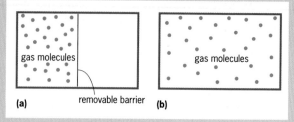

Fig 15.12 **Molecules of a gas in one half of a container (a) rapidly diffuse throughout the container (b) when the central barrier is removed**

- Suppose we start with 10 molecules only.
- When the molecules have had a chance to spread out, each molecule has a 50 per cent chance of being on the left side, that is, probability $\frac{1}{2}$.
- For all the molecules to exist on the left-hand side there is a probability of:

$$\frac{1}{2}\times\frac{1}{2}\times\frac{1}{2}\times\frac{1}{2}\times\frac{1}{2}\times\frac{1}{2}\times\frac{1}{2}\times\frac{1}{2}\times\frac{1}{2}\times\frac{1}{2}, \text{ or } \left(\tfrac{1}{2}\right)^{10}.$$

This is 1 in 1024 – not a likely event, but one that can easily happen for a brief moment as the molecules speed around.

- But now suppose we increase the number of molecules to 10^{23} (less than the Avogadro number). We replace 10 as used previously by 10^{23}. The chance of the molecules being on the left-hand side becomes:

$$\left(\tfrac{1}{2}\right)^{10^{23}} \text{ or } 1 \text{ in } 1024^{22} \text{ or } 1 \text{ in } 1.7\times10^{66}.$$

(To show that $1024^{22} = 1.7\times10^{66}$, use the x^{y} key on your calculator.)

The likelihood of all the molecules being on the left side is therefore very small indeed:

- Suppose we want all the molecules to exist on the one side for a time t.
- For this to happen for total time t, there must be a time of $1.7\times10^{66}t$ when the molecules are dispersed on both sides.
- On average, we expect to wait for half of this time before a given situation occurs.
- So we can expect to have to wait for a time of the order of magnitude $10^{66}t$ for the atoms to all be on the left side.
- So for all the molecules to exist on the left side for a time t of one microsecond we must wait $(10^{66}\times10^{-6})$ seconds or approximately 10^{60} seconds. This is over 3×10^{52} years, which is many billions of times longer than the age of the Universe.

All this arises because there are approximately 1.7×10^{66} ways of arranging the molecules between the two sides of the container. For all of them to be on one particular side is one special arrangement.

Relation between entropy and number of arrangements

We say that entropy is a measure of the number of ways in which a system can exist. Whereas there is only one way of choosing all the molecules to be on one side, there are many ways in which we can select out half the molecules to exist in one half of the container, leaving the remainder in the other half. So a distribution with the molecules arranged either side of the barrier has a high entropy (lots of ways of choosing the molecules), and all molecules on one side has a low entropy (one way).

This was understood by Boltzmann, and the idea was developed much later by Planck. It became known as the Boltzmann–Planck hypothesis, expressed as:

$$S = k \ln W$$

where W is the number of ways in which an arrangement can be set up (for instance an arrangement of molecules in a box), k is Boltzmann's constant and S is the value of the entropy measured in the same units as k (that is, J K^{-1}).

As a system becomes more disordered, there is an increase in the number of ways its molecules or components can be arranged; the equation shows directly how the entropy increases. The equation eventually made an important contribution to quantum theory, which is covered in Chapter 17.

?

D Show that $(\frac{1}{2})^{10^{23}}$ is the same as 1 in 1024^{22}.

LIFE ON EARTH AND ENTROPY

AT THIS POINT, you might wonder about the implications of entropy for biological life and its complexity. For instance, DNA is a very complex, ordered biological molecule, built from atoms and molecules that were arranged in a more random way. This seems to represent a *reverse* of the process of increasing entropy. Is it an example of overall negative entropy?

Certainly there is a *local* decrease in entropy as the arrangement of a biological molecule builds up. However, the entropy version of the second law makes it clear that there must be an overall increase in entropy of the molecule's universe – its surroundings. So where exactly is the expected *increase* in entropy, an increase that should more than balance the local biological decrease?

On the wide scale, a rapid increase in entropy is taking place in the Sun, the source of energy that sustains Earth's life. The Sun's nuclear reactions produce huge amounts of thermal energy as random movements of particles and radiation. More locally, many biological processes involve a breakdown (reduction in complexity and order) in the material of their surroundings and, in so doing, give off thermal energy as a by-product.

It is rather as if the Sun and the Earth are acting as components of a heat engine (which is discussed next). Just as in a heat engine, so, in life processes, energy is dissipated and the overall entropy of the surroundings is increased.

?

E Explain why living matter would be unable to develop on Earth without a source of energy arising from the formation of the Solar System.

Fig 15.13 **A Honda 750 cc motorcycle engine –** an example of an internal combustion engine

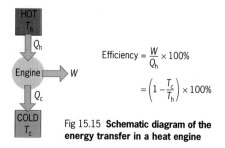

Efficiency $= \dfrac{W}{Q_h} \times 100\%$

$= \left(1 - \dfrac{T_c}{T_h}\right) \times 100\%$

Fig 15.15 **Schematic diagram of the energy transfer in a heat engine**

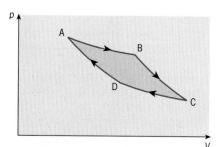

Fig 15.16 **The Carnot cycle for an ideal gas consisting of two isotherms AB and CD and two adiabatics BC and DA**

?

F Devise closed cycles on a pV diagram using isothermal, adiabatic, constant pressure or constant volume changes. Restrict yourself to either three or four sides for your cycles. Even with this restriction, there are a number of possibilities.

4 HEAT ENGINES

The internal combustion engine

Engines are designed to do work. The type of engine with which we are most familiar is the car engine, which uses the thermal energy produced by igniting petrol (Fig 15.13). It is called an **internal combustion engine**.

Air and fuel are mixed during the intake stroke (Fig 15.14). After the intake valve is closed, the air and fuel mixture is compressed as the piston goes through the compression stage of the cycle. The spark plug ignites the mixture and this produces the expansion, which is the part of the cycle producing the power. Finally, spent gases are expelled through the engine's exhaust.

The motion is kept going by having a number of pistons connected together at different stages in the cycle. The straight-line piston motion becomes a rotation in the wheels.

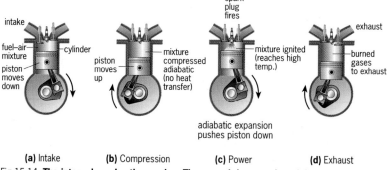

(a) Intake   **(b)** Compression   **(c)** Power   **(d)** Exhaust

Fig 15.14 **The internal combustion engine. The up and down motion of the piston is converted to rotatory motion to drive the car**

The internal energy of the engine is unchanged at the end of the cycle compared with that at the beginning. However, energy is lost to the surroundings and in the exhaust gases. So the energy supplied thermally by igniting the petrol equals the work done plus the thermal losses that heat the surroundings.

If Q_h = 'heat' input (hot source), W = work done and Q_c = 'heat' output (thermal losses; cold sink), then:

$$Q_h = W + Q_c$$

The amount of work done in a cycle equals the amount of energy actually used. However, it is not possible to do this work without a thermal loss of energy. That is, there must be a **thermal sink**. In Fig 15.15, we show transfer of energy thermally to a thermal sink.

The Carnot cycle

We can put together two isotherms and two adiabatics (the names given to the changes) to produce a closed cycle. This is called a **Carnot cycle** (Fig 15.16). In a closed cycle we finish at the same pressure, volume and temperature at which we started. The Carnot cycle is not the only cycle we can produce on a pV diagram, since we can also have changes at constant volume and constant pressure or other changes, as well as isothermal and adiabatic changes. If we follow the cycle in the clockwise direction shown, we see that the gas does net work on its surroundings equal to the area enclosed by the cycle.

The Carnot engine

Thermodynamics limits the amount of work that can be extracted between the hot source and the cold sink. An engine operating using the Carnot cycle (Fig 15.16) consisting of two adiabatics and two isothermals is the most efficient engine possible. It has an important feature: it can be operated in reverse. It is called a **Carnot engine** and is the standard by which other engines are compared. A Carnot engine cannot actually exist – it is just an imaginary ideal engine in which there are no energy losses, such as through friction and conduction.

The temperatures of the hot source and cold sink are crucial to the amount of work that can be extracted. We might expect this as we know from everyday experience that when engineers are designing an internal combustion engine or the operation of the turbines of a power station they try to achieve high temperatures on the input side.

It turns out that we can define the source temperature T_h and the sink temperature T_c in terms of the ratio of the thermal energies Q_h and Q_c transferred:

$$\frac{T_h}{T_c} = \frac{Q_h}{Q_c}$$

Note that the temperatures T_h and T_c are measured on the kelvin scale – this is *very important*. Although we cannot go into details here, this relationship between the temperatures and energy inputs and outputs can be used for defining the thermodynamic scale of temperature, and the scale agrees completely with the scale defined by using a constant gas thermometer and the triple point.

Efficiency

The efficiency of an engine is defined by:

$$\frac{\text{the work done by the engine}}{\text{the energy supplied to the engine}} \times 100\%$$

which we can express as: $\frac{W}{Q_h} \times 100\%$

But the work done is the difference between the energy supplied and the energy given to the sink (conservation of energy), that is, $Q_h - Q_c$. So the efficiency of the engine is given by:

$$\frac{Q_h - Q_c}{Q_h} \times 100\% = \left(1 - \frac{Q_c}{Q_h}\right) \times 100\%$$

$$= \left(1 - \frac{T_c}{T_h}\right) \times 100\%$$

G Internal combustion engines operate at high input and exhaust temperatures. Calculate the efficiency of an engine operating between 3000 °C and 1000 °C. How might the efficiency of the engine be improved? Why is 100 per cent efficiency not attainable?

See questions 7 and 8.

EXAMPLE

Q A power station uses superheated steam (at high pressure) at approximately 500 °C. The cold sink corresponds to the temperature at which steam condenses at atmospheric pressure, that is, 100 °C. What is the maximum theoretical efficiency for the power station?

A Using the equation for efficiency: $\left(1 - \frac{T_c}{T_h}\right) \times 100\%$,

$$\text{efficiency} = \left(1 - \frac{373}{773}\right) \times 100\%$$

$$= (1 - 0.48) \times 100\%$$

$$= 52\%.$$

Refrigerators and heat pumps

We mentioned that a Carnot engine is reversible. In real life, we operate cycles reversibly to remove energy from a cold source and get rid of this energy at a higher temperature. Refrigerators and freezers do this (Fig 15.17). Here the cooling cycle involves the evaporation of a low boiling point liquid in the refrigerator. The energy for this evaporation comes from the internal energy of the contents of the refrigerator, but we need a source of work.

The vapour is condensed using a compressor. This compressor, of course, does work using energy from the electrical mains. As the vapour is compressed and condensed, it gets hotter and loses (dissipates) its energy thermally through a long metal pipe at the back of the refrigerator. So, overall, internal energy extracted from the food is dissipated by the element which becomes warm at the rear of the cabinet. Hence, a refrigerator will warm the room it is in.

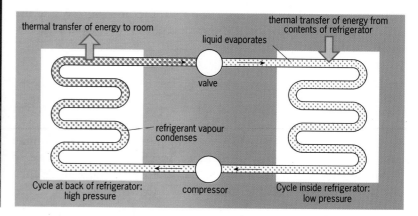

Fig 15.17 **The back of a domestic refrigerator, showing the compressor at the bottom, and the looped pipe in which the refrigerant circulates**

We show the operation of a refrigerator in Fig 15.18(a) in the same way as we did for the heat engine. The efficiency, or more appropriately the **coefficient of performance**, as values can be greater than 100 per cent, is defined by:

$$\frac{\text{energy extracted from the cold source}}{\text{work done by compressor}} = \frac{Q_c}{W} \times 100\%$$

$$= \frac{Q_c}{(Q_h - Q_c)} \times 100\%$$

$$= \frac{T_c}{(T_h - T_c)} \times 100\%$$

A **heat pump**, as shown in Fig 15.18(b), operates in a similar way to a refrigerator by extracting energy thermally from a cold source such as a river and delivering it into a body such as a house at a higher temperature. Here we are interested in the amount of internal energy added to the house rather than the amount extracted from the river, so we define the percentage coefficient of performance by:

$$\frac{\text{energy added to the hotter body}}{\text{work done by pump}} = \frac{Q_h}{W} \times 100\%$$

$$= \frac{Q_h}{(Q_h - Q_c)} \times 100\%$$

$$= \frac{T_h}{(T_h - T_c)} \times 100\%$$

H Obtain appropriate temperatures found in a typical kitchen and in a kitchen refrigerator, and estimate the theoretical coefficient of performance of the refrigerator.

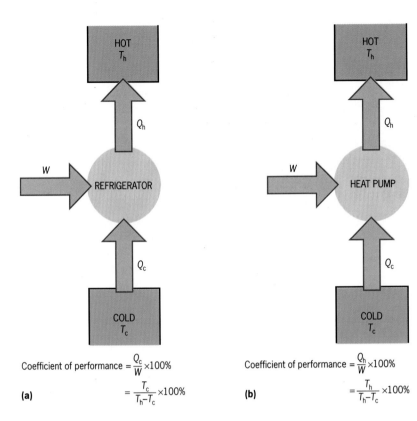

Fig 15.18 **The energy transfer in** (a) **a refrigerator and** (b) **a heat pump**

Coefficient of performance $= \dfrac{Q_c}{W} \times 100\%$

$= \dfrac{T_c}{T_h - T_c} \times 100\%$

(a)

Coefficient of performance $= \dfrac{Q_h}{W} \times 100\%$

$= \dfrac{T_h}{T_h - T_c} \times 100\%$

(b)

On the basis of this equation, heat pumps look very attractive. A small amount of work can be used to transfer a large amount of thermal energy. In practice, real heat pumps turn out to have a much lower efficiency than the efficiency of an ideal heat pump based on the Carnot cycle. Heat pumps also have high capital cost, partly because they are not made in large numbers.

In addition, although energy can be continually extracted thermally from a neighbouring river (assuming the river does not freeze up), extracting internal energy from other sources such as neighbouring ground can lead to permanent frozen conditions.

> **?**
>
> I Given that a refrigerator and a heat pump operate between the same hot source and cold sink, show that the coefficient of performance for the heat pump is equal to 1 + coefficient of performance of the refrigerator.

Heat pumps and entropy

We have already seen that entropy is a measure of disorder. When energy is transferred thermally to a body, this increases the disorder.

We can measure the entropy change best if the temperature remains constant. For example, suppose we add energy thermally to a solid at its melting point. The resulting liquid is more disordered than the solid and the molecules move around randomly with a range of speeds. The **change of entropy** ΔS can be shown to be related to the change of thermal energy ΔQ by:

$$\Delta S = \frac{\Delta Q}{T}$$

where T is temperature measured on the kelvin scale.

We can also assume constant temperature when we extract from, or put energy into, a large reservoir. When the temperature is not constant, working out the entropy change becomes more difficult (and usually involves calculus).

We can consider the change of entropy that occurs when a heat engine operates. Here we can assume that the hot source and the cold sink remain at constant temperature.

See question 9.

EXAMPLE

Q A heat engine operates between 373 K and 300 K and has an efficiency of 18 per cent. What is the change of entropy when work of 1000 J is done by the engine?

A First we find the transfer of energy from the hot source.

1000 J of work is done with 18 per cent efficiency. Therefore, $(100/18 \times 1000)$ J must be transferred from the hot source. That is:

$$Q_1 = 5560 \text{ J}$$

Loss of entropy of the hot source

$$= Q_1/T = 5560/373 \text{ J K}^{-1}$$

$$= 14.9 \text{ J K}^{-1}$$

Gain of energy by the sink

$$= Q_2 = Q_1 - W = (5560 - 1000) \text{ J}$$

$$= 4560 \text{ J}$$

Gain of entropy of the heat sink

$$= Q_2/T = 4560/273 \text{ J K}^{-1}$$

$$= 16.7 \text{ J K}^{-1}$$

Net change in entropy

$$= 1.8 \text{ J K}^{-1}$$

As we have come to expect, there is a net increase in entropy.

5 POSTSCRIPT – THE THIRD LAW

There is a **third law of thermodynamics**. It states:

> **The entropy of a system approaches a constant value (zero in a simple system) as the temperature approaches absolute zero.**

To achieve zero temperature requires an infinite number of steps – we have to extract a bit of energy, and a bit more, and a bit more and so on. A practical statement of the law is:

> **It is impossible to achieve absolute zero temperature for a system by a finite number of steps.**

At the start of this chapter we saw that a temperature as low as 200 billionths of a kelvin has been reached. This statement of the third law tells us that we shall never get to 0 K.

SUMMARY

After studying this chapter you should be able to:

■ Provide a statement of the zeroth law and indicate its application to temperature measurement.

■ Provide a statement of the first law and apply the equation $\Delta U = \Delta Q + \Delta W$.

■ Carry out calculations for isothermal (constant temperature) and isobaric (constant pressure) changes of an ideal gas.

■ Use the relationship between C_V and C_p for an ideal gas in calculations.

■ Use (not derive) the relationship $pV^\gamma = $ constant in calculations for an adiabatic change of volume of an ideal gas.

■ Provide a statement of the second law and explain its importance.

■ Have an understanding of the concept of entropy, $S = k \ln W$, and change of entropy, $\Delta S = \Delta Q/T$, and calculate ΔS values for simple applications.

■ Understand the principles underlying heat engines, refrigerators and heat pumps and carry out simple calculations, including those for efficiencies (or coefficients of performance).

QUESTIONS

For some questions you will need also to refer back to Chapter 14.

1 The internal energy of a system can be increased by heating it or by working on it. Give an example of each. For heating to occur, a temperature difference is necessary. What else is necessary for heating to take place?

Fig 15.Q1 shows a motor-driven belt rubbing slowly over the surface of a copper cylinder. The cylinder gets hot. Student A says, incorrectly, that the motor is heating the belt which is heating the cylinder. Student B says, correctly, that the motor is working on the belt which is working on the cylinder. State the measurements you could take to show that Student B is correct.

copper
cylinder

motor

Fig 15.Q1

The belt travels at 70 mm s^{-1} across the surface of the cylinder and the tension difference between the two sides of the belt is 14 000 N. Explain why the rate of working is 980 W.

The cylinder has a mass of 300 g and a specific heat capacity of 390 J kg^{-1} K^{-1}. Calculate how long it would take for the temperature to rise by 5 K. What assumption have you made in this calculation?

How far will the belt have moved in the time taken for this temperature rise? Use this answer to give a further reason why Student B's description of the energy transfer is preferred to Student A's.

[ULEAC: June 1993, A level Physics, Paper 2, Section II, Q8(c)]

2 Five moles of an ideal gas at volume 3.0 m^3 and temperature 300 K expand to a volume of 6.0 m^3. Draw on graph paper a plot of pressure p versus volume V for the gas and estimate the work done by the gas in expanding. Use the expression given in the Extension box on page 321 for ΔW to calculate the work done. Compare your two values.

3 Fig 15.Q3 [next column] shows the idealised and practical forms of indicator diagram for a four-stroke internal combustion engine.

State and explain three differences between the two diagrams.

[NEAB: June 1993, Physics Advanced, Syllabus B, paper 1, section B, Q9]

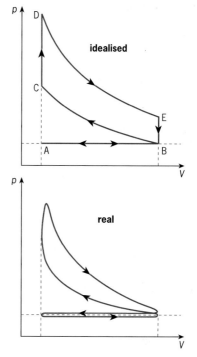

Fig 15.Q3

4 An ideal gas at 300 K is adiabatically compressed to half its original volume and then cooled at constant volume until the pressure is restored to its initial value. What is the final temperature?

5 An ideal gas is contained in a hollow cylinder, sealed at one end, with a frictionless piston at the other. Initially the gas occupies a volume of 1.0×10^{-3} m^3 at a pressure of 200 kPa and at a temperature of 300 K.

a) The gas is heated isothermally. How is the thermal energy Q supplied related to the change in internal energy ΔU and the work W done by the gas? Explain your answer. Does the gas expand or must the piston be pushed in?

b) How many moles of gas does the cylinder contain?

c) The cylinder is used as a Stirling engine by suitable coupling to the piston. The gas undergoes the following cycle of changes:
   (i) compressed isothermally to half its initial volume,
   (ii) heated at constant volume to 450 K,
   (iii) expanded isothermally back to the original volume,
   (iv) cooled at constant volume to the initial condition.
   Calculate the pressure at the end of each stage. Draw a pV diagram showing the gas undergoing these four processes.

d) Explain how you could use the pV diagram to find the thermal efficiency of the Stirling engine, which can be defined as:

$$\frac{\text{(work done by the gas)} - \text{(work done on the gas)}}{\text{work done by the gas}}$$

[OCSEB: June 1993, A level Physics, Physics 3/4, Q8]

6

a) An ideal gas, with initial properties p_1, V_1 and T_1, expands adiabatically to a state with properties p_2, V_2 and T_2.

   (i) Write down an equation relating p_1, V_1 and T_1 to p_2, V_2 and T_2.

   (ii) Write down an equation relating p_1, V_1, p_2 and V_2.

b) The above equations can be combined in the form:

$$\frac{T_1}{T_2} = \left(\frac{p_1}{p_2}\right)^{[(\gamma-1)/\gamma]}$$

where γ is the ratio of the specific heat capacities of the gas. A pump raises the pressure of air for pneumatic road construction equipment from atmospheric pressure of 1×10^5 Pa to an absolute pressure of 4×10^5 Pa (Fig 15.Q4). The pump acts almost adiabatically. External temperature is 280 K. Estimate the air temperature immediately after compression.

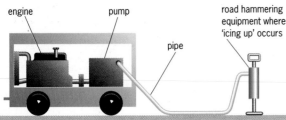

engine   pump   pipe   road hammering equipment where 'icing up' occurs

Fig 15.Q6

c) Assuming that the air in the pipes feeding the pneumatic road construction equipment falls to ambient temperature before reaching the equipment, explain why the operators sometimes experience 'icing-up' of the equipment when the air is released.

(Ratio of the principal specific heat capacities of air $(\gamma = 1.4$.)

[NEAB: June 1994, Physics Advanced, Syllabus B, paper 1, section B, Q10]

7 A heat engine operating between two reservoirs has an efficiency of 25 per cent and does 3500 J of work on the surroundings in each cycle. Calculate the amount of energy absorbed in each cycle from the hot reservoir and the amount each cycle given out to the cold reservoir.

8 An inventor claims to have constructed a cyclic engine operating between two reservoirs at temperatures of 490 K and 340 K respectively. In each cycle, the engine extracts 8000 J of energy from the high-temperature reservoir and rejects 5200 J of energy to the low temperature reservoir, while performing 2800 J of work on the surroundings. Would you be prepared to give financial support to the project? Justify your answer with calculations and remember that the most efficient engine is a Carnot engine.

9 2400 J of energy are transferred from a heater at a constant temperature of 600 K into oil at a temperature of 400 K. What is the entropy change of the heater in J K^{-1}?

[NEAB: June 1993, Physics Advanced, Syllabus B, paper 1, section A, Q39]

Assignment

pV DIAGRAM FOR AN INTERNAL COMBUSTION ENGINE

This Assignment concerns the Otto cycle, which is used as an approximation to the cycle used in an internal combustion engine.

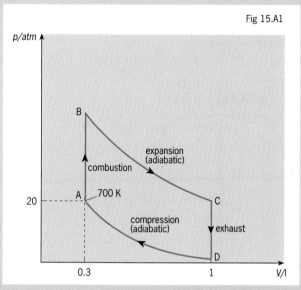

Fig 15.A1

The cycle has a combustion stroke AB which takes place at constant volume (see Fig 15.A1 for the pV diagram shown schematically). There is an adiabatic stroke BC in which the power is transmitted, an exhaust stroke CD at constant volume and an adiabatic compression DA to the starting point A of the cycle. You will use a calculator and graph paper to look at the operating conditions and performance of the engine undergoing this cycle.

At point A on the cycle the volume of the chamber of the engine is 0.3 litre, the temperature is 700 K, and there is a gas pressure of 20 atmospheres. At point B after ignition the pressure is 200 atmospheres, and after expansion (points C and D) the volume of the chamber is 1 litre. Note that 1 atmosphere is approximately 1×10^5 Pa and 1 litre is 10^{-3} m^3, but *for the moment* we shall use these everyday units of atmosphere and litre.

1 Knowing the positions of A and B on the pV diagram, you need to obtain data to plot lines BC and AD. Along BC, $pV^\gamma = $ constant$_1$; and along AD, $pV^\gamma = $ constant$_2$. For the mixture of gases used in a combustion engine, γ may be taken as 1.4.
a) Using the x^y button on your calculator, obtain numerical values for constant$_1$ and constant$_2$.
b) Continuing to use your calculator and knowing the values of constant$_1$ and constant$_2$, obtain values of pressure

corresponding to volumes of 0.6, 1.0, 1.5 and 2.1 (litres) for the adiabatic expansion BC and the adiabatic compression DA. (Having obtained values for BC, you may find it is easy to write down values for AD.)
c) Now draw a scaled graph for the indicator diagram ABCD.

2 Assuming that the ideal gas equation $pV = nRT$ applies, use the conditions:
a) at A to calculate the temperature at B,
b) at A to calculate the temperature at D,
c) at D to calculate the temperature at C.

Lines BC and AD represent adiabatics. By thinking how isothermal lines might appear on the graph, consider whether your calculated temperatures T_B, T_D and T_C seem sensible.

3 Using the ideal gas equation $pV = nRT$ at a fixed point (A, say), calculate the number of moles of gas in the chamber. You will have to convert atmosphere and litre to SI units for this. This can be done by remembering what volume is occupied by 1 mole of an ideal gas at STP.

4 You are given that the specific heat C_V of the gas is 20.8 J mole^{-1} K^{-1}. The energy thermally gained is $Q_{in} = nC_V(T_B - T_A)$ and the energy thermally transferred from the gas is $Q_{out} = nC_V(T_C - T_D)$. Calculate Q_{in} and Q_{out}.

5 The percentage efficiency of the engine is given by (net work/Q_{in}) × 100, where the net work equals $Q_{in} - Q_{out}$. Calculate the efficiency of the Otto engine. Explain why the efficiency of a real internal combustion engine is much lower.

6 Calculate the efficiency for an ideal (Carnot) engine operating between the highest temperature (T_B) and the lowest temperature (T_D). Compare the value with that obtained for the Otto engine in question **5** and comment on the difference.

7 The net work done by the engine equals the area enclosed by your pV diagram. Estimate this area. Do this by estimating the number of squares enclosed by your graph and check the size of the squares in units of pV (Pa × m^3). Check your value for work done with that used in question **5**.

8 By choosing an appropriate cycle rate, estimate the power produced by this ideal Otto engine.

ENERGY, TEMPERATURE AND THE LAWS OF THERMODYNAMICS

This Chapter Map brings together the ideas you have covered in your study of **14 Energy and temperature** and **15 The laws of thermodynamics**. The main concepts and equations are included, with linked boxes showing how the ideas are interconnected.

Refer to the syllabus you are following and use the map as a guide to the key information you need to know and understand. The map will help you identify areas you are not confident about and may wish to study further.

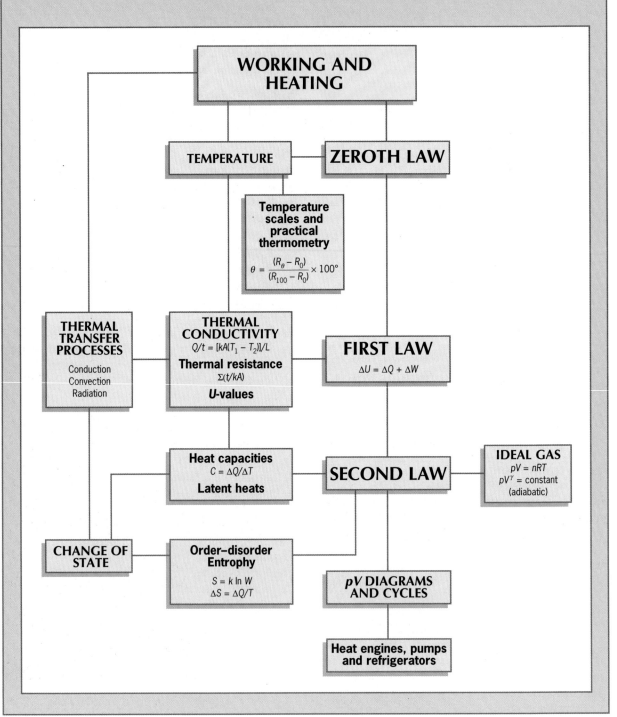

WORKING AND HEATING

TEMPERATURE — **ZEROTH LAW**

Temperature scales and practical thermometry
$$\theta = \frac{(R_\theta - R_0)}{(R_{100} - R_0)} \times 100°$$

THERMAL TRANSFER PROCESSES
Conduction
Convection
Radiation

THERMAL CONDUCTIVITY
$Q/t = [kA(T_1 - T_2)]/L$
Thermal resistance
$\Sigma(t/kA)$
U-values

FIRST LAW
$\Delta U = \Delta Q + \Delta W$

Heat capacities
$C = \Delta Q/\Delta T$
Latent heats

SECOND LAW

IDEAL GAS
$pV = nRT$
$pV^\gamma = \text{constant}$
(adiabatic)

CHANGE OF STATE

Order–disorder Entropy
$S = k \ln W$
$\Delta S = \Delta Q/T$

pV DIAGRAMS AND CYCLES

Heat engines, pumps and refrigerators

WAVES, QUANTA AND ATOMS

I T MAY NOT be obvious that light, X-rays and radio waves are all the same kind of radiation – electromagnetic waves – and not obvious what they have to do with the electromagnetism that is referred to in the second section. So we cover the connections here.

Electromagnetic radiation travels as waves, and we use its wave properties to explain refraction, diffraction, interference and polarisation. We then see that, strangely, electromagnetic radiation can also behave as particles that we call photons. These ideas are linked with the nature of the atom and the quantisation of energy.

When atoms are disturbed (excited) they emit radiation having definite patterns called spectra, which can tell us a great deal about atomic structure. Where nuclei are radioactive, the atomic nucleus can be a source of radiation. The study of radioactivity led not only to nuclear power and nuclear medicine (covered later in 24 Medical physics) but also to a better understanding of the nature of matter (dealt with more fully in 26 Deep matter).

Our eyes can detect light only in the narrow visible spectrum – our brain interprets the images produced on the retina. Now we have devices that can use almost any part of the electromagnetic spectrum to create information-carrying images – of atoms in an alloy to quasars which emitted radiation billions of years ago. We end with a review of the ways images are produced, stored and processed.

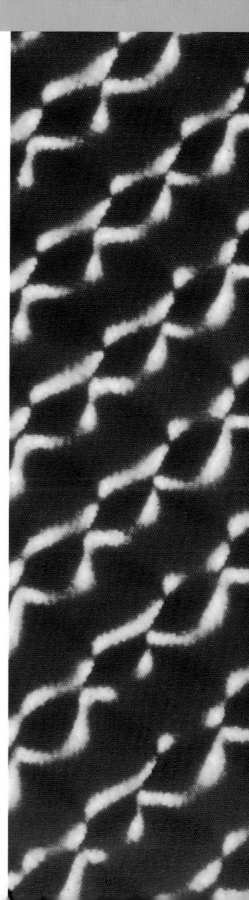

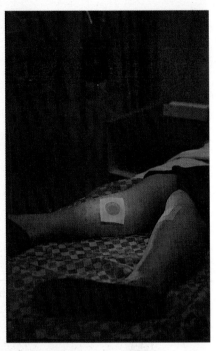

Photodynamic therapy treatment: the cancerous tissue in the patient's leg has accumulated the dye that is being illuminated with laser light. As a result, the cancer cells are killed

LIFE ON EARTH depends on one chemical in particular – chlorophyll. It can interact with red light in sunlight; in a photochemical reaction, its electrons become more energetic. A chain of reactions forms sugar from water and carbon dioxide in the surroundings, with oxygen as a useful by-product.

A photochemical reaction is now being applied to the treatment of cancers, in a procedure called photodynamic therapy (PDT). Its key process parallels the action of chlorophyll. A patient is given a harmless drug designed to build up in cancerous tissue. The drug is a small part of the very complex chlorophyll molecule and acts as a light-absorbing dye. Dyes absorb certain frequencies from white light directed at them, so the light they reflect lacks those frequencies, which is why a dye looks coloured: leaves look green because green light is, roughly, white minus the red.

In PDT, light comes from a laser tuned to a frequency absorbed by the drug. The laser produces a low-energy dose of red light which penetrates quite deeply into the body and is no harm to ordinary tissue. The important thing is what happens to the energy carried by the absorbed radiation.

In typical laser treatment, absorbed radiation heats up target tissue enough to kill the cells, both ordinary and cancerous. In PDT, the less energetic radiation that the drug absorbs triggers a photochemical reaction which releases a poison consisting of single atoms of oxygen. In contrast to the harmless diatomic oxygen molecules we breathe from the air, single atoms are highly reactive and are able to kill living cells. Photodynamic therapy is effective because, since the poison only forms at the point where the dye accumulates, only the diseased cells are killed.

The electromagnetic part of the Universe

We can say that the Universe consists of matter and electromagnetic radiation. Light is just one small (visible) part of the range of electromagnetic radiation, and the radiation itself has the characteristics of **wave** and **particle** behaviour. The particle – **photon** – aspect of electromagnetic radiation appears in Chapter 17, which deals with quantum theory; Chapter 26 also features the photon, as one of the family of particles making up the universe of matter.

This chapter covers the **wave** aspect of the range of radiations observed in the **electromagnetic spectrum**, the **speed** of the radiations, how they are produced and detected, and how we make use of the wave nature of the radiations. So we deal with **diffraction, interference** and **polarisation**. (21 Communications covers the use of electromagnetic radiations as information carriers.)

1 THE ELECTROMAGNETIC SPECTRUM

A very hot object such as a star produces a range of radiation like that of the Sun, see Fig 16.1. **Light** is the tiny fraction that we detect with our eyes. White light can be separated into a range of colours to form a **spectrum** of white light, from violet with a wavelength of about 4×10^{-7} m, to red of about 7×10^{-7} m wavelength.

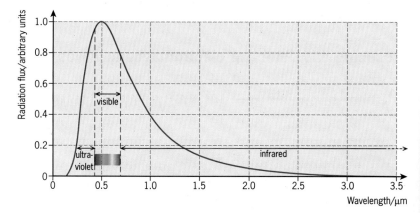

Fig 16.1 **Graph showing the continuous spectrum of solar radiation**

Fig 16.2 **A prism splits white light into the colour spectrum that we can see**

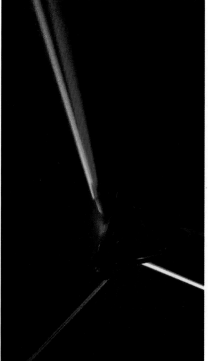

Why 'electromagnetic'?

The radiations are produced and detected by the *acceleration or sudden movement of electrons* (or occasionally of other charged particles). For example, a **radio signal** is produced by electrons as they oscillate to and fro in an aerial, as in Fig 16.3. The result is an interlocked pair of fields, electric and magnetic, oscillating at the frequency of the electron current. We can think of the **electric field** as being produced by the *charge* on the electrons.

A **magnetic field** is produced whenever charges move, and is proportional to the size of the current (see page 246).

What is surprising (and hard to explain) is that the interlocking fields move away from the aerial at a speed which in a vacuum is the same for all of them, whatever their frequency. This is the **speed of light, *c*.** These moving fields *behave like waves*: they can be *diffracted*, they *interact* with each other to show *interference*, and they can be *polarised*. These terms are explained below.

When the waves reach another metal rod, the electric field component exerts a varying *electric* force on the electrons in the metal which oscillate in time with the variations of the field.

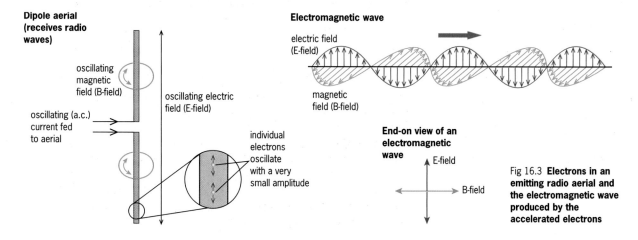

Fig 16.3 **Electrons in an emitting radio aerial and the electromagnetic wave produced by the accelerated electrons**

(a) Loop aerial B-field

B-field

loop in section

oscillating B-field

(b) Electron moves between orbitals in atom

visible light

A loop-shaped metal aerial – as in Fig 16.4(a) – encloses the varying magnetic field of the waves and, by the laws of **electromagnetic induction,** an e.m.f. is induced in the loop.

Infrared radiation, light and the shorter wavelengths beyond the visible spectrum are also made by electrons moving within atoms, see Fig 16.4(b). There is more about this in Chapter 17.

Fig 16.4 (a) **A loop aerial detects electromagnetic waves because the varying B-field includes an electric current in the loop.** (b) **Electrons in atoms move from higher to lower energy orbits and produce light**

Producing and detecting electromagnetic waves

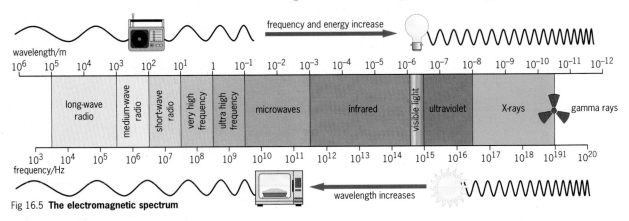

frequency and energy increase

wavelength/m

10^6 10^5 10^4 10^3 10^2 10^1 1 10^{-1} 10^{-2} 10^{-3} 10^{-4} 10^{-5} 10^{-6} 10^{-7} 10^{-8} 10^{-9} 10^{-10} 10^{-11} 10^{-12}

long-wave radio | medium-wave radio | short-wave radio | very high frequency | ultra high frequency | microwaves | infrared | visible light | ultraviolet | X-rays | gamma rays

10^3 10^4 10^5 10^6 10^7 10^8 10^9 10^{10} 10^{11} 10^{12} 10^{13} 10^{14} 10^{15} 10^{16} 10^{17} 10^{18} 10^{191} 10^{20}

frequency/Hz

wavelength increases

Fig 16.5 **The electromagnetic spectrum**

Radio waves

Radio waves are produced by the *accelerated* motion of free electrons as explained above, and are used mainly in communication systems – see Chapter 21.

A process was developed in the 1930s to provide short-wavelength radio waves – **microwaves** – to help in the medical treatment of damaged tissue. Most school laboratories now have low power microwave sources for experiments on the nature of electromagnetic waves. Microwaves are still used in physiotherapy – and in cooking!

?

A How do the two common designs of TV aerials (loop and rod) support the idea that TV signals are somehow both *electric* and *magnetic*?

Radar

The potential of these very short radio waves for the **ra**dio **d**etection **a**nd **r**anging of wartime aircraft was quickly exploited with the development of **radar**. In radar transmitters (Fig 16.6), electrons oscillate in small metal cavities called *magnetrons* which contain a strong magnetic field.

The waves produced have a frequency determined by the size of the cavity and the strength of the magnetic field, depending how long it takes an electron to complete an oscillation. The waves are led out of the cavity through a metal tube called a *waveguide*. The metal walls reflect the waves and keep them moving in the right direction, towards a small aerial at the focus of a metal mirror. They are then emitted as a beam. The transmitting aerial and mirror can also act as the main *receivers* of the radiation, with a detector at the focus.

Microwave ovens

Similar magnetrons are used in **microwave ovens**. The frequency of the microwaves is selected to match the resonance frequency of water molecules so that energy from the waves is transferred efficiently to the kinetic energy of the molecules. This raises the temperature of any food containing water.

Fig 16.6 **A radar aerial at Heathrow airport for ground–air control**

Infrared radiation

Fig 16.7(a) **Rescuers use body-sensing equipment at a fire or earthquake**   Fig 16.7(b) **The infrared image of a victim**

Infrared radiation (IR) overlaps with short microwaves in the electromagnetic spectrum (Fig 16.5), but in practice the term describes the (invisible) hot body emissions which begin just beyond the visible red. Infrared is readily absorbed by matter (it excites movement in molecules) and raises its temperature.

Being sensitive to near infrared, the human skin is a simple detector – we feel the 'heat'. A thermometer with a blackened bulb (or sensor surface) will show a rise in temperature when placed in the red region of a white-light spectrum. But it shows a greater temperature rise when placed in the dark region just outside the red – as was first shown in 1800 by the astronomer Sir William Herschel (Germany/Britain, 1738–1822).

Special film takes photographs in infrared, but is difficult to use and has largely been replaced by electronic detecting methods. Modern infrared detectors use solid state (electronic) detectors which act rather like TV cameras. In industry, they monitor processes which cause temperature differences. They are used in Earth satellites both for military purposes and to observe the growth of crops, measure surface temperature, etc (see page 68 and Chapter 23). Infrared detectors can be made sensitive to a narrow wavelength band: in the rescue cameras used to find fire or earthquake victims, the detectors are most sensitive to the infrared radiation that is characteristic of body temperature.

Electronic devices (eg semiconductor light-emitting diodes) also emit infrared, and are widely used in the remote switches of household electronic systems such as TV sets, video recorders and hi-fi systems.

B Suggest reasons why **(a)** food in metal containers cannot be cooked in a microwave oven, **(b)** an empty glass or plastic container doesn't get hot in a microwave oven.

See question 1.

Fig 16.8 **A hand-held remote control directs an IR signal at the equipment**

Light

Visible radiation is detected by both human and animal eyes and also by a range of devices including ordinary photographic film, photoelectric cells of various types and by very sensitive charge-coupled devices (CCDs). There is more about light-sensitive devices and the control of light in Chapter 19.

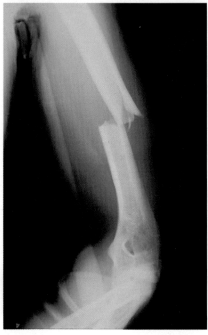

Fig 16.9(a) **Ultraviolet lighting is used to create special effects**

Fig 16.9(b) **X-ray image of a broken arm bone**

Short wave and ionising radiation

Ultraviolet radiation (UV) is produced by changes in the energy levels of atomic electrons, as we shall see below, and by very hot bodies. UV was discovered in 1801 during early photochemical experiments in which light blackened silver chloride. A type of radiation beyond the visible violet was discovered to have more effect on silver chloride than visible light. As explained below, *the shorter the wavelength the more energetic the radiation*, so UV is very effective at exposing photographic film. UV is good at making chemicals called *phosphors* glow, and low-intensity UV is used for special effects: Fig 16.9(a). It is also energetic enough to ionise atoms and so can harm living tissue, causing sunburn and skin cancers.

X-rays have shorter wavelengths than ultraviolet. They are produced when electrons decelerate very rapidly, such as when high-speed electrons are stopped by colliding with a metal target. They were first discovered, accidentally, by William Röntgen (Germany, 1845–1923). He was experimenting with cathode rays (electron streams in an evacuated tube) and noticed that they caused a phosphor screen some metres away to glow. He soon found that the rays could pass through soft tissue but were selectively absorbed by denser material such as metal and bones. Within weeks of the discovery, X-rays were being used in hospitals for diagnosis and treatment – rather too soon and too dangerously.

X-rays are used for medical diagnosis (see Chapter 24) and to inspect metal objects for flaws. X-rays also played a significant role in the science of X-ray crystallography which, amongst other things, led to the discovery of the helical structure of DNA.

Gamma (γ) radiation is produced by changes in the internal energy of an atomic nucleus in radioactive decay (see Chapter 18). Gamma rays are very energetic and penetrating: photographic film and ionisation detectors such as Geiger counters will detect them.

Damage caused by ionising radiation

Because of the high energy associated with photons of such short wavelength radiation (see below), UV, X- and gamma radiation can all damage living tissue. These radiations **ionise** atoms and molecules in living cells and disrupt their biochemical processes. The cells may die, or worse, they may become cancerous. Small changes to the DNA in sperm or ovum cells are particularly dangerous since they may cause *genetic mutations* to appear in offspring.

2 THE CHARACTERISTICS OF AN ELECTROMAGNETIC WAVE

Electromagnetic waves have a frequency f and a wavelength λ related by the simple wave formula:

$$c = f\lambda$$

The amplitude of the wave is measured by either its electric or its magnetic component, depending on the instrument used. As with all waves, the energy carried by the wave is proportional to the square of the amplitude. As we shall see in Chapter 17, when radiation has a wavelength of comparable size to atoms and molecules, we need to think of the energy as being carried as packets or **quanta**, in the form of particles or **photons**.

The speed of electromagnetic waves

The speed of electromagnetic waves travelling in a vacuum is the same for all frequencies and is *defined* as $2.997\ 924\ 58 \times 10^8$ m s^{-1} *exactly*. In 1983 it was found that measurements using radio waves gave the best results and consequently the speed of light was defined as 299 792 458 m s^{-1}. The same year, this value was confirmed by a very careful experiment which used a stabilised laser working with red light. It no longer makes sense to measure the 'speed of light' (see page 8).

What decides the speed of electromagnetic waves?

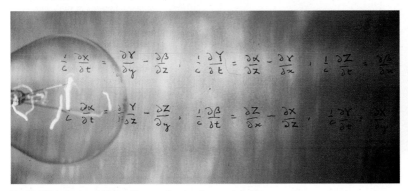

Fig 16.10(a) **Some of Maxwell's equations as he recorded them**

Fig 16.10(b) **The Scottish physicist James Clerk Maxwell (1831–1879) whose equations brought together electricity and magnetism**

In 1864, James Clerk Maxwell was considering the speed of the alternating electric and magnetic fields that define a wave. From theory, he worked out that the speed of these fields is determined by the electric and magnetic constants of the medium in which the waves travel. These are the *force constants* ε and μ (see pages 237 and 268).

In a vacuum, the electric field constant is ε_0 and is called the **permittivity of vacuum**. If one charged particle is moving relative to the other, there is an additional force on both particles, the *magnetic* force. The magnetic field (force) constant for a vacuum is μ_0 and is called the **permeability of vacuum** (see page 268). Air is near enough a vacuum for everyday calculations. Maxwell was able to prove from theory both that electromagnetic waves other than light should exist, and also that the speed of all such waves in empty space is given by the Maxwell formula:

$$c = (\varepsilon_0 \mu_0)^{-\frac{1}{2}} \quad \text{or} \quad c^2 \varepsilon_0 \mu_0 = 1$$

In a vacuum, the constants have the values:

$$\mu_0 = 4\pi \times 10^{-7} \text{ H m}^{-1} \text{ (by theoretical definition)}$$

$$\varepsilon_0 = 8.854\ 187\ 82(7) \times 10^{-12} \text{ F m}^{-1} \text{ (calculated from } c \text{ and } \mu_0)$$

?

C Use the Maxwell formula to check that these values agree with the quoted value for c.

See questions 2 and 3.

Electromagnetic waves in matter

The speed of electromagnetic waves is a maximum in a vacuum and less in materials which are 'transparent' to the waves. As a general rule, electromagnetic waves cannot travel at all through 'opaque' materials containing *free* electrons (eg metals) as the waves lose so much energy to the electrons. *Bound* charged particles (including electrons) may also absorb energy from the waves, but do so at

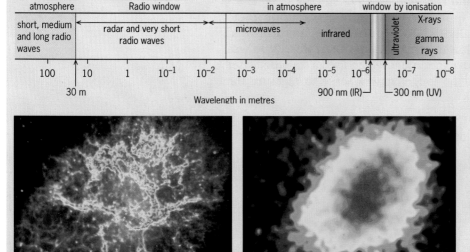

Absorption and reflection by ionised layers in upper atmosphere	Radio window	Absorption by molecules in atmosphere	Optical absorption window by ionisation	Atmospheric

Fig 16.11(a) **The atmosphere acts as a filter for some electromagnetic frequencies. Radiation reaches the Earth's surface in two main bands: optical (with some UV and IR) and short-wave radio. There are also a few narrow windows in the near infrared and sub-millimetre microwave bands**

Fig 16.11(b) **Astronomers use radiation that penetrates the atmosphere to produce images of stars. Images of the same objects differ according to the radiation used. For instance, the Crab Nebula is shown on the left as an optical telescope image and on the right as a radio telescope image**

definite frequencies or bands of frequencies that depend on the atoms or molecules of the medium. Fig 16.11 shows, for example, that the transparency of the Earth's atmosphere varies in different regions of the electromagnetic spectrum.

Refraction and the speed of electromagnetic waves

We have noted that electromagnetic waves travel more slowly in a transparent medium than in a vacuum. Fig 16.12(a) shows what happens as waves enter a transparent medium in which their speed is c_m. Their frequency stays the same but their wavelength gets less such that:

$$c_m = f\lambda_m$$

This change in speed also causes the **refraction** effect – the wavefronts change direction when they enter or leave the surface of the material at other than 90° (at an angle to the normal).

Fig 16.12(b) shows a set of parallel wavefronts of single frequency radiation entering a transparent medium. As the leading edge enters the medium, the wave slows down but the 'outside' section of the front does not, so it catches up on the inside section. Inside the medium the distance between successive fronts is smaller and the direction of travel of the wave has changed. Snell's law of refraction follows directly from this effect.

?

D Sound waves travel more quickly in glass than in air. What should be the shape of a 'sound lens' that is meant to focus a beam of light to a spot?

Fig 16.12 **A change in wave speed produces a change in wavelength, but not frequency, and causes refraction**

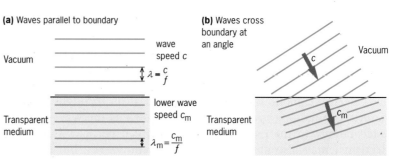

(a) Waves parallel to boundary

Vacuum

wave speed c

$\lambda = \dfrac{c}{f}$

Transparent medium

lower wave speed c_m

$\lambda_m = \dfrac{c_m}{f}$

(b) Waves cross boundary at an angle

Vacuum

c

Transparent medium

c_m

Snell's law

Fig 16.13 shows the geometry when the leading edge of a particular wavefront travels for a time t inside the medium before the rest of the front enters. Inside the medium the leading (left) edge A travels a distance $c_m t$, whilst the right edge B travels a distance ct. As the diagram shows, simple geometry gives the result that:

$$\frac{\sin i}{\sin r} = \frac{c}{c_m} = n \text{ (a constant)}$$

Constant n is the **refractive index** and is a property of *both* media since it involves the speeds of electromagnetic radiation in both media. Table 16.1 shows the refractive indices of *light* for some ordinary transparent materials with reference to air. The **absolute refractive index** is the value for light entering the medium from a *vacuum* ('free space'), and is the value quoted in most reference books.

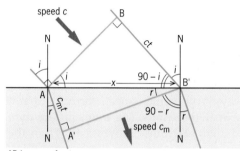

$$\sin i = \frac{ct}{x} \quad \sin r = \frac{c_m t}{x}$$

$$\text{Refractive index } n = \frac{c}{c_m}$$

$$= \frac{x \sin i}{t} \div \frac{x \sin r}{t}$$

$$= \frac{\sin i}{\sin r}$$

AB is a wavefront:
t seconds later it has reached A'B'

As B goes to B',
so A goes to A' in the same time t

Fig 16.13 **Snell's law of refraction: the geometry of wave refraction and proof of the sine formula**

Table 16.1 **Refractive indices for light travelling between air and materials**

Material	n (at λ = 589 nm)
diamond	2.419
fused quartz (SiO_2)	1.458
bottle glass	1.520
borosilicate (glass beakers)	1.474
ice	1.309
polystyrene	1.6
polythene (low density)	1.51
acrylic	1.49
Optical glass:	
light crown glass	1.541
dense crown glass	1.612
light flint glass	1.578
dense flint glass	1.613
Liquids:	
ethanol	1.361
glycerol	1.473
water	1.333
Gases (at 0° C, 1 atm):	
air	1.000 293
carbon dioxide	1.000 45

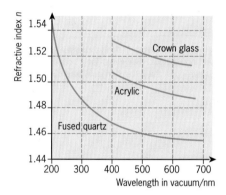

Fig 16.14 **Graph of refractive index n against wavelength. Dispersion occurs because refractive index varies with wavelength (after Serway)**

Refractive index varies with wavelength

The speed of light in a given transparent medium is also likely to vary with frequency – the refractive index is different for different frequencies. Fig 16.14 shows how the refractive index of fused quartz and crown glass varies with the vacuum wavelength of radiation between short ultraviolet wavelengths (~200 nm) and near infrared (~750 nm). Fused quartz is widely used in optical devices as it is transparent over a wide range of wavelengths.

Note that glass has a higher refractive index for light of shorter wavelengths (higher frequencies) than for longer wavelengths: light of shorter wavelength is refracted more. This is why **prisms** produce a spectrum from white light, with blue light deviated more than red light (Fig 16.2).

Dispersion and chromatic aberration

Dispersion is a serious problem that the makers of optical instruments with lenses have to solve. Dispersion means that red light is brought to a focus further away from a positive lens than blue light is (Fig 16.16). This blurs images, an effect called **chromatic aberration**. Newton solved the problem for telescopes by designing one in which the light was focused by a curved mirror (see Chapter 19).

An **achromatic** lens can be made – a combined *double-lens* using two different types of glass (eg crown and flint glass). One lens is positive and stronger than the other, negative, lens. The overall combination is positive, but the negative lens is made from a more dispersive type of glass so that the total dispersion of the combination can be made very small.

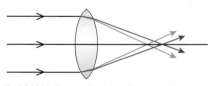

Fig 16.15(a) **Chromatic aberration occurs because blue light moves more slowly than red light, and so is refracted more**

Fig 16.15(b) **Chromatic aberration: the image has blurred colours at it edges**

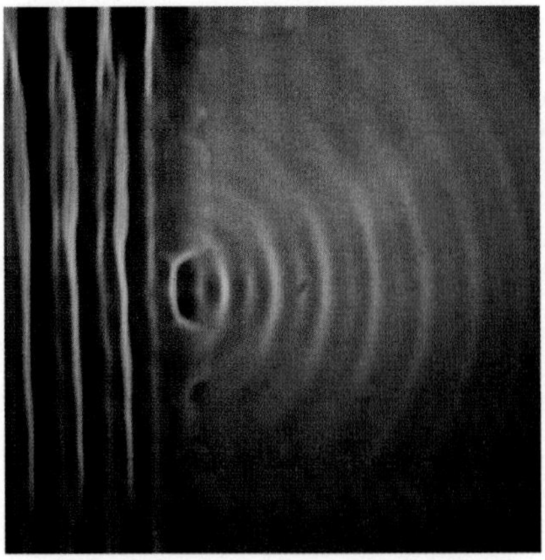

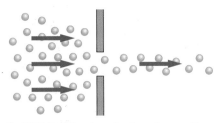

Fig 16.16 **Water waves passing through a gap and diffracting**

Fig 16.17 **Particles streaming through a gap. They don't spread out. (But could they bounce off the edges? – see the Extension box, page 354)**

3 ELECTROMAGNETIC RADIATION AS A WAVE

So far in this chapter we have thought of electromagnetic radiations as waves. But physics in fact uses three main models of light - a straight line 'ray' model (see Chapter 19), a particle (photon) model (see Chapter 17), and the wave model which we shall develop in this chapter.

The diffraction of waves

Fig 16.16 shows a set of parallel water waves passing through a small gap in a barrier. The waves spread out after passing through. This spreading out is called **diffraction** and is a key to distinguishing *wave* behaviour from *particle* behaviour.

How waves move forward

We see in Fig 16.16 that a wave crest such as a single ripple moves forward and keeps its direction. At the crest of the water wave there is a volume of water temporarily lifted above its normal level. Imagine raising just a small part of a water surface (eg by touching it and lifting your finger away). You would expect, correctly, that it would immediately collapse in all directions. The water surface then oscillates up and down and a circular pulse moves away from the site of the initial disturbance.

Why does the ring of raised water in the pulse (a circular crest) keep going with the same shape? To answer this – and to explain diffraction properly – we have to use the idea of the **superposition** of waves, an idea first used by Jan Christian Huygens (Holland, 1629–1695) to support his theory that light was in fact a wave.

Huygens' principle

The principle of superposition

When two water wave pulses meet, they pass through each other. However, at the meeting point the wave amplitudes add: two crests

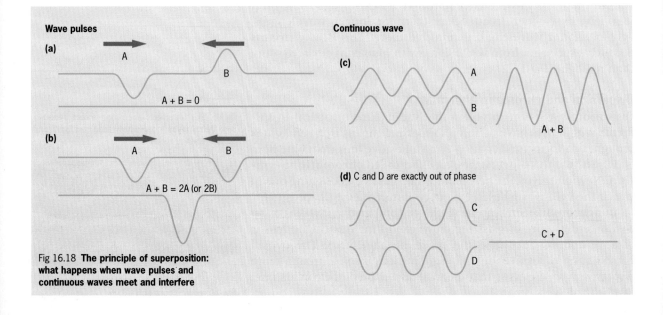

Fig 16.18 **The principle of superposition: what happens when wave pulses and continuous waves meet and interfere**

produce a larger crest, and similarly, two troughs make a deeper trough. A crest meeting an equal-sized trough produces an instant of completely flat water. See Figs 16.18(a) to (d) which illustrate the **principle of superposition** by which the amplitudes of two waves which occupy the same space at the same time simply add together. (See page 134 to compare with the superposition of mechanical waves.)

Huygens used this principle to explain how *light* waves propagate. He imagined that each point on a straight wavefront (equivalent to the water wave crest) was the source of another wave (ie as it 'collapsed').

But the waves (or *wavelets*) from each of these point sources **interfered** with the wavelets from the other points in such a way that the only places where they added together was in the forward direction. Thus a plane (straight-fronted) wave would move forward as a plane wave and therefore *light would travel in straight lines* (see Fig 16.19).

But note that at the edges of the plane wavefront, the wavelets would have no supporting wavelets to interfere with. The waves here could propagate sideways, in other words *diffraction* would occur.

Fig 16.19 **Huygens' model of light wave propagation**

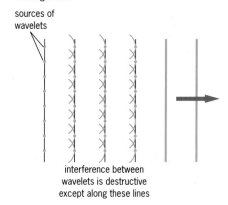

(a) Each point on a wavefront produces 'secondary wavelets'. For clarity, only a few points and wavelets are shown. For a very long wavefront, a plane wave reproduces itself as a plane wave, that is, 'light travels in straight lines'

sources of wavelets

interference between wavelets is destructive except along these lines

(b) But when the wave passes through a small hole, it spreads out: **diffraction**

4 INTERFERENCE AND WAVELENGTH

Young's two-slit interference experiment

Fig 16.20(a) shows a parallel beam of light (eg from a distant point object) reaching an opaque screen with two narrow slits in it. Light is diffracted at each slit and the spreading beams overlap. This means that at any point along a line such as XY, light is received from both slits A and B. The contributions from the slits reaching a point such as P may be **in phase** or **out of phase**, as in Fig16.20(c), and the resulting intensity at P could be anything from zero to double the contribution from any one slit.

See question 4.

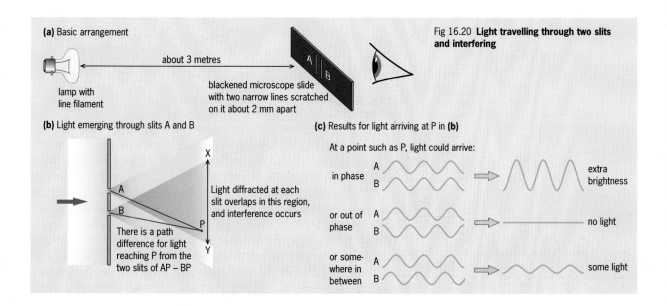

(a) Basic arrangement

about 3 metres

lamp with line filament

blackened microscope slide with two narrow lines scratched on it about 2 mm apart

Fig 16.20 **Light travelling through two slits and interfering**

(b) Light emerging through slits A and B

Light diffracted at each slit overlaps in this region, and interference occurs

There is a path difference for light reaching P from the two slits of AP – BP

(c) Results for light arriving at P in **(b)**

At a point such as P, light could arrive:

in phase — A B — extra brightness

or out of phase — A B — no light

or somewhere in between — A B — some light

What exactly happens at P depends on the geometry, in fact, on the *difference* in the path lengths AP and BP in Fig 16.21(b):

- If AP − BP is zero, a whole wavelength or a whole number of wavelengths, the contributions are in phase, and we have a brightness at P.

- If AP − BP is a half wavelength or an *odd* number of half wavelengths, the contributions are exactly out of phase, destructive interference occurs and there is a darkness at P.

You can imagine that as the point P in Fig 16.20(b) moves along XY, the path difference changes and P will be alternately in bright and dark zones, called **interference fringes**. These fringes are easy to see when you look at a distant lamp (say, 3 metres away) through two slits scratched about a millimetre apart in a blackened microscope slide. This is **Young's two-slit interference experiment**, and was one of the first experiments that led physicists to accept that light was indeed a wave. This experiment can be used to measure the wavelength of light.

Measuring the wavelength of light

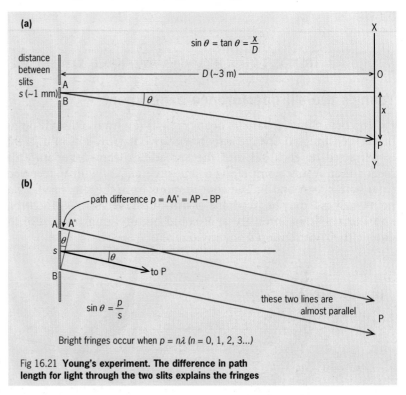

Bright fringes occur when $p = n\lambda$ $(n = 0, 1, 2, 3...)$

Fig 16.21 **Young's experiment. The difference in path length for light through the two slits explains the fringes**

Fig 16.21 shows Fig 16.20(b) with angles drawn to a more realistic scale. The slit separation *s* is a millimetre or two, and the distance *D* from the slits to the plane of observation (XY) is several metres. O is a point equidistant from the slits A and B, so the path difference is zero and O is in the centre of a bright fringe. P is distance *x* from O, chosen so that P is in the centre of another bright fringe.

Look at the triangle in Fig 16.21(a). The angle θ is very small, so that $\sin \theta = \tan \theta$, and we can write $\sin \theta = \tan \theta = x/D$. Fig 16.21(b) shows the detail nearer the two slits. The path difference between light reaching P from A and B is *p* (AA′). The two lines AP and BP are almost parallel, since P is so far away. Thus the angle ABA' is also θ, and we can write:

$$\sin \theta = \frac{p}{s}$$

Bright fringes occur when the path difference $p = n\lambda$. So, combining the two expressions for $\sin \theta$ gives:

$$\frac{x}{D} = \frac{n\lambda}{s} \text{ or } x = \frac{n\lambda D}{s}$$

Bright fringes are separated by values of x corresponding to one extra wavelength of path difference, that is, the fringe separation is:

$$\Delta x = \frac{\lambda D}{s}$$

The fringes produced by the arrangement shown in Fig 16.20(b) are viewed through either a microscope eyepiece with a micrometer scale (marked in millimetres) or a movable (travelling) microscope mounted on a vernier scale. The fringe separation is measured by counting a whole number of fringes – as many as possible – and measuring the value of x on Fig 16.21(a).

D is measured with a good steel tape. A travelling microscope (or some optical imaging method) measures the small distance s. The accuracy of the final result is likely to depend most critically on this measurement.

Coherence

We do not observe interference effects if the path difference between two interfering wavelets is longer than about 30 cm. Neither do we see the effects when light is combined from two *different* sources (two lamps or two stars, say). This is because the waves are emitted randomly, in short bursts, as in Fig 16.22(a). Interference still occurs, but the effect is so random and on such a short time scale (about 10^{-8} s) that it is impossible to observe. For all practical purposes, we can only observe interference when light from the same 'burst' is combined: such light is described as **coherent** (meaning 'belonging together'), see Fig 16.22(b). **Lasers** can emit continuous coherent light – particularly useful for applications of interference effects, such as making holograms.

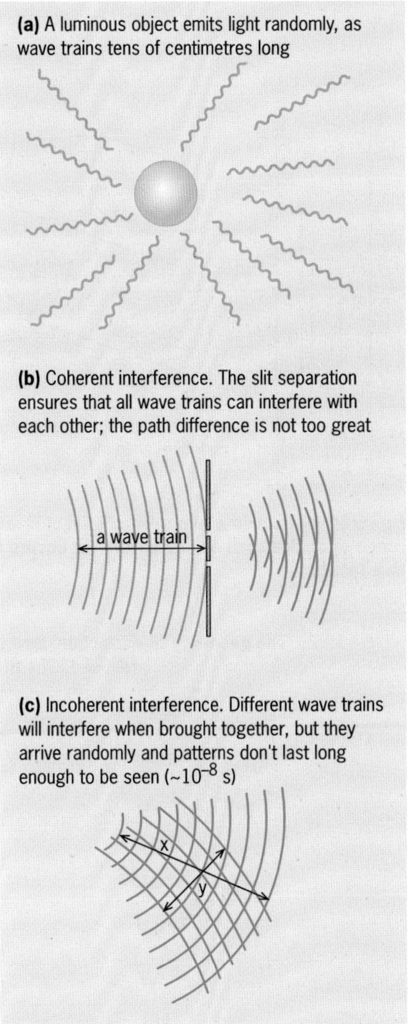

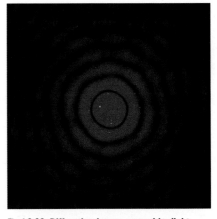

(a) A luminous object emits light randomly, as wave trains tens of centimetres long

(b) Coherent interference. The slit separation ensures that all wave trains can interfere with each other; the path difference is not too great

a wave train

(c) Incoherent interference. Different wave trains will interfere when brought together, but they arrive randomly and patterns don't last long enough to be seen ($\sim 10^{-8}$ s)

x

y

Fig 16.22 **Random light emission, coherence and incoherence**

See questions 5, 6 and 7.

5 DIFFRACTION AND IMAGE FORMATION

Diffraction becomes more noticeable when waves pass through a small hole (aperture). A greater fraction of the incident wave energy moves 'sideways', and interference effects occur which produce a **diffraction pattern** of the kind shown in Fig 16.23. The image of a small source of light (effectively a point source) is no longer a point, and the diffraction pattern is a more or less inaccurate version of the point object.

All optical instruments direct light to pass through apertures (or to reflect off mirrors) of finite size. So some diffraction will occur, and the images produced will tend to lack clarity. They lose fine detail as light from separate points on the object reaches the image as overlapping diffraction patterns. Diffraction also affects the images made by telescopes, which is one reason why they have to be so large (see page 352).

Fig 16.23 **Diffraction image caused by light passing through a small hole. Waves interfere constructively in the bright regions and destructively in the dark regions**

Diffraction through a slit

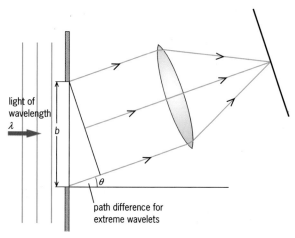

Fig 16.24 Diffraction at a narrow slit. Light is diffracted through a range of angles such as θ. Interference occurs if the light is brought to a focus

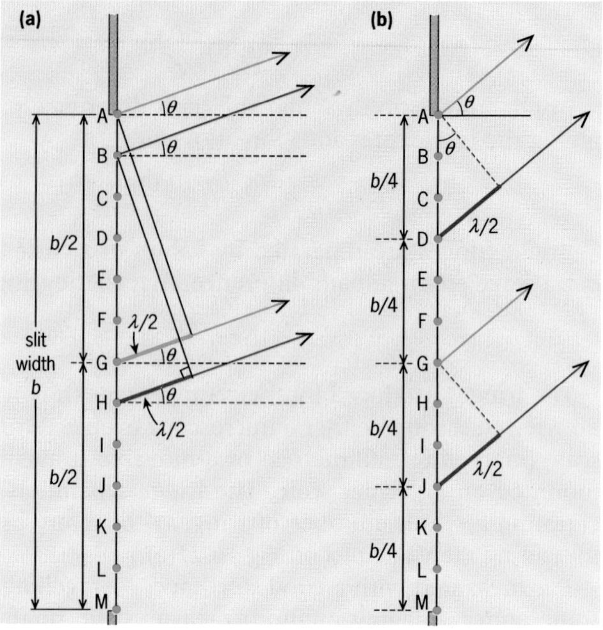

Fig 16.25 **The geometry of diffraction: how a pattern of fringes is produced. See the text for explanation**

Although most optical devices have circular apertures, we can deal more easily with the geometry of a rectangular slit, and it gives a result like that for circular apertures. In Fig 16.24, incident light of a single wavelength λ arrives from a distant object at a narrow slit of width b. Look back at Fig 16.19: using Huygens' principle, we can think of the wavefront that fills the slit as a set of small point sources which produce circular wavelets spreading out in all directions. Consider the light which travels at an angle θ to the direction of the incident light. This light contains waves from all the point sources – and we can imagine as many of these as we like. Ultimately, this light will be brought to a focus by a lens, say.

In Fig 16.25(a), light from point G of the slit has further to travel to this final focus than light from points A to F (similarly for points H to M). When light through the whole of the slit is combined in an image, there will be interference effects as the light from each wavelet is in or out of phase with others. As described below, at certain values of θ, no light will be focused on the screen.

Look again at Fig 16.25(a). The slit of width b is drawn in two halves. There is an arbitrary number of point sources labelled A to G in the top half of the slit, and an equal number labelled H to M in the bottom half. Angle θ is such that the path difference between the wavelets of light from the point sources A and G is exactly half a wavelength. Similarly, the pair of points B and H produce light with a path difference of half a wavelength, so does pair C and I, and so on. When combined, light from each of these pairs of points will interfere *destructively*. We can make the number of wavelet sources as large as we like – they will still pair off to provide destructive interference in the direction θ. There will be no light diffracted in directions when:

$$\sin \theta = \frac{\lambda/2}{b/2} = \frac{\lambda}{b}$$

Similarly, other pairs of points give zero intensity at another value of θ, as shown in Fig 16.25(b). This time, the slit is divided into four equal sections such that the light from the point source A destructively combines with light from the point D. As before, light from B will combine destructively with that from E, and C with F. In the lower half of the slit, light from points G and J, H and K (etc) will do the same. Thus there will be no light at the image when the light travelling in the direction shown is combined. This direction θ is given by:

$$\sin \theta = \frac{\lambda/2}{b/4} = \frac{2\lambda}{b}$$

We could go on to find 'no light' directions at other angles, by dividing the slit into sixths, eighths, tenths etc. In general, a diffraction pattern from a rectangular slit has zero intensity at angles given by the relation:

$$\sin \theta = \frac{n\lambda}{b}$$

where n is a whole number. This relation is usually written in the form:

$$b \sin \theta = n\lambda$$

Fig 16.26 is a graph of the intensity of a diffraction pattern from a narrow rectangular slit. The **central maximum** contains most of the light. It has a width of 2θ for the value of θ when $n = 1$. The peak of the central maximum corresponds to $n = 0$, that is, all light reaching it comes from pairs of points in the two halves of the slit which have zero path difference.

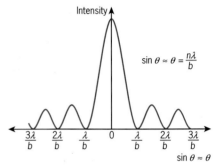

Fig 16.26 **The diffraction pattern of parallel light passing through a narrow slit (width b). The central maximum has greater intensity than the side fringes**

Diffraction by a circular aperture

The geometry of a circular aperture is more complex, but the equivalent result for the angular width of the central diffraction maximum is given by:

$$\sin \theta = \frac{1.22\lambda}{D}$$

Here, D is the *diameter* of the aperture.

▨ See question 8.

EXAMPLE

Q
a) What is the angular width of the central maximum of the image of a star when photographed in blue light of wavelength 450 nm? The image is produced by a telescope system with a circular aperture of diameter 0.2 m.

b) The image is formed 0.5 m from the prime focusing lens. What is the size of the central maximum on the photographic plate?

A
a) Using the diffraction formula for a circular aperture:

$$\sin \theta = \frac{1.22\lambda}{D}$$

$$= \frac{1.22 \times 450 \times 10^{-9}}{0.2}$$

$$= 2.75 \times 10^{-6}$$

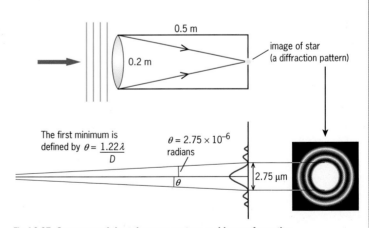

The first minimum is defined by $\theta = \frac{1.22\lambda}{D}$

$\theta = 2.75 \times 10^{-6}$ radians

2.75 μm

Fig 16.27 **Geometry of the telescope system and image formation**

The angle is so small that we can write $\theta = 2.75 \times 10^{-6}$ *radians*, so the image has a central maximum of 2θ, equal to 5.5×10^{-6} radians.

b) The angle calculated in **a)** produces a central maximum of width:

$$0.5 \times 5.5 \times 10^{-6} \text{ m} = 2.75 \times 10^{-6} \text{ m (2.75 μm)}$$

Comments on the Example

Stars are so far away that each star is effectively a point and its image on a photographic plate or other detector is a diffraction pattern.

Astronomers need to make out the diffraction pattern of any one star from those of its nearest apparent neighbours, as in Fig 16.28(b). So, for good *resolution* between images, the diffraction image should be made as small as possible. Astronomers do this by making *b* as large as possible for a given wavelength, that is, they use as large a telescope as possible. Also, light of smaller wavelength gives a smaller diffraction pattern. In practice, this tends to be blue light because the atmosphere absorbs the shorter UV wavelengths. The largest optical telescope (at the observatory of Mauna Kea in Hawaii, Fig 19.38) has a diameter of 9.82 m. A radio telescope uses much longer wavelengths – typically 6 cm compared with 6×10^{-7} cm for optical telescopes – and would need to be 10 million times larger to get the same kind of resolution!

Resolving power: the Rayleigh criterion

Resolving power is the measure of how good an optical system is at clearly distinguishing detail, for example, of two stars close together, or two structures in a living cell. The standard measure of the resolving power of any image-making device is the **Rayleigh**

Fig 16.28(a) **The Hubble space telescope: it detects objects 50 times fainter and 7 times further away than a ground-based telescope**

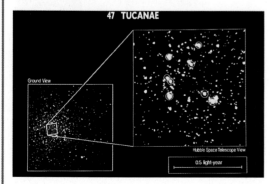

EXAMPLE
The Hubble space telescope

Q The mirror aperture of the Hubble space telescope has a diameter of 2.40 metres.

a) What is its theoretical resolving power in green light of wavelength 5.20×10^{-7} m?

b) At its closest approach Mars is 7.83×10^{-10} m from Earth. What is the closest possible distance between two small objects on the Martian surface that would allow them to be distinguished as separate by the Hubble telescope?

A **a)**   Resolving power $\theta = 1.22\lambda/D$

$$= \frac{1.22 \times 5.20 \times 10^{-7}}{2.40}$$

giving: $\theta = 2.6 \times 10^{-7}$ radians

b) Let the separation be *x* metres. The angle subtended by this distance at the telescope must be equal to or greater than θ, so at the limit:

$$\frac{x}{7.83 \times 10^{10}} \geq 2.64 \times 10^{-7}$$

giving: $x \geq 20\ 697$ m, or 20.7 km

Fig 16.28(b) **Comparing star pictures taken by a ground-based telescope (small box), and the Hubble telescope which has far better resolution and has led to the discovery of new stars**

criterion (due to Lord Rayleigh, Britain 1842–1919). This states that it is possible to separate the images of two point objects if the centre of the central maximum of one image lies on the first dark fringe of the other. This is illustrated in Fig 16.29, and shows that when the criterion is satisfied the central maxima are separated by an angle of θ given by the relationship:

$$\sin\theta = \frac{1.22\lambda}{D}$$

In most optical instruments the angles are so small that the criterion is usually written, with θ in radians, as:

$$\theta = \frac{1.22\lambda}{D}$$

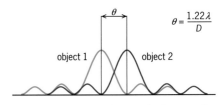

Fig 16.29(a) **Graph of light intensity for two point sources with central maxima separated by angle θ**

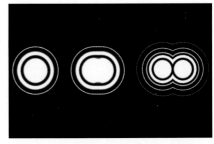

Fig 16.92(b) **Illustrating the Rayleigh criterion: diffraction images for pairs of point sources. Left: 1 source; middle: two unresolved sources; right: two sources completely resolved**

6 THE DIFFRACTION GRATING AND SPECTRA

The study of the electromagnetic spectrum of radiation in and near the visible has produced some of the most important discoveries in science. The element helium is named after the Greek for Sun as it was first identified in the spectrum of sunlight before being discovered as a gas. Many other examples are given in Chapters 27 and 28.

Prisms were first used to produce spectra but glass absorbs ultraviolet and some infrared radiation. Prisims are useless for microwave and radio studies.

Modern spectroscopy relies heavily on the principle of the **diffraction grating**. A diffraction grating is a glass or metal sheet with a large number of very fine parallel lines ruled on it. These lines diffract radiation in such a way that a series of spectra is formed (described below). If the grating lines are close enough together, the spectrum is spread over a very large angle compared to that produced by a prism – see Fig 16.30(a) – and much more detail can be observed. In 1882, the American Henry Rowland made the first grating, ruling lines with the accuracy and constant separation needed.

A typical grating used in school laboratories today might have 5 000 lines per centimetre. *Transmission* gratings are ruled on good optical glass and the radiation passes through the series of slits between. Cheaper gratings are made by copying these on to a plastic film. *Reflection* gratings are ruled on good metal mirrors and do not suffer from absorption problems. The mirrors may be curved, to focus the spectra without the need for lenses.

?

G Human eyes are quite good at resolving images formed in light of about 500 nm. Estimate how much bigger your eyes would have to be to get equivalent resolution if they worked at UHF TV wavelengths (*l* about 1 m).

H Two radio telescopes on different continents are linked to provide data on a radio star. The telescopes are separated by 4000 km and observe on a wavelength of 21 cm. What is the angular resolution (resolving power) of this arrangement?

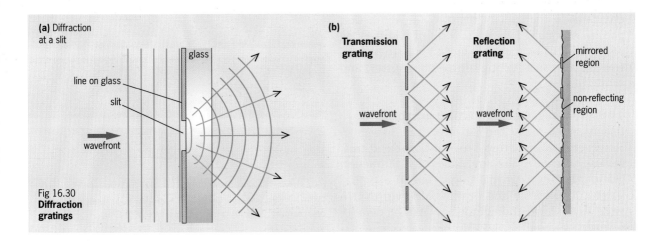

Fig 16.30 **Diffraction gratings**

(a) Diffraction at a slit

glass

line on glass

slit

wavefront

(b)

Transmission grating

wavefront

Reflection grating

mirrored region

non-reflecting region

wavefront

The diffraction grating

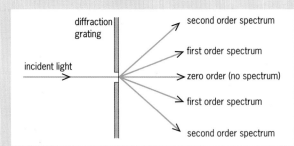

Fig 16.31 **A diffraction grating produces several spectra**

We can analyse the principle of the diffraction grating like the way we considered the formation of a diffracted image by a narrow slit on page 350. Fig 16.31 shows what happens when a plane beam of parallel light (eg from a distant point) passes through one slit in a diffraction grating.

Light is diffracted and travels in circular waves from each slit, as in Fig 16.30(a). Suppose the slits are a distance d apart and the light from each successive slit has to travel an extra distance λ. For light of a given wavelength λ, there is a series of angles at which the waves reinforce each other by travelling paths that differ by whole numbers of wavelengths before they are combined by the lens. This will happen for light diffracted at an angle such that $\lambda = d \sin \theta$. This causes the first order spectrum. Spectra will also occur at other angles for which the path difference for light from adjacent slits is 2, 3, 4 or more wavelengths. First-order spectra arise at angles given by:

$$n\lambda = d \sin \theta$$

where n = 2, 3, 4 etc; n = 2 defines the second order spectrum.

The envelope of the grating diffraction pattern

The width b of *each slit* is also important. The spectra must obey the basic diffraction rules for waves passing through a narrow slit: this means that the envelope of the spectra fits into the diffraction pattern produced by a narrow slit of width b. Fig 16.33 shows the effect produced for a single narrow slit, two slits (Young's fringes of course!) and more slits.

The advantages of diffraction gratings

If d is small, sin θ can be large and so can θ. However, if the spectrum is to be detectable, θ must be less than 90º. The advantage of a diffraction grating, however, is that it can spread a typical spectrum of visible and near visible light over a wide angle, large enough to show its fine detail.

The sharpness of the spectrum

As in Fig 16.34, maximum brightness occurs at an angle of θ for a given wavelength λ of light. Consider light travelling at an angle very close to θ. Suppose light in this just-off-θ direction produces a path difference between the first slit and the second slit of just 1/100th of a wavelength less than a whole wavelength. (Each slit, of course, sends light in all directions, so at all angles.)

What is the effect of having so many slits? Slit 3 will send light at this angle-just-close-to-θ with a path length which is 2/100th of a wavelength away from a whole number of wavelengths, slit 4 will have a path difference 3/100ths of a wavelength – and so on. Slit 51 will produce light at this angle which is 50/100ths of a wavelength different from a whole number of wavelengths, meaning that *it will be exactly out of phase with the light from slit 1*. Light from slits 1 and 51 will cancel each other. Similarly, so will light from slit 2 and 52, 3 and 53... and so on. If there are enough slits – and there will be at 5000 or so per centimetre – the bulk of the slits sending light in this close-to-θ direction will cancel each other out.

The result is that for any diffraction grating with many slits, light emitted at angles close to θ will be cancelled out and the intensity graph of light of a single wavelength will be very sharp, as in Fig 16.33(e). Compare this with (b), the graph of intensity for two slits.

Now, the light has to go somewhere – its energy is not destroyed on passing through the grating. Not only does the grating therefore produce sharp maxima, they are also very bright since they contain all the energy in the incident waves.

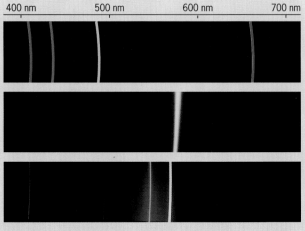

Fig 16.32 **Line emission spectra of hydrogen (top), sodium (middle) and mercury (bottom) produced by a typical grating**

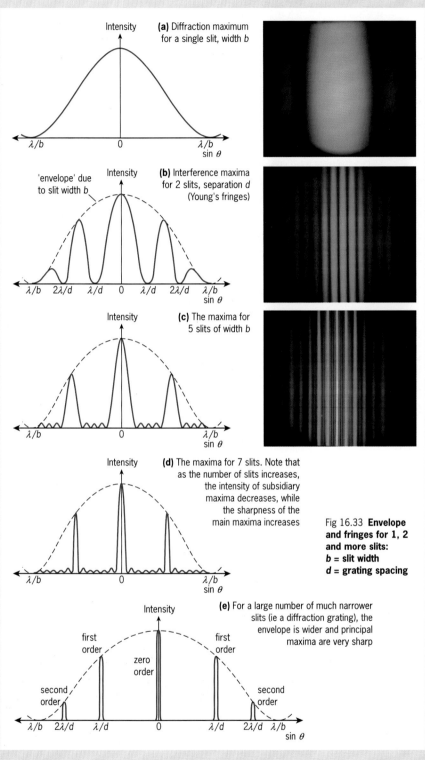

Fig 16.33 **Envelope and fringes for 1, 2 and more slits:**
b = slit width
d = grating spacing

The light in the direction of A, A', A'' etc will always be in phase. But light B will be destroyed by light B' ... and so on: see text

Fig 16.34 **Explanation of why a diffraction grating produces such sharp peaks (see text). Note that the angles are exaggerated and that the diagram is not to scale**

See questions 10, 11 and 14.

The sharpness of the peaks in Fig 16.33(e) means that the grating can distinguish between wavelengths that are very close together. In theory, if a grating has N slits it will produce a line at $\sin \theta$ for light of a given wavelength, such that the intensity is zero for values which differ from $\sin \theta$ by $\sin \theta / N$. This formula assumes perfect gratings. In practice, however, gratings cannot be ruled to the accuracy that would allow the results to be as good as the formula predicts.

7 POLARISATION

Electromagnetic waves are transverse: both the electric and magnetic components oscillate at right angles to the direction of travel (look back at Fig 16.3). If all the electric oscillations are made to lie in the same plane, the waves are said to be **plane polarised** (Fig 16.35).

Fig 16.35 **Polarisation: the diagrams show the electric component of an electromagnetic wave only, as 'seen' end on**

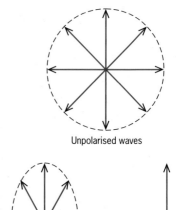

Unpolarised waves

Partly polarised   Plane polarised

The easiest way to polarise a beam of light is to pass it through a sheet of polarised film, manufactured as Polaroid. This contains a large number of small aligned crystals which absorb light whose electric component is aligned along the long axes of the molecules. The emerging beam is thus plane polarised (Fig 16.36). This light will not pass through a Polaroid filter whose crystals are aligned perpendicularly to the first.

Polarised light is now a common feature of everyday life. Many people use Polaroid sunglasses, and the liquid crystal displays (LCD) on watches, calculators and portable computers involve the use of polarised light. Polaroid sunglasses and filters reduce reflection glare because they absorb the polarised component.

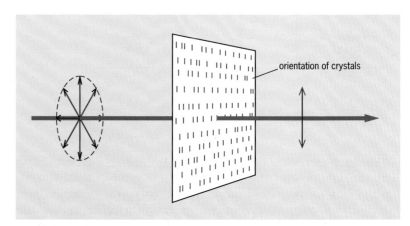

orientation of crystals

Fig 16.36 **A Polaroid sheet transmits light waves with the electric field oscillating in one direction only**

Polarisation by reflection

When light is partially reflected by a transparent medium, such as a sheet of glass, both the reflected and refracted light is partly polarised (Fig 16.36). When the reflected and refracted beams are at right angles to each other, all the reflected light is polarised. This occurs at an angle of incidence called the **Brewster angle**, **B**, where $\tan B = n$, the refractive index of the reflecting medium.

■ See question 12.

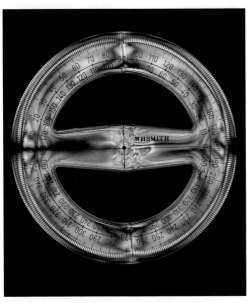

Fig 16.37 **The appearance of a protractor viewed through crossed polaroids**

Fig 16.38 **A Polaroid sheet removes reflection glare from the screen**

SUMMARY

Having studied this chapter, you should be able to do the following:

■ Understand the nature and production of electromagnetic waves.

■ Know the properties of the main sections of the electromagnetic spectrum.

■ Know what determines the speed of electromagnetic waves and use the formula:

$$c = \frac{1}{\sqrt{\varepsilon_0 \mu_0}}$$

■ Understand why a change in speed may produce **refraction** and **dispersion** of light.

■ Understand the principle of **superposition** for waves.

■ Understand the meaning of **diffraction** and **interference** and what happens when light passes through a single narrow slit and two narrow slits (Young's experiment).

■ Know and be able to use Young's two-slit formula:

$$\Delta x = \lambda D / s$$

■ Derive the formula for Young's two-slit interference:

$$n\lambda = d \sin \theta$$

and know how the experiment may be used to measure the wavelength of light.

■ Know the conditions for interference between two wave trains, especially the concept of **coherence**.

■ Understand the meaning of resolution for an optical instrument and the significance of the Rayleigh criterion: $\sin \theta = 1.22\lambda / b$.

■ Understand how a diffraction grating works, and why it is an improvement on prisms for use in spectroscopy.

■ Know the meaning of **polarisation** for light, and how polarised light is produced.

QUESTIONS

1 Describe briefly three ways of detecting infrared radiation.

2 The electric and magnetic fields in electromagnetic radiation are at right angles to each other. Describe another physical situation or effect in which electricity and magnetism seem to show this perpendicular relationship.

3 The electromagnetic waves in a TV aerial cable actually travel in the plastic insulator between the central copper wire and the cylindrical copper sheath.

a) Why is the speed of the electromagnetic waves in the cable different from their speed in a vacuum?

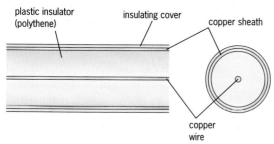

Fig 16.Q3 **A simple diagram of a TV cable**

b) A typical plastic insulator has a permittivity ε of 2.2×10^{-11} F m^{-1}, and a magnetic permeability μ of 1.2×10^{-6} H m^{-1}. Use this data to calculate the expected speed of the TV signal in a cable using this material.

c) Such plastic may be used to make a lens for focusing electromagnetic waves. What is the refractive index of the plastic?

4 When two waves meet, interference may take place, due to an effect called *superposition*. Draw annotated diagrams to illustrate this effect.

5 A detector can distinguish the presence of a certain kind of invisible radiation to a linear resolution of 2 cm. It is suspected that the radiation consists of waves which might have wavelengths between 1 and 5 cm.

a) Suggest an experimental arrangement which would demonstrate conclusively that the radiation is wave-like.

b) Design an experiment, or modify the one you have described in **a)**, to measure the wavelength of the radiation (accurate to, say, 10%).

6 Fig 16.Q6 shows the fringe pattern produced in a Young's double slit experiment.

Fig 16.Q6 **Young's fringes using a double slit and showing diffraction minima**

a) Explain why the fringe pattern seems to disappear at the two angles shown, and then reappears again at greater angles.

b) Take measurements from the photograph so that you can estimate the ratio of slit separation to slit width.

7 Two loudspeakers emitting sound of the same frequency produce an interference pattern of loud and quiet regions. Why don't two torches shining on a screen produce an interference pattern?

8 A small coin is glued to a glass sheet and placed in the beam from a laser. The shadow of the coin has light and dark fringes surrounding it and a bright spot in the centre. Explain in general terms why these effects occur.

9 The Jodrell Bank Lovell radio telescope has a diameter of 76 m and typically works at wavelengths of 8 cm. The eye is most sensitive to green light of wavelength about 5×10^{-7} m.

a) Estimate (or measure using a mirror and a ruler) the diameter of the pupil of your eye.

b) Calculate which of these two detectors of electromagnetic radiation has the better resolution. By what factor?

10 A diffraction grating for 29 mm microwaves consists of a set of parallel copper bars with a separation of 75 mm. At what angles will the first order and second order diffraction maxima occur?

11 The sodium spectrum contains two yellow lines with very similar wavelengths : 5.896×10^{-7} m and 5.890×10^{-7} m. They are viewed using a diffraction grating with 6000 lines per cm.

a) What is the angular separation of these two lines in the first order spectrum?

b) Your dark adapted eye has a pupil diameter of about 3 mm. Would you be able to see these two lines as separate with the unaided eye?

12 Light reflected from a sheet of clear ice at an angle of 53° is found to be completely plane polarised. What is the refractive index of ice?

See page 375 for the CHAPTER MAP that covers Chapters 16 and 17

13 This question is about the quality of vision of eyes of different sizes.

a) A human has an eye in which the diameter of the pupil is about 4 mm. Estimate the minimum angle (in radians) for resolution by this eye at a wavelength of 600 nm.

b) A mouse has an eye in which the diameter of the pupil is only 1 mm. Calculate the following ratios:
  (i) (the amount of light entering the mouse's eye per second)/(the amount of light entering the human's eye per second)
  (ii) (the minimum angle for resolution by the mouse's eye)/(the minimum angle for resolution by the human's eye)

c) In fact, a mouse can only just distinguish two objects 0.1 m apart at a distance of 2 m.
  (i) Calculate the angle subtended by these objects at its eye.
  (ii) Suggest a physical reason which might account for this, the ability to distinguish, being so much poorer than the theoretical resolution.

[OCSEB: 1990, Nuffield, Short Answer paper]

14

a) Describe an experimental arrangement to display the diffraction pattern from two narrow parallel slits of spacing d when illuminated by light of wavelength λ. Sketch and account for the variation in intensity with angle in the diffraction pattern. Why is a similar pattern not observed from the headlights of a distant car?

b) Sketch the diffraction pattern seen using a large number of slits of the same spacing d, and explain why the pattern differs from that in **a)**.

c) Sketch the diffraction pattern when a broad slit of width a is used. Explain the reasons for the shape of the pattern as carefully as you can.

d) The pattern of an array of many *broad* slits, of width a and repeat spacing d, can be obtained by taking a combination of the two relevant diffraction patterns, that is, those in **b)** and in **c)**. Sketch the result when $d = 2a$. What would be the advantages and disadvantages of constructing a diffraction grating in this way, rather than using narrow slits?

[OCSEB: 1994, Special paper]

17 Atoms, spectra and quanta

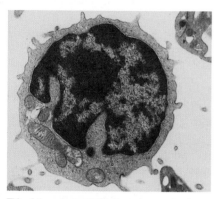

This picture, from an electron microscope, shows a white blood cell which helps the body fight disease. It is taken with a beam of electrons, which we normally think of as particles. Here, they are acting as waves

THE PICTURES ON THIS PAGE were taken using modern techniques which rely on the wave nature of particles, such as electrons, and the particle nature of light which Chapter 16 should have convinced you 'is a wave'. This is the world of *quantum physics* – probably the most successful theory in physics at providing explanations and accurate predictions.

The pictures were made by extremely useful imaging devices that work because of the strange ability of the very small – atoms, electrons, light waves – to show quantum effects. These effects mean we have to think again about the meaning of 'real' – and be prepared to set aside some of the ideas which have worked so well in explaining and predicting the behaviour of everyday objects like cars, tennis balls and planets.

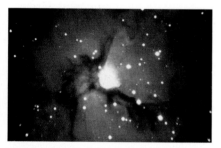

This picture of the Trifid Nebula. was taken with a charge-coupled device (CCD) – an electronic light detector that relies on the fact that light sometimes behaves as a particle – although we normally think of it as a wave

An overview

Physicists have developed quantum theory during the twentieth century. Physics before 'the quantum' is called **classical physics**; after the quantum, it has become **modern physics**. Quantum theory originated from two apparently quite different areas of physics.

Firstly, there was a puzzle that seemed just a small fault in the otherwise superb theory developed in the nineteenth century to explain the nature of **electromagnetic radiation**. The spectrum of radiation emitted by a hot object – so-called **black body** (or thermal) radiation – didn't match the theory.

Secondly, there was the (accidental) discovery of **radioactivity**, described in Chapter 18. This led physicists to discover the nucleus, and to propose a new model of the atom. According to the theories at the time, a nuclear atom surrounded by orbiting electrons would immediately self-destruct as its orbital electrons spiralled into the nucleus. At the same time, they would radiate energy as electromagnetic waves. This doesn't happen, a fact which proved to be the link between the two unsolved problems.

1 THE CONTINUOUS SPECTRUM OF A HOT BODY

Hot objects emit electromagnetic radiation and its intensity varies with wavelength in a pattern that depends on the temperature of the body. This radiation pattern is called by various names: **temperature radiation, thermal radiation** or **black body radiation**. Fig 17.1 shows the intensity of the radiation plotted

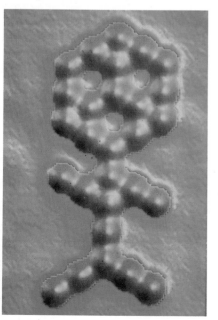

The small mounds in this picture are the outlines of atoms: they are images of the whirling cloud of particles that we call electrons. The picture was taken by a scanning tunnelling microscope (STM) which not only relies on the wave properties of electrons but also on the strange fact that electrons can 'tunnel' through an apparently impassable energy barrier

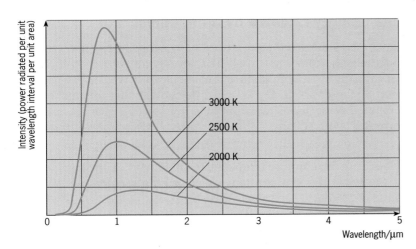

Fig 17.1 **The graphs show how the distribution of the intensity of radiation from a hot object changes as the temperature is increased: the peak shifts to shorter wavelengths. Check that $\lambda_{max}T$ is a constant. The total energy emitted is represented by the area under the graph**

against wavelength for a generalised object at several temperatures. The graphs are based on experimental results.

Real materials may not emit radiation in precisely these distributions, particularly at low temperatures, hence the notion of an ideal or 'perfect' radiator called a **black body radiator**. An argument (which we won't go into here) suggests that a perfect *radiator* must also be a perfect *absorber* of radiation. Anything that perfectly absorbs all the radiation falling on it will look black – hence the term *black body* (Fig 17.2).

The graphs have a similar pattern. The intensity reaches a peak at a particular wavelength, say, λ_{max}. The higher the temperature, the smaller λ_{max} becomes. It obeys a law discovered experimentally in 1893 by Wilhelm Wien (Germany, 1864–1928), known as **Wien's displacement law**:

$$\lambda_{max}T = \text{constant}$$

where T is the absolute temperature of the body (in kelvin). The constant has a value of 2.898×10^{-3} m K. The graphs also increase in area as the temperature increases, showing that the hotter the body the more energy it radiates per second.

Cool objects emit entirely in the infrared. As they first become red hot, they start to emit visible radiation (at about 700 K). They then become white hot, then blue hot, as shorter wavelengths begin to dominate at even higher temperatures.

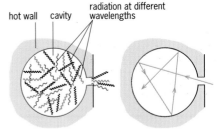

On entering, radiation is reflected so many times, losing energy each time, that it is eventually totally absorbed

Fig 17.2 **Cavity or black body or temperature radiation. The radiation in the cavity of a mass of hot material settles down into a steady distribution of intensity at different wavelengths. The exact distribution depends only on temperature. A small hole lets out a sample of radiation to study**

■ See questions 1 to 4.

EXAMPLE

Q At what wavelengths do **a)** the Sun and **b)** the human body emit maximum intensity of radiation? (Assume Wien's constant is 2.9×10^{-3} m K, the surface temperature of the Sun is 6000 K, and the human body has a skin temperature of about 35°C.)

A a) The Sun has a surface temperature of 6000 K, so by Wien's law:

$$\lambda_{max} \times 6000 = 2.9 \times 10^{-3} \text{ m K}; \qquad \lambda_{max} = 4.8 \times 10^{-7} \text{ m}$$

b) The skin temperature of the body is 308 K, thus:

$$\lambda_{max} \times 308 = 2.9 \times 10^{-3} \text{ m K}; \qquad \lambda_{max} = 9.4 \times 10^{-6} \text{ m}$$

Comment on the Example

The wavelength of maximum energy emission for the Sun is 480 nm, which is in the visible range (green). At a λ_{max} of about 10 μm, the human body emits most energy in the infrared, hence the popularity of infrared burglar alarms and body-sensor light switches.

Total energy radiated by a hot object

Experiments showed that for a black body (a perfect radiator) emitting radiation, the total energy emitted per second (power P) increases very rapidly with temperature. This behaviour was investigated by Josef Stefan (Austria, 1835–1893). He showed that the total energy P emitted *per second* as radiation by a hot object is given by a simple formula:

$$P = \sigma A T^4$$

where T is the absolute temperature (in kelvin), A the surface area of the body and σ a constant called **Stefan's constant**. It has a value of 5.6696×10^{-8} W m^{-2} K^{-4}.

This relationship was later deduced from theory by Stefan's student, Ludwig Boltzmann, and is now known as the **Stefan–Boltzmann law**.

2 PLANCK AND THE FIRST QUANTUM THEORY

The ultraviolet catastrophe

The laws about the continuous spectrum of radiation emitted by hot objects were discovered during the nineteenth century. The relationship between temperature and intensity was the **Rayleigh–Jeans law**. This related the intensity I at a particular wavelength λ to the absolute temperature T of a black body by a formula:

$$I = \frac{2\pi ckT}{\lambda^4}$$

where c is the speed of light, k is a thermal constant (the Boltzmann constant, see page 160. I is the power emitted per unit area having wavelength λ in a small wavelength interval $\Delta\lambda$. The equation was based on the idea that electromagnetic radiation is emitted by oscillating charged particles (see page 339). The oscillators were taken to be the atoms in the wall of the black body.

But the formula had a serious flaw – it didn't work! The villain is the term λ^4. It means that the Sun, with $T = 6000$ K, should emit ever-increasing quantities of energy as the wavelength gets smaller (Fig 17.3). The energy emitted in the ultraviolet should be immense

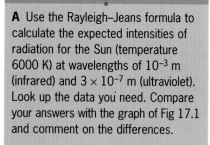

?

A Use the Rayleigh–Jeans formula to calculate the expected intensities of radiation for the Sun (temperature 6000 K) at wavelengths of 10^{-3} m (infrared) and 3×10^{-7} m (ultraviolet). Look up the data you need. Compare your answers with the graph of Fig 17.1 and comment on the differences.

Fig 17.3 **The Rayleigh–Jeans formula fitted observations only at longer wavelengths for very hot objects**

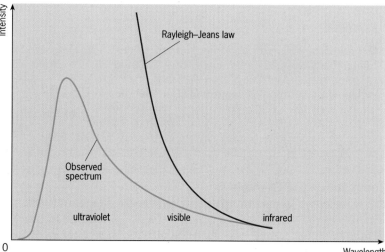

– and become infinite at even shorter wavelengths. In reality, as Fig 17.3 shows, the curve bends over at the Stefan maximum and less energy is radiated as wavelength decreases. This failure of nineteenth century classical physics to account for the short-wavelength decrease became known as the **ultraviolet catastrophe**. ▨ See question 5.

Short-wavelength energy emission and Planck's constant

The solution was eventually found in 1900 by the German physicist Max Planck (1858–1947). He saw the problem – it was an assumption that seemed to everyone else at the time to be perfectly sensible. As explained in Chapter 16, electromagnetic waves are produced when electrons or charged particles oscillate. The incorrect formula assumed that the charged particles in matter could oscillate at any frequency. But Planck thought that if the oscillators were more like the strings of a musical instrument, they would oscillate *only at definite fixed frequencies*. Fig 17.4(a) shows the natural modes of vibration of a stretched string – which can have an infinite number of frequencies, but they cannot vary continuously.

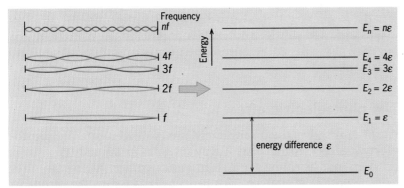

Fig 17.4(a) **The natural vibrations of a stretched string can have only certain definite frequencies. Frequency is 'quantised**

Fig 17.4(b) **The energy of a simple oscillator is also quantised with equal differences between the 'energy levels' or energy states**

When Planck fed this idea into the mathematics, there turned out to be a natural limit to the short wavelength (high frequency) radiation. He produced a formula giving an intensity–wavelength curve that matched the experimental evidence.

He also showed that, just like in a vibrating string, the energy E of an oscillator is proportional to its frequency f:

$$E = hf$$

where h is **Planck's constant** with a value of 6.626×10^{-34} J s.

As for a stretched string, the frequency of an oscillator cannot change continuously, and neither can the frequency of Planck's oscillating charges. The frequency can have values of f, $2f$, $3f$... nf, where f is a fundamental frequency and n is a whole number, now called a **quantum number**. Planck's formula relating energy to frequency means that the energy of an oscillator cannot change continuously either – it must change in small, discrete steps, called **quanta**. In other words,

$$\text{oscillator energy} = hf, 2hf, 3hf \ldots nhf$$

Note the similarity to Fig 17.4. Planck received the Nobel Prize for his theory in 1919.

Einstein and the quantisation of light

Planck did not himself take the next step, which was to recognise that the *radiation itself is also quantised*, meaning it is in small packets. The step was made by an unknown young civil servant working in the Patent Office in Berne, Switzerland, the 26-year-old Albert Einstein. His argument was simple:

- An oscillating charge can accept or lose energy only in small quantities of value ΔE ($= h\Delta f$ where Δf is its change in frequency).
- It gains or loses the energy as electromagnetic radiation.
- Therefore the radiation must also be emitted in small packets, each carrying energy ΔE.

Einstein suggested that the emitted radiation should also obey the Planck rule, and have its own frequency f_{rad} such that the energy E_{rad} it carries is related by:

$$E_{rad} = hf_{rad}$$

This meant that the radiation from a hot object was no longer thought continuous. In fact, it consists of a set of radiation packets each carrying a bit of energy – *a quantum of energy*. This was difficult to imagine: radiation was a packet of energy but was also a wave – it had a frequency! It is what we now call a **photon**.

Einstein used this idea to explain another small effect that was puzzling physicists at the turn of the twentieth century – the photo electric effect.

See questions 6 and 7. ◼

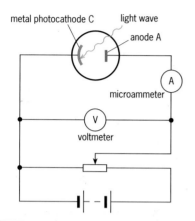

Fig 17.5(a) **Positively charged anode A collects all the electrons emitted from C**

$$E_{K,max} = eV_s$$

where $E_{K,max}$ is the maximum kinetic energy and e is the electronic charge.

Brightness does not change V_s, but colour (frequency) does:

$$E_{K,max} = hf - W$$

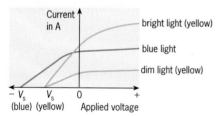

Fig 17.5(b) **As the voltage of A is gradually decreased, the current in A falls. V has a negative reading (V_s) when the current falls to zero. It is just negative enough to stop the fastest (most energetic) electrons emitted**

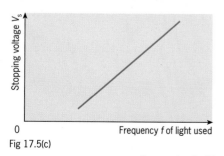

Fig 17.5(c)

See question 8. ◼

The photoelectric effect

Fig 17.5(a) shows the arrangement used to investigate the **photoelectric effect**, together with typical results in (b) and (c).

When light is shone on a clean metal surface, the surface may become positively charged. This happens because energy carried by light waves is absorbed by free electrons in the metal, giving them enough kinetic energy to 'jump' out of the metal. But there are a number of puzzling features in the photoelectric effect which can't be explained by the simple wave model of light, see Fig 17.6(a).

One is that electrons are not emitted unless the frequency of the incident light is greater than a certain threshold value. For example, blue light produces a photoelectric effect in sodium, but red light doesn't. Perhaps red light doesn't carry enough energy? The simple wave model suggests that even if the waves of red light are less energetic than blue light waves, eventually the beam of red light will transfer enough energy to liberate electrons from the metal atoms. But this does not happen.

Another strange fact is that when even a very weak beam of blue light is shone on sodium metal, electrons are released almost immediately. It is hard to explain why this happens so quickly using a wave theory in which the light is spread evenly all over the metal surface. How could it be concentrated enough to give even just a few electrons the energy to escape?

A third observation is also difficult to explain. The electrons leave the metal surface with kinetic energy, which must have been given to them by the light. Some have more kinetic energy than others, and experiments show that the *maximum* kinetic energy $E_{K,max}$ posssessed by the emitted electrons depends on the *frequency* of the light, not on its *intensity*. Weak ultraviolet light produces electrons with greater energy from sodium than, say, very bright blue light does.

Einstein's theory that light arrived in packets with energy proportional to frequency gives a simple explanation of all this. In the photoelectric effect, all of the energy carried by one light-packet (one photon) is given to one electron. There are forces which normally hold the electron inside the metal, and it needs some minimal energy to do work against these forces before it escapes from the metal. Suppose this minimal escape energy is W, as in Fig 17.6(b). If the electron gets energy E from a photon, then the *maximum* kinetic energy that an escaped electron can have is $E_{K,max}$ such that:

$$E_{K,max} = E - W$$

Einstein's formula gave $E = hf$, so we have:

$$E_{K,max} = hf - W \quad \text{or} \quad hf = E_{K,max} + W$$

This model explains the experimental results:

- If hf is smaller than W, the electron cannot escape.
- There is a **threshold** frequency f_0 such that $hf_0 = W$.
- A graph of $E_{K,max}$ against f will be a straight line of slope h with the intercept on the energy axis being $-W$.

W is called the **work function** of the metal (see Table 17.1).

Einstein's theory is illustrated in Fig 17.6(b). Compare this with Fig 17.5(c).

Einstein's model also explains why there is a such a small delay between irradiating the metal and the escape of electrons. Just one photon might be enough to cause an electron to be emitted – there is no need to wait for the energy to 'build up'.

The model also explains why increasing the *intensity* of the radiation does not increase the maximum energy of the electrons emitted. *No electron can gain more than one quantum of energy*:

- Brighter light means more photons, not more energetic photons.
- Brighter light means more electrons emitted, not electrons with more energy.

For many physicists, the Planck–Einstein quantum theory remained 'just a theory'. But in 1922, Arthur Compton (USA, 1892–1962) did experiments which showed that X-ray quanta behaved like particles when they bounced off electrons (**Compton scattering** – see Chapter 24). The photons not only had *energy*, they also had *momentum*. This seems very odd, since momentum is defined as mass × velocity, and photons have no mass. But see Chapter 25 for an explanation of how all energy has an equivalent mass.

3 QUANTUM THEORY AND THE NUCLEAR ATOM

There are two main ways of finding what atoms or other small particles are like. One is to fire something even smaller at them and see either how they break up or how the projectile bounces off them. The other is to shake them about (give them some energy) and see what comes out! Both these methods were used to give physicists their first idea of what an atom is like.

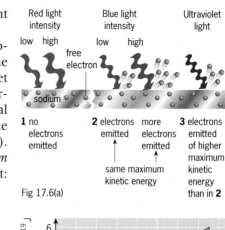

Fig 17.6(a)

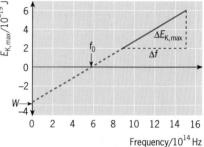

Fig 17.6(b) **Graph of the Einstein photoelectric equation:**
$$E_{K,max} = hf - W$$
No electrons are emitted for frequencies of light less than f_0.
At intercept value f_0, $W = hf_0$.
The slope of the graph is the Planck constant h:
$$h = \Delta E_{K,max} / \Delta f$$

Table 17.1 **Work functions of some metals**

Metal	W/eV	W/10⁻¹⁹ joules
sodium	2.28	3.65
aluminium	4.08	6.53
copper	4.7	7.52
zinc	4.31	6.9
silver	4.73	7.57
platinum	6.35	10.2
lead	4.14	6.62
iron	4.5	7.2
potassium	1.81	2.9

■ See question 9.

?

B Use Fig 17.6(b) to do the following.

(a) From its work function, identify the metal involved.

(b) Estimate a value for the Planck constant.

Probing the atom with alpha particles

The first method was used in 1909. Alpha particles from a radioactive source were fired at a thin film of metal atoms: see the Geiger–Marsden experiment on page 391. Some particles bounced back at angles that meant they had hit something small and with mass. From this, Rutherford worked out in 1911 that an atom has a positive nucleus surrounded by negative electrons. He suggested that the electrons could be orbiting round the nucleus like planets round the Sun. But such an atom would not be stable. An orbiting electron, like an orbiting planet, has an acceleration directed towards the attracting object. An accelerating electron continuously radiates electromagnetic waves, so should (he thought) lose energy and spiral into the nucleus.

See question 10. ▥

Rutherford's model was saved in 1913 by the Danish physicist Niels Bohr. He used the new quantum ideas of energy, saying that the electron could have only certain 'allowed' energy states, with definite energy gaps between them and corresponding to definite orbits. So electrons could not lose energy continuously and spiral into the nucleus. Electrons could only move between orbits by gaining or losing definite, fixed quanta of energy.

This seemed a very far-fetched idea at the time – but Bohr backed it up with calculations of how much energy an atom could gain or lose, and matched this with the energy of the light quanta it emitted.

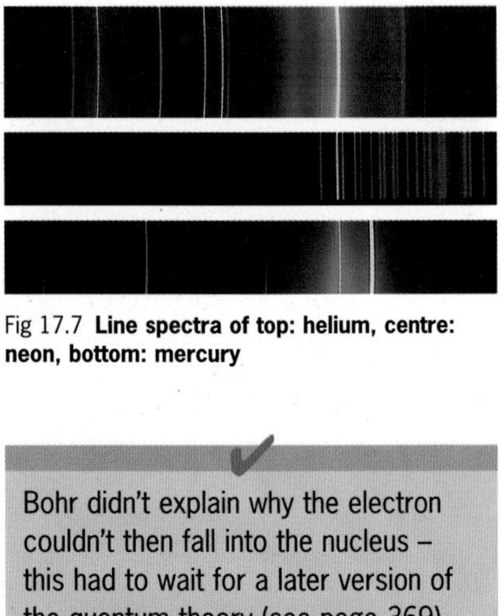

Line spectra

The spectrum of the radiation emitted by a hot body is continuous because there are very many different kinds of oscillators in any real lump of matter, so that in practice quanta exist at all frequencies.

But there are discontinuous spectra called **line spectra** – the mysterious pattern of radiation emitted by *pure* elements when they are heated or electrically 'disturbed'. See Fig 17.7 and Fig 16.32 for examples of line spectra. The lines are differently coloured images of the slit that was illuminated by the original light. The arrangement for producing a line spectrum is shown in Fig 17.8.

Fig 17.7 **Line spectra of top: helium, centre: neon, bottom: mercury**

Each element has a unique spectrum. The spectrum of the smallest atom, hydrogen, has four visible lines (see page 354), and a large number of invisible ones in the ultraviolet and infrared. Niels Bohr was able to explain the pattern of the spectra – and to calculate their wavelengths.

> Bohr didn't explain why the electron couldn't then fall into the nucleus – this had to wait for a later version of the quantum theory (see page 369).
>
> 'Orbital' is the term that mainly chemists use to describe the modern quantum-theory fuzziness of an electron in an atom.

The single electron in the hydrogen atom could exist in a set of definite orbits, each associated with a definite quantity of energy. It was normally in the 'lowest orbit'. When you supply energy to a hydrogen atom, as in a discharge tube, the electron can move

See question 11. ▥

to an orbit further from the nucleus. Such *excited* atoms are

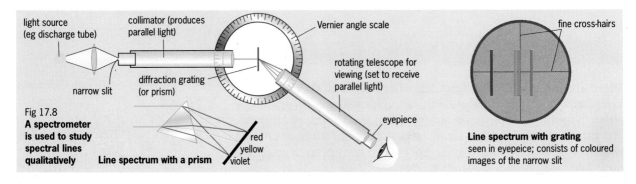

light source (eg discharge tube)

collimator (produces parallel light)

Vernier angle scale

narrow slit

diffraction grating (or prism)

rotating telescope for viewing (set to receive parallel light)

eyepiece

fine cross-hairs

Fig 17.8 **A spectrometer is used to study spectral lines qualitatively**

Line spectrum with a prism

red
yellow
violet

Line spectrum with grating seen in eyepeice; consists of coloured images of the narrow slit

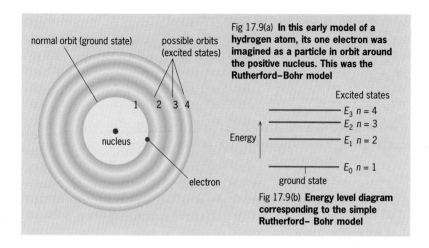

Fig 17.9(a) **In this early model of a hydrogen atom, its one electron was imagined as a particle in orbit around the positive nucleus. This was the Rutherford–Bohr model**

normal orbit (ground state)

possible orbits (excited states)

nucleus

electron

Excited states

E_3 $n = 4$
E_2 $n = 3$
E_1 $n = 2$

Energy

E_0 $n = 1$

ground state

Fig 17.9(b) **Energy level diagram corresponding to the simple Rutherford– Bohr model**

unstable, and the electron quickly falls back to its normal orbit. As it does so, it emits a quantum of electromagnetic radiation. This is illustrated in Fig 17.9, which shows a simple (early) model of a hydrogen atom and a more modern **energy level diagram**.

For an electron to make the transition from level 1 to level 2 (Fig 17.10), the hydrogen atom needs to be given exactly 10.2 eV (1.63×10^{-18} J) of energy. An electron in orbit 2 is not stable: it quickly falls back to orbit 1. In this transition, it then emits the energy difference ΔE as a photon of light, with energy 1.63×10^{-18} J. Planck's formula allows us to calculate the frequency of this light:

$$\Delta E = E_1 - E_0 = hf$$

giving:

$$f = \frac{1.63 \times 10^{-18}}{6.63 \times 10^{-34}} = 2.46 \times 10^{15} \text{ Hz}$$

This is the frequency of a line in the spectrum of hydrogen with a wavelength of 122 nm, which is in the ultraviolet range.

Ionisation

If the atom gains enough energy for the outermost electron to leave it completely, the atom becomes **ionised**. The energy values shown in Fig 17.11 are in fact measured from the state in which the electron is a long way from the nucleus: the electron has escaped, and the atom is ionised. The energy required to ionise a hydrogen atom is 2.18×10^{-18} J (13.6 eV).

The success of the quantum theory – both to *explain* the simple Rutherford model of an atom and the pattern of lines in the hydrogen spectrum, and also to *predict* the values of the wavelengths – led to its acceptance.

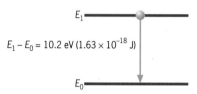

E_1

$E_1 - E_0 = 10.2$ eV (1.63×10^{-18} J)

E_0

Fig 17.10 **For a hydrogen atom, a photon of energy 1.63×10^{-18} J is emitted when an electron moves to the ground state from the first excited state**

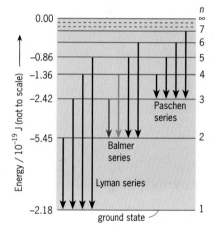

Energy / 10^{-19} J (not to scale)

0.00
−0.86
−1.36
−2.42

−5.45

−2.18

n
∞
7
6
5
4
3

Paschen series

2

Balmer series

Lyman series

ground state

1

Fig 17.11 **Hydrogen emits several series of lines in its spectrum. Each series is named after the scientist who discovered it. Each arrow corresponds to a different transition, and it is the transitions from higher excited states (energy levels) to lower ones that produce the spectral series**

■ See question 12.

4 WAVY ELECTRONS

'I suddenly got the idea, during the year 1923, that the discovery made by Einstein in 1905 should be generalised by extending it to all material particles and notably to electrons.'

Louis de Broglie, PhD thesis, 1924

Prince Louis de Broglie (1892–1987) first studied history and then turned to physics. In his PhD thesis he suggested that the formula $E = hf$ should apply to electrons as well as photons. There ought, he said, to be a 'fictitious wave' associated with electrons. If so, it should be possible to diffract electrons, just as you could diffract light.

The concept of negative potential energy is discussed in 3 Newton's universe, in terms of the gravitational field.

✓

See 25 Spacetime physics.

In general, $\Delta E_K = v\Delta p$ for a particle.

$$\Delta E = c\Delta p$$

So, by substitution:

$$E = pc \qquad [1]$$

But $\qquad E = hf = hc/\lambda$

giving $\qquad \lambda = hc/E = h/p$ from [1].

✓

See page 349 for diffraction grating theory.

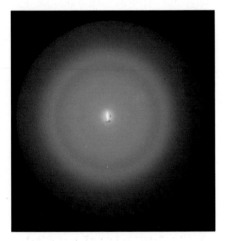

See questions 13 to 16. ▨

The wavelength of an electron

Relativity theory shows that a photon, although without mass, nevertheless has a momentum p related to its wavelength λ by the relation:

$$\lambda = \frac{h}{p}$$

By analogy with the photon, an electron should have a wavelength given by:

$$\lambda = \frac{h}{mv}$$

where m is the mass of an electron and v its speed. The wavelength of an electron therefore depends on its speed: the faster it goes, the smaller its wavelength.

The predicted diffraction of electrons was first shown to happen in 1925. C. J. Davisson was investigating the effect of bombarding a nickel target with low-energy electrons (54 eV). But the surface had been specially cleaned, causing the growth of large crystals of nickel on the target surface. The rows of atoms in the crystals acted as a diffraction grating so that the reflected beams showed regions of electrons or no electrons. Ordinary diffraction grating theory ($n\lambda = d \sin \theta$) allowed the wavelength of electrons to be calculated and showed that de Broglie's theory was correct.

Fig 17.12 shows the diffraction pattern of rings produced when a beam of electrons passes through a 'grating'. The grating consists of the regular rows of carbon atoms in a graphite crystal.

The pattern is exactly the same as is produced by light passing through a similar type of grating – although the grating spacing for electron diffraction is very much smaller. This is because the wavelength of an electron is far smaller than the wavelength of visible light. Now we shall calculate how much smaller.

Fig 17.12 **Diffraction rings formed by standard laboratory diffraction apparatus**

EXAMPLE

Q Calculate the wavelength of an electron accelerated by a potential difference (V) of 8 kV (as in a TV tube). The mass m of an electron $= 9.11 \times 10^{-31}$ kg, the charge e on an electron $= 1.60 \times 10^{-19}$ C.

A Kinetic energy gained by electron:

$$\tfrac{1}{2}mv^2 = \text{change in electrical potential energy}$$

$$= eV$$

Thus: $\qquad v = \sqrt{\dfrac{2eV}{m}} \;=\; \sqrt{\dfrac{2 \times 1.6 \times 10^{-19} \times 8000}{9.11 \times 10^{-31}}}$

$$= \quad 5.30 \times 10^7 \text{ m s}^{-1}$$

(This is about 18 per cent of the speed of light, but we ignore relativistic effects!)

Wavelength of electron: $\quad \lambda = \dfrac{h}{mv}$

$$= \frac{6.63 \times 10^{-34}}{9.11 \times 10^{-31} \times 5.30 \times 10^7}$$

$$= 1.37 \times 10^{-11} \text{ m}$$

For comparison, the wavelength of light is of the order of 10^{-7} m.

What does the electron wave actually mean?

De Broglie called the electron's wave 'fictitious'. The modern view is that the wave is linked to the **probability** of finding an electron at a certain point in space. We can write a *wave equation* for an electron, as for any other wave. And, just with any other wave, we see that the amplitude varies – with distance, say.

Theory and experiment both suggest that:

The probability of finding an electron at a given point is proportional to the square of the amplitude of the wave at that point.

The amplitude is given by a rather complicated wave formula, the **Schrödinger wave equation**, which was put forward in 1926 by Erwin Schrödinger (Austria, 1887–1961). There is more about this on pages 370–371).

Now *we know why atoms do not self-destruct!*

De Broglie's idea was the key to unlocking the last secret of the atom – why it didn't immediately collapse in a dramatic burst of electromagnetic radiation. The simple answer is this:

● The electron is a wave whose length decreases with energy.

● If you try to squash an electron wave closer to the nucleus, the wavelength must get smaller.

● When its wavelength is as small as a nucleus, its energy becomes so great that the attractive force of the nucleus isn't big enough to keep it there.

The (simple) mathematics of this is given on page 371.

Fig 17.13 shows how an electron wave fits into a hydrogen atom, together with the probability function which is simply the square of the electron wave.

Solutions of the Schrödinger wave equation can exist only for certain definite values of the total energy of the electron: bear in mind that the energy determines the momentum, which in turn determines the wavelength. In principle, any valid solution must reach a zero value at some reasonable distance from the nucleus – otherwise the electron has a finite probability of escaping completely from the atom.

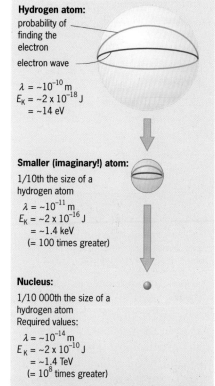

Hydrogen atom:
probability of finding the electron
electron wave

$\lambda = {\sim}10^{-10}$ m
$E_K = {\sim}2 \times 10^{-18}$ J
    $= {\sim}14$ eV

Smaller (imaginary!) atom:
1/10th the size of a hydrogen atom
$\lambda = {\sim}10^{-11}$ m
$E_K = {\sim}2 \times 10^{-16}$ J
    $= {\sim}1.4$ keV
    (= 100 times greater)

Nucleus:
1/10 000th the size of a hydrogen atom
Required values:
$\lambda = {\sim}10^{-14}$ m
$E_K = {\sim}2 \times 10^{-10}$ J
    $= {\sim}1.4$ TeV
    (= 10^8 times greater)

Fig 17.13 Squashing a wavy electron. As the wave gets smaller, its wavelength decreases, and its momentum and energy increase.

Energy is proportional to $1/\lambda^2$, so the increase is great.

Even nuclear forces are too small to keep an electron in the nucleus: its escape velocity would be too great.

▨ See question 17.

Electron tunnelling

One of the unexpected consequences of the wave theory is that the successful wave functions do *not* become zero at the edge of the atom. This effect would be forbidden in classical theory, because the edge of the atom is the place where the electron has zero kinetic energy, so going further means that its kinetic energy is negative, which is a physical impossibility. However, it seems to be allowed in quantum wave theory, so that there is a definite probablility that electrons can exist outside the atom without breaking any energy rules. This gives rise to the phenomenon called **tunnelling**, which is used in the electron scanning tunnelling microscope (see photo at foot of page 360).

A brief look at the real hydrogen atom – electron clouds and the Schrödinger wave equation

Fig 17.15 shows how the potential energy of the electron–proton system varies with separation of the particles in a hydrogen atom. Remember that the potential energy is negative. When the electron is at the edge of the atom its kinetic energy is zero. Closer to the nucleus it speeds up as the potential energy gets less. The total energy of the system stays constant at 2.18×10^{-18} J (13.6 eV).

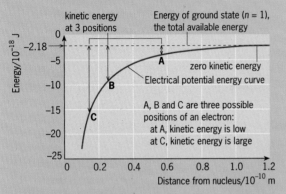

Fig 17.15 **The 'potential well' of a hydrogen atom. As the electron gets closer to the nucleus, electrical potential energy gets less and kinetic energy increases**

We can no longer assume that the electron is a simple particle, of course. What wave mechanics tells us is that the electron can be anywhere around the nucleus, with a probability of being at any particular *distance* which is given by a wave formula. The *probability* of the electron being at a particular *place* is given by the square of the electron-wave's amplitude at that place. What is the shape of this electron wave?

We know that it depends on the electron's momentum, since we have $p = h/\lambda$ (de Broglie). But as the electron moves nearer to the nucleus, its speed and momentum increase, so its wavelength gets smaller. Fig 17.16 shows a possible shape for the wave of an electron in a hydrogen atom. It shows that the wave is curved more steeply nearer the nucleus as it would have a smaller wavelength there. We can't show it as a whole wavelength because it varies with distance, but the dotted lines show the wavelength that fits the curve at a particular point.

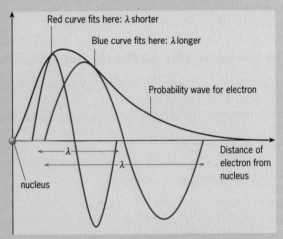

Fig 17.16 **The shape of the electron wave could be fitted to a set of sine waves of changing wavelength, with shorter and shorter wave-lengths the closer they get to the nucleus. As the wave-lengths get shorter, their contributions reach a peak, but quickly go to zero as the distance from the nucleus approaches zero**

Fig 17.14(a) **There is a finite probability of the electron existing outside its potential well**

Fig 17.14(b) **Electrons can 'tunnel' from A to B because the probability curve extends through a narrow insulator. There is more about this on page 402**

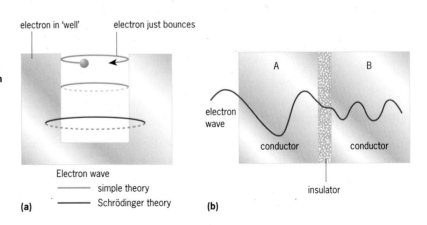

The wave nature of electrons is thus more than just a convenient model for giving right answers about atoms. It is vital to the understanding of semiconductor materials (now so important in practical microelectronic devices), the nature of chemical bonds and the shapes of molecules.

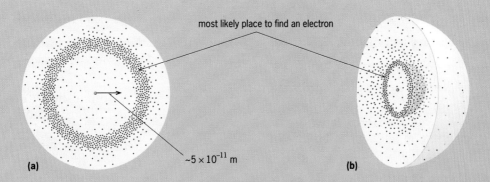

most likely place to find an electron

~5×10^{-11} m

(a)

(b)

Fig 17.17(a) **The electron moves so fast that all we observe is a 'cloud' of varying electron density, which is densest at about 0.5×10^{-10} m from the nucleus**

Fig 17.17(b) **The electron occupies three-dimensional space**

Fig 17.18 shows probability waves, which are simply the electron wave squared. The diagram in Fig 17.17(a) shows the atom as a simple two-dimensional object, but of course it is three-dimensional, and Fig 17.17(b) gives a better picture. Chemists tend to think of the space the electron occupies as a sort of cloud and draw an atom with orbits showing the region where there is the probability of finding an electron.

The Schrödinger wave equation

The wave equation for an ordinary wave, such as a wave on a string, tells us how the displacement (amplitude) of any disturbance at a position x on the string varies with time t, and/or how the displacement varies with position along the string at a given instant of time. The equation for a sine wave on a string is developed on page 125:

$$y = A \sin\left(\frac{\omega x}{v}\right)$$

where y is the amplitude of a wave travelling at speed v along a string; y varies with ω (where $\omega = 2\pi f$). The wave has a maximum amplitude of A when $x = 0$. The wave equation can also be written as a differential equation:

$$\frac{d^2 y}{dx^2} = \frac{1}{v^2}\frac{d^2 y}{dt^2}$$

For electron waves we can relate speed v to the momentum and hence the kinetic energy of the electron. Schrödinger did this and produced a form of the above wave equation which is usually written as:

$$\frac{d^2 \Psi}{dx^2} = -\frac{8\pi^2 m}{h^2}(E - v)\,\Psi$$

where we use ψ instead of y as the amplitude at a point x, E is the total energy of the electron, and V is its potential energy at x. This equation does not involve time and has been simplified to one

dimension (x). When we apply the equation to a hydrogen atom, we define x in terms of r, the distance of the electron from the nucleus, and introduce the fact that V depends upon r:

$$V = \frac{e^2}{4\pi\varepsilon_0 r}$$

This gives:

$$\frac{d^2 \Psi}{dr^2} = -\frac{8\pi^2 m}{h^2}\left(E - \frac{e^2}{4\pi\varepsilon_0 r}\right)\Psi$$

Fig 17.18 shows in orange the resulting one-dimensional curve produced by plotting the square of this function against r for a value of total energy $E_1 = -2.18 \times 10^{-18}$ J. This is the total energy of the atom in its normal (ground) state when the quantum number n is 1. The curve drops to zero, showing that the electron is enclosed inside the atom. In fact, we can get similar closed solutions only for values of E given by E_1/n^2. This satisfies Balmer-type formulae for energy levels in the hydrogen atom.

Fig 17.18 gives curves for the energies shown. In each case, the main peak corresponds to the most likely radius of the 'orbit' of the electron in the excited atom.

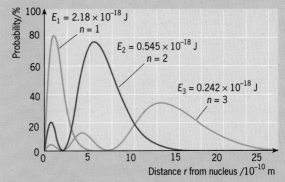

Fig 17.18 **Probability curves for the electron in three energy states in a hydrogen atom**

See question 18.

SUMMARY

After studying this chapter you should be able to do the following.

■ Understand the nature of black body (temperature) radiation.

■ Know Wien's and Stefan's laws relating wavelength, temperature and intensity in black body radiation.

■ Understand how classical theory had to be changed to a quantum theory in order to account for the experimental data.

■ Understand that Planck's quantum theory explained black body radiation.

■ Know about the photoelectric effect and how Einstein explained its strange results using quantum theory and the idea of the photon ($E = hf = W + E_K$).

■ Know that Rutherford's nuclear model of the atom helped to explain how quantum theory worked.

■ Understand how quantum theory explains the emission of light from an excited atom, using the idea of energy levels, and that it was able to explain the patterns in the line spectrum of hydrogen ($hf = E_1 - E_2$).

■ Understand the idea that electrons had a wave nature, with a de Broglie wavelength $\lambda = h/p$.

■ Know how wavy electrons produce a better model of the atom than particle electrons can – with the waves as an indicator of the probability of finding an electron.

■ As an extension, be familiar with the Schrödinger wave equation as applied to the hydrogen atom.

QUESTIONS

Data which you may need to answer some of the questions
Planck constant h: 6.6×10^{-34} J s
mass of electron m: 9.1×10^{-31} kg
speed of light c: 3.0×10^{8} m s^{-1}
charge of proton/electron: $\pm1.6 \times 10^{-19}$ C
Wien's constant in $\lambda_{max}T$ = constant: 2.9×10^{-3} m K
Stefan's constant: 5.7×10^{-8} W m^{-2} K^{-4}

1 Write short explanations of the following:

a) Colour changes occur when a metal is heated to a high temperature.

b) The grill of a cooker has to be at a high temperature whilst a microwave cooker is quite cool.

2 An electrically heated radiator is kept at a temperature of 40°C. It has a surface area of 2 m² and is rated at 2 kW.

a) What is the maximum wavelength at which it radiates?

b) How much energy does it radiate per second?

You should find that it radiates at a rate significantly less than 2 kW. Suggest what happens to the remainder of the energy transferred to the radiator each second.

3

a) One of the brightest stars in the sky is Vega. It emits radiation with a peak wavelength of 240 nm. Use the Wien displacement law to calculate its surface temperature.

b) What would be the peak wavelength of the radiation emitted by a red giant star with a surface temperature of 3000 K?

4 The eye is most sensitive to light with a wavelength of 560 nm. At what temperature would a (black) body radiate most strongly at this temperature?

5 What was the *ultraviolet catastrophe*? Briefly describe how it was overcome.

6

a) Calculate the energy of a photon of electromagnetic radiation with a wavelength of:
(i) 200 nm, **(ii)** 600 nm **(iii)** 2 μm **(iv)** 10 mm.

b) In which part of the electromagnetic spectrum is each of the above photons?

7 Estimate the number of photons emitted per second in the visible region by an ordinary 60 W electric lamp. These lamps are about 2 per cent efficient – most of the energy is used to heat the surroundings.

8 Red light, however bright it is, cannot produce the emission of electrons from a clean zinc surface. But even weak ultraviolet radiation can do so. Explain why this is.

9 The work functions of three metals are shown in the table:

	eV	J/10^{-19}
potassium	1.81	2.9
sodium	2.28	3.65
zinc	4.31	6.9

a) Which metal shows a photoelectric effect with the lowest frequency of radiation?

b) Calculate the minimum (threshold) frequency that can cause the emission of electrons from zinc.

c) What other property/ies of sodium suggest that it is fairly easy for its atoms to lose electrons?

d) Monochromatic light of wavelength 500 nm shines on a clean potassium surface. What is the maximum kinetic energy of the electrons emitted?

10 A simple model of the atom is of a small massive nucleus surrounded by electrons in orbit.

a) In this model, what keeps the electrons in orbit?

b) Why does this model contradict the classical theory that accelerated charges emit electromagnetic radiation?

c) Explain how a model in which an electron behaves as a wave can overcome this contradiction.

11 Elements may be identified by their *line spectra*.

a) Describe an experiment to observe the line spectrum of an element and measure the wavelengths of its lines.

b) Briefly explain the physical reasons why the line spectra of different elements are unlikely to be identical.

12 The table below gives the wavelengths of the first four lines in the Lyman series of the line spectrum of hydrogen.

Wavelength/nm	95	97.3	102.6	121.6

a) These lines are formed when the atom changes its energy state to a value of -2.18×10^{-18} J from four higher levels. Draw an energy level diagram to illustrate the formation of these lines. Label the changes in energy states with the wavelength each change produces. Number the levels on your diagram from $n = 1$ to $n = 5$, starting with $n = 1$ as the lowest.

b) Use the relation $E = hf$ to calculate the energies associated with the four higher levels. Plot a suitable graph to check whether the higher levels are related to the lowest level by the relation $E_n = E_1/n^2$.

13 This question compares the diffraction of light (see Chapter 16 Electromagnetic radiation) with the diffraction of electrons.

a) A diffraction grating is set up to form a spectrum of white light. Which is diffracted through a larger angle, blue light or red light?

b) A typical diffraction grating has a grating spacing d of 5×10^{-6} m. By what angle is the first order maximum produced for light of wavelength 5.0×10^{-7} m? (Use $n\lambda = d \sin \theta$.)

c) A student considers using the grating to produce the diffraction of electrons. He intends to place the grating in a vacuum tube and fire a beam of electrons into it.

The emerging electron beam would be displayed using a fluorescent screen as used in cathode ray tubes. The arrangement would be good enough to detect a diffraction angle of 2°.

(i) The detecting screen only works if the electrons hitting it have been accelerated by a potential difference of 2 kV. What is the wavelength of electrons that have been accelerated by this voltage?

(ii) If the grating is capable of diffracting electrons, what would be the angle of the first order maximum for these electrons? Is the experiment feasible?

(1 eV = 1.6×10^{-19} J, for electrons momentum $p = h/\lambda$, kinetic energy $E_K = p^2/2m$.)

14 A follow-on from q.13: Electron diffraction is more easily seen by firing a beam of electrons at a crystal in which rows of atoms act as the grating. Graphite may be used for this: the carbon atoms are in rows with a spacing of 1.23×10^{-10} m.

a) In an experiment, a beam of electrons was diffracted by graphite so that the second order maximum was at an angle of 0.167 radian. Use the grating formula to calculate the wavelength of the electrons.

b) The electrons were accelerated by a potential difference of 5000 V. Calculate the kinetic energy and hence the momentum of the electrons.

Use the relation $p = h/\lambda$ to obtain a value for the Planck constant h.

15 Free electrons in a metal wire carrying an electric current have have an average speed between collisions with metal atoms of about 2×10^6 m s^{-1}.

a) Show that the wavelength of such electrons is about four times the size of an atom (an atom is about 10^{-10} m in diameter).

b) Modern semiconductor microchips may contain components as layers only a few atoms thick. What problems might this cause?

16 A neutron has a mass of 1.67×10^{-27} kg. Slow neutrons ('thermal' neutrons) are used in nuclear power stations to initiate the fission of uranium fuel. Such neutrons have an average kinetic energy of 0.04 eV. (1 eV is 1.6×10^{-19} J.)

a) Calculate the momentum of neutrons which have this value of kinetic energy.

b) Use the de Brogie formula to calculate the wavelength of neutrons with this energy.

17 A singly ionised helium atom consists of a nucleus with charge +2e and an electron of charge –e. Would this ion be larger or smaller than a hydrogen atom?

18 An electron in an atom may be modelled very simply by thinking of it as a standing wave in a box. Fig 17.Q19 shows such a wave. An atom is about 10^{-10} in diameter. The nucleus consists of one proton.

Fig 17.Q19

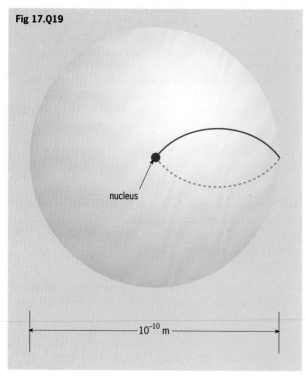

nucleus

10^{-10} m

a) What is the longest wavelength of a standing wave that could fit into this box?

b) Use the de Broglie formula to calculate the momentum of an electron with this wavelength.

c) Calculate the kinetic energy of that electron.

d) The electrical potential energy between an electron and a proton distance r from it is given by
$E_p = e^2/4\pi\varepsilon_0 r$

How much energy would have to be transferred to electrical potential energy to move the electron a long way from the proton?

e) Does the electron whose energy you calculated in part c have enough kinetic energy to escape from the proton?

f) Suppose we had an electron with a wavelength that allowed it to fit into a one-proton atom with a diameter 10^{-11} m. Repeat the above procedures to see if this electron–proton combination could be stable.

ELECTROMAGNETIC RADIATION; ATOMS, SPECTRA AND QUANTA

This Chapter Map brings together the main ideas from Chapters **16 Electromagnetic radiation** and **17 Atoms, spectra and quanta**. It shows the main concepts and equations that you have studied in these chapters, and the way they are interlinked and depend on each other.

The map is a guide to the key information you need to know and understand for the syllabus you are following, and you can use it as a revision checklist. You can look through the chart to identify the areas you are not confident about and may need to study further.

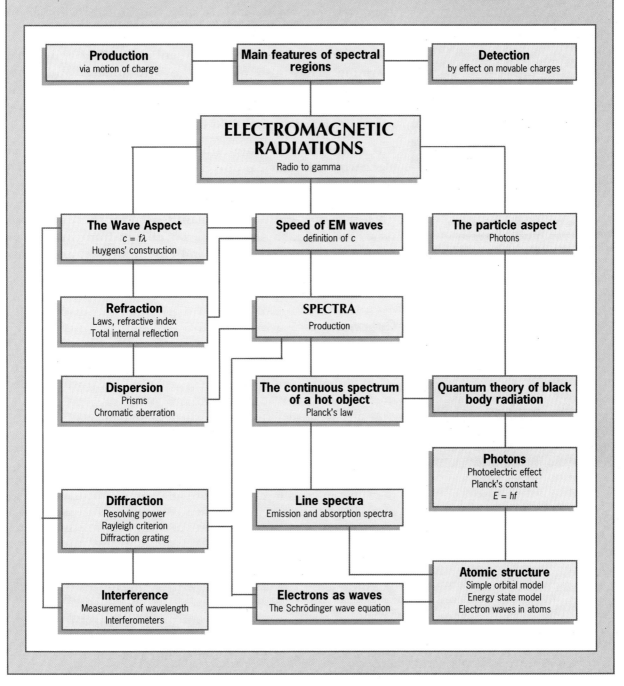

Production
via motion of charge

Main features of spectral regions

Detection
by effect on movable charges

ELECTROMAGNETIC RADIATIONS
Radio to gamma

The Wave Aspect
$c = f\lambda$
Huygens' construction

Speed of EM waves
definition of c

The particle aspect
Photons

Refraction
Laws, refractive index
Total internal reflection

SPECTRA
Production

Dispersion
Prisms
Chromatic aberration

The continuous spectrum of a hot object
Planck's law

Quantum theory of black body radiation

Photons
Photoelectric effect
Planck's constant
$E = hf$

Diffraction
Resolving power
Rayleigh criterion
Diffraction grating

Line spectra
Emission and absorption spectra

Atomic structure
Simple orbital model
Energy state model
Electron waves in atoms

Interference
Measurement of wavelength
Interferometers

Electrons as waves
The Schrödinger wave equation

18 The atomic nucleus

Above: The nuclear power station at Creys-Malville in France: any radioactive pollution from nuclear power stations is invisible.

Below: The Drax coal-fired power station in North Yorkshire. Its cooling towers give off steam which appears as clouds of water vapour, but the smoke from the central chimney includes sulphur and nitrogen oxides from burning the coal

FRANCE PRODUCES 75 per cent of its electrical power from nuclear power stations and only 12 per cent from fossil fuels. This compares with just 28 per cent nuclear power in Britain, and 70 per cent from fossil fuels.

Fossil fuels contribute to the greenhouse effect and global warming. They also pollute the atmosphere, and most developed countries are searching for ways to reduce their dependence on fossil fuels. This suggests that making more use of nuclear power should be a good thing.

But in the United States (the world's largest producer of electricity from nuclear power), electricity suppliers decided in 1987 not to build any more nuclear power stations. Britain, too, cancelled plans in 1995 for building two large nuclear power stations. The main reason for these decisions was economic). In the debate on energy from nuclear or fossil fuel sources, economic and environmental arguments compete.

Overview

The atomic nucleus was not known about before the early twentieth century. Yet, thirty years on, in 1945, two nuclear bombs were to kill an estimated 200 000 people in Japan, an event which ended the Second World War in a matter of days.

As a general rule, the nucleus provides useful energy through a process that occurs only with certain rare, very large nuclei – **nuclear fission. Radioactivity** itself is a minor source of energy, but always accompanies fission reactions.

Life on Earth depends on the energy from nuclear reactions in the Sun known as **nuclear fusion**, in which two hydrogen nuclei combine to form a helium nucleus. But attempts to use nuclear fusion as a steady, controlled source of cheap and almost pollution-free energy have so far been unsuccessful.

1 THE NUCLEAR ATOM

Atoms are mostly empty space. We picture a hydrogen atom, for example, as a very small central blob of matter surrounded by a single rapidly moving electron. The central **nucleus** contains 99.95 per cent of the atom's mass but only a ten-billionth part of its volume. The space surrounding the nucleus is occupied by the single electron, moving so rapidly that we can think of it as being everywhere in the space at once (see page 371).

This chapter deals with the evidence we have about the nature and structure of the nucleus, and why some nuclei break up into two smaller nuclei (**fission**) and others are **radioactive.**

Nuclear patterns

Practically all the mass of an atom is the mass of its nucleus, which consists of particles called **nucleons**. There are two kinds of nucleon with very slightly different masses (see Table 18.1). The **proton** is electrically charged: its positive charge is equal in size to the charge on an electron. It is slightly less massive than its partner, the uncharged nucleon called the **neutron**.

The Periodic Table of the chemical elements, shown in Table 18.2, arranges the elements in the order of the number of protons in the nucleus. This number is the **proton number** or the **atomic number**, symbol **Z**, (Fig 18.2).

Adding the number of protons and neutrons gives the **mass number** or **nucleon number** of an element, symbol **A**. For the actual mass of a nucleus, see page 383. Fig 18.3 shows the conventional way of writing the proton and mass numbers of a nucleus.

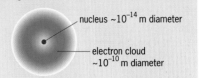
Fig 18.1(a) **A modern view of an atom. It has a very definite size, but most of its volume is filled with 'non-localised' electrons forming a fuzzy cloud**

nucleus $\sim10^{-14}$ m diameter

electron cloud $\sim10^{-10}$ m diameter

■ See question 1.

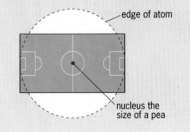
Fig 18.1(b) **A nucleus is very small compared with an atom. If it were the size of a pea, an atom would be as big as a football pitch**

edge of atom

nucleus the size of a pea

Fig 18.2 **Simple models of atomic nuclei**

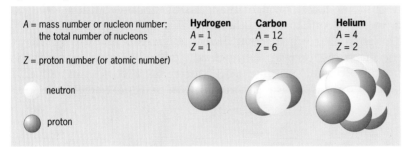

A = mass number or nucleon number: the total number of nucleons

Z = proton number (or atomic number)

○ neutron

● proton

Hydrogen A = 1 Z = 1
Carbon A = 12 Z = 6
Helium A = 4 Z = 2

Fig 18.3 **Describing the carbon nucleus**

A: mass (nucleon) number — **12**
Z: proton (atomic) number — **6** **C**

Table 18.1

Nucleon	proton	neutron
Mass/10^{-27} kg	1.672 623	1.674 929
Charge/e	1.00	0.00

A Use the data given in the Periodic Table to find the number of neutrons in the nuclei of calcium (Ca, Z = 20) and tellurium (Te, Z = 52).

Table 18.2 **The Periodic Table**

groups	I	II												III	IV	V	VI	VII	0
periods																			
1									1 H hydrogen 1										2 He helium 4
2	3 Li lithium 7	4 Be beryllium 9												5 B boron 11	6 C carbon 12	7 N nitrogen 14	8 O oxygen 16	9 F fluorine 19	10 Ne neon 20
3	11 Na sodium 23	12 Mg magnesium 24												13 Al aluminium 27	14 Si silicon 28	15 P phosphorus 31	16 S sulphur 32	17 Cl chlorine 35.5	18 Ar argon 40
4	19 K potassium 39	20 Ca calcium 40	21 Sc scandium 45	22 Ti titanium 48	23 V vanadium 51	24 Cr chromium 52	25 Mn manganese 55	26 Fe iron 56	27 Co cobalt 59	28 Ni nickel 59	29 Cu copper 64	30 Zn zinc 65		31 Ga gallium 70	32 Ge germanium 73	33 As arsenic 75	34 Se selenium 79	35 Br bromine 80	36 Kr krypton 84
5	37 Rb rubidium 85.5	38 Sr strontium 88	39 Y yttrium 89	40 Zr zirconium 91	41 Nb niobium 93	42 Mo molybdenum 96	43 Tc technetium 98	44 Ru ruthenium 101	45 Rh rhodium 103	46 Pd palladium 106	47 Ag silver 108	48 Cd cadmium 112		49 In indium 115	50 Sn tin 119	51 Sb antimony 122	52 Te tellurium 128	53 I iodine 127	54 Xe xenon 131
6	55 Cs caesium 133	56 Ba barium 137	57 La lanthanum 139	72 Hf hafnium 178.5	73 Ta tantalum 181	74 W tungsten 184	75 Re rhenium 186	76 Os osmium 190	77 Ir iridium 192	78 Pt platinum 195	79 Au gold 197	80 Hg mercury 201		81 Tl thallium 204	82 Pb lead 207	83 Bi bismuth 209	84 Po polonium 210	85 At astatine 210	86 Rn radon 222
7	87 Fr francium 223	88 Ra radium 226	89 Ac actinium 227	104 Db dubnium 261	105 Jl joliotium 262	106 Rf rutherfordium	107 Bh bohrium	108 Hn hahnium	109 Mt meitnerium										

58 Ce cerium 140	59 Pr praseodymium 141	60 Nd neodymium 144	61 Pm promethium 147	62 Sm samarium 150	63 Eu europium 152	64 Gd gadolinium 157	65 Tb terbium 159	66 Dy dysprosium 162.5	67 Ho holmium 165	68 Er erbium 167	69 Tm thulium 169	70 Yb ytterbium 173	71 Lu lutetium 175
90 Th thorium 232	91 Pa protactinium 231	92 U uranium 238	93 Np neptunium 237	94 Pu plutonium 242	95 Am americium 243	96 Cm curium 247	97 Bk berkelium 247	98 Cf californium 251	99 Es einsteinium 254	100 Fm fermium 253	101 Md mendelevium 256	102 No nobelium 254	103 Lr lawrencium 257

metal / non metal

atomic no. symbol name mass no.

Forces in the nucleus

The electrical force

The positively charged protons exert an electrical force of repulsion on each other. The force between two protons, of equal charge $+e$, can be calculated from the formula:

$$F_E = \frac{ke^2}{r^2}$$

where k is the electric force constant (9×10^9 N m² C⁻²) and r is a typical nuclear distance, about 10^{-14} m. (See also equation page 238.) So the force between the two protons in a helium nucleus is 2.3 N. This is a repulsive force and is immense on the nuclear scale because protons are so close together. The force would, on its own, cause the protons to fly apart at high speeds in a nuclear disintegration.

The strong nuclear force

But nuclei do exist, and are quite hard to break apart. So there must be another force, stronger than this electrical repulsion force, to attract nucleons to each other and so keep the nucleus together. This is the **strong nuclear force**, about 100 times as strong as the electrical force in the nucleus. Unlike electrical and gravitational forces, the strong nuclear force has a very short range: it reaches only as far as from one nucleon to its next neighbours (Fig 18.4).

?

B Check that the force between two protons in a helium nucleus is about 2.3 N. Use the data:

$r = 10^{-14}$ m;
force constant $k = 9 \times 10^9$ N m² C⁻²;
$e = 1.6 \times 10^{-19}$ C.

Table 18.3 **Comparing the ranges and sizes of forces**

Force	Relative strength	Range
Strong nuclear	1	about 10^{-15} m
Electromagnetic	10^{-2}	infinite (inverse square law)
Gravitational	10^{-38}	infinite (inverse square law)

See question 2. ■

1 An atom of an element is electrically neutral

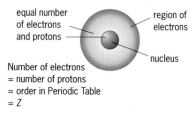

equal number of electrons and protons — region of electrons — nucleus

Number of electrons
= number of protons
= order in Periodic Table
= Z

2 Electrons are arranged in shells

shells have fixed numbers of electrons

shells (usually) fill up before the next shell is occupied

3 Electrons are added to the outer shell, one at a time ≡ sequence in Periodic Table, row by row

Number of electrons:

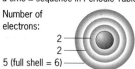

2
2
5 (full shell = 6)

This is fluorine, a halogen - which is very reactive

4 Elements in vertical columns of the Periodic Table have the same number of outer electrons

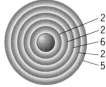

2
2
6
2
5

This is chlorine, also a very reactive halogen

Fig 18.5 **Connection between atomic structure, chemical behaviour and Periodic Table position**

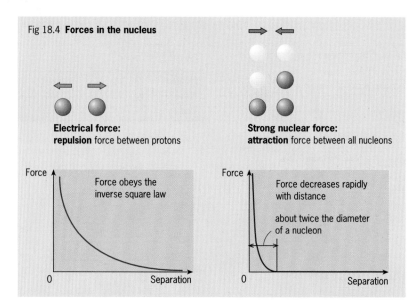

Fig 18.4 **Forces in the nucleus**

Electrical force:
repulsion force between protons

Strong nuclear force:
attraction force between all nucleons

Force — Force obeys the inverse square law — Separation — 0

Force — Force decreases rapidly with distance — about twice the diameter of a nucleon — Separation — 0

Chemistry and atomic structure

Fig 18.5 shows how the chemical behaviour of an element relates to its place in the Periodic Table. More and more nucleons join together to make larger and larger nuclei, and patterns emerge linking atomic structure and chemical behaviour.

In chemical reactions, the outer electrons of atoms interact. Elements with the same number of outer electrons react similarly, and are arranged in columns in the Periodic Table called **groups**. Examples of groups include the alkali metals (lithium, sodium etc), and the halogens (fluorine, chlorine etc), and the noble gases (helium, neon etc).

Isotopes

All nuclei except the hydrogen nucleus contain both protons and neutrons. For the smaller nuclei, the number of neutrons is about equal to the number of protons – but this is a very rough rule because of **isotopes**.

All atoms of an element have the same number of protons, but they may have different numbers of neutrons and are then called **isotopes** of that element. The mass of an atom of one isotope of an element will then be different from the mass of another isotope. For example, the nucleus in the most common atom of carbon has 6 protons and 6 neutrons. But we can find seven forms of carbon nuclei with different masses. So carbon has seven isotopes – as in Table 18.4. They all have the same number of protons (and electrons), so they occupy the same place in the Periodic Table and have identical chemical properties.

Only carbon-12 and carbon-13 nuclei are **stable**; the rest are radioactive (see below).

C Look at the Periodic Table on page 377 and identify four elements which have an equal number of protons and neutrons.

Table 18.4 **Isotopes of carbon**

Nucleon no. A	Proton no. Z	No. of neutrons $(A - Z)$
9	6	3
10	6	4
11	6	5
12 (98.89%)	6	6
13 (1.11%)	6	7
14	6	8
15	6	9

Relative atomic mass

The mass of an atom is very small. So it is usually given as a multiple of a standard unit, the **atomic mass unit, u**. This multiple is called the **relative atomic mass** or **RAM**. This is the mean of the relative atomic masses of the naturally occurring isotopes.

> **The atomic mass unit is defined as one-twelfth of the mass of the commonest isotope of carbon (carbon-12).**

D Draw diagrams to illustrate the structure of the nuclei of the following isotopes of carbon:

carbon-9, carbon-13, carbon-15.

The problems of being a large nucleus

Fig 18.6 is a graph of the elements, with proton number Z (on the x-axis) plotted against (nucleon) number A, for the most common isotope of each element. It has a slight curve which shows that A (protons + neutrons) increases more rapidly than Z (protons only). For example, the nucleus of lead (Pb-206) has 82 protons and 124 neutrons, whilst calcium (Ca-40) has 20 of each. We see this increase in the neutron–proton ratio more clearly in the graph of Fig 18.7.

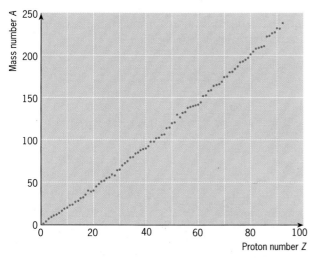

Fig 18.6 **Graph of proton number against mass (nucleon) number**

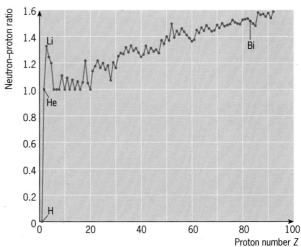

Fig 18.7 **Graph of neutron–proton ratio**

Nuclei larger than those of bismuth (Bi) with more than 83 protons and 1.5 times as many neutrons, are unstable. They are all **radioactive** – which means that they break down (disintegrate) to emit:

● a group of 2 protons plus two neutrons as an **alpha particle**, *OR*

● an electron (as a **beta particle**), *OR*

● a photon (a **gamma ray**).

The largest nuclei can often break down in another way, splitting into two smaller, roughly equal sized nuclei, plus some spare neutrons. This is **nuclear fission.**

Repulsion versus attraction in the nucleus

The **liquid drop model** is a simple model of the atomic nucleus that likens the nucleons in a nucleus the molecules in a drop of water. The molecules in the drop are moving around – they have kinetic energy – but forces of attraction between the molecules hold them together. This is what causes *surface tension*: molecules at the surface experience a net inward force and the drop forms into a sphere. A sphere is the shape with the smallest number of particles in its surface for a given volume.

Of course, molecules with enough kinetic energy can escape – they evaporate. Again, when the drop becomes too large, the attracting forces may not be strong enough for it to keep its shape under the action of gravity, and it may even split up into smaller drops.

Now think of an atomic nucleus. Like water molecules in a drop, its particles are attracted by each other (the nucleus too has a kind of surface tension). The nucleus holds together because nucleons exert a strong force of attraction on each other – the **strong nuclear**, but *short-range*, force. An important fact is that the range is so short that any one nucleon can only attract its adjacent neighbours.

Gravity, which disrupts a liquid drop, is too weak to affect nuclear particles. But a nucleus contains its own disruptive force – the electrical force whereby protons repel each other. In a stable nucleus, the attractive strong nuclear force is easily large enough to counteract the disruptive **electrical** force. But as the nucleus gets larger, this stops being true because the repelling force is *long range*: the nucleons are held in a nucleus only by their nearest neighbours, but any proton is repelled away by *all* the other protons. The more protons we add, the bigger is this repulsive force – but the attracting force per particle stays the same. See Fig 18.8.

Fig 18.8(a) **An unstable sphere: repelling particles have positive potential energy. This is rapidly converted to kinetic energy when the sphere splits apart (surface particles shown only)**

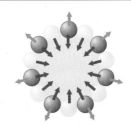

Fig 18.8(b) **A mixture of attracting and repelling particles may or may not be stable**

So protons in the nucleus have a tendency to make the nucleus fall apart. This happens in fact with just two protons! There is no nucleus with just two protons. The helium nucleus is stable only because of its two neutrons: they add a strong nuclear force without adding an electric repulsion force. We can think of neutrons as having the role of adding glue to the nucleus, or as diluting the repulsion forces.

But adding too many neutrons can also make a nucleus unstable (look back at the graph of Fig. 18.6). Nuclei with excess neutrons are **radioactive,** tending to emit an alpha particle; or, if they are large enough, they will split into two roughly equal parts in nuclear **fission**. These effects are hard to explain on a simple liquid drop model, but happen when a nucleus has more than 83 protons. Explaining alpha decay needs quantum theory (see page 402). Fission occurs when a large nucleus is given some extra energy, for example, by absorbing another neutron. The nucleus 'wobbles', changing its shape into a dumbbell, which allows the repulsive force to break it up into two parts, as shown in Fig 18.9.

See question 3.

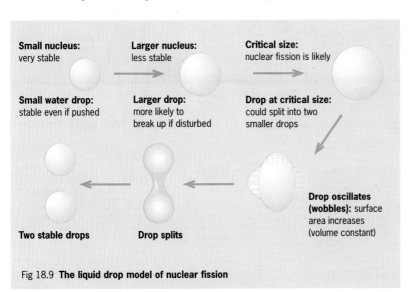

Fig 18.9 **The liquid drop model of nuclear fission**

2 NUCLEAR FISSION

The most common isotope of uranium is U-238. It has 238 nucleons, and is quite stable against fission (although it is radioactive). The uranium isotope with 235 nucleons (U-235) has three fewer neutrons and so a larger ratio of protons to neutrons which makes it more likely to split. We can artificially split the U-235 nucleus by adding a neutron to it. This probably disturbs the delicate balance of forces just enough to cause the nucleus to disintegrate. See Fig 18.17 on page 387.

The nucleus splits into two smaller nuclei, plus two or three neutrons. These neutrons will trigger further disintegrations if they are absorbed by other U-235 nuclei, and can lead to a **chain reaction**, in which induced nuclear fission spreads through a mass of U-235. An uncontrolled chain reaction is used in nuclear bombs and a controlled reaction is used in nuclear electrical power stations. Nowadays, other fissile nuclei tend to be used as much as the naturally occurring U-235.

The size of a nucleus

A simple theory
Protons and neutrons have almost exactly the same mass.

Assuming that they also occupy the same space, volume v, the volume of a nucleus with A nucleons is Av.

If the nucleus is spherical with radius r, then we can write:

$$\text{approximate volume of a nucleus} = Av$$

$$= \frac{4}{3}\pi r^3$$

This result is usually written as:

$$r = r_0 A^{\frac{1}{3}}$$

where r_0 is an experimental constant of value 1.2×10^{-15} m.

The experimental evidence
When a metal is bombarded with a beam of charged particles, some of the particles bounce back almost exactly along the approach path. The first such experiment used alpha particles as in Fig 18.10, and led to the discovery of the nucleus. It is described on page 390.

What happens is shown in Fig 18.12(a). The approaching particle is repelled by the nucleus. Its kinetic energy E_K is converted to electrical potential energy E_P.

The point of closest approach occurs when all the E_K becomes E_P, that is:

$$\frac{1}{2}mv^2 = k\frac{qQ}{d}$$

$$= k\frac{2Ze^2}{d}$$

where k is the electric force constant (which is 9×10^9 N m^2 C^{-2}) and the other quantities are as shown in Fig 18.10.

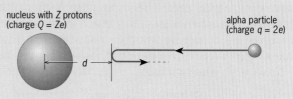

nucleus with Z protons
(charge $Q = Ze$)

alpha particle
(charge $q = 2e$)

Fig 18.10

The energy of the approaching particle is known and so are the charges of the particle and the nucleus. This means that we can calculate d. A missile with high energy gets very close to the nucleus, so d is very nearly the same as the nuclear radius r.

Experiments have shown that most nuclei are approximately spherical with a constant r_0 as given above.

?

E The mass number of platinum is 195. Calculate the size of a platinum nucleus.

F An alpha particle ($Z = 2$) collides with a platinum nucleus of $Z = 78$. The alpha particle has kinetic energy 8×10^{-13} J (5 MeV) and the charge on a proton is 1.6×10^{-19} C. Calculate how close the alpha particle gets to the centre of the nucleus. Is this value consistent with the size of nucleus you calculated in **E**?

---

3 BINDING FORCE AND BINDING ENERGY

A mass on the Earth's surface is held there by the force of gravity. When the mass is lifted above the ground, the Earth–mass system gains energy. Work has been done to separate the two bodies. This energy is the **gravitational potential energy** covered in Chapter 4. In that chapter, we calculated the gravitational potential energy per kilogram of a mass on the Earth's surface. Since the force is attractive, and since the potential energy when the mass is at infinity is zero, the potential energy at the surface is therefore negative.

Although this value is calculated for the mass m, it is in fact the energy of the Earth–mass *system*, not of the movable mass alone. We could call it the **binding energy** of the Earth–mass system, which Chapter 4 shows to be:

$$-G\frac{Mm}{r}$$

where G is the gravitational constant, M the mass and r the radius of the Earth.

In the same way, we can think of the particles in the nucleus as making up a system which contains binding energy. The attracting force here is not gravity, but the net sum of the *strong nuclear force* and the *electric force*, which act in opposite directions. The effect is much the same for the Earth–mass system: it takes a supply of energy to break up the nucleons of a normally stable nucleus. You cannot work out the binding energy of a nuclear system by summing the forces for the individual nucleons. The two main forces involved obey different laws, unlike the simple one-force gravitational system. But we can do it easily using a very simple, but quite amazing idea:

Mass and energy are equivalent.

See questions 4 and 5.

4 THE MASS DEFECT

Physicists have measured the masses of nuclei, and the individual protons and neutrons, to a high degree of accuracy, as we saw in Table 18.1.

A helium nucleus has 2 protons and 2 neutrons, and so we would expect it to have a mass of $2m_p + 2m_n$ totalling $6.695\,104 \times 10^{-27}$ kg. But this is *greater* than the actual mass of a helium nucleus. We seem to have lost $0.048\,322 \times 10^{-27}$ kg! This lost mass is called the **mass defect**.

Mass of proton m_p
  $1.672\,623 \times 10^{-27}$ kg

Mass of neutron m_n
  $1.674\,929 \times 10^{-27}$ kg

Mass of helium nucleus
  $6.646\,782 \times 10^{-27}$ kg

G Use the Einstein mass–energy relationship to check that the binding energy of the helium nucleus is 4.349×10^{-12} J.

Einstein to the rescue

We explain this lost mass by saying that it represents the loss of energy that occurred when the particles combined to form helium. Just as a lump of matter loses potential energy in falling to Earth, so nucleons lose potential energy when they fall together to make helium nuclei. The energy loss appears as a mass loss according to the Einstein relation $\Delta E = c^2\Delta m$. Using the value of the speed of light c as 3×10^8 m s^{-1}, we can calculate the mass defect for helium as a **binding energy** of 4.349×10^{-12} J.

To convert electronvolts to joules, multiply the value by the electronic charge e, 1.6×10^{-19} C.

Binding energy per nucleon

It is often more useful to consider the **binding energy per nucleon** (BEPN) in a particular nucleus. This is the energy needed to take one nucleon out of a nucleus, doing work against the net attracting force. For helium-4, the binding energy per nucleon is a quarter of the total binding energy calculated above from the mass defect: 1.0873×10^{-12} J.

Physicists also prefer to use the **electronvolt** as the unit of energy. The electronvolt is explained on page 232. The value of the binding energy per particle for helium-4 is –7.07 MeV.

H (a) How many joules are equivalent to **(i)** 2 eV, **(ii)** 4 MeV?

(2 eV is the typical energy required to ionise an atom; 4 MeV is the typical energy acquired by an alpha particle when ejected from a nucleus.

(b) Convert the following to electronvolts: **(i)** 1.09×10^{-12} J, **(ii)** 25 J.

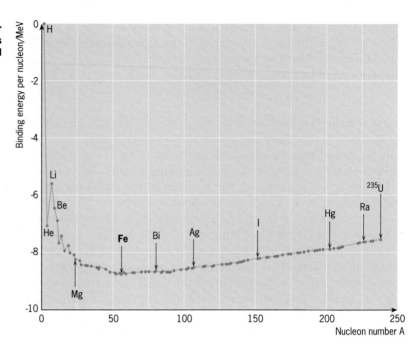

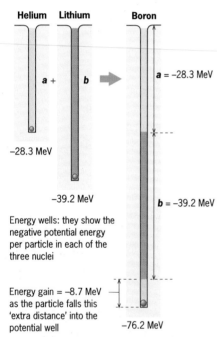

Energy wells: they show the negative potential energy per particle in each of the three nuclei

Energy gain = –8.7 MeV as the particle falls this 'extra distance' into the potential well

Fig. 18.12 **Fusing helium and lithium to make boron releases 8.7 MeV of binding energy**

The **binding energy per nucleon** is useful because it tells how strongly bound the nucleons are in different nuclei. Fig 18.13 is a graph of BEPN against proton number. The nucleus in which the particles are most strongly bound is that of iron ($Z = 26$, $A = 56$). This is because iron has the lowest value of BEPN – when they enter iron, nucleons have 'further to fall' than for any other nucleus. It means that more energy has to be supplied to get them out again. As the next section shows, the position of nucleons on this curve decides whether or not we can get energy by their fusion or fission. This means that iron (Fe) has the most stable nucleus.

Energy from nuclear fusion

In theory, any two nuclei could be combined together to form a larger one. Consider adding a helium nucleus ($Z = 2$, $A = 4$) to a lithium nucleus ($Z = 3$, $A = 7$). We would end up with an isotope of boron ($Z = 5$, $A = 11$). Breaking up the smaller nuclei into nucleons would cost their binding energies (in MeV):

Binding energy for 4 nucleons of He/MeV	–28.3
Binding energy for 7 nucleons of Li/MeV	–39.25
Total binding energy for 11 nucleons/MeV	–67.55

If we could join these two small nuclei together to form boron, we would create a nucleus with a binding energy of –76.21 MeV. This would produce an energy *gain* of 8.66 MeV (76.21 – 67.55) per nucleus of boron produced – in forming boron, the nucleons have fallen further into the energy well, so have less potential energy. This difference in energy would leave the nucleus as radiation, say. But lithium is a rare element, so this fusion is not a cost-effective source of nuclear energy.

The formation of helium from four protons (by a complex route, see page 387 and Chapter 27) releases 26.7 MeV per helium nucleus formed. Look at the graph in Fig 18.11: the fall from hydrogen (1 proton) to helium is large, and suggests a large energy release when hydrogen is converted to helium. The falls from helium + lithium to boron are smaller, so less energy is released.

See questions 6 and 7.

COULD WE MAKE GOLD BY NUCLEAR FUSION?

THE VALUE PUT ON GOLD has tempted people down the ages to try and find a way of making it from other chemicals. Is it practicable, now that we can fuse smaller (light) nuclei, to make larger nuclei, including gold?

Binding energy of iron/MeV	-492.5
Binding energy of iodine/MeV	-1072.6
Total binding energy/MeV	-1565.1
Binding energy of gold/MeV	-1559.4

Energy is only released up to the minimum of the curve at iron ($Z = 26$, $A = 56$). Any process that fuses nuclei near the bottom of the curve to create a nucleus larger than iron needs *more* energy to create the new nucleus than we get from breaking up the smaller ones.

For example, suppose we could create a gold nucleus ($Z = 79$, $A = 197$) by combining iron ($Z = 26$, $A = 56$) with iodine ($Z = 53$, $A = 127$).

So we would have to supply energy to the process because the final state is at a higher energy level than the sum of the two initial states. There is a net energy *input* of 5.7 MeV. It is more economical to dig gold out of the ground.

It is fusions that take in energy which eventually stop energy production in stars and lead to their 'death'. What exactly happens is described in Chapter 27.

Even with smaller nuclei, it is not easy to produce fusion. The electric repulsion between positively charged nuclei is a long-range force and produces a barrier stopping nuclei getting close enough to each other for the strong nuclear force of attraction to come into play. So, for nuclear fusion, nuclei need to collide at speeds high enough to overcome the repulsion barrier. There can be nuclear fusion in stars because the very high temperatures in their cores (over 10 million kelvin) mean that the average speed of the particles is very high.

> ✔ See 160 for the link between temperature and mean kinetic energy in a gas (and plasma):
>
> $$\tfrac{1}{2}m\overline{c^2} = \tfrac{3}{2}kT$$

5 MEASURING NUCLEAR MASSES

The mass spectrograph

Accurate measurements of the masses of nuclei are made using the **mass spectrograph**, see Fig 18.13.

This device vaporises atoms and then ionises them with a beam B of electrons. The ions are accelerated and enter the spectrograph chamber D which separates ions of different mass into different paths.

The *separation* is done using a magnetic field directed into the plane of the diagram. The magnetic field exerts a force at right angles to the direction of motion of the ions – so providing a centripetal force to make the ions move in a circular path. A **centripetal force** F makes a mass m move in a circle of radius r at a speed v such that:

$$F = \frac{mv^2}{r}$$

F is a magnetic force of size BQv, for a field of strength B acting on charge Q travelling with speed v:

$$F = BQv = \frac{mv^2}{r}$$

and so the radius of the path is given by:

$$r = \frac{mv}{BQ} \qquad [1]$$

This simple theory assumes that all the atoms are ionised to the same degree, so that the charge Q is the same for them all, namely 1.

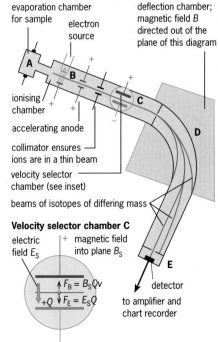

Fig 18.13 **The mass spectrograph**

The velocity selector

Note that equation 1 involves the speed v. If we want all ions of the same mass and charge to follow the same path, they must also have the same speed. This is achieved by the *velocity selector* labelled C in Fig 18.13. Ions enter the selector with a wide range of speeds and are in both a magnetic field B_s and an electric field E_s (see diagram inset). These fields are at right angles to each other and to the direction of the ion beam, with the result that the electric force ($E_s Q$) and the magnetic force ($B_s Qv$) on the ions act in opposite directions. When these forces are equal, the ions are undeflected and carry on to go through the exit aperture into the *deflection chamber* D. But the forces only balance out at a particular value of v, when:

$$B_s Qv = E_s Q$$

so that:

$$v = E_s/B_s \qquad [2]$$

Ions with different speeds are deflected, so do not pass into chamber D.

The strength of the deflecting field B in chamber D is varied, so bringing, in turn, ions of different mass into the detector E. The detector counts individual ions and so measures the relative quantities (abundances) of ions of different mass in the sample. By combining formulae 1 and 2 above:

$$r = \frac{mv}{BQ} \text{ and } v = \frac{E_s}{B_s}$$

we get:

$$r = \frac{mE_s}{BQB_s}$$

and so:

$$m = \frac{QrBB_s}{E_s} \qquad [3]$$

That is, $m = kB$, where k is a constant.

A typical readout from a mass spectrometer is shown in Fig 18.14.

See questions 8 and 9. ▨

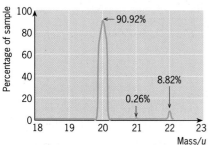

Fig 18.14 **Readout from a mass spectrometer as a graph, showing the isotopes of neon**

?

I An ion enters a mass spectrometer with a speed of 800 m s^{-1}. It passes between two plates, 5 mm apart, with a potential difference of 1000 V between them. What value of B-field (in tesla) must be applied in order that the ion is not deflected?

The unified atomic mass constant, *u*

It is hard to measure k accurately (equation 3), so values of m for different ions are measured by comparison with a *standard nuclear mass*. This is the nucleus of carbon-12, which is defined to have a mass of exactly 12 **unified atomic mass constants** (12 u), where u is $1.660\ 540\ 2 \times 10^{-27}$ kg.

On this scale, the mass of a proton is $1.007\ 276\ 5$ u.

6 NUCLEAR FUSION ON EARTH

The hydrogen bomb and the tokamak

Nuclear fusion was achieved artificially in the 'hydrogen bomb', first exploded in 1952. This used the uncontrolled chain reaction of nuclear *fission* (a fission bomb) to create a high enough temperature for isotopes of hydrogen (normal hydrogen and deuterium $^2_1 H$) to fuse together to form helium and release enormous energy. This is called a **thermonuclear process.** Such bombs are many thousands of times more destructive than the first nuclear bombs used on Japan in 1945.

CONTROLLED FUSION – THE ULTIMATE ENERGY SOURCE?

RUSSIA, THE UNITED STATES and the European Union have for many years tried to develop a working version of a controlled thermonuclear energy source. The main problems are to produce and maintain a high enough temperature for the process to begin, whilst at the same time keeping the very hot particles close enough together for fusion to continue, and in a controlled way.

The working gases are a mixture of two hydrogen isotopes, deuterium and tritium ($^2_1$H and $^3_1$H). At high temperature, the gas molecules separate into atoms and then into electrons and nuclei, so forming a **plasma**. The plasma is heated to a temperature of over 10^7 K to trigger a fusion reaction. Then, the nuclei have enough kinetic energy on average for a significant fraction of them to fuse on collision.

$$^2_1H + ^3_1H \rightarrow ^4_2He + ^1_0n \quad \text{plus } 17.59 \text{ MeV}$$

Induction heating brings the gas to high temperatures: electric coils round the chamber act as the primary coils of a transformer and the plasma nuclei act as the secondary. Alternating current at high frequency in the primary coils induces a current of ions, which means the ions gain kinetic energy. The high temperature can also be reached by firing powerful lasers into the gas.

At these high temperatures, the ordinary solid matter of a container would melt or evaporate. So the plasma has to be held in a 'magnetic bottle': a strong field makes the high-speed nuclei move in circles or tight spirals in the shape of a hollow doughnut – a torus – well away from the wall of the chamber. Fig 18.15 shows details of the equipment used, called a tokamak.

By 1996, the best results were from the experimental fusion reactor at the Joint European Torus at Culham, UK. In 1991, a 1.7 MW pulse of power was obtained for a period of 2 seconds. Even so, more energy was supplied to the machine to produce the result than was obtained from fusion.

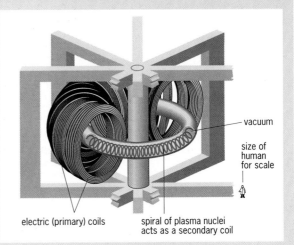

vacuum

size of human for scale

electric (primary) coils

spiral of plasma nuclei acts as a secondary coil

Fig 18.15 **A tokamak nuclear fusion reactor**

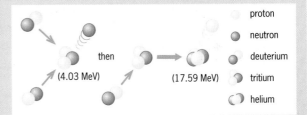

proton

neutron

then

deuterium

(4.03 MeV)

(17.59 MeV)

tritium

helium

Fig 18.16 **A likely fusion reaction in a tokamak**

The advantages of fusion as a source of energy are that the raw material is deuterium which is easily extracted from the sea, and the end product is non-radioactive helium. However, highly radioactive tritium is formed during the reaction. Also, neutrons released during fusion are captured by the chamber walls, so making new, radioactive isotopes. It is therefore unlikely that controlled fusion will become a practical energy source until well into the twenty-first century.

7 NUCLEAR REACTORS

The binding energy curve of Fig 18.11 shows that for nuclei larger than the iron nucleus at the low-point of the curve, energy might be generated by splitting a nucleus into two parts. For this to work, both new nuclei must be near the bottom of the curve. The best known fissionable nucleus is of uranium-235. The nucleus can split spontaneously, a very rare event, but one that can be triggered by the nucleus absorbing a neutron.

The fission (disintegration) products shown in Fig 18.17, barium and krypton, are just two of many possibilities. Note that two neutrons are also released. This means that a nuclear disintegration may cause a **chain reaction** in which more nuclei split as they absorb the released neutrons.

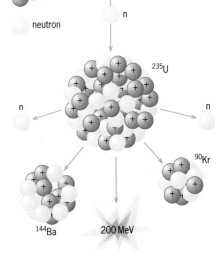

proton

neutron

n

^{235}U

n

n

^{90}Kr

^{144}Ba

200 MeV

Fig 18.17 **Nuclear fission. Adding a neutron to a potentially unstable nucleus induces fission, a process used in nuclear fission reactors**

?

J A nucleus of uranium-235 splits spontaneously into two smaller nuclei and three neutrons. Identify two of the smaller nuclei which could be produced, apart from barium and krypton.

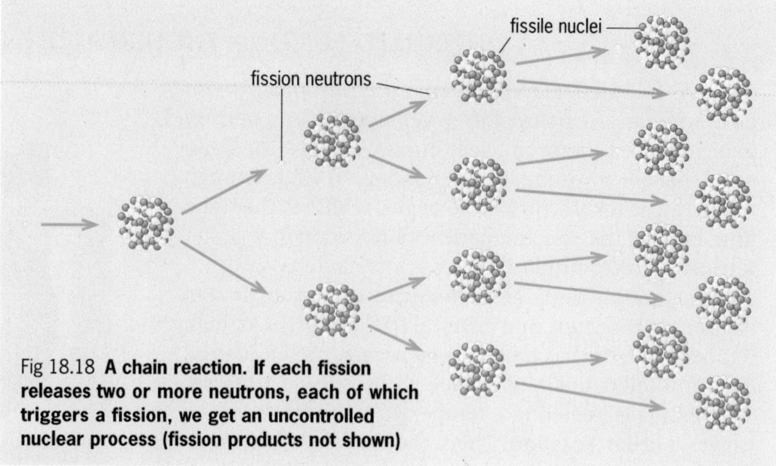

Fig 18.18 **A chain reaction. If each fission releases two or more neutrons, each of which triggers a fission, we get an uncontrolled nuclear process (fission products not shown)**

Critical mass

In a small mass of fissile material – such as uranium enriched with uranium-235 – some natural (spontaneous) fission occurs all the time, producing neutrons. Most of them escape from the mass completely, and so do not cause fission of other nuclei. This is because any lump of matter is mostly empty space, and the chance of a neutron colliding with the tiny nucleus of an atom is very small.

But suppose the mass, and so the size of the fissile material, is increased. There are more neutrons released and there is also an increase in the total 'target area' of nuclei for each neutron. So there will be more *induced* fission, and more energy will be released in the mass – it will get hotter.

Explosion or meltdown?

This is a classic case of **positive feedback**: the more fission, the more neutrons are released, so there is more fission. In a large enough mass of the fissile material, induced fission produces a runaway chain reaction: a very large number of nuclei will disintegrate in a very short time. There is a sudden very large release of energy – an explosion. For a given composition of material, the mass which will occur depends on its shape. But for a sphere, it is called the **critical mass**.

In practice, simply putting lumps of enriched uranium in a container will not have this effect. Before the critical mass is reached, the energy generated will melt (indeed, vaporise) the uranium. This is a 'meltdown', as happened in the worst nuclear reactor accident of all time at Chernobyl in 1986.

?

K The total binding energies of the nuclei in the nuclear reaction described are:

uranium-235	–1736.8 MeV
barium-144	–1161.5 MeV
krypton-90	–754.6 MeV

(a) Show that the energy released in this example of fission is about 180 MeV per event.

(b) One gram of uranium-235 contains 2.5×10^{21} nuclei. Estimate how much energy (in joules) would be released when 1 gram of U-235 is split in a nuclear reactor.

Structure of a nuclear reactor

In a nuclear reactor, energy is generated in **fuel rods** contained in thin non-corrosive tubes inserted into the reactor **core**. Around the core is a gas or water under pressure. The fuel rods contain either plutonium or uranium oxide, made with an 'enriched' mixture of 97 per cent uranium-238 and 3 per cent uranium-235. The core of a typical reactor contains about 80 tonnes of uranium oxide. Only the uranium-235 nuclei are likely to undergo fission, the ultimate energy source in the fuel rod.

Fig 18.19 **Sizewell B power station**

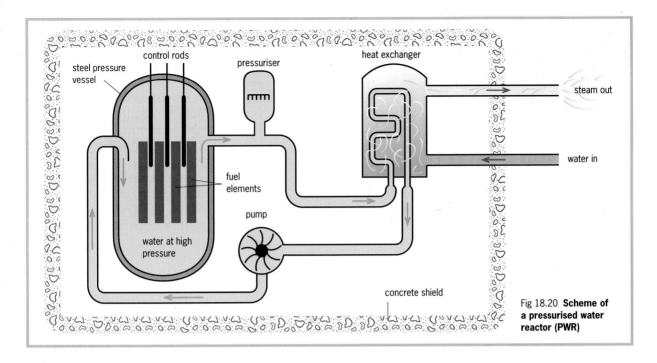

Fig 18.20 **Scheme of a pressurised water reactor (PWR)**

Increasing induced fissions by using slow neutrons

The neutrons from spontaneous random fission of U-235 are emitted at high speeds. Fast neutrons have a much smaller probability of entering a nucleus than slow ones; they tend to bounce off.

To slow down the neutrons, the fuel rods are surrounded by a **moderator.** This used to be graphite, but modern pressurised water reactors use the water which also acts as a coolant. The fast neutrons collide with coolant molecules and gradually lose kinetic energy until they are slow enough to have a much greater chance of causing **induced fission** when they reach the next fuel rod.

Controlling the reactions

Once induced fission starts, too many neutrons could be produced and the reaction could start 'running away'. Therefore, excess neutrons must be removed, so that the fission rate is controlled and energy is released at a steady rate. To do this, cylindrical **control rods** are inserted into the reactor core near the fuel rods. The material of the control rods, for example boron or cadmium, absorbs neutrons.

Using the energy from a nuclear reactor

The fission products are large nuclei which separate at high speed. They collide with other nuclei and their kinetic energy is shared out, so that the reactor core containing the fuel rods, moderator and control rods becomes very hot. Nowadays, the reactor core contains and is surrounded by a fluid, most commonly water under pressure, as in **pressurised water reactors** (PWRs), see Fig 18.20.

The energy is carried from the core by superheated water at a temperature of 325 °C kept from vaporising by a pressure of 150 atmospheres. This water heats a secondary loop of water in a 'heat exchanger', and produces steam to drive a turbine that produces electricity.

The trouble with nuclear power

It seems that nuclear power avoids many of the problems we associate with the most common UK energy source, fossil fuel. Nuclear power does not add to the greenhouse effect, which is mainly due to carbon dioxide from burnt fuel, and its raw material will not run out as quickly as, say, oil reserves. Nor does it produce the gases from burning coal and oil which affect health and cause acid rain.

But nuclear power has its own problems. There is the risk of accidents in which radioactive materials can escape into the environment. Also, even for perfectly managed systems, there are the problems of nuclear waste and of closing down (decommissioning) power stations at the end of their useful life. These are problems because they may expose people (and other forms of life) to harmful ionising radiations, as described below.

L What costs are ignored in the current use of fossil fuels, compared with nuclear fuels?

Nuclear waste

Nuclei of the non-radioactive material in the core may become radioactive isotopes when they absorb neutrons. Also, the spent fuel rods contain the fission products which tend to be radioactive, as well as the unused uranium-238. The uranium can be recycled, but eventually the main components of the reactor core become radioactive waste and have to be taken out.

Radioactive waste includes both short-lived isotopes which are highly active, and isotopes with very long half-lives (such as plutonium-239, half-life 24 000 years, formed from uranium-238). The waste is kept in storage tanks for years to allow the 'high level' wastes to decay to less active or stable isotopes.

Nuclear power station operators would like to store all wastes deep underground but it is difficult to get agreement on where to do this. The other problem is how to decommission a nuclear power station when it becomes obsolete and is no longer economical to run. A large, slightly radioactive mass of building material has to be made safe – and this could involve burying the whole site in concrete.

Dealing with waste is so costly that nuclear power may be no cheaper than power from conventional power stations using fossil fuels. But, of course, the true environmental cost of using fossil fuels is not yet paid. For example, the consequences of acid rain and the more distant and unknown effects of global warming are yet to be seen.

8 HOW WE KNOW ABOUT THE NUCLEUS

The very first nuclear probe – alpha scattering

The first evidence for the existence of the nucleus was the alpha-scattering experiment of Geiger and Marsden in 1909 at the University of Manchester under the leadership of Ernest Rutherford.

At that time, most physicists' idea of an atom was of a round ball completely filled with a positive 'jelly' in which the newly discovered electrons were embedded, like currants in a currant bun. Rutherford told his assistants, the experienced Hans Geiger and a young undergraduate, Ernest Marsden, 'See if you can get some effect of alpha particles directly reflected from a metal surface.' A diagram of the apparatus they used is shown in Fig 18.21. It shows the main features of all such collision or scattering experiments:

- a source of high-speed particles,
- a vacuum to avoid unwanted collisions,
- a target, or more precisely the atomic or subatomic components of the target,
- a detector of the scattered particles or any fragments produced,
- a means of measuring the paths of scattered particles or fragments.

A radioactive source, contained in a lead box with a small hole in it, emitted alpha particles.

The targets included thin gold foil only a few atoms thick. Most alpha particles went straight through the foil, and were detected by the faint glow (scintillation) that each particle produced when it hit a detecting glass screen coated with a phosphorescent chemical. The scintillations were so faint that Geiger and Marsden had to allow half an hour before each observation session for their eyes to become 'dark adapted'.

The detecting screen was fitted to a pivoted arm moved along a scale marked in degrees. About 1 alpha particle in 8000 was reflected back through a large angle. A typical distribution of particles by angle is shown in the graph of Fig 18.22.

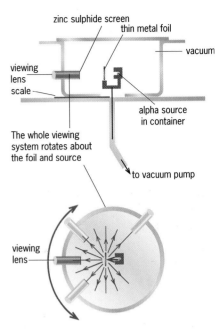

Fig 18.21 **Geiger and Marsden probe the atom – and discover the nucleus**

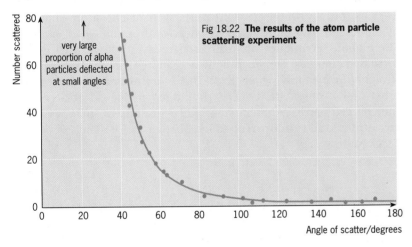

Fig 18.22 **The results of the atom particle scattering experiment**

An alpha particle is deflected by repulsion from another positively charged object. The 'positive jelly' model would have provided a repulsion force far too small to deflect the massive alpha particle through a large angle – it predicted a deflection of 2° at most. Rutherford deduced that there must be a very much stronger electric field inside an atom, which could only be produced by a very high charge density, such as if all the charge were locked into one small volume.

For the fast and quite massive alpha particle to bounce back at all, most of the mass of the atom would also have to be squashed into this small space. Rutherford named this small space, in which all the positive charge and most of the mass of an atom is concentrated, the atomic **nucleus**. By comparison, the electrons had negligible mass but occupied most of the space.

Fig 18.23 illustrates Rutherford's idea of what was happening in the gold foil experiment. Alpha particle **A** passing some distance from the nucleus is hardly deflected at all – the electric field is too

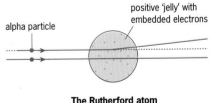

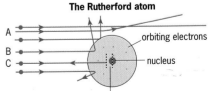

Fig 18.23 **Alpha scattering: Rutherford saw that the only model that would produce large scattering angles required a massive, positive nucleus**

EXAMPLE

Q An alpha particle with kinetic energy 5 MeV makes a head-on collision with a gold nucleus. How close to the centre of the nucleus does it get?

The charge Q on a gold nucleus is $79e$, where e is the electronic charge 1.6×10^{-19} C, an alpha particle has charge $q = 2e$, the electric force constant k ($1/4\pi\varepsilon_0$) is 9×10^9 N m^2 C^{-2}.

A The alpha particle stops at a distance r from the centre of the nucleus, when all its kinetic energy has been converted to electrical potential energy in the field of the gold nucleus. So:

$$E_K = \frac{1}{4\pi\varepsilon_0} \frac{Qq}{r}$$

Rearranging gives:

$$r = \frac{1}{4\pi\varepsilon_0} \frac{Qq}{E_K}$$

so, inserting values:

$$r = 9 \times 10^9 \times \frac{79 \times 1.6 \times 10^{-19} \times 2 \times 1.6 \times 10^{-19}}{5 \times 10^6 \times 1.6 \times 10^{-19}}$$

giving:

$$r = 4.6 \times 10^{-14} \text{ m}$$

small. A closer approach such as B's produces a larger deflection. A (rare) head-on 'collision', **C**, stops the alpha particle dead in its tracks and repels it back along the path it came.

Rutherford wasn't very keen on mathematics, and it took him nearly two years to produce a calculation which related his nuclear model of the atom to the distribution of the number of deflected particles and the angles they were deflected through.

See questions 10 and 11.

Particle accelerators

Early experiments used naturally energetic particles to probe the nucleus, but by the 1920s physicists were using 'atom-smashing machines' to accelerate charged particles in a controlled manner. Their main techniques are still used today, though with greatly increased energies. One technique accelerates charged particles in a straight line by applying an electric field – a **linear accelerator**. The other uses a combination of electric and magnetic fields to accelerate the charged particles in a circular path – leading to such devices as 'cyclotrons' and **synchrotrons.** See Chapter 26 for details of particle accelerators.

See question 12.

9 UNSTABLE NUCLEI AND RADIOACTIVITY

Nuclei are **radioactive** when there is a chance that changes will take place in the nucleus that result in the emission of **ionising radiation**. The radiation is energetic enough to knock electrons out of other atoms and molecules to produce charged particles, **ions**, meaning that it **ionises** them.

The discovery of radioactivity

Radioactivity was discovered in 1896 by Henri Becquerel (France, 1852–1908). He was investigating the property that some natural minerals have of glowing in the dark – fluorescence. He detected the faint light emitted by fluorescent substances using photographic plates. One day he found (by accident!) that a salt of uranium emitted *invisible* radiation which blackened the photographic plates.

Soon, three types of radiation were discovered, and because they were then unidentified they were simply labelled **alpha** (α), **beta** (β) and **gamma** (γ) rays, after the first three letters of the Greek alphabet. Table 18.5 lists the nature and properties of the radiations.

Radiation	Alpha	Beta	Gamma
Nature	helium nucleus (2n + 2p)	electron	photon of electromagnetic radiation
Symbol	He	e	γ
Charge	+2e	−e or +e	none
Range in air	1–5 cm	10–100 cm	infinite – obeys the inverse square law
Range in matter	stopped by eg a sheet of paper	stopped by eg a thin (1 mm) sheet of aluminium	intensity reduced to half by eg 1 or 2 cm of lead
Mass in atomic mass units	4 u	5×10^{-4} u	zero
Ionising ability	very high 1	high 1/100	low 1/10000

Table 18.5 **Properties of ionising radiations**

Marie Curie (Poland/France, 1867–1934) and her husband Pierre (France, 1859–1906) established that the source of radiation in their experimants was the element uranium. They discovered other elements that also emitted energetic radiation, such as the new elements polonium and radium, by detecting their radioactive properties. Later, the physicist Ernest Rutherford (New Zealand/Britain, 1871–1937) and the chemist Frederick Soddy (Britain, 1877–1956) worked together to establish the nature of the radiations and track down the changes to the elements that were linked to them. Their results began to make sense when Rutherford's team discovered the nucleus in the years 1909–1911.

Detecting the radiations

Not surprisingly, the instruments used to detect the ionising radiations from radioactive materials do so by making use of the ionisation process.

The earliest, simplest and cheapest is the **leaf electroscope**, shown in Fig 18.24. When charged, the leaf (usually made of gold) is repelled by the like charges on the central rod and is deflected outwards. The metal case is earthed. Ionising radiation that passes from a nearby source through the electroscope case will ionise molecules in the air. The rod and leaf attract ions of opposite charge to themselves, and eventually all their charge is neutralised. As this happens, the leaf angle decreases. The rate of decrease is proportional to the rate of ionisation produced by the radiation.

The cloud chamber

The path of the ionising radiations was first made visible by the **cloud chamber.** Fig 18.25 shows the tracks of alpha particles in a cloud chamber that contains damp, cold air. The track is a small cloud of tiny water droplets that form around the ions produced from air atoms by the radiation. Water drops will form only when there is some kind of small particle or ion (called a 'condensation nucleus') to start the condensation process. You can see similar effects in the condensation trails formed at the wing tips or propeller tips of high-flying aircraft when the air is cold and damp enough.

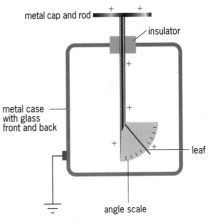

Fig 18.24 **A leaf electroscope**

Fig 18.25 **Alpha particle tracks in a cloud chamber**

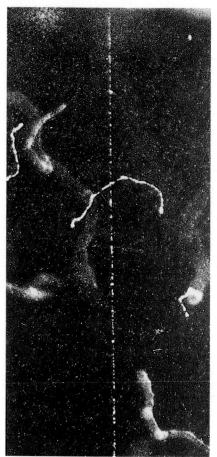

Fig 18.26 **Travelling up the centre is the track of a fast moving beta ray emitted during radioactive decay. The squiggly tracks are of less energetic electrons knocked from atoms by X-rays**

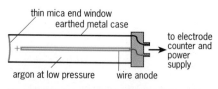

Fig 18.28(a) **An end-window Geiger–Müller tube for the detection of alpha particles**

Fig 18.28(b) **A Geiger counter: its scale shows the count for the luminous watch**

Fig 18.26 shows some tracks made by beta particles, which have a longer range in air than alpha particles. They also produce a much thinner track, showing that they are less effective at causing ionisation. These two effects are linked: the more ionisation produced, the greater is the energy taken from the ionising particle. Alpha particles are much more massive than beta particles (helium nuclei compared with electrons) and also carry double the charge. So alpha particles interact with more molecules and lose more energy per centimetre of path than beta particles.

Fig 18.27 shows a small cloud chamber used in schools. It works continuously as long as the base is kept cold enough for the vapour (water or ethanol) in the chamber to be supercooled. Cloud chambers are now obsolete for serious research into radioactivity or energetic subatomic particles. They were first overtaken by **bubble chambers** (now also obsolete) and then a variety of **particle detectors** using thin metal plates or wires. These are described in Chapter 26.

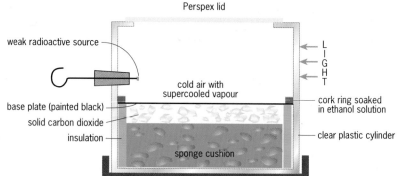

Fig 18.27 **A simple 'continuous' cloud chamber. Ionising radiation creates condensation nuclei, so producing visible trails of the paths of the radiation. This works best for alpha particles**

The Geiger counter

The most common device – and one of the oldest – for detecting and monitoring ionising radiation is the **Geiger counter**. The key part of the counter is the **Geiger–Müller tube**, the GM tube, as in Fig 18.28(a). A sealed metal tube has a thin wire down the middle and is filled with a non-reactive gas (eg argon) at low pressure. The wire is kept at a voltage of about +400 V with respect to the metal case, which is earthed. The radiation enters the tube through a thin window in the front. Alpha particles are easily absorbed by matter so the window in an alpha detector is a thin film of mica – but strong enough to withstand the pressure difference between the atmosphere and the gas inside the tube.

Radiation entering the tube ionises argon atoms. The positive argon ions are attracted to the metal case and electrons to the central wire. These charged particles are accelerated by the voltage and will collide with other argon atoms. If they are moving fast enough, the collisions will ionise the atoms. The new particles are in turn accelerated and will cause further ionisation, producing an avalanche effect. The sudden large flow of charge in the tube causes a pulse of current in the central wire which is amplified electronically. The amplified pulses may be counted by a **scaler** or averaged out to give a current reading on a **ratemeter**. The pulses may also be fed into an audio circuit to give a series of clicks, so providing an audible indication of the radiation level.

The charge avalanche takes time to be cleared out of the gas, so there is a **dead time** (of about 200 ms) in which the tube cannot react to any incoming radiation.

Beta and gamma detectors are similar: windows for beta detectors can be made of glass, and gamma detectors can be made entirely of metal.

Working voltage of a GM tube

The voltage on the central rod has to be set high enough to accelerate ions and electrons and cause the avalanche effect. But the voltage should not be so high as to produce a continuous discharge, as at B in Fig 18.29. This happens when a continuous supply of ions hitting the metal walls have enough energy to eject electrons from it, which are then accelerated to cause further ionisation of the argon atoms.

Fig 18.29 is a plot of the count rate, which is proportional to the mean current in a GM tube, against voltage when the ionising radiation enters the tube at a constant rate. The **operating plateau** at A shows the range of applied voltage for which the tube works well. At plateau voltages, all ionising particles of the same type will produce the same size pulse of charge.

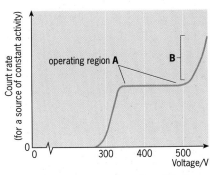

Fig 18.29 **Characteristic curve for a GM tube**

Solid state detectors

In industry and research laboratories, the most common kind of radiation detector is 'solid state'. The energy of the incoming radiation (particle or photon) liberates electrons from semiconductor material. This produces a pulse of current and so indicates the arrival of the radiation. Such detectors can both count individual events and also measure the energy of the radiation.

Photographic detection

Photography works because photons of light have an ionising effect on certain chemicals – generally salts of silver such as silver nitrate. **Radiation badges**, described in Chapter 24, use this effect. Such a badge has to be worn by anyone who works in an area where they might be exposed to ionising radiation, such as a nuclear power plant, an X-ray department in a hospital, or a radioactivity laboratory.

Background radiation

A Geiger counter set up anywhere on Earth will always register a count. This count is due to **background radiation,** produced by tiny fragments of radioactive elements present in all rocks and soil, the atmosphere and even in living material itself. Also, the Earth is continuously bombarded by high-speed particles from outer space and from the Sun (called **cosmic rays**). Cosmic rays smash up the nuclei of atmospheric molecules and produce other high-speed particles which can cause ionisation. Many of these reach ground level.

The background count varies considerably from place to place on Earth. As detected by a Geiger counter, it is typically one or two counts a second. It is a lot more where there are igneous rocks such as granite, which are relatively rich in minerals containing natural radioactive elements. Also, the radioactive decay of these elements produces a radioactive gas, radon, which may accumulate in buildings, so increasing the local background count.

See question 13.

M A GM tube has a detection area of about 3 cm². It shows an average background count of 1.5 per second in your home. Estimate how many ionising particles are likely to be hitting your body per second.

10 ACTIVITY AND HALF-LIFE

The **activity** of a radioactive source is the number of ionising particles it emits per second. Each emission corresponds to a change in the nucleus of one atom and is also called the **decay rate**. The SI unit of activity (and decay rate) is the **becquerel** (Bq):

$$1 \text{ Bq} = 1 \text{ decay per second}$$

See question 14.

What we actually record is the reading on any radiation detector we use to monitor the effect of the radiation. Usually this is the **count rate** as measured by a Geiger counter. The count rate is simply the number of counts recorded per second, and in simple experiments it is taken to be directly proportional to the activity. Note that it is very unlikely that all the decays of a source will be recorded. Usually, some of the radiation is trapped inside the source, and in most measurements only a small fraction of the radiations actually

See question 15.

enter the detector to be counted.

As time goes on, the activity of a source (and so the count rate) decrease in a consistent manner. Fig 18.30 shows a typical plot of the alpha particle count rate against time for a small sample of the

See question 16.

radioactive gas radon ($^{220}_{86}$Rn).

See question 17.

Fig 18.30 **Activity curve for radon-220**

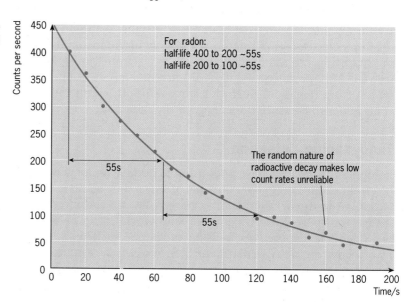

For radon:
half-life 400 to 200 ~55s
half-life 200 to 100 ~55s

55s

55s

The random nature of radioactive decay makes low count rates unreliable

The graph has been drawn as a 'best fit' through actual numbers recorded by a counter. The count rate in counts per second is calculated from the total number of counts measured at 10-second intervals.

There are two significant features of the graph. One is that the values aren't all on the line, so the recording has some inbuilt uncertainty. Secondly, see the red and blue sets of lines. They tell us that the time taken for the count rate to fall from 400 s⁻¹ to 200 s⁻¹ is the same as the time for the count rate to fall from 200 s⁻¹ to

See question 18.

100 s⁻¹. That is, the graph suggests that:

**The time taken for the activity of a radioactive sample
to decrease to a half of any starting value is the same,
whatever the starting value of the count rate.**

This time is called the **half-life** of the radioactive substance. We return to this when looking at the shape of the curve.

Randomness of decay

Both these features of radioactive decay have the same cause: *The time at which any given radioactive nucleus emits its radiation cannot be predicted*. It is impossible to tell which one of a hundred radon nuclei will be next to emit its alpha particle. All the nuclei are identical, and all are equally liable to decay. The breakdowns are completely *random*. But if we have a sample of several hundred billion radon nuclei, a pattern would emerge. Bear in mind that ten billion (10^{10}) radon nuclei have a mass of a few thousandths of a billionth of a gram (about 4×10^{-12} g).

What is constant about the radioactive process is that each nucleus has a **definite probability** of undergoing decay. Starting with a large enough number of radon nuclei, we can be reasonably certain that a predictable number will emit an alpha particle in the next ten seconds. Similarly, with a large enough sample, the values of count rate we measure will coincide closely with the smooth graph we draw between the values. Our estimate of half-life will be more reliable.

N The Avogadro constant is 6×10^{23} per mole. Radon has a relative atomic mass of 220. Show that 10^{10} radon atoms have a mass of about 4×10^{-12} g.

The shape of the half-life curve

We associate an **exponential change** with any system where the change is proportional to the quantity that is changing, either increasing or decreasing. An example of an *increase* is a biological population growth, where the number of young born is proportional to the number of organisms in the population.

The graph plotted in Fig 18.30 is a **decay curve** for an exponential change to a radioactive material, and here the *decrease* in the population of nuclei is exponential: the decay rate *decreases by equal fractions in equal times*. The fraction that we use for a radioactive material is a half, so we refer to its *half-life* (page 396). This property is characteristic of many physical changes: on page 221 we saw it for the charging and discharging of capacitors.

Where ΔN is the number of nuclei that decay in a small time Δt, then:

$$\Delta N = -kN\Delta t \qquad [1]$$

The minus sign is because ΔN is a *reduction* in N: as time goes on N decreases. The **decay constant** k is a measure of the *probability of a nucleus decaying in the following second*. (Each radioactive nucleus has its own value of k.)

The Extension box on page 398 gives the calculus version of deriving the exponential form of radioactive decay.

The Assignment to this chapter asks you to use a spreadsheet program to explore this relationship and show that it leads to the half-life property.

The decay constant and half-life

For a radioactive isotope, the decay constant k indicates the probability that a nucleus will undergo decay. Its value is in terms of the *fraction* of the potentially active nuclei in a sample that do actually decay in any second of time. So, the larger the value of k, the lighter the number of nuclei decaying in a given time – and the shorter the half-life. The Extension box shows that the relation between half-life and the decay constant is:

$$\text{half-life (in seconds)} = \frac{\ln 2}{k} \qquad [2]$$

where ln 2 is the logarithm to base e of 2 (0.693).

O The half-life of radon-220 is 55.5 s. Use equation 2 to show that the decay constant for radon-220 is 0.0125 s^{-1}.

Random decay and the exponential rule

Equation 1 on page 397 can be written in calculus notation as:

$$\frac{dN}{dt} = -kN$$

Rearranging this gives: $\frac{dN}{N} = -kdt$

which can be integrated to give:

$$\ln N = -kt + A \qquad [3]$$

where A is a constant and ln is the logarithm to base e. Assume that we start with a number N_0 of radioactive nuclei. When $t = 0$, equation 3 gives:

$$\ln N_0 = A$$

so that we can rewrite equation 3 as:

$$\ln N = -kt + \ln N_0 \qquad [4]$$

or: $\qquad kt = \ln N_0 - \ln N$

$$kt = \ln \frac{N_0}{N} \qquad [5]$$

In equation 5, the number of nuclei has changed from N_0 to N in time t.

Equation 5 gives rise to a simple connection between half-life and decay constant k: if $N = N_0$ when $t = 0$, then after a time $t_{\frac{1}{2}}$ (the half-life), the number has decreased to $N_0/2$. So inserting values $t = t_{\frac{1}{2}}$ and $N = N_0/2$ in equation 5, we get:

$$kt_{\frac{1}{2}} = \ln \frac{N_0}{0.5N_0} = \ln 2$$

which can be rearranged to give the half-life:

$$t_{\frac{1}{2}} = \ln 2/k \qquad [6]$$

Also, equation 4 can be rewritten as:

$$\ln N - \ln N_0 = -kt$$

or: $\qquad \ln(N/N_0) = -kt \qquad [7]$

and by converting logarithms to exponentials:

$$N/N_0 = e^{-kt} \quad \text{or} \quad N = N_0 e^{-kt}$$

showing the *exponential* form of the relationship.

Fig 18.31 shows a logarithmic plot of a decay curve: $\ln N$ is plotted against time. As indicated in equation 7, the intercept on the $\ln N$ axis gives $\ln N_0$, and the slope of the line is $-k$.

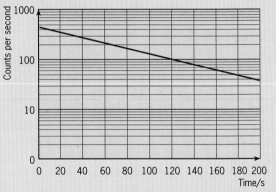

Fig 18.31 **Log graph for the decay curve for radon-22 gas**

Nuclides

The word **nuclide** is the name for atoms which have identical nuclei; that means they have the same proton number (atomic number) and the same mass number, so they have the same number of neutrons. Compare this with 'isotopes' (see page 379), atoms whose nuclei have the same proton number but *different* mass numbers (because they have different numbers of neutrons).

Table 18.6 gives some values for the half-lives of selected nuclides.

Table 18.6 **Half-lives and types of emission of some nuclides**

Nuclide	Comment	Emission	Half-life
uranium-238	Most common isotope of uranium	α, γ	4.51×10^9 y
thorium-232	100% of normal thorium found on Earth	α, γ	1.4×10^{10} y
radon-220	Radioactive gas – a decay product formed when thorium-232 decays	α	55.5 s
plutonium-239	Found naturally but in minute quantities; made artificially by bombarding uranium with neutrons. Fissile: used in nuclear reactors and nuclear bombs. The most dangerous component in nuclear waste: highly toxic and difficult to store, with a long half-life	α	24 360 y
americium	Formed by bombarding plutonium with neutrons; often a laboratory source of alpha particles	α, γ	433 y
carbon-14	Formed from nitrogen-14 by the action of cosmic rays; useful for radioactive dating of organic materials	β⁻	5730 y
potassium-40	Naturally occurring isotope (0.12% abundance), a significant source of internal (body) background radiation	β⁻	1.3×10^9 y
radium-226	The first element to be discovered due to its radioactivity (Pierre and Marie Curie, 1898)	α, γ	1622 y

EXAMPLE

Q The half-life of bismuth ($^{212}_{83}$Bi) is 60.6 minutes. What is **a)** its decay constant, **b)** the activity of 1 g of bismuth-213?

The Avogadro constant is 6.02×10^{23}.

A **a)** Half-life of bismuth = $(\ln 2)/k$, so:

decay constant $k = 0.693/(60.6 \times 60) = 1.91 \times 10^{-4}$ s^{-1}

b) 212 g of bismuth-212 contains 6.02×10^{23} atoms. Therefore number of atoms in 1 g:

$$N = (6.02 \times 10^{23})/212 = 2.84 \times 10^{21}$$

The probability of decay for bismuth-212 is k per second. Thus we would expect kN atoms to decay per second:

activity $kN = 1.91 \times 10^{-4} \times 2.84 \times 10^{21}$ Bq

$= 5.4 \times 10^{17}$ Bq

▮ See question 19.

11 HOW NUCLEI CHANGE WHEN THEY DECAY

When a nucleus emits an alpha particle it loses two neutrons and two protons. Its mass number decreases by 4 and its proton (atomic) number by 2. This means that it has become a *different element*. For example, thorium-232 emits an alpha particle and becomes an isotope of radium:

$$^{232}_{90}\text{Th} \rightarrow {}^{228}_{88}\text{Ra} + {}^{4}_{2}\alpha$$

The radium-228 then emits a negative beta particle (an electron), leaving its mass unchanged since the electron has negligible mass compared with a nucleon. What has happened is that a neutron has decayed and emitted an electron, so changing to a proton. The proton number increases by one to 89. This means that the new nucleus has approximately the same mass but becomes a new element – actinium:

$$^{228}_{88}\text{Ra} \rightarrow {}^{228}_{89}\text{Ac} + {}^{0}_{-1}\beta$$

We have represented these changes as two nuclear equations. Note that they are doubly balanced: both the mass numbers and the proton numbers have the same totals on each side of the equation.

Many of the natural radioactive nuclides on Earth are the result of similar alpha and beta emissions from larger nuclei that were formed during the final explosion of a dying star, billions of years ago. This event formed some nuclides with such long half-lives that they still exist. Examples, and their half-lives, are:

thorium-232: 1.41×10^{10} y
uranium-238: 4.51×10^{9} y
uranium-235: 7.1×10^{8} y

The decay of these nuclides and their radio-active 'daughter' products (Fig 18.32) provide a large amount of energy which helps to keep the interior of the Earth hot enough to be molten (see Chapter 23).

P Explain why:

(a) the emission of an alpha particle from thorium-232 results in a nucleus of radium-228,

(b) the emission of a beta particle (β⁻) then changes the radium into an isotope of actinium.

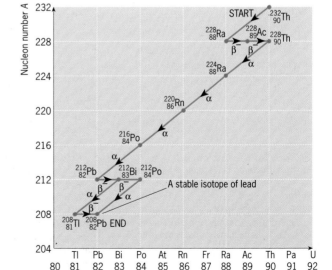

Fig 18.32 **The decay series of thorium**

Gamma emission

Gamma rays are photons of very short-wave electromagnetic radiation. Their wavelength is about 10^{-13} m, about a thousandth of the wavelength of X-rays. They have a photon energy some ten million times that of a photon of light.

The following is a simple model to explain why gamma rays are emitted. A nucleus that has emitted an alpha particle or a beta particle tends to have its nucleons arranged in an 'excited' energy state higher than its normal 'ground' state. In emitting the gamma ray, it reaches that ground state (Fig 18.33).

This is similar to the model used to explain why low-energy photons are emitted when electrons in an atom are in an excited state and move between energy levels so that the atom reaches its ground state. The energy of a gamma photon from the nucleus is so much greater than a photon emitted by an excited atom because the nuclear forces between nucleons are far greater than the electromagnetic forces between the nucleus and the electrons.

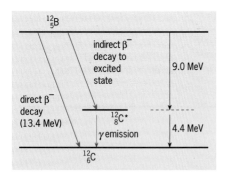

Fig 18.33 **Gamma decay occurs when a nucleus is created with its nucleus in an excited state. This nucleus is $^{12}_{6}C^{*}$**

Nuclear stability – or what causes radioactive decay?

This is not an easy question to answer! The two forces in the nucleus oppose each other. One is the **electric repulsion** force between protons. The other is the short-range **attractive** force, the **strong nuclear** force, which acts on both protons and neutrons. For a nucleus to be stable, the net attractive force for any nucleon must be greater than the net repulsive force. The puzzling thing is that when this is not the case, the resulting unstable nucleus does not break up immediately. But uncertainties and probabilities are a feature of the quantum nature of matter and forces.

Fig 18.34 shows the nuclear structure of the lightest nuclei. Note that some of them are unstable.

The neutron adds to nuclear stability by providing an attraction force without adding to the repulsion force. So, as Fig 18.35 shows, when the proton number of stable nuclei increases, we get even more neutrons.

Artificial nuclides can be made by adding neutrons to the nucleus, but most artificial nuclides are radioactive, often with half-lives measured in seconds. It seems that having too many neutrons in a nucleus is as perilous as having too few – in both cases the nucleus tends to become unstable.

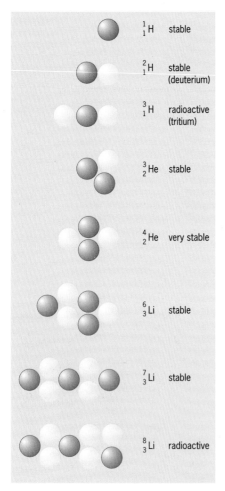

Fig 18.34 **Nuclear composition of light elements**

Fig 18.35 **A graph of nucleon number plotted against proton number for beryllium to neon. Each vertical row contains the isotopes of the same element**

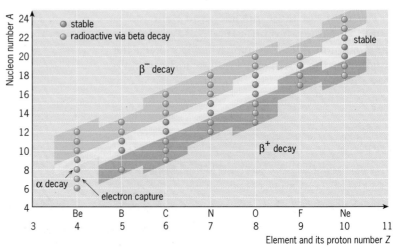

Beta decay

In nearly all nuclear decays, a beta particle is emitted. In some decays, the beta particle is an electron, β⁻. But other unstable nuclei undergo what is called **positive beta decay**, emitting a positive beta particle, β⁺, which is a positive electron or **positron**.

To summarise: when there are too many neutrons, a neutron turns into a proton by β⁻ decay; too few neutrons, and a proton turns into a neutron by β⁺ decay:

$$\beta^- \text{ decay:} \quad {}^{14}_{6}\text{C} \rightarrow {}^{14}_{7}\text{N} + \beta^-$$
$$\beta^+ \text{ decay:} \quad {}^{15}_{8}\text{O} \rightarrow {}^{15}_{7}\text{N} + \beta^+$$

There must be a delicate balance between the numbers of protons and neutrons in a stable nucleus. Look back to see Fig 18.7 (page 379) which shows the neutron–proton ratio for the most common isotope of each of the stable elements. After a rather jagged start, the ratio increases slowly with increasing nuclear size. For the heaviest elements, the nucleus is so large that another kind of instability sets in – a tendency to split into two smaller ones, and nuclear fission occurs (see page 381).

A new force in nature – the weak force

To explain the tendency for nuclei with a proton–neutron ratio outside the stability zone to become radioactive by beta decay, nuclear physicists proposed another natural force, the **weak nuclear force**. This fourth force of nature completed a quartet of fundamental forces, adding to the gravitational, electromagnetic and the strong nuclear force. There is more on this in Chapter 26.

Alpha decay

We can think of a particle in a nucleus as being at the bottom of a potential energy well. This is similar to picturing a mass on the Earth's surface at the bottom of the Earth's 'gravitational well', see Fig 18.38. In both cases, there has to be some energy input for the object to reach an 'escape speed' and get away from the well.

An alpha particle is identical to the particularly stable helium nucleus, and alpha decay tends to occur with the larger unstable nuclei. But how does the alpha particle 'escape' from the well? Fig 18.36(a) suggests that it needs to be given kinetic energy to get

Fig 18.36 **Potential energy wells for a nucleus and the Earth. On classical theory, the alpha particle does not have enough kinetic energy to escape**

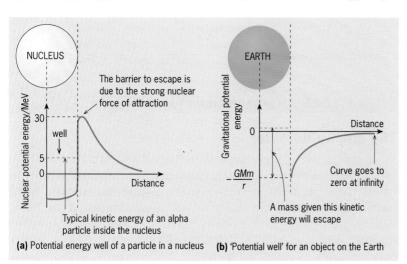

(a) Potential energy well of a particle in a nucleus   (b) 'Potential well' for an object on the Earth

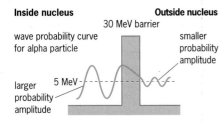

Fig 18.37 **The quantum explanation for alpha decay: Tunnelling occurs, so it must be possible for an alpha particle with a kinetic energy of only 5 MeV to exist outside the nucleus. This is not a *certainty*, hence radioactive decay is random**

✓

See 17 Atoms, spectra and quanta for more about quantum physics and probability waves.

Table 18.7 **Forces and emissions**

Process	Dominant force
α decay	strong nuclear
β decay	weak nuclear
γ decay	electromagnetic

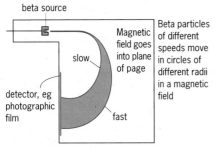

Fig 18.38 **Experiments on the deflection of beta particles from a pure source show that they are emitted with a range of speeds**

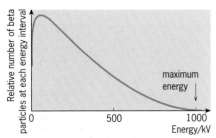

Fig 18.39 **The beta particle spectrum of bismuth-210. The neutrino was 'invented' to explain why so many particles had less than the maximum energy**

Fig 18.40 **Every household should have a smoke detector**

away from the other particles in the nucleus. It does have some kinetic energy – the nuclear particles are imagined as joggling about inside the nucleus like gerbils in a bag – but not enough to allow a particle to escape.

To explain alpha emission we need to use the quantum theory of matter. This says that, like a gerbil that eats through the side of the bag, an alpha particle could 'tunnel' its way out. A quantum view of the alpha particle is shown in Fig 18.37.

In this theory, the alpha particle is represented by a probability wave of a size that fits into the nucleus. But, as Fig 18.37 shows, the wave doesn't end *inside* the nuclear potential energy barrier; there is a small but finite probability that the alpha particle is outside the nucleus, where the electric repulsion force alone acts and gives the alpha particle its kinetic energy as it accelerates away.

To sum up, the forces involved in radioactive decay are linked to the emissions as shown in Table 18.7.

Energy conservation and beta decay – the neutrino

In radioactive decay, both alpha particles and gamma radiation are emitted from a given nuclide with a definite energy which is characteristic of the nuclide. This is not a property of beta particles. Measurements show a range of energies for the beta particles emitted from a given radioactive substance (Figs 18.38 and 39).

This spread of energies posed a serious problem to nuclear physicists. Why did some emerge with less energy than others? It seemed to contradict the principle of the conservation of energy. To avoid this appalling prospect, Wolfgang Pauli suggested in 1930 that the missing energy was being carried away by a *new* subatomic particle. The particle had to have a very small mass, and, because no one had yet detected such an energetic particle, it must carry no charge. This particle has been named the 'little neutron', or **neutrino** (symbol ν, the Greek letter 'nu'). We now know that there is more than one kind of neutrino, each with a mass very much smaller than the mass of an electron – possibly zero. Thus the beta reactions on page 401 should be written as:

$$^{14}_{6}C \rightarrow \, ^{14}_{7}N + \beta^- + \bar{\nu}$$

$$^{15}_{8}O \rightarrow \, ^{15}_{7}N + \beta^+ + \nu$$

In these reactions ν is the 'ordinary' neutrino, and $\bar{\nu}$ its antiparticle, called the **antineutrino**. When a neutron decays to form a proton and a negative beta particle, the antineutrino is involved:

$$^{1}_{0}n \rightarrow \, ^{1}_{1}p + \beta^- + \bar{\nu}$$

Particles and antiparticles are discussed fully in Chapter 26.

13 USES AND DANGERS OF RADIOACTIVITY

The smoke detector

Most radioactive sources used in industrial and other applications are radioactive isotopes of stable elements that are manufactured by using the stream of neutrons from a nuclear reactor. Being uncharged, neutrons can penetrate the atomic nucleus quite easily. When an ordinary stable nucleus gains extra neutrons, it becomes unstable with a probability of decaying by either negative or positive beta emission, as explained above.

Most applications of radioactivity are in industry or medicine, but the most common is found in almost every home – the **smoke detector** (Fig 18.41). It uses a small amount of the nuclide americium-241 ($^{241}_{95}\text{Am}$) with a half-life of 433 years. There are tiny amounts of it in mineral ores of uranium, but commercially it is produced by irradiating plutonium with neutrons in a nuclear reactor.

As a nuclide with a comparatively massive nucleus, americium-241 decays by the alpha process. In the smoke detector, alpha particles in a steady stream collide with molecules in the air and ionise them. The ions move towards the electrodes, maintaining a small continuous current. Smoke particles from a fire are much bigger than air molecules and are more likely to intercept the alpha particles, absorbing them and so decreasing the ionisation current. When the current drops, an alarm is triggered via an electronic amplification system.

See Chapter 24 for medical uses of radioactivity.

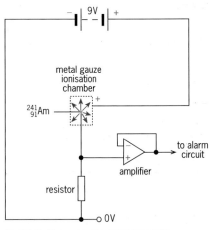

Fig 18.41 **How a smoke detector works**

CARBON DATING

THE CARBON DIOXIDE in the atmosphere is used by plants to build their main structures: roots, stems, trunks, branches, leaves and seeds. Atmospheric carbon is a mixture of the main isotope carbon-12 (98.89%), carbon-13 (1.11%); also tiny traces of carbon-14 which is radioactive with a half-life of 5730 years. It is being formed continuously as high energy cosmic rays from space collide with atmospheric nuclei. They break up, producing neutrons and other fragments. The neutrons then enter nitrogen nuclei which are disturbed enough to lose a proton and so form carbon-14. The process is:

$$^{14}_{7}\text{N} + ^{1}_{0}\text{n} \rightarrow ^{14}_{6}\text{C} + ^{1}_{1}\text{p}$$

Carbon-14 eventually combines with oxygen to form carbon dioxide, CO_2. Both $^{12}CO_2$ and $^{14}CO_2$ in the atmosphere are taken up by plants. Therefore, carbohydrates in all plants contain radioactive carbon in the same ratio as carbon-14 to carbon-12 in the atmosphere at any time. So do carbohydrates in animals and other organisms whose tisues contain carbon derived from plants.

When a plant dies, its store of carbon-14 stops being recycled, so that as radioactive decay occurs the proportion of carbon-14 decreases. Thus the ratio of carbon-14 to carbon-12 in a piece of wood, say, many hundreds of years old, will give a measure of the time that has elapsed since that piece of wood died.

A modern piece of wood has an activity per kilogram (its specific activity) due to carbon-14 of $250\ \text{s}^{-1}$. Say the specific activity of wood from an old building is $125\ \text{s}^{-1}\ \text{kg}^{-1}$, then the wood must be one half-life old: 5730 years. More generally, if the zero age specimen has a specific activity A_0 and the specimen of age t has a specific activity A then $A = A_0 e^{-kt}$ where k is the decay constant of carbon-14.

As an example, suppose A was measured to be $50\ \text{s}^{-1}\ \text{kg}^{-1}$. Then we can say:

$$50 = 250 \times e^{-kt} = \frac{250}{e^{kt}}$$

or: $e^{kt} = 5$

giving: $kt = \ln 5 = 1.61$

A half-life of 5730 years means a decay constant k of $\ln 2/5730$ per year, so the age of the specimen is:

$$t = \frac{1.61}{k} = 5730 \times \frac{1.61}{0.693} = 13\ 300 \text{ years}$$

Carbon dating is not an exact method. The energy of the beta particles emitted is low and some would get absorbed inside the material, so small samples are vaporised to avoid this. But with such small samples, the error in counting these random events is significant, especially with older materials of low activity.

The method also assumes that cosmic ray activity has been constant over many thousands of years, though it has not. To correct for this, the radioactivity in the wood of tree rings of living trees (bristlecone pines in California) many thousands of years old, has been used to calibrate radio carbon dates much more accurately. Now, the accuracy of dating is about ±200 years in the range 600 to 10 000 years.

Fig 18.42 **Examples of the bristlecone pine tree, famous worldwide for its use in setting the scale for radiocarbon dating**

Fig 18.43 **A carbon-containing sample is burnt and the amounts of carbon-12 and carbon-14 it produces are analysed, with the results displayed on screen**

Carbon dating is the best-known dating technique. It has revolutionised the work of archaeologists, geologists and planetologists. At the same time, they are also using longer-lived radioactive nuclides to date materials over far greater time spans. For more, see Chapter 27.

■ See question 21.

The dangers of ionising radiations

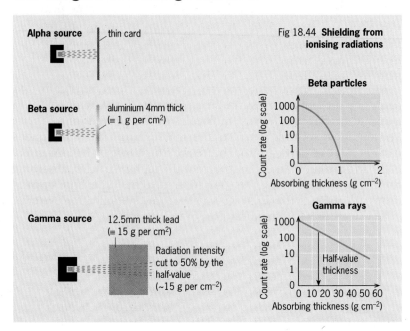

Fig 18.44 **Shielding from ionising radiations**

Living cells contain complex molecules which are easily damaged by ionising radiation. The key molecules likely to be damaged are DNA and RNA. They are involved in the production of proteins, which include the enzymes that control the synthesis of all biochemicals needed for the growth and reproduction of cells.

Alpha radiation is heavily ionising and can be most damaging to cells, so alpha emitters are dangerous if they enter the body. But alpha radiation is easily absorbed: its range in air is only a few centimetres and a sheet of paper is enough to act as a shield (Fig 18.44).

See question 22. ■

Beta particles are less ionising but have a longer range. So they can penetrate deeper into the body from external sources. Anyone working with beta emitters must wear protective clothing.

Gamma radiation is the least ionising but far more penetrating, so is considered the most dangerous.

Measuring the biological effect

The variety of radiations and their differing penetration and ionising effects means that simply measuring the activity of a source doesn't tell us much about the effect the emissions may have on the body. What matters is a combination of the energy released in the body by the ionising radiation and the relative ionising effect on living tissue.

The energy absorbed in the tissue is measured as the **dose,** in units of joule per kilogram, called the **gray** (Gy):

$$1 \text{ Gy} = 1 \text{ J kg}^{-1}$$

To allow for the different ionising abilities of the radiation, we multiply the energy absorbed per kg in grays by a **quality factor** which depends on the radiation: for X-rays, gamma rays and beta particles the factor is 1; for alpha particles is 20. The result is a quantity called the **dose equivalent**, and this is measured in **sieverts** (Sv).

Older units still in use:
curie = 3.7×10^7 Bq
rad = 0.001 Gy
rem = 0.001 Sv

Table 18.8 **Units related to radioactivity**

Quantity	Unit	Definition
activity	becquerel	(Bq) decay rate in decays per s^{-1}
dose	gray (Gy)	energy absorbed in tissue from radiation, in J kg^{-1}
dose equivalent	sievert (Sv)	a measure of likely tissue damage, calculated as: dose × quality factor

THE ATOMIC NUCLEUS

There are many ideas in this chapter about the nature of the nucleus, its component parts and how they and their characteristics were discovered. The different kinds of radioactivity of nuclei help to explain the properties of the nucleus. These ideas are brought together in the Chapter Map, which shows how they link. In your studies, cross-match the entries in the map with the syllabus you are following, and check that you understand the ideas.

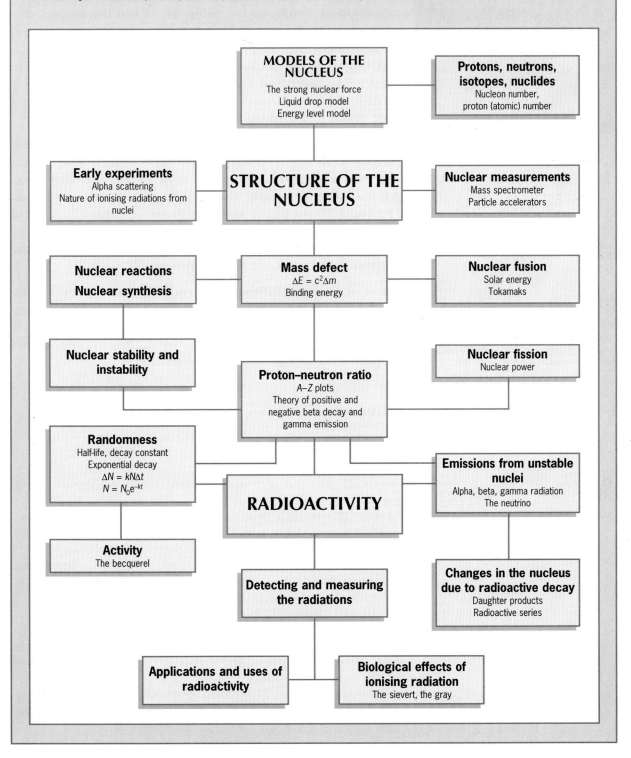

MODELS OF THE NUCLEUS
The strong nuclear force
Liquid drop model
Energy level model

Protons, neutrons, isotopes, nuclides
Nucleon number,
proton (atomic) number

Early experiments
Alpha scattering
Nature of ionising radiations from nuclei

STRUCTURE OF THE NUCLEUS

Nuclear measurements
Mass spectrometer
Particle accelerators

Nuclear reactions
Nuclear synthesis

Mass defect
$\Delta E = c^2 \Delta m$
Binding energy

Nuclear fusion
Solar energy
Tokamaks

Nuclear stability and instability

Proton–neutron ratio
A–Z plots
Theory of positive and negative beta decay and gamma emission

Nuclear fission
Nuclear power

Randomness
Half-life, decay constant
Exponential decay
$\Delta N = kN\Delta t$
$N = N_0 e^{-kt}$

RADIOACTIVITY

Emissions from unstable nuclei
Alpha, beta, gamma radiation
The neutrino

Activity
The becquerel

Detecting and measuring the radiations

Changes in the nucleus due to radioactive decay
Daughter products
Radioactive series

Applications and uses of radioactivity

Biological effects of ionising radiation
The sievert, the gray

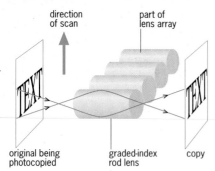

An insect's eye and a photocopier both use an array of rod-shaped lenses to produce an image. Each rod provides part of the image, and all the parts contribute to a complete picture

OPTICAL LENSES, such as the lenses in a camera viewfinder, a telescope or a microscope, allow us to see the images of a wide range of objects. In these instruments, a lens (or several lenses) takes the image to the eye, which itself has a lens, too. But an image can be formed by many lenses packed side by side in a layer. The eyes of insects have this arrangement, with hundreds of small rod-shaped lenses, each directing part of the image to the creature's brain.

A photocopier has a similar imaging system. Up to several million cylindrical 'microlenses', each only a few micro-metres across, form a flat array, which scans the item being copied a line at a time. The light from particular points on the copied item passes through the microlenses, and on to the detector, to build up a complete image. The refractive index of the microlenses is graded, which means the microlenses don't have to be 'lens-shaped'.

Microlenses are also used in producing the image on the liquid crystal display (LCD) screen of a portable laptop computer. The brightness and clarity of the image is steadily being improved and, at the same time, the energy required to power the equipment is being reduced.

Introduction

In today's world, images have become as important as writing – or even speech – as a means of communicating ideas. Images are rich in information: we say that 'a picture says more than a thousand words'. Fig 19.1 shows some symbols that clearly convey ideas more rapidly and simply than words.

Every day we experience the communicating power of images, particularly through television and advertising. Also, current physics and technology convert complex data into images – from particle tracks in accelerators (see Fig 19.52) to ozone concentrations measured by satellites. In this chapter, we will be looking at the basic science and the techniques used to produce images.

First, the chapter deals with the working of a simple thin lens, and how lenses work in imaging devices such as the human eye, cameras, telescopes and microscopes. The chapter goes on to show how images are displayed and recorded. (The way they are trans-mitted is left to 21 Communications.) Finally, we look briefly at some modern methods of imaging which rely on advances in physics and involve complex technology.

Fig 19.1 **Symbols we see regularly and which instantly convey information more rapidly than words**

1 HOW A CONVERGING LENS FORMS AN IMAGE

We are familiar with convex glass lenses. They have equal surfaces that we can think of as parts of a sphere, and are called **spherical** (or **simple**) lenses. Figs 19.2–5 show what happens to the wavefronts of light waves when they pass through a simple glass lens. A lens like this, which alters a plane wavefront to make the light waves pass through a point, or **focus**, is called a **converging** or **positive** lens.

Fig 19.2 **Plane light waves passing through a simple lens**

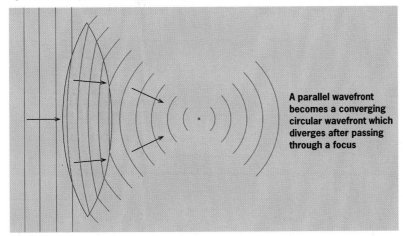

A parallel wavefront becomes a converging circular wavefront which diverges after passing through a focus

Fig 19.3 **Beam diagram**

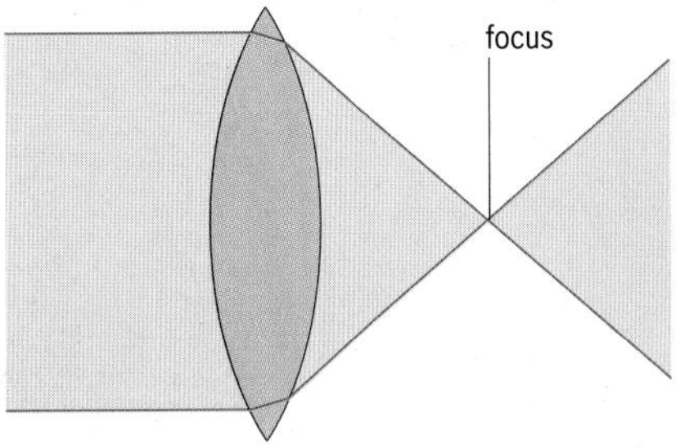

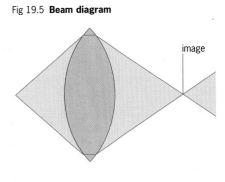

focus

Fig 19.4 **Diverging light waves passing through a simple lens**

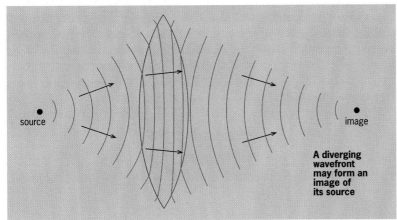

source

image

A diverging wavefront may form an image of its source

Fig 19.5 **Beam diagram**

image

In Figs 19.2–5 you can see that the wavelengths of the light are shorter after they pass across the air–glass boundary. This is because the speed of the waves is less in glass than in air, and the waves don't go as far in each period of the wave motion. This drop in speed causes **refraction**, meaning that the light changes direction wherever the wavefront is not parallel to the boundary. (See page xxx for more about refraction.)

Wave diagrams like those in Figs 19.2–5 are hard to draw, and they also hide some of the features of the light paths. Here, it makes sense to revert to an old but very useful model of light which assumes that light travels in straight lines. Where this model breaks down (as it will!), we shall use the more advanced wave model.

Figs 19.6(a) and (b) show just some of the light rays we could draw. The rays show the *direction* of the waves, which means that the rays are at right angles to the wavefront. When a ray hits the glass surface, it is refracted as shown. Notice in Fig 19.6 that the light ray hitting the centre of the lens does so at 90° and that light rays increasingly far from the centre of the lens hit the surface at smaller and smaller angles. We will return to this soon.

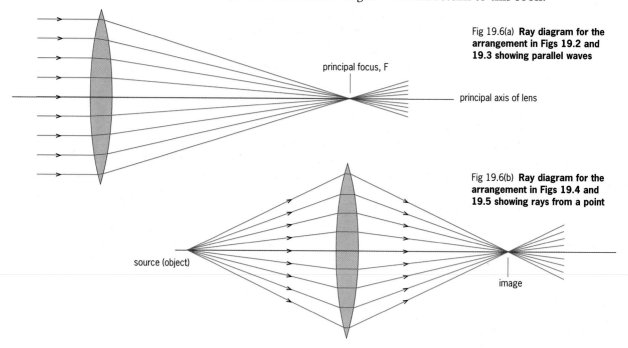

Fig 19.6(a) **Ray diagram for the arrangement in Figs 19.2 and 19.3 showing parallel waves**

Fig 19.6(b) **Ray diagram for the arrangement in Figs 19.4 and 19.5 showing rays from a point**

Let's now look at Fig 19.7(a), which shows light passing through a plane surface, such as the surface of a glass block. The line at right angles to the boundary is the **normal**, and we can see that the angle *i* made with the normal by the incident ray is greater than the angle *r* of the refracted ray. At the same time, as *i* increases, so *r* increases, while remaining less than *i*, see Fig 19.7(b).

The light rays are obeying **Snell's law**, the **law of refraction**:

$$\frac{\text{sine of angle of incidence}}{\text{sine of angle of refraction}} = \frac{\sin i}{\sin r} = \text{constant}, n$$

The constant *n* is the **refractive index** (air to glass for glass lenses).

Fig 19.7 **The angles which define the law of refraction. The angle of incidence is larger in (b) than in (a)**

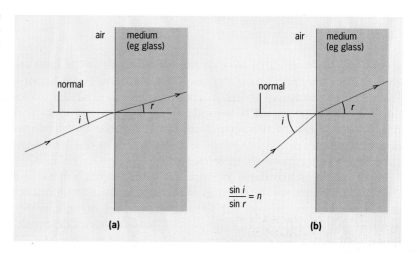

We can now look at Fig 19.8, which is part of Fig 19.6(b) and shows rays from a point source passing through a lens. The radius lines of the curved surface are shown extended beyond the surface as normal lines. We can see again that the law of refraction applies to the rays as they cross the air–glass boundary.

The spherical geometry of the simple lens makes the rays converge (but not accurately) to one point or **focus**. For parallel rays (that form a plane wavefront), this point is called the **principal focus**, as in Fig 19.9 and also Fig 19.6(a). The distance of this point from the centre of the lens is called the **focal length** of the lens, CF in Fig 19.9. A lens has *two* principal focuses, one on each side of the lens, at equal distances from it, F and F′ in Fig 19.9.

In practice, the focus for simple spherical lenses is only at a point if the diameter of the lens is very small compared with the radius of curvature of its surfaces. Otherwise, the lens forms a partly blurred image. A lens defect like this, caused either by the shape or the material of the lens, is called an **aberration**. For a lens with spherical geometry, it is called **spherical aberration**. Removing aberrations is technically difficult, which explains why good optical instruments, such as camera lenses, are expensive.

Predicting the image

Fig 19.9 shows how a simple lens forms an image of an object. We can *predict* the position and size of an image either by drawing or by using formulae.

Drawing to find the image

We can draw any number of rays to help us find the position and size of an image, but the three rays that are most helpful to draw are shown in Fig 19.9, with the labels we use when drawing ray diagrams.

Fig 19.9 shows an upright object OX close to a lens. In ray diagrams, the rays are assumed to change direction at a line that represents a plane in the centre of the lens. The object could be anything, but by convention it is drawn simply as an arrow. As Fig 19.6(a) showed, any ray that is parallel to the **principal axis** of the

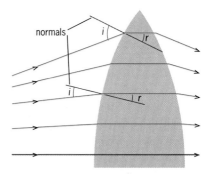

Fig 19.8 **Refraction of light by a simple lens**

B Light of frequency 5.0×10^{14} Hz enters some glass.

(a) Light travels at 3.0×10^8 m s^{-1} in air. What is the wavelength of the light in air?

(b) The frequency of the light stays the same in the glass, but the light slows down to a speed of 2.0×10^8 m s^{-1}.
(i) What is its wavelength in the glass?
(ii) What is the refractive index of the glass?

See question 8 (a and b)

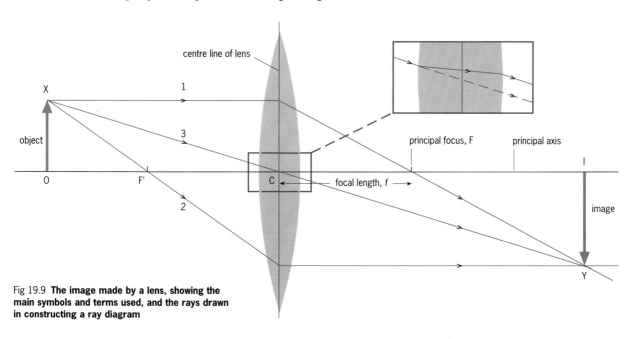

Fig 19.9 **The image made by a lens, showing the main symbols and terms used, and the rays drawn in constructing a ray diagram**

lens is refracted to pass through the principal focus, F. In this way, we can predict the direction of the ray labelled 1 in Fig 19.9.

We use the same idea for ray 2. Going through F′, it emerges from the lens parallel to the principal axis.

The third useful line, for ray 3, goes straight through the centre of the lens – it does not deviate. But see the close-up drawing. At the centre of the lens, the two faces are parallel to each other. If the lens is thin compared with the distances of object and image, the slight sideways displacement of the ray is not significant. All three rays pass through the same point, Y. They all started at X, and it is clear that a screen placed at Y would catch them all together again: Y is a *focused image* of X. Similarly, any point on OX would give a focused image of itself somewhere between I and Y. IY is a **real image**.

These construction rays can be drawn to scale to find the position and relative size of the image of any object placed in front of the lens. Objects viewed through a lens, and their images, are usually much smaller than the distances they are from the lens. In such cases, it is best to make the vertical scale larger than the horizontal one.

C Check that you have understood Fig 19.9 and followed the description for constructing a ray diagram. Show that an object, 5 cm tall, placed 10 cm from a lens, of focal length 6 cm, produces an inverted image 15 cm from the lens. The image should be 7.5 cm high. By drawing on A4 graph paper, you can use a scale of 1:1 for both vertical and horizontal distances.

Finding the image by formula

It is usually quicker and more accurate to find the position and size of an image by using the **lens formula**:

$$\frac{1}{u} + \frac{1}{v} = \frac{1}{f}$$

where u is the distance of the object from the lens centre, the *object distance*; v is the distance of the image from the lens centre, the *image distance*; and f is the *focal length* of the lens. Fig 19.10 gives the geometry required to prove this formula for a thin converging lens.

You can now find the position and size of any image simply by inserting the other given values in the formula.

D Use the formula to check the results for image distance and size that are given in question C.

Fig 19.10 **Proof of the thin lens formula**

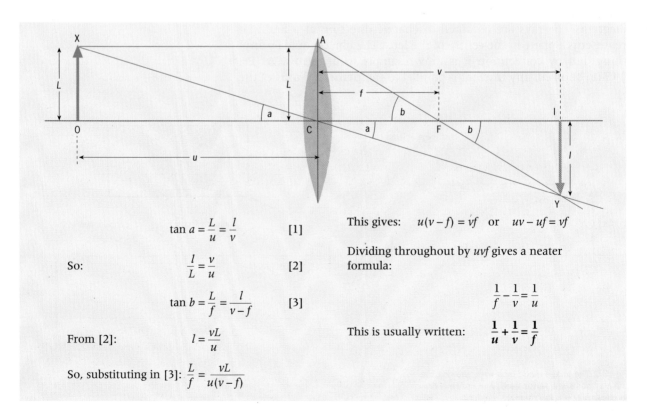

$$\tan a = \frac{L}{u} = \frac{l}{v} \quad [1]$$

So: $$\frac{l}{L} = \frac{v}{u} \quad [2]$$

$$\tan b = \frac{L}{f} = \frac{l}{v-f} \quad [3]$$

From [2]: $$l = \frac{vL}{u}$$

So, substituting in [3]: $$\frac{L}{f} = \frac{vL}{u(v-f)}$$

This gives: $$u(v-f) = vf \quad \text{or} \quad uv - uf = vf$$

Dividing throughout by uvf gives a neater formula:

$$\frac{1}{f} - \frac{1}{v} = \frac{1}{u}$$

This is usually written: $$\frac{1}{u} + \frac{1}{v} = \frac{1}{f}$$

2 THE IMAGE IN A DIVERGING LENS

Fig 19.11 shows a lens with *concave* spherical surfaces and what happens to parallel light rays (that is, a plane wavefront coming from a distant object) when they pass through it. The rays are refracted so that they seem to **diverge** from a single point, F. This point is the **principal focus** of a diverging lens. As light doesn't actually come from the point, or pass through it, it is a **virtual focus** (see the description of the plane mirror on page 420 for more about the word 'virtual').

Fig 19.12 shows how a diverging lens forms an image of an object. As for a converging lens, you would see the image by looking at the object through the lens. But light doesn't actually pass back through the diverging lens to the image, so you cannot catch the image on a screen. It is therefore a **virtual image**.

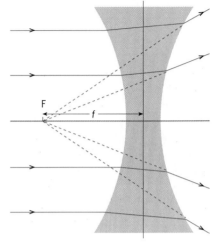

F is a virtual principal focus

Fig 19.11 **A diverging lens has a virtual principal focus and a negative focal length**

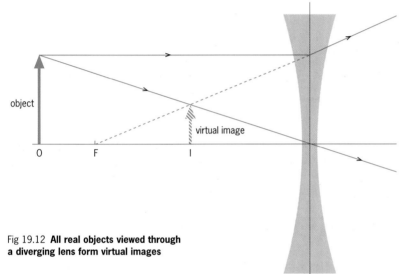

Fig 19.12 **All real objects viewed through a diverging lens form virtual images**

See question 1.

The sign convention

The lens formula works for all simple optical devices (even mirrors, as explained on page 422). But we have to know whether the images, principal focuses – and even objects – are real or virtual. Where they are virtual, the convention is to give *negative* values to distances measured from them to the lens or mirror. For example, the principal focus of a diverging lens is virtual, so its focal length is given a negative sign.

Suppose we place a real object 20 cm from the diverging lens as shown in Fig 19.12. The principal focus of the lens is 10 cm from the lens, so its focal length is –10 cm. The lens formula gives:

$$\frac{1}{20} + \frac{1}{v} = -\frac{1}{10}$$

So:

$$\frac{1}{v} = -\frac{1}{10} - \frac{1}{20} = -\frac{3}{20}$$

and:

$$v = -\frac{20}{3} = -6.7 \text{ (cm)}$$

So the distance of the image from the lens centre is 6.7 cm, and the negative sign tells us that the image is also virtual.

Converging lenses have positive focal lengths and are often called *positive* lenses. By contrast, diverging lenses are called *negative* lenses.

Fig 19.13 **Optician fitting a lens into a spectacles frame. Is the lens convergent or divergent?**

3 THE POWER OF A LENS

In optics, lenses are usually described in terms of their **power**. The more powerful the lens, the closer to the lens is the image that the lens forms of a distant object. The power of a lens is defined as the reciprocal of its focal length measured in metres:

$$\textbf{power} = \frac{1}{f}$$

The unit of power is called the **dioptre**, symbol D, and so a lens of focal length +10 cm (0.1 m) has a power of +10 D. A diverging lens of focal length −5 cm (0.05 m) has a power of −20 D. This way of describing lenses lets us work out what happens when two lenses are used together. The combined power of the lenses is simply the sum of the power of each lens, bearing in mind their signs, as shown in Fig 19.14.

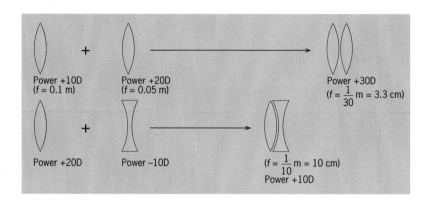

Fig 19.14 **Lens combinations: (a) two positive, (b) positive plus negative. In combination, the powers of lenses are added**

The camera lens as an expensive hole

We have seen that the quality of the image made by a lens depends on its shape (see spherical aberration, page 415). Image quality also depends on the **aperture** of the lens, that is, the part of the lens through which light is allowed to pass. (Though the lens diameter is fixed, the aperture can be varied.)

A lens of large diameter captures more light and produces a brighter image than a smaller diameter lens. However, it is difficult and expensive to correct for aberrations in large aperture lenses. So, for example, in ordinary cameras with a single lens, the lens is usually 'stopped down' to make the picture sharper. The light used to form the image passes through the central region only of the lens (see Fig 19.15), and this reduces spherical aberration.

Fig 19.15 **'Stopping down' to reduce lens aperture and so correct spherical aberration. A 'stop' excludes the outer rays which would have been brought to a focus closer to the lens than the inner rays**

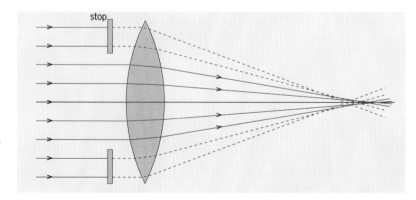

The quality of the image is also affected by **diffraction**. The image is blurred as points on the object become circular diffraction patterns. Points on the image subtending an angle smaller than a value of θ are not seen as separate; their **resolution** is not possible. θ is determined by wavelength λ and aperture d in the **Rayleigh criterion**:

$$\theta \approx \frac{\lambda}{d}$$

Diffraction effects are less when the aperture is large, so accurate optical instruments have a design conflict. Large aperture means better **resolution** (see Chapter 16) but more spherical aberration; small aperture reduces aberration but lowers resolution.

See question 8c.

?

E A camera lens is made of two components, a diverging (negative) lens of focal length 69 mm, and a converging (positive) lens of focal length 29 mm. What are:
(a) the powers of each component,
(b) the power of the combination,
(c) the focal length of the combination?

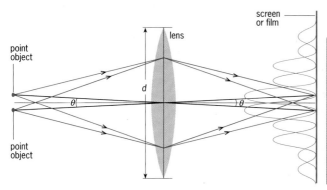
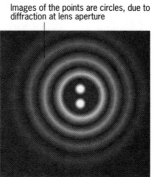
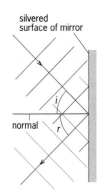

Fig 19.16 **Diffraction: the images of two points can just be 'resolved', meaning that they can be seen as separate, when the points subtend an angle at the lens such that** $\sin \theta \approx \theta > \lambda/d$

Fig 19.17 **Plane waves reflected in a plane mirror. The angle of incidence *i* equals the angle of reflection *r***

4 MIRRORS

Plane mirrors are the most common optical devices we come across. The ordinary household mirror is a sheet of flat glass, 'silvered' on the back with a layer of metal paint (the metal is usually aluminium), which is then protected by a coat of ordinary paint.

Light passes through the glass and is reflected by the silvering. A plane wavefront is reflected as a plane by a flat mirror, and the geometry results in the well-known rule:

The angle of incidence equals the angle of reflection.

The angles are shown in Fig 19.17, and are measured from the **normal** line drawn at right angles to the surface.

As shown in Fig 19.18, the image of an object reflected in a plane mirror is behind the mirror and as far from the mirror as the object is. Fig 19.19 shows the geometry to prove this.

See question 1.

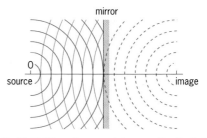

Fig 19.18 **Waves from a point source O reflected in a plane mirror. Circular waves are reflected so that they appear to come from the image, which is as far behind the mirror as O is in front**

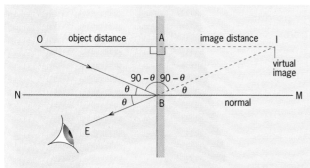

Fig 19 19 **Using rays to prove that the image is as far from a plane mirror as the object**

1. By the law of reflection:
  angle OAB = angle IAB = 90°
  angle OBN = angle EBN (θ)

  By equal angle theorem (straight line properties):
  angle OBN = angle IBM (θ)
  and so: angle OBA = angle IBA ($90 - \theta$)

  Triangles OAB and IAB similar, line AB common.
  So: OA = IA

2. Alternatively, $\tan(90-\theta) = \dfrac{OA}{AB} = \dfrac{IA}{AB}$
  So: OA = IA

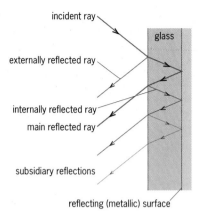

incident ray

glass

externally reflected ray

internally reflected ray
main reflected ray

subsidiary reflections

reflecting (metallic) surface

Fig 19 20 **Ordinary 'silver'-backed mirrors produce blurred images because multiple reflection occurs**

Notice that light does not pass through the mirror, and the image is really an optical illusion. It isn't really 'there' – you couldn't, for example, catch it on a photographic film placed at the image position. The image is therefore a **virtual** image. Compare this image with the one produced by the lens in Fig 19.6(b) which is formed from rays of light and can be caught on a photographic film. Our brain perceives ('sees') virtual images by making them into 'real' ones using another optical device – our eyes.

An ordinary glass mirror produces a blurred image because of multiple reflections – see Fig 19.20. Such a mirror is not suitable for optical instruments which need to give sharp images. There are two main solutions to this problem.

Front-silvered mirrors

These have the metallic layer on the front surface. Very fine metal particles are deposited on glass either from solution or from metal vapour in a vacuum. The large mirrors used in reflecting telescopes (see page 427) are made in this way. The metal surface is easily damaged, by touching or corrosion, and has to be treated with care.

Reflecting prisms and Snell's law

A glass (or clear plastic) **reflecting prism** provides a much cheaper and more practical *plane* mirror by using the effect of **total internal reflection**.

Fig 19.20 shows light rays which are partly internally reflected. Fig 19.21 shows how total internal reflection occurs when the angle of incidence of a ray exceeds the **critical angle** c for the medium. When a ray reaches the inside of a plane boundary (say, between air and glass) at an angle of incidence greater than the critical angle, it cannot pass through the surface. Fig 19.21 also shows the relationship, by **Snell's law**, between the refractive index of a medium and its critical angle.

See question 2. ■

See question 3. ■

Fig 19.21 **How total internal reflection occurs**

Total internal reflection occurs at angles of (internal) incidence greater than c, the critical angle. By Snell's law:

$$\frac{\sin 90}{\sin c} = n \text{ or } n = \frac{1}{\sin c}$$

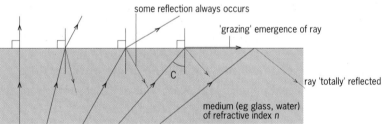

some reflection always occurs

'grazing' emergence of ray

c

ray 'totally' reflected

medium (eg glass, water) of refractive index n

light rays at increasing angles to normal

Fig 19.22 **A reflecting prism – a prism used as a mirror. Angle a must be greater than the critical angle c for the material used**

The action of a prismatic mirror is shown in Fig 19.22. Such mirrors are used in some cameras and binoculars.

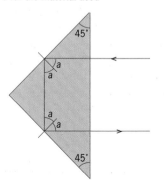

45°

a
a

a
a

45°

Fig 19.23 **The use of prisms allows binoculars to be compact**

5 OPTICAL INSTRUMENTS: SINGLE LENS DEVICES

The magnifying glass

A single converging lens can be used as a **magnifying glass** – sometimes called a **simple microscope**. The object is placed closer to the lens than the principal focus. As shown in Fig 19.24, the lens produces a magnified virtual image which is upright. The eyepiece lens of a (compound) microscope or telescope may act as a magnifying glass in this way.

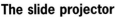

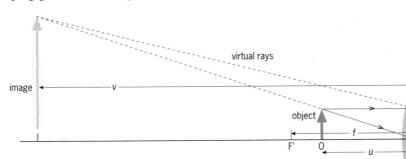

Fig 19.24 **Ray diagram for a converging lens used as a magnifying glass. The image is upright, magnified and virtual**

F An object is placed 4 cm from a converging (positive) lens of focal length 6 cm. Show (by calculation or by drawing) that the image produced is virtual, 12 cm from the lens and 3 times as large as the object.

Projectors

A **projector** uses a single converging lens to produce a real, enlarged image on a distant screen. Everyday examples include cinema projectors, slide projectors and overhead transparency projectors. In each case, the design problem is to get enough light on to a small image so that the final enlarged image is bright enough to be seen. The light source is often a filament lamp operating at high temperature, emitting a great deal of infra-red radiation which can damage the slide or film by heating it.

■ See questions 4 and 5.

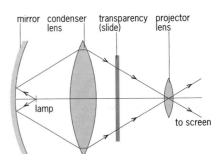

Fig 19.25 **The components in a slide projector**

The slide projector

Fig 19.25 shows a typical arrangement for a **slide projector**.

The first lens (the **condenser** lens) is there to collect as much light as possible to illuminate the slide. A special glass filter absorbs infrared wavelengths. It, and a fan, help to keep the slide cool. The slide is placed very close to the principal focus of the projector lens, so that the image distance – and therefore the image size – is large.

EXAMPLE

Q A slide has to be magnified to become 300 times larger when projected on a screen. The focal length of the projector lens is 3.00 cm.
a) How far from the lens must the screen be placed?
b) How close to the lens must the slide be placed?

A
a) Magnification $= \frac{v}{u} = 300$, so $\frac{1}{u} = \frac{300}{v}$

Using the lens formula gives:

$$\frac{300}{v} + \frac{1}{v} = \frac{1}{0.03}$$

This simplifies to $v = 0.03 \times 301 = 9.03$ m $= 903$ cm. So the screen must be placed 903 cm from the projector lens.

b) From the magnification ratio we get the object distance:

$$u = \frac{v}{300} = \frac{903}{300} = 3.01 \text{ cm}$$

So the slide has to be just a tenth of a millimetre outside the principal focus of the lens.

Comment: The slide is usually in a fixed holder, and the position of the lens is adjusted to provide a clear image. For accurate focusing, the lens is mounted in a holder which has a good screw thread with a very small pitch. Have a look at a slide or film projector to see this arrangement.

The overhead projector

The most common projecting system you are likely to see is the **overhead projector**. This works like a slide projector with the addition of a mirror so that the transparency can be horizontal. The transparency is much larger than a slide, so the condenser lens is usually a **Fresnel lens**, shown in Fig 19.26.

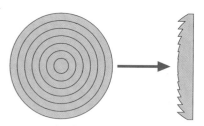

Fig 19.26 **A Fresnel lens (left) enlarges print for people with impaired sight. Used as a condenser in an overhead projector, such a lens (right), it saves space, weight and cost**

Spherical mirrors

Concave mirrors

A spherical mirror has much the same image-forming properties as a glass lens. As shown in Fig 19.27, the **concave mirror** focuses parallel rays to a point – its principal focus, F. To find the image of an object placed in front of a mirror, we can use the same kind of ray-drawing technique as for a lens (see Fig 19.28).

Similarly, we can use the same formula relating object distance, image distance and focal length as we use for the lens, namely: $1/u + 1/v = 1/f$.

A concave mirror can be compared with a positive (converging) lens, and can produce both real and virtual images. Concave mirrors are used as shaving and make-up mirrors in their virtual, magnifying mode. As with a lens used as a magnifying glass (see above), the object has to be nearer the mirror than the principal focus. Concave mirrors are also used as the *objectives* in large astronomical telescopes (see the description of the Newtonian telescope on page 427).

The simple geometry of the circle in Fig 19.27 shows that the focal length f of the mirror is half the radius of curvature of the sphere of which the mirror is a section, so that PC = 2f. This focal length applies only if the angle made between the edges of the mirror and its sphere's centre is very small, in other words, that the diameter of the mirror is small compared to its radius. Otherwise, we find that the image is blurred, a form of spherical aberration like that shown in Fig 19.15.

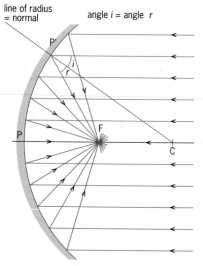

Fig 19.27 **A concave spherical mirror. Parallel rays are focused at F. PF = f, radius PC = 2f (spherical aberration not shown)**

See questions 6 and 7. ▓

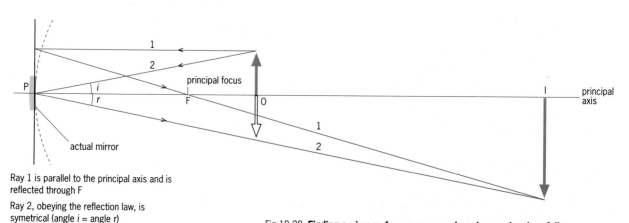

Ray 1 is parallel to the principal axis and is reflected through F

Ray 2, obeying the reflection law, is symetrical (angle *i* = angle *r*)

The image is real, inverted and larger than the object

Fig 19.28 **Finding an image for a concave mirror by ray drawing. A line perpendicular to the principal axis defines the position of the reflecting surface**

Convex mirrors

Convex mirrors act like concave (diverging) lenses. Their focal lengths are negative and they form virtual images of real objects (Fig 19.29). They are used to give a wide field of view in car rear-view mirrors, in blind exits or entrances and as security mirrors in shops (Figs 19.30 and 19.31).

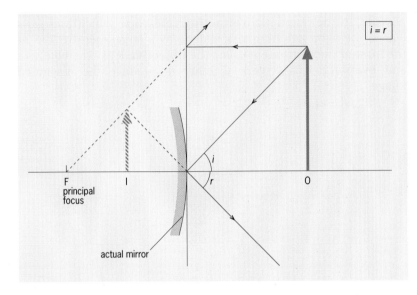

actual mirror

Fig 19.29 **Ray diagram for the image formed by a convex mirror. The image is upright, virtual and smaller than the object**

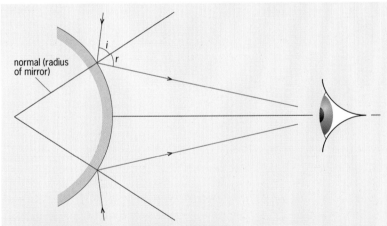

Fig 19.30 **A convex mirror gives a wide field of view. It produces a diminished picture of a large area**

Fig 19.31 **Convex mirrors are used as security mirrors in shops and on roads at sharp bends and concealed entrances**

?

G Find the position and size of the image of a child 1.4 m tall who stands 2 m from a convex (diverging) mirror of focal length 4 m. Where would such a mirror be useful?

?

H (a) Draw a scale diagram (use graph paper) to find the position and size of an image made in a concave mirror. Data: radius of curvature of mirror 80 cm; distance of object from mirror 20 cm; height of object 4 cm.

(b) Check the results obtained in **(a)** by calculation using the formula $1/u + 1/v = 1/f$.

?

I Find by drawing or by calculation the position and size of the image produced when a slide 24 mm by 32 mm is placed 40 mm from a concave mirror of focal length 36 mm. Is the image real or virtual?

The astronomical refracting telescope

An astronomical telescope is designed not only to *magnify* the apparent size of distant objects but also to *gather as much light* as possible. In the traditional **refracting telescope** which uses glass lenses, the light gathering task is done by a large-aperture lens at the front of the instrument called the **objective** lens. The objective and **eyepiece** lenses work together to produce the magnified image.

Fig 19.32 **A telescope concentrates light so that images look brighter, and separates the light coming from distant objects – it magnifies. Angles α and β have been exaggerated for clarity**

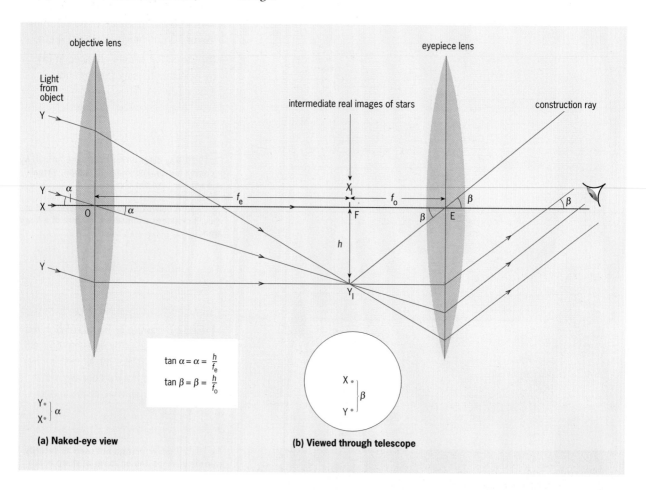

$$\tan \alpha = \alpha = \frac{h}{f_e}$$

$$\tan \beta = \beta = \frac{h}{f_o}$$

(a) Naked-eye view

(b) Viewed through telescope

Fig 19.32 shows the light rays from distant objects forming a 'pencil' of light that passes through the telescope. In optics, a distant object producing almost perfectly parallel rays (a plane wavefront) is said to be at **infinity**.

Follow the main diagram of Fig 19.32 from left to right. The objective lens forms a real image of the object in the plane of its principal focus, F. The eyepiece is placed so that F is also at the position of the plane of *its* principal focus. Accordingly, the final pencil of light leaving the eyepiece and entering the eye is composed of parallel rays. The image is virtual, upside down (inverted) and also at infinity. The telescope set up like this is in **normal adjustment**, the usual arrangement for viewing distant objects. The position of the eyepiece can be altered to produce a real image, on a photographic plate, for example.

See questions 8b and 9. ■

Magnifying power

Fig 19.32 shows that the emerging set of rays from the eyepiece makes a larger angle, β, with the central axis of the telescope than angle α made by the incoming rays. What this means is shown in Figs 19.32 (a) and (b).

Imagine that we are studying a system of two stars, X and Y. Without the telescope, we would see the system as quite small, as in Fig 19.32(b). Suppose star X is at the centre of the picture. It sends light along the main axis of the telescope. Star Y is off-centre and sends light at angle α. When the light emerges from the telescope we see that Y appears to be further apart from X, as shown in Fig 19.32(b). In other words, the image of the system appears larger than with the naked eye. With a telescope, stars appear brighter and further apart, and we are able to see faint objects that were previously invisible.

The **magnifying power** of the telescope is simply the ratio of the image size seen with the telescope and the image size seen without it. The ratio is also the ratio ($\tan \beta / \tan \alpha$) which, when the telescope is in normal adjustment, is also the ratio:

$$OF/FE \quad \text{or} \quad f_o/f_e$$

where f_o is the focal length of the objective and f_e is the focal length of the eyepiece.

The angles α and β are usually very small, so we can write (using radians):

$$\text{magnifying power} = \beta/\alpha$$
$$= f_o/f_e$$

Where do you put your eye?

A simple telescope is designed for use with the human eye. This means that all the light gathered by the objective has, eventually, to form a beam that just fits the pupil of the eye – nominally taken to be 5 mm in diameter. Fig 19.33 shows the extent of a beam of light entering the objective than can reach the eyepiece and so be used to form the final image.

R is the position of the **eye ring,** the hole you look through in telescopes and binoculars. The eye is best placed at R because this is where the emerging beam is smallest, ensuring that:

- the light is most concentrated, so the image is **brightest;**

- all of the useful light entering the objective is seen, so we have the **greatest field of view.**

J An astronomical telescope has an objective lens of focal length 1.2 m. What should the focal length of the eyepiece be to produce a magnifying power of 150?

K A typical reflecting telescope used by amateur astronomers has a mirror of focal length 1.2 m. What is its magnifying power when used with an eyepiece of focal length 40 mm? It is possible to buy eyepieces of focal length 2.4 mm. Can you think of any disadvantages in using an eyepiece of such short focal length?

For angles θ smaller than 5° or so, we have $\sin \theta \approx \tan \theta = \theta$ radians.

See questions 8c, 10 and 11.

Fig 19.33 **The eye ring of a telescope at position R**

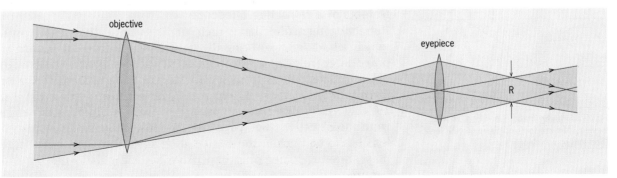

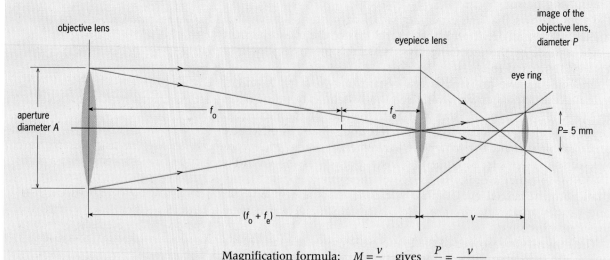

Magnification formula: $M = \dfrac{v}{u}$ gives $\dfrac{P}{A} = \dfrac{v}{f_o + f_e}$

Lens formula:

$$\frac{1}{u} + \frac{1}{v} = \frac{1}{f}$$

Referring to the eyepiece lens, since $u = f_o + f_e$ and $f = f_e$:

$$\frac{1}{f_o + f_e} + \frac{1}{v} = \frac{1}{f_e}$$

$$\frac{1}{v} = \frac{1}{f_e} - \frac{1}{f_o + f_e} = \frac{f_o + f_e - f_e}{f_e(f_o + f_e)}$$

This gives:

$$v = \frac{f_e(f_o + f_e)}{f_o}$$

Combining these gives aperture diameter of objective:

$$A = \text{diameter of eye ring (in mm)} \times \frac{f_o}{f_e}$$

That is:

$$A = 5\frac{f_o}{f_e} \text{ (mm)} = 5M$$

where M is the magnifying power of the telescope, and the diameter of the eye ring equals the diameter of the pupil, that is, 5 mm.

The position of the eye ring is given by v, ie it is a distance L/M from the objective, where L is the separation of the lenses.

Fig 19.34 The eye ring of a telescope, its magnifying power and use of formulae

It can be shown that the position and size of the eye ring is the same as the *image of the objective lens made by the eyepiece*. Fig 19.34 shows how this fact is related to the separation of the lenses and hence the magnifying power of the telescope.

Defects of a refracting telescope

Refracting telescopes have their name because they use lenses, which refract light, to bring the light rays to a focus. It is hard to make large refracting astronomical telescopes because the large objective lens suffers from several kinds of aberrations. The most inconvenient of these is **chromatic aberration**, shown in Fig 19.35. Rays of different colours are focused at different distances from the objective. (See page 345 for more about this.) Also, as a lens has to be thick in the middle, a large aperture lens has a large mass, making it not only expensive but difficult to mount and control.

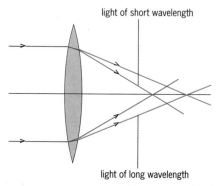

Fig 19.35 In chromatic aberration, shorter wavelengths are focused nearer to the lens than longer wavelengths

See questions 10 and 11. ■

GALILEO EXPLOITS THE TELESCOPE'S POTENTIAL

AT THE GREAT annual fair at Frankfurt-am-Main in Germany in 1608, a wandering trader was offering a novelty for sale – a tube with a lens at each end, one concave and the other convex. When you looked through the tube, it magnified distant objects seven times. In October, a Dutch spectacle maker applied for a licence to make and sell similar 'telescopes' (meaning 'far seers') but was refused because so many other people were already making them. By the following year, they were available for sale in London, Paris and in Italy.

Galileo Galilei was then one of the most famous 'natural philosophers' in Europe – the word 'scientist' was not yet in use. Galileo is still revered as one of the greatest physicists of all time, but unlike some physicists, he was also smart. In August 1609 he went to his employers, the Senate of Venice, and offered them a device – a telescope – which would enable their trading fleets to see enemy ships long before the enemy could see them. The grateful senators immediately doubled his salary, just a few days before the local spectacle makers learned about telescopes and started making them for sale.

This kind of commercial smartness made the senators of the great trading Republic of Venice quite proud. A professor of physics and mathematics who could outsmart *them* was worth keeping, and they gave him a contract for life.

In fact, Galileo was (as far as we know) the first person to explain how the telescope worked. His improved design is named after him (while the man at the Frankfurt Fair was never known). Fig 19.36 shows how it produces an image. Unlike the later – and better – astronomical telescope, it produces its image the right way up. The only place you are likely to see one is in theatres as the binoculars called 'opera glasses'.

Galileo used his telescope to look at the mountains on the Moon and the satellites of Jupiter. His observations convinced him that the old ideas of the nature of the Universe were wrong: the heavens were *not* 'perfect' – and the Sun *was* at the centre of a Copernican solar system of moving planets. It was these ideas that got him into trouble with the Catholic Church in Italy.

Fig 19.36 **The Galilean telescope is shorter than an astronomical telescope of the same power and produces an upright image. The image would be formed at X, but the eyepiece straightens out the rays. The 'virtual object' for the eyepiece is at its principal focus, so rays are made parallel and the image is at infinity**

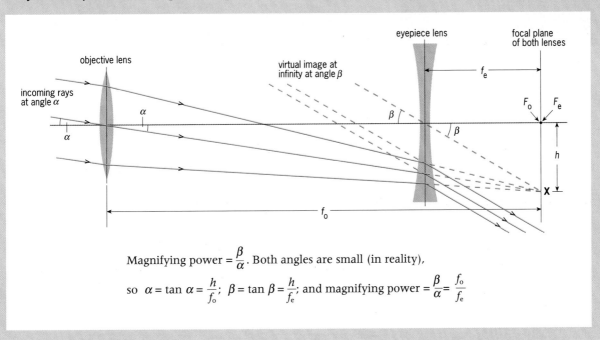

Magnifying power $= \dfrac{\beta}{\alpha}$. Both angles are small (in reality),

so $\alpha = \tan \alpha = \dfrac{h}{f_o}$; $\beta = \tan \beta = \dfrac{h}{f_e}$; and magnifying power $= \dfrac{\beta}{\alpha} = \dfrac{f_o}{f_e}$

Reflecting telescopes

Some of the defects of refracting telescopes can be avoided by using a curved mirror as the objective, as in the **Newtonian reflecting telescope**.

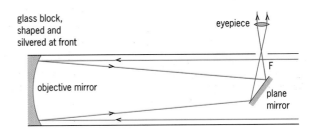

Fig 19.37 **The Newtonian reflecting telescope uses a concave mirror instead of a lens as the objective, so avoids chromatic aberration**

L Why doesn't a Newtonian telescope suffer from chromatic aberration?

See question 12. ■

Fig 19.38 **The world's largest reflecting telescope at Mauna Kea in Hawaii. Its mirror consists of 36 smaller hexagonal mirrors controlled by computer to work together in making an image. (The largest single-mirror telescope, at Zelenchukskaya, Russia, is 6 m in diameter)**

While the largest refracting telescope has an objective lens 1 m in diameter, the largest reflecting telescope in Hawaii (Fig 19.38) has an objective mirror with an aperture of 10 m diameter. The mirrors in such telescopes do not suffer from chromatic aberration, and spherical aberration may be removed by making the shape slightly parabolic rather than perfectly spherical. Also, telescope apertures can now effectively be increased with the aid of computers, as in the Mauna Kea telescope, Fig 19.38.

The objective mirror in Fig 19.37 produces a real image of a distant object at its principal focus, F. This is usually offset by using a plane mirror so that the eyepiece can be mounted conveniently at the side.

The Cassegrain telescope

We have seen that large magnifications require a long focal length. This means that the tube of a Newtonian telescope is inconveniently long, especially for amateur users. A Cassegrain system reduces the length considerably, as shown in Fig 19.39. It uses a convex *secondary mirror* and extends the focal length, forming the image via a small hole in the main mirror. It is technically much easier to make a spherical mirror than a parabolic one to the accuracy required, and a Cassegrain telescope uses simple spherical mirrors with a glass *correction plate*. Its complex lens is shaped to counteract the defects of the mirrors.

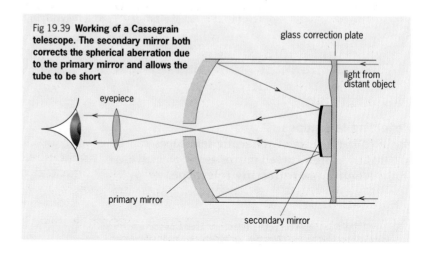

Fig 19.39 **Working of a Cassegrain telescope. The secondary mirror both corrects the spherical aberration due to the primary mirror and allows the tube to be short**

glass correction plate

light from distant object

eyepiece

primary mirror

secondary mirror

The compound microscope

A compound microscope consists of two lenses, a lens combination which reduces aberration. It is designed to produce an enlarged virtual image of a small object. Light gathering is not as important as it is in a telescope, since the object can be brightly illuminated if necessary. Fig 19.40 shows how a microscope forms its image.

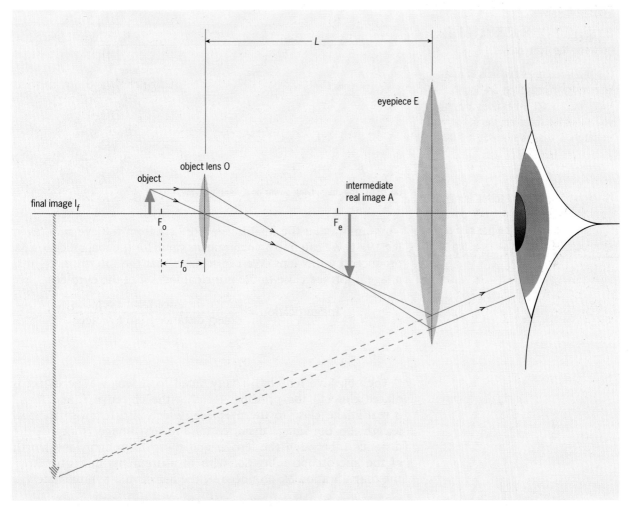

Fig 19.40 **Formation of an image in a compound microscope. In practice, the final image is roughly 25 cm from the eye and eyepiece, with I$_f$ very close to F$_e$**

The objective lens O forms a slightly magnified (intermediate) real image of the object at A. A is just inside the focal length of the eyepiece E, so that the final image I is virtual and highly magnified. It is usually arranged so that this final image is formed about 25 cm from the eye. This is the *near point* of the eye, where we normally place something that we want to see as large as possible (see below).

Magnification in a compound microscope

The magnification produced by the compound microscope depends on where the final image is actually formed. The microscope is focused by moving the object closer to or away from the objective lens, a high power lens (with a small focal length).

We can find an approximate value for its magnification which is good enough for everyday purposes as follows.

M Telescopes are usually made so that different magnifying powers are produced by changing the eyepiece (see marginal question K). Microscopes are made so that it is easy to change the objective lens, for example by having three on a swivelling mount.

(a) Suggest a reason for this difference between the instruments.

(b) The three objective lenses of a laboratory microscope are labelled ×10, ×20, ×40. What does this tell you about the focal lengths (or powers) of the three lenses?

N A compound microscope is often used to produce a real image on a screen. Explain how the position of the eyepiece lens has to be adjusted to do this.

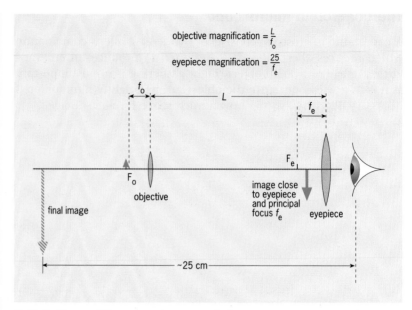

Fig 19.41 **The magnifying power of a compound microscope**

Assuming that the final image is 25 cm from the eyepiece, as in Fig 19.41, we can find an *approximate* value for the magnification M_e produced by this lens. We can assume that the intermediate, real image is formed close to the principal focus f_e of the eyepiece:

$$\text{magnification} = \frac{\text{image distance from eyepiece}}{\text{object distance from eyepiece}}$$

$$M_e = \frac{25}{f_e}$$

The objective lens has a very small focal length. The object is placed close to the principal focus of the objective f_o and forms a real image close to the eyepiece, which also has a small focal length. So the image distance for the first image is almost the distance L between the lenses, and therefore roughly the length of the microscope tube. So, with measurements in centimetres, the magnification M_o produced by the objective is approximately:

$$M_o = \frac{L}{f_o}$$

The total magnification of the compound microscope is therefore:

$$M = M_o M_e = \frac{25L}{f_o f_e} \text{ (measurements in cm)}$$

Binoculars

Binoculars are a pair of telescopes, one for each eye. The magnifying optical principles are the same as in the astronomical telescope described above. The practical difference is that the tube is much shorter, and so the final image is upright.

Both these effects are produced by using reflection prisms (see Fig 19.22). The light travels along the tube three times, so that the optical path is roughly three times the length of the binoculars. The prisms are arranged so that they invert the telescope image, which means it is finally the right way up.

See question 13. ■

7 CAPTURING IMAGES

The images formed by optical (and other imaging) devices can be 'caught' and stored in a variety of ways, ranging from the biochemical method using eyes and brain, through the traditional photochemical processes of film and plate cameras, to the newer systems, which include video cameras, charge-coupled devices (CCDs), magnetic discs and compact discs.

While the imaging process itself (using lenses and mirrors) depends on the *wave* aspect of light, image-capturing systems generally rely on the *particle* aspect of light. It is the energy of the photon that determines how the devices work and are designed.

The human eye

Details of the eye are shown in Fig 19.42(a). Light entering it is focused by two lenses. First, the light meets the **cornea**, a curved layer which is transparent and fairly hard. It acts as a fixed focus lens and in fact does most of the refracting of the light entering the eye. Then the light reaches the **lens**. It refracts the light only slightly, mainly because it is surrounded by substances of almost the same refractive index as itself, and so the refraction (change in speed and direction) of light at the lens surfaces is quite small.

The job of the lens is to adjust the focal length of the cornea–lens combination, allowing for a range of object distances between about 25 cm at the **near point** and infinity at the **far point** for the normal eye.

The focal length of the eye lens is adjusted by changing its shape: the radii of curvature of its surfaces are changed by a ring of muscle round the lens called the **ciliary muscle**, and Fig 19.42(b) shows how it alters the shape of the lens.

?

O Why does the cornea produce more of a focusing effect than the eye lens?

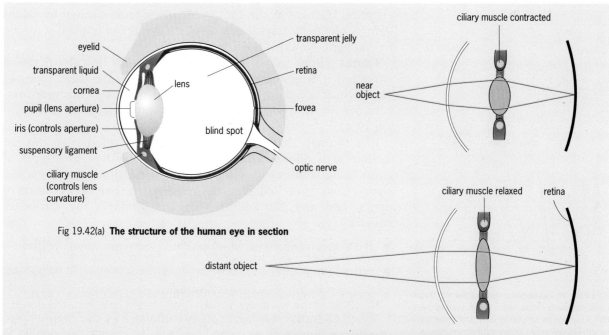

Fig 19.42(a) **The structure of the human eye in section**

Fig 19.42(b) **Accommodation, showing how the shape of the lens is adjusted by the ciliary muscle (dimensions not to scale)**

When the eye is relaxed, the radial suspensory ligament keeps the lens in tension. The eye forms clear images of objects at the far point (infinity, rays parallel), and the focal length of the cornea–lens combination is the length of the eye – about 17 mm. When the eye looks at a near object (rays diverging), the ciliary muscle contracts, the lens becomes fatter, and the focal length is shortened, in order to refract the rays to form a focused image on the retina. Rapid change in lens shape relies on swift feedback between brain and eye, and is called **accommodation**.

The quantity of light entering the eye depends on the size of the **pupil**, the hole in the **iris**. Its size is controlled by the muscles of the iris. The light travels on to reach the **retina** which contains two types of light-sensitive cells, called **rods** and **cones** because of their shape. It is the functioning of these cells that enables us to 'see'. Briefly, this is the mechanism. The rods and cones contain light-sensitive pigment molecules. Photons of particular energies (frequencies) reach the pigment molecules. These molecules absorb energy and a reversible photochemical reaction is triggered. The molecules then rapidly lose energy in a process that generates an electrical potential, and a signal is transmitted to the brain via the **optic nerve**.

The **fovea** is the point on the retina directly in line with the lens. The fovea is where an image is resolved in the finest detail and is the part of the retina most crowded with cones. The rods are more numerous away from the fovea. The average human eye contains about 125 billion rods but only 7 billion cones.

Rods can 'distinguish' between different intensities of light, but not between light of different frequencies (photon energies). They are more sensitive at low light intensities, and are particularly effective for seeing in the dark. Since there are more of them away from the centre of the retina, at night, faint objects are most easily seen 'out of the corner of the eye'. If we suddenly move from bright light to darkness, or a light is switched off, it takes our eyes time to adjust to the dark and make out our surroundings. In this time, the photo-chemical pigment in the rods, which was broken down by bright light, is gradually resynthesised. When the retina is again sensitive to dim light, we say it is **dark-adapted**.

See question 16. ■

Cones allow us to see colour. Theories of colour vision vary, but it is generally accepted that there are three types of cone which are sensitive to overlapping regions of the spectrum, centring on the **red**, **blue** and **green** wavelengths, as shown in Fig 19.43.

It is thought that the enormous range of colours we can see around us is due to different combinations of the three types of cone being stimulated, and to the way the brain interprets the signals from them. Red, blue and green are the **primary colours** of optics, and any pair of these colours produces the secondary colours:

- red + green (or minus blue) produce the sensation of **yellow**,
- red + blue (or minus green) produce the sensation of **magenta**,
- green + blue (or minus red) produce the sensation of **cyan**.

When all three types are strongly stimulated by the energy range of photons found in daylight, the brain interprets the signals as 'white'. (In reality, colour vision is much more complicated than this simple description suggests.)

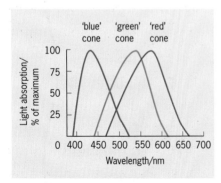

Fig 19.43 **The absorption spectra for the three cone pigments in the retina. Each type of cone has a maximum sensitivity at a particular wavelength (at the peak) where light absorption is taken as 100 per cent for that type. (Cones for blue light are in fact much less sensitive than for green and red light)**

Defects of the eye

The near point is the closest distance at which we can see objects clearly, and this depends largely on the strength or weakness of the ciliary muscles – look back to Fig 19.42(b). The distance of the near point increases with age: at 10 years, it is about 18 cm; in a young adult, 25 cm; and by the age of 60, it may be as much as 5 metres.

This age-related defect is called **far sight** or **hypermetropia**. The eye lens is too weak, so that near objects are blurred because their images are formed behind the retina. Fig 19.44 illustrates these defects and how spectacles with positive lenses can correct them.

> **P (a)** Estimate the focal length of your cornea–eye lens combination when you look at an object 25 cm from your eye (and see it clearly in focus).
>
> **(b)** Look back at Fig 19.13. Is the lens for someone who is long or short sighted?

Fig 19.44 **Long sight and its correction**

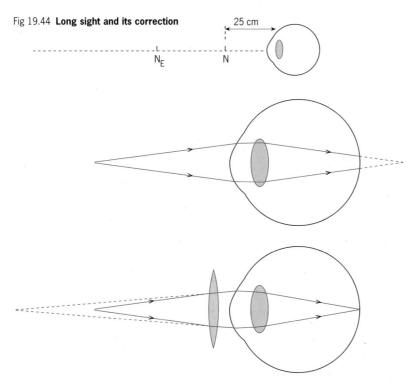

(a) A weak eye lens means that the effective near point N_E is more than 25 cm from the eye at N, the normal near point for a young adult. The far point is still infinity unless the lens is too weak when relaxed (if lens can still accommodate enough)

(b) Long sight (hypermetropia): The eye lens cannot be made strong (convex) enough to focus close objects on the retina

(c) Long sight corrected: A positive lens helps to reinforce the strength of the eye lens

A person may have **short sight** or **myopia**. In this defect, the eye's ciliary muscles are too strong and, even when relaxed, the lens is too powerful. This means that distant objects are focused in front of the retina. Negative lenses correct this defect, as shown in Fig 19.45.

Fig 19.45 **Short sight and its correction**

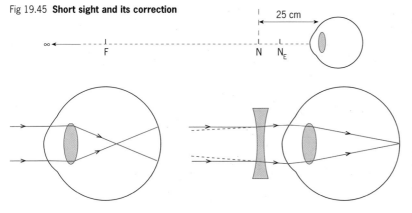

(a) Range of accommodation for short-sighted eye The near point N_E is closer than the usual 25 cm at N. The eye's far point of F should be at infinity - but could be less than a metre away

(b) Short sight (myopia)
Left. The eye lens is too strong, so distant objects are focused in front of the retina
Right. A negative lens produces a 'virtual object' closer to the eye, within its range of accommodation

The lens in the ageing eye tends to get less and less flexible, so that it becomes practically a 'fixed focus' device. Then, a person may need two pairs of spectacles – one for seeing close objects clearly ('reading glasses', with positive lenses) and one for seeing distant objects clearly (with negative lenses or weaker positive ones). Some people wear **bifocals**, spectacles with lenses that have a 'reading' part and a 'long distance' part.

At any age, the eye lens may have a condition called **astigmatism**. The eye cannot focus at the same time on vertical lines and horizontal ones, and a point source tends to be seen as a line. This is because the lens is not spherical and, for example, the radius of curvature in the horizontal direction is not equal to the radius in the vertical direction. Astigmatism is corrected by lenses with curvatures that counterbalance the eye lens curvatures.

Cameras

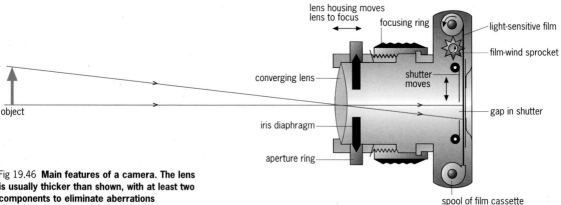

Fig 19.46 **Main features of a camera. The lens is usually thicker than shown, with at least two components to eliminate aberrations**

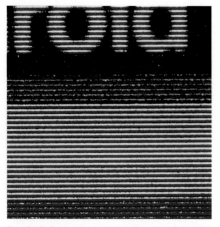

Fig 19.47 Enlarged part of a transparency film. This is Polaroid film which has colours in narrow strips, as on a television screen. Combinations give the range of visible colours

Q Name the main features that ordinary photographic cameras and eyes have in common. Name a main aspect of their working in which they differ (apart from the fact that one is alive and the other is not).

The design of the average camera has many similarities with the eye's design, including automatic focusing and the control of the amount of light entering. But focusing is done by moving the camera, instead of changing the shape of the focusing lens, towards or away from the light-sensitive film. The lens itself is usually compound (to reduce chromatic aberration) and great care is taken to reduce other lens aberrations. The pupil of the eye is replaced in the camera by the **aperture** (or **stop**), and its size is controlled by a mechanical version of the iris, also called an **iris diaphragm**.

Whether the camera is an instant snapshot camera, or a video, film or telescope camera, the optical process which results in an image is the same. The difference is in how the image is then recorded and made visible.

Photographic colour film is made from layers of plastic impregnated with photosensitive dyes. Light of appropriate photon energy triggers off chemical changes in the different dyes. The changes are made visible by *developing* the film, using other chemicals. The dyes used in colour slides reproduce as closely as possible the colours of the light entering the camera to form the image. The full range of colour can be produced by using dyes of just three colours – red, yellow and blue.

A **cine camera** produces its images on a moving sheet of photographic film, at a typical rate of 24 frames (pictures) per second. When projected, we see the images moving smoothly and continuously. This is due to an effect in the retina called **persistence of vision**. It takes time for the response of the eye cells to a stimulus to die away. The image in the eye persists effectively for about a twentieth of a second. A new image appears about every fiftieth of a second, before the old one fades, and we don't notice the 'join'.

A **video camera** works on the same principle as the cine camera, but the image is sensed electronically rather than chemically, with a series of electrical signals being recorded on magnetic tape.

Camcorders and charge coupled devices

Camcorders are small video cameras with **charge coupled devices**, known as CCDs. CCDs are also used as the sensors which capture the images in astronomical telescopes. A CCD consists of a large number of very small picture element detectors, called **pixels**. A typical small CCD has an array of 385 by 578 pixels, tiny electrodes built on to a thin layer of silicon.

When a photon hits the silicon layer, an electron-hole pair is created. A pixel electrode collects the charge produced, which depends upon the intensity of the light falling on the nearby silicon. After a brief time, the electrodes are made to release the charge they have collected in a programmed sequence as the pixels are scanned electronically. The scan produces a digital signal which encodes the light pattern on the CCD, to give the image.

Fig 19.48 **A CCD (held over a printed circuit board). Many thousands of photon-sensitive pixels occupy the central rectangle**

Fig 19.49 **A CCD image of the spiral galaxy NGC 3184. The CCD collects photons over a period of time to build up the fine detail of the image**

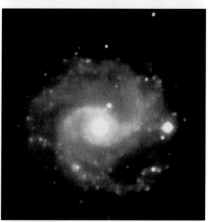

A nanometre, symbol nm, is 10^{-9} metres.

Colour pictures require three sets of sensitive areas with colour filters (red, green, blue) which produce a more complicated output containing colour as well as brightness information.

CCDs are sensitive to light of wavelengths 400 nm (violet) to 1000 nm (infrared). Their main advantage in both scientific and TV applications is their *efficiency*. Photographic film uses less than 4 per cent of the photons reaching it. The 'wasted' photons simply miss the sensitive grains in the film, or fail to trigger a chemical change. CCDs can make use of about 70 per cent of the incident photons, partly because they cover a larger area but mainly because the pixels are more sensitive than the chemicals in a film. (The eye is only about 1 per cent efficient at sensing photons.)

In astronomy, the collection time can be made very much longer than in a TV camera, allowing detail in a faint object to build up. Even so, the detection efficiency of CCDs means that the time of exposure is much less than is needed for a photographic image. Also, the output of current is proportional to the number of photons arriving, which means that the intensity of spectral lines and the brightness of stars can be measured more accurately than is possible with photographic methods.

But photographic plates do have the advantage that the light-sensitive grains are smaller than CCD pixels, so they can produce images with better resolution. Also, the largest CCDs available commercially are about 7 mm square, which means that the image has to be small. But CCDs can be 'butted' together to make larger image areas, and CCD technology is likely to be even more useful and more widely available in the future.

See question 14.

Display screens

We already see a great deal of the world on our TV screens, and developments in information technology seem likely to make everyday communication through electronic mailing systems as common as the printed word. There are two main kinds of video screen to display the information: the **cathode ray tube** (CRT), with a phosphor screen which is made to glow by high speed electrons, and **liquid crystal display** (LCD). The image is produced by a scanning process.

The cathode ray tube

Cathode ray tubes (CRTs) with phosphor screens are the oldest: 'cathode rays' was the name given to streams of what were later discovered to be electrons. They were emitted from the negative electrode (cathode) of a high voltage discharge tube. Nowadays, electrons are produced by heating a metal oxide. Energy transferred from the electrons as they reach the screen makes the phosphor coating emit light.

CRTs are not only the most common tubes for TV receivers, radar devices and computer monitors but are also in such useful laboratory equipment as cathode ray oscilloscopes. CRTs waste energy by having to heat the electron emitter, and they need high voltages (over 2 kV) to accelerate the electrons to energies capable of exciting the phosphor. However, the technology is well established and CRTs are likely to be in use for some time.

See question 15.

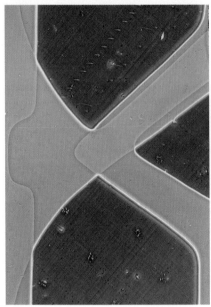

Fig 19.50 **Part of the liquid crystal display of a calculator, magnified 160 times**

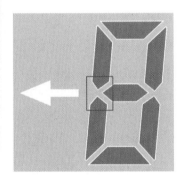

Liquid crystal display screens

Many portable devices, such as calculators, watches and portable computers use **liquid crystal displays** (LCDs). The use of liquid crystals to produce images is described on page 356.

Briefly, the crystal is able to polarise light, and so can cut out light polarised in certain directions. The polarising effect in liquid crystals can be switched on and off by applying an electric field to it. In an LCD screen an array of small liquid crystal picture elements (pixels) is back-lit by a bright screen. A pixel can be changed from bright to dark electronically, and the array of pixels can be controlled by a data signal just like the screen of the cathode ray tube in a TV set.

Each pixel is connected individually, to a **thin film transistor**, many thousands of these making up a large single block – a kind of microchip. This individual control of each pixel means that the pixels can be brighter, and the brightness can also be changed faster, so that the screen is 'refreshed' more often. As these types of screen become cheaper and larger, they may eventually take over from the cathode ray tube in televisions.

?

R Compare the advantages and disadvantages of using liquid crystal screens and cathode ray screens for video display.

8 BEYOND THE VISIBLE SPECTRUM

An image is a record of *information*. Increasingly, the information is being coded digitally (see page 470). For example, a scanner used to convert a painting into a stream of data for storage in a computer digitises its continuously varying light intensities and colour.

The object used as the basis of an image need not be visible to the eye, that is, it need not be a source or reflector of visible light waves. The whole range of radiation in the electromagnetic spectrum may be used to make a visible image. For example, X-rays, ultraviolet and infrared have been used for a long time to make visible images on photographic film. Images are now also made using sound, electrons (as waves), and even changes in the tiny magnetic fields of spinning nuclei, as described below.

Fig 19.51 **A satellite image of farmland. The red and infrared radiation reflected from the land varies according to the crops growing on it, and a computer converts these variations into different colours**

Ultrasound

Ultrasound scanners are now a commonly used in medicine to investigate and diagnose disorders. Ultrasound consists of sound waves of very high frequency and correspondingly small wavelength. Body tissues of different types absorb and reflect these sound waves to varying degrees, and images can be made of organs deep within the body. You can read more about ultrasound on page 539. Note that sound is not part of the electromagnetic spectrum.

Magnetic resonance imaging

Hospitals commonly run appeals to raise funds for a 'scanner'. Often, this is a scanner that can probe body systems in even more detail than ultrasound, using a technique that relies on the magnetic property of atomic nuclei and is known as magnetic resonance imaging, or MRI. (See page 546 for more about this.) The low intensity electromagnetic waves used (at radio frequency) do not affect biological tissues.

See question 17.

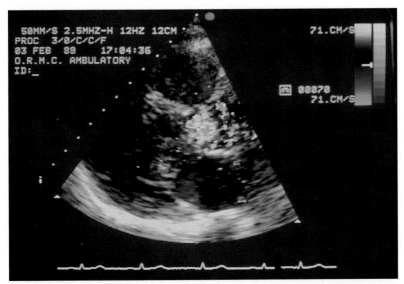

Fig 19.52 **The ultrasound scan of a healthy heart. Inside the scan triangle, the three round dark areas show the chambers of the heart. The orange area is of blood flowing between two chambers**

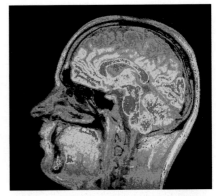

Fig 19.53 **An MRI image of a section through a human head, showing structures of the brain, airways and facial tissue (false colour is added)**

?

S The momentum p of an electron is related to its wavelength λ by the de Broglie formula $p = h/\lambda$, where h is Planck's constant.

(a) Explain why this relationship means that in the electron microscope, shorter wavelengths require higher accelerating voltages than longer wavelengths.

(b) Estimate **(i)** the momentum and **(ii)** the kinetic energy of the electrons in a typical electron microscope.

(c) Hence calculate the accelerating voltage required for the microscope. Data: mass of electron 9.1×10^{-31} kg, $h = 6.6 \times 10^{-34}$ J s, elementary charge $e = 1.6 \times 10^{-19}$ C.)

9 IMAGING THE ULTRA-SMALL

The electron microscope

Quantum physics produced the greatest surprise of all when it suggested that, just as light waves have a particle aspect, so material particles should have a wave aspect (see page 367). The realisation that electrons – as waves – could be used to produce images in much the same way that light waves do, led to the development of the **electron microscope**.

The problem with making images of any kind with waves, is one of diffraction (see page 349). Diffraction affects the *resolution* of an image, that is, how far apart objects or details in an object must be before they can be seen as separate in an image. The limit of resolution for a microscope or telescope is defined by the Rayleigh criterion (see page 419).

The clarity of the image is determined by the wavelength – the shorter the better. Blue light has the shortest wavelength in the visible range, though ultraviolet may be used with photomicrography. The visible wavelengths are of the order of a few hundred nanometres – about 3 to 5×10^{-7} m.

In an electron microscope, electrons are accelerated to a high speed before hitting the object being studied. The higher the speed, the shorter the wavelength (see page 368). The wavelength of an electron in an electron microscope is typically about 10^{-11} m – about 10 000 times shorter than the wavelength of light. Therefore resolution is 10 000 times better.

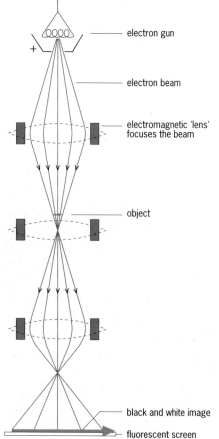

Fig 19.54 **The operation of the transmission electron microscope. The electron 'rays' are focused by powerful coil magnets, as in a TV tube. Their wave property allows detail in an object to be imaged**

electron gun

electron beam

electromagnetic 'lens' focuses the beam

object

black and white image

fluorescent screen

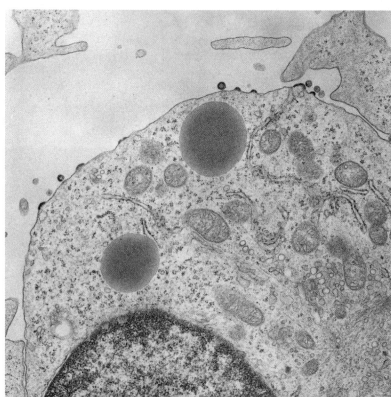

Fig 19.54 **A transmission electron micrograph: part of a white blood cell known as a T-cell (its dark nucleus is bottom left). The cell's role is to combat infection, but it has itself been infected with HIV. The AIDS virus particles multiply inside it, and can be seen as small dark spheres budding off its outer surface at the top of the picture**

The scanning tunnelling microscope

The images in Fig. 19.56 were produced by a scanning tunnelling microscope and show the actual atoms at the surfaces of objects. The images are formed by exploiting a strange property of the electron, the fact that it can behave as if it is in several places at once: by quantum theory, an electron could be at any place defined by its wave equation (see section 4, Chapter 17).

In the scanning electron microscope, a very fine needle passes at a constant tiny distance over the surface of an object, like the needle of a record player over the groove of a disk, but not quite touching it. The needle has a probability of 'capturing' electrons on the surface, even though physics suggests that the electrons do not have enough energy to escape. The electrons are imagined as **tunnelling** through a potential energy barrier.

The greater the number of electrons there are at a position on the surface – at the outer boundaries of atoms – the more the electrons will 'tunnel'. The variation in current they produce can be used to build up an image at the atomic level of detail. All this is done with just a small positive voltage on the needle.

The scanning tunnelling microscope has the following advantages over the electron microscope: it is very useful indeed for studying surfaces that would be damaged by high-voltage electron bombardment, and it does not require a vacuum to produce the images.

See questions 18 and 19.

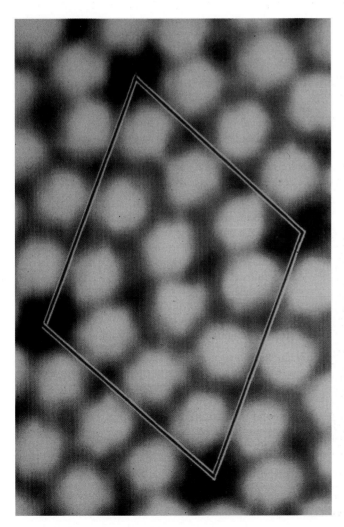

Fig 19.56 The scanning tunnelling microscope gives us images of individual atoms at surfaces

Above: A spot of gold atoms (yellow to red) about 3 atoms thick on a graphite base (green)

Left: In this image of a silicon crystal magnified 45 million times, the orange spots are each a silicon atom, and the faint extensions between some atoms represent the electrons binding atoms together. The diamond indicates the repeating pattern in the crystal

SUMMARY

As a result of studying the topics in this chapter you should:

■ Understand the importance of images and imaging in modern life.

■ Understand how lenses form images, and be able to distinguish between real and virtual images.

■ Understand the meaning of power in dioptres for a lens, as $1/f$ (in metres).

■ Know how to find the position and size of images in single lenses and mirrors both by drawing and by calculation, using the lens formulae:
$$\frac{1}{u} + \frac{1}{v} = \frac{1}{f} \quad \text{and} \quad M = \frac{v}{u}$$

■ Understand resolution and its calculation by the Rayleigh criterion:
$$\theta = \lambda/d$$

■ Understand how a range of optical devices work (magnifying glass, projector, camera, refracting and reflecting astronomical telescopes, compound microscope), including the eye, its possible defects and their correction.

■ Be able to use the formulae for magnifying power:
$$\text{telescope:} \frac{f_o}{f_e} \quad \text{microscope:} \frac{25L}{f_o f_e}$$

■ Know how images are 'captured' by film and other image-recording devices.

■ Understand the basic physical principles underlying some modern imaging systems such as CCDs, display screens, the ultrasound scanner, magnetic resonance imaging, the electron microscope and the scanning tunnelling microscope.

QUESTIONS

1 Do you need a mirror as tall as you are in order to see all of your body? Draw a ray diagram to explain your answer.

2 Explain the difference between a *real* image and a *virtual* image.

3 Explain why a prismatic mirror that uses total internal reflection gives a sharper image than an ordinary back-silvered mirror.

4 Find the size and position of the magnified virtual image formed by a magnifying glass of focal length 4 cm when the object is placed 3 cm from the lens.

5 A positive (converging) lens of focal length 50 mm is used in a slide projector.

a) How far from the lens must the slide be placed in order to form a clear image of it on a screen 12 m away from the mirror?

b) The image on the slide is 36 mm wide. How wide is the image on the screen?

6 Why are shaving and make-up mirrors often concave mirrors?

7 Why are convex mirrors used as rear-view mirrors in cars?

8

a) Explain what is meant by the *principal focus* and the *focal length* of a converging lens.
Indicating the position of the principal focus in each case, draw ray diagrams to show how a converging lens can produce **(i)** an upright magnified image of a real object, **(ii)** a clearly focused image of the Moon on a screen.

b) The Moon subtends an angle of 9.2×10^{-3} rad at the centre of a converging lens of focal length 0.80 m. The lens produces a focused image on a screen, and this image is viewed by an observer whose eye is 0.25 m from the screen.
Calculate:
(i) the diameter of the image on the screen,
(ii) the angle subtended at the eye by this image,
(iii) the angular magnification produced.

c) The lens in **b)** is used with another lens of power +20D to form an astronomical telescope. Draw a ray diagram showing the telescope used in normal adjustment to view the Moon. Show the passage through the instrument of **two** rays which are not parallel to the axis. Calculate:
(i) the separation of the lenses,
(ii) the angular magnification,
(iii) the angle subtended at the eye by the final image.

[NEAB: 1993 Syllabus A]

9 Give two reasons why astronomical telescopes have as large an aperture as possible.

10 An astronomical refracting telescope consists of two lenses, the objective and the eyepiece. The objective lens has a focal length of 60 cm. What is the magnifying power of the telescope when it is used (in normal adjustment) with eyepiece lenses of focal length **a)** 25 mm, **b)** 45 mm?

11 The Mount Palomar reflecting telescope has as its objective a parabolic mirror of focal length 80 m. A pair of stars has an angular separation of 0.001 radian when viewed with the unaided eye. What is their apparent separation when viewed through this telescope when the eyepiece used has a focal length of 250 mm?

12 Why do large astronomical telescopes use curved mirrors rather than lenses as their objectives?

13 Why are binoculars shorter than telescopes of the same magnifying power?

14 The information in this chapter about the use of charge coupled devices (CCDs) is likely to be out of date by the time you read it. Use a library or other source of information to gather information which will enable you to answer the following questions.

a) What is the physical size of the CCD used in a typical amateur video recorder (camcorder)? How many pixels does it have?

b) What other uses for CCDs are there **(i)** in industry, **(ii)** in scientific applications?

15 What decides the sharpness (resolution) of an image on a TV screen – the wavelength of the electrons used to make it or the pixel size of the screen? Give as numerate an answer as you can.

16 Why does a colour TV screen need only three colours (red, green, blue)?

17 Suggest reasons for using imaging devices which use techniques like magnetic resonance imaging and ultrasound scanning, rather than X-rays in medical diagnosis.

18 Write a short illustrated magazine article of about 500 words, explaining to non-physicists the importance of physics in creating the images they see in everyday life.

19 We need a wave theory to explain how images are formed, but a particle theory to explain how images are sensed or captured. Is this statement correct? Discuss it by considering the action of any optical imaging device.

Assignment

OPTICAL FIBRES

All long-distance telephone calls in Britain are now carried by optical fibres rather than copper wires. The radiation used as a carrier signal has a frequency of about 10^{14} Hz which allows optical fibres to carry very many simultaneous telephone conversations or data transmissions. Quality and carrying capacity is good enough for data-hungry TV films to be sent down optical fibre cables. There is more about this in Chapter 21 Communications.

Unlike copper wire carrier systems, optical fibre systems are not affected by electrical interference. Also, they are light and thin, so that a cable of given size can hold many more fibres than it could copper wires.

But there are two main problems in using optical fibres: attenuation of the signal, and overlap of information due to differing path lengths for the same data.

Attenuation

The signal enters an optical fibre as an infrared beam with the information coded digitally, as in Fig 19.A1. Some of this radiation is lost through scattering by molecules in the fibre, and some energy is lost by absorption processes. A typical value for the total power loss in a cable is 2.0 decibels per kilometre.

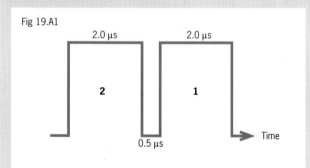

Fig 19.A1

2.0 µs   2.0 µs

2   1

0.5 µs   Time

Signal overlap

A simple optical fibre has a central glass core 60 µm in diameter. It is surrounded by a protective glass sheath 32.5 µm wide with a refractive index lower than that of the core itself (Fig 19.A2(a)). Such fibres are called step-index or single mode fibres.

The lower refractive index of the sheath allows total internal reflection to occur, so that once radiation is fed into the central core at a suitable angle, it travels down the core, bouncing from side to side, as shown in Fig 19.A3. Notice that two paths are shown. This is because, in this

design, the signal can travel by many different paths.

Suppose path 1 is taken by a single pulse of a digital signal and path 2 by the next consecutive pulse. Clearly, path 2 is longer than path 1, and eventually the second pulse will overtake and overlap with the first. This means that information is lost.

Another design of fibre uses glass in which the refractive index decreases gradually from the core outwards (Fig 19.A2(b)). Such fibres are called graded index fibres. They also allow total internal reflection. The advantage over single mode fibres is that rays which move in a region of lower refractive index can travel faster, yet the change in refractive index from centre to edge is designed so that all paths in the fibre have the same transit time.

However, today's optical fibre systems use single mode fibres that are thinner than shown in Fig 19.A2(a), thin enough for good signal transmission while avoiding the expense of graded index fibres.

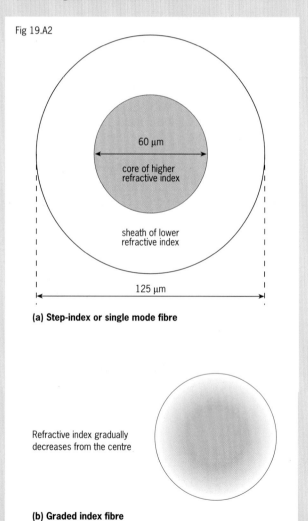

Fig 19.A2

60 µm

core of higher
refractive index

sheath of lower
refractive index

125 µm

(a) Step-index or single mode fibre

Refractive index gradually
decreases from the centre

(b) Graded index fibre

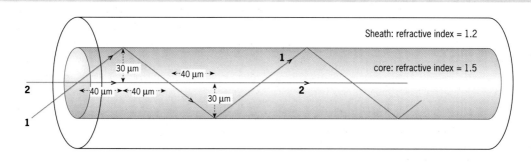

Fig 19.A3 **Paths of radiation moving in an optical fibre**

Data

The relation between attenuation A in decibels (dB) to power in watts is:

$$A = -10 \log_{10} \frac{P_o}{P_i}$$

where P_o is the power output and P_i is the power input.

Speed of light in a vacuum = 3.0×10^8 m s⁻¹.

$$\text{Refractive index} = \frac{\text{speed of light in a vacuum}}{\text{speed of light in a medium}}$$

1 A laser puts a signal at 10 mW power into a fibre optic cable of attenuation 2 dB km⁻¹.

a) What attenuation is produced by a cable of length 40 km?

b) Show that the output signal power of the signal at this distance is 10^{-10} W.

c) Signal power has to be significantly greater than random noise in the system, which is 1.0×10^{-18} W. The signal-to-noise ratio must be at least 20 dB. What is the lowest acceptable signal power in the cable?

d) Weak signals can be amplified to increase the signal-to-noise ratio. What is the maximum length of this cable that can be used between the laser and the receiver without having to use amplifiers between them?

2 The receiver for the fibre illustrated in Fig 18.A2 is 1 km from the laser.

a) How far does the first pulse actually travel as it goes from laser to receiver?

b) How far does the second pulse travel?

c) Calculate the times taken for each pulse to travel through the cable.

3

a) The signal data is carried by square pulses each 2.0 μs long, separated by an interval of 0.50 μs. Sketch the two-pulse signal on entry and at the receiver. Comment on the meaning of this.

b) Suppose the two pulses had entered a graded-index fibre in the same way. Sketch their possible paths and mark on where the radiation is travelling **(i)** at its fastest, **(ii)** at its slowest.

IMAGING

The chapter map below draws together the main concepts and areas of information covered in this chapter. The weight of type indicates the level of information and how the main concepts can be subdivided. The map also shows the connections between several main concepts and other related chapters.

You can use the map as a revision checklist by referring to the syllabus that you are following and identifying on the map the key information that you need to know and understand.

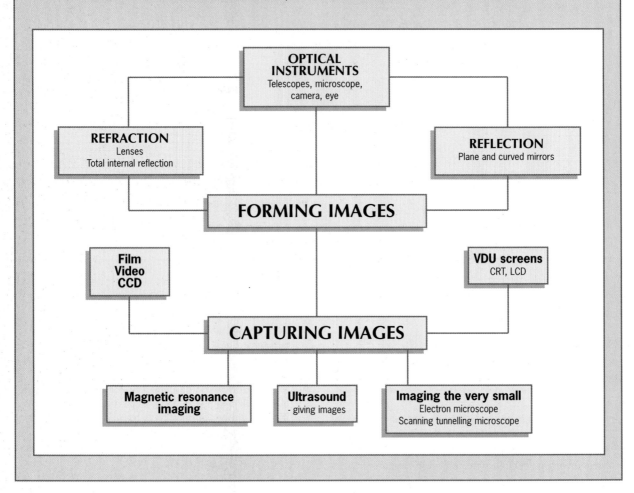

PHYSICS FOR CONTROL AND COMMUNICATION

WITH ITS COMMUNICATIONS revolution, the twentieth century may well be remembered as the age of communications. Modern communications rely both on the electromagnetic radiations which carry information, and also on the materials and techniques we use to control, store, transmit and receive the information.

We begin with the properties of the materials, essentially their electronic and magnetic properties, studied under the heading of condensed matter (sometimes referred to as solid state physics): the useful properties of conductors, insulators and semiconductors depend on the structures that arise when atoms are close together – 'condensed' – and the way that electrons are shared between the atoms. Some of these materials also have magnetic properties: they can be magnetised, either temporarily or permanently, and used in data storage devices.

You also learn how information for communications systems may be coded and transferred over long distances, and the electronic devices that make this possible.

20 Condensed matter

Computer-generated model of a high temperature superconductor $YBa_2Cu_3O_{6.5}$. Yttrium atoms are marked in grey, barium in green, copper in blue and oxygen in red

IN 1948, WILLIAM SHOCKLEY filed a patent for his new invention, the junction transistor. He had combined the theory of quantum mechanics from earlier in the century with the fast-growing knowledge of new materials and their uses. Ten years later, the electronic revolution began with the appearance of transistor radios and the first modern computers.

Since then, improved theory and technology have led to miniaturised devices of silicon, germanium and, more recently, gallium arsenide. Some devices have only a few layers of atoms, so that conduction is effectively within a two-dimensional surface. Electronic equipment containing these tiny devices fills our homes and workplaces. It is not often in history that a single technological breakthrough has such far-reaching effects.

What development will have such an impact in the future? It could be a superconductor that operates at room temperature. With such a device, we could in theory transmit electrical power over any distance with practically no energy losses.

Physics will continue to have the potential for changing our world when applied with the right materials.

1 INTRODUCTION: ATOMS AND THEIR OUTER ELECTRONS

You see that this chapter is titled 'Condensed matter', but it might have been called 'Solid state physics'. So why wasn't it? Partly because 'Condensed matter' is more fashionable. But more importantly, 'solid' obscures the fact that atoms and electrons in all materials are in constant motion, and 'physics' hides the need for an input from chemistry and the Periodic Table.

All atoms consist of a nucleus of protons and neutrons surrounded by electrons. A simple model of the atom has electrons moving round the nucleus in a series of orbits like planets orbiting the Sun.

But the theory of this simple model could not justify precise orbits with electrons having fixed energies. As explained in Chapter 17, the atom was better understood only with quantum mechanics.

Then, electrons could be treated as waves confined around a circular path or, more accurately, as three-dimensional waves filling the near space around the nucleus. Just as we can set up standing waves on a string attached at one end (Fig 20.1) with the other end fixed, so the waves associated with the electrons become standing waves set up as if confined within a spherical box.

✔ It requires slightly different amounts of energy to set up the different modes of vibration of a guitar string.

?

A Standing waves are set up in a string of length 4 m. Calculate the wavelengths of the first four harmonics. Show that they correspond to lengths of $8/n$ metres where n is an integer.

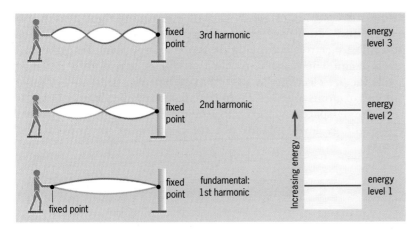

Fig 20.1 **Standing waves on a string. Each mode has an associated energy which can be shown as an energy level**

Many different modes of vibration become possible, and each of these modes involves a different amount of energy. The different modes of vibration of the electron wave lead to a series of **energy levels** which we can think of as like the energies associated with a series of planets moving around a central body.

The properties of an atom (excluding properties due to the nucleus) rely on the possible energy levels that the electrons can occupy in their wave-like form. So we represent the atom as a series of available electronic energy levels, see Fig 20.2(a). The number of electrons available to fill up the energy levels depends on the atomic number of the particular atom. If the atom is in its ground state, that is if it has not been excited, the electrons fill the levels one by one, starting from the bottom level. The bottom level corresponds to the level available to an electron which on average is nearest to the nucleus. (If we go back to thinking of the electron as a particle, then it moves around very rapidly and we cannot be sure where it is at any one time, but we can say some are nearer to the nucleus than others.) Only two electrons can occupy each level; this comes from quantum mechanics. Even these two electrons themselves must have different properties. We can imagine them as spinning tops: the two electrons in any single energy level must spin in opposite directions. This is shown in Fig 20.2(b) by tops spinning upright and upside-down.

When atoms come together to form compounds, it is the outer **valence** electrons in the highest filled energy levels that interact. As in Fig 20.3 we can picture that above this are empty levels and

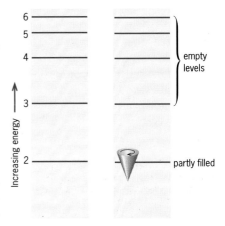

Fig 20.2 **Electron energy levels in an atom and adding electrons with spin to the levels. Electrons are represented by tops spinning upright or upside-down**

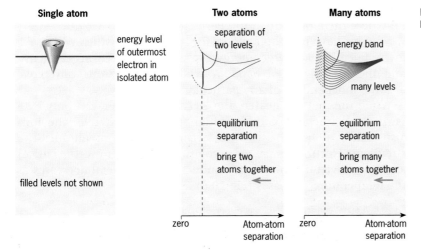

Fig 20.3 **Creating electron energy bands by bringing atoms close together**

below there are inner electrons which are more tightly bound. If two identical atoms are brought very close together, each with its outermost electron occupying the same energy level, then the two levels of equal height split in value, producing very slightly different levels for the electrons to sit within. When you bring a very large number of atoms together, then the same very large number of levels very closely spaced will exist for the electrons. We say that such a single energy level spreads out (in magnitude) to form **energy bands**.

2 METALS, INSULATORS AND SEMICONDUCTORS

Conduction in a metal

Conduction in a metal arises because free electrons can circulate among a lattice of ions (Fig 20.4). The ions are the original atoms minus their 'freed' outer electron which belongs to no particular atom. There may be more than one outer electron; it would be two for an atom of an element from Group 2 of the **Periodic Table** (see page 377).

The electrons move very fast: their speed is about 10^5 m s^{-1}. But their velocities are in all different directions so the electrons make no net progress in one direction round a circuit. However, when an e.m.f. is applied to the ends of a metal conductor, the electrons do move slowly round the circuit, forming a current. We say slowly as they move with a **drift velocity** which is typically 10^{-5} m s^{-1}. The drift velocity can vary by orders of magnitude depending on the cross-sectional area of the conductor and the current carried.

This model fits our band model. The electrons in the current are moving within an energy band. Just as people in a crowd cannot move around unless there is an area to move into, so electrons cannot move around within the energy bands unless there is a vacant *energy* space in which to move. If all energy states are filled, the electrons cannot change their energy.

To move round a circuit easily, an electron needs to be able to move within the energy bands as well as spatially through the material. So for good conduction in the metal, it is best to have a large number of filled and a large number of unfilled energy states. This can easily happen if *filled* and *unfilled* levels in isolated atoms spread apart and start to overlap when the atoms are brought close together (Fig 20.5). Below these overlapping levels, there are likely to be other completely filled levels. The number of levels depends on where the atoms are in the Periodic Table (page 377).

Now we have built up a model for the metal in which we have a lower filled band, a large energy gap where no electrons can exist, and the partially filled band in which the electrons can move around, independent of any particular atom.

When the metal is heated, the electrons go faster, but only by a very small fraction of their initial speed. Meanwhile, the ions also warm up and vibrate with larger and larger amplitude. The circulating electrons make more collisions with the ions (the lattice) and are slowed down. So the conductivity of a metal decreases as the temperature increases (Fig 20.6).

?

B Briefly, explain electronic differences between atoms from different columns of the Periodic Table.

C Assume that there is an electrical circuit around your room and that it includes a light bulb. Estimate how long, on average, it will take for a particular electron to complete the circuit from the moment you switch on. Approximately what *total* distance will the electron have travelled during this time? Why does the light go on instantaneously?

See question 1. ■

+ ion   ● electron

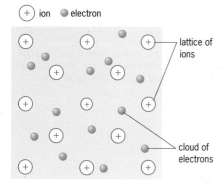

lattice of ions

cloud of electrons

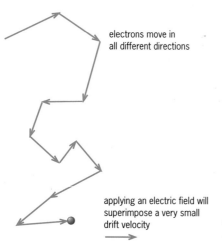

electrons move in all different directions

applying an electric field will superimpose a very small drift velocity

Fig 20.4 **A metal consists of electrons dispersed and moving freely within a lattice of ions. If an e.m.f. is applied, the electrons drift slowly round the circuit**

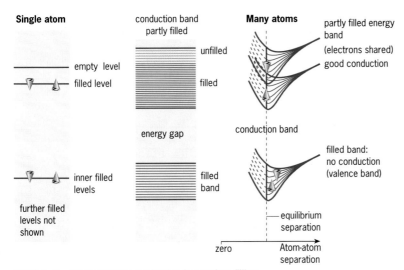

Fig 20.5 **Overlapping bands can lead to incomplete filling**

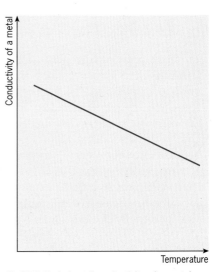

Fig 20.6 **Variation of conductivity of a metal with temperature**

Conduction in an insulator

An insulator has a lower, *filled* electron band, a bandgap that is quite large, and then another band which is empty, as in Fig 20.7(a). The electrons cannot move around in the filled band because there are no vacant energy states to move to, nor can electrons jump the gap to the unfilled band. Consequently the insulator has a very low electrical conductivity. Even if we heat the insulator, we cannot excite electrons to move from the filled to the unfilled band because it requires too large an energy jump.

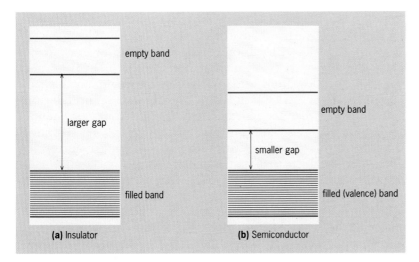

Fig 20.7 **Band structure of** (a) **an insulator,** (b) **a semiconductor at 0 K**

Conduction in a pure semiconductor

Semiconductors have had a massive impact on the design and usefulness of electronic equipment. Computers are amongst the best known and most useful equipment that contain them. Computers are only of a manageable size and speed of operation because semiconductor devices can be made very small and so electrons have to move only tiny distances. As the name suggests, a semiconductor carries current less easily than metal, but much better than an insulator. This is because it contains free charges which carry current, but fewer than in a metal.

Conduction electrons in an intrinsic semiconductor

Suppose an electron needs energy E to escape from its surroundings, and that this energy is supplied thermally. The temperature T (in kelvin) multiplied by the Boltzmann constant k indicates the amount of energy available to the electron. It can be shown that the number of electrons which acquire enough energy E to jump the energy gap is proportional to $\exp(-E/kT)$.

In fact, for N atoms, each with an electron available to break free, the number of electrons n which actually do so is given by:

$$n = N\,e^{-E/kT}$$

The number n rises rapidly as the temperature T increases (Fig 20.8). It is these electrons which reach the conduction band and allow electrical conduction in a semiconductor. The resistance of a semiconductor decreases as the number of carriers increases: double the number of carriers, and the resistance halves, and so on. It is not difficult to see that the resistance will vary proportionally to $e^{E/kT}$.

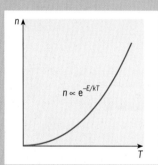

Fig 20.8 **How the number of electrons moving into the conduction band varies with temperature**

Additionally, by escaping from the lower valence band, each electron leaves behind a gap called a hole. We can liken this effect to a passenger

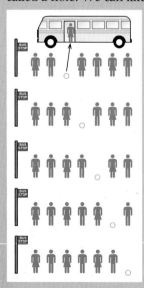

moving from a bus queue on to the bus. In the queue, as a person leaves, the rest move up, one by one, to fill the gap left by the previous person. You will see a gap move down the queue in the *opposite* direction to the actual movement of the passengers (Fig 20.9).

Fig 20.9 **The movement of people to fill a space in a bus queue is similar to the movement of electrons to fill a hole in the valence band of a semiconductor**

Similarly, in the valence band of the semiconductor, once a gap or hole is left, an adjacent electron can jump into it. Applying an e.m.f. to the semiconductor makes electrons jump like this. With repeated jumps of electrons, one after another, holes move through the semiconductor in the opposite direction to the electrons.

Electrons move under the influence of the e.m.f. towards the positive potential (corresponding to the bus-stop sign), while the holes in the valence band move the opposite way. It is *as if* the holes have a positive charge. So we say that such a semiconductor has electrons in the conduction band and holes in the valence band and that both act as carriers of charge. The electrons move *against* the electric field as they have negative charge, the holes move *with* the field as they are of positive charge.

We get the overall conductivity by adding together the contributions from both types of charge. Their contributions must be added as, although they have opposite charge, they also move in opposite directions. When electrons that are excited into the conduction band leave behind an *equal* number of holes, we describe the semiconductor as intrinsic and the conduction as intrinsic conduction.

Because far fewer charges move in a semiconductor than in a metal, a semiconductor has a lower electrical conductivity. This lower conductivity is slightly balanced by the mobility of the electrons (their ability to move in an applied e.m.f.) which is much higher for a semiconductor than for a metal: the drift velocity of semiconductor electrons is usually many metres per second.

As the temperature of a semiconductor is increased, the mobilities of the electrons (and holes) decrease, just as they do in a metal. But the increased temperature enables more electrons to be excited from the valence to the conduction band. This effect happens rapidly as the temperature effect is exponential and it swamps any variation of the mobilities. Overall, as temperature rises, the increased number of carriers increases the conductivity of the semiconductor (Fig 20.10).

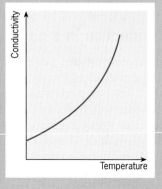

Fig 20.10 **Variation of conductivity of an intrinsic semiconductor with temperature**

At 0 K, the band structure of a semiconductor looks similar to that of an insulator; filled band, bandgap and unfilled band, as in Fig 20.7(b). The difference between the semiconductor and the insulator is that the semiconductor's bandgap is small. Electrons in the filled band require only a little energy to excite them into the higher band.

We can estimate whether electrons will be excited thermally by comparing the energy required to jump up through the energy gap with the magnitude of kT where k is Boltzmann's constant and T is the temperature of the semiconductor measured in kelvin. Typically, the bandgap energy is quite a lot larger than kT, but sufficiently small for the thermal energy to excite a significant number of electrons to make the jump. Once electrons are excited into the upper band, they can move freely and so the semiconductor can carry electricity. This upper band is called the **conduction band**.

Fig 20.11(a) **Gallium metal rods and arsenic nuggets are fused at high temperature to form this lump of gallium arsenide**

Examples of semiconductor materials

Typical semiconductor materials are the elements silicon and germanium, with atoms in a structure like that of diamond (page 147). The atoms are *covalently* bonded, meaning that adjacent atoms *share* outer electrons and come from Group 4 of the Periodic Table (see page 377).

For silicon the bandgap is 1.1 eV and for germanium it is 0.67 eV. kT measured in eV at room temperature is 0.026 eV, that is, $(8.6 \times 10^{-5} \times 300)$ eV.

Another commonly used semiconductor material is the compound gallium arsenide with a bandgap of 1.43 eV. Gallium is from Group 3 and arsenic from Group 5. The atoms take up alternate sites in the diamond structure and are also covalently bonded: adjacent atoms share electrons (unlike ions which lose or gain electrons).

Compound semiconductors are similar to single-element conductors. For example, gallium arsenide is a more complicated material to prepare and use than single element semiconductors, yet it is popular. The reason is that the electrons move through gallium arsenide very much faster than they do in silicon or germanium, so devices operate more rapidly. Other, less used, semiconductor materials contain atoms of three different elements.

By altering the proportions of elements and compounds, semiconductors can be made with the right bandgaps for their use. For example, semiconductors used in infrared camera can be made to detect light of infrared frequency. The incoming photons of light each have an energy equal to the semiconductor bandgap energy and are absorbed as they excite electrons up across the gap. In reverse, light is emitted by semiconductor lasers that are designed to produce light of a particular frequency and colour. They are used in CD players and for optical communications.

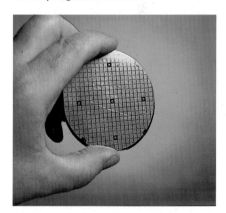

Fig 20.11(b) **A silicon sheet: each square is doped and becomes an integrated circuit**

✔ The energy required to excite electrons from the valence to the conduction band is always stated in eV rather than in J.

?
D What is 1 eV when measured in J?
E Estimate the fraction of electrons which are excited from an energy band through an energy gap of 1.1 eV at a temperature of 300 K. What is the fraction at 600 K?

Thermistors and light-dependent resistors

Thermistors detect variations in temperature and so are rather like thermometers, while LDRs, light-dependent resistors, detect variations in light level. Varying the temperature of thermistors changes the number of available electrons for conduction, while incoming light alters the number in LDRs.

A thermistor is made of a semiconductor material such as nickel oxide. Its resistance very rapidly decreases exponentially (not linearly) as its temperature rises because the number of free carriers increases exponentially. As we have seen, increasing the temperature, and hence thermal energy, excites more electrons into the conduction band.

A typical variation of resistance for a thermistor is 4.7 kΩ at room temperature and 270 Ω at 100 °C. So thermistors are used in electronic circuits where big changes of output signal are required for small variations in temperature (page 210). An example is a computer used for monitoring a process that produces thermal energy, such as a chemical reaction. Being sensitive to a tiny energy change, the thermistor gives rise to a large input signal that is used to record and control the process.

LDRs are usually made of cadmium sulphide which has a bandgap of 2.6 eV – larger than the bandgap for a typical semiconductor. For LDRs, it is photons of light which provide the energy to excite the electrons from the valence into the conduction band. The dark resistance of an LDR is typically 1 MΩ, and this decreases with increasing light intensity to 1 kΩ or less, depending on the magnitude of intensity. LDRs are also easily incorporated into electronic circuits (pages 209 and 212).

?

F Sketch the variation of resistance of a thermistor with temperature.

Conduction in a doped semiconductor

The precise conducting properties of semiconductors used for transistors (see page 455), thermistors and LDRs are controlled by adding very small amounts of 'impurity' which is a very tiny quantity of another element. The process of adding the impurity is called **doping** and the extra material is the **dopant**. Dopant atoms go into positions in the original semiconductor lattice to replace a few of the atoms of the pure material. Their size must be similar to the size of the atoms they are displacing. Assuming we start with a Group 4 semiconductor such as silicon, then we choose either a Group 3 element such as boron or a Group 5 element such as arsenic or phosphorus. Choosing to replace with a Group 3 or a Group 5 element leads to quite different changes.

Arsenic-doped silicon

Starting with a lattice of silicon atoms as in Fig 20.12(a), each atom shares its outer four electrons with four other atoms and shares a further four electrons from its neighbours to establish a filled shell of eight electrons. Conductivity occurs when some of the electrons are excited thermally into the conduction band. If we now replace a few silicon atoms with arsenic, each arsenic atom brings with it an *extra* electron, as in Fig 20.12(b). This extra electron in arsenic is easily excited into the conduction band. For the bulk material, the band structure diagram on the right of Fig 20.12(b) shows these extra arsenic electrons within the energy band diagram just below the conduction band.

At room temperature they are thermally excited up into the conduction band. By controlling the number of dopant atoms we can control the number of current-carrying electrons and hence the conductivity. The dopant atoms are referred to as **donors** because they are giving up electrons and the semiconductor is called **n-type** because it has *negative* type carriers.

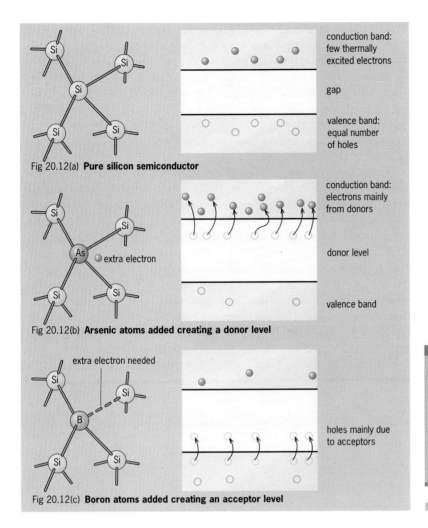

Fig 20.12(a) **Pure silicon semiconductor**

conduction band: few thermally excited electrons

gap

valence band: equal number of holes

Fig 20.12(b) **Arsenic atoms added creating a donor level**

As   ● extra electron

conduction band: electrons mainly from donors

donor level

valence band

Fig 20.12(c) **Boron atoms added creating an acceptor level**

extra electron needed

B

holes mainly due to acceptors

?

G How would you expect the conductivity to vary with temperature for an extrinsic n-type semiconductor in which the carrier concentration is due to donors? (Remember that the conductivity will depend on the carrier concentration and the carrier mobility.)

■ See question 2.

Boron-doped silicon

We can instead add boron atoms as a dopant. They have one electron *less* than the silicon atoms and the lattice can easily take up electrons from the valence band. This leaves holes in the valence band to carry current. The boron atoms are called **acceptors** as they are taking up electrons and the semiconductor is called **p-type** (*positive* carriers). As in Fig 20.11(c), we represent the acceptors on the energy band diagram at a level just above the valence band.

A semiconductor in which the conduction is dominated either by electrons from donor atoms or holes from acceptor atoms is called an **extrinsic** semiconductor.

We can either have an *n-type* semiconductor in which the carriers are predominantly electrons or if we have a *p-type* semiconductor in which the carriers are holes. For either case, we can find out the number of carriers by measuring the Hall voltage (see page 254).

Note that: $I = neVA$
and: $V_H = BI/nte$
(See page 255.)

Hall voltage V_H

carrier density n

current I

t

area A Magnetic field B

Remember the Hall effect

■ Try questions 3 and 4 at the end of the chapter.

The pn junction – the junction diode

A pn junction consists of a piece of p-type semiconductor joined to a piece of n-type semiconductor. The lattice of atoms in which the p-type and n-type impurities are inserted is often the same material. Holes exist in the p-type material and extra electrons in the n-type material and it is these which can carry current.

?

H Explain how the Hall voltage can tell us which type of carrier exists in an extrinsic semiconductor.

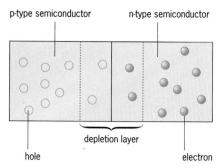

p-type semiconductor   n-type semiconductor

hole   electron

Fig 20.13 **A pn junction showing the depletion layer**

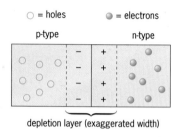

○ = holes   ● = electrons

p-type   n-type

depletion layer (exaggerated width)

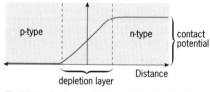

p-type   n-type   contact potential

depletion layer   Distance

Fig 20.14 **Contact potential within the depletion layer**

Very close to the junction, the holes and electrons cancel each other out, so that there are very few free carriers here. This region is called the **depletion layer** (Fig 20.13). Because there is now a shortage of holes on the p side of the junction, it is as if there is a negative charge there relative to the rest of the material, and the shortage of electrons on the n side makes this a positively charged region. This gives rise to a contact potential (Fig 20.14) maintained by the distribution of holes and electrons; there are no charge carriers, so there is no current.

Let us apply a *reverse bias* to the junction. This means that we apply a voltage difference across the junction with a positive voltage going to the n-type side and a negative voltage to the p-type side. This increases the contact potential, see in Fig 20.15(a), and attracts even more free carriers (both holes and electrons) away from the contact region.

If we apply the voltage difference the other way round, in forward bias, the applied potential difference will cancel the contact potential, see Fig 20.15(b). The depletion layer is narrowed and then removed and the pn junction conducts normally. Hence the junction diode passes current in forward bias but not reverse bias, see Fig 20.15(c), and can be used for **rectification**, meaning that current is allowed to pass in one direction only (as described on page 287).

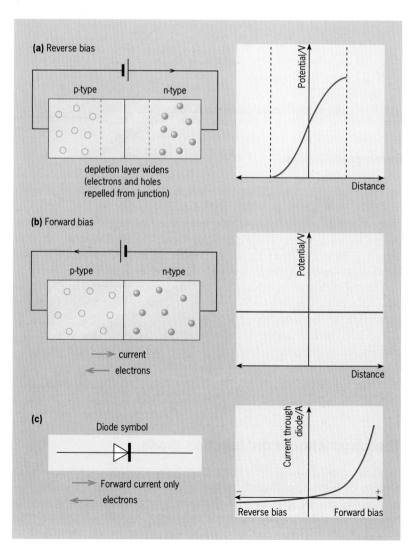

(a) Reverse bias

p-type   n-type

depletion layer widens
(electrons and holes
repelled from junction)

Potential/V

Distance

(b) Forward bias

p-type   n-type

→ current
← electrons

Potential/V

Distance

(c)

Diode symbol

→ Forward current only
← electrons

Current through diode/A

Reverse bias   Forward bias

Fig 20.15 **A junction diode in verse and forward bias.**

(a) **In reverse bias, electrons and holes are attracted away from the junction and there is a large potential to be overcome, preventing a current.**

(b) **In forward bias, the potential is removed.**

(c) **The diode symbol, and the overall variation of current as the potential is varied both positive and negative**

A light-emitting diode (LED) is made from a junction of two semiconductors, gallium phosphide and gallium arsenide. There is a current only one way, as with an ordinary junction diode, but as current passes through the junction, light is emitted as electrons drop between energy levels.

Red, yellow and green LEDs have been in use from some time, but it is only recently that blue LEDs have become practicable. For blue light ($\lambda = 450$ nm), a forward voltage drop of 2.8 V is needed. Gallium nitride (GaN) LEDs have been developed with forward voltages of 3.4 eV ($\lambda = 360$ nm) and aluminium nitride (AlN) LEDs with ultraviolet outputs (6.2 eV). LEDs can now be made as bright as a 500 W traffic light whilst using just one tenth of the energy. They also respond (light up) more quickly and should soon take over from filament bulbs in, for example, car stoplights as well as traffic lights. Ultraviolet LEDs could also replace the gas discharge tubes used in fluorescent lighting.

The pnp transistor

The pnp transistor was invented in 1951 John Bardeen, Walter Brattain and William Shockley, at the Bell Research Laboratories in the USA; it is used very widely as a controllable switch.

It consists of a thin layer of n-type material sandwiched between two thicker regions of p-type material. The transistor turns on and controls a large current through one of a pair of pn junctions (the high-resistance collector-base junction) by passing a small current through the other (low-resistance base-emitter junction). Transistors are particularly suited for controlling the on–off binary logic used in computers.

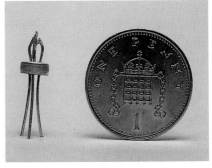

Fig 20.16 **The pnp transistor**

?

I Show that for an LED to emit yellow light ($\lambda = 600$ nm), electrons of charge e must fall through a voltage difference of 2.0 V. (Remember the energy of a photon is *hf*.)

J A car stoplight becomes visible some time after the brakes have been put on. A bright red LED lights up 0.25 s faster than a filament lamp. Make a quantitative estimate of the effect this advantage has in traffic travelling at motorway speed.

3 SUPERCONDUCTIVITY

In 1911, a Dutch physicist, Heike Kammerlingh-Onnes, was investigating what happened to the electrical resistivity in metals as he reduced their temperature down to that of liquid helium, 4.2 K. He found that the resistivity of platinum levelled out to a constant low value. Impurities in the platinum, however small in number, scattered the electrons and added a finite resistance.

The great surprise came when he used purer mercury. Below 4.2 K it appeared to have no resistivity at all, yet on an increase in temperature, the resistivity shot up at 4.2 K (Fig 20.17). Some important change was happening at this temperature.

The critical temperature at which mercury becomes a superconductor is called the **transition** temperature. A metal which can 'go superconducting' has a transition temperature below which resistivity drops to zero and it becomes a superconductor. In addition, both the current density in the superconductor and any external magnetic field adjacent to it must be below critical values.

An important application of superconductivity is *superconducting magnets* which are held at low temperature using liquid helium. The current through the coils of the magnets keeps flowing indefinitely and there is no heating of the wire. The magnetic fields produced are exceptionally stable and in medicine are used in scanners for magnetic resonance imaging (MRI) of the human body (see page 546).

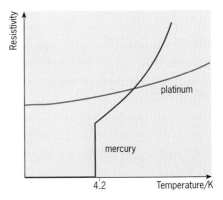

Fig 20.17 **Resistivity of platinum and mercury at very low temperatures**

?

K What would happen if the temperature suddenly went up or the current density increased significantly through the superconducting coils of the magnet in a magnetic resonance imager?

The *TjB* surface of a superconductor

The critical conditions for superconductivity can be shown on a *TjB* diagram, such as Fig 20.18. This is a three-dimensional plot with temperature *T* along the *x* axis, current density *j* within the superconductor up the *y* axis, and the magnetic field *B* (adjacent to the superconductor) along the *z* axis.

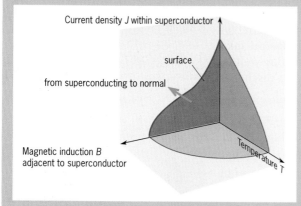

Fig 20.18 **Critical *TjB* surface for a superconductor**

The surface on the *TjB* plot identifies all the points at which the critical conditions apply. As soon as any one of *T*, *j* or *B* increases, and we move in three dimensions out through the critical surface, the superconductor changes from superconducting to normally conducting. We say that a change of phase has occurred. (It is the same as if we plotted a critical surface on a *pVT* diagram to show the *p*, *V*, *T* conditions when a gas changes to liquid.) Table 20.1 shows the transition temperatures for some metals and alloys without any magnetic field present. It includes the value for YBCO, the high temperature superconducting alloy $YBa_2Cu_3O_{6.5}$, pictured at the start of this chapter.

Table 20.1 **Transition temperatures for some superconducting materials in zero magnetic field**

Material	T_c/K
Al	1.20
Hg	4.15
Pb	7.19
Nb	9.26
Nb_3Ge	23.2
YBCO	93

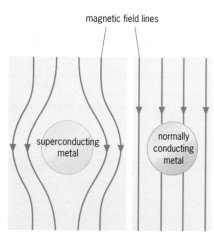

Fig 20.19 **A magnet levitating above a superconductor**

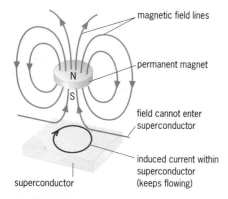

Fig 20.20 **A superconductor excludes magnetic field lines, but a normally conducting metal does not**

Magnetic levitation and the Meissner effect

A superconductor can keep a permanent magnet suspended above its surface (Fig 20.19). As the magnet starts to fall it induces an e.m.f. and hence a current in the superconductor, obeying Faraday's law (see page 264): charge flows around at the surface of the super-conductor totally unresisted. And in agreement with Lenz's law (page 265), the effect of this current is to oppose the motion of the magnet, so it remains levitated.

To explain more fully, the current rises to the value that produces a magnetic field which exactly matches and opposes the magnet's field. The internal 'supercurrent' at the superconductor surface can flow continuously and so it behaves as another magnet. Its poles are in the opposite direction to those of the permanent magnet, and the magnets repel. The separation adjusts until the force of repulsion exactly balances the gravitational field.

The field lines of the permanent magnet are excluded from entering the surface of the superconductor, and this is named the **Meissner effect**. When a superconductor is placed in parallel lines of magnetic induction *B*, the magnetic field lines are pushed out and bend round it (Fig 20.20). This is not the case for an ordinary metal or when the superconductor goes normal. Only advanced quantum mechanics can explain superconductivity, so we shall not explain it here!

High temperature superconductors

Interest in superconductors was reinforced in 1987 when a super-conductor with a transition temperature as high as 93 K was dis-covered. This is the material illustrated at the start of the chapter. It has the approximate formula $YBa_2Cu_3O_{6.5}$ and is often called YBCO because it contains yttrium, barium, copper and oxygen. Other materials with high transition temperatures were discovered and, like YBCO, they are compounds with imprecise ratios of elements.

These superconductors are likely to be more useful than the low temperature superconductors. In particular, their transition temperature is above that of liquid nitrogen which is relatively cheap and can be used to cool the superconductor. Unfortunately, it has so far proved difficult to make these 'high temperature' superconductors in bulk form with tensile strength. It may be some years before useful superconductors at room temperature are made.

See question 5.

L List three possible applications for room-temperature superconductors (if and when they become available). Explain the advantage of a superconductor over a conventional conductor in each case.

4 DIELECTRICS

We have already talked about the large bandgap in insulators (page 451). A material which cannot conduct electric charge is often called a **dielectric** (the use of dielectrics in capacitors is discussed on page 237). The dielectric constant, or relative permittivity, of an insulator is defined as:

$$\varepsilon_r = \frac{\text{capacitance of capacitor with dielectric between plates}}{\text{capacitance of capacitor with vacuum between plates}}$$

ε_r is always greater than 1. It is 1.0006 for air, 2.1 for Teflon, a polymer used for insulation, and 310 for strontium titanate, a special ceramic used to produce high capacitance layers inside small electronic devices. Inserting a dielectric between capacitor plates allows more charge to be stored than if there is just air, and it increases capacitance.

The electric field between the capacitor plates causes the molecules within the dielectric to *polarise*. What we mean by this is that the positive and the negative charges become separated within the molecules.

As you can see from Fig 20.21(b), this polarisation of the molecules causes a slight excess of positive charge on one surface of the dielectric (the surface furthest from the positive plate) and an equal excess of negative charge on the other surface. Only a few excess charges are enough to produce a sufficiently large internal electric field to cancel out most of the applied field. (The charge on the surface of the dielectric balances much of the charge on the plates).

While the capacitor is connected to a constant voltage source, this source supplies more charge in order to maintain the electric field at its original strength, and to maintain constant voltage across the plates. So a lot more charge is retained on the plates with a dielectric than with air. The more that molecules polarise in the dielectric, the more the extra charge that can be added to the plates. For a material to have a large dielectric constant, it must polarise easily.

There is a limit to the voltage that can be applied across the capacitor plates before breakdown and sudden discharge of the capacitor. Capacitors are always marked with the highest safe voltage which can be applied to the plates. Dry air at atmospheric pressure breaks down at 3000 V mm^{-1} (think of the discharge from a Van de Graaff machine or the lightning in a thunderstorm).

Modern, purer dielectric materials can withstand much higher electric fields. (Impurities in the dielectric usually cause breakdown at lower fields than in the pure material.) At breakdown, the dielectric acts as a conductor. A small number of mobile carriers are so accelerated that molecules ionise, more charge carriers are produced, and the situation avalanches into total conduction. (You can compare

M Name a material other than strontium titanate which is called a ceramic.

$$C = \frac{\varepsilon_0 \varepsilon_r A}{d} \text{ from page 237.}$$

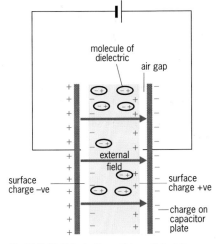

Fig 20.21(a) **A capacitor cut across to show the very thin metal foil that is wound round (with insulator layer between) to increase the area**

molecule of dielectric

air gap

external field

surface charge –ve

surface charge +ve

charge on capacitor plate

Fig 20.21(b) **Polarisation within a dielectric**

N Assuming a breakdown voltage of 8 kV mm^{-1} for a ceramic insulator, calculate the minimum thickness of insulator necessary for a 500 kV transmission line.

See question 6.

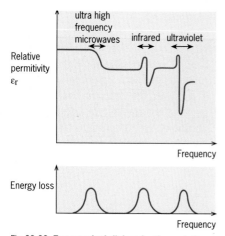

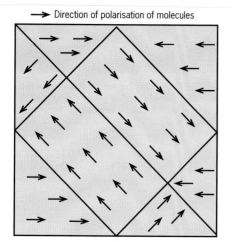

Fig 20.22 **For a typical dielectric, the variation with frequency of** (a) **its relative permittivity and** (b) **the energy loss within it**

this to just a small movement of snow on a mountain which can magnify into a large-scale avalanche.)

So far, we have looked at what happens when we apply a static field to the dielectric. When an a.c. electric field is applied, the response of the material depends on how easily the polarisation charges can reverse their orientation. This in turn depends greatly on the type of bonding between the molecules in the dielectric material.

The permittivity of the dielectric usually varies considerably with frequency (Fig 20.22). As the electric field rapidly changes direction, the polar molecules rotate relative to their neighbours. The interaction between neighbouring molecules causes a loss of energy; this energy is provided by the a.c. source and is lost as thermal energy, in giving motion to the particles.

Also, any charged particles tend to move to follow the alternating field. How successfully and at what frequency they can do this depends on their mass. For a dielectric not containing polar molecules, electrons move relative to the nuclei and can keep up with the field, even at high frequencies. For a dielectric made up of polar molecules, electrons cannot keep pace with a rapidly changing field, so the dielectric constant falls in value at high frequencies.

Ferroelectrics

Some materials polarise spontaneously, that is, there is separation of charges without the presence of an applied electric field. Such materials are called **ferroelectrics**. This does not mean that they are made of iron (ferro) but they can possess permanent electric fields similar to the permanent magnetism in alloys containing iron. Quartz and barium titanate are ferroelectrics.

A recent example is PZT. It is an alloy of oxides of lead (**P**b), **z**irconium and **t**itanium, and its properties can be adjusted by varying the proportions of the components. For instance, it can be used to detect ultrasonic sound waves of particular frequencies in underwater detection systems.

The direction of polarisation may not be the same throughout a ferroelectric. Instead, the polarisation may have a **domain** structure (Fig 20.24) in which the internal electric field changes from domain to domain (i.e. from region to region).

Domain structure is particularly important for the **piezoelectric effect** in which energy applied mechanically is directly transferred to give energy electrically within the material.

When polarisation is established in a sample of piezoelectric material in one direction only, then the material becomes slightly *longer* in that direction than it would be if the domains were arranged in different directions and summed to zero. Therefore, by altering the overall polarisation, it is possible to set up or to release a strain, that is, it is possible to lengthen the sample or to release an extension.

Just as importantly, the reverse can occur. An applied mechanical stress can set up a strain within the specimen that can alter the overall polarisation.

One practical advantage of a piezoelectric material is that it can respond very quickly – at megahertz frequency. That is why, used in the field-to-strain mode, piezoelectrics are used as vibrators in watches and clocks. Quartz crystals, for example, vibrate with a very precise frequency, like a pendulum but with much greater accuracy. In the strain-to-electric-field mode, the piezoelectric is used in the spark igniter, see the Feature box.

Fig 20.24 **Domain structure within a ferroelectric**

See question 7. ■

ELECTRIC FIELDS IN THE KITCHEN

THE ABSORPTION OF ENERGY that gives motion to particles is put to good use in the **microwave oven**. Here, the interaction is between the electric field component of the microwaves and the water molecules in the food. Microwave energy becomes thermal energy for cooking the food.

Most of us have used a **spark igniter** that doesn't need a battery to light the jets of a gas cooker: we press a knob which presses on a small ceramic block and this produces a spark.

Pressing the knob applies a force to one face of the block and the force is resisted at its opposite face where it is fixed. The force sets up an electric field because the molecules in the crystalline grains, and hence their charges, are compressed and realigned. The ceramic is called a **piezoelectric** (piezo = pressure) and it is part of an electric circuit with a small gap. The field in the piezoelectric sets up a high voltage at one side of the gap, and the voltage is discharged when the gas-igniting spark jumps across the gap.

Fig 20.23 (a) **A microwave oven,** (b) **a spark igniter**

5 MAGNETIC MATERIALS

We have discussed ferroelectric materials in which polarisation of the molecules can maintain permanent electric fields, and we have mentioned permanent magnetism in iron. Materials which can be made magnetic in a similar way to iron alloys are called **ferromagnetic**. Like ferroelectrics, the property is associated with the component atoms. Also like ferroelectrics, they may show a domain structure.

We usually connect magnetic fields with moving charges (electrons) within a wire and, as a current passes through the wire, we detect the associated magnetic field round the wire (see page 269). But many atoms themselves, such as those in ferromagnets, have so-called magnetic moments. That is, they behave rather like very small bar magnets. As in a current-carrying wire, the magnetic field which we detect in a ferromagnetic material must arise from moving charges: in the material, the field arises from the motion of certain unpaired outer electrons in its atoms. Whereas many atoms have magnetic moments, once these atoms are in combination as ions or molecules, the magnetic moments usually combine and cancel. It is only certain elements such as iron, nickel and cobalt (transition elements, see page 377) that have magnetic moments which align in such a way as to maintain the magnetism in the bulk material.

To form a permanent magnet from ferromagnetic material, the magnetic moments of its atoms are lined up in the same direction by inserting the specimen into a solenoid (Fig 20.25). Current through the turns of the solenoid is gradually increased and the magnetic field along the axis of the solenoid increases in proportion.

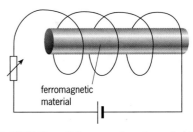

ferromagnetic material

Fig 20.25 **Magnetising a specimen in a solenoid**

Remember (from page 269):
$$B = \mu_0 nI$$
(where n is the number of turns per unit length), so the magnetising field is proportional to the magnetising current.

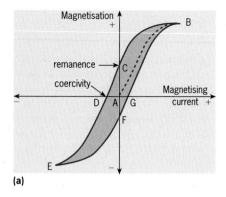

(a)

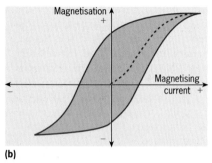

(b)

Fig 20.26 **Hysteresis loops for** (a) **a soft magnetic material and** (b) **a hard magnetic material**

?

O In a magnetic material the magnetic moments of atoms are aligned. If you used a solenoid to demagnetise a specimen would you use d.c. or a.c. current? Can you think of another way of demagnetising the specimen without using a solenoid?

See question 8. ■

The magnetisation of a specimen follows the magnetising current, but not always linearly. Look at Fig 20.26(a): starting from zero magnetisation, the magnetisation curve follows AB, when at B the specimen has reached magnetic **saturation**. At this stage, all the atoms with magnetic moments have become aligned with the magnetic field. If the current in the solenoid is now decreased, the magnetisation in the specimen also decreases, but lags, following curve BC. When the current through the solenoid is zero and the magnetising field is also zero, there remains a residual value of the magnetic induction at C called the **remanence**. This means that the material remains partly magnetised.

We now reverse the current in the solenoid. The reverse magnetic field from the solenoid must reach a particular value, the **coercivity,** at D before the field in the specimen has been eliminated. At this stage, the magnetic moments of the atoms in the material will be randomly aligned. (Alternatively, the magnetic moments may be aligned *within* domains, but the domains will then be randomly aligned.) Further increase in the reverse magnetic field now starts to produce a reverse magnetisation in the specimen that ultimately saturates at E.

As we continue to cycle the current, the field in the specimen follows the outside loop BDEGB. The area enclosed within the loop represents the energy lost (ultimately heating the magnet). If this area is small, then the specimen consists of a **soft** magnetic material. If the area within the loop is large and coercivity is large, then it is a **hard** magnetic material, as in Fig 20.26(b). This makes sense as it needs a large amount of reverse current to drive a hard material into its reverse magnetisation.

A hard material is suitable for a permanent magnet where we want it to be difficult to reverse the direction of magnetisation. A soft material, that gives little opposition to change of magnetisation so that energy loss is small, is suitable for a transformer core. Steel cannot be easily magnetised and is a *hard* magnetic material; iron can be easily magnetised and is *soft*.

Let us look again at the curves in Fig 20.26. Choose a particular magnetising current, and you will see that it has two possible maximum values of magnetisation for the specimen. (This is apart from high current in the solenoid, when the magnetisation has saturated.) Which value applies in practice depends on whether we are increasing or decreasing the current in the solenoid, since the value of the specimen's magnetisation depends on its previous magnetisation history, shown as the path taken on the magnetising current–magnetisation graph. And so we say that the value for the specimen is path dependent, a property we call **hysteresis**.

When an electromagnet is used to make measurements or obtain data such as magnetic resonance in a varying magnetic field, the hysteresis is kept small. This is done by using a very soft magnetic material, when the solenoid current–magnetisation variation will be close to a single curve. Then, the energy loss within the magnetic material during the cycle is negligible.

SUMMARY

By studying this chapter you should have learned the following:

■ When atoms come close together, electron energy levels split and spread out to form energy bands.

■ Conduction in a metal occurs because electrons can move among partially occupied energy levels of the conduction band.

■ Insulators cannot conduct electricity because they have a filled valence band and an empty conduction band separated by a large energy gap.

■ A pure semiconductor has a small energy gap between valence and conduction bands, and so electrons can be thermally excited into the conduction band at finite temperature, leaving an equal number of holes in the valence band.

■ Semiconductors can be doped with donor and acceptor impurities to produce n-type and p-type semiconductors respectively.

■ The electrical conductivity of a metal decreases with temperature, whereas that of an intrinsic semiconductor increases with temperature.

■ The Hall effect can be used to find the type and number of carriers in an extrinsic semiconductor.

■ Superconductors require temperature, current density and external magnetic field to be below critical values, in order to have negligible resistance.

■ Superconductors exclude magnetic field; this is the Meissner effect.

■ Dielectrics exhibit polarisation of charge. Compared with air, a dielectric between capacitor plates increases the ability of the capacitor to store charge.

■ Soft and hard magnetic materials exhibit hysteresis loops, with small and large enclosed areas respectively.

QUESTIONS

1 Explain the term drift velocity in the context of electrical conduction of a metal. A wire has radius $r = 1.00$ mm and carries a current of $I = 5.0$ A. It is made of a metal with 9.0×10^{28} free electrons m^{-3}. Estimate the drift velocity of the electrons.

2 You are supplied with three specimens:
(i) pure germanium, **(ii)** germanium doped with antimony (Group 5) and **(iii)** germanium doped with indium (Group 3).

a) What is the nature of the charge carriers in each case?

b) Sketch the variation of resistivity with temperature of the pure germanium specimen. Would you expect the variation to be different for the other two specimens? Explain.

3 In measuring the Hall effect in a metal, why is it necessary to use a thin foil of metal, whereas in measuring the effect in a semiconductor it is suitable to use a sample of a few mm thickness?

4 A steady d.c. current of 45 mA is passed through a rectangular slice of an n-type semiconductor of size 12 mm × 5 mm × 2 mm, as shown in Fig 20.Q4. A uniform magnetic field of 100 mT is applied perpendicular to the slice and a potential difference of 4.1 mV is measured across the sample.

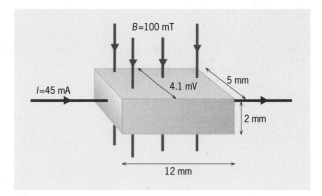

Fig 20.Q4 **A Hall sample**

Explain why this potential difference develops and calculate the electron carrier density in the semiconductor.

5 A conducting wire at low temperature consists of superconductor filaments mixed in a matrix of copper filaments. By mistake, the superconductor is allowed to enter its normal (resistive) state.

a) What conditions might have changed for this to happen?

b) Discuss what now happens to the current flow, and hence say why the superconductor has been mixed with the copper.

c) Assuming that the copper now conducts away excess heat, indicate any further change that might occur.

6

a) A parallel plate capacitor consists of plates with area $A = 36$ mm^2 and separation $d = 3$ mm, and is connected to a potential difference $V = 3$ kV. Calculate the charge on the plates.

b) A sheet of insulating plastic material of dielectric constant 3.00 is inserted between the plates to fill the space completely. A quantity of charge flows around the circuit as the sheet is inserted. Explain why. Calculate the amount of charge which flows.

7

A spark igniter contains a block of piezoelectric ceramic measuring 2 mm square by 1 mm thick, with electrodes on its square faces. A load of 30 N is applied to the square faces by thumb pressure on a lever. The piezoelectric generates a surface charge of 30 C m^{-2} per unit strain, and the charge generated is proportional to the strain.

a) The Young modulus (see page 94) of the ceramic is 70 GN m^{-2}. Calculate the strain in the ceramic.

b) Hence calculate the charge on the surfaces of the ceramic.

c) The ceramic acts as a parallel plate capacitor.

Use: $\qquad C = (\varepsilon_0 \varepsilon_r A)/d$

to calculate the capacitance and hence the voltage established across the ceramic. The relative permittivity ε_r of the ceramic is 68.

d) Can the block produce enough voltage to produce a 1.5 mm spark in dry air? (Breakdown voltage in air is 30 kV cm^{-1}.)

8

Explain why an iron bar retains some magnetism when removed from a solenoid carrying a d.c. current, but retains no magnetism when the solenoid carries an a.c. current. Use ideas of domain theory in your arguments.

CONDENSED MATTER

The main ideas that relate to charge conduction and distribution in metals, semiconductors and other materials are brought together in this Chapter Map, which shows how these ideas are linked. You can use the map to cross-match with the needs of the syllabus you are following. The map should also help you to identify areas that you may need to study further.

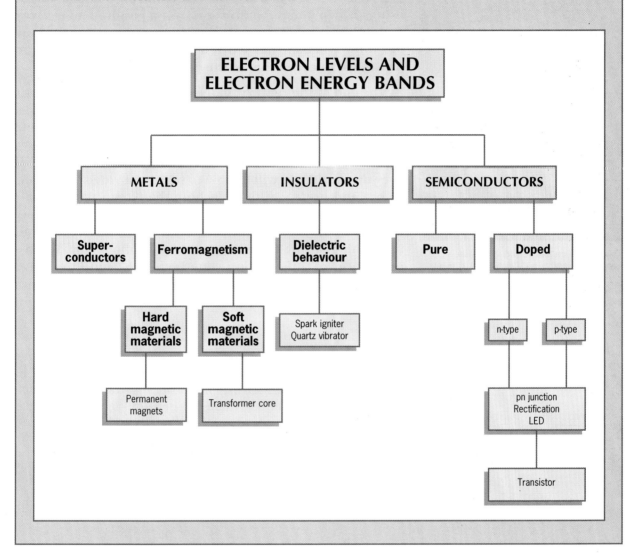

21 Communications

The information transmitted down a single telephone line from a videophone like the one in the photograph would have taken up the entire capacity of a transatlantic telephone cable twenty years ago

THE GIRL IN THE PICTURE is in school having a lesson. She is using a video phone. Her tutor is over a hundred miles away at another school but, with the video phone, she can show her tutor what she has written on the screen, and they can pass work to each other at the press of a key.

By the end of the century, video phones like this will be just one of several facilities available over the telephone line. Many people will have them in their homes, and portable versions will be commonplace. We will be able to shop over the phone line and have access to libraries and databases all over the world. We will be able to select the video film of our choice and use a washing machine that contacts the repair engineer itself when it breaks down – before we even know it's broken ourselves! In addition, it will be possible to do all these things at the same time on one telephone line. We live in a world of 'information superhighways' and interactive multimedia. Though we may not use them yet, these things are possible now.

Introduction

All the examples described above are part of the revolution in communications which has happened since Alexander James Bell first made a successful telephone call in 1876. Satellite and radio communications have added to this progress, and have contributed to giving us information in an instant from the furthest parts of the world.

In this chapter, we will be answering the question: How is it possible to transmit so much information so easily? But first, we need to understand some important aspects of radio and telephone communication.

1 RADIO COMMUNICATION

Sound offers a flexible means of communication between people. We come equipped with our own transmitter and receiver – our voice and our ears. It does have some disadvantages, though: sound does not carry far, and if several people 'transmit' at the same time, the 'receiver' finds it very difficult to sort the messages out.

Radio waves offer one answer because they 'carry' the sound (or any other information we send) much further – as electromagnetic waves, travelling at a speed of 3×10^8 metres per second in a vacuum. Fig 21.1 shows where radio waves fit into the electromagnetic spectrum, and some of the features of the **radio spectrum**.

Radio waves have frequencies ranging from tens of kilohertz to thousands of megahertz. The sound waves we hear on the radio and telephone are within the range of human hearing; that is, from

Remember:
kilohertz 10^3 Hz
megahertz 10^6 Hz

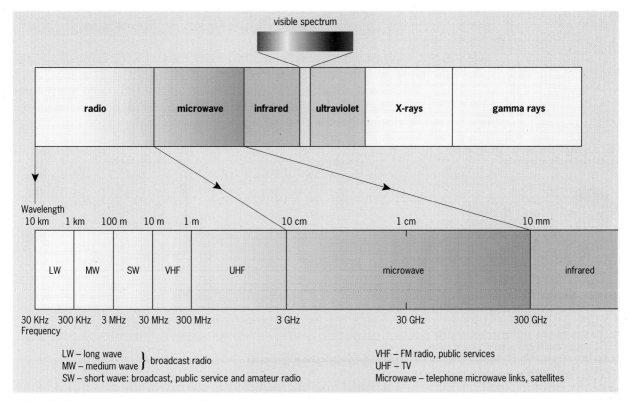

Fig 21.1 **Radio waves, shown in relation to other parts of the electromagnetic spectrum, with their frequencies and main uses in radio broadcasting, services, television and satellite**

about 20 Hz to 20 000 Hz. Generally, radio waves have frequencies far higher than the frequency of the sound information that they 'carry'.

The transmitted signal

Communication systems using radio waves can transmit a variety of information which can be received at its destination as sound, video pictures or data of various types. But before we look at the signal that is received, let's see what a transmitted signal consists of.

There are two signals that go to make the signal being transmitted. First, there is a radio signal of constant **amplitude** and **frequency**, which is often called the **carrier wave**. This carries the information from the second signal, the **information signal**, which is combined with the carrier. The information signal fluctuates (varies), as the information changes. It can also be intermittent, meaning it may stop and start. The two signals – the carrier signal and the information signal – are combined in a process called **modulation**.

The modulation process

There are several different ways of achieving modulation. In the simplest type of modulation used to transmit simple codes such as Morse code, the radio wave is merely switched on and off.

More complex methods of modulation are needed to transmit sound information such as speech or music. The two methods, namely **amplitude** modulation (AM) and **frequency** modulation (FM), which are used by broadcasting stations (such as Radio 1, Virgin, Channel 4 television, and so on), are described below.

?

A Using the equation:
speed = frequency × wavelength,

(a) Calculate the wavelength of the radio signals with these frequencies:
(i) 200 kHz, (ii) 1 MHz, (iii) 100 MHz, (iv) 500 MHz, (v) 10 GHz.

(b) Calculate the frequency of radio signals with these wavelengths:
(i) 200 m, (ii) 80 m, (iii) 2 m, (iv) 20 cm, (v) 10 mm.

✔

Both sound and radio travel as waves (see Chapter 6).

When a microphone converts sound into a changing electrical voltage, we refer to it as a *sound signal*: the information carried by the sound wave is transferred to the signal.

Similarly, in a radio, the received radio *wave* is converted to a *signal* that is then processed by the radio: the signal is a fluctuating voltage which then causes a fluctuating current to flow along a wire.

2 AMPLITUDE MODULATION (AM)

See question 1. ■

When the information signal is combined with the carrier wave, the amplitude of the carrier signal is altered. Fig 21.2 is a block diagram for a simple **amplitude modulated** radio transmitter designed to transmit sound signals. Read through the notes on the diagram to see how the signal is built up.

Fig 21.2 **Block diagram for an amplitude modulated radio transmitter**

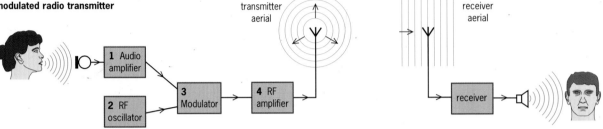

1. Audio amplifier: A microphone converts the sound into an electronic signal, which is usually very small, so an amplifier makes it larger.

4. Radio frequency (RF) amplifier: The signal produced by the modulator is a more complex radio frequency signal. A special amplifier makes this signal stronger before it is fed into the aerial which 'radiates' the signal.

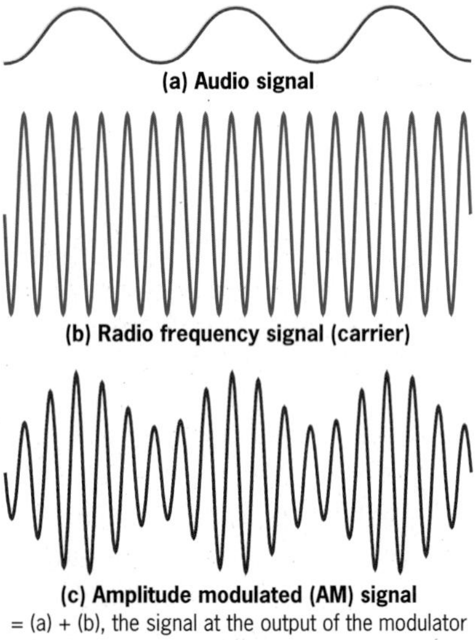

2. Radio frequency oscillator: An oscillator is a circuit that generates an electronic 'carrier' signal, in this case, a radio frequency signal, of constant amplitude and frequency.

3. Mixer or modulator: The circuit in which the information signal, in this case the signal produced from the sound, and the carrier are combined.

The modulator signal

The signal from the modulator is quite complex, as Fig. 21.3 shows. It is a mixture of signals fed into the modulator. It is easier to understand if we think about a transmitter sending a simple sound signal such as a pure tone – a note of single frequency. Let's call it f_0, as in Fig 21.3(a). The radio frequency carrier signal has a frequency of f_c. As in Fig 21.3(b), f_c would be a much higher frequency than f_0.

The mixer combines these two signals to produce two new frequencies as well as the two signals we started with. The two new signals are the sum and the difference of f_0 and f_c (see Fig 21.3):

$$(f_c + f_0) \text{ and } (f_c - f_0)$$

The output of the mixer contains all four frequencies:

$$f_0, f_c, (f_c + f_0) \text{ and } (f_c - f_0)$$

Three of the frequencies at the output of the modulator are roughly the same; they are all radio frequency signals. The odd one out is f_0. As an audio frequency signal, its frequency is much lower than the other three. The last stage in our simple transmitter is an amplifier. It is a radio frequency amplifier which only amplifies the three radio frequency signals. f_0 is effectively filtered out; it doesn't reach the aerial.

Fig 21.4 shows a useful way of displaying the three radio frequency signals that are transmitted. It is a frequency spectrum and shows the carrier and the two other frequencies either side of it.

(a) Audio signal

(b) Radio frequency signal (carrier)

(c) Amplitude modulated (AM) signal
= (a) + (b), the signal at the output of the modulator

Fig 21.3 **Graphical representation of audio, radio and amplitude modulation (AM) signals**

?

B A carrier frequency signal of 1 MHz is modulated with an audio signal of 10 kHz. Calculate the sum and difference frequencies produced by the modulator.

See question 11. ■

Fig 21.4 **The three radio frequency signals transmitted by an amplitude modulated radio transmitter, shown as a frequency spectrum**

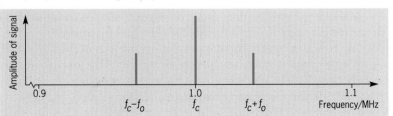

Bandwidth

In most real radio transmissions, the sound signal will be more complex than the single tone mentioned earlier. Instead of a single pure frequency, it will be a range of audio frequencies representing, say, a person's voice or the music of a group. For telephone communication and most radio, the sound is transmitted in a **band of audio frequencies** ranging from 300 Hz to 4 kHz. When this band of frequencies is used to **amplitude modulate** the carrier, the resulting output is the **carrier**, f_c, and two **sidebands**.

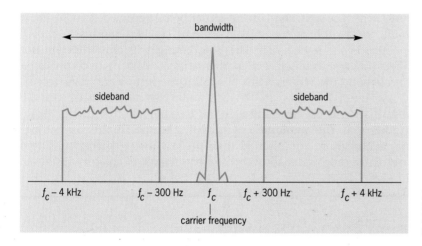

Fig 21.5 **Frequency spectrum of a typical medium wave band station with an audio bandwidth of 300 Hz to 4 kHz. Notice the upper and lower sidebands**

In amplitude modulation, the signal transmitted includes all the frequencies shown in Fig 21.5. The sidebands are the important part because they carry the information. (It would be pointless to just send the carrier signal!)

The signal transmitted covers a range of frequencies, and we say that the transmitted signal has a **bandwidth**. The bandwidth of the signal just described and shown in the diagram is 8 kHz. That means that this radio station occupies 8 kHz of frequency space. Notice also that:

$$\text{bandwidth} = 2 \times \text{maximum audio frequency}$$

Radio transmissions are organised into bands by international agreement. They are given names relating to their wavelengths, for example, the 'medium wave band' and 'long wave band'. Each band itself covers a particular range of frequencies. The medium wave band covers a range from 500 kHz to about 1.6 MHz.

If radio stations on a band wanted to use amplitude modulation to transmit signals over a range similar to that in Fig 21.5, they could each be allowed 10 kHz of frequency space. This would avoid any overlap of adjacent stations, and they could operate without interfering with each other.

See question 2.

?

C What is the bandwidth of the medium wave band described in the text?

?

D What is the maximum number of radio stations, each allowed a bandwidth of 10 kHz, that could operate on the medium wave band without interfering with each other?

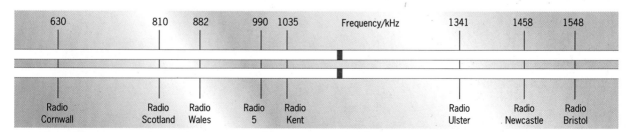

Fig 21.6 **The medium wave band showing the carrier frequency of some of the radio stations broadcast in the United Kingdom**

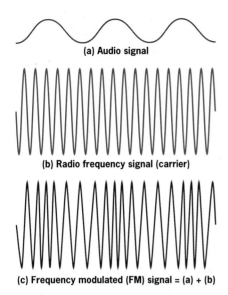

(a) Audio signal

(b) Radio frequency signal (carrier)

(c) Frequency modulated (FM) signal = (a) + (b)

Fig 21.7 **Graphical representation of audio, radio and frequency modulated (FM) signals**

3 FREQUENCY MODULATION (FM)

Frequency modulation is another very common form of modulation. FM gives a far better quality signal than AM because AM is very easily affected by 'noise' and FM is not.

Noise means the random signals that are present in all circuits, and the atmospheric noise in the signal picked up by the aerial in a radio. Noise can also be produced by electrical machinery, such as electric drills or vacuum cleaners.

Whatever the source, the noise is a signal which affects and interferes with the *amplitude* of the signal being received, and so, on a radio, you will hear a signal distorted by crackles and hiss.

Since FM is a *frequency* variation, changes in amplitude do not affect it nearly as much. The mathematics of FM is more complicated and beyond the scope of this book, but the results are important.

As Fig 21.7 shows, in FM the audio signal which carries the information is used to modulate the frequency of the carrier signal (rather than the amplitude, as in AM). The spectrum of an FM signal is also more complicated than the AM signal spectrum. There are many more sidebands and the bandwidth is greater, as shown in Fig 21.10.

THE WAY RADIO WAVES TRAVEL

RADIO WAVES TRAVEL at the speed of light. The Voyager spacecraft sent messages straight back to Earth from the outer edge of the Solar System over 5000 million km away in just over four and a half hours. But the distance and paths followed by radio waves on Earth are affected by the Earth itself and on the state of the upper atmosphere which itself changes with time of day, season and the level of solar activity.

At low frequencies, radio communication depends on the wave travelling (being propagated) in contact with the surface of the Earth; this is called **surface or ground wave propagation**. During the daytime, broadcasts from medium waveband stations can travel nearly 200 km like this. Above 2 MHz, a surface wave weakens rapidly with distance (it is **attenuated**).

At frequencies between 2 MHz and about 20 MHz, radio waves are reflected off the ionosphere, a layer of ionised molecules which reaches from about 40 km to about 300 km above the Earth's surface. Using this mode of propagation, radio waves can travel 4000 km in one 'hop' and can easily travel round the Earth with several hops. When a radio wave travels like this, it 'skips' over large areas where the signal strength will be very weak.

The upper limit of **ionospheric propagation** varies with the sunspot cycle, which goes through eleven-year cycles of activity. At periods of high sunspot activity, the upper frequency can reach 30 MHz.

Above 30 MHz, radio waves travel mainly by **space wave**, which is a line-of-sight wave. Both Earth-bound and satellite TV use this mode of propagation. Range is then limited by the curvature of the Earth. At these frequencies, the ionosphere has little effect.

Fig 21.9 **Radio brings us the voice of people from the most remote parts of the world**

Fig 21.8 **The three main modes of propagation of radio waves**

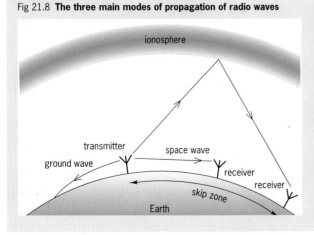

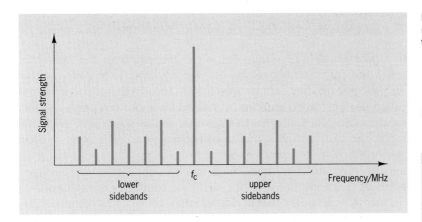

Fig 21.10 **Frequency spectrum for an FM signal. Notice the larger number of sidebands than in AM**

■ See question 3.

?

E (a) Why are stereo signals not transmitted on the medium wave band?

(b) What is the largest number of stereo stations that could use the British FM band?

(c) Capital Gold transmits on 1548 metres. Why would the station prefer to be FM?

FM is used for stereo sound transmissions where high quality reproduction is important. Because FM requires a very big bandwidth, FM stations in Britain are allowed a bandwidth of 150 kHz. This means that the signals need to be transmitted on a frequency band which is much higher than for AM. The FM radio stations in Britain fit into a band from 88 MHz to 108 MHz.

4 AM AND FM AS ANALOGUE METHODS OF MODULATION

Radio and telephone systems

So far, we have concentrated on radio signals. Telephone signals can also be amplitude modulated or frequency modulated. Many telephone signals can be sent down a cable – or, today, an optical fibre – in a band like a band of radio signals. We will look at optical fibres later on page 478.

Amplitude modulation and frequency modulation are **analogue** methods of modulation. By 'analogue' we mean that there is a direct relationship between the information signal and the electronic signal that is transmitted. In the case of radio, the loudness and pitch of the sound is used to change the *amplitude* and *frequency* of the transmitted signal. Analogue signals change continuously, covering a range of possible values.

As shown above, such signals take up a slice of frequency space, occupying a small part of the electromagnetic spectrum. Both AM and FM signals travel a limited distance, and beyond this they are so weak that they are difficult to detect.

Limitations of AM and FM

We have seen that AM and FM are widely used in radio and telephone communication. Some television systems use AM for picture information and FM for the sound. So AM and FM are certainly effective means of communication, but they cannot provide the flexibility and reliability required for some modern communications.

They are also very open: once a signal is transmitted, anyone with a suitable receiver can listen in. They are both 'real time' systems, which means that the signal can be received only at the time it is transmitted. As radio airways become more congested, this is a disadvantage.

Fig 21.11 **A videophone allows us to see the person we are talking to**

5 DIGITAL COMMUNICATIONS

It is now possible to send a complete message over the radio that takes only milliseconds to transmit. Signals can also be coded so that only the person they are intended for can receive them. This system is called **packet radio** and has been used for about ten years by radio amateurs and for longer by military and commercial users. In packet radio, the transmitted signal is converted to a series of very short pulses which are compressed together into a very short period of time, usually only milliseconds. These signals are computer messages, pages of text compiled and decoded by a computer at each end.

Personal computer-based videophones allow us to communicate face to face. We can have video conferences with each person seen in their window on the monitor screen. We can also send any information that can be displayed on a computer monitor by transferring a file of written or diagrammatic information, for example.

These are just some of the possibilities of **digital systems**. In using them, communications technology has surprisingly reverted to the first effective electrical method used to transmit information, that is, Morse code, which consists of a series of short and long pulses.

A pulse is made by switching the signal 'on' for a short time. In Morse code, long pulses ('dashes') are three times as long as short ones ('dots'). Combinations of pulses are used to represent letters, and these are strung together to make words and sentences. Samuel Morse devised his code to take advantage of the new wire telegraph that he invented in 1837. Messages could be sent over very long distances – the first transatlantic cable carrying Morse came into use in 1858, while the first transatlantic telephone line wasn't laid until 1956! Digital transmissions are less affected by noise interference, they can travel further and still be detected, and they takes up less frequency space, so that more stations can use the same band.

What is a digital code?

We have seen that Morse code (see Fig 21.12) is a simple digital code. Electronic digital codes and digital signals are also very simple: they have only two values. They are either 'on' or 'off'. A digital signal is made up from combinations of 'on' and 'off' pulses. 'Off' is given the code '0' and 'on' the code '1'.

Fig 21.12 **Morse code for the letters A (· —) and B (— · · ·)**

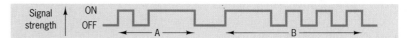

The binary code

A simple code in common use is the **binary code**. The binary number system is based on just two digits, 0 and 1 (as compared to the ten used by the decimal system – 0 to 9). Table 21.1 shows what signals equivalent to the binary codes for decimal numbers 1 to 15 look like.

The third column of the table shows each binary number represented as a 'four-bit' binary word. Each digit of the binary number is called a **bit** (from **bi**nary dig**it**). Zeros are added as extra bits to the left of any number that is less than four bits. Digital signals are made up from four-bit words or multiples of them: 8 bit (called a **byte**), 16 bit, 32 bit, or even 64 bit.

Decimal	Binary	Four-bit binary
0	0	0000
1	1	0001
2	10	0010
3	11	0011
4	100	0100
5	101	0101
6	110	0110
7	111	0111
8	1000	1000
9	1001	1001
10	1010	1010
11	1011	1011
12	1100	1100
13	1101	1101
14	1110	1110
15	1111	1111

Table.21.1 **Decimal and binary equivalents**

Digital signals

Images and sounds are changed into **digital codes**, and the codes are transmitted as pulses. Then the receiver equipment changes the pulses back into the original sounds or images. Let us look further at this process.

Digital signals are easy to receive. All the receiver needs to do is to detect whether a pulse is high or low. Digital signals, transmitted either as a radio signal or down a telephone line, tend to change after they have travelled some distance. Fig. 21.13 shows how a very regular pulse changes as it travels a long distance. But as long as the receiver can still distinguish between 'high' and 'low', the original signal can be reproduced accurately. Analogue signals do not have this abrupt 'high'–'low' pattern. They change continually and become corrupted over much shorter distances.

Older undersea telephone systems use special equipment that amplifies and reshapes the analogue signal. They are called **repeaters** and are required every 3 nautical miles. Modern digital systems which use optical fibres have repeaters every 30 nautical miles! The digital repeaters change the corrupted pulses, like that on the right in Fig 21.13, back into perfect pulses, as shown on the left.

?

F (a) How many bits are needed to represent the following decimal numbers as binary numbers? (i) 16, (ii) 32, (iii) 64

(b) Each of the decimal numbers in (a) can be written as 2^n, where n is a whole number. Show that the answers to (a) can all be written as $(n + 1)$.

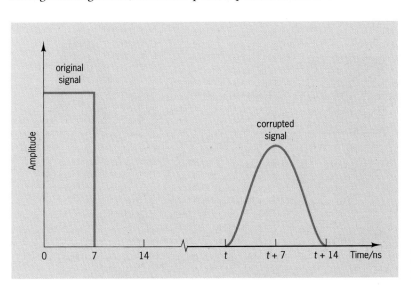

Fig 21.13 **After travelling a long distance, a digital signal loses its original shape as it becomes corrupted**

6 SIGNAL CONVERSION

Many signals that we wish to send start out being analogue signals. Examples are speech or music which change continually. To benefit from the advantages of a digital system, the analogue signal must first be changed into digital signals. This process is called **analogue to digital conversion**. The digital signal is then transmitted, and at the receiver it is turned back into an analogue signal. This is called **digital to analogue conversion**. Ideally, this signal should be a perfect reproduction of the analogue signal we began with.

Pulse code modulation (PCM)

There are several ways that signals can be converted into digital pulses. We will look at one of them, **pulse code modulation** or **PCM**, which is the method commonly used in communication systems.

In particular, let's see what happens inside a digital telephone line using PCM, as shown in Fig 21.14.

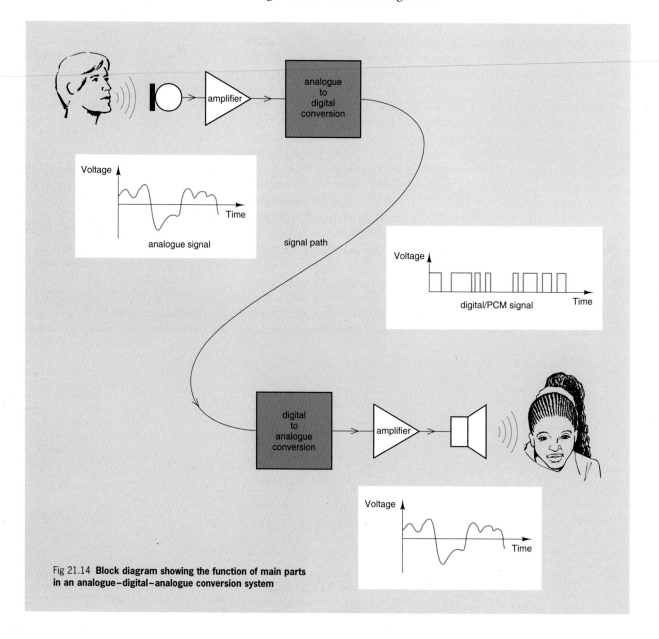

Fig 21.14 **Block diagram showing the function of main parts in an analogue–digital–analogue conversion system**

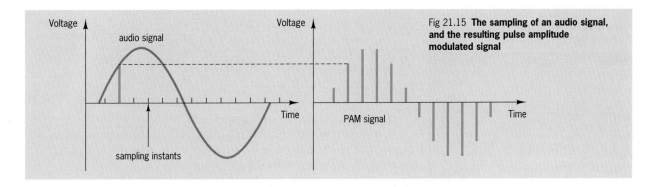

Fig 21.15 **The sampling of an audio signal, and the resulting pulse amplitude modulated signal**

Analogue to digital conversion

First, the microphone changes the sound of a voice into an electronic analogue signal which is an alternating voltage which varies over a range of voltages and frequencies. The signal travels in this form to a local telephone exchange where it is changed into a digital signal using PCM.

The voltage of the analogue signal is measured – it is **sampled** – several times during each cycle of the signal. With a typical audio frequency signal, this would mean that samples are taken at intervals of about 125 microseconds, that is, 8000 samples every second. The samples are very short pulses of varying amplitude and taken together they show the same shape as the original analogue signal, as in Fig 21.15. The number of pulses per second is called the **sampling rate**. At this stage, the signal has become **pulse ampltude modulated** (PAM).

The next step is to convert the varying amplitude pulses into a binary code. The maximum analogue voltage, which depends on the microphone and the amplifier used, is subdivided into a fixed number of levels. This process is called **quantisation** and the levels are called **quantisation levels**. For example, if the maximum signal voltage is 1 volt, this would be the highest level. If sixteen quantisation levels were used, then the first level above zero would represent 1/16 volt. Each sampled pulse is then measured and matched to these levels and is given the value of the closest level.

As shown in Fig 21.16, each amplitude pulse is given a digital code which is a binary number. The size of the number, that is, how many bits it has, depends on the number of quantisation levels. If there are eight levels (including zero), as in Fig 21.16, then a three-bit code will do.

> ✓
> The sample rate of 8000 times per second depends on the highest frequency to be transmitted. Digital telephone systems use the same practical rules as for AM, with a maximum frequency of 4 kHz. 8000 times per second is twice that maximum frequency.

See questions 4, 5 and 6.

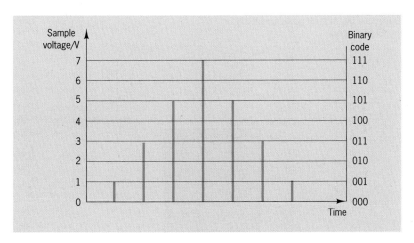

Fig 21.16 **Amplitude pulses are converted into a binary code**

Table 21.2 **The voltage and corresponding binary code of each successive sample taken of the signal shown in Fig 21.15**

Sample	Level/voltage	Code
1	1	001
2	3	011
3	5	101
4	7	111
5	5	101
6	3	011
7	1	001

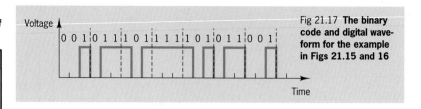

Fig 21.17 **The binary code and digital waveform for the example in Figs 21.15 and 16**

To summarise, the number of quantisation levels determines the number of bits in the code for each pulse. Each pulse is represented by a binary number with the same number of bits, no matter what its level, and each binary number is called a **word**.

Each binary code is then transmitted in sequence, starting with the first sample. If the binary signal were displayed on an oscilloscope, it would look like a series of pulses, each having the same amplitude, but these pulses would now have different lengths. The digital signal for our example is shown in Fig 21.17.

Each binary digit is allocated a fixed period of time. The length of the pulse is the time for which it is high or low. Binary zeros (0) are represented by zero volts, and binary one (1) by a fixed voltage (often, +5 V is used). A series of several consecutive 1's or 0's represents a longer pulse. The signal is now **pulse code modulated**.

In a telephone system, these digitised pulses are transmitted along a coaxial cable, or they are converted into pulses of infra-red radiation and sent along an optical fibre. They may even be transmitted as pulses of radio waves at microwave frequencies. The information from the Voyager space probes was sent back to Earth as a PCM radio signal (see page 472).

Digital to analogue conversion

At the receiver, the process is reversed. As it arrives, each binary word is decoded by the receiver as the voltage of the relevant quantisation level. As the samples are decoded, they are reassembled to make the original analogue signal. Fig 21.18 shows the digital signal changed back into a pulse amplitude modulated (PAM) signal. This is further processed to give the original analogue signal.

The quality of reproduction depends on the number of quantisation levels, that is, how close together they are within the range used. More levels means bigger binary codes. Larger binary codes can be sent within the same time by giving each binary digit a shorter time.

?

G Why would 16 quantisation levels need a four-bit number?

H Why is it better to send pulses of the same amplitude rather than pulses of varying amplitude?

?

I Pulse code modulation is also used for digital audio recording. Some compact discs use a 32-bit code. How many quantisation levels would this give?

Fig 21.18 **The digital signal arrives at the receiver, which decodes it to reproduce the original pulse amplitude modulated signal**

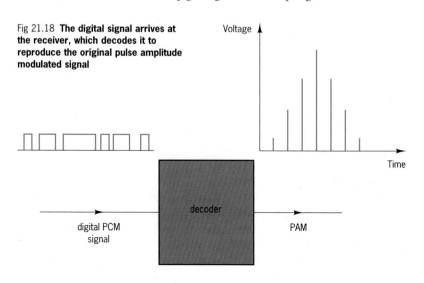

UPDATING TELEPHONE SYSTEMS

COPPER WIRES HAVE traditionally carried telephone calls as an analogue signal from our homes to the local exchange. To replace this system by optical fibres would be very costly. But our phones can be easily updated: the new handsets themselves now convert calls to a digital signal which can travel down the copper wire.

This system, called the Integrated Digital Services Network (ISDN), has made our telephones much more versatile. An ISDN telephone with one line can carry data at the rate of 144 kbit s^{-1}. This allows two channels of 64 kbit s^{-1} each, and a 16 kbit s^{-1} channel for sending other signals. The larger channels can carry voice or data such as computer data or fax.

A videophone uses one channel for voice and the other for picture data, so it's like using two telephone lines (and it costs twice as much). The 16 kbit signal channel could carry **telemetry**, too. This includes data from metering devices such as electricity and gas meters, or it could be the message to a service engineer from the faulty washing machine mentioned at the start of this chapter.

As electronic communications technology improves, we can expect the speed of systems like this to increase, and in the next few years, our home telephone lines are likely to improve from 144 kbit systems to circuits that can handle data rates of over 100 Mbits s^{-1}.

Digital signals and bit rates

An analogue signal would be described by its 'frequency range', such as that for a radio station. This term has no meaning for a digital signal. Instead, we describe a digital signal by the number of bits – binary digits – per second.

To calculate the **bit rate**, we need to know the sampling rate and the number of quantisation levels. The sampling rate is the number of samples every second. The levels determine the binary code and the number of 'bits per sample' required.

So the bit rate is given by:

(sampling rate) × (number of bits per sample)

J A 64 kbit channel used on ISDN lines allows the same quality of audio described earlier in the chapter, that is, up to 3.4 kHz. Suggest some of the advantages that a faster bit-rate would give.

EXAMPLE

Q Calculate the bit rate for a signal which has a sampling rate of 8 kHz and where 16 quantisation levels have been used.

A The sampling rate is 8000 per second.
16 quantisation levels can be coded by a 4-bit binary number

because: $16 = 2^4$.
So bit rate = (sampling rate) × (number of bits/sample)
= 8000 × 4

The bit rate for the signal = 32 000 bits per second

The **medium** through which a signal travels will place a limit on the maximum frequency and on the maximum bit rate. We will look at this in more detail on page 477.

7 MULTIPLEXING

We have seen different ways of transmitting information. The next question is: How can we send more than one signal along a telephone cable at the same time? Alternatively, how do modern communication satellites handle thousands of telephone calls at a time? **Multiplexing** is the process that achieves this, by combining many individual signals and sending them together in one signal.

Different radio stations can share the same band because each has its own carrier frequency. The carrier frequencies are spaced out so that the stations do not overlap and interfere with each other. AM stations transmit using a bandwidth of 8 kHz.

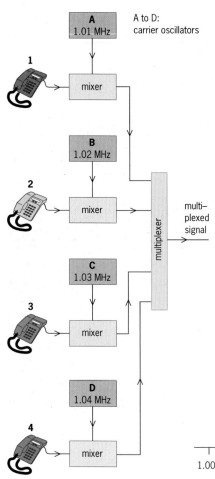

A to D:
carrier oscillators

A
1.01 MHz

1

mixer

B
1.02 MHz

2

mixer

C
1.03 MHz

3

mixer

D
1.04 MHz

4

mixer

multiplexer

multi-
plexed
signal

Frequency division multiplexing

Analogue signals can be assembled to produce bands rather like a band of radio signals. A band containing several signals is called a **group**. This grouping together of analogue signals is called **frequency division multiplexing**.

Let's see how four telephone lines can be multiplexed together to form a small group. This is shown in Fig 21.19, with each telephone line representing a line with one telephone connected to it.

As each telephone call comes in, it is amplitude modulated with a carrier signal. The first carrier signal is at 1.01 MHz, the second at 1.02 MHz, the third at 1.03 MHz and the fourth at 1.04 MHz. The band is made up from four channels; each channel is allowed 10 kHz of bandwidth. At the multiplexer, all the channels are combined to give a **multiplexed signal**.

The cable from the telephone exchange therefore carries a band of frequencies from 1.00 MHz to 1.05 MHz. This is our 'group'. At the destination exchange, the process is reversed. The band of frequencies is filtered to split it into separate channels. Each channel is then demodulated to recover the original telephone call. (Demodulation of AM signals is described on page 501.)

Fig 21.19 **Block diagram for four multiplexed telephone lines**

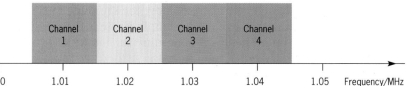

Channel 1	Channel 2	Channel 3	Channel 4

1.00   1.01   1.02   1.03   1.04   1.05   Frequency/MHz

Time division multiplexing

Digital signals are multiplexed using a different method, called **time division multiplexing**.

In the example of pulse code modulation in the calculation above, the analogue signal is sampled at a rate of 8 kHz. That means that samples are taken every 125 μs (125 microseconds). But each sample lasts for only 2 or 3 μs, and that leaves over 120 μs free between samples – time in which the system would not be carrying any information. For the system, this is wasted time.

In time division multiplexing, this 'free' time is used to fit in other signals. In practical systems, at least 30 channels are multiplexed. Just three channels are shown in Fig 21.20.

kilo = thousand:
    8 kHz = 8000 Hz
micro = millionth (μ):
    125 microseconds (μs)
    = 0.000125 seconds (s)

See questions 6–10. ■

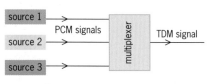

source 1

PCM signals

source 2

multiplexer

TDM signal

source 3

Fig 21.20 **Three channels which are time division multiplexed**

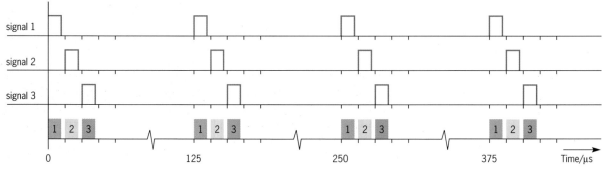

signal 1

signal 2

signal 3

| 1 | 2 | 3 | | 1 | 2 | 3 | | 1 | 2 | 3 | | 1 | 2 | 3 |

0   125   250   375   Time/μs

8 MEDIA – HOW SIGNALS TRAVEL

Radio waves can travel through free space or along a coaxial cable. Electrical signals can travel down a wire, and light pulses can be sent along an optical fibre. Free space, coaxial cable, wire, optical fibre – all these are referred to as **media** through which a signal can be sent. Each has its uses, and when you make a telephone call, it is possible that your message may be travelling through every one of these media. Let's follow a typical long-distance call from a caller in Britain to her uncle in California, as shown in Fig 21.21(a).

Fig 21.21(a) **Simple circuit diagram for a transatlantic telephone call**

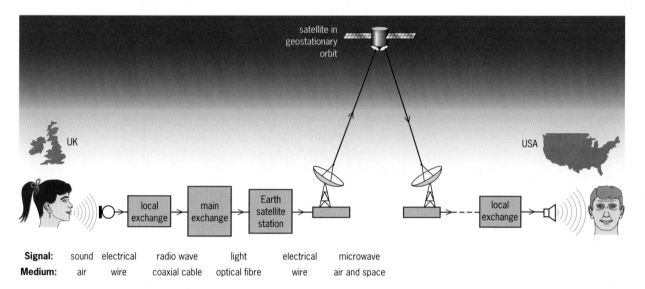

| Signal: | sound | electrical | radio wave | light | electrical | microwave |
| Medium: | air | wire | coaxial cable | optical fibre | wire | air and space |

Our call would first travel as an electrical signal from the telephone to the local exchange along a **wire** (wires are in pairs for a complete circuit). These wires are likely to carry only one call at a time – but that's all right because that's all a single phone can handle.

At the local exchange, our call would be digitised and multiplexed with other calls from the area, and sent to a bigger exchange at the nearest large town. All the calls could travel along a coaxial cable as a radio wave.

A **coaxial cable** is a cylindrical wave guide: the radio wave is guided along the cable between the inner and outer conductors which are shown in Fig 21.22. The space between the conductors is filled with polythene or a similar dielectric material (see page 457 for more on dielectrics). Coaxial cables can carry radio signals with frequencies of tens of megahertz or tens of megabits per second.

Fig 21.21(b) **The satellite receiving dish at British Telecom's station at Madley in Herefordshire, where international telephone calls are relayed**

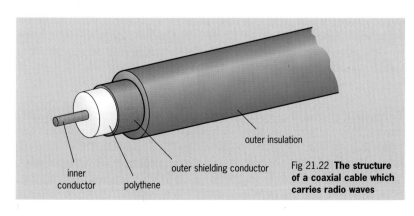

inner conductor, polythene, outer shielding conductor, outer insulation

Fig 21.22 **The structure of a coaxial cable which carries radio waves**

Optical fibres

Our long-distance call, and many other transatlantic calls like it, would be directed from the large exchange to a satellite ground station. The link between the exchange and the ground station is by **optical fibre** and, to use it, the signal has to be converted from electrical pulses into pulses of light. Infrared is used in high-speed optical systems. The frequency of infrared is of the order of 10^{14} Hz, a million times higher than the upper frequency limit of coaxial cables. With bit rates approaching 10^{14} bits per second, we can see that an optical fibre is capable of handling vast amounts of information. Here, it means a vast number of phone calls all at the same time.

How an optical fibre works

An optical fibre is made from a fine cylinder of glass about 5 mm in diameter. This is the core, and is enclosed by cladding which is made from a less dense glass. If light enters the core at a large enough angle, it will be trapped in the core because the beam is totally internally reflected at the boundary between the core and the cladding. The beam travels down the fibre after multiple reflections, as shown in Fig 21.23.

Signals travelling down a fibre are not affected by electrical interference, and they are very secure because they do not create a magnetic field which a current flowing down a wire would create.

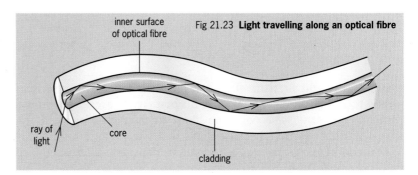

inner surface of optical fibre

ray of light

core

cladding

Fig 21.23 **Light travelling along an optical fibre**

At the satellite ground station, the signal is converted back into an electrical signal (it continues to be digitised). This is then transmitted up to a geostationary satellite, now as a **microwave** signal. The microwaves have frequencies of about 10 GHz (10 gigahertz or 10^{10} Hz). Microwaves don't have the signal-handling capacity of optical signals, yet they allow thousands of telephone calls and several television channels to be transmitted at the same time.

The satellite re-transmits the signal down to another ground station in America. Our call may then be directed to its destination by optical fibre or coaxial cable, or both. Its last lap to the telephone in California will probably be by wire – unless our caller's uncle is using a mobile telephone.

Transatlantic telephone calls now travel by undersea optical fibre cables. It is the vast potential of optical fibre systems that makes the 'information superhighway' possible. The optical fibre systems being installed now are handling only a small fraction of what they are capable of carrying, and there is plenty of room for expansion over the next decade or two.

?

K What happens to the signal during the last lap of this call?

THE INTERNET AND OPTICAL FIBRES

THE INTERNET HAS come of age with digital communications. What started as small networks linking computers in US research establishments and colleges has grown into a world-wide system linking countless different networks together at little cost.

One of the Internet's most popular features is the World Wide Web. This part of the system gives users access to a range of multimedia resources – stereo sound, graphics and pictures, including video. All these require the high bit speed of fibre optics to make it cheap enough for general use. Optical fibre cables have fibres in pairs carrying signals in opposite directions. Six or twelve pairs are usual in a cable, and each individual fibre can cope with data at about 500 megabits per second.

Data transmission has been around for a long time, and this chapter has shown how different kinds of signals can be digitised. With the flexibility and reliability that digital modulation allows, the huge capacity of optical fibres has made the videophone and the Internet possible. Now it is easy to transmit many different types of information together along a single fibre, and still have room for thousands of other signals. With time division multiplexing, too, all this information can travel along one fibre. Currently, a single fibre can have 7680 speech channels, and TAT 12/13, the latest transatlantic optical fibre pair, can carry 300 000 telephone conversations all at the same time.

Fig 21.24 **Optical fibres, the growth of data flow on the Web over two years (1993–4 and fibre optic cable**

SUMMARY

Having studied this chapter, you should be able to understand the following:

■ Electromagnetic waves, in particular radio waves and light waves, can be used to carry information.

■ The process of adding information to a carrier wave is called modulation. Two types in common use are amplitude modulation and frequency modulation.

■ The bandwidth is the range of frequencies from a broadcasting station or in a signal.

■ Bandwidth = 2 × audio bandwidth. Bit rate = sampling rate × number of bits per sample.

■ Noise is a random signal that is always present in circuits. Atmospheric noise and other interference from electrical machinery can also affect signals.

■ Analogue and digital signals can be converted from one to the other.

■ Digital signals travel further without corruption than analogue signals.

■ Pulse code modulation is a digital modulation method used for communications and audio recording which is capable of very high quality reproduction.

■ Multiplexing is the process of combining many signals into one.

■ Electromagnetic waves can travel through space, along a wire or along an optical fibre.

■ Optical fibres offer a huge bandwidth and are not affected by electrical interference.

QUESTIONS

1 Fig 21.Q1 shows a block diagram of a complete model of a communications system.

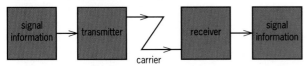

Fig 21.Q1

Four communications systems are listed in **a)** to **d)**. For each one, name the transmitter, the receiver and the carrier of the message.

a) Two people talking to each other across a room.

b) A telephone conversation between two people.

c) An FM radio station transmitting music.

d) A message sent by e-mail from one computer to another over the Internet.

2 Fig. 21.Q2 shows a block diagram of an amplitude modulated transmitter and the frequency spectrum that results from a 1 MHz carrier being modulated by a 1 kHz pure tone.

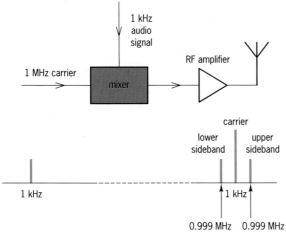

Fig 21.Q2

a) The lowest frequency is not amplified by the radio frequency amplifier. What is the bandwidth of the transmitted signal?

b) The pure tone is replaced by an audio signal covering the frequency range 300 Hz to 3.4 kHz. List the frequencies produced by the mixer for **(i)** the 300Hz signal, **(ii)** the 3.4 kHz signal.

c) Draw a frequency spectrum diagram of the signal transmitted from the aerial.

d) What is the bandwidth of the signal in **d)**?

e) Estimate how many stations transmitting similar signals could fit into a band covering the frequency range 800 kHz to1200 kHz.

f) Improved quality audio can be transmitted by using an audio signal with a range of 20 Hz to 20 kHz. How many stations with hi-fi audio would fit into the same band?

3
a) What is frequency modulation?

b) Why are AM broadcasts more likely to be affected by noise than FM broadcasts?

c) In the UK, FM broadcasts use a band of frequencies from 88 MHz to 108 MHz. Hi-fi quality broadcasts require a bandwidth of 1 MHz.
    (i) How many such radio stations could use the 88–108 MHz band?
    (ii) The number of stations using this FM band in the UK is many more than your answer in **(i)**. Why don't these stations interfere with each other?

4 Fig 21.Q4(a) shows an analogue signal drawn on graph paper. The horizontal scale represents time, measured in milliseconds. The vertical scale gives the voltage of the signal. Copy this diagram on to a sheet of millimetre graph paper.

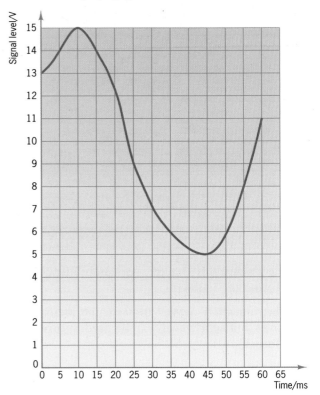

Fig 21.Q4(a)

a) Draw vertical lines every 5 milliseconds, starting at 0 ms.

b) Read off the length of each line in volts and tabulate them.

c) Convert each voltage into a four-bit binary code.

The signal above is always positive. Fig 21.Q4(b) shows a very similar signal. This signal alternates – it has positive parts and negative parts. How do we encode this signal so that we can distinguish between positive and negative?

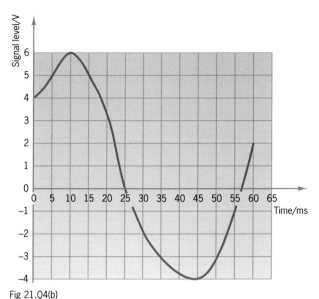

Fig 21.Q4(b)

The answer is to use a 5-bit code; that is, to add a fifth bit to the 4-bit code. The extra bit is added to the left so that it becomes the most significant bit. If the signal is positive, a '1' is added, if the signal is negative, a '0' is used.

d) Repeat parts **a)**, **b)** and **c)** for the signal in Fig 21.Q4(b), using a 5-bit code.

5 The analogue signal shown below is sampled every millisecond with 16 sampling levels.

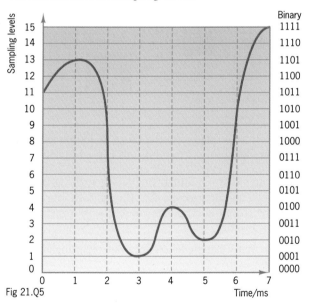

Fig 21.Q5

Complete the table below showing the binary code and digital output of this signal at the sample times shown. What is the highest frequency for an analogue signal that can be properly converted at this sampling rate?

Time/ms	0	1	2	3	4	5	6	7
Binary code	1011							
Digital output								

[Nuffield Advanced Physics question]

6 This question is about sampling rates. The sampling rate used for telephone systems is 8 kHz–8000 samples per second.

a) What is the time interval from one sample to the next?

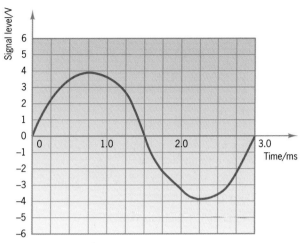

Fig 21.Q6(a)

b) Fig 21.Q6(a) shows a low frequency audio signal. Sample this signal at the time intervals you calculated in **a)**.
Carefully draw a diagram representing each sample as a vertical line along a horizontal time axis with each sample equally spaced out.

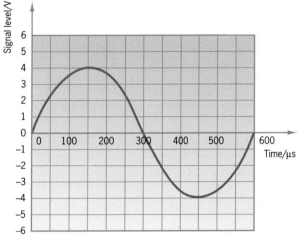

Fig 21.Q6(b)

c) Fig 21.Q6(b) shows a similar signal but of higher frequency.
Sample this signal at the same time intervals and draw a diagram showing the samples as in **b)**.

d) The maximum frequency at which this sample rate is used so that the reconstructed signal resembles the original signal is 4 kHz.
How many samples would be taken every cycle at this frequency?
How faithfully would these samples enable the signal to be reconstructed?

7 Read the passage below and then answer the questions.

British Telecom uses a sampling frequency of 8 kHz to convert audio signals in the range 300 Hz to 3.4 kHz into digital form. An 8-bit code is used to define the sample levels. Several signals are sent on one carrier by time division multiplexing (TDM), where each signal is allocated a 25 microsecond slot, using a clock rate of 1 MHz. Synchronisation pulses are sent within each 25 microsecond slot.

a) How many sampling levels are available when using an 8-bit code?

b) Calculate the sampling period.

c) Explain why a sampling frequency of 8 kHz is suitable for this application.

d) Describe how TDM would allow five signals to be transmitted on one carrier.

[Nuffield–Chelsea Curriculum Trust, 1994]

A telephone system samples at a frequency of 8 kHz, and each sample is encoded as an 8-bit number.

a) How many samples are taken each second?

b) Calculate the bit rate of the transmission.

Time division multiplexing (TDM) is a method of sending several channels down the same cable or fibre at the same time.

c) What will the new bit rate be if 32 telephone channels are multiplexed?

d) Main telephone trunk lines between major centres in Britain carry information at a bit rate of 140 Mbit s^{-1}. Use this information in Fig 21.Q9 to find the number of telephone channels that a line can carry at the same time.

e) The figure of 140 Mbit s^{-1} is a digital rounding up of the number involved. What is the actual number represented?

f) TAT 12/13 is a new transatlantic optical fibre telephone system. It operates at 5 Gbit s^{-1}. Estimate the number of telephone channels it can carry.

[Nuffield–Chelsea Curriculum Trust, 1994]

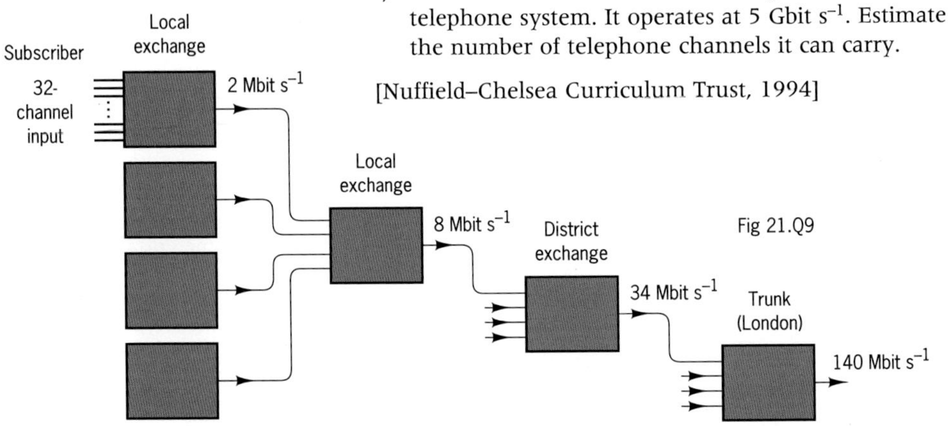

Fig 21.Q9

8
a) The samples in question 6 are coded using a 4-bit code. How many quantisation levels will be available?

b) A CD received as a Christmas present is marked 'High definition 20-bit sound'. The sleeve notes gives the following description:
'The 20-bit recording process is true to every nuance and inflection of musical performance. 20-bit, with its significantly higher resolution captures far more detail than 16-bit, the current CD standard. 20-bit is more responsive with up to 16 times the resolution of conventional 16-bit recordings.'
The notes state that the current CD standard is 16-bit recording. **(i)** Describe what this means. How many sampling levels does this allow? **(ii)** Why does 20-bit recording allow 'up to 16 times the resolution' of the standard 16-bit recording?

9 Digital systems describe the rate of information transfer as the bit rate. This is the number of binary digits transmitted each second. The units used for bit rate are: bit s^{-1}, kbit s^{-1}, Mbit s^{-1} and so on. The bit rate in a single channel is calculated using the relationship:

bit rate = number of digits × sampling frequency

10 British Telecom use a sampling frequency of 8 kHz to convert audio signals in the range of 300 Hz to 3.4 kHz into digital form. An 8-bit code is used to define the sample levels. Several signals are sent on one carrier using time division multiplexing where each signal is allocated a 25 microsecond slot using a clock rate of 1 MHz. Synchronisation pulses are sent within each 25 microsecond slot.

a) How many sample levels are available using an 8-bit code?

b) Why is a sampling frequency of 8 kHz suitable in this application?

c) Calculate the sampling period.

d) Why are synchronization pulses sent?

11 This exercise should help you understand how amplitude modulation is produced by mixing different frequencies. You will need to use a spreadsheet which has the facility to plot graphs.

Amplitude modulation is achieved by adding two signals of very different frequencies. For the sake of this exercise, we will use a simple modulating frequency – a single pure tone of frequency f_0. When this is added to a high frequency carrier (frequency f_c).

The amplitude of the carrier is represented by the formula:

$$a = a_0.\sin(2\pi.f_c.t)$$

where a_0 is the maximum amplitude.

The two sideband frequencies are similar but have slightly different frequencies.

$$b = b_0.\sin[2\pi(f_c - f_0)t]$$
and
$$c = c_0.\sin[2\pi(f_c + f_0)t]$$

b_0 and c_0 are the amplitudes of the two sidebands; these are usually the same.

You can use a spreadsheet to add these three signals together and plot a graph of the result. The example below uses Pipedream, but most spreadsheet programs will do. You should use a machine with at least 4 megabytes of RAM.

The spreadsheet uses five columns:
The first column is time in microseconds. In the example shown, time increases in steps of 5.

spreadsheet to give an amplitude modulated signal					
Time/us	Sum	Carrier	LSB	USB	
0.00000	0.00000	0.00000	0.00000	0.00000	
5.00000	0.83388	0.52125	0.14930	0.16333	
10.00000	1.42267	0.88967	0.25899	0.27401	
15.00000	1.59358	0.99725	0.29999	0.29634	
20.00000	1.29700	0.81245	0.26141	0.22313	
25.00000	0.62093	0.38945	0.15350	0.07798	
30.00000	−0.23518	−0.14773	0.00487	−0.09231	
35.00000	−1.01950	−0.64160	−0.14506	−0.23284	
40.00000	−1.50217	−0.94736	−0.25650	−0.29830	
45.00000	−1.54286	−0.97537	−0.29991	−0.26759	
50.00000	−1.13177	−0.71741	−0.26377	−0.15060	

Fig 21.Q11(a) **Start of spreadsheet**

The three signals are represented by the following functions (note that 'π' appears as 'pi' on some spreadsheets):

$$\text{carrier} = \sin[\text{rad}(2*\pi*t)]$$

The amplitude, a_0, and the carrier frequency are both set at 1.

Sidebands:
$$\text{LSB} = 0.2\sin[\text{rad}(1.9*\pi*t)]$$
$$\text{USB} = 0.2\sin[\text{rad}(2.1*\pi*t)]$$

b_0 and c_0 are both given the value of 0.2, and $(f_c - f_0)$ is 1.9, while $f_c + f_0$ is 2.1.

The Sum column is the sum of the last three columns:

$$\text{Sum} = \text{carrier} + \text{LSB} + \text{USB}$$

Fig 21.Q11(b) shows the graph produced from the spreadsheet. Time in microseconds is plotted horizontally. The vertical axis is the amplitude of the modulated signal.

a) Use the information from the graph to show that:
   (i) the carrier frequency is about 17.5 kHz,
   (ii) the modulating frequency is about 440 Hz.

b) The values of all the variables can be easily changed using a spreadsheet. Try changing the following.

   (i) b_0 and c_0: It is best to keep them the same. These control the depth of modulation which is about 66 per cent – see Fig 21.Q11(b). Try out different values to see the result.

   Find the value required to produce 100 per cent modulation. This happens when the carrier amplitude just reaches zero at some times. Modulation over 100 per cent causes distortion.

   (ii) f_c, $(f_c + f_0)$ and $(f_c - f_0)$: These change the carrier and modulating (audio frequencies.

   (iii) **Time:** Making the increments of time smaller produces smoother graphs but it needs more memory on your computer.

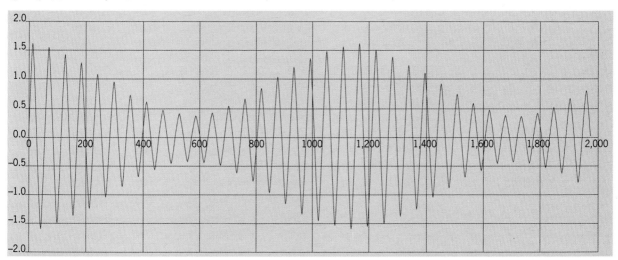

Fig 21.Q11 (b)

See page 506 for the CHAPTER MAP that covers
Chapters 21 and 22

Assignment

COMMUNICATIONS IN PRACTICE

Communications systems allow information to be sent and received as signals travelling through media such as a copper wire, a glass fibre or space. As the signal travels away from its source it becomes weaker: it is being **attenuated**. When a signal is sent over a long distance, it is necessary to boost it by **amplification**.

Measuring amplification and attenuation

The distances some signals travel are huge, and so is their attenuation. A signal transmitted from a communications satellite at a power of 30 watts will arrive at a receiving dish on Earth at a power of about 3 picowatts (3×10^{-12} watts). That means the signal has been attenuated by 10^{13} times. A range as large as this is easier to deal with using a logarithmic scale, and the decibel, dB, a unit which is defined as a logarithmic function.

The ratio of two powers, P_1 and P_2 (given in the same units, decibels or dB) is defined as:

$$\text{ratio} = 10.\log_{10}(P_1/P_2)$$

If the ratio is positive, then the ratio is said to show **gain** – it has been amplified. If the ratio is negative it is called **loss** – the signal is attenuated.

For example:
A signal of 150 mW is amplified and the output is 600 mW. Calculate the gain of the amplifier.

$$\text{gain} = 10.\log_{10}(600/150) = 10.\log_{10}4$$

$$= 10 \times 0.602 = 6.02 \text{ decibels (or 6.02 dB)}$$

1

a) Calculate the gain when a signal of 40 mW is amplified to **(i)** 80 mW, **(ii)** 160 mW, **(iii)** 240 mW, **(iv)** 400 mW, **(v)** 4 W.
b) Calculate the loss when a signal of 500 mW ia attenuated to **(i)** 250 mW, **(ii)** 125 mW, **(iii)** 62.5 mW, **(iv)** 50 mW, **(v)** 500 µW.

It is worth remembering that a change of 3 dB means a gain or loss of two times. A logarithmic scale has the advantage that successive gains and losses can be found by simple addition or subtraction.

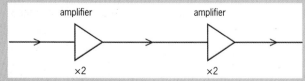

Fig 21.A1 **Two-stage amplifier**

The overall gain of the two amplifiers in Fig 21.A1 can be found in two ways. Quite simply, the overall gain is four times ((2×2) or, if the dB gains are added, 3 dB + 3 dB = 6 dB. (6 dB is equivalent to a gain of 4 – see the example above.)

To find the overall effect on a signal, just add together all the gains and losses in decibels.

Some maths to remember:
$$\log_{10}(2 \times 2) = \log_{10}2 + \log_{10}2$$
$$= 0.3010 + 0.3010 = 0.6020$$
In general, for multiplication, $\log_{10}(a.b) = \log_{10}a + \log_{10}b$ and for division, $\log_{10}(a/b) = \log_{10}a - \log_{10}b$

c) What is the overall gain or loss of the arrangements in Fig 21.A2?

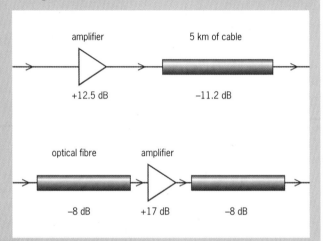

Fig 21.A2

Noise

Noise is the major problem of all communications systems. The everyday meaning of 'noise' is any unwanted sounds. The background noise of lots of people having separate conversations in a room is a good model of noise in an electrical system. Somebody entering the room just hears a noise – none of the conversations is clear enough to follow. A person in the room will probably have difficulty listening to their conversation: the other conversations are just background noise, but they are about as strong as the 'signal' being listened to.

In communications systems, the noise comes from various sources:

In any circuit or cable, the electrons are in random motion, and this random movement of electrons causes the noise. The noise increases with the length of the cable. It also increases with temperature – known as 'thermal' noise.

'Atmospheric' noise affects radio transmissions and microwave satellite transmissions. Molecules in the atmosphere absorb energy and re-radiate it as electromagnetic radiation. Natural effects such as lightning can affect radio signals around the Earth.

2 Explain why PCM signals are less affected by noise than analogue signals.

Signal-to-noise ratio. When a signal travels down a cable the level of the noise is important in determining the power needed for the signal to reach the other end. The signal must be distinguishable above the noise. This is measured by the signal-to-noise ratio, measured in decibels.

Signal-to-noise ratio

$$= 10.\log_{10} \text{(signal power)/(noise power)}$$

The value of the signal-to-noise ratio depends on the method of modulation. For telephone systems a ratio of 20 dB is used – that is, the signal must be 20 dB stronger than the noise.

3 If the noise in a telephone line is measured at 100 mW, what must the power of the signal be in watts?

Long-distance cables. The attenuation of a coaxial cable is measured in dB km^{-1}. A typical value is 14 dB km^{-1}.

4

a) A signal starts at 100 dB above the noise level which is fairly constant along the cable. If an acceptable signal-to-noise ratio is 40 dB, how many kilometres can a signal travel before the signal-to-noise ratio becomes unacceptable?

Repeaters are used at regular intervals to amplify the signal. This ensures the signal-to-noise ratio never falls below a predetermined value.

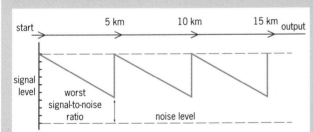

Fig 21.A3 **Repeaters amplifying the signal**

b) Fig 21.A3 shows a typical system using three repeaters. If the lowest acceptable signal-to-noise ratio is 40 dB, estimate in dB **(i)** the attenuation per kilometre of the cable, **(ii)** the gain of each repeater.

Repeaters are very expensive, especially ones designed to operate on the sea bed.

c) Why not make the signal stronger to begin with, and do without repeaters?

d) If the output signal in Fig 21.A3 is 100 mW, how big would the input signal have to be reach the output without using repeaters?

Optical fibre communications
Coaxial cables are being replaced by optical fibres. The fibres are made from glass and the signal is converted into pulses of infra-red light of wavelength 1.55 μm. Light passing down a fibre is absorbed and scattered, and 1.55 μm is the wave-length at which these factors are the smallest. In fact, at this wavelength the attenuation is as low as 0.3 dB km^{-1}.

5

a) If this fibre were used to replace the cable in questions **4a)** and **b)**, what would the repeater spacing be?

One other advantage of optical fibres is the huge bandwidth. Coaxial cables can be used efficiently up to frequencies of about 500 MHz.

b) What is the upper frequency limit for an optical fibre?

The term 'bandwidth' doesn't have the same meaning in digital transmissions, although it can be related to length of each bit. The duration of each bit requires at least half of a cycle of a sine wave.

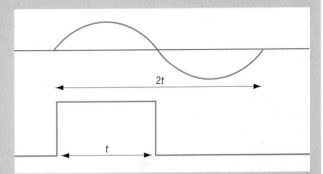

Fig 21.A4

The period of the sine wave will be twice the duration of the pulse)

c) Estimate the minimum possible duration of a bit and the bit rate from the frequency calculated from **5b)**.

The signal sent down a fibre is **pulse code modulated**. Electrical pulses are converted into pulses of infra-red laser light. The aperture of the fibre is so small (about 3 to 8 μm for monomode fibres) that a semiconductor laser is used as the light source. The 'electronics' switches the laser on and off many millions of times a second. (The availability of electronic circuits that can switch fast enough is one of the factors limiting the bit rate in use on optical fibres.)

6

a) Bit rates of 140, 280 and 560 Mbits s^{-1} have been used for several years. What is the duration of one bit at these speeds?

Dispersion is another factor which limits the maximum bit rate (see page XXX).

b) Explain how dispersion limits the bit rate.

Above: A stage cluttered with wiring restricts the musicians

Below: Madonna can move freely around the stage wearing a radio microphone and earpiece

WHEN POP GROUPS PERFORM on stage, they no longer have to put up with a clutter of wires and speakers because of the progress made in electronic sound engineering.

Singers and musicians cannot hear directly what other members of the band are playing. Up until recently, they have had to rely on monitor speakers on stage to play back the sound of the band, to get their own timing and sound levels correct (with other speakers directed at the audience).

Now, thanks to miniaturised electronic circuits, singers and players can wear a tiny earpiece instead of listening to the monitor speakers. Sound engineers mix the sound from all the instruments and voices and send it to a radio transmitter about 100 metres from the stage. Each musician carries a battery-powered receiver that picks up the signal from the transmitter and sends it along a wire to the earpiece. The receivers have volume controls, so each musician can adjust the sound to a comfortable level.

This system has several advantages. The stage is safer without stray speakers and wires and the musicians need not suffer from hearing damage because they can control their own sound level. The dreadful howling noise which used to come from sound feedback into amplifiers is a thing of the past. And the audience hears a better sound because it comes only from speakers directed at them.

Introduction

This chapter describes the behaviour of electronic circuits and systems. Electronics is a practical subject, so it will be best if you try the circuits out and check that they behave in the way described here. Only then will you fully understand the ideas in this chapter.

Electronic devices perform a vast range of tasks. For example, an electronic system in washing machines allows different types of material to be washed at the right temperature for the right length of time. The system controls water flow in and out of the machine, and heats the water if necessary. It also controls the rinsing and spinning.

As another example, some car radios tune themselves automatically to the strongest signal. Though the driver may not know it is happening, on a long journey, the radio retunes as the car travels between the range of one transmitter and the next. This ensures that the sound output remains as clear as possible.

In this chapter we will look at the principles of some of the most useful electronic circuits and systems, including some of those that can be used in washing machines.

1 ELECTRONIC SYSTEMS

Input, process and output

Any electronic system can be broken down into three parts as shown in Fig 22.1. Electronic systems respond to **input** signals. A signal is something that varies with time and carries information. For example, some of the signals we human beings respond to are changes in temperature or light intensity, changes in sound level or the weight of a bag of sugar. These signals could also apply to electronic systems.

The input block converts the input signal – the information – into an electronic signal. This will be a voltage which changes to reflect the changing information. The electronic signal is passed into the **process** block. For the time being, we will simply say that this block 'processes' the signal. The rest of this chapter describes some different types of process device.

The signal coming out of the process block is still an electronic signal but it has been changed, that is, it has been processed. The last block, the output block, responds to this processed electronic signal and changes it into some other form – such as a sound or light signal, or perhaps it turns a motor on or off.

The input and output blocks use devices that convert signals from one form to another. These are called **transducers**.

Fig 22.1 **Block diagram of any electronic system**

Transducers

Transducers can be divided into **input transducers** and **output transducers**. Input transducers are devices that convert some physical quantity, such as temperature or light level, into a voltage or some other electrical quantity. Input transducers are sometimes referred to as **sensors**.

A microphone converts the changes in air pressure of sound into a changing voltage. A light-dependent resistor responds to changes in light intensity by changing its resistance. This change of resistance can be used in a simple circuit to produce a changing voltage (Fig 22.2).

A What physical quantity could these input transducers change into electronic signals?

(a) a thermistor,

(b) a strain gauge,

(c) a Hall probe.

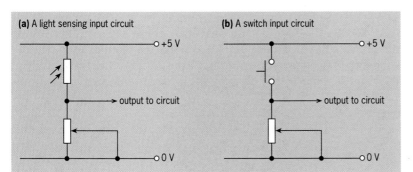

(a) A light sensing input circuit

(b) A switch input circuit

Fig 22.2 **Two kinds of input circuit**

Output transducers change electronic signals back into some physical quantity. For example, a loudspeaker converts an electronic signal into changes in air pressure – that is, sound.

Electronic systems can be divided into two kinds, although there are many examples where they are used together to make useful systems. These two parts are **digital** systems and **analogue** (or **linear**) systems.

B What physical quantities do these output transducers change electronic signals into?

(a) a motor,

(b) a bulb,

(c) a heater.

+5 V high (1)

+0 V low (0)

Fig 22.3 **A 'square' wave with a 'peak' voltage of 5 V, typical for a digital signal**

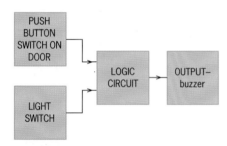

Fig 22.4 **Block diagram of a car door alarm system**

Table 22.1 **Combinations of switches for the lights-on alarm in a car**

Light switch on/off	Door switch	Audible warning
off	open	off
off	closed	off
on	open	off
on	closed	on

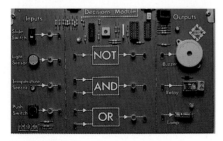

Fig 22.5 **Examples of logic gates on a 'decisions' board**

2 DIGITAL SYSTEMS

Digital systems process digital signals. A digital signal is one which has discrete values (that is, separate as opposed to continuous, possibly varying values). The switch in Fig 22.2(b) produces a digital signal. The switch is either open or closed, so the voltage output of the circuit will be either 0 V or 5 V, which is the usual voltage range of the power supply. In simple digital systems there are only two possible signal states. These two states are called **low** and **high**.

Some circuits are designed to make digital signals. An **astable** produces a square wave – it has a frequency and amplitude just like the sine wave more typical of alternating signals, but the signal alternates sharply between a high and low value (Fig 22.3).

The square wave is a digital signal that changes at a regular rate. If the period (time for one cycle) is constant, circuits like this can be used for timing.

3 DECISION CIRCUITS – LOGIC GATES

Most new cars have an audible alarm that warns the driver that the lights are on when the car door is opened. The electronic system that does this uses two inputs:

- A push-button switch which is *closed* when the door is opened.
- The light switch – off when the lights are off and on when the lights are on!

There is one output (Fig 22.4), which activates the buzzer that produces the audible warning. Table 22.1 shows the possible combinations. This way of summarising the behaviour of a digital system is called a **truth table**. But how do we design a digital system that will behave as described in the truth table?

The solution to this problem is quite simple if we use one of a group of very useful digital circuits called **logic gates**. Logic gates are simple electronic elements (Fig 22.5) which make decisions about an output based on one or more inputs. The simplest logic gate, the **NOT gate** or **inverter**, has only one input. Other logic gates have two or more inputs.

We have seen that truth tables summarise the behaviour of logic gates and that the inputs and outputs to these gates will depend on the power supply (often 0 to 5 V); also that an input or output at 0 V is described as 'low', and an input or output at or near 5 V is 'high'. Low and high therefore describe the voltages at inputs and outputs and are given the binary codes 0 and 1.

Fig 22.6 shows a simple circuit for testing a two-input logic gate. The two switches allow each input to be set at either low or high.

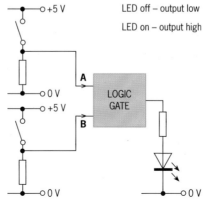

LED off – output low
LED on – output high

Fig 22.6 **A simple circuit for testing a two-input logic gate. When the switch is open the input A or B into the gate is low. The switch is closed to make the input go high. The LED monitors the output**

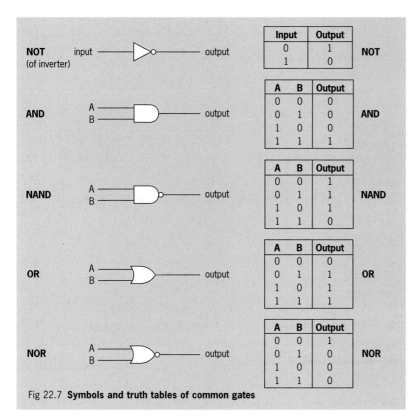

Fig 22.7 **Symbols and truth tables of common gates**

?

C Which gate from Fig 22.7 could be used to make the decisions necessary in the car sidelights warning system?

Fig 22.8 **NOR and NAND can be used as NOT gates by connecting the two inputs together**

Fig 22.9 **An AND gate made from two NANDs**

?

D (a) Explain how to make a two-input NOR gate from two input NANDs.

(b) Explain how to make a two-input NAND gate from two input NORs.

Fig 22.7 shows the symbols and truth tables for several useful gates. Both NOR and NAND can be used as NOT gates. If both inputs are connected together they become NOT gates (Fig 22.8). Two NAND gates can be put together to make an AND gate (Fig 22.9).

■ See questions 1 to 4.

Control system for a washing machine

Fig 22.10 shows the block diagram for a slightly more complex system, which controls the programmes of a washing machine. At each point in a programme, the control system has to make a decision. For example, there are only some conditions which will turn the washing machine motor on, and these are summarised in Table 22.2.

The motor switches on only when the door is closed *and* when the temperature of the water is higher than 55 °C (2 and 3 in Table 22.2), *or* when the dial is set to allow the load to be washed at temperatures of less than 55 °C, in which case the wash switch is on (1 in Table 22.2).

Table 22.2

Water temp. sensor	Wash switch	Door	Motor
<55 °C	Off	Open	OFF
<55 °C	Off	Closed	OFF
<55 °C	On	Open	OFF
<55 °C	On	Closed	ON [1]
>55 °C	Off	Open	OFF
>55 °C	Off	Closed	ON [2]
>55 °C	On	Open	OFF
>55 °C	On	Closed	ON [3]

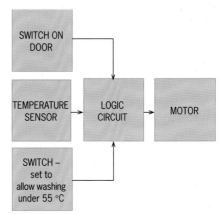

Fig 22.10 **Block diagram of the control system for a washing machine**

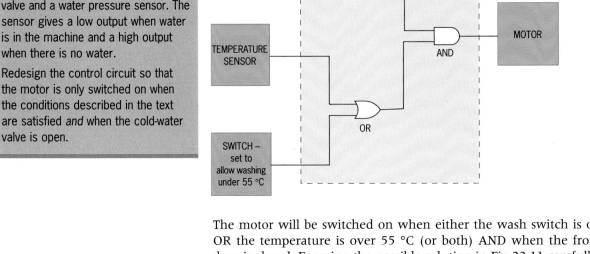

?

E The washing machine indicated a problem because the cold-water inlet valve kept sticking. The manufacturers designed a second model with a new valve and a water pressure sensor. The sensor gives a low output when water is in the machine and a high output when there is no water.

Redesign the control circuit so that the motor is only switched on when the conditions described in the text are satisfied *and* when the cold-water valve is open.

Fig 22.11 **A washing machine control circuit**

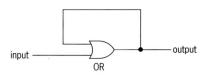

Fig 22.12 **A simple latch circuit. When the input changes from low to high, the output changes to high and stays high, even if the input goes low again**

The motor will be switched on when either the wash switch is on OR the temperature is over 55 °C (or both) AND when the front door is closed. Examine the possible solution in Fig 22.11 carefully. Does it satisfy the truth table on page 489?

Latch circuit and bistable

Once the machine is running, the temperature of the water sometimes falls just below 55 °C. This turns the motor off (unless a wash at below 55 °C is selected). Now, suppose that tests on different fabrics show that a drop of a few degrees does not affect the quality of the wash as long as the temperature is 55 °C at the start of the wash. The designer decides to adapt the circuit so that the motor will stay on even if the temperature does drop. A **latch circuit** (Fig 22.12) is added after the temperature sensor to ensure that it stays shut (on) while the machine is operating.

When the output of the temperature sensor goes high, the output of the latch should go high. If the temperature drops below 55 °C, the output of the temperature sensor will go low. However, the output of the latch should stay high. It should be possible to 'reset' the latch when the next washing cycle is started.

A circuit that behaves like this is called a **bistable**. Fig 22.13 shows a bistable made from two NOR gates and Table 22.3 shows its truth table.

The two inputs are called SET and RESET for reasons which should become clear when you examine the behaviour of the circuit. An important feature of this circuit is that the output of each gate is connected to one of the inputs of the other gate. This is an example of **feedback**.

If the switches in Fig 22.13 are push-button switches, they will be spring loaded. That is, when they are pressed they go from low to high, but as soon as they are released they return to low. It is possible to hold both switches down, but it is not a 'natural state'. When thinking about the behaviour of the circuit, we should treat the switches as though they are pressed briefly and then released.

Let us summarise Fig 22.13. When the SET input is changed, the outputs both change. If the SET input is changed again, nothing happens – not until the RESET input is changed. The outputs are 'latched'. The two outputs are always different (Table 22.4).

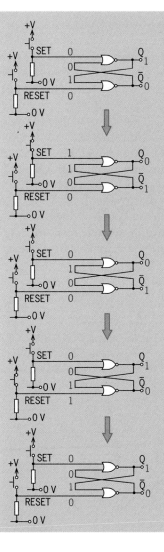

To begin with, both inputs are low. It is a matter of chance which state the two outputs are in. Suppose Q is high and Q̄ is low. This means that both inputs to the top NOR gate are low – which is what we would expect if its output is high! It also means that one of the inputs into the lower gate is high. This makes sure its output is low.

Now, suppose the SET input goes high. This makes its output go low and, in turn, makes both inputs to the lower gate low, which changes the output of that gate to high.

If SET is returned to low, then nothing changes at the outputs. The other input of the top gate is already high, which keeps its output low. This keeps the inputs to the lower gate both low.

Nothing changes until the RESET input is made high. This makes its output go low. Both inputs to the top gate are now low – its output now goes high. We are back where we started.

If RESET is released so that it returns to low, the two outputs will not change. The next change will happen only when SET is made high.

Fig 22.13 **A bistable made from two NOR gates. The two outputs of a bistable are called Q and Q̄. Q̄ means 'not Q'. Notice from Table 22.4 that this is always true. The truth table for the NOR gate is given in Table 22.3**

Table 22.3 **NOR gate truth table**

A	B	Output
0	0	1
0	1	0
1	0	0
1	1	0

Table 22.4 **Truth table for the bistable of Fig 22.13**

SET	RESET	Q	Q̄
0	0	1	0
1	0	0	1
0	0	0	1
0	1	1	0
0	0	1	0

?

F Suppose the output Q starts low and Q̄ is high. Work out the truth table showing the sequence of changes. (After the first stage the sequence should be the same!)

(When this is used in the washing machine circuit, the Q̄ output would be connected to the input of the OR gate.)

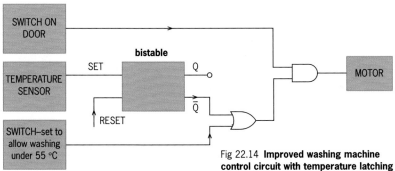

Fig 22.14 **Improved washing machine control circuit with temperature latching**

■ See question 5.

The washing machine control circuit including the temperature latch is shown in Fig 22.14. The bistable latch is also a simple 'memory'. It 'remembers' what happens to it!

Now let us turn our attention to the temperature sensor itself. Temperature is a physical quantity which varies over a wide range. The water in a washing machine might have a temperature at some time of anything between about 10 °C and 100 °C. There are various kinds of sensor which respond to changes in temperature (see Table 14.1, page 308). One example of a temperature sensor is a **thermistor**, a semiconductor device whose resistance varies *continuously* with temperature, so it is an *analogue* device.

4 ANALOGUE SYSTEMS

Many electronic circuits are designed to respond to signals that change over a wide range of values, like the verying temperature of the water in the washing machine. These are known as **analogue systems**.

A circuit very similar to those in Fig 22.2 can be used as a temperature sensing input (Fig 22.15). The thermistor is joined to a variable resistor to make up a **potential divider**. The voltage at the output of this circuit will change with temperature. The output voltage range can be adjusted by changing the variable resistor since the variable resistor changes the sensitivity of the circuit.

This simple analogue circuit can be used with a logic gate to change an input from low to high. To understand how this happens, we must examine the behaviour of a logic gate in more detail.

Fig 22.15 **A 'temperature sensing' input board, with a graph showing how the resistance of the thermistor varies with temperature**

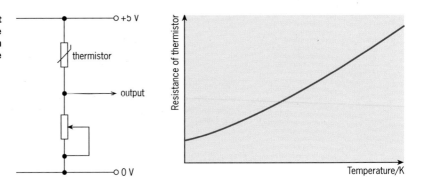

The logic gate in more detail

We need to know the answers to two questions:

● How big must the voltage at the input be in order to be 'high'?

● How low is low?

The Feature box gives some information about different kinds of logic families. However, the circuit in Fig 22.16 helps to answer our two questions.

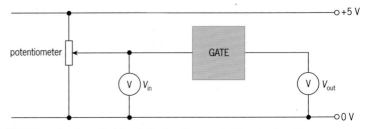

Fig 22.16 **Transfer characteristics (output voltage against input voltage) for different logic gates can be plotted using this circuit**

The potentiometer can be used to vary the input voltage from zero up to the maximum value of the supply. The voltmeters monitor the input and output voltages. A graph of output voltage against input voltage answers both our questions. This graph is called the **voltage transfer characteristic**.

Fig 22.19 in the Feature box shows two typical graphs. The answers to our two questions are not exactly the same for the two logic families shown. However, this does not matter so long as we know the characteristics of the gate we are using.

INTEGRATED CIRCUITS

LOGIC GATES COME as parts of integrated circuits (ICs) or chips (Fig 22.17).

Fig 22.17 **A memory chip**

The two input gates dealt with in this chapter are available in IC packages containing four of each gate. If a NAND gate is needed, then you have to use a 'quad NAND gate' (Fig 22.18). You cannot get such simple gates on their own!

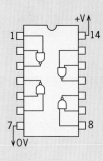

Fig 22.18 **A 'pin' diagram of a CMOS 4011 quad NAND gate chip. The chip contains four NAND gates with common power supply pins (7 and 14)**

There are also several different types or 'families' of logic gates available. The two most common ones are TTL, short for transistor transistor logic, and CMOS, short for complementary metal oxide semiconductor.

TTL is a very fast logic family which operates from a 5 V power supply. CMOS is slow but much more tolerant of different power supplies. It will operate from 3 V to 18 V. The graphs in Fig 22.19 show the transfer characteristics for TTL and CMOS. To make comparison easier, both have been measured using a 5 V power supply.

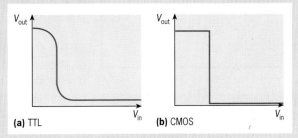

(a) TTL   **(b)** CMOS

Fig 22.19 **Transfer characteristics for** (a) **TTL and** (b) **CMOS**

If the temperature-sensing circuit is used with a TTL gate, then anything below 1.2 V will be low. If the input increases a bit above this, then it is high.

The variable resistor in the sensing circuit is set so that the output voltage is just below 1.2 V. As the temperature rises, the resistance of the thermistor increases and the voltage across it also increases. As this voltage crosses the 1.2 V hurdle, the input changes from low to high. The output of the gate also changes.

5 THE OPERATIONAL AMPLIFIER

This circuit works quite well, but there are better ways of doing the job. There is a family of amplifiers available in integrated circuit packages which are very useful in all sorts of applications. These are called **operational amplifiers**. It will be useful first to look at the properties of a simple amplifier.

The simple amplifier

We are all familiar with audio devices such as radios and CD systems. An audio amplifier is an important part of all of these systems. The amplifier takes a small, weak signal and makes it stronger. A good amplifier preserves the characteristics of the audio signal such as frequency, and just increases the amplitude of the signal, as in Fig 22.20.

The amplitude of the signal going into and out of the amplifier could be measured in volts, and so the amplifier may be described as a **voltage amplifier**. The output voltage will be bigger than the input voltage: the ratio of the output to the input is called the **gain** of the amplifier. The gain may be expressed as a number (such as 10 times) or may be measured in decibels (dB), see page 494.

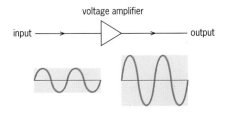

Fig 22.20 **A simple voltage amplifier showing input and output signals**

EXAMPLE

Q The input to an amplifier has a peak voltage of 20 mV. The output has a peak voltage of 2 V. What is the gain of the amplifier?

A Gain = $\dfrac{\text{output voltage}}{\text{input voltage}} = \dfrac{2\text{ V}}{20\text{ mV}}$

= 100 times (or × 100)

Or, in dB:

Gain = $20 \log_{10} V_{out}/V_{in}$

= $20 \log_{10} 100 = 20 \times 2$

= 40 dB

?

H A signal of 5 mV is fed into an amplifier with a gain of 20 dB. Calculate the voltage of the output signal.

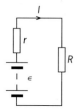 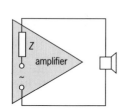

(a) A battery of e.m.f. ϵ with internal resistance r connected to a load resistor R

(b) The output of the amplifier – a source of alternating e.m.f. in series with the 'output impedance' of the amplifier connected to a load (a loudspeaker)

Fig 22.21 **Analogy between a battery with internal resistance and a signal source**

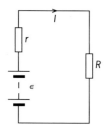

Fig 22.22

The decibel

Very often in electronics we wish to compare the relative amplitude of two signals. The signals could be the input and output from an amplifier, and the ratio could be very large, sometimes as big as millions. For this reason it is easier to use a logarithmic scale:

Linear scale, N	1	10	100	1000	10 000	100 000
Logarithmic scale, $\log_{10} N$	0	1	2	3	4	5

We use a unit called the **decibel**. By definition, the ratio of two signals in decibels is:

$$\text{dB} = 20 \log_{10}(A_2/A_1)$$

where A_1 and A_2 are the amplitudes of the two signals. We would usually measure the amplitudes as voltages.

When a signal passes into an amplifier, there is a current. The size of the current depends on the signal voltage at any instant, but it also depends on the **input impedance** of the amplifier. The impedance is measured in ohms (see page 292).

The signal coming out of the amplifier is also a current and has a voltage (usually both are fluctuating) and is passed into another device or **load**. In the case of an audio amplifier, the load is usually a loudspeaker. The output of the amplifier behaves as a signal source. The current that this source drives through the load depends on the source voltage, the impedance of the load and the **output impedance** of the amplifier.

It helps to think of the output of the amplifier as a source of alternating e.m.f. with an internal resistance – rather as a battery is made up of a source of e.m.f. ϵ in series with its internal resistance r, as in Fig 22.21.

When the signal source is connected to a load, it operates most efficiently when the load impedance is the same as the output impedance of the amplifier.

Maximum power theory

This is a useful idea from d.c. circuits that helps in the design of efficient electronic systems.

How does the energy transferred from a battery to a load depend on the resistance of the load (Fig 22.22)?

Consider the two extremes of the load – that is, zero resistance and very high (infinite) resistance.

When the load has zero resistance ($R = 0$), the current I will be limited only by the internal resistance r of the battery and will therefore have a maximum value. However, the p.d. across zero resistance is zero. No energy is transferred in the load – but the battery will get very hot!

When the load resistance R is infinite, the current in the load will be very small. Again, the energy transferred in the load (IV) will be very small.

It is reasonable that for some value of load resistance in between, the energy transferred will be a maximum. That value can be found using calculus, as follows.

Now:
$$\epsilon = Ir + IR$$

or:
$$I = \frac{\epsilon}{r + R}$$

The power W in the load is given by:

$$W = I^2 R = \frac{\epsilon^2 R}{(r + R)^2}$$

As R changes, so does W. W will be a maximum when dW/dR is zero:

$$\frac{dW}{dR} = \frac{d}{dR}\left(\frac{\epsilon^2 R}{(r + R)^2}\right) = 0$$

That is: differential $= \dfrac{(r + R)^2 \epsilon^2 - 2\epsilon^2 R(r + R)}{(r + R)^4} = 0$

Therefore: $(r + R)^2 \epsilon^2 - 2\epsilon^2 R(r + R) = 0$

This simplifies to: $(r + R) - 2R = 0$

which gives the solution: $r = R$

That is, the maximum energy transfer happens when the load resistance is the same as the internal resistance of the battery.

Properties of operational amplifiers

The operational amplifier or **op-amp** is nearly the perfect amplifier. Fig 22.23 shows the basic characteristics of an op-amp.

Op-amps have some important properties:

1 They are **differential** amplifiers. They have two inputs. The output of the amplifier is given by:

> output = gain × *difference* between the two inputs.

One input is called the **non-inverting** input and the other the **inverting** input (Fig 22.23).

$$V_{out} = A(\epsilon_1 - \epsilon_2)$$

For example,
if: $A = 10^6$, $\epsilon_1 = 0$ V and $\epsilon_2 = 2$ μV,
then: $V_{out} = 10^6(0 - 2)$ μV
$= -2$ V

(Notice that when $\epsilon_1 = 0$, the output is negative – the amplifier is said to 'invert' the input.)

2 They have a very high gain A, typically 10^5 to 10^6, with some having gains up to 10^9. This gain is called the **open loop gain**.

3 All inputs and outputs are measured relative to earth (0 V). Most op-amps use a 'split rail' power supply which means that Earth is taken as the mid-point of the power supply. The supply voltage limits the output voltage of the amplifier. V_{out} cannot be bigger than $\pm V$ (Fig 22.24).

4 The inputs draw very tiny currents – so small that we can consider that they draw no current at all. That is, they have a very high input impedance. (A typical op-amp will draw a current much smaller than a micro-amp, and often as low as nano-amps.) A voltage of 1 mV at the input will drive a current of 1 pA into the op-amp (Fig 22.25).

5 The output impedance is very low. This means that the output can deliver quite a high current, up to 20 mA!

?
I How many times bigger than a pico-amp is a milliamp?

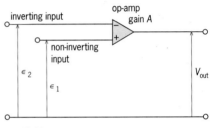

Fig 22.23

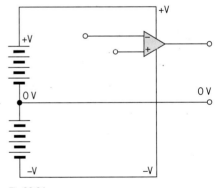

Fig 22.24

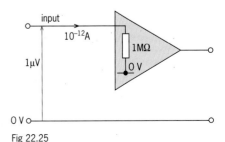

Fig 22.25

Op-amps in use

Op-amps are not used by themselves. The simplest circuits use resistors, in particular – one between the input and the output. This provides **feedback** between the output and the input and makes some very useful circuits possible. In the examples described, the feedback is **negative feedback**. This means that a fraction of the output is 'fed back' to the *inverting* input.

The inverting amplifier

In Fig 22.26, R_f is the value of the feedback resistor between the output and the inverting input. R_{in} is the input resistor value. The inverting input is earthed, that is, at 0 V.

You will find it easier to understand this circuit if you try it out. If the values of the two resistors R_f and R_{in} are equal, the output voltage will be the same size as the input but it will also be inverted. That is, an input voltage of +1 V will give an output of –1 V. To analyse how this circuit behaves, we must start at the output.

Fig 22.26 **An inverting amplifier**

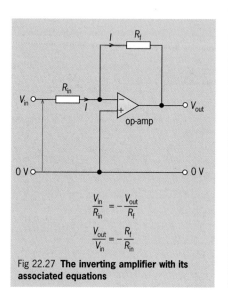

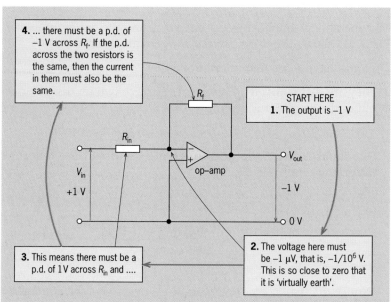

4. ... there must be a p.d. of –1 V across R_f. If the p.d. across the two resistors is the same, then the current in them must also be the same.

START HERE
1. The output is –1 V

3. This means there must be a p.d. of 1 V across R_{in} and ....

2. The voltage here must be –1 μV, that is, –1/10⁶ V. This is so close to zero that it is 'virtually earth'.

$$\frac{V_{in}}{R_{in}} = -\frac{V_{out}}{R_f}$$

$$\frac{V_{out}}{V_{in}} = -\frac{R_f}{R_{in}}$$

Fig 22.27 **The inverting amplifier with its associated equations**

The two resistors are effectively in series. The current drawn by the inverting input of the op-amp is very small compared with the current through the two resistors. This will still happen when the two resistors are not the same size since resistors in series have the same current in them. So the mathematics of this circuit is fairly simple (Fig 22.27).

The output voltage can be a precise multiple or fraction of the input – it depends only on the two resistors. Notice that the output voltage does not depend on any of the characteristics of the op-amp, yet the circuit would not work without it.

The circuit carries out a precise mathematical operation, in this case multiplication by a constant, the gain of this circuit (V_{out}/V_{in}), which is called the **closed loop gain**.

One way of looking at the circuit is to say that the output will become whatever value is necessary to make the difference between the two inputs equal zero. In this case, the output goes to a value that makes the non-inverting input zero. (The other input has been fixed at zero.)

The voltage of the power supply limits the maximum output voltage from the amplifier. No matter what the theoretical gain of a circuit, it is impossible for the output to be bigger than the supply voltage. Real op-amps saturate at one or two volts short of the power supply voltage.

For example, if an op-amp is operating from a ±15 V supply, then the output will never be bigger than ±13 V, and so ±13 V is called the **saturation voltage**. This can be measured by plotting the **transfer characteristic** of the op-amp (Fig 22.28).

The slope of the central part of the graph is the gain of the inverting amplifier, R_f/R_{in}.

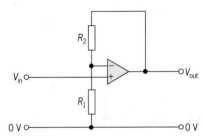

J What will be the overall gain of the circuit in the drawing?

See question 6.

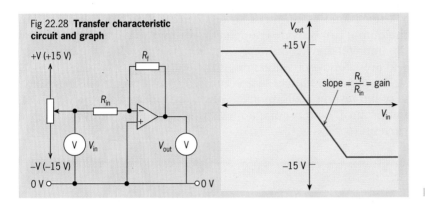

Fig 22.28 **Transfer characteristic circuit and graph**

$$\text{slope} = \frac{R_f}{R_{in}} = \text{gain}$$

The non-inverting amplifier

The op-amp can be used to make an amplifier that does not invert the input signal. The circuit is shown in Fig 22.29. This circuit also operates so that the difference between the two inputs is so close to zero that it can be taken as zero.

This time, neither input is connected to earth (0 V). If the input to the non-inverting input is V_{in}, then we assume that the voltage at the inverting input is also V_{in}.

R_1 and R_2 form a **potential divider** (see page 209). The voltage at the non-inverting input V_{in} of Fig 22.29 is given by:

$$V_{in} = \frac{R_1}{R_1 + R_2} \times V_{out}$$

This is equal to the input voltage, V_{in}, so that:

$$V_{in} = \frac{R_1}{R_1 + R_2} \times V_{out}$$

The gain is:

$$\frac{V_{out}}{V_{in}} = \frac{R_1 + R_2}{R_1} = 1 + \frac{R_2}{R_1}$$

Again, this is the closed loop gain and it depends only on the two resistors.

There is one version of the non-inverting amplifier that is particularly useful. This is the **voltage follower**.

The voltage follower

Fig 22.30 shows a simple version of the non-inverting amplifier. The output is connected directly to the inverting input, which means that R_2 is zero. The other resistor in Fig 22.29, R_1, is removed. This is equivalent to making its size infinite. Therefore:

$$\text{gain} = 1 + 0/\infty = 1 + 0 = 1$$

So what good is an amplifier with a gain of 1?

Fig 22.29 **A non-inverting amplifier**

K If R_1 is 10 kΩ, what value of R_2 would give a gain of 10?

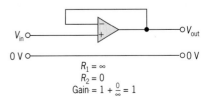

Fig 22.30 **The voltage follower (compare this circuit with Fig 22.29)**

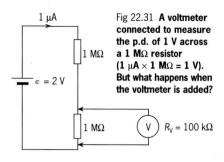

Fig 22.31 **A voltmeter connected to measure the p.d. of 1 V across a 1 MΩ resistor** (1 μA × 1 MΩ = 1 V). But what happens when the voltmeter is added?

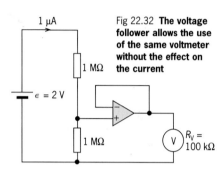

Fig 22.32 **The voltage follower allows the use of the same voltmeter without the effect on the current**

The advantage of this circuit lies in the basic properties of the op-amp. The input draws a very small current and the output can provide a much larger current. The technical explanation is: 'It has a very high input impedance and a very low output impedance.' The following example illustrates the use of the voltage follower.

Fig 22.31 shows a voltmeter connected to measure the p.d. of 1 V across a 1 MΩ resistor. A typical moving coil voltmeter might have a resistance of about 100 kΩ (see the Assignment for Chapter 9), so this is not a very good way of measuring the p.d. since the voltmeter will draw most of the current and give an incorrect reading.

Fig 22.32 shows a better way of measuring the p.d. using a voltage follower.

The op-amp input draws far less current than the voltmeter, but the output can provide enough current for the voltmeter. A typical op-amp will draw about 1 pA with a voltage of 1 V at the input, but will provide several milliamps at the output.

The voltage follower is a very useful little circuit and has some very interesting applications in physics. The examples in Fig 22.33, for instance, both require a measuring device that draws very little current.

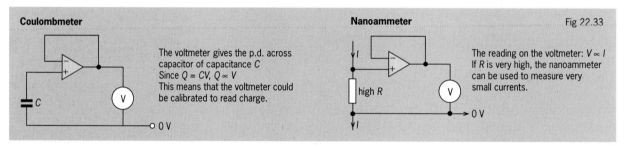

Coulombmeter

The voltmeter gives the p.d. across capacitor of capacitance C
Since $Q = CV$, $Q \propto V$
This means that the voltmeter could be calibrated to read charge.

Nanoammeter   Fig 22.33

The reading on the voltmeter: $V \propto I$
If R is very high, the nanoammeter can be used to measure very small currents.

?

L How much current would the voltmeter in Fig 22.31 draw with a p.d. of 1 V across it? Estimate the reading on the voltmeter.

M In the circuit of Fig 22.32, where does the 'extra' current come from?

The comparator

The op-amp can be used as a switch. There is no feedback, so the two inputs do not have to be the same.

The output depends on the difference between the two inputs. In Fig 22.34, an op-amp with an open loop gain of 10^6 has a voltage of 15 μV at the non-inverting input. (The inverting input is held at 0 V.) The output will be at +15 V.

If the input is reduced by 15 μV to zero, the output also becomes zero. If the input is further reduced to –15 μV, the output shoots down to –15 V. A change of 30 μV (0.000 03 V) is enough to switch the output of the op-amp from positive saturation to negative saturation.

The op-amp can be used as a voltage comparator. Fig 22.35 shows a typical application of this.

The thermistor and the fixed resistor R form a potential divider which fixes the voltage at the non-inverting input of the op-amp.

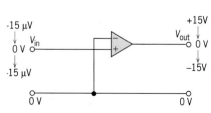

Fig 22.34 **A basic comparator – the op-amp as a switch**

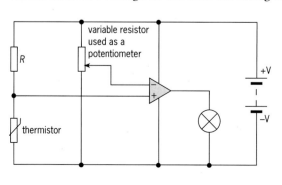

Fig 22.35 **A temperature-sensitive comparator**

As the temperature rises, this voltage increases as the resistance of the thermistor increases. (If the temperature falls, the voltage will decrease.) The voltage at the inverting input is set by the variable resistor which is used as a potentiometer.

The potentiometer is set so that the voltage at the inverting input is the same as the voltage at the non-inverting input. The difference between the two voltages will then be zero and the lamp at the output will be off. As soon as the temperature increases, the voltage difference will be greater than zero and the lamp will light.

Comparators are very sensitive switches. The temperature need only change by a very small amount before the voltage difference exceeds 15 μV.

■ See questions 7 to 10.

The integrator

Integration is a mathematical procedure that 'sums'. That is, it is used to add up a series of values. A good example is the use of integration to find the area under a graph (see Appendix 2). Op-amps make almost perfect integrators. Fig 22.36 shows the circuit.

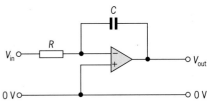

Fig 22.36 **An op-amp as an integrator**

Theory of the integrator

The capacitor of capacitance C provides the feedback between the output and the inverting input. The input voltage is passed via the resistor of resistance R, to the inverting input (hence the negative sign in equations below). If the current at some point is I and the charge on the capacitor at the same time is Q, then:

$$V_{out} = -\frac{Q}{C} \quad \text{or} \quad Q = -CV_{out}$$

and:

$$I = \frac{V_{in}}{R} = \frac{dQ}{dt} = \frac{d(-CV_{out})}{dt}$$

That is:

$$\frac{V_{in}}{R} = -C\frac{dV_{out}}{dt}$$

Rearranging, we get:

$$dV_{out} = -\frac{1}{RC}V_{in}\,dt$$

We now integrate to get V_{out}:

$$V_{out} = -\frac{1}{RC}\int V_{in}\,dt$$

This is the output of the op-amp.

One application of the integrator is the **ramp generator** (Fig 22.37). The input is a steady voltage – one which does not change with time. The output increases gradually: the slope depends on the input voltage.

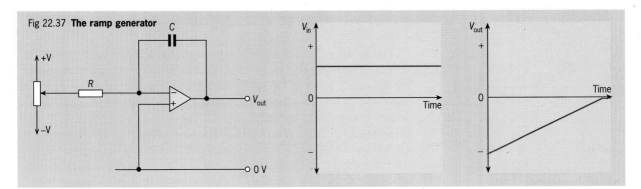

Fig 22.37 **The ramp generator**

aerial

TUNED
CIRCUIT → DETECTOR → AUDIO
CIRCUIT

select
the signal

processes
the audio
signal

unmixes the
RF and AF
components

earth

Fig 22.38 **AM receiver block diagram**

6 RADIO RECEPTION

We shall finish this chapter by looking at a simple radio designed to receive amplitude modulated signals. (See Fig 21.2, page 466, for a block diagram of an AM transmitter.)

A transmitter adds an audio frequency signal to a radio frequency carrier to produce the amplitude modulated signal which is transmitted.

A radio receiver receives this signal and carries out the reverse process, eventually 'un-mixing' the radio signal to produce the original audio frequency signal. Fig 22.38 is a block diagram showing the important parts of an AM receiver.

The radio signal is transmitted as an electromagnetic wave and the signal is 'picked up' by an aerial. A length of wire makes a very effective simple aerial; many radios use a telescopic metal rod.

Electromagnetic waves and aerials

Electromagnetic waves have an electric field component and a magnetic field component. Different kinds of aerial respond to one or other of these components (Fig 22.39).

In the presence of a radio signal, free electrons in a wire or telescopic aerial oscillate to produce an alternating current at the same frequency as the radio signal. The oscillation of the electrons in the wire aerial is induced by the *electric field* component of the radio wave. See Fig 16.3, page 339.

Most radios designed to receive medium and long wave signals use a ferrite rod aerial, which consists of a coil of wire wound on a ferrite rod. This aerial responds to the *magnetic field* component of the radio wave. The changing magnetic field induces a current in the coil. The ferrite rod concentrates the local field, making it stronger inside the coil.

Fig 22.39 **Left: A radio with an aerial uses the electric field component of the radio wave**

Right: A radio with a ferrite rod uses the magnetic field component

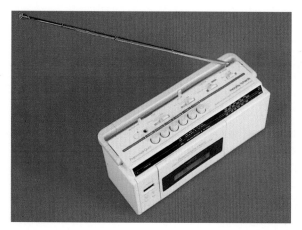

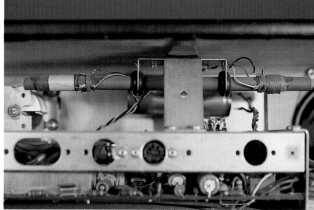

The problem is that there are many radio signals, each with a different frequency and each inducing a current of the same frequency as the radio wave. The aerial carries all these alternating currents into the radio. The **tuned circuit** selects one signal out of the many present in the aerial (Fig 22.40).

The tuned circuit is a parallel *LC* circuit. That is, it is a capacitor in parallel with an inductor (see page 293). Most tuned circuits use a variable capacitor so that the circuit can be 'tuned' to cover a range of frequencies. The parallel *LC* circuit has a resonant frequency.

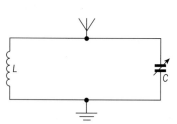

Fig 22.40 **Tuned circuit**

If the resonant frequency matches a signal from the aerial, this signal will be passed on. The tuned circuit behaves like a filter, blocking some signals and allowing a narrow range of frequencies through (Fig 22.41).

The 'ideal' tuned circuit allows only one signal to pass through, so that the signal produced by the radio is free from interference from other signals. However, the tuned circuit must allow a band of frequencies through that is equal to the bandwidth of the AM signal (see page 467).

Simple radios also need a good earth wire in contact with the ground to allow unwanted signals to be carried away. Note that this literally means the earth, not the mains power supply earth.

The next stage is the 'un-mixing' or **demodulation** of the selected signal. A very simple and effective demodulator can be made using a single diode and a capacitor. The diode actually rectifies the signal, that is, it allows current in only one direction. The rectified signal still contains the lower frequency signal and the radio frequency carrier.

The capacitor is chosen so that its reactance is low at the frequency of the radio signal. The radio frequency part of the rectified signal then passes through the capacitor to earth (Fig 22.42).

The lower, audio frequency part of the rectified signal is not strong enough to drive a loudspeaker, but it can be amplified by an audio amplifier.

Remember from Chapter 13 that the resonant frequency of a parallel LC circuit is:

$$f = \frac{1}{2\pi}\sqrt{\frac{1}{LC}}$$

N A variable capacitor can be varied between 200 pF and 500 pF. It is used in a tuned circuit intended to cover the medium wave band, that is, from 0.5 to 1.5 MHz. Calculate the size of a suitable inductor to be used with the capacitor to make a tuned circuit.

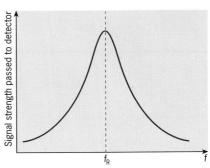

Fig 22.41 **Resonance curve for parallel *LC* circuit**

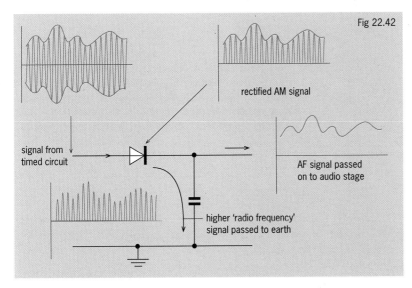

Fig 22.42

rectified AM signal

signal from timed circuit

AF signal passed on to audio stage

higher 'radio frequency' signal passed to earth

As its name suggests, the audio amplifier is designed to amplify audio frequency signals. A good amplifier should amplify equally over the whole of the audio frequency range (20 Hz to 20 kHz), although for most simple radios a much narrower range still gives acceptable results. Fig 22.43 shows a simple 'crystal set' radio.

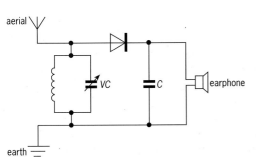

aerial

VC
C
earphone

earth

Fig 22.43 **The simple crystal set will drive a high impedance earphone when tuned to a strong signal. A good aerial and earth are very important to this simple radio. Note that there is no power supply!**

SUMMARY

After studying this chapter you should:

■ Know that any electronic system can be broken down into three parts: input, process and output.

■ Understand the terms transducer and sensor.

■ Know that there are two kinds of electronic system: digital and analogue.

■ Be able to use truth tables to analyse the behaviour of digital systems.

■ Understand the behaviour of logic gates and how they are combined to make more complex circuits.

■ Understand the terms op-amp, inverting amplifier and non-inverting amplifier.

■ Understand the principles of radio reception.

QUESTIONS

1 Describe the function of the logic circuits in Fig 22.Q1. Hint: Draw up a truth table for each circuit.

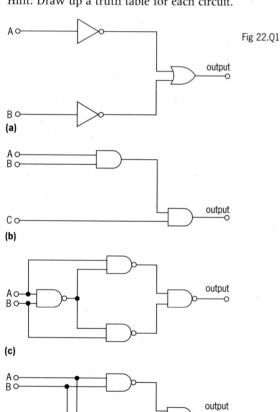

Fig 22.Q1

(a)

(b)

(c)

(d)

2 The truth table for the exclusive OR gate is given in the table.

exclusive OR gate

Inputs		Output
A	B	
0	0	0
0	1	1
1	0	1
1	1	0

a) Design an exclusive OR gate from two input AND and NOR gates. (This should take three gates to solve!)

b) Redesign your solution, but this time only using NOR gates. What is the smallest number of gates that are needed?

c) Repeat part **b)**, this time using only NAND gates. How many gates are needed?

3

a) Data pulses passing between commercial banks are often transmitted simultaneously by two different routes and are monitored at intervals to ensure that they remain identical. Fig 22.Q3 shows such a monitoring system using a logic gate as a comparator.

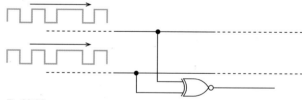

Fig 22.Q3

(i) Draw up a truth table comparing inputs and output of the comparator.

(ii) Sketch the output from the comparator when the trains of pulses shown in Fig 22.Q3 are identical.

b) Show how an array of NAND gates could perform the same logic function as the comparator in Fig 22.Q3.

[UCLES: Specimen Paper 1996]

4 You are asked to design a circuit that adds together two binary numbers. The rules of binary arithmetic are:

$0 + 0 = 0$
$0 + 1 = 1 + 0 = 1$ carry 0
$1 + 1 = 10$, that is 0 carry 1

The circuit can be represented by a truth table and a block diagram as shown in Fig 22.Q4.

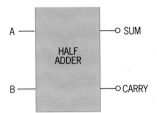

Fig 22.Q4

What gates will be needed to make this circuit?

5

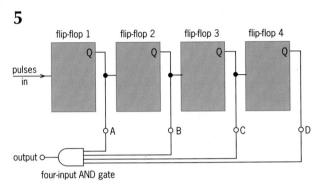

Fig 22.Q5(a)

A square wave is fed into the input of the first flip-flop as shown in the first line of Fig 22.Q5(b). Copy this diagram on to graph paper and add the four signals coming out of each flip-flop.

The Q outputs are also fed into a four-input AND gate. Add a fifth signal to your diagram to show the output from the AND gate.

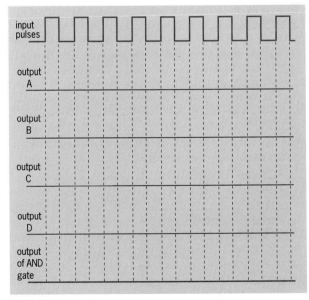

Fig 22.Q5(b)

6

An electrical device monitoring conditions in a manufacturing plant has an internal resistance of 10 kΩ and is generating an e.m.f. of 1.0 mV.

a) The transducer is connected directly to a moving coil meter of resistance 1.0 kΩ as shown in Fig 22.Q6(a).

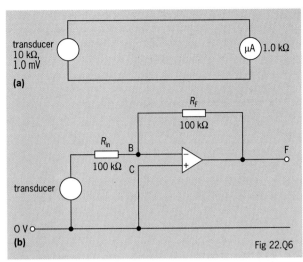

Fig 22.Q6

Calculate the current in the circuit and the p.d. across the meter.

b) The transducer is now connected to the operational amplifier circuit shown in Fig 22.Q6(b). Assume that the op-amp behaves 'ideally' in this circuit.
   (i) What is the 'voltage' at B?
   (ii) What is the current through the input resistor of value R_{in}?
   (iii) What is the output voltage at F?

c) As the gain of the circuit is −1, how can the circuit be modified so that the output voltage at F is −1.0 mV?

7

The operational amplifier in Fig 22.Q7 saturates at ±13 V.

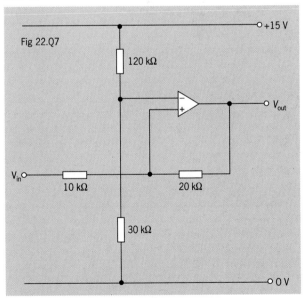

Fig 22.Q7

a) What is the voltage at the inverting input of the operational amplifier?

b) When the output of the amplifier is −13 V, what value of input voltage will make the output switch from −13 to +13 V?

c) What input voltage will make the output change back from +13 to −13 V?

8 The input voltage, V_A, to the integrating circuit shown in Fig 22.Q8(a) varies with time in the manner shown in Fig 22.Q8(b).

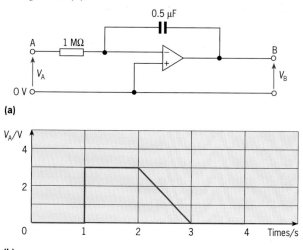

(a)

(b)

Fig 22.Q8

a) Initially the output, V_B, at B is zero. Show that the magnitude of V_B at $t = 2$ s is 6 V.

b) Copy and complete the table by calculating values of the potential, V_B, at B at the indicated times.
Assume that the operational amplifier does not reach saturation at any time.

Time/s	0	1	2	3	4
V_B/V	0		6		

c) Copy Fig 22.Q8(c). Using the calculated points, sketch on it the way in which V_B varies throughout the four seconds.

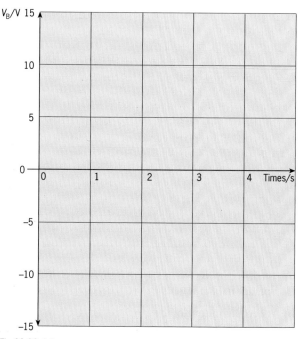

Fig 22.Q8 **(c)**

[OCSEB: 1989, Nuffield Science, Physics, short-answer paper, Q8]

9

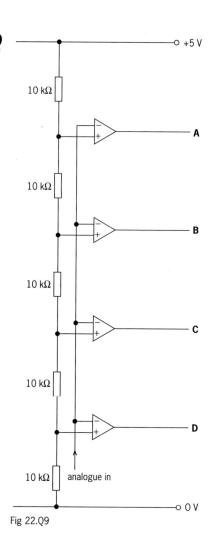

Fig 22.Q9

Fig 22.Q9 shows a simple analogue-to-digital converter (sometimes called a flash converter). It is based on operational amplifiers used as comparators. The circuit uses a 0 to 5 V power supply. The analogue signal is fed into the inverting amplifier of each amplifier.

a) What is the voltage at the non-inverting input of each amplifier?

With a 5 V power supply, the output of each amplifier will be either high (1) or low (0).

b) Copy and complete the truth table showing the digital output for different voltages.

Analogue voltage	Amplifier output			
	A	B	C	D
0				
1				
2				
3				
4				

c) What will the output of the amplifiers be for an analogue input of **i)** 1.5 V, **ii)** 4.5 V?

10 An operational amplifier can also be used to make a digital-to-analogue converter. Fig 22.Q10(a) shows a four-bit digital-to-analogue converter.

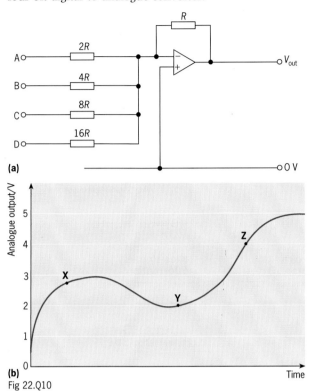

(a)

(b)   Time
Fig 22.Q10

Each of the inputs A to D can be high (5 V) or low (0 V).

a) What will the output voltage be if each input in turn is high, while the three others are low? Copy and complete the table.

Input high	Output
A	
B	
C	
D	

b) Copy and complete the table by calculating the output voltage for each input combination.

Input				Output/V
A	B	C	D	
0	0	0	0	
0	0	0	1	
0	0	1	0	
0	0	1	1	
0	1	0	0	
0	1	0	1	
0	1	1	0	
0	1	1	1	
1	0	0	0	
1	0	0	1	
1	0	1	0	
1	0	1	1	
1	1	0	0	
1	1	0	1	
1	1	1	0	
1	1	1	1	

This circuit converts a digital signal into an analogue signal with 16 different levels.

c) The signal in Fig 22.Q10(b) is obtained at the output of the circuit in Fig 22.Q10(a). What was the digital input at the points labelled X, Y and Z?

d) A 'smoother' analogue output could be obtained by converting the digital signal into twice as many 'levels', that is, 32 levels. Draw a circuit showing how the circuit in Fig 22.Q10(a) should be changed. Calculate the value of any extra resistors you use.

ELECTRONICS

The Chapter Map includes important concepts and applications from Chapters **21 Communications** and **22 Physical electronics**, and shows the way they interlink and depend on one another.

It is important that you relate the areas covered in these chapters to your own experience of practical electronics, so that you understand the theory behind the applications.

Use the map to match the requirements of the syllabus you are following, and to identify the areas you may need to study further.

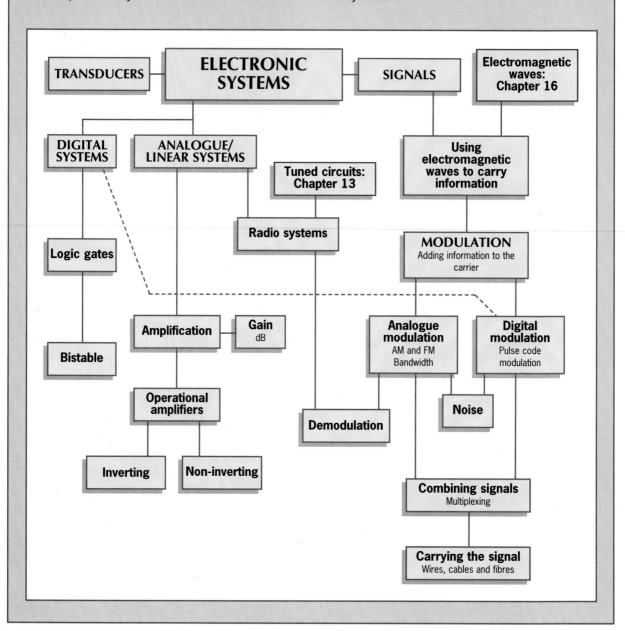

HUMAN PHYSICS

THE PHYSICAL CHARACTERISTICS of the Earth have
determined the nature of life that has evolved on it.
We humans continually use the Earth's resources to our
advantage, and we need to be aware that the living
world and the physical world could be irreversibly
altered by our activities, for example, in our use of fuel,
land and water or in waste disposal.

We have important choices to make. We could carelessly
exploit finite resources in the short term. Or we could
take account of ecological systems and develop a long-
term balance between our needs and the use we make
of the Earth's resources. The aim ought to be for a
healthy population in a self-sustaining environment.

Here, we look at the basic physical character of the
Earth and its systems, and how we can monitor change
in order to make informed choices. And through
medical physics we see how modern physical techniques
and discoveries are used to diagnose and treat disease.

23 Earth and atmosphere

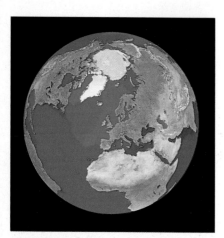

Clues to possible life elsewhere include substances that resemble biochemicals in space and the discovery of planetary systems round stars like our Sun

THOUGH LIFE MIGHT once have existed on Mars, and may still survive in some primitive form below the planet's surface, most scientists think that the Earth is the only planet in the Solar System that sustains life now. However, the Sun is not the only star known to have a planetary system and, since the Universe is very large, there may be many Earth-like planets where life has evolved.

The search for life and 'intelligence out there' engages many scientists and stimulates the public imagination. It cannot be proved that there isn't life beyond the Solar System, but positive proof is yet to be found.

Table 23.1 **Data on the Earth**

Age	4.6×10^9 years
Dimensions	
Shape	a flattened spheroid
Polar radius	6356.779 km
Equatorial radius	6378.164 km
Mean radius	6371.03 km
Mass	5.976×10^{24} kg
Volume	1.0832×10^{21} m^3
The ocean and crust	
Mass of ocean	1.45×10^{21} kg
Fraction of surface covered by ocean	70%
Mean crust thickness: under oceans	8 km
Mean crust thickness: under continents	~ 45 km (from 30 km to 70 km approx)
Energy, temperature and pressure	
Solar constant	1.353 kW m^{-2}
Albedo	0.3
Mean surface temperature	288 K
Standard atmospheric pressure	1013 millibar = 1.013×10^5 Pa
Energy input from Sun	1.7×10^{17} W
Gravity	
Surface (standard)	9.80665 N kg^{-1}
At latitude 45°	9.80612 N kg^{-1}

1 THE AGE AND EVOLUTION OF THE EARTH

Some 5 billion years ago, there was a massive star, a supernova, in our arm of the Milky Way galaxy which exploded and triggered a wave of compression in the nearby interstellar gas. One result was the formation about 4.6 billion years ago of the Solar System. For more about its origins, see Chapter 27.

% abundance Element	Crust	Mantle	Outer core	Inner core
oxygen	46.4	~20		
silicon	27.7	~10		
aluminium	8.1			
iron	5	~30	~90	100
calcium	3.6			
sodium	2.8			
potassium	2.6			
magnesium	2.1	~30		
nickel			~10	
other	<2	~10		

Table 23.2 **Elements of the Earth and their percentage abundance**

The accepted model suggests that the Earth and the inner planets were formed by **accretion**, a process in which progressively larger particles collide and clump together, as described in Chapter 27. As accretion progressed, small 'protoplanets' or **planetesimals** attracted each other and smaller objects. Potential energy of the many objects changed to kinetic energy as they accelerated towards each other. On collision, this energy was transferred to random molecular movements, making the growing planets hot. The heavy elements that formed the inner planets include long-lived radioactive elements such as uranium and potassium.

Radioactive decay releases energy, and this made a large contribution to heating the early Earth. See Table 23.3. The longest lived of the radioactive nuclides are used to measure the ages of rocks, both on Earth and in meteorites. These measurements confirm the currently accepted value for the age of the Earth: 4.6 billion years (4.6 **aeons**).

Table 23.3 **The long-lived radionuclides in the Earth's crust**

Nuclide	Half-life (10^9 y)	Mean abundance in crust (g per tonne)	Heating effect (10^{-7} W per tonne of rock)
uranium-238	4.9	2.2	} 1.6 (combined)
uranium-235	0.7	0.02	
potassium-40	1.3	0.24	0.7
thorium-232	14	8	1.6

See question 1.

Why the Earth has a layered structure

The energy gained in the accretion process, together with the energy released in radioactive decay, made the early Earth molten. The denser material – metals like iron and nickel – sank to the centre, whilst the lighter material – mostly oxides and silicates – floated to the surface. This cooled to form the Earth's crust. As a result, the main chemical components of the Earth were separated into layers (see page 517).

The early atmosphere was probably formed by **outgassing** (bubbling up to the surface) of oxides of hydrogen and carbon including H_2O and CO_2. Any lighter gases such as hydrogen, neon and even methane (CH_4) would have escaped since their mean molecular speeds in a hot atmosphere would be close to the escape speed from Earth (see pages 66 and 532).

Measuring the age of the Earth

Radioactive elements can be used to determine the age of rocks, provided that the radioactive elements were pure and associated with no 'daughter products' when the rock was formed. This may not be true for sedimentary rock (formed from eroded older rocks) but is likely to be true for igneous rocks (those solidifying from the molten state).

To be useful for measuring times of geological interest, the nuclides must have long half-lives. The uranium–lead method illustrates the principles.

Uranium–lead age measuring

The half-life of uranium-238 is known to be 4.51×10^9 years as it decays to thorium-234. This is the start of a decay chain of radioactive nuclides that finally produces a stable isotope of lead, lead-206 (Fig 23.1). However, the decay of thorium-234 and its daughter products to lead takes the short geological time of only a few hundred thousand years, so it is accurate enough to take the half-life of the whole chain of uranium to lead as being 4.51×10^9 years.

The method requires very careful chemical sampling. Suppose that in a rock sample we find a mass of 2.0×10^{-6} g of uranium-238 and 3.4×10^{-7} g of lead-206.

Fig 23.1 **As uranium-238 decays, lead-206 builds up. The dashed line shows how the ratio of their masses changes with time**

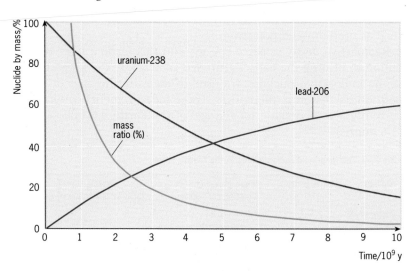

The decay formula for N_0 nuclei decaying to N nuclei in time t, with a decay constant λ, is:

$$N = N_0 e^{-\lambda t}$$

This can usefully be rearranged as:

$$\frac{N_0}{N} = e^{\lambda t} \quad \text{or} \quad \lambda t = \ln \frac{N_0}{N}$$

giving:

$$t = \frac{\ln (N_0/N)}{\lambda} \qquad [1]$$

We can relate the decay constant to half-life as follows, knowing that $N = 0.5 N_0$ (ie $N_0 = 2N$) for $t = $ half-life T:

$$e^{\lambda t} = 2$$

so that:

$$\lambda T = \ln 2$$

and:

$$\lambda = (\ln 2)/T = 0.693/T$$

Combining this with equation 1, we get:

$$t = \frac{\ln (N_0/N)}{0.693} T \qquad [2]$$

Now we have to convert the masses of the elements to the equivalent number of nuclei. The Avogadro constant 6.0×10^{23} gives the number of atoms in 238 g of uranium-238 and in 206 g of lead-206. So, the number of uranium atoms in the rock sample is:

$$\frac{2.0 \times 10^{-6}}{238} \times 6.0 \times 10^{23} = 5.0 \times 10^{15}$$

The number of lead atoms $= \dfrac{3.4 \times 10^{-7}}{206} \times 6.0 \times 10^{23} = 9.9 \times 10^{14}$

The lead atoms were originally uranium atoms, so the number of uranium atoms at the start is the sum of these:

$$N_0 = (5 + 0.99) \times 10^{15} = 5.99 \times 10^{15}$$

and the ratio: $\qquad N_0/N = 5.99/5.0 = 1.2$

Now we can calculate the age of the rock using equation 2:

$$t = \frac{\ln 1.2}{0.693} \times 4.51 = 1.2 \text{ billion years}$$

In practice, this kind of measurement probably has errors. It is highly unlikely that there was no lead in the sample when the rock was formed. Also, lead may have been leached out. But uranium-238 is always associated in a fixed ratio with the isotope uranium-235. This also decays to lead, but to lead-207. This means that the ratio of the masses of the two lead isotopes is also fixed, and this fact can be used to check the reliability of the lead-206 actually measured in the sample.

> **?**
> **A** Show that the abundance of potassium-40 in the Earth's crust 4.6 billion years ago was about 0.95 g per tonne (see Table 23.3).

■ See question 2.

2 EARTH'S TEMPERATURE AND ENERGY BALANCE

The temperature range necessary for life

Life on Earth depends on chemical processes which on the whole do not occur at temperatures above about 35 °C (308 K). Below 0 °C (273 K) biochemical reactions are very slow. Also, pure water freezes at 273 K. The very many processes involving water as a solvent or as a reactant are impossible at temperatures much lower than this. So the normal range of temperature for life is between 0 °C and 35 °C.

What decides the surface temperature of the Earth?

These are the three main factors which determine the surface temperature of a planet in the Solar System:

● The radiation energy received from the Sun.

● The energy released internally in the planet (mostly due to radioactivity).

● The energy radiated away from the planet into space.

> **?**
> **B** Give two reasons why the interior of the Earth is hotter than its surface.

Energy from the Sun

The energy radiated from the Sun peaks in the visible part of the spectrum, and its total quantity is 3.8×10^{26} W. Of this, the Earth receives a tiny fraction, 1.353 kW m^{-2}, a quantity called the **solar constant**, S. But the Earth is not a perfect absorber: about 30 per cent of the visible radiation is reflected – by clouds, ice, the sea and land. This reflected proportion is called the **albedo**, a, of the Earth: $a = 0.30$.

The radiation energy reaching the Earth heats the ground, the sea and the atmosphere. It evaporates water and powers the water cycle. However, the atmosphere is too 'transparent' to solar energy to be significantly warmed directly by solar radiation; it is warmed by contact with the ground. The heated, moving atmosphere then shares out the absorbed energy more evenly across the Earth: air heated near the ground in the tropics moves in convection currents towards latitudes near the poles. Convection in the oceans also helps redistribute energy over the Earth. The Gulf Stream, for instance, brings warm water from the Caribbean area to western Europe.

Fig 23.2 **The energy balance of the Earth**

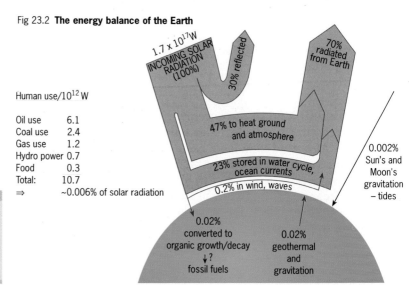

Human use/10^{12} W

Oil use	6.1
Coal use	2.4
Gas use	1.2
Hydro power	0.7
Food	0.3
Total:	10.7
⇒	~0.006% of solar radiation

C Explain why the continuous supply of radiant energy from the Sun does not produce a continuous rise in the temperature of the Earth.

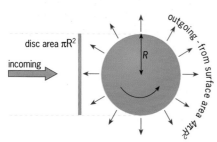

Fig 23.3 **Energy exchange for a bare Earth**

Fig 23.2 shows what happens to the incoming energy on Earth. The warm Earth also radiates energy into space. An equilibrium is established of a steady mean surface temperature because the power radiated equals the power delivered by the Sun's radiation:

$$P_{absorbed} = P_{reradiated}$$

We can estimate the effect of the atmosphere on the temperature of the Earth's surface in a calculation in which we assume that the Earth is 'bare', acting as a 'black body' (page 305), without an atmosphere, as in Fig 23.3.

Energy absorbed per second (in watts):

= solar constant $S \times (1 - \text{albedo}, a) \times$ area of Earth disk

so: $\qquad P_{absorbed} = 1.353 \times 10^3 \times (1 - 0.3) \times \pi R^2 = 947 \pi R^2$

where R is the mean radius of the Earth.

The radiation emitted is calculated using Stefan's law:

The radiation energy emitted per second per unit area by a black body at temperature T is σT^4

where σ is the **Stefan constant**, 5.57×10^{-8} W m^{-2} T^{-4}.

Total power radiated by Earth $= A\sigma T^4$

where A is the area of the whole Earth, that is, a *sphere* of radius R. So the power radiated is:

$$P_{\text{radiated}} = 4\pi R^2 \times 5.67 \times 10^{-8} \times T^4$$

Equating absorption and radiation gives:

$$4 \times 5.67 \times 10^{-8} \times T^4 = 947$$

From this, the temperature of the bare Earth's surface is:

$$T = 254 \text{ K or } -19 \text{ °C.}$$

But the Earth is not bare, and the actual mean surface temperature is 288 K or +15 °C. A small part of the extra 34 degrees is due to an outflow of energy from the interior of the Earth, *but most of it is a result of the Earth having an atmosphere.*

> **?**
> **D** Check the calculated value of −19 °C for the mean surface temperature of a bare Earth.
>
> ▨ See question 3.

3 THE ATMOSPHERIC BLANKET

Molecules in the atmosphere contain energy in various ways.

- They have **translational kinetic energy**, as they move through space.

- They may spin and so have some **rotational kinetic energy**.

- They also have **vibrational energy**, alternately potential and kinetic as the atoms in molecules oscillate to and fro.

> ✔
> Remember: $\frac{1}{2}m\overline{c^2} = \frac{3kT}{2}$
>
> See page 160.

As a rough rule, the temperature of a gas is proportional to the mean translational kinetic energy of its molecules.

However, the energy that is important in making the atmosphere a 'blanket' for radiation from the Earth is the *vibrational* energy of molecules. Molecules can increase their vibrational energy by absorbing photons carrying exactly the right amount of energy to increase the vibrational energy by one quantum (Fig 23.4).

The energy required by a carbon dioxide molecule to jump up a quantum level happens to have a frequency that is close to the peak of the radiation emitted by the warm Earth, at 2×10^{13} Hz (infrared, at $\lambda = 15$ μm). Other gases can absorb energy in much the same way: ozone (O_3) requires energy at a photon frequency of 3×10^{13} Hz ($\lambda = 10$ μm), and water at 4 to 5×10^{13} Hz ($\lambda = 5$ μm).

> ✔
> Ozone is formed from atmospheric oxygen (O_2) which is split up by ultraviolet radiation from the Sun and then recombines as O_3. There is too little ozone to have a thermal effect as great as carbon dioxide's, but as it also absorbs in the ultraviolet region, it cuts down this dangerously ionising radiation from the Sun.

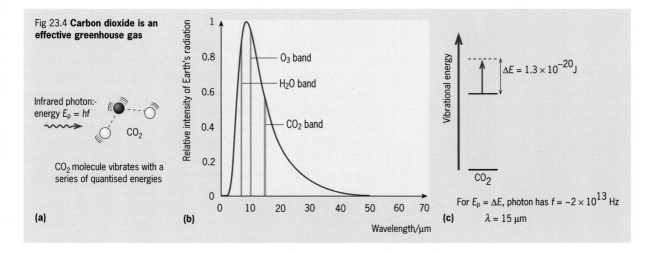

Fig 23.4 **Carbon dioxide is an effective greenhouse gas**

(a) Infrared photon:- energy $E_p = hf$

CO_2

CO_2 molecule vibrates with a series of quantised energies

(b) Relative intensity of Earth's radiation vs Wavelength/μm
O_3 band
H_2O band
CO_2 band

(c) Vibrational energy

$\Delta E = 1.3 \times 10^{-20}$ J

CO_2

For $E_p = \Delta E$, photon has $f = \sim 2 \times 10^{13}$ Hz
$\lambda = 15$ μm

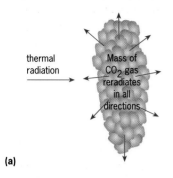

(a)

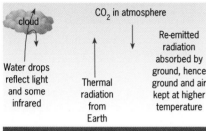

(b)

Fig 23.5 **The atmosphere as a blanket, helped by clouds: the energy that carbon dioxide reradiates is much more than for other gases of the atmosphere**

?

E Work through the calculation to check the value of 16 °C for the mean surface temperature of the Earth.

See questions 4 and 5. ■

The greenhouse effect

The energy absorbed by a carbon dioxide molecule is re-emitted almost immediately as an infrared photon. Now, all the photons absorbed by carbon dioxide were travelling *outwards* from the Earth. But roughly half of the re-emitted photons (Fig 23.4(b)) are moving *back to Earth*.

Remember that the incoming radiation from the Sun has an energy peak in the visible part of the spectrum. The atmosphere is mostly transparent to this part. Because energy *absorption* by atmospheric gases is at the peak of the radiation emitted by the warm Earth, the atmosphere acts as a blanket. This is like the way that air in greenhouses is kept warmer than air outside: sunlight and short-wave infrared can get through the glass, but the longer wave infrared that is reradiated from the interior is absorbed within. Hence, energy absorption in the atmosphere is called the **greenhouse effect**. See Fig 23.5.

The result is that only about 60 per cent of the thermal radiation from Earth gets through into space, that is, the **transmission factor** $b = 0.6$. The energy balance in watts can now be recalculated:

$$\text{incoming power} = \text{outgoing power}$$

From page 512:

$$\text{incoming power} = S(1 - a)\pi R^2 = 947\pi R^2$$

$$\text{outgoing energy rate} = b \times 4\pi R^2 \times \sigma T^4$$

Equating these gives:

$$T = 289 \text{ K or } 16 \text{ °C}$$

The prediction from this simple model is close to the measured value for the mean surface temperature of Earth, which is 288 K or 15 °C.

More about the greenhouse effect, and the carbon dioxide balance

Carbon dioxide is just 0.03 per cent of atmospheric gases. Four billion years ago the atmosphere was probably 98 per cent carbon dioxide, rather like the atmosphere of Venus today. The Earth was then cool enough for water to exist as a liquid. Carbon dioxide is water soluble, so rain washed it out of the atmosphere, and chemical processes trapped it as carbonate in rocks such as chalk, limestone and dolomite.

The oceans still contain a large mass of carbon dioxide dissolved in them. Animals in plankton, the sea organisms with the largest total mass, make their shells from carbon dioxide. These eventually fall to the sea floor and become sedimentary rocks. The oceans are therefore a **sink** for carbon dioxide – they remove the gas from the atmosphere. Over a long time, some of this is recycled as the ocean crust goes deeper into the Earth by the process of **subduction**, see Fig 23.26(b). There, it is melted, heated carbonates release carbon dioxide, and so subduction is a **source** of carbon dioxide. The gas emerges into the atmosphere through volcanoes.

Fig 23.5(b) shows the main stores of carbon dioxide as a gas, in solution or trapped in solid compounds on the Earth's surface. A much greater quantity is trapped in rocks of the sub-surface crust.

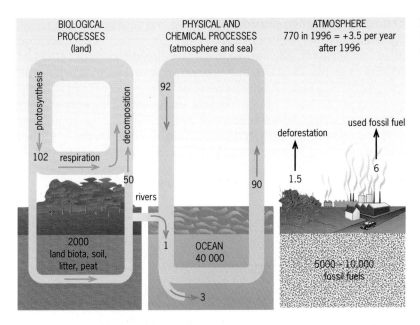

Fig 23.6(a) **Flows and reservoirs of the Earth's carbon dioxide. Figures are in gigatonnes (10^9 tonnes) of carbon**

A quicker recycling process than plate tectonics involves organisms. Plants take carbon dioxide out of the atmosphere, use the carbon and release oxygen. The oxygen is used by most organisms in respiration and re-emerges combined with carbon as carbon dioxide.

Dead plants and animals are a sink for the carbon from carbon dioxide if they become fossilised without oxidation. This has happened many times on the Earth, providing large stores of carbon combined in the **hydrocarbons** of coal, gas and oil.

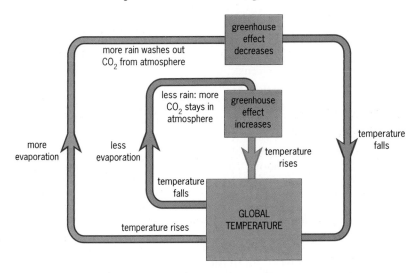

Fig 23.6(b) **Feedback control of global temperature via carbon dioxide. This assumes that CO_2 inputs and losses from other effects are a net zero**

The carbon dioxide balance

About 30 per cent of the carbon dioxide is continuously moving in and out of the atmosphere by various processes, but over long periods of the Earth's history, oxygen and carbon dioxide have been in balance, in a self-regulating feedback system. Suppose there is an excess of carbon dioxide. This is good for plant growth, so plants take up more carbon dioxide and release more oxygen, restoring the balance. If an excess of oxygen is produced, fires burning vegetation kill plants and put more carbon dioxide into the atmosphere as a combustion product. Fig 23.6(b) shows one of the feedback processes that helps maintain a stable temperature on Earth, via atmospheric carbon dioxide.

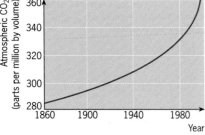

Fig 23.7 **Buildup in atmospheric carbon dioxide concentration since the nineteenth-century industrial revolution**

Atmospheric carbon dioxide levels are increasing

We are concerned nowadays about the greenhouse effect because there has been a steady *increase* in carbon dioxide levels over the last century or so (see Fig 23.7). This is largely due to human activity – the burning of fossil fuels and the increase in the number of farm animals. Both result in more carbon dioxide emission. Farm animals also produce methane, which is an even more effective greenhouse gas.

The consequence of increased carbon dioxide levels is **global warming**. There has been both global warming and cooling in past ages, with warm (interglacial) periods or cool periods (ice ages) lasting for thousands to tens of thousands of years (Fig 23.8).

Fig 23.8 **Temperature variations for the past 15 000 years and prediction for the next 25 000 years.**

The causes of ice ages are not precisely known. There may be several: for example, increased volcanic activity not only produces more carbon dioxide but also more dust, so that the Earth's surface gets less solar radiation. The output of solar energy may vary with time, and the inclination of the Earth's spin axis relative to its orbit also varies. The relative positions of oceans and continents affect ocean and air currents. These currents can shift, and such changes can trigger cooling or warming processes

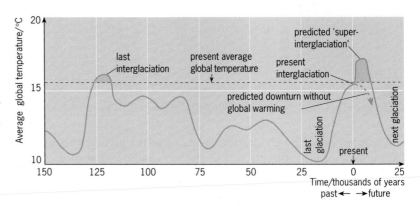

4 THE ENERGY SOURCES INSIDE THE EARTH

Radioactive elements have a chemistry which tended to make them oxidise in the young Earth: they formed compounds whose density made them float to the upper layers of the molten Earth, to be concentrated in the crust. Even so, since the mantle (see Fig 23.11) contains so much more material, the mantle produces much more radioactive heating than the crust.

The rate of flow of this energy due to radioactive decay through surface rocks is about 80 mW per square metre. Only about a fifth of this comes from the crust, with the rest from sources deeper in the Earth. The energy is about 10 times as much as all human power sources, but 4000 times less than the Earth receives from the Sun.

Thermal conduction and convection

Energy released in the core and mantle moves upwards by both conduction and convection. In the rigid lithosphere (crust and outer mantle), conduction is the main steady mechanism. The movement of tectonic plates and the associated volcanic activity are other significant ways in which the hot interior loses energy.

The rate at which energy flows thermally, by conduction, through a solid is given by a simple 'heat flow' formula:

$$\text{energy flow} = \frac{\text{temperature difference}}{\text{thermal resistance}}$$

This is analogous to Ohm's law for charge flow (current), see Fig 23.9, but in thermal calculations we tend to use **thermal conductance** K rather than resistance, so we have:

$$\text{energy flow} = \text{temperature difference} \times \text{conductance} \qquad [1]$$

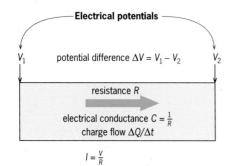

Fig 23.9 **The analogy between electrical flow and thermal flow**

As with electricity, we define conductance as:

$$\text{conductivity} \times \frac{\text{area of cross section}}{\text{length}} \qquad [2]$$

Thermal conductivity λ depends on the nature of the material and has units $\text{W m}^{-1}\text{ K}^{-1}$.

EXAMPLE

Q The crust is mainly composed of the igneous rocks granite and basalt. These have a mean thermal conductivity λ of 2.5 $\text{W m}^{-1}\text{ K}^{-1}$. Estimate the temperature at a depth of 4 km.

A Thermal energy flow = 8×10^{-2} W m^{-2} = $\Delta T \times \lambda A/L$ where $A = 1$ m^2 and $L = 4 \times 10^3$ m.

Then, rearranging: $\quad \Delta T = \dfrac{8 \times 10^{-2} \times 4 \times 10^3}{2.5} = 128$ K

Temperature difference $\Delta T = 128$ K. So, the temperature in a deep mine should be about 144 °C.

Comment on the Example

Note that the result represents a temperature gradient of about 3×10^{-2} K per metre. The measured temperature gradient is less, at 2×10^{-2} K m^{-1}. We assumed that the energy source was at a depth of 4 km. In fact, sources are spread fairly uniformly throughout the rock, some being therefore closer to the surface (Fig 23.10).

5 THE INTERNAL STRUCTURE OF THE EARTH

In the early history of the Earth, its material formed layers on a very large scale, making it **stratified**. Fig 23.11 shows the main layers. We get only indirect evidence for the structure and even the composition of layers deep inside the Earth, from measurements of earthquake waves. The study of these waves is called **seismology**.

F Check that the units for thermal conductivity match the quoted relationships [1] and [2].

G The Earth has a radius of 6.4×10^6 m. The energy flowing through surface rocks is 8×10^{-2} W m^{-2}, and is due mainly to radioactivity.

(a) Calculate how much energy is conducted per second for the whole Earth.

(b) The average energy released per second, due to earthquakes is 10^{11} W. Express this as a percentage of the energy generated inside the Earth.

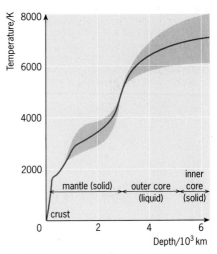

Fig 23.10 **A possible geotherm for the internal temperature of the Earth. This is based on theory, and the shaded regions indicate uncertainty**

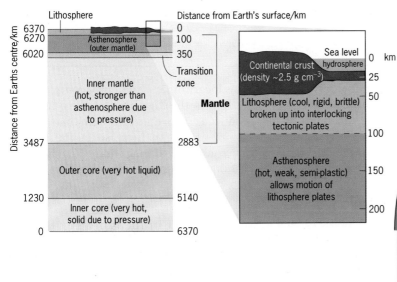

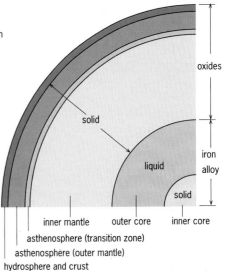

Fig 23.11 **The main internal zones of Earth**

Earthquakes and seismology

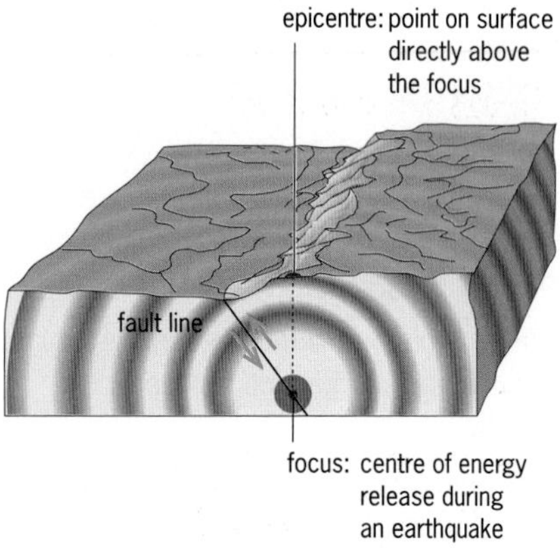

There is an earthquake when one large mass of rock suddenly moves against another. This often happens along a zone of weakness called a **fault line** where the rock has already broken (Fig 23.12). Before that happens, large adjacent blocks of rock may become distorted at the fault line without any slippage. Energy is stored elastically and eventually, when blocks slip past each other, the enormous energy is released as waves of deformation. Waves can only propagate in parts of the Earth's crust where the rocks are cool enough to be *elastic* rather than *plastic*. (Plastic behaviour allows materials to distort gradually without the build-up of stored energy.

Fig 23.12 **The focus of an earthquake is the centre of energy release – where the first movement on a fault occurs. The epicentre is the surface point vertically above the focus**

The Richter scale

A fairly strong earthquake that shakes houses and causes slight damage is graded 5 on the Richter scale (see Table 23.4). It has a typical energy release of 6×10^{10} J. The nearer such an earthquake occurs to the surface, the more damage it does. The damage is caused because the ground suddenly accelerates sideways: buildings are damaged because the structures can't keep up with the motion of their foundations.

The energy is transmitted as **earthquake waves**. A typical Richter scale 5 earthquake produces a *series* of accelerations due to the wave nature of the earthquake shock, with a maximum of about 5 m s^{-2} (0.5g).

The most devastating earthquakes, at Richter scale 8.6 or so, release about 10^{18} J of energy and produce a maximum acceleration of 10g. On average, there are ten such earthquakes a century.

Table 23.4 **The Richter scale for earthquakes**

Richter scale no.	No. of earthquakes per year	Typical effects of this magnitude
<3.4	800 000	Detected only by seismometers
3.5–4.2	30 000	Just about noticeable indoors
4.3–4.8	4 800	Most people notice them, windows rattle
4.9–5.4	1 400	Everyone notices these: dishes may break, open doors swing
5.5–6.1	500	Slight damage to buildings: plaster cracks, bricks fall
6.2–6.9	100	Much damage to buildings: chimneys fall, houses move on foundations
7.0–7.3	15	Serious damage: bridges twist, walls fracture, buildings may collapse
7.4–7.9	4	Great damage, most buildings collapse
>8.0	one every 5 to 10 years	Total damage, surface waves seen, objects thrown in the air

The Richter scale formula is logarithmic – and fairly complex. The magnitude value is proportional to the logarithm of the amplitude of the strongest wave produced. The energy in an earthquake increases by a factor of 30 for each unit increase in the Richter magnitude. This means that, compared to magnitude 5, magnitude 8 involves $30 \times 30 \times 30 = 27\ 000$ times more energy.

Seismic waves

The energy is carried away from the site or **focus** of an earthquake as a set of waves. The nearest point on the surface to an earthquake is usually directly above the focus, and is called its **epicentre** (Fig 23.12). There are three main types of earthquake wave, shown in Table 23.5 and Fig 23.13. The waves are detected by a **seismometer**.

A typical seismometer has a mass suspended by springs in a way that prevents it from being affected by the motion of the ground. See Fig 23.14.

Table 23.5 **Seismic waves**

Wave	Description	Speed in upper crustal rock
P wave	Primary (pressure; push-pull) wave: a body wave that travels through rock as a longitudinal compression wave	6 km s^{-1}
S wave	Secondary (shear; shake) wave: also a body wave, travels through rock as a transverse shear wave	3.5 km s^{-1}
Surface wave (L)	Surface waves are transverse waves that make the ground undulate – like sea waves; they do not travel *through* the Earth but can reach depths of several hundred kilometres	depends on depth: 3.2 km s^{-1} at the surface

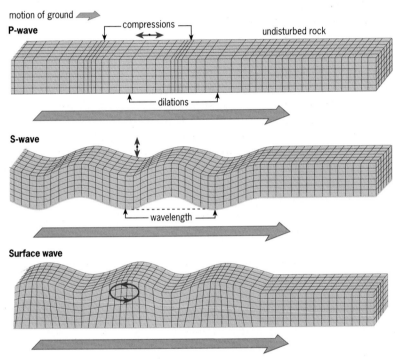

Fig 23.13 **P waves, S waves and surface waves**

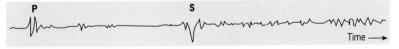

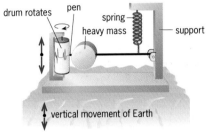

(a) Measuring vertical movement

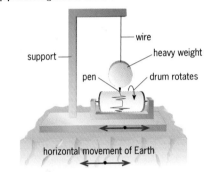

(b) Measuring horizontal movement

Fig 23.14 **The working parts of seismometers for measuring vertical and horizontal movement of the ground**

Fig 23.15 **A seismogram showing the arrival of P and S waves of an earthquake near Sumatra in the Indian Ocean**

The mass in a seismometer is connected to a pen which draws a trace on a sheet of moving paper. When the ground moves, so does the paper, but the suspended mass and its pen stay still. The trace is a record of how the ground moved. Fig 23.15 shows a typical trace called a **seismogram** produced by an earthquake.

Seismic waves as deep Earth probes

Fig 23.16 shows the expected paths of seismic waves through an imaginary planet made of a single material. The paths curve because the increase of pressure and temperature with depth alters the properties of the material: it becomes denser and also more elastic. The speed v of a **body wave** (a P or S wave that goes through a body) depends on both these properties:

$$v = \sqrt{\frac{E}{\rho}}$$

where E is an **elastic constant** and ρ is the **density** of the material.

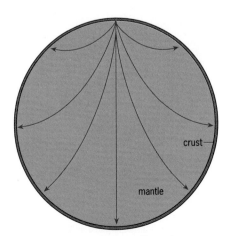

Fig 23.16 **Waves through a uniform planet**

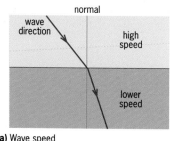

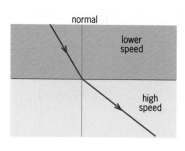

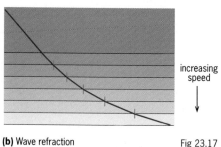

(a) Wave speed

(b) Wave refraction

Fig 23.17

E increases more rapidly with depth than ρ does, so the waves speed up the deeper they go. When waves change speed in a medium, *refraction* occurs. The steady change in speed causes continuous refraction, so there is a gradual change in direction (see Fig 23.17).

Establishing the speed of seismic waves

From data collected at seismic stations all over the Earth, the paths of seismic waves have been worked out as follows.

The times of arrival of P and S waves from an earthquake of known location were recorded (by at least three nearby stations), establishing surface wave speeds. More remote stations then calculated the paths of the P and S waves from the times of their arrival. The speeds of the waves were found to vary in a definite way for these long distances, so such data was used to measure the directions (Fig 23.18) and speeds (Fig 23.19) of the waves at different depths.

The calculations have been supported by laboratory experiments and theory about the behaviour of minerals and rocks at high temperatures and pressures. Now, the epicentre and focus of an earthquake can be precisely pinpointed, however far from the nearest station.

See the end-of-chapter Assignment. ■

Shadow zones

Two main effects are noted for waves from a distant earthquake (Fig 23.18). First, there is a large **shadow zone** for **S waves** on the opposite side of the Earth from the focus, a region that S waves do not reach. The *liquid* outer core causes this shadow: S waves are shear waves and do not pass through liquids which cannot support a shear stress.

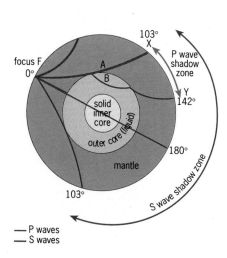

— P waves
— S waves

Fig 23.18 **Some possible paths of P and S waves from an earthquake at F**

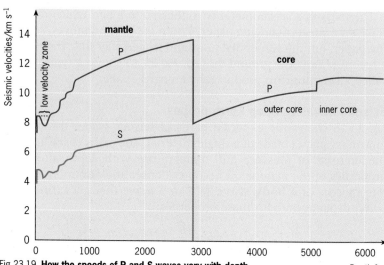

Fig 23.19 **How the speeds of P and S waves vary with depth**

A smaller shadow zone for **P waves** surrounds the S wave shadow zone. P waves are longitudinal and can pass through liquids, but the sudden change in density at the liquid–solid boundary causes considerable refraction. Fig 23.18 shows two paths for P waves from a focus F, travelling very close together. Wave A just scrapes past the boundary and arrives at X. Wave B just enters the liquid zone and is refracted to Y, producing a shadow zone XY.

See question 6.

Artificial 'earthquakes' used to check data

For use in calibrating recording instruments, earthquakes are not ideal: they are unpredictable and they release differing amounts of energy. More precise data can be gathered from man-made explosions in the Earth's crust. Small ones are frequently set off for research purposes, for instance, to help in oil and gas exploration.

Many larger, nuclear explosions were set off in the 1950s to 1980s for military research, and one country would make careful measurements of nuclear weapons tested by another country. These explosions were used to calibrate seismometers, and the data collected helped improve our understanding of the Earth's structure.

Seismic wave data provides the main evidence that the solid *inner core* is made of almost pure iron. The surrounding liquid layer seems to be a mixture of iron, nickel and sulphur. The mantle consists of compounds of lighter elements, mainly iron and magnesium, with silicon and oxygen. They make up complex silicates, similar to the mineral olivine which is a ferromagnesium silicate. Look back at Table 23.2.

6 THE EARTH'S CRUST: ISOSTASY AND PLATE TECTONICS

The temperature and pressure of the material in the outer mantle (asthenosphere) makes it behave as a semi-plastic: rapid increases in force cause brittle fracture, but the material is very slowly distorted under gradual forces. Such forces are due to **gravity** and **convection**.

Isostasy and gravity

The crust of the Earth consists of several continental blocks, varying from 30 to 70 km in thickness (average 35 km). Between them is a thinner crust 8 to 9 km thick which forms the floor of the oceans and extends under the edges of continents. The greater weight of the continental blocks pushes them into the lithosphere, as shown in Fig 23.20(a).

Both ocean and continental blocks rest on the asthenosphere, which is a soft, partially molten layer in the outer mantle. The blocks can move up or down as erosion or accumulation of their material makes them lighter or heavier. This process of floating balance is called **isostasy**, see Fig 23.20(b).

Eroded material may collect in shallow ocean basins between continents. Its weight depresses the basin so that, as more and more material is collected, the ocean doesn't become measurably much shallower. Some 10 to 30 million years ago such a process took place in the shallow sea between the Eurasian continent and a then separate Indian subcontinent. The two blocks eventually collided and pushed up the sea bed, which is now part of the Himalayas.

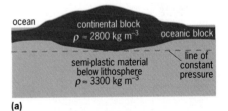

(a)

(b)

Fig 23.20 **The principle of isostasy: less dense material sinks into denser material until pressures are stabilised**

See question 7.

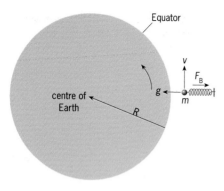

Force due to gravity on mass m is mg newtons.

This provides:

1 the force F_B read by the spring balance
2 the centripetal force F_C needed to keep the mass 'in orbit': $\dfrac{mv^2}{R}$
ie $mg = F_B + F_C$
What we measure in an experiment is F_B, not mg.

Fig 23.21 **Centripetal force lowers the measured value of g**

?

H A point on the Equator is 6.3782×10^6 m from the centre of the Earth.

(a) What is its rotational speed?

(b) What force is required to give a kilogram of matter at that point a centripetal acceleration (v^2/R)?

(c) What fraction of the standard value of the Earth's field (9.8067 N kg^{-1}) is this centripetal force?

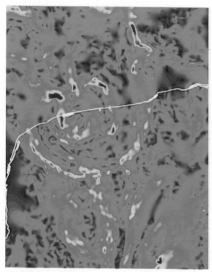

Fig 23.22 **A gravitational anomaly in the Chicxulub region of the Yucatan Peninsula, Mexico, led geologists to discover that an asteroid collided with Earth 65 million years ago (see also page 57)**

Gravitational anomalies

The gravitational field is not the same everywhere at the surface of the Earth. There are five main reasons why the value of the field strength g varies from place to place.

1 Variation of g due to the shape of the Earth

The Earth is not a perfect sphere but is flattened at the poles: the North and South poles are nearer to the centre than the equator by about 10 km. Earth's gravitational field is therefore not perfectly radial (though in Chapter 3 we assumed it was), and the surface field strength g varies with latitude: the fields at the poles (latitude 90°) are stronger than at the Equator (latitude 0°).

2 Variation of the measured value of g due to the spin of the Earth

Because the Earth spins, the net force measured on an object at the Equator is less than the actual gravitational force on it (Fig 23.21). This is because some of the gravitational force keeps the object in its circular path, that is, to provide a *centripetal force*. The speed of rotation decreases with distance towards the poles, so the centripetal force decreases, resulting in an increase of g with latitude.

3 Variation of g with height

The value of g decreases with height according to the inverse square law of gravity:

$$g = \frac{GM}{R^2}$$

where M is the mass of the Earth and R the distance of the point from its centre.

4 Variation in rock density, and mascons

The Earth is not perfectly symmetrical in its composition. Rocks of the crust are less dense than mantle rocks and, since the thickness of the crust varies, so does the underlying gravitational mass and similarly the local gravitational force. The crust itself is not uniform: there are local increases in density which are observed as mass concentrations or **mascons**.

5 Isostatic effects

A continent depresses the rigid outer mantle and the asthenosphere (semi-plastic mantle) underneath it. When the materials are in balance, the extra continental volume is balanced by the lower density of its crustal rock: Fig 23.23(a). But **plate tectonic** movements tend to destroy the balance: some parts of the Earth's plates are sinking and other parts rising. This produces gravitational anomalies as shown in Fig 23.23(b) and (c).

Fig 23.22 shows a gravitational anomaly round Chicxulub. Gravitational anomalies provide the main evidence for variations in the density and thickness of crustal layers.

Gravimeters (Fig 23.24(b)) are instruments for measuring very small differences – as small as one part in 10^8 – in the local gravitational field. They are used to identify the differences in density of regions in the crust. Oil-bearing rocks are less dense than solid rock, so the low gravitational field of a region might indicate the presence of oil-bearing rock. Gravity surveys are also used to locate minerals; for example, metallic ores are usually denser than other rock.

(a) Original, stable system

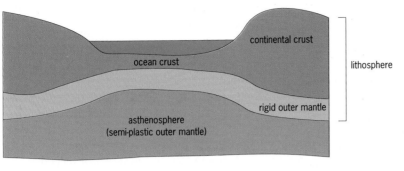

gravity value

continental crust

ocean crust

lithosphere

rigid outer mantle

asthenosphere
(semi-plastic outer mantle)

(b) Erosion produces sediments of low density, pushing away higher density material. Value of *g* decreases

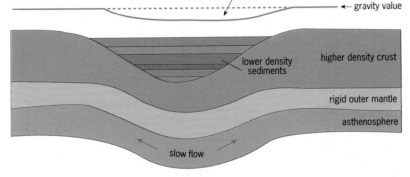

anomaly

gravity value

lower density sediments

higher density crust

rigid outer mantle

asthenosphere

slow flow

(c) Long after erosion stops, balance is restored and gravity returns to normal

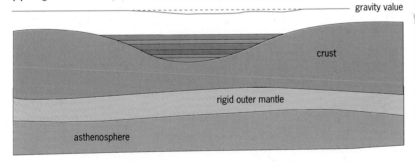

gravity value

crust

rigid outer mantle

asthenosphere

Fig 23.23 **Isotasy and gravitational anomalies (thickness of crust exaggerated for clarity)**

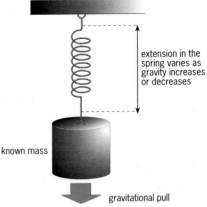

extension in the spring varies as gravity increases or decreases

known mass

gravitational pull

Fig 23.24(a) **The principle of the gravimeter**

(Fig 23.24(b)) **The gravimeter**

Tectonic plates and convection forces

The lithosphere is rigid material in seven main **tectonic plates** and several smaller ones (Fig 23.25). These plates move over the Earth's surface, carrying both continental and oceanic crust with them.

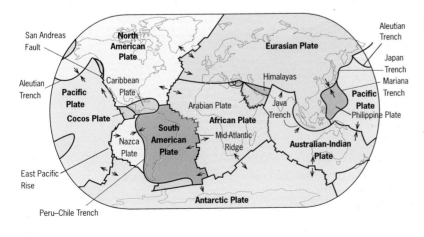

Fig 23.25 **The tectonic plates of the Earth**

Mid-ocean ridges: where magma wells up and ocean floor spreads

Direction of plate movement

The source of the energy for this very large-scale slow movement is believed to be the decay of radioactive nuclides deep in the Earth's interior. This energy in the materials of the core and mantle is thought to produce large convection cells (currents) in the inner, liquid mantle. The circulating material heats the asthenosphere and keeps it more fluid than the inner mantle. Smaller convection cells in the asthenosphere itself are thought to play a role in moving the tectonic plates of the rigid lithosphere (see Fig 23.26(a)).

See questions 8 and 9. ■

We do not fully understand the forces that move tectonic plates around on the Earth's surface. Fig 23.26(b) shows three mechanisms proposed for the plate movements. It is assumed that convection cells in the asthenosphere produce a drag force, but this is unlikely to be great enough to explain the motions observed.

Tectonic plates are known to move at speeds that vary from 1 cm per year to 18 cm per year: the Atlantic Ocean is widening at an average speed of 3 cm per year (this movement broke the Atlantic telephone cable in 1927), and the Pacific Plate is moving away from the Nazca Plate at between 16 and 18 cm per year.

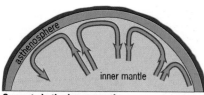

Currents in the inner mantle

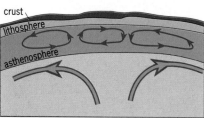

...and in the asthenosphere

Fig 23.26(a) **Convection currents in the mesosphere and the asthenosphere are thought to exert forces on the layers above, to make the tectonic plates move relative to one another**

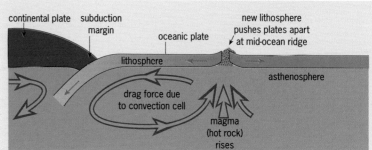

1 Sideways force from convection cells in asthenosphere and accumulating new lithosphere

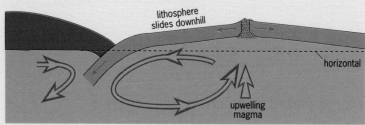

2 Forces in **1**, with additional downsliding of oceanic plate

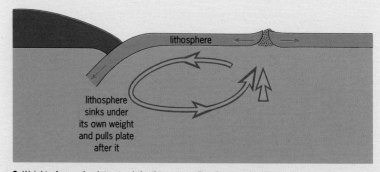

3 Weight of oceanic plate at subduction zone splits plate at mid ocean

Fig 23.26(b) **Three possible mechanisms of plate movements**

7 GEOMAGNETISM

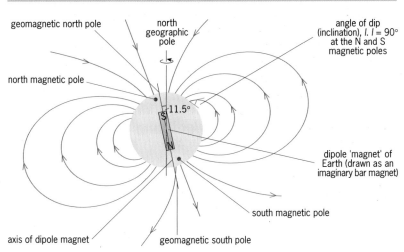

Fig 23.27
The Earth's magnetic field

The magnetic field of the Earth is shown in Fig 23.27. On a large scale, it is similar to the field of a bar magnet. The axis of this 'bar magnet' is not through the geographic poles but cuts the Earth at the **geomagnetic north pole** in Northern Canada at 79°N, 71°W, and at the **geomagnetic south pole** in the south Pacific at 79°S and 109°E. The **north magnetic** and **south magnetic poles** are the places where the field dips down into the Earth at an angle of 90°. The geomagnetic north pole attracts the *north* end of a compass needle, so the imaginary 'bar magnet' must be south at the geographic North Pole end, as labelled in Fig 23.27.

The positions of the poles are slowly changing: the magnetic axis moves around the Earth's geographic poles – its axis of spin – at about 0.18° per year. Over geological time the magnetic poles seem to wander in loops close to the geographic poles.

Magnetic reversal

Even more mysteriously, the direction of the Earth's field reverses every few hundred thousand years or so. Sometimes, the new direction lasts for up to a million years, sometimes for just a few thousand years, see Fig 23.28. This **magnetic reversal** provides the main evidence for **continental drift**, the movement of continents on their plates.

On cooling down from the molten state, hot rocks containing a magnetisable material such as iron or nickel are magnetised in the direction of the Earth's magnetic field at that time. Geomagnetic surveys across the North Atlantic in the 1950s produced the mirror-like patterns of magnetism either side of the Mid Atlantic Ridge, as shown in Fig 23.29. In 1963, this pattern was interpreted as a record of the widening of the Atlantic due to continental drift.

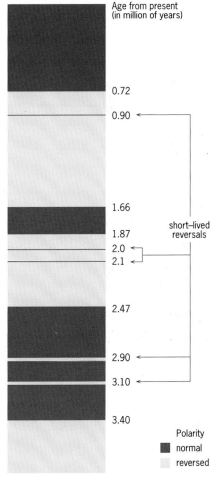

Fig 23.28 **How the polarity of the Earth's magnetic field has changed during the last 4 million years**

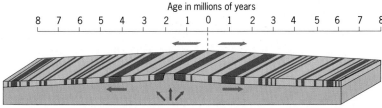

Black stripes = normally magnetised crust (N)   White stripes: reversely magnetised crust (S)

Fig 23.29 **Rocks either side of an ocean ridge show symmetric patterns of magnetic reversal**

Liquid magma emerges continuously from the Mid Atlantic Ridge (see Fig 23.25). As it cools its iron and nickel are magnetised in the direction of the Earth's field, so that as the ocean floor widens, a giant 'magnetic tape recording' is made. The further the rock is from the Ridge, the earlier it was formed from the magma. Geologists already knew the dates of the main field reversals from dating surface rocks (see Fig 23.28), and were able to measure the rate of movement (about 2 to 3 cm per year).

The pattern of field reversals is now known for the past few hundred million years. This has helped geologists to plot the motion of continents and to reconstruct their likely positions on the Earth throughout its history.

See questions 10 and 11. ■

Why is the Earth a magnet?

Iron, nickel and their compounds lose their ability to become magnetised at temperatures above 600 K or so, the **Curie temperature**, so the hot core and mantle cannot be permanent magnets.

Current theory is based on the nature of the semi-plastic layer in the outer mantle and the liquid outer core, and suggests that the Earth's magnetic field is generated in the liquid core. The field here is probably ten to a hundred times stronger than in the mantle, which is in turn up to ten times stronger than the surface field.

Radioactive heating of the liquid in the core produces *convective* flow. The liquid is a conductor of electricity and its motion produces a dynamo effect: the core acts as a self-excited dynamo, producing an electromotive force and an associated magnetic field. As the Earth spins, the core currents are dragged round by the magnetic linkage. This is not a strong link, so the effect is a slow drift of the magnetic axis, and it results in what is at present a slow westward movement of the magnetic poles. This movement is the 'secular variation' marked on maps (about 1° in 6 years).

In this model, the field can be reversed quite easily – but it can't, so far, explain why a reversal occurs.

8 THE ATMOSPHERE: CLIMATE AND WEATHER

The atmosphere is a mixture of gases, see Table 23.6.

The pressure of the atmosphere at the Earth's surface is about 10^5 Pa (1 bar or 1000 millibars, as given on weather forecasts). Atmospheric pressure varies locally and depends on the temperature and humidity of the mass of gas below a height of about 15 km. This region is the **troposphere**, and is where 'the weather happens'. The pressure variation is quite small, at ±30 mbar or so, and is caused by the differing densities of tropospheric air: cool dry air is denser than warm damp air. Fig 23.30 shows the main features of the atmosphere.

Air pressure decreases rapidly with height, following (roughly) an exponential decay law:

$$p = p_0 e^{-kh}$$

where p is pressure at height h, p_0 is pressure at sea level and k is a constant. See the Extension box on page 528 for a proof of this relationship.

The pressure formula assumes that the air is at a constant temperature throughout, but this is not the case. See Fig 23.30. The air gets

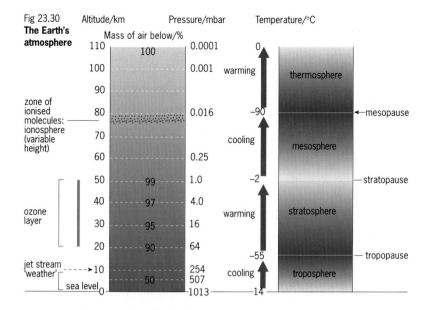

Fig 23.30
The Earth's atmosphere

Gas	Composition of the atmosphere/% by volume	Composition of the atmosphere/% by mass
N_2	78.08	75.52
O_2	20.95	23.14
H_2O	0.1–2.8	0.06–1.70
Ar	0.93	1.29
CO_2	0.03	0.05

Table 23.6 **The gases of the atmosphere**

cooler up to the top of the troposphere, then its temperature rises in the **stratosphere**, cools in the **mesosphere** and warms up again in the **thermosphere**. This complex temperature pattern is partly due to the way that the atmosphere absorbs solar and terrestrial radiation at different levels, and partly because of a gradual cooling with height.

The overall cooling is what you would expect if a volume of gas drifted upwards into regions of *lower pressure*. Near the ground, air is heated by contact with the Sun-warmed land and seas. It expands, gets less dense and rises, carrying energy to higher levels. The gas expands, does work against its surroundings, loses energy and so cools: this process is called **adiabatic** cooling, meaning that the gas is not heated or cooled by external means.

Global circulation of air

If the Earth spun vertically in its orbit, the Sun would always be overhead at the Equator. The equatorial zone would be the warmest and the poles the coolest. Thermal convection would produce a permanent circulation pattern as shown in Fig 23.34 on page 529.

Warm air rises above the Equator, cools, then falls down at about 30° latitude. This air then moves towards the Equator to replace the uplifted air, causing a more or less steady air flow called the **trade winds**.

In fact, the Earth is tilted on its axis, so that the warmest zone (and its trade winds) moves up and down with the seasons. A hemisphere gets its summer when the warm zone is furthest into it (Fig 23.31).

Fig 23.31 **The spin axis of the Earth is tilted, so that climatic zones oscillate about the Equator as its position in orbit changes**

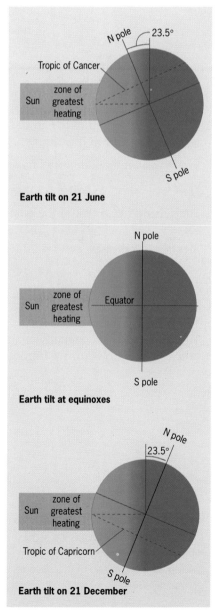

Earth tilt on 21 June

Earth tilt at equinoxes

Earth tilt on 21 December

Pressure variation with height in the atmosphere

First balance the forces

Fig 23.32 shows the cross-section (area ΔA) of a small cylinder of a still column of air in the atmosphere; Δh is the thickness of the cylinder. There is a small pressure difference Δp between the top and bottom of this cylinder.

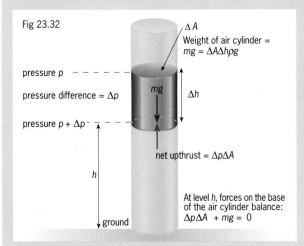

Fig 23.32

Weight of air cylinder = $mg = \Delta A \Delta h \rho g$

pressure p

pressure difference = Δp

pressure $p + \Delta p$

ΔA

mg

Δh

net upthrust = $\Delta p \Delta A$

h

At level h, forces on the base of the air cylinder balance:
$\Delta p \Delta A + mg = 0$

ground

The atmosphere is stable, so the pressure difference must provide an upward thrust which just balances the weight of the air in the cylinder. The weight of the air in it is simply:

$$\text{volume} \times \text{density of air } (\rho) \times g$$

That is: $\text{weight} = \Delta A \Delta h \rho g$

There is no net thrust at the base of the cylinder, so:

$$\text{net upward thrust} + \text{downward force (weight)} = 0$$

$$\Delta p \Delta A + \Delta A \Delta h \rho g = 0$$

$$\Delta p = -\Delta h \rho g \qquad [1]$$

Then bring in the gas laws

The density of air depends on its pressure p; to take this into account, we introduce the gas law:

$$pV = nRT \qquad [2]$$

R is the gas constant and n the number of moles in the volume V, see page 156. Air is a mixture of gases, so n is the total of moles for all the gases in volume V.

The density $\rho = \text{mass/volume}$. For a volume V of gas containing n moles, its mass is nm where m is the mean molar mass of its molecules, so:

$$\rho = nm/V \qquad [3]$$

From equation 2: $V = nRT/p$

so: $$\rho = nm\,\frac{p}{nRT} = \frac{pm}{RT} = Kp \qquad [4]$$

where K is a constant. In this equation, m and R are constants, and we assume that temperature T is also constant.

Then put them together

Substituting for density in equation 1, we get:

$$\Delta p = -\Delta h \rho g = -\Delta h K p g$$

which we rewrite as $$\frac{\Delta p}{p} = -k\Delta h$$

where k is another constant $(= Kg)$; we assume that g doesn't vary much with height in the atmosphere.

Finally, integrate the equation

To find an expression for the pressure p at any height h we integrate both sides of equation 5.

$$\int_{p_0}^{p} \frac{dp}{p} = -k \int_{0}^{h} dh$$

which integrates to $\ln p - \ln p_0 = -kh$.

So: $$\ln \frac{p}{p_0} = -kh$$

which we can rewrite as: $\dfrac{p}{p_0} = e^{-kh}$ or $p = p_0 e^{-kh}$

?

I Mount Everest is about 8 km high. What is the atmospheric pressure at the top as a proportion of the value at ground level?

The Coriolis effect

The Earth is spinning eastwards under the moving air. So that, seen from the ground at points north of the Equator, winds from further north appear to come from the north-east, rather than straight down from the north (Fig 23.34(a)). This spin also affects the apparent direction of wind from the South Pole and winds moving north and south from the Equator (Fig 23.33).

The explanation is that a point on the ground near the poles rotates at a slower speed than a point nearer the Equator: it travels a smaller circle in the same time. So, for an air mass moving at constant speed from pole to Equator, we find a gradual change in

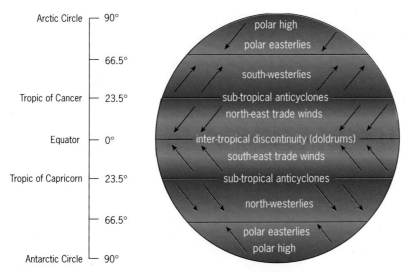

Arctic Circle — 90°
— 66.5°
Tropic of Cancer — 23.5°
Equator — 0°
Tropic of Capricorn — 23.5°
— 66.5°
Antarctic Circle — 90°

Fig 23.33 **The main wind belts at the Earth's surface**

the difference between ground speed and the speed of the air mass. The result is that the air moves in a curve relative to the ground, as in Fig 23.35.

Blocks of air moving away from the Equator spin relative to the Earth's surface – *clockwise* in the northern hemisphere, *anticlockwise* in the southern. These blocks of warm air collect water vapour from the oceans, creating the familiar 'lows' or depressions that provide much of the rain that falls on the British Isles.

The Earth doesn't actually provide a force to produce these changes in direction; the apparently curving movement of air is due to the relative motion of the air and Earth, and is known as the **Coriolis effect**.

Friction

The Earth does affect the wind's movement since there is a frictional force between moving air and the ground. As a result, winds are faster even a few tens of metres above ground than at ground level. Hoisting a flag will show that the wind is faster at the top of the flag-pole than on the ground.

Atmospheric gases

The composition of the atmosphere is given in Table 23.6 (page 527). The two most common elements of the Solar System, hydrogen and helium, are missing from the table because their amounts in the air are so small. One reason for the absence of hydrogen is that it is a very reactive gas: much of it is combined with oxygen in water. It also forms a large fraction of fossil hydrocarbons and of carbohydrates in living matter.

Helium is continually being created on Earth as a result of radioactive decay: alpha particles are helium ions. But both hydrogen atoms and helium atoms are continually lost to space from the atmosphere simply because the atmosphere is too hot. An atmosphere is held to a planet by its gravitational pull. But any object travelling above a certain speed is able to escape from the planet (hydrogen and helium for example). This speed is called the planet's **escape speed** (see page 67 for an explanation of this). The escape speed for Earth is 1.1×10^4 m s⁻¹.

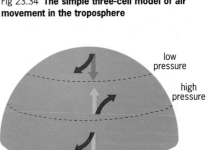

Fig 23.34 **The simple three-cell model of air movement in the troposphere**

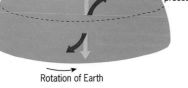

Rotation of Earth

Direction of pressure gradient: air would move this way on the Earth's surface without the Coriolis effect

Wind direction as observed on the ground on a spinning Earth

Fig 23.35 **The Coriolis effect. Compare with surface winds in Fig 23.33**

I The speed of a satellite in orbit is related to its centripetal acceleration:

$$a = mv^2/R$$

Take $a = 9.8$ N kg⁻¹ for a near-Earth satellite in polar orbit at a height of 800 km, and the radius of the Earth as 6400 km. Estimate the speed and deduce the orbital period of the satellite.

MONITORING THE EARTH BY SATELLITE

EARTH SATELLITES CONTINUOUSLY monitor the atmosphere and the Earth's surface. As described in Chapter 4, satellites may be in equatorial or in polar orbit.

Equatorial orbiting satellites are geostationary (geosynchronous) satellites, in orbit at a height of 36 000 km above the Equator. Their orbital period matches the rotation of the Earth, so that they stay above the same point on the Earth. It takes several such satellites evenly spaced in the orbit to monitor the whole Earth. For example, there are six weather satellites: METEOSAT (Europe), GOES E, GOES W, TIROS (USA); Himawari (Japan), INSAT (India). Each of these photographs a quarter of the Earth's surface every 30 minutes, in the visible, the mid infrared and the thermal bands (see the last column in Fig 23.36).

The visible radiation shows cloud cover. The thermal infrared wavelength used is at the peak of the natural thermal (black body) radiation from the Earth, and so can measure surface temperature. The near infrared can penetrate cloud and photograph the surface, so it shows such things as flooding, ice and snow.

Not all wavelengths can be used since the atmosphere strongly absorbs some of them. The **atmospheric window** of wavelengths is shown in Fig 23.38.

Fig 23.37 **A NOAA satellite view of half the Earth's weather**

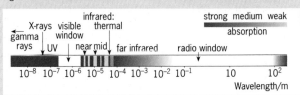

Fig 23.38 **The 'windows' for electromagnetic transmission through the atmosphere**

Polar satellites orbit in north–south paths at heights of between 700 and 1000 km. At these heights, satellites orbit the Earth in a period of 100 minutes or so. Their sensors pick up data from a track of the Earth's surface as it passes beneath them. The Earth spins under the satellite, so in time all the surface can be monitored. How long this takes depends on the period of the satellite and the width of the track covered by the sensors. For example, the LANDSAT satellites which monitor land use (crops, vegetation) use a track width of 185 km and can photograph the whole Earth in 16 days. This process of data collection is called **Along Track Scanning Radiometry**.

Image resolution

The resolution of the final images depends on the number of picture cells (pixels) used. The larger the area photographed, the less detail there is in each pixel; the smallest area a pixel in a weather satellite covers is 2.5 km square, good enough to show major meteorological features.

Polar satellites produce more detailed pictures. In each overflight or 'pass', they cover tracks ranging from 100 km to 185 km, with pixels of areas from 10 m square (SPOT) to 30 m square (LANDSAT TM). Stereoscopic pictures can be taken by angling the detectors differently on two passes over the same area. The greater the detail required, the longer it takes to build up the image for a large area.

The data is digital, and so of high quality, in a range of wavelengths at regular and frequent intervals. The cost is less than that for normal aerial photography.

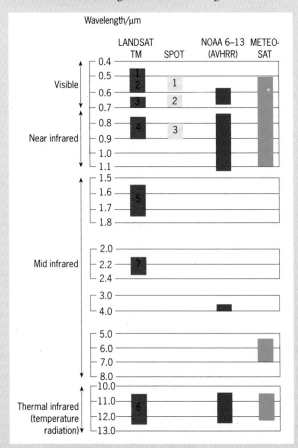

Fig 23.36 **Wavelength bands monitored by the main satellite programmes. The numbers 1–7 refer to bands, see section: How satellites collect information**

See questions 12 and 13.

There are two main monitoring programmes using satellites: the US National Oceanographic and Atmospheric Administration (NOAA) programme and the US Landsat programme. There are several others, such as the French SPOT satellites, the European Space Agency's Earth Resources Satellites – ERS1 (1991) and ERS2 (1994) – and the Coastal Zone Colour Scanner (CZCS).

How satellites collect information

Look again at Fig 23.36 showing the wavelength bands used by different satellite programmes. Electronic sensors produce a signal for each of the numbered bands. The signals are transmitted to Earth and may be combined to give various types of image. For example, a **true colour image** may be produced by combining signals in bands 1 (blue), 2 (green) and 3 (red) in the Landsat set (Fig 23.39).

Fig 23.39 **Left: LANDSAT satellite true-colour image of the British Isles. Right: False-colour image from Landsat satellites. red = pasture, orange-brown = arable land, blue-grey = urban areas, light blue = costal sand and silt, olive green = moorland and conifers, pale blue-green = bog and alkaline soil area**

A **false colour image** may emphasise other types of radiation – from crops, for instance. Fig 23.39 shows an image of southern England made from a visible component Band 3 (red) and bands 4 and 5 (near and far infrared). On a TV monitor, the signals are directed so that band 3 produces a blue image, band 4 produces red and band 5 green. Band 4 is characteristic of radiation reflected from leaves and so distinguishes between built-up areas and crops or wooded areas (see Using the data, below). Different combinations of band signals distinguish between different types of rock and soil, dry from damp areas, and so on. It is even possible to identify different species of trees and types and growth of crops.

The image can be **enhanced** using computer techniques. Curently the best images are obtained from the NOAA's Advanced Very High Resolution Radiometer, which uses the visible and the near and thermal infrared. The mid infrared is monitored by Landsat so that the two monitoring programmes complement each other.

Passive detectors make use of the *natural* radiations reflected from or emitted by surface features. Some recent satellites send out radar signals to monitor the surface; they are **active** systems and need to be very sensitive as the reflected power is so small. The technique they use is called **synthetic aperture radar**: the small quantity of data collected on each overflight is stored and combined later to give a clearer image. This system is very good at measuring distances and is used to monitor ocean levels to check any possible effects of global warming: the mean height of the ocean can be measured to within a centimetre or so.

Fig 23.40 **An image using Aperture Synthesis Radar by ERS1 of the Cote d'Azure. The magenta 'threads' in the sea indicate an oil slick**

Using the data

The data from equatorial weather satellites produces the kind of satellite image that we are most familar with, while polar orbiting satellites keep an eye on land use. Crops are identified by the infrared spectrum signature of the radiation they reflect. The SPOT satellite is specially designed for this. For example, crop growth is monitored well enough to be able to predict yield several months before harvest. In Europe, farmers are paid *not* to grow certain crops on particular areas of 'set aside' land. LANDSAT and SPOT keep a continuous record of land use, checking that crops are not grown on subsidised land.

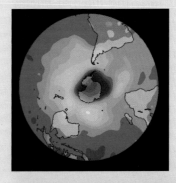

Fig 23.41 **The Antarctic 'ozone hole' (dark central area) in 1996. The image is from the Ozone Vertical Sounder instrument of the US TIROS waether satellite**

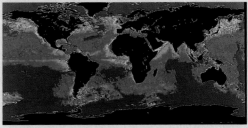

Fig 23.42 **The distribution of phytoplankton of the Earth's oceans from NASA's Nimbus 7 research satellite.**
   Levels of phytoplankton (tiny plants) indicate the health of the oceans and the circulation of currents. Red = densest, violet = least dense (grey = data gaps)

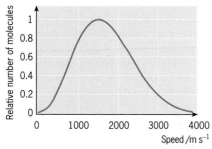

Fig 23.43 **The distribution of speeds of molecules in the atmosphere at 273 K**

?

H Use equation 1 in the text to check the quoted value for the root mean square speed of a hydrogen molecule for gas at 273 K.

Molecules in the atmosphere move with a range of speeds that depends on all their random collisions, but with a mean value of speed determined by the temperature of the gas. Fig 23.43 shows a typical distribution.

As explained on page 160 the average kinetic energy E_k of molecules in a gas at temperature T is:

$$E_K = \tfrac{1}{2}m\overline{c^2} = \frac{3kT}{2} \qquad [1]$$

Here, 'c squared bar' (also written $<c^2>$) is the root mean square velocity of the molecules, m the molecular mass in kilograms, k the Boltzmann constant and T the kelvin temperature of the gas. Note that this relationship applies to all the molecules in the gas, so that lighter molecules must move faster, on average, than more massive ones.

Surprisingly, the temperature at the top of the atmosphere is quite high: 273 K. The mean speed of a hydrogen molecule (mass 3.3×10^{-27} kg) at 273 K is about 1.9×10^3 m s^{-1}. But the high speed tail of the distribution curve (Fig 23.36) means that a significant fraction of hydrogen molecules are at speeds greater than the escape speed for Earth, of 1.1×10^4 m s^{-1}, so hydrogen atoms will escape. Helium is also a light gas and will tend to 'leak' away from the gravitational pull of the Earth.

SUMMARY

After studying this chapter, you should:

■ Know that radio-dating gives the age of rocks, hence information on the origin and age of the Earth.

■ Understand the energy balance between incoming and outgoing radiation, affected by the atmosphere and solar radiation.

■ Understand the way that the internal energy sources of the Earth link with tectonic movements and outward heat flow.

■ Know how the Earth's internal structure, plate tectonics and isostasy are studied from seismic waves and gravitational and geomagnetic variations.

■ Know that the atmosphere has four main regions and that there is an exponential decrease of pressure with height: $P = P_0 e^{-kh}$.

■ Understand the absorption and transmission of radiation in the atmosphere, and know how it is monitored by satellite and has significance for the greenhouse effect.

■ Know how the atmosphere circulates by convection currents and winds.

■ Know that the Earth Satellites in the Landsat and NOAA programmes carry out remote sensing of the atmosphere and the Earth's surface.

QUESTIONS

1 There is a theory that the Earth was formed by the collision and accretion of large numbers of small metallic and stony objects.

a) Explain why this process would have made the early Earth hot.

b) Calculation shows that to disperse all the mass in the Earth to infinity against its own gravitation would require energy of 2.5×10^{32} J. The Earth is mostly iron, with perhaps a mean specific heat capacity of about 500 J kg^{-1} K^{-1}. The mass of the Earth is 6×10^{24} kg. Assuming that the Earth was formed by very cold objects falling from a great distance away, estimate its temperature soon after formation.

c) Do you think it was likely that the early Earth became as hot as your estimate? Justify your answer.

2 An isotope of potassium-40 decays by *electron capture*, in which an inner orbital electron is taken into the nucleus:

$$^{40}\text{K} + \text{e}^- \rightarrow {}^{40}\text{Ar} + \nu \text{ (neutrino)}$$

The half-life of this process is 1.3×10^9 years.

The argon usually stays trapped in the crystal lattice of the potassium mineral, provided the mineral is kept cool. Argon is a non-reactive element.

Analysis of a potassium feldspar crystal in a sample of igneous rock showed that for every 24 000 atoms of ^{40}K present, there were 12 000 atoms of argon.

a) How many potassium-40 atoms were present when the rock crystallised? State any assumptions you make in arriving at this value.

b) Calculate the age of the rock sample. (The decay constant = 0.693/T, where T is the half-life.)

3 The planet Venus is 1.08×10^{11} m from the Sun and has a diameter of 1.2×10^7 m.

a) The Sun radiates energy at the rate of 3.8×10^{26} W. Calculate the power received at the orbit of Venus per square metre.

b) This energy warms Venus. At what temperature will it radiate enough energy to balance the incoming energy from the Sun?

c) The actual surface temperature of Venus is about 460 °C. How can this be reconciled with the result obtained in part (b)? (The Stefan constant = 5.7×10^{-8} W m⁻² K⁻⁴.)

4 The greenhouse effect is due to certain gases present in the atmosphere.

a) What effect do they have on the mean temperature of the Earth's surface?

b) Explain why this effect is called the *greenhouse* effect.

Questions 5, 6 and 12 are adapted from a specimen question set by the University of London Examinations and Assessment Council.

5

a) Explain what is meant by the *greenhouse effect*. Name the most abundant greenhouse gas in the Earth's atmosphere.

b) Why is it essential that the Earth's atmosphere contains some greenhouse gases?

c) Give two different sources and sinks of atmospheric carbon dioxide.

6 Fig 23.Q16 is a simplified cross-section showing the paths of some seismic waves through the Earth.

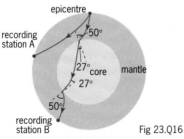
Fig 23.Q16

a) Explain why the waves follow curved paths.

b) How can the seismogram at station A be used to determine its distance from the epicentre?

c) Why can the signal received at station B *not* be used in this way?

d) Using the information in the diagram, calculate the speed of the waves in the core if their speed in the mantle is 13.6 km s⁻¹.

7 What is meant by *isostasy*?
A continent is mostly made of granite: density ρ_g = 2800 kg m⁻³. It is supported on a bed of a rock called peridotite: density ρ_p = 3300 kg m⁻³.

a) Draw a simple sketch showing a rectangular block of granite 'floating in a sea' of peridotite. Label the depth of granite under the peridotite as **X**, and the height of the granite above the peridotite as **Y**.

b) Think about the pressure at the level of the base of the granite block. Show that the following relationship is true:

$$\rho_p X = \rho_g (X + Y)$$

c) A typical continent rises 5 km above the level of peridotite. How thick is the continent?

8 Describe some differences between *continental* crust and *oceanic* crust.

9 What are the main differences between the *lithosphere* and the *asthenosphere*?

10 What role has the study of geomagnetism played in support of the theory of plate tectonics?

11 The Deep Sea Drilling Project of the 1960s took samples of igneous rocks at various distances from the Mid Atlantic Ridge. Their ages were measured by reference to both magnetic anomalies and by radioactive dating. The results for some drilling sites are shown in Table 23.Q11.

Table 23.Q11

Site	14	15	16	17	18	19	20	21
Distance from Mid Atlantic Ridge/km	730	420	220	710	500	1010	1280	1680
Age/10⁶ y	39	24	10	32	27	48	67	75

a) Plot a graph of rock age against distance from the Mid Atlantic Ridge.

b) Comment on the significance of this graph.

c) Use the graph to estimate the rate at which the Atlantic Ocean is widening.

12 Geostationary satellites orbit at a height of 36 000 km above the Earth's surface. The radius of the Earth is 6400 km. Calculate the period T of a **polar** orbiting satellite such as NOAA at a height of 800 km above the Earth's surface, given that:

$$(\text{period of orbit})^2 = \text{constant} \times (\text{radius of orbit})^3$$

In what way do polar orbiting and geostationary satellites complement each other for weather forecasting purposes?

13 Look at the range of wavelengths used by orbital satellites shown in Fig 23.37. For any one satellite system, explain the choice of wavelengths used.

Assignment

FINDING THE EPICENTRE OF AN EARTHQUAKE

Fig 23.A1 shows a typical trace from a seismometer. The P waves arrive before the S waves because they travel faster. Suppose P waves travel at 6 km s^{-1} and S waves at 3.5 km s^{-1} in the upper crust. The seismogram can only give the *difference* in arrival times.

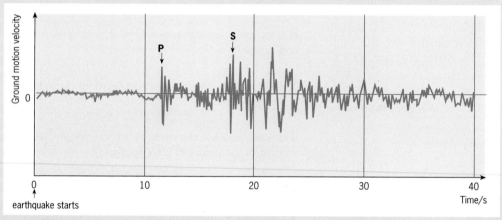

Fig 23.A1 **Seismogram, showing when P and S waves first arrive**

1

a) Copy the table and complete it to show how far P and S waves have travelled at the times shown after the earthquake.

Time/s	2	4	6	8	10	12	14	16	18	20
P distance /km										
S distance /km										

b) Draw a graph of distance versus time showing both types of wave.

c) Use the graph to determine the distance of the earthquake from the seismic station for the P–S time interval in Fig 23.A1.

2

a) Explain how you would find the position of an earthquake for different seismometers that give different P–S wave separation times. What is the minimum number of seismograms that are needed to pinpoint the earthquake?

b) Show that the distance D of an earthquake from a seismometer is given by the formula:

$$D = T \frac{V_S V_P}{V_P - V_S}$$

where T is the time gap between P and S waves travelling at speed V_p and V_s respectively.

(Hint: Begin by writing down the times for each wave to cover a distance D, then subtracting the times to give the time gap.)

Note

The graphical method is easiest to use even professionally, since the speed of the waves varies with distance from the earthquake when the distance is large. This is because the waves travel through layers of different types of rock. The result is that the graphs are curves rather than straight lines like the one that you have drawn.

EARTH AND ATMOSPHERE

The Chapter Map includes the main ideas on the Earth's structure and its energy balance related to the effect of the atmosphere. Use the map to check that you understand the ideas that are required by the syllabus you are following.

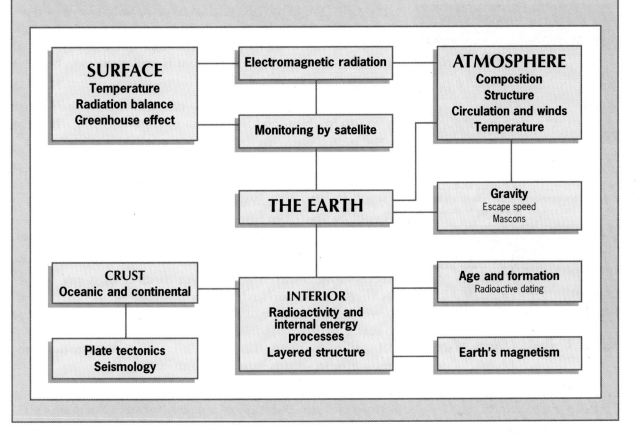

SURFACE
Temperature
Radiation balance
Greenhouse effect

Electromagnetic radiation

ATMOSPHERE
Composition
Structure
Circulation and winds
Temperature

Monitoring by satellite

THE EARTH

Gravity
Escape speed
Mascons

CRUST
Oceanic and continental

INTERIOR
Radioactivity and internal energy processes
Layered structure

Age and formation
Radioactive dating

Plate tectonics
Seismology

Earth's magnetism

24 Medical physics

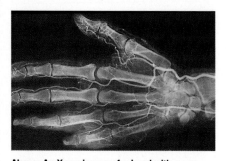

Above: An X-ray image of a hand with a contrasting dye used to highlight the vein structure

Below: An infrared image showing variation in haemoglobin density as yellow where it is at a low level, and red and blue at higher levels

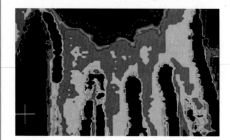

MOST CHILDREN PLAYING with an electric torch have noticed the red glow that comes through their hand when they put it over the lens. Light passes through quite easily, but it doesn't give a clear picture of what is under the skin because the soft tissue scatters light.

An X-ray image of a hand is a fine-tuned version of the picture using torchlight. Blood vessels can be shown clearly when the blood is injected with a dye that strongly absorbs photons.

A different picture is obtained when a very short pulse of infrared light from a laser is used. The light that penetrates the hand and comes straight out the other side is detected in an equally short time, while scattered photons, which take longer to pass through the hand because they travel further, are not recorded. So the resulting image is sharp.

Images from infrared are not yet as clear as X-ray images because resolution with this technique is still poor. However, they can be useful because the colours indicate the quantity of haemoglobin (the oxygen-carrying compound) in blood. An advantage is that infrared radiation does not ionise atoms, so it is safer to use than X-rays.

With improvements such as computer enhancement giving better resolution (detail) of images, infrared imaging could become a very safe technique for both patients and medical workers.

This chapter deals in Part A with the role of physics in **medical diagnosis** (finding out what is wrong), and in Part B with **therapy** (providing treatment), through the use of ultrasound, X-rays, radioactive nuclides, lasers and magnetic resonance imaging (MRI). All these methods can be used in diagnosis, mostly to produce images, but only some of them can be used in therapy.

PART A: PHYSICS IN MEDICAL DIAGNOSIS

A trained observer can find out a great deal about the state of health of the human body by looking at it, examining by touching it, studying substances from it, and listening to its internal noises. Usually, the patient can also describe the nature and site of any pain. But until the end of the nineteenth century, finding out more required **invasive techniques**, such as cutting an organ open in an exploratory operation. This carried the risk of trauma (damage and shock) and infection. But in 1895 Wilhelm Konrad Röntgen discovered X-rays. He saw that they cast a shadow of a hand on a luminous screen. One of his X-ray images revealed the bones and the ring on a finger of his wife's hand. Within weeks of the discovery, doctors were using home-made X-ray machines to look at broken bones and other internal structures. They did so with dangerous enthusiasm, not knowing that X-rays were ionising radiations that could harm living tissue.

X-rays are an example of a **non-invasive** method for probing deep into the body without cutting into it. Doctors now use a wide range of non-invasive (and often less risky) techniques, both as probes for diagnosis, and also to provide treatment.

1 IMAGING USING ULTRASOUND

Ultrasound waves are longitudinal pressure waves at a frequency well beyond the upper limit of human hearing (>20 kHz). Low intensity ultrasonic waves pass through tissue without causing harm, and are reflected at the *boundaries* between different biological structures. These *reflections* allow images (body scans) of internal organs to be created by an **ultrasound scanner**.

The piezoelectric transducer

See questions 1 and 2.

Typical diagnostic frequencies are in the range 1 to 5 MHz. Such high frequencies are produced by an electromechanical effect known as **piezoelectricity**. Certain crystals, such as quartz (SiO_2), produce an electric charge on their surfaces when compressed (page 460). The effect is used in the everyday piezoelectric gas igniter: squeezing the handle compresses a crystal and the charge produced makes a spark to ignite the gas.

The effect is used in reverse to produce ultrasound. A high frequency alternating voltage applied to the crystal surface makes the crystal compress and expand at the same frequency, and the crystal's vibrations generate the ultrasound waves. Vibrations are best at one of the crystal's natural **resonant frequencies** – determined by its size and how it has been cut – and so the applied voltage is tuned to match one of the crystal's resonant frequencies.

A Explain what resonance is, and give two examples of a resonating effect or system.

Fig 24.1 shows a typical ultrasound emitter. It uses a piezo-crystal of lead zirconate-lead titanate, PZT. It is a more efficient energy converter than quartz, and so can produce higher power output. The emitter also acts as a receiver: incoming waves alter the crystal's vibrations which, in turn, generate electrical signals. The piezoelectric element can be thought of as a **transducer**: as an emitter it converts electric potentials to mechanical vibrations, and as a receiver it converts the energy of mechanical vibrations into varying electric potentials.

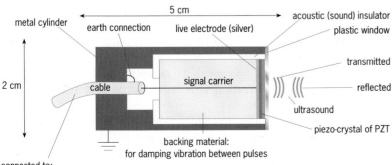

connected to:
power when used as a transmitter;
an amplifier and cathode ray tube
when a receiver

Fig 24.1 **A typical piezoelectric transducer for medical use**

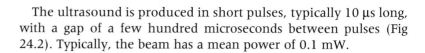

Fig 24.2 **Ultrasound pulses used in diagnostic scans**

The ultrasound is produced in short pulses, typically 10 μs long, with a gap of a few hundred microseconds between pulses (Fig 24.2). Typically, the beam has a mean power of 0.1 mW.

Building an ultrasound image

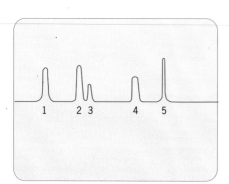

Fig 24.3 **An A-scan: peaks identify surfaces 1 to 5 at increasing depths in the body**

With the transducer in the 'receive' mode, the reflections from different layers in the body return before the next pulse is transmitted. The result is displayed on the screen of a cathode ray oscilloscope as a set of line peaks as in Fig 24.3. Each peak shows the position of a reflecting surface and the height of the peak shows how reflective the surface is. This is called an **A-scan**.

A **B-scan** produces an image that is easier to interpret. The probe is scanned across the body in a series of lines. The strength and position of the return signal is stored electronically, then transferred to produce an image on a TV screen. The signal strength now controls the brightness – and even the colour – of a spot on the screen, so building up a two-dimensional image.

Ultrasound waves do not enter the body easily; they are strongly absorbed in air. So a liquid is used to couple the transmitter with the skin – either a thin film of oil or water-based cellulose jelly.

Ultrasound image quality

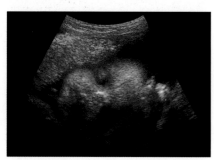

Fig 24.4 **Ultrasound scan of the face of a full-term (9 month) fetus**

As we have seen, the ultrasound image is formed from a set of reflection pulses from material at different depths in the body section scanned. The sharpness of each reflection depends on the duration of the transmitted pulse: if the pulse is too long, then echoes will overlap. But if the pulse is too short, the energy carried is small and the reflections may be too weak to be detected above the background 'noise' of the detection system. Thus, the best pulse length is a balance between the two.

Another problem is multiple reflections: the reflected pulse from a boundary may be bounced back into the patient to be reflected once again. This gives multiple overlapping signals which reduce clarity. Also since the image of an area of the body is built up over time using a scanning procedure, any movement of the patient or internal organ means that linear images taken at different times do not match, and the compound image will be blurred.

See question 3.

Resolution

The image of an object in the body is made up of reflections from small details. **Diffraction** will occur if the wavelength used is too large. Then, the image will lack clarity, making **resolution** poor. As a general rule, ultrasound will just resolve details of the same size as its wavelength. This means that if we need to resolve detail to a level of 1 mm, the ultrasound must have a wavelength equal to or smaller than 1 mm. Diagnostic ultrasound uses frequencies in the range 1 to 15 MHz, resolving details as small as 0.1 mm.

✓
See page 349 for more about diffraction and resolution.

See question 4.

Blood flow measurement

One useful application of ultrasound is the measurement of the rate at which blood flows in blood vessels. The method uses the **Doppler** principle. The ultrasound is reflected by blood particles (red blood cells) and, as the particles are moving, the reflected waves are shifted in frequency by an amount determined by the speed of the blood.

See Appendix 4 for more about the Doppler effect in sound. Higher ultrasound frequencies, of 5 to 10 MHz, are used to make images. The beam enters the blood vessel at an angle θ (Fig 24.5). The frequency shift Δf is measured and related to blood speed v by the formula:

$$\Delta f = \frac{2fv \cos \theta}{c}$$

where c is the speed of the ultrasound.

B The speed of ultrasound in soft tissue is 1540 m s^{-1}. What frequency of ultrasound has a wavelength that can resolve detail to 0.1 mm? (Take the speed of sound to be 340 m s^{-1}.)

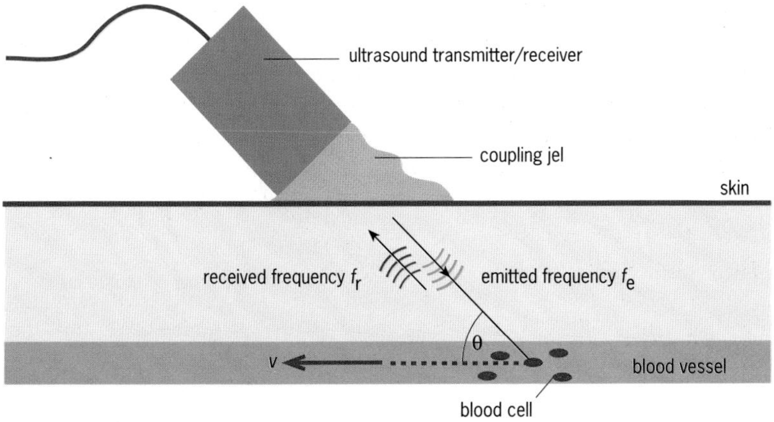

Relative blood flow speed in receiver direction is $v \cos \theta$
Doppler shift $\Delta f = f_r - f_e$

Fig 24.5 **Using ultrasound to measure blood flow**

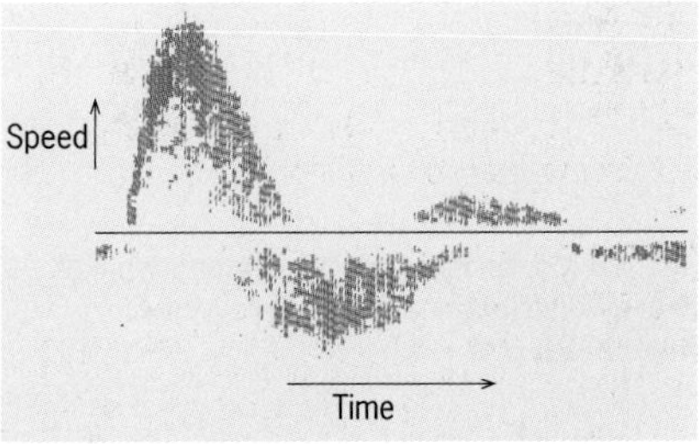

Fig 24.6 **Typical Doppler scan for an artery using an ultrasound scanner. This shows how the speed of blood changes with heartbeat**

Dangers of ultrasound

Ultrasonic waves carry energy. Some of it is absorbed in tissue and causes heating. Bones are particularly energy absorbent. At some frequencies, small objects in the body could resonate and literally be shaken to pieces. So care is taken to keep a low combination of exposure time and intensity. Ultrasound can also cause **cavitation**, the production of small gas bubbles which absorb energy, expand and may damage surrounding tissue. Damage is unlikely at the frequencies and intensities used for diagnosis, but is useful in some kinds of treatment (see page 549).

2 IMAGING WITH X-RAYS

X-rays are ionising electromagnetic radiations (photons) with short wavelengths (of about 10^{-8} to 10^{-12} m) and correspondingly high photon energies (of about 100 eV to 1 MeV). Diagnostic X-rays give best results at energies of about 30 keV, and are produced by bombarding a tungsten anode with electrons accelerated through potential differences of 60 to 125 kV. Fig 24.7 shows the arrangement in a typical diagnostic X-ray machine.

X-rays are produced when electrons are rapidly decelerated as they strike the anode. It becomes very hot, so the usual material is tungsten which has a very high melting point. The electrons also disturb (excite) tungsten atoms which then emit more high frequency photons at particular wavelengths. These photons add a line spectrum of **K** and **L** lines to the continuous spectrum produced by the decelerating electrons.

The spectra in Fig 24.8 show that the distribution of photon energies depends on the target anode and the tube voltage and current, as explained in the next sections.

Fig 24.7 **Plan of the X-ray machine. The anode is strongly heated by the electron impact. It must be cooled, and may also be rotated to reduce wear on the target metal**

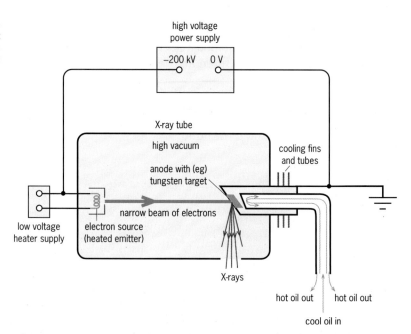

Fig 24.8 **Continuous spectra showing typical energy distributions for an X-ray tube. Note that E = hf**

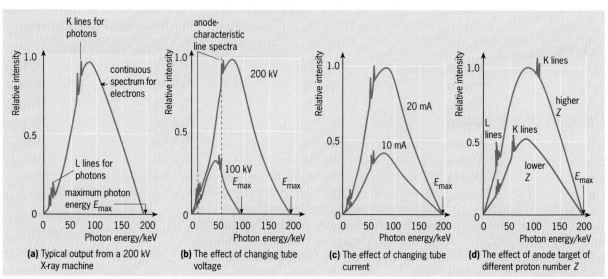

(a) Typical output from a 200 kV X-ray machine

(b) The effect of changing tube voltage

(c) The effect of changing tube current

(d) The effect of anode target of different proton number Z

Tube voltage

The higher the potential difference through which the electrons move, the more kinetic energy E_K they gain, and so the higher the frequency f of the X-ray photons produced:

$$\text{maximum energy of photon } hf = E_K = eV$$

where h is the Planck constant, e is the electronic charge and V is the accelerating voltage. Most electrons lose energy in heating the anode, and only a few have this maximum energy. Fig 24.8(b) shows the effect of increasing tube voltage.

Tube current

Increasing the tube current, which means increasing the number of electrons moving from cathode to anode, increases the number of X-ray photons produced:

$$\text{beam intensity} \propto \text{tube current}$$

This effect is shown in Fig 24.8(c).

Target anode material

Increasing the proton number Z of the anode material increases the likelihood that electrons produce X-ray photons:

$$\text{output beam intensity} \propto Z$$

A change in Z also changes the frequency (energy) of the line spectra, which are characteristic of the target atoms. This effect is shown in Fig 24.8(d).

See Chapter 17 for more about photon energies.

C (a) Explain why high-energy X-rays have short wavelengths.

(b) Explain why increasing the electron current produces more X-rays rather than X-rays with more photon energy.

See questions 5, 6 and 7.

Radiography: how X-rays produce images

X-rays interact with matter in various ways. In all of them, the material removes photons (absorbs energy) from the direct beam and so causes **attenuation**, meaning that the energy of the beam is diminished.

Radiography, the term given to producing images with X-rays, relies on the fact that different types of tissue cause differing attenuations. An X-ray image is really a *shadowgraph* and the darkest shadows are cast by the strongest absorbers (attenuators) of the X-rays. There are four main processes that can reduce the intensity of an X-ray beam:

- **Simple scattering** occurs when X-ray photons bounce elastically off the nuclei of atoms. They do not lose energy but change direction so that they do not reach the detector.

- A photon may instead **ionise** an atom, transferring all or most of its energy in doing so. This is essentially the **photoelectric effect** (see page 364) in which the photon energy liberates an electron from an atom. X-rays have high energy and tend to knock out inner orbital electrons in this ionising process. (Ions usually result from loss of outer orbital electrons.)

- Sometimes a photon will collide with an *outer* electron in an atom. The photon acts as a particle with a particular momentum, which it shares with the electron. The photon goes off at an angle after losing energy to the electron. This process is called the **Compton effect**.

- A photon with very high energy that travels very close to the nucleus of an atom may disappear completely. Its energy is enough to produce a pair of particles – an electron and a positive electron (a positron). The photon's energy has been converted to matter in a process called **pair production** (see page **584**). The energy has to be high enough to satisfy Einstein's relationship $E = mc^2$, where m is the sum of the masses of the two particles produced.

See Chapter 26 for more about pair production.

Measuring the total attenuation

Each of the four processes produces attenuation which depends on the mass of matter interacted with, and this is measured in terms of the **mass attenuation coefficient** μ_m. Table 24.1 shows how μ_m for the four processes depends on the photon energy E and the nuclear charge, that is, the number of protons Z in the nucleus.

Table 24.1 **Attenuation processes for X-rays in matter**

Process	How μ_m depends on photon energy E	How μ_m depends on Z	Photon energy range in which the process is important in soft tissue
simple scatter	$\propto 1/E$	$\propto Z^2$	1–20 keV
photoelectric effect	$\propto 1/E^3$	$\propto Z^3$	1–30 keV
Compton scatter	falls very gradually as E increases	does not depend on Z	30 keV–20 MeV
pair production	increases slowly as E increases	$\propto Z^2$	above 20 MeV

D The average values of Z for muscle and bone are 7.4 and 13.9 respectively. Estimate the ratio of attenuation due to the photoelectric effect between equal masses of bone and muscle.

As a general rule, attenuation gets less as photon energy increases, so the higher the X-ray energy, the more the photons penetrate matter. In diagnostic radiography, an optimum photon energy of about 30 keV produces the best contrast between different types of tissue. This is because at 30 keV energy the main attenuation process is the *photoelectric effect*, with absorption proportional to the cube of the proton number Z. This means that bones, which are mainly calcium with $Z = 20$, produce significantly more attenuation per unit mass than soft tissue (mostly water with hydrogen: $Z = 1$ and oxygen: $Z = 16$).

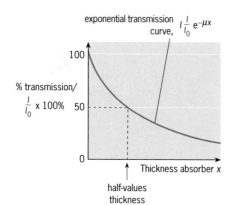

Fig 24.9(a) **Ideal graph for attenuation of a near-monoenergetic X-ray beam**

See question 8. ■

Homogeneous beams

It is not easy to obtain an intense beam of X-rays containing photons of just one energy – a **homogeneous** or **monoenergetic** beam. A near-monoenergetic beam can be obtained by **filtering** it: the beam passes through a metal sheet which absorbs some X-ray photons, more of the low energy photons than the high energy ones – as you should be able to deduce from the second column of Table 24.1. This means that when an X-ray beam is filtered, the beam becomes more penetrating and is described as *harder*.

In the ideal case, a near-monoenergetic beam is attenuated in matter to give the percentage transmission curve shown in Fig 24.9(a). The shape of the graph should be familiar: AB is an **exponential fall**. This is because each small distance Δx in the material produces a small attenuation $-\Delta I$, which is proportional both to Δx and to the beam intensity I:

$$-\Delta I = \mu I \, \Delta x \qquad [1]$$

μ is a constant for a given X-ray wavelength in a given attenuating material, and is called the **linear attenuation coefficient**.

In any situation where the *change* in a quantity is proportional to the (varying) quantity itself, the result is an exponential change. We can rewrite 1 as:

$$\frac{\Delta I}{I} = -\mu \Delta x$$

or in calculus notation:
$$\frac{dI}{I} = -\mu dx \qquad [2]$$

Integrating equation 2 gives:

$$\ln I = -\mu x + C \qquad [3]$$

where I is the intensity at a depth of penetration x. C is a constant which we can identify by the fact that when x is zero, the beam has its starting unattenuated value I_0, so:

$$\ln I_0 = C$$

Putting this value for C in equation 3 gives:

$$\ln I - \ln I_0 = -\mu x$$

which simplifies to:
$$\frac{I}{I_0} = e^{-\mu x} \qquad [4]$$

Fig 24.9(a) above shows that we can define a **half-value thickness** (compare *half-life* in radioactivity) which is the thickness of a material that cuts the X-ray intensity by a half. We can use equation 4 to state the half-value thickness $x_{1/2}$ in terms of the linear attenuation coefficient μ as follows.

$$I = \tfrac{1}{2}I_0$$

so:
$$e^{-\mu x_{\frac{1}{2}}} = \tfrac{1}{2}$$

or:
$$e^{\mu x_{\frac{1}{2}}} = 2$$

giving:
$$x_{\frac{1}{2}} = \ln 2 / \mu \qquad [5]$$

Note that filtering the beam makes it harder, that is, more penetrating, which means that the value of μ for a given material increases.

The inverse square law

Like light, X-rays spread out from the source according to the inverse square law. This means that the intensity decreases as $1/r^2$ where r is the distance from the source. This effect adds to attenuation.

X-ray image quality

The X-ray image or shadowgraph is usually produced on special photographic film. The sharpness of the image is affected by the *size of the X ray source*, known as the **focal spot**, and the *scattering* effect as photons pass through the object.

As an example of exponential change, think of radioactive decay: the amount of a radioactive material that decays in a given time depends on the amount of material there is, and that amount varies with time.

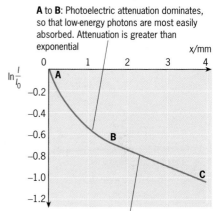

A to B: Photoelectric attenuation dominates, so that low-energy photons are most easily absorbed. Attenuation is greater than exponential

B to C: The beam is now 'hard', with attenuation obeying an exponential law: $I = I_0 \, e^{-\mu x}$

Fig 24.9(b) **A logarithmic plot of attenuation in a typical X-ray beam after passing through different thicknesses: total x (mm) of a metal**

See question 9.

Fig 24.10 **The sharpness of an X-ray image is affected by the size of the source**

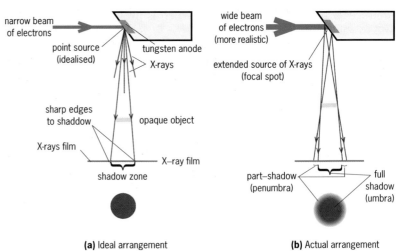

(a) Ideal arrangement

(b) Actual arrangement

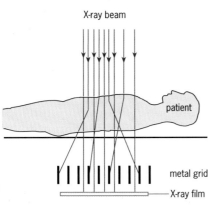

Fig 24.11 **A metal grid is used to remove scattered X-rays. This improves contrast (the size of the grid spacing is exaggerated)**

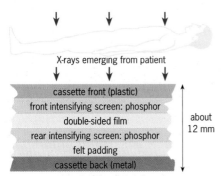

Fig 24.12 **Arrangement of film and phosphors in an intensifying screen cassette (patient and cassette not to scale)**

?

E (a) Suggest why ordinary light-sensitive film is more sensitive to light than X-ray film is to X-rays.

(b) Explain the advantage of using two layers of phosphor as shown in Fig 24.12.

A point source produces perfectly sharp shadows, see Fig 24.10(a). But X-rays originate in a small spot of finite size on the tungsten anode as in Fig 24.10(b), and so the shadow contains an edge effect – a **penumbra**. The penumbra can be reduced by placing the film as close to the object (part of the patient) as possible.

Photons scattered by nuclei in the object carry no information and merely blur the final image, reducing *contrast* between the darker and lighter areas. To minimise this effect, a filter grid is used as shown in Fig 24.11. Only unscattered photons can reach the film.

Clearer pictures would be produced if higher energy (harder) beams were used and the exposure time increased. But this would increase the risk of damage to the patient because atoms in living cells would be ionised, and that increases the risk of cancer.

Improvements in detection systems allow better images with quite low beam intensities. For example, a fluorescent (phosphor-coated) screen placed in front of and behind light-sensitive film will absorb X-rays and re-emit the energy as light in a pattern matching the X-ray image. Fig 24.12 shows how the phosphor and light-sensitive films are arranged. The film is much more sensitive than ordinary X-ray film, so images can be produced using low intensity X-ray beams.

When an X-ray image of the digestive system is required, the patient swallows a harmless solution of barium sulphate (a 'barium meal'). This enhances image contrast, since barium atoms have a high Z value. Similarly, harmless high-Z dyes are injected into blood (see the photo for the Opener to this chapter).

3 MAGNETIC RESONANCE IMAGING (MRI)

This technique gives images of tissues deep in the body by using radio waves and a rather obscure property of nuclei, their **nuclear magnetic resonance**, or **NMR**. The process is now generally called **magnetic resonance imaging**, **MRI**.

The nucleus of an atom spins. It is also charged, and a spinning charge generates a magnetic field. Just as one magnet becomes aligned in the presence of another (eg a compass needle in the Earth's magnetic field), so hydrogen atoms are aligned in a magnetic field. The field has to be very strong, and a hydrogen nucleus can align itself in one of two ways, which correspond to two different quantised energy states.

Suppose that all the nuclei are in just the lower state, with their magnetic fields aligned the same way. A small extra oscillating magnetic field added to the first can cause the hydrogen nuclei to flip from the lower to the higher energy state. In this flip, the hydrogen nuclei take energy from the oscillating field, so altering its strength. It is this change in strength of the oscillating magnetic field that is detected in MRI.

In practice, the oscillating field is the magnetic component of a high radio frequency electromagnetic wave, and the alteration in the signal corresponds to the number of hydrogen nuclei present. The molecules of biological tissues contain plenty of hydrogen nuclei, especially carbohydrates and water. The number varies with the chemical composition, so different tissues extract different amounts of energy from the applied signal.

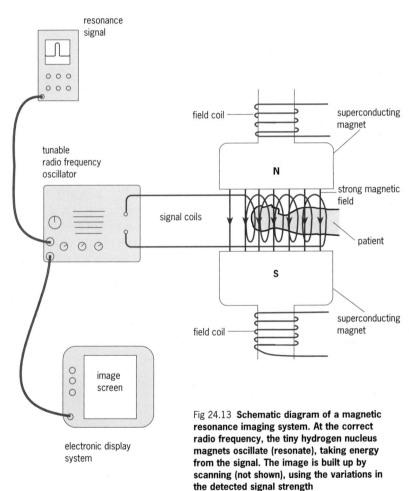

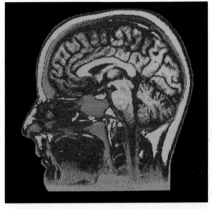

Fig 24.14 **A scan from an MRI through the head of a woman, showing the structure of the brain, spine, head bones and facial tissues**

Fig 24.13 **Schematic diagram of a magnetic resonance imaging system. At the correct radio frequency, the tiny hydrogen nucleus magnets oscillate (resonate), taking energy from the signal. The image is built up by scanning (not shown), using the variations in the detected signal strength**

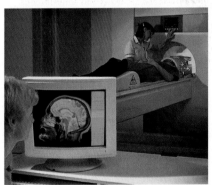

Fig 24.15 **An MRI scanner in use: the patient lies in the MRI detector, while the image of her head appears on the screen**

The signal is sampled spatially, as in ultrasound, so its variation gives a map of the tissues under study. The variation is converted to a screen image (Figs 24.14 and 15).

Magnetic resonance imaging needs expensive equipment, but it is a particularly useful technology for probing delicate areas of the body such as the brain. This is because the energy carried by the radio signal is very small and at a frequency far from the frequencies at which molecules of the body vibrate, so it does no damage. Lower frequencies (such as those in microwave ovens) might provide information – but at the expense of cooked tissue!

4 IMAGING USING LIGHT

Images can be made of the inside of hollow organs by sending a beam of light into them through optical fibres. This is the principle of the **endoscope** (Fig 24.16).

Fig 24.16 **The main features of an endoscope**

For more about fibre optics, see pages 442 and 478.

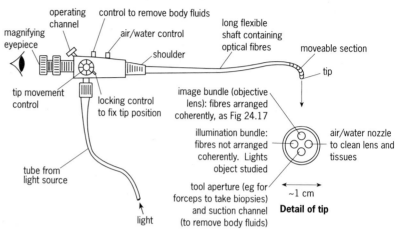

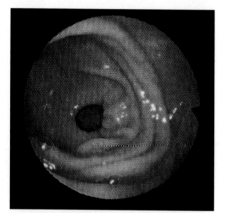

Fig 24.17 **An image of the exit from the stomach into the duodenum, taken with an endoscope. Each tiny facet of the image corresponds to the end of an optical fibre**

In an endoscope there are two bundles of very narrow optical fibres. The illumination bundle carries light to the object being studied, and the image bundle carries back reflected light to provide the image. The image fibres are aligned *coherently* (Fig 24.18) so that the mosaic image formed best matches the object. This image is viewed or photographed through a magnifying eyepiece.

The bundle of fibres is in a flexible probe that is inserted into the body, for example at either end of the digestive system, or through blood vessels to view the heart. The tool aperture is used in treatment, for example, to introduce a powerful laser that can be focused on a small area of unhealthy tissue to burn it away.

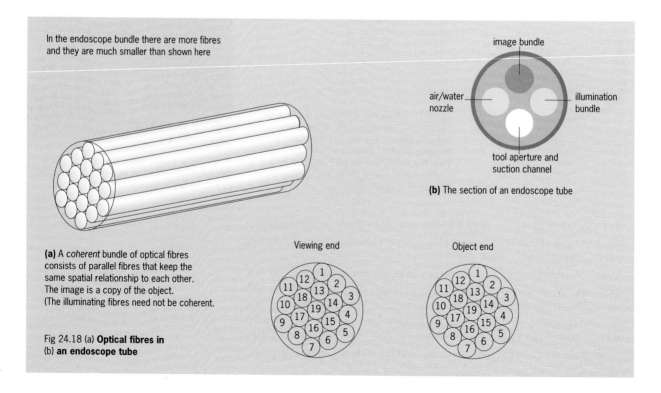

In the endoscope bundle there are more fibres and they are much smaller than shown here

(a) A *coherent* bundle of optical fibres consists of parallel fibres that keep the same spatial relationship to each other. The image is a copy of the object. (The illuminating fibres need not be coherent.

Fig 24.18 (a) **Optical fibres in** (b) **an endoscope tube**

(b) The section of an endoscope tube

5 IMAGING USING RADIOACTIVE TRACERS

Radioactive nuclides emit ionising radiation. This can be detected by film, by a Geiger counter or by a scintillation counter.

In diagnoses, radioactive nuclides are injected into the bloodstream and their passage through an organ is followed using one of various detection instruments.

Alternatively, specific areas can be **radiolabelled**. Biochemical processes involving a series of chemical reactions can be followed using a radioactive isotope of an element in a compound in the series. For example, the uptake of iodine in the thyroid gland is checked by using radioactive iodine.

Table 24.2 shows some of the medical uses for a range of radioactive tracers.

> See Chapter 18 for basic information about radioactivity, half-life, isotopes and radiation detection methods.

■ See question 13.

Table 24.2 **Some radioactive tracers used in medicine**

Organ/tissue	Tracers	Uses
General body composition	^{3}H, ^{24}Na, ^{42}K, ^{82}Br	Used to measure volumes of body fluids and estimate quantities of salts (eg of sodium, potassium, chlorine)
Blood	^{32}P, ^{51}Cr, ^{125}I, ^{131}I, ^{132}I	Used to measure volumes of blood and the different components of blood (plasma, red blood cells) and the volumes of blood in different organs. Also used to locate internal bleeding sites
Bone	^{45}Ca, ^{47}Ca, ^{85}Sr, ^{99m}Tc (see page 548)	Used to investigate absorption of calcium, location of bone disease and how bone metabolises minerals
Cancerous tumours	^{32}P, ^{60}Co, ^{99m}Tc, ^{131}I	Used to detect, locate and diagnose tumours, ^{60}Co is used to treat tumours
Heart and lungs	^{99m}Tc, ^{131}I, ^{133}Xe	Used to measure cardiac action: blood flow, volume and circulation. Labelled gases used in investigations of respiratory activity
Liver	^{32}P, ^{99m}Tc, ^{131}I, ^{198}Au	Used in diagnosing liver disease and disorders in hepatic circulation
Muscle	^{201}Tl	Diagnosis in organs; in particular, heart muscle
Therapy	^{32}P, ^{131}I	^{131}I used in treatment of cancerous thyroid; ^{32}P used in treatment of certain blood cancers
Thyroid	^{99m}Tc, ^{123}I, ^{125}I, ^{131}I, ^{132}I	Used to investigate thyroid function. ^{132}I is especially useful for pregnant women and for children

The quantity of tracer used must be as small as possible to minimise harmful ionising radiation. Exposure time is reduced if the substance that is labelled with a tracer is quickly eliminated from the body, or when the isotope has a short half-life.

The lifetime of the tracer must be matched to the time scale of any process being studied. Often, a short half-life is useful, as in monitoring blood flow which can be studied in a short period of time. The tracer should be easy to detect and its position identified accurately.

The best type of emission is gamma (γ) radiation because gamma rays travel easily through matter and cause little ionisation. But low energy beta (β) radiation is also useful. The isotopes in Table 24.2 are a mixture of beta and gamma emitters.

Alpha (α) particles are heavily ionising. This means that alpha emitters are very dangerous and are therefore not useful for producing images.

F Why are alpha particles not useful for producing images?

Technetium-99m

One of the most useful tracers is an isotope of the artificial element **technetium**: $^{99m}_{43}\text{Tc}$. The 'm' indicates that it is a **metastable** nuclide, meaning that the nucleons in its nucleus are at an energy level higher than in stable technetium. Such nuclei return to normal with a half-life of 6 hours, emitting gamma rays of 140 keV, an energy that makes them particularly easy to detect. The decay produces ordinary technetium-99 which is a naturally occurring radioactive material but has a half-life of 216 000 years, and so it is practically stable, emitting very little radiation.

The production of artificial isotopes

Most of the radioactive materials used in medicine do not occur naturally but are produced in nuclear fission reactors at nuclear power stations from natural, often stable, isotopes. Usually, a sample of the natural isotope is irradiated with neutrons obtained as a by-product in the nuclear power reactors (see page 388 for more about nuclear fission reactors).

One reaction involves the capture of a neutron by the nucleus, which then immediately emits a gamma ray and also becomes a radioactive isotope of the original nuclide. This is the n,γ reaction. For example: ordinary sodium-23 is converted to radioactive sodium-24:

$$^{23}_{11}\text{Na} + ^{1}_{0}\text{n} \rightarrow ^{24}_{11}\text{Na} + \gamma$$

In this process the element stays the same. Not all the nuclei will be changed. It is chemically impossible to separate the two isotopes, so that you cannot get a pure sample of the active isotope. Ordinary sodium is the carrier for radioactive sodium. Phosphorus-32 and potassium-42 are other examples of tracers that can be produced by neutron irradiation.

A pure sample of the radioactive isotope phosphorus-32 can be made using another reaction: ordinary (stable) sulphur is irradiated by neutrons and the irradiated nuclei emit protons in the n,p reaction:

$$^{32}\text{S} + ^{1}\text{n} \rightarrow ^{32}\text{P} + ^{1}\text{p}$$

The new nuclide is chemically different from the irradiated nuclide, and so it can be separated chemically. This process is also used to produce carbon-14 from nitrogen-14, and sulphur-35 from chlorine-35. These pure and entirely radioactive substances are called carrier-free.

Other processes include the n,α reaction, which also produces pure samples. Phosphorus-32 is an example:

$$^{35}_{17}\text{Cl} + ^{1}_{0}\text{n} \rightarrow ^{32}_{15}\text{P} + ^{4}_{2}\alpha$$

The phosphorus is then separated chemically from the chlorine. This process is also used to produce hydrogen-3 (tritium) from lithium-6.

The very useful tracer isotope metastable technetium-99 is produced when the radioactive isotope of molybdenum, $^{99}_{42}\text{Mo}$, decays by beta emission:

$$^{99}_{42}\text{Mo} \rightarrow ^{99m}_{43}\text{Tc} + ^{0}_{-1}\text{e} + \bar{\nu}$$

(Inside the nucleus, $\text{n} \rightarrow \text{p} + \text{e}^- + \bar{\nu}$.)

This process has a half-life of 67 hours, and the technetium can be produced using a 'molybdenum-cow', a column of alumina into which the molybdenum has been absorbed. When required, the technetium can be flushed out with a saline solution, leaving the insoluble molybdenum behind, although the solution must then be cleaned of aluminium and other impurities.

Fig 24.19 **Production line for technetium-99m sources used in medicine to produce gamma ray scintigrams, see page 550**

■ See question 11.

PART B: PHYSICS IN MEDICAL TREATMENT

Many physical diagnostic processes can also be adapted to provide treatment, known as **therapy**.

1 THERAPY USING ULTRASOUND

Ultrasound is used to heat small volumes of tissue and destroy small tumours, malignant groups of (cancerous) cells. This requires more intense ultrasound beams than are used in diagnosis, but causes little or no harm to surrounding tissue when a wide beam is accurately focused, or when several beams meet at the point being treated.

Bladder stones can be shattered by the resonance effect when the ultrasound frequency matches their natural frequency of vibration. We have seen (page 539) that **cavitation**, produced when small air bubbles absorb energy and expand, can be harmful. But used carefully, cavitation also promotes wound healing and the repair of damaged bones.

2 THERAPY USING IONISING RADIATIONS

Ionising radiations can kill living cells and both **X-rays** and **radiations** from radioactive nuclides are used to treat malignant (cancerous) tumours in **radiotherapy**.

For treating tumours deep within the body, higher energy X-ray photons are required. Lower energy photons are more easily absorbed in soft tissue (see Table 24.2): they fail to reach deep into the body and may also cause damage to the tissues that absorb them.

High energy X-rays are produced by using high voltages, of up to 2 MV compared with the 120 kV used to produce diagnostic X-rays. The output beam of high voltage tubes contains a wide range of photon energies and the lower energy photons are removed by metal filters (aluminium, tin, lead and gold). The X-rays are delivered by several beams which converge on the site of the malignant cells (Fig 24.20(a)). This reduces the harm to surrounding tissue. Alternatively, a single beam can be used while the patient is rotated about the target point.

In some cases, the therapy requires even more energetic photons. These are actually 'artificial' gamma rays, produced in **supervoltage devices** – linear or circular electron accelerators (described in Chapter 26). These instruments accelerate electrons using potential differences between 4 MV and 42 MV, and the electrons give rise to the high energy photons.

Teletherapy uses high energy gamma rays from radioactive nuclides. ('Tele' means 'at a distance'.) These gamma rays bridge the gap between X-rays and photons from supervoltage devices. The nuclides commonly used are caesium-137 and cobalt-60 (Fig 24.20(b)). Both have long enough half-lives for the output from the source to be roughly constant during a course of treatment. The source is placed in a lead-lined steel container which emits the gamma rays through a small aperture. The source has the advantage of being small, and simpler to use than X-ray machines. However, it cannot be turned off, and is a small but significant radiation hazard for the medical staff working with it.

See 404 for more about the harmful biological effects of ionising radiation.

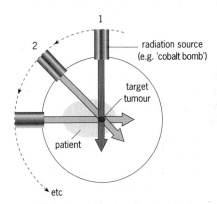

Fig 24.20(a) **Multiple beam therapy. Short doses of radiation are given at each of a number of positions (1, 2, 3 etc). Only the tumour receives the dose each time. The same effect can be obtained by rotating the patient about the axis of the tumour**

Fig 24.20(b) **The 'cobalt bomb' used in teletherapy. Gamma rays from the cobalt-60 source have an energy of 1 MeV (compared with X-rays which have energies of 8–10 MeV), often used to treat secondary cancers**

RADIATION DETECTORS IN MEDICINE

THE BASIC METHODS of detecting ionising radiations are covered in Chapter 18, pages 393–395. Most have been adapted to medical diagnosis and for monitoring in radiotherapy.

A **Geiger counter** is used to measure beta radiation to detect contamination or spillage of radioactive materials. The counter is battery driven and is usually portable. It may include a warning buzzer to show when contamination has reached a dangerous level. Miniature Geiger counters are small enough (20 mm square by 2 mm) to be used inside the body.

An **ionisation chamber** is sometimes used to monitor exposure to radiation and is more accurate than the ordinary film badge described on page 551. The ionisation chamber measures the small current or charge collected when the gas inside the chamber is ionised.

Gamma radiation is of great use in medical diagnosis, since it produces very little damaging ionisation. For the same reason, it is difficult to detect with a Geiger counter or ionisation chamber. The preferred method is to use a series of photomultiplier tubes arranged to form a **gamma camera** (part is shown in Fig 24.21).

A **photomultiplier** works in the following way. A gamma ray (emerging through a patient, say) reaches an ionic crystal substance, such as sodium iodide, which emits light when hit by the gamma ray: the crystal *scintillates*. Because of this effect, the gamma camera scan is called a **scintigram**. The number of light photons emitted depends upon the gamma ray's energy. The photons then hit a photocathode, a material that ejects electrons when bombarded by photons in the *photoelectric effect*. The number of electrons is multiplied using a sequence of charged metal plates called **dynodes**. Electrons collide with the plates and have been accelerated enough to liberate secondary electrons from the plates. There may be ten dynodes, each producing four secondary electrons for each incident electron. The electron current is thus multiplied by a factor of 4^{10}, that is, by about 10^6. The whole system is called a **photomultiplier tube**.

The gamma camera consists of an array of photomultiplier tubes connected to a recording and display system. An image is formed which matches the distribution of radioactive emissions from the patient (Fig 24.21(a)).

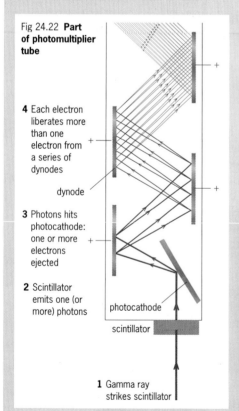

Fig 24.22 **Part of photomultiplier tube**

4 Each electron liberates more than one electron from a series of dynodes

dynode

3 Photons hits photocathode: one or more electrons ejected

2 Scintillator emits one (or more) photons

photocathode

scintillator

1 Gamma ray strikes scintillator

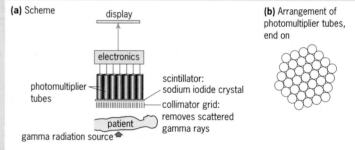

(a) Scheme

display

electronics

photomultiplier tubes

scintillator: sodium iodide crystal

collimator grid: removes scattered gamma rays

patient

gamma radiation source

(b) Arrangement of photomultiplier tubes, end on

Fig 24.22(a) **The gamma ray scintillation camera**

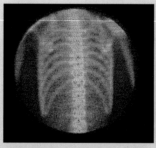

Fig 24.22(b) **Scintigram of a healthy person's torso. The radiotracer, technetium-99m, is concentrated in bone. Cancerous bone would appear as 'hot spots' on the scan**

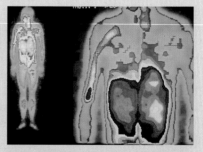

Fig 24.22(c) **Scintigram of a healthy person. The radiolabel thallium-201 shows the liver and stomach (false colour: purple, left and right)**

■ See question 12.

?

G Explain why an alpha emitter produces only localised effects in implant therapy.

Implanted radiation sources are also used to treat malignant tumours. These deliver a small but continuous dose of radiation at the site of the tumour, which kills the cells. 'Needles' contain a radionuclide (such as radium-226 or gold-198). The use of small sources can produce very localised effects. This is especially important with radium which is an alpha emitter.

Unsealed sources can be injected or ingested (swallowed), and are used when they accumulate selectively in malignant tissue. For example, radioactive iodine accumulates in the thyroid gland and so is used to treat cancer of that gland. Colloidal suspensions of gold-198 target malignant cells carried in fluids lining the lungs and in the abdominal cavity.

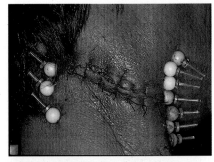

Fig 24.23 **Implant needles, which contain radioactive iridium, used to treat a lymphatic cancer. The needles remain in place for from 24 hours to one week**

3 THERAPY USING LASER TREATMENT

Malignant tissue (for instance, a cancer tumour) absorbs **laser radiation** more strongly than healthy tissue: the output from lasers is usually pulsed and each pulse contains a definite quantity of energy. The wavelength of radiation, its intensity and the time of exposure are chosen to suit the absorption properties of the tumour.

Darkly pigmented tumours absorb best and are most easily destroyed. This is why lasers are also used to decolour skin blemishes, such as 'port wine' birthmarks, see Fig 24.24(a) and (b). In the eye, a detached retina can be spot welded – *coagulated* – back on to the wall of the eyeball by laser light, as in Fig 24.24(c).

Lasers can produce light with enough energy to cut through tissue. As it does so, it 'heat seals' blood vessels and so there is less bleeding than during scalpel surgery. With this technique, diseased parts of the liver can be removed, whereas using a knife might lead to a life-threatening loss of blood.

Fig 24.24(a) **Laser eye surgery**

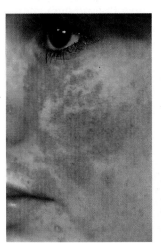
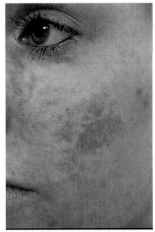

Fig 24.24(b) **A 'port wine' birth-mark: Left: before and Right: after laser treatment**

4 MONITORING RADIATION TO HEALTH WORKERS

Medical workers and patients are protected as far as possible from unnecessary exposure to ionising radiation, by careful monitoring of the **radiation dose** they receive.

The most common monitoring device for medical workers is the **film badge**, see Figs 24.25 and 24.26.

The film badge contains two types of film: one is fast (sensitive), the other slow (less sensitive). The films are in a light-tight box and radiation enters through three windows. One is 'open': the light-tight cover does not include an absorber and lets all the radiation through. The other two areas are covered with several absorbing filters which have two functions: to indicate the *degree* (total exposure) of irradiation, and to identify the *type* of radiation:

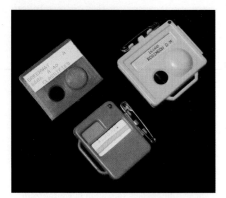

Fig 24.25 **The type of film badge worn by a medical worker to monitor exposure to radiation**

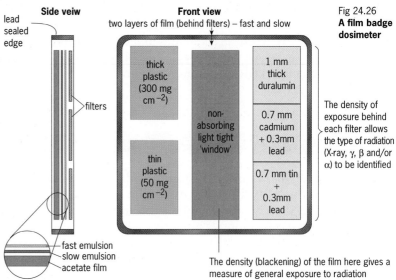

Side veiw

lead sealed edge

filters

fast emulsion
slow emulsion
acetate film

Front view
two layers of film (behind filters) – fast and slow

thick plastic (300 mg cm^{-2})

thin plastic (50 mg cm^{-2})

non-absorbing light tight 'window'

1 mm thick duralumin

0.7 mm cadmium + 0.3mm lead

0.7 mm tin + 0.3mm lead

Fig 24.26
A film badge dosimeter

The density of exposure behind each filter allows the type of radiation (X-ray, γ, β and/or α) to be identified

The density (blackening) of the film here gives a measure of general exposure to radiation

- Irradiated film shows darkening when it is developed, and the **image density** depends on the total exposure to radiation.

- Low energy radiation darkens the part of the film with no absorber or a low absorber in the way, but has little effect in other parts. The higher the energy of the photons and particles, the more likely they are to pass through the filters of increasing thickness or density, and so the more likely they are to reach and affect the underlying film. Blank areas show that no high energy radiation has been received. In general, the contrast between the different areas of the film indicates the range of radiations the user has been exposed to.

The filters and film speeds are chosen to simplify analysis of the type and degree of exposure. The badges are cheap, reliable and sufficiently accurate for their monitoring task, measuring exposure dosage to about 20 per cent.

SUMMARY

After studying this chapter, you should understand and be able to describe the following main topics and techniques related to medical physics.

▨ Non-invasive techniques are important for diagnosis and treatment.

▨ Ultrasound is generated and detected by piezolectric crystals, and is used to produce images.

▨ Ultrasound is used for diagnosis, body scanning and blood flow measurements; there is a relationship between the frequency used and the resolution obtained.

▨ X-ray beams of different types are generated and detected to produce images.

▨ X-rays interact with matter: they become filtered and attenuated ($I = I_0 e^{-\mu x}$), and so may be controlled to produce images (shadowgraphs).

▨ X-rays for treatment are generated by high voltages. X-rays reach the patient as multiple beams or as a single beam in rotational treatment.

▨ Magnetic resonance imaging is a non-hazardous imaging technique.

▨ Ingested radioisotopes are used for diagnosis and therapy.

▨ Medically useful radioisotopes are produced (eg technetium-99m) from naturally occurring isotopes.

▨ The following are instruments used to detect and monitor ionising radiations and to make images: film, Geiger and scintillation counter, photomultiplier tube and gamma camera.

▨ Safety aspects are involved in the use of ultrasound and ionising radiations.

QUESTIONS

Data: speed of light $c = 3 \times 10^8$ m s^{-1}
Planck constant $k = 6.6 \times 10^{-34}$ J s
charge on an electron $e = 1.6 \times 10^{-10}$ C

1 Why is ultrasound diagnosis called a *non-invasive* technique?

2 State some advantages that ultrasound has compared with X-rays as a diagnostic (imaging) system.

3 A beam of ultrasound with a frequency of 4 MHz travels at a speed of 1500 m s^{-1} in soft tissue. The reflection from the moving surface of an internal organ is measured to have a Doppler shift of 400 Hz. How fast is the surface moving?

4 What factors decide the resolution of detail in an image produced by ultrasound? What is the smallest object that can be 'seen' using ultrasound of frequency 10 MHz?

5
a) Sketch a diagram of the energy distribution of photons from an X-ray tube. Label the two distinct features of the spectrum. What decides the maximum photon energy?

b) Draw diagrams to illustrate the effect on the distribution you have drawn in part **a)** of:
  (i) increasing the tube voltage,
  (ii) increasing the tube current,
  (iii) Using a target metal of higher proton number Z.

6 Calculate the minimum wavelength of X-rays produced by an accelerating voltage of 150 kV.

7 Some X-ray tubes use a rotating anode. Explain why this is done.

8 In X-rays used in radiography, the beam is first passed through a *filter*, usually a sheet of aluminium.

a) What two main effects does this have on the beam?

b) What is the medical purpose of filtering the beam in this way?

9 The mass attenuation coefficient μ_m of a material for X-rays is related to the linear attenuation coefficient μ by the relation:

$$\mu_m = \mu/\rho$$

where ρ is the density of the material (see page 542). Aluminium has a mass attenuation coefficient of 0.012 m^2 kg^{-1}, and a density of 2700 kg m^{-3}.

a) Calculate the linear attenuation coefficient for aluminium.

b) Calculate the thickness of aluminium required to reduce the intensity of a beam of X-rays to 10 per cent of the incident value.

10
a) What is the principle of the use of radioactive isotopes as tracers in diagnosis? Explain properties that it is desirable for a suitable tracer isotope to have.

b) Name **two** radioactive isotopes which are commonly used for tracer studies on several different organs or parts of the body. For *each* isotope:
  (i) state its approximate half-life, the type(s) of radiation it emits and how this radiation is monitored in a medical application,
  (ii) describe how it is used in one particular medical study.

[OCSEB: 1993, A-level Physics]

11
a) Describe two ways in which radioactive isotopes for medical use may be produced by a nuclear reactor.

b) What is meant by a *carrier-free radioisotope*? Do the methods you have described in **a)** produce a carrier-free isotope?

12 An ordinary Geiger counter may be used to monitor ionising radiation. Explain in what ways a gamma ray scintillation camera is different from a Geiger counter, and what advantages it has compared with a simple Geiger counter in medical diagnosis.

13 ^{132}I is a gamma ray emitter which can be used as a tracer in the human body. Two identical samples of ^{123}I are prepared for use in a thyroid uptake test. Explain how these are used to measure thyroid function.

[NEAB: 1994 A Level Physics (part question)]

MEDICAL PHYSICS

The chapter you have studied covers the range of techniques for diagnosis and therapy that are summarised in this Chapter Map. Use it to check the requirements of the syllabus you are following.

DIAGNOSIS

THERAPY

RADIATIONS
Ultrasound
X-rays
Ionising
Gamma rays
Light (laser)
Radio (MRI)

SOURCES

DETECTORS

MONITORING
of users, patients

PHYSICS AT THE EDGE

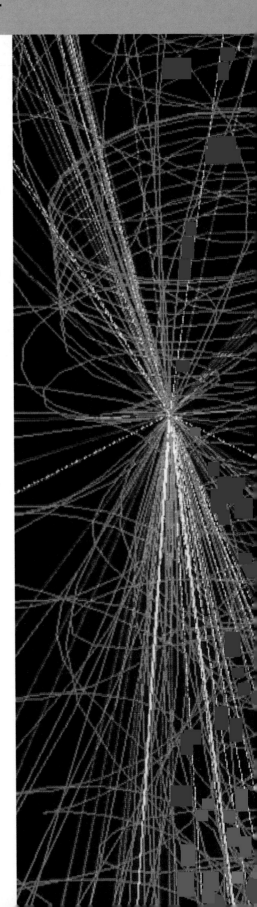

AS WELL AS the technological advances that continually affect our day-to-day lives, the twentieth century has seen a revolution in our understanding of the Universe. Scientists have pondered on its origins and extent, its various parts and how they behave. What we see and touch around us may give us clues. But far away in time and space, things are so very different that our Earth-bound experience of physics could hinder our understanding of the Universe.

Each chapter in this section deals with topics that are at the edge of our knowledge of physics, so that some of the theories and models may be temporary and may be altered later.

The key ideas of relativity and the quantum nature of matter and energy were discovered just at the beginning of the twentieth century and were exploited only within the past 50 years. So it is no surprise that the new ideas on the nature of matter, and the picture of the Universe that came from these ideas, have yet to affect the way we think in everyday life about the nature of matter.

Here we build on earlier topics: 2 Moving in space and time underpins the new relativity theories of Einstein described in 25 Spacetime physics. 26 Deep matter relies on an understanding of the nature of atoms, nuclei and radiation which is dealt with in 17 Atoms, spectra and quanta. Information in 27 Astrophysics and 28 Cosmology deals with stars and the Universe at large, and links closely with the physics of the first two chapters in this section.

'TO QUICKLY GO...'

THE VOYAGER 2 SPACECRAFT was launched by NASA in 1977. Its mission was to visit each of the outer planets and take close-up photographs and other measurements. It took 12 years to reach Neptune, its final port of call in the Solar System. It is now travelling in the direction of the star Sirius, leaving the Solar System at a speed of about 10 km per second. Sirius is a comparatively near star, 8.6 light years away, and Voyager 2 should be near it after a journey lasting 300 000 years or so.

The laws of relativity forbid any object that has mass from travelling faster than c, the speed of light, which is about 3×10^8 metres per second. But the laws also tell us that a starship could get to Sirius in a shorter time than you would calculate in the usual way.

If we could break the laws of physics and travel at the speed of light we could reach Sirius and get back in 17.2 years. Then, relativity theory shows that our starship clock would record a time interval of zero!

But how quickly could we do it, at a speed *close* to the speed of light, say? The answer depends not only on how *fast* the starship goes, but also on where the clock is that measures the time. The maths is very simple – although the ideas are very strange. Suppose NASA wants a starship to get to Sirius 8.6 light years away in just 9 years. To do this it would have to travel at a speed of $0.956c$, that is, $8.6c/9.0$. Relativity treats distance and time as almost interchangeable, and a version of Pythagoras' theorem is used to calculate the time:

$$(\text{time elapsed on ship's clock})^2 = (\text{time elapsed by NASA clock})^2 - (\text{Sirius distance in light units})^2$$

Putting in the values, we get:

$$(\text{time elapsed on ship's clock})^2 = (9)^2 - (8.6)^2$$

This gives a ship's time interval of 2.6 years. Mission control back at Houston will have aged 9 years. For a round trip, the astronauts would be 5.3 years older, the NASA ground team 18 years older.

This is the famous 'twin paradox': a twin in the spacecraft would come back younger than her brother back at NASA. It is a paradox because it is hard to explain why the effect is not symmetrical, that is, why each twin isn't younger than the other!

Departure: 0 years — Starship's travelling calendar clock

Starship leaves Earth

8.6 light distance years

travel time on starship's clock 2.6 light years — Arrival Sirius

Earth clock
Start: 0 years
Arrive: 9.0 years

Travelling calendar clock reads 2.6 years

The trip to Sirius: traveller's elapsed time is 2.6 years; stay-at-home's elapsed time is 9 years.
   Another example: light measured on Earth takes 120 000 years to cross the diameter of our Galaxy. A starship travelling at almost the speed of light could cross it in a few days – as measured by its own clock. As it did so, the Earth would get more than 120 000 years older.

1 SPACE AND TIME

Chapter 2 (Motion in space and time) brought out the close relationship that exists in modern physics between distance and time, and we saw that the main link is the *speed of light*. Distance is measured in the unit of the metre, itself based upon the *distance light travels in a unit of time*. This chapter extends this basic and simple relationship more fully, and deals with the consequences of ideas first put forward by Albert Einstein in 1905.

The constancy of the speed of light

The basis of Einstein's ideas is that the speed of light (in empty space) is a constant for all observers. However fast you travel, light still shoots away from you – or towards you – at the same speed of $2.997\ 924\ 58 \times 10^8$ m s^{-1}. If you drive a car faster, you can catch up with the one in front. But however fast your spacecraft goes you can never even *gain* on the light wave in front. Light doesn't obey the ordinary laws of *relative* motion. This led Einstein to call his theory a **principle of relativity**.

The consequence of the principle is that *distances*, *time intervals* and the *time at which an event occurs* will be different for observers who move at a *steady velocity* with respect to each other. This chapter covers such simple situations, with objects moving at constant velocities.

In **time dilation** we observe that processes take longer to happen in objects that are moving relative to us. The effect is most dramatic at relative speeds close to the speed of light. Time dilation leads to other effects: objects *shrink* in the direction of travel – the **Lorentz contraction** – and their *mass increases*. From these ideas we can then move on to what is certainly the most widely known (but not the best understood) formula in all physics: $E = mc^2$.

Einstein produced his **special principle of relativity** in 1905. Its consequences in cosmology, particle physics and nuclear physics are immense and are discussed in Chapters 26 to 28. This chapter develops the principle in detail, both in words and by some mathematics involving the **Lorentz factor**.

We begin by considering the idea of time.

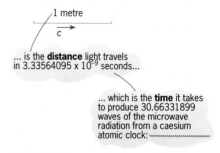

... is the **distance** light travels in 3.33564095 x 10^{-9} seconds...

... which is the **time** it takes to produce 30.66331899 waves of the microwave radiation from a caesium atomic clock:

Fig 25.1 **Light in the form of microwave radiation from caesium defines both time and distance**

See question 1.

2 A (VERY) BRIEF HISTORY OF TIME

The idea of time turns out to be fundamental to (Einsteinian) relativity. So we begin by recalling some aspects of time and its measurements that you probably take for granted.

Time and clocks

Clocks do two things. They tell us the *time of day*, and they can be used to define a *time interval*. As mechanical clocks were improved and became more widely available in the early seventeenth century, the idea grew that they measured something that flowed steadily, as described by Isaac Newton:

Absolute, true and mathematical time, of itself, and from its own nature, flows equably without relation to anything external...

Newton had defined a kind of clockless time. But it doesn't work.

0.4 ms

Fig 25.2 **Comparing the time on a stationary clock and watch**

Whose time is it?

A **muon** is a short lived small particle of matter created when there is an energetic collision, for example, a collision in a particle accelerator or caused by cosmic rays. The muon is highly unstable and very fast, with a half-life of about a millionth of a second. But when particle physicists study muons, their lifetime is seen to be many times longer. So the external observer and the high-speed muon seem to experience time at different rates.

Setting a clock

A clock is anything that can be used to measure time. Ordinary clocks tend to use oscillating systems such as a pendulum or a vibrating quartz crystal. But very long times, such as the age of the Earth, can be measured using the half-lives of radioactive materials. The half-life of a beam of muons can also 'tell the time'. The radioactive material and the beam of muons are sorts of clocks. *But how do we actually compare clocks?*

Imagine that watches and clocks could easily be set to microsecond accuracy. Suppose you check your watch by looking across the town square at the town clock. If its says noon exactly and your watch agrees exactly, you are happy. But the town clock is 120 metres away; when you *see* the town clock at noon it is already noon + $(120/c)$ seconds, where c is the speed of light. Your wristwatch will be slow by 0.4 microseconds.

If you set your watch by the radio time signal beeps, it might be slower by far more. The national master clock is at Rugby, Warwickshire. Its signal for noon has to travel to the radio station, then to you – say 400 km altogether. Timed exactly according to this signal, your watch would be slow by 1.3 milliseconds. This is a very long time by the standard, for example, of the clock in a computer running at 100 MHz which easily measures time to an accuracy of 10^{-8} s.

But it's no problem to a person seeking maximum accuracy. If you know how far you are from the master clock, you can easily work out the correction to ensure that your watch tells the same time of day. Also, however fast or slow the 'time carrier' (the signal or process that 'carries' the time), your watch could still be an accurate measurer of *time intervals*.

The problems occur when clocks start moving, *relative to each other…*

3 MOVING CLOCKS, AND THE LAWS OF PHYSICS IN MOVING REFERENCE FRAMES

A reminder about everyday relativity

Chapter 2 explained *frames of reference* and 'everyday' relativity. A reference frame can be simply you, your laboratory and your measuring instruments.

Four hundred years ago, Galileo realised that he could not prove that the Earth moved – or stayed still – simply by making measurements of objects moving about on Earth. His example was a goldfish bowl in a ship's cabin. The fish move about the bowl in exactly the same way whether the ship is moving or not (as long as the ship is sailing at a steady speed on a calm sea). Movement of the goldfish or water in the bowl gives you no clue about the movement of the ship. For that you would have to look out through the porthole at the shore – and then you would have to assume that it is the ship and not the shore that is actually moving.

Imagine you are sitting at a desk and drop your pen on to it from a height of a few centimetres. It should fall directly below where you let it go. Exactly the same would happen if you were sitting in an aircraft travelling at a steady speed where you and the cabin are the reference frame (look back at Fig 2.19, page 19). But from a frame of reference outside yours, dropping the pen would be seen as moving sideways as well as falling down (Fig 25.3).

The motion of the aircraft is adding a sideways motion to the pen. This is simply in accordance with the laws of mechanics: speeds add up.

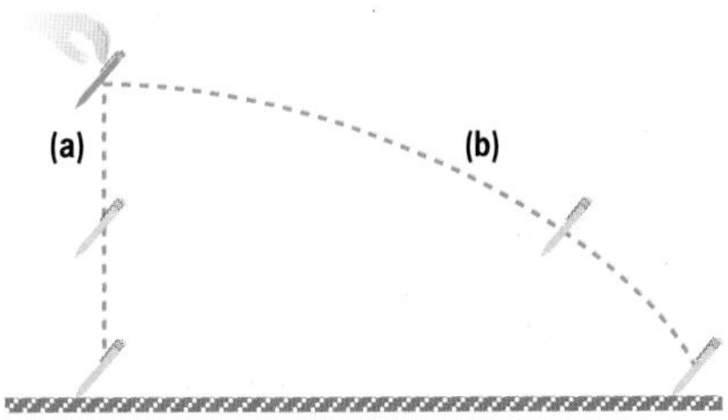

Fig 25.3 **Dropping a pen in two frames of reference:** (a) **the passenger's and** (b) **someone hovering outside as the plane passes – a stationary observer**

Light doesn't behave like pens – or any form of matter

The strange fact that emerged towards the end of the nineteenth century was that light doesn't behave like falling pens, with an observed movement that differs according to the reference frame. Nor does light behave like bullets fired from a moving gun. For example, if a gun is fired while moving towards a fixed target, the bullets move faster to the target than if both gun and target are stationary. This agrees with the ideas of Galileo and Newton, that any object in a moving frame of reference has the speed of the reference frame (the gun, in the case of the bullets) as well as any extra speed of its own. Experimenters thought that light would behave like objects and gain *extra* speed when the light source and the observer were moving towards each other relatively. But in spite of very careful measurements, the experiments all failed to show that this happens.

Albert Michelson and Edward Morley carried out the experiments described in the Extension box on the behaviour of light. The implication of the null result (a result that disproves the starting assumption) is as follows. If you are travelling in a rocket at half the speed of light and send out a light beam ahead of you, you would measure the speed of the light as c, as you would expect. But c would also be the speed measured *by any other observer moving at any other steady speed relative to your rocket*. Compare this with throwing a pen forwards in a train.

For pens: pen speed v_p plus train speed $v_t = (v_p + v_t)$

For light: light speed c plus rocket speed $0.5c = c$

This is shown in Fig 25.4.

The meaning of this null result was profound: its explanation was one of the great turning points in physics.

> If you are finding all this difficult, work again through pages 18 to 20 in Chapter 2.

■ See question 2.

■ See question 2.

Fig 25.4 (a) **Objects obey Galileo's ideas and Newton's laws: speeds add up.** (b) **Light doesn't go any faster or slower, however hard you 'throw' it**

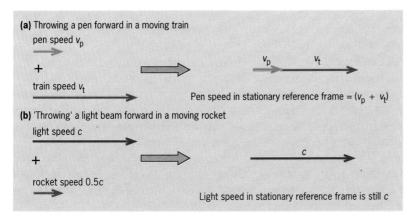

(a) Throwing a pen forward in a moving train

pen speed v_p

+

train speed v_t

Pen speed in stationary reference frame $= (v_p + v_t)$

(b) 'Throwing' a light beam forward in a moving rocket

light speed c

+

rocket speed $0.5c$

Light speed in stationary reference frame is still c

The Michelson–Morley experiment

Fig 25.5 **Albert Michelson was born in Strelno, Poland, but emigrated to the USA and became a physicist serving in the US navy. He was an instructor at the US Naval Academy when he began his research, but then left to become a professor of physics at Cleveland, Ohio. There, he was joined by the chemistry professor Edward Morley. The famous experiments that they carried out produced null results, providing the first crack in the well-built structure of nineteenth-century physics**

It seemed obvious in the late nineteenth century that, if light is a wave, there must be a medium to carry it: scientists thought that, since light travelled through the vacuum of space, a vacuum should have at least some of the properties of a material. Just as sound waves are carried by air, so light waves were imagined as disturbances in what was called a **luminiferous** (or light-carrying) **aether** (pronounced ether) occupying all space (and all matter too) in the Universe. Light and all bodies were thought to travel through it.

Starting in 1880, the American physicist and naval officer Albert Michelson (Fig 25.5) carried out experiments to measure the speed of the Earth through the aether. He found it a harder task that he had expected. When he moved to the University of Cleveland, Ohio, he was joined by chemistry professor Edward Morley.

The two scientists assumed that light travels at a constant speed c in the aether. Their aim was to measure the speed of the Earth as it moved through the aether, relative to the speed of light. If, for example, the Earth was moving through the aether in the *same* direction as the light beam of their experiment, then (they thought) the light should appear to be travelling *slower* than c because of a sort of Doppler effect. Think of a road (the aether) with a cyclist (the Earth) moving along it. A car (light) travelling at constant speed comes up from behind the cyclist and overtakes. The car will approach and move away from the cyclist at a speed that is slower than its actual speed.

So when the Earth moves at speed v in the same direction as a beam of light, Michelson and Morley expected a light speed of less than c,

namely $(c - v)$. And a light beam moving in the opposite direction should have a speed of $(c + v)$.

Not knowing in which direction the Earth was travelling through the aether, Michelson and Morley arranged their apparatus to measure light speeds coming from different directions. Since every 6 months the Earth's direction changes by 180°, they also hoped that any change of the Earth's direction relative to this aether would affect their light speed measurements, too. The interferometer they used is shown in Fig 25.6.

Simply, Michelson and Morley measured the time taken for light to travel along a pair of equal-length arms placed at right angles to each other. If at one time the Earth was stationary relative to the aether, beam 1 and beam 2 would take the same time to cover the equal distances. But if the Earth on its orbit then moved in the aether in the direction of one of the arms – so the experimenters thought – the time taken would be different for the two light beams.

Michelson and Morley measured time in a very modern way for over a hundred years ago – in terms of the number of complete waves made by the light along each arm and observed as an interference pattern at the screen. The light was monochromatic and therefore of constant frequency and wavelength. So, when the apparatus was rotated through 90°, if light took longer to travel along one arm than before rotation, the number of waves would be greater in that direction. The apparatus was sensitive enough to detect a time difference equivalent to a hundredth of the time period t of one wavelength. They expected to see a difference of up to four hundredths of t, estimated from the known speed of the Earth in its orbit. A simple account of the mathematics they used is given below Fig 25.6.

But after many experiments there was never any difference between sets of results: there was always a null result. Compared with the aether, the Earth appeared to be at rest – even when it swung to the other side of the Sun and was travelling in the opposite direction!

This null result – finally confirmed in 1887 – led first to the theory that electrical forces (ie connected with the electromagnetic waves of light) will shrink

mirror M₁

beam 1 — arm 1, length L

mirror M₂

beam 2

light source   half-silvered mirror M

arm 2, length L

Beams 1 and 2 combine to produce interference fringes on a screen

Screen

The whole apparatus floats on a pool of mercury so that it is easily rotated through 90°

Expected light speed to M₁

M₁

c

from Pythagoras $\sqrt{(c^2 - v^2)}$

M v

Direction of Earth's motion relative to light source: speed v

Expected light speeds to M₂

v c

$(c + v)$

v c

$(c - v)$

The beam is split by the half-silvered mirror M.
Beam 1 goes along arm 1, length L, and beam 2 along same-length arm 2.
For beam 2, relative to Earth, the light speed to the right is $(c - v)$.
But when it travels left after reflection at M_2, it is $(c + v)$.
So it takes a total time T_2 to travel from the half-silvered mirror to M_2 and back of:

$$T_2 = \frac{L}{c - v} + \frac{L}{c + v} = \frac{2Lc}{c^2 - v^2} = \frac{2Lc}{c^2\left(1 - \frac{v^2}{c^2}\right)} = \frac{2L}{c}\left(1 - \frac{v^2}{c^2}\right)^{-1}$$

Beam 1 travels to and from mirror M_1 at the same speed of $(c^2 - v^2)^{\frac{1}{2}}$, so

that it takes a time of $T_1 = \dfrac{2L}{(c^2 - v^2)^{\frac{1}{2}}}$ to do the double journey.

This can be rearranged as: $T_1 = \dfrac{2L}{c}\left(1 - \dfrac{v^2}{c^2}\right)^{-\frac{1}{2}}$

Thus the time difference $T_2 - T_1 = \Delta T = \dfrac{2L}{c}\left[\left(1 - \dfrac{v^2}{c^2}\right)^{-1} - \left(1 - \dfrac{v^2}{c^2}\right)^{-\frac{1}{2}}\right]$

Now, v^2/c^2 is very small, so that we can expand the brackets according to the
binomial theorem: $(1 - x)^n = 1 - nx$ (for $x \ll 1$)

You can do this to show that the expression for ΔT simplifies to Lv^2/c^3.
Michelson and Morley determined the value of ΔT by swinging the whole
apparatus through 90°, so swapping one arm for the other. They expected this
to cause the interference pattern to change by moving the fringes side-ways by
an amount equal to a path difference due to a time difference *twice* ΔT.

Path difference = $c \times 2\Delta T$

which corresponds to a shift of n fringes such that:

path difference = $n\lambda = 2c\Delta T$

so that: $n = \dfrac{2c\Delta T}{\lambda} = \dfrac{2Lv^2}{\lambda c^2}$

Fig 25.6 **Above: The interferometer that Michelson and Morley used to measure the velocity of the Earth through the aether (space), and the mathematical explanation of their results. Below: What Michelson and Morley expected to find**

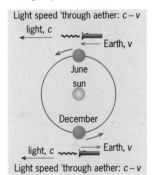

Light speed 'through aether: $c - v$

light, c

Earth, v

June

sun

December

light, c Earth, v

Light speed 'through aether: $c - v$

?

A In the apparatus of Fig 25.6, L was 10 m and the light had a wavelength of 5×10^{-7} m. Show that this corresponds to a fringe shift of about 0.4 if the Earth moves at a speed of 30 km s⁻¹.

any object when it moves through the aether. This was the **Lorentz–FitzGerald contraction** theory: it explained the result – but otherwise led nowhere!

Then in 1905, Einstein, who probably didn't even know about Michelson and Morley's experiment,

produced a much more radical theory based on two simple assumptions about the nature of physics and of light. Not only did it explain the null result; it also predicted the other 'relativistic' effects that had not then been observed.

4 EINSTEIN'S THEORY

Albert Einstein produced the theory that explained the null results of the Michelson–Morley experiment. The Lorentz–FitzGerald theory was seriously flawed since it was invented to explain just one effect (see the Extension box). Einstein produced a formula for contraction with a simpler (if at first unbelievable) assumption about the constancy of the speed of light.

Einstein's two assumptions

Einstein took account of the nature of light and the fact that it was an electromagnetic effect. His theory was based on two simple assumptions (or 'postulates'):

- Physical laws – mechanical, optical and electromagnetic – are the same in all uniformly moving frames of reference.
- The speed of light in a vacuum is the same for all observers, in all uniformly moving frames of reference.

To recap: your frame of reference is the set of objects like tables, chairs, clocks or metre rules that happen to be at rest relative to you. You, and these things, could well be moving with respect to other objects – think of doing experiments in a train or an aircraft. An experimenter in another train travelling past you would be in a different frame of reference.

Frames of reference are called **inertial frames** when they are not accelerating in a straight line or rotating. In these frames, Newtonian relationships like $F = ma$ apply simply, whereas in rotating frames we experience imaginary forces like centrifugal and Coriolis forces.

5 STRETCHING TIME – TIME DILATION

The first and very surprising consequence of Einstein's principle of relativity is called **time dilation**. This means that:

> **A process that takes a certain time to occur in a moving system is observed to take a *longer* time by someone outside that system than by someone moving *with* the system.**

As in Fig 25.7, the observer outside the system could see a clock in a moving spacecraft move its second hand, say, 10 seconds, and also see that this took 11 seconds measured by her own watch. Meanwhile, an observer in the moving system would also see that while his watch counted 11 seconds, the outsider's watch changed by 10 seconds. Both would say that the other person's watch counted seconds too slowly. The situation is *symmetrical*: either could say that the other is moving; neither has the right to declare that they are at rest relative to the Universe at large.

A mathematical proof of time dilation is very simple and needs no more than an understanding that distance = speed × time, and a knowledge of Pythagoras' theorem. Other proofs use even less mathematics.

There are two difficulties: one is believing the final result, the other is in setting up a scenario in which time dilation is important. In his popular explanations, Einstein used the situation of a moving train being struck by lightning, with the guard on the train and the station master some distance away arguing about exactly when the strike occurred – hardly the first thing either would worry about under the circumstances!

Fig 25.7 **Time measured from inside and outside systems**

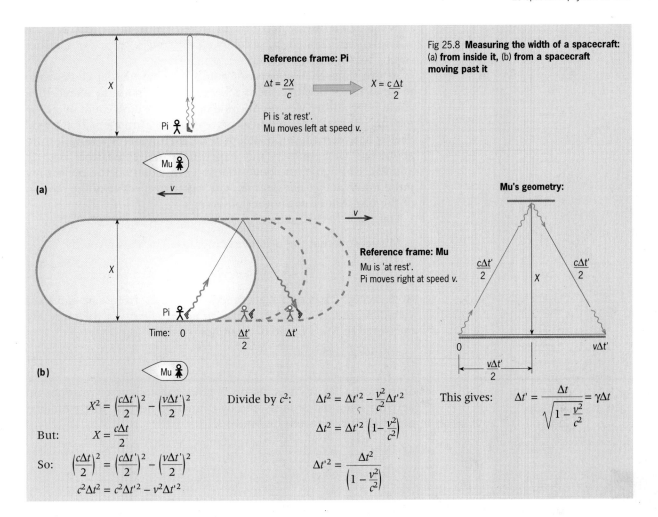

Fig 25.8 **Measuring the width of a spacecraft: (a) from inside it, (b) from a spacecraft moving past it**

Reference frame: Pi

$$\Delta t = \frac{2X}{c} \quad \Longrightarrow \quad X = \frac{c\Delta t}{2}$$

Pi is 'at rest'.
Mu moves left at speed v.

Reference frame: Mu

Mu is 'at rest'.
Pi moves right at speed v.

Mu's geometry:

$$X^2 = \left(\frac{c\Delta t'}{2}\right)^2 - \left(\frac{v\Delta t'}{2}\right)^2$$

But: $\quad X = \frac{c\Delta t}{2}$

So: $\quad \left(\frac{c\Delta t}{2}\right)^2 = \left(\frac{c\Delta t'}{2}\right)^2 - \left(\frac{v\Delta t'}{2}\right)^2$

$$c^2\Delta t^2 = c^2\Delta t'^2 - v^2\Delta t'^2$$

Divide by c^2:

$$\Delta t^2 = \Delta t'^2 - \frac{v^2}{c^2}\Delta t'^2$$

$$\Delta t^2 = \Delta t'^2\left(1 - \frac{v^2}{c^2}\right)$$

$$\Delta t'^2 = \frac{\Delta t^2}{\left(1 - \frac{v^2}{c^2}\right)}$$

This gives: $\quad \Delta t' = \dfrac{\Delta t}{\sqrt{1 - \dfrac{v^2}{c^2}}} = \gamma\Delta t$

We shall imagine a scenario that would be more common nowadays: a measurement made by astronauts in 'deep space'. The mistake to avoid here is to think that relativity works only in outer space, and is only of significance to astronauts. Instead, bear in mind that one of the consequences of relativity and time dilation is the equivalence of mass and energy, according to the well-known formula $E = mc^2$. This relationship describes the source of the energy that powers stars – and so makes life possible on Earth.

Back to our scenario: A large, strange, deserted spacecraft is found in space, and is boarded by an astronaut, Pi, who is given the task of measuring its internal dimensions. He does this using an infrared gun, which measures distance X in terms of the time taken for an infrared pulse to go to and from an internal wall, as shown in Fig 25.8(a).

As Pi does this, his twin, Mu, sees him through the window of a scout vehicle from the space station where they both work. She decides to check the measurements which he has radioed in. Mu is doubtful: measuring with the same kind of infrared gun, she does not agree that the pulse travelled through space as shown in (a).

When Pi and Mu return to the space station, Mu tells Pi that, as the spacecraft was moving, the pulse actually travelled a greater distance than X, along two sides of a triangle, as in Fig 25.8(b). She suspects that there is something wrong with Pi's timing system. Pi points out that his infrared gun is exactly the same as Mu's. They check them by measuring the same distance inside the space station, and the results agree. Pi also asserts that the deserted spacecraft was actually at rest, and that it was the scout vehicle that was doing the moving. Therefore, when Mu measured the distance through the window, her infrared pulse followed the longer path – so that her timing was wrong. They argue inconclusively.

At this stage their supervisor points out that neither could claim to be at rest – because there is no third fixed point to which they could refer. Either viewpoint could be true – or both could in fact be moving. There is no physical process or measurement that could decide between these possibilities. He then delivers a long lecture about the Michelson–Morley experiment and shows that both Pi and Mu can agree, provided that in each situation the measuring pulse *travels at the same speed through space*, irrespective of the motion of the transmitter or another observer.

This leads to the following conclusions (refer to Fig 25.8).

Mu accepts that the distance X measured by Pi is given by:

$$2X = c\Delta t, \text{ ie that } X = \frac{c\Delta t}{2} \tag{1}$$

Both accept that Mu's geometry is correct from her point of view, and that this implies a *different* time of flight for the pulse $\Delta t'$, measured in Mu's frame of reference, so that:

$$X^2 = \left(\frac{c\Delta t'}{2}\right)^2 - \left(\frac{v\Delta t'}{2}\right)^2 \tag{2}$$

Substituting for X in equation 2 gives the following relationship between the time Δt measured by Pi and the time $\Delta t'$ that would be observed (or deduced) by Mu:

$$\Delta t' = \frac{\Delta t}{\sqrt{1 - \frac{v^2}{c^2}}} = \Delta t \left(1 - \frac{v^2}{c^2}\right)^{-\frac{1}{2}} \tag{3}$$

This result is worked out in detail below Fig 25.8, and is the effect known as **time dilation**. It is a consequence of the fixed and unvarying speed of light, and the result applies to any time interval $\Delta t'$ that we measure, using *our* clocks, for a process taking a time Δt in a body moving at speed v relative to us. The process could be a pendulum swinging, a body ageing, a muon decaying or a light wave oscillating.

See question 3. ■

The quantity $\dfrac{1}{\sqrt{1 - \frac{v^2}{c^2}}}$ or $\left(1 - \frac{v^2}{c^2}\right)^{-\frac{1}{2}}$ is called the **Lorentz factor.**

B Calculate the value of the Lorentz factor for a speed of 2×10^8 m s^{-1}.

It is awkward to write but keeps turning up in relativistic expressions, so it is often written as γ, that is, the time dilation effect becomes $\Delta t' = \gamma \Delta t$, as shown in Fig 25.8.

Time dilation is symmetrical

We could work out that time dilation is symmetrical from Pi's point of view, which would assume that Mu was doing the moving and would produce an identical result, except that Pi would deduce that it was Mu's time that was going slow. As we saw in Fig 25.7, the (Einsteinian) relativistic situation is fundamentally symmetrical.

Thinking about time dilation

Cosmic rays enter the upper atmosphere of the Earth at speeds very close to the speed of light. Most of the cosmic rays are protons,

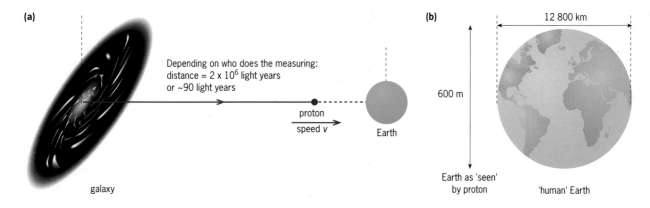

Fig 25.9 **The journey to Earth and the diameter of Earth from a high-speed proton's viewpoint**

accelerated to high speeds in intense electric and magnetic fields linked with supernova explosions. How long do they take to reach Earth?

Suppose a supernova exploded in a galaxy some 2 million light years away and released a proton in our direction travelling at almost the speed of light – at $0.999\,999\,999c$. By our measurement, the proton should take just over 2 million years to reach us, as in Fig 25.9(a). What would be the time of flight of the proton, as read by its own 'watch'? We can find this using the equation $\Delta t' = \gamma \Delta t$:

$$\text{time measured by us} = \gamma \times \text{proton time}$$

So: $$\text{proton time} = \text{our time}/\gamma$$

In this example: $\gamma = \left(1 - \dfrac{(0.999\,999\,999)^2}{1}\right)^{-\frac{1}{2}}$

A reasonably good calculator can just about cope with this calculation to give $\gamma = 22\,400$.

Thus: $$\text{proton time} = \frac{2\,000\,000}{22\,400} = 89.3 \text{ years}$$

So, as far as the proton is concerned, it has spent less than 90 years on the journey. In practice, many cosmic ray protons travel faster than this and would take less than a second – in proton time – to complete the journey.

How far has the cosmic ray travelled?

The answer seems obvious – the galaxy it came from is 2 million light years (1.9×10^{22} m) away. But from the point of view of the proton, it travelled for about 90 years at a speed of (near enough) 3×10^8 m s^{-1}. It has thus travelled just 90 light years, that is, 8.5×10^{17} m. It has collapsed the distance between its galaxy and Earth by a factor of 22 400.

From the point of view of this rather slow cosmic ray, the Earth itself would only be about 600 m across, as Fig 25.9(b) shows. A typically much faster cosmic ray would 'see' the Earth as just a few centimetres in diameter. The cosmic ray, of course, 'thinks' that the Earth is doing the moving. We know that it is all relative. These results reinforce the fact that time and space (distance) are interdependent.

See questions 4 and 5.

See question 6.

C A muon is created in the upper atmosphere by a collision between a cosmic ray and an oxygen nucleus. The muon moves Earthwards at a speed of $0.998c$. In its own reference frame, the muon has an expected lifetime of 2×10^{-6} s.

(a) Show that its expected lifetime in a laboratory on Earth is about 30 microseconds.

(b) How far will it travel before decay, **(i)** in its own reference frame, **(ii)** as observed in the laboratory?

The Lorentz–FitzGerald contraction

The account on page 565 shows that time dilation also shrinks distances for moving objects. This explains the **Lorentz–Fitzgerald contraction** outlined on page 561. With the usual notation, the contraction is written as:

$$x_0 = \gamma x$$

Here, x is the length *that we measure* (in the same direction as v) for any object moving relative to us at speed v; x_0 is the length the moving object would measure *itself* to be – think of it as a moving metre rule. In our reference frame, it would be 1 metre long when at rest next to us. Note that when it is moving with respect to us we see it to be shorter, so to get our observed value x we *divide* by γ, which is always greater than 1:

$$x = x_0/\gamma$$

or: $$x = x_0\left(1 - \frac{v^2}{c^2}\right)^{\frac{1}{2}} = \textbf{length contraction}$$

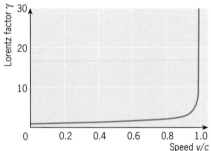

D Centaurian spacecraft are spherical. What shape would a Centaurian spacecraft appear to have as it passes you at a relative speed of 0.5c?

Fig 25.10 **The graph shows that the Lorentz factor γ increases very rapidly to an infinite number at speeds over 95 per cent of the speed of light**

Lorentz factor γ / Speed v/c

E State some ways in which everyday life would be different if light travelled at 13 m s^{-1} (that is, at about 30 miles an hour).

See question 7. ■

A closer look at the Lorentz factor γ

At the heart of the Lorentz factor is the quantity $(1 - v^2/c^2)$. As the size of v gets closer and closer to that of c, v/c and therefore v^2/c^2 get closer and closer to 1. Thus the quantity in the brackets gets closer and closer to zero, and so does its square root.

The Lorentz factor γ is the inverse of this square root, so as the square root gets smaller and smaller, γ gets larger and larger. The result of *multiplying* by γ, as in $t' = \gamma t$ and $m = \gamma m_0$ (see below), is therefore a very large value as v approaches c. When v equals c, γ becomes an infinite number.

When we *divide* a quantity by the Lorentz factor, the result approaches zero for high values of v. This is equivalent to *multiplying* the quantity by $\sqrt{1 - \frac{v^2}{c^2}}$ or $\frac{1}{\gamma}$.

It will help you to look now at Fig 25.10, which shows how the Lorentz factor γ varies with v up to values close to the speed of light. Nothing changes very much until $v = 0.6c$ or thereabouts. Then, for $v > 0.995c$, the Lorentz factor increases very rapidly indeed for small fractional increments of speed.

6 MASS AND SPEED

One of the more surprising and significant consequences of Einstein's theory of relativity is that the *mass* of a moving body increases with speed. The mass of the object will be m, compared with the mass m_0 that it has at rest in our reference frame. So:

$$m = m_0\gamma$$

The **rest mass** m_0 of the body is especially significant in particle physics. The 'mass' of an object such as a proton or electron quoted in tables is its rest mass. At speeds up to about $0.2c$ the actual mass of a particle is very close to its rest mass (Fig 25.11). The particle's observed mass increases very rapidly at speeds close to that of light as it is multiplied by the Lorentz factor.

A simple proof of the relativistic mass formula is given in the next Extension box on page 568.

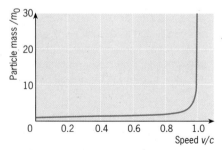

Particle mass $/m_0$ / Speed v/c

Fig 25.11 **The mass of a moving particle increases as the Lorentz factor increases. At speeds over 95 per cent of light speed, there is a rapid mass increase to an infinite value**

See questions 8 and 9. ■

The significance of relativistic mass

The mass acquired by a moving particle due to its motion is just as 'real' as what we call its 'ordinary' mass. The gravitational force on it increases in accordance with Newton's gravity law, and it becomes harder for a force to accelerate it, exactly in accordance with Newton's laws of motion.

The **relativistic mass formula** (see Extension box on page 568) explains why no object that has mass can travel faster than the speed of light in a vacuum. An accelerating force does work which appears as the kinetic energy of the object. As the speed approaches c, the mass increases and so does kinetic energy. If it could reach the speed c, it would have infinite mass and so infinite kinetic energy, and to get this result the force would have to be either infinite or act for an infinite time.

F Estimate the mass your body would have if you travelled at $0.999c$.

Bearing in mind that you are actually travelling at this speed relative to some distant galaxy, explain why you don't in fact feel this extra massiveness.

See questions 10 and 11.

7 MASS AND ENERGY

The best known equation (and one of the shortest) in all physics is:

$$E = mc^2$$

A simple proof of this equation is given in the Extension box on page 570. Don't make the mistake of thinking that the Einstein mass–energy relationship applies only to high-speed particles and nuclear reactions. We can justify it in words by considering what happens when you heat a gas. Most of the energy put in to make the gas 'hotter' goes to make its molecules move faster. As they move faster they gain more mass as shown by the equation:

$$m = m_0 \gamma$$

When a cup of coffee cools down, it loses energy – and also loses the mass-equivalent of that energy. *Any gain in energy is a gain in mass.*

The mass–energy of an atom

The mass of an atom is made up of the rest masses of its particles (nucleons and electrons), together with mass due to both their kinetic and their potential energies. As the nucleons move around in the region of the nucleus and the electrons move around outside the nucleus, kinetic energy is continually being interchanged with potential energy.

Since we are concerned with changes of energy relating to changes of mass, the formula is often written:

$$\Delta E = c^2 \, \Delta m$$

This relation applies to such things as the burning of a fuel, where the final mass of the chemical products is always less than the initial masses of the reactants.

The mass change in a normal chemical reaction is tiny. It is far too small to worry about when doing ordinary chemical calculations. More mass gets converted to energy in nuclear reactions like fission and fusion; but even so, there are practical physical reasons that make it impossible for all the energy equivalent of the rest mass of the nuclei to be converted to, say, kinetic energy. Ultimately, any process would end with everything in a stable form of matter: protons, electrons and neutrinos (see Chapter 26).

The relativistic mass formula

The derivation of the relativistic mass formula is based on what is probably the most fundamental law in all physics – the **law of conservation of momentum**. We shall apply this law to a simple, perfectly elastic collision between objects of equal mass travelling towards each other at equal speed. Thus the total momentum is zero.

Imagine a strange futuristic game in which two players move at high speed and attempt to hit each other with (perfectly elastic) rubber balls.

The expected scenario is shown in Fig 25.12 from the points of view of both players, again Pi and Mu. They are moving towards each other on close parallel paths at a very high relative speed *v*. They are in deep space, so no other forces are involved, and they have no frame of reference against which to see whether either or both of them are moving in their rapid approach towards each other. But a referee is there to keep an eye on them.

In Fig 25.12(a) Pi throws his ball at right angles to his direction of motion, calculating that, by the

time Mu gets opposite, she will just be hit by the ball. He throws the ball at speed *u*, which is very much smaller than *v*, to avoid injury!

(There is an approximation hidden in this proof: Pi and Mu *should* move directly towards each other with speed *v* – and so would collide! We put them slightly apart and make *v* very much bigger than *u* so that we can ignore the extra mathematics needed to calculate exactly.)

From Mu's point of view, as in (b), we get an exactly symmetrical picture: Mu throws her ball at speed *u* and at right angles as well, thinking that Pi will just get opposite to Mu as her ball reaches Pi's path.

However, the referee sees that the balls fail to hit either player but meet halfway and bounce off each other, as in (c).

Did both players throw their missiles at the same time? Yes! says the referee. *No!* say the players. Pi claims that Mu threw her ball before he did, and Mu insists that Pi threw first. Let's keep with Pi to see why they make such claims.

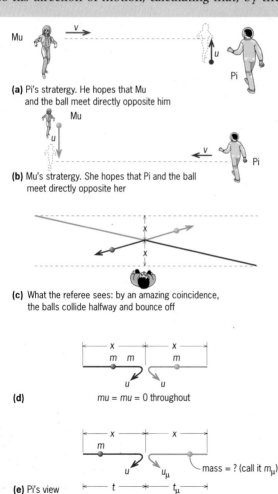

(a) Pi's stratergy. He hopes that Mu and the ball meet directly opposite him

(b) Mu's stratergy. She hopes that Pi and the ball meet directly opposite her

(c) What the referee sees: by an amazing coincidence, the balls collide halfway and bounce off

(d) $mu = mu = 0$ throughout

(e) Pi's view — mass = ? (call it m_μ)

For Pi, Mu's processes run slower than his: for any interval such as time between throwing and collision:

$$t_\mu = \gamma t \qquad [1]$$

Conserving momentum means:

$$m_\mu u_\mu = -mu \quad \text{or} \quad \frac{m_\mu}{m} = \frac{u}{u_\mu} \qquad [2]$$

Also:

$$u_\mu = -\frac{\text{distance } (x/2)}{t_\mu}$$

and:

$$u = \frac{\text{same distance } (x/2)}{t}$$

So:

$$\frac{u}{u_\mu} = \frac{t_\mu}{t} = \frac{\gamma t}{t} = \gamma \text{ (from equation 1)}$$

Thus, equation 2 becomes:

$$\frac{m_\mu}{m} = \gamma$$

or:

$$m_\mu = \gamma m$$

which we generalise to:

$$m = \gamma m_0$$

Fig 25.12 **Diagrams to illustrate the proof of** $m = m_0 \gamma$

For Pi, Mu is moving towards him (almost directly) at a high speed v. Remember that, in this game, speed u is much smaller than v. Pi observes that *any* time interval relating to Mu is *longer* than any equivalent process in his (Pi's) time frame by the time dilation factor γ. But both balls met (obviously at the same time!). From this fact, Pi asserts that Mu's ball must have been in flight for a *longer* time than his own ball was. If Pi's ball was in flight for t seconds, then he judges that Mu's was in flight for a time t_μ seconds:

$$t_\mu = \gamma t$$

Both balls travelled an equal *sideways* distance x, as shown in (c). The two players agree about this, since distance only shrinks *in line with* the direction of the players' motion.

So Pi concludes that Mu's ball travelled more slowly (sideways) to the collision point – it covered the *same* distance (x) in a *longer* time. Pi asserts that, compared to the speed of his own ball, the sideways speed of Mu's ball is:

$$u_\mu = u/\gamma$$

Momentum must be conserved!

Fig 25.12(c) shows the referee's view of the collision looked at sideways. Balls with equal mass and equal but opposite speed collide and move away. Momentum must always be conserved. Both before and after the collision, (d), it is zero.

Pi's view is different, see (e). One ball (his) comes in at a greater speed than the other (Mu's). But the laws of physics still hold and their momenta still add up to zero. This can only be true of the reduction in the sideways speed of that ball is compensated for by an increase in mass by exactly the same factor (γ). Thus:

$$m_\mu = m\gamma$$

where m_μ is the mass of the ball as observed in Pi's frame of reference and m is the mass of the same ball in Mu's reference frame. For Mu, her ball is moving very slowly; for Pi it is moving very quickly at a speed very close to v.

If we developed the argument from Mu's point of view, we would obtain the same result, only this time Mu asserts that Pi fired first, and that Pi's ball travelled more slowly and so gained in mass. This relationship is more general in the form:

Relativistic mass m of a body moving at speed v relative to observer = rest mass m_0 × Lorentz factor.

or:
$$m = m_0\gamma = \frac{m_0}{\sqrt{1 - \dfrac{v^2}{c^2}}}$$

The rest mass is the mass of an object moving with zero speed *relative to the observer*, that is, it is stationary in the observer's reference frame.

The mass of a massless particle

A photon of light has zero *rest* mass. But it has energy E and so it must have an equivalent relativistic mass given by:

$$m = E/c^2$$

Photon momentum

This means that a photon also has momentum p:

$$p = mc = E/c$$

de Broglie again

The energy of a photon is given by the Planck–Einstein relation:

$$E = hf$$

where h is the Planck constant and f is the frequency of the radiation. We can write this in terms of light speed and wavelength λ:

$$E = hf = hc/\lambda$$

Now, as the momentum p of a photon is E/c, we can also see that:

$$p = \frac{E}{c} = \frac{hc}{c\lambda} = \frac{h}{\lambda}$$

which is the de Broglie relationship you first met for electrons on page 368.

> **?**
> **G** Which would have a greater force exerted on it in a gravitational field, a photon of red light or a photon of blue light? Would their motion be different in the field? Explain your answer.

A proof that $E = mc^2$

Fig 25.13 shows an idealised arrangement. It is a closed box with side walls of negligible mass and with equal masses at each end. We consider what happens when a photon is emitted from the left end of the box, as (a), and is then absorbed at the right end. This is equivalent to transferring mass, energy and momentum from left to right.

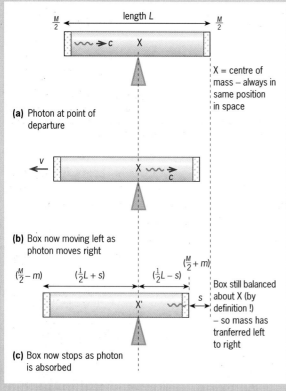

Fig 25.13 **Photons carry energy and mass such that $E = mc^2$**

We observe the box from outside, and have to imagine and calculate what happens inside. How would we *know*, from observation, that a photon had been emitted and reabsorbed?

First, we might notice that the box starts to move to the left, owing to the recoil effect as momentum is carried away by the emitted photon travelling right. Then the box would stop moving, (c), as the photon hit the right end with exactly the same quantity of momentum it had carried away from the left end.

We can quantify this as follows:

The box is L metres long, with a mass M, shared equally between the two ends. The photon has a mass-equivalence m and travels at speed c, giving it momentum p.

As usual, we can put $m = E/c^2$ and so momentum p of the photon is $mc = E/c$.

Conserving momentum as the photon is emitted gives:

> momentum of box moving left at speed v
> = photon momentum p

That is:
$$(M - m)v = E/c$$

But m is very very very small compared to M, so this simplifies to:
$$Mv = E/c$$

so that the speed of the box is:
$$v = E/Mc$$

The photon hits the right end of the box t seconds later, where:
$$t = L/c$$

During this time, the box has moved s metres to the left, where:
$$s = vt$$

We substitute first for v and then for t to give:
$$s = Et/Mc = EL/Mc^2$$

8 RELATIVITY EXPLAINS ELECTROMAGNETISM

The magnetic effect of a current (of moving charges) appears as a force. The force is always caused by the *relative* motion of charged particles – the electrons drifting along in a wire, say, and the electrons in a detector such as a Hall probe. Magnetic forces are a consequence of *relativistic time dilation*.

Consider a mass M on a spring (Fig 25.14). It will oscillate if disturbed with a time period T:

$$T = 2\pi \sqrt{\frac{M}{k}}$$

where k is the spring constant (force per unit extension).

When such an oscillating system moves relative to us *sideways* at a speed v, T changes by the usual dilation factor γ. In effect, this

Now comes the really tricky idea. It is to do with the **law of moments**. The box is completely isolated. Its centre of mass to start with is shown as X in Fig 25.13(a). When the photon moves from left to right, the mass distribution is altered, so it is now shown as X' in (c). *But from our point of view, outside the box, X and X' are the same place.* The box has moved, not its centre of mass.

This is one of the simple, subtle bits of physics that you have to think hard about. It is generally true that the spatial position of the centre of mass of a completely isolated system is unaffected by any internal processes. We need this idea for the next step – applying the law of moments. Look at Fig 25.13(c) to confirm that, taking moments about the unchanged centre of mass:

left-hand moment = right-hand moment
$$(\tfrac{1}{2}M - m)\,(\tfrac{1}{2}L + s) = (\tfrac{1}{2}M + m)\,(\tfrac{1}{2}L - s)$$

This expression simplifies nicely to:

$$m = Ms/L$$

and since $s = EL/Mc^2$, it simplifies further to:

$$m = E/c^2$$

Thus the mass associated with a quantity of energy E is equal to E/c^2.

Comment

This derivation has flaws. First, the set-up is highly impractical. Take the box: no body is rigid enough to start moving quickly enough to get anywhere by the time the photon crosses the box. Also, we have proved $E = mc^2$ only for a photon – a massless particle – and a fuller treatment is required for a mathematical proof that the relation holds for a lump of matter.

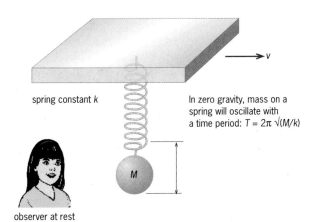

spring constant k

In zero gravity, mass on a spring will oscillate with a time period: $T = 2\pi \sqrt{(M/k)}$

M

observer at rest

Fig 25.14 **An oscillating system viewed sideways. T increases when the system moves at speed v: M gets larger and k gets smaller**

appears to us as either M getting larger or k getting smaller. Some advanced mathematics shows that it is a combination of both effects: the mass increases, and the force decreases by a factor of order v^2/c^2.

We have imagined a force due to a spring, but any force will be affected in this way when the masses move relative to us, as in Fig 25.15.

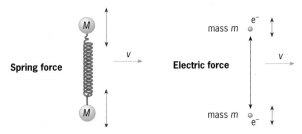

Spring force

Electric force

Fig 25.15 **A pair of electrons can oscillate like two masses on a spring. If they move relative to a stationary observer, the period of oscillation increases: again, M increases and the force decreases. The force (electric) decreases in *all* directions**

Now consider two wires with electrons free to move among fixed positive ions, as in Fig 25.16. When there is no current in the wires there is no net force between them. The forces exist but balance out, as Fig 25.16 shows:

attraction between opposite charges
= repulsion between like charges

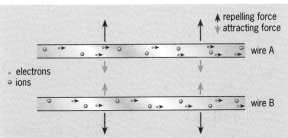

repelling force
attracting force

electrons
ions

wire A

wire B

There are two pairs of attracting forces and two pairs of repelling forces:

Attracting: 1. electrons in A and ions in B
2. electrons in B and ions in A

Repelling: 3. ions in A and ions in B
4. electrons in A and electrons in B

With no current, all forces are equal and cancel out. With a current of moving electrons, force 4 gets less. There is a net attraction between the wires: this is electromagnetism.

Fig 25.16 **Magnetism is a consequence of relativity**

But when the wires carry a current, electrons are moving with a mean speed v relative to us (and the fixed positive ions). We have seen that as v increases, M increases, and so the force decreases. Sothe force of repulsion between the electrons gets less. The net force of attraction is now greater than the net repulsion force. As a result, the wires are pulled together. This is the effect we call **electromagnetism**.

9 OTHER EFFECTS DUE TO RELATIVITY

Relativistic momentum

The momentum of a moving body is given by $p = m_0 v$. If v is very large then the increase in mass means we have to write:

$$\text{momentum } p = \gamma m_0 v$$

The total energy of a moving mass

The work done in accelerating a body is transferred to its kinetic energy. At speeds very much smaller than c, the kinetic energy E_K is given by the formula $E_K = \frac{1}{2} mv^2$. However, for higher speeds we need to use the relativistic form of the equation, which allows for the relativistic increase in mass. The formula becomes:

$$E_K = \gamma m_0 c^2 - m_0 c^2$$

where m_0 is the rest mass of the body. Rearranging gives:

$$\gamma m_0 c^2 = E_K + m_0 c^2$$

The quantity $\gamma m_0 c^2$ is the **total energy** E of the moving body, consisting of its kinetic energy E_K plus its rest mass energy $m_0 c^2$.

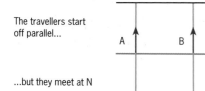

See question 18 at the end of Chapter 26 for a simple proof of this formula.

The 'momenergy' formula

In many cases, such as in particle physics, the momentum of the particle is better known than its speed, so this relationship is often written in terms of momentum p:

$$E^2 = p^2 c^2 + (m_0 c^2)^2$$

See Chapter 26 for examples of the use of this relationship.

At very high speeds, say at $0.999c$, the rest mass energy is negligible compared with the body's kinetic energy, so that we have simply:

$$E = pc$$

– the same formula as for a photon on page 569.

10 GENERAL RELATIVITY

Spacetime grips mass, telling it how to move: mass grips spacetime, telling it how to curve.

This single sentence carries the seeds of general relativity. It began with Einstein thinking of a simple 'relative' situation in which he imagined being in a closed box, like a lift. He would feel a force at his feet, interpreted as 'weight'. When he dropped a pen it would fall to the floor with a constant acceleration.

But were these effects caused by a downward all-pervading force (gravity), or by the fact that the box was being accelerated upwards by some unknown agency? From experiments done *inside* the lift only, noone could tell! He then applied his *theory of special relativity* to this situation and borrowed some advanced geometry (from his teacher, Hermann Minkowski) to extend it to the Universe at large.

First, Einstein abolished gravity as a 'force that acts through space'. He replaced it with geometry – a simple consequence of the *curvature of spacetime*. Fig 25.17 provides an analogy. Imagine two

The travellers start off parallel...

...but they meet at N

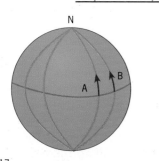

Fig 25.17
Q: What mysterious force draws A and B together?
A: No force – it's just geometry

people flying aeroplanes due north at a steady speed from two points on the Equator. They both travel in straight lines – but will inevitably find themselves getting closer and closer together – ultimately colliding at the North Pole. Is there some mysterious force pulling them together? No. It is all because the Earth is a sphere: its surface is curved.

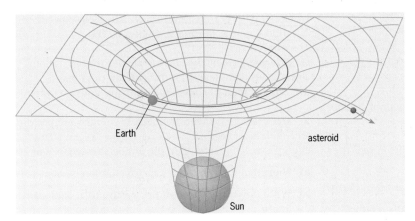

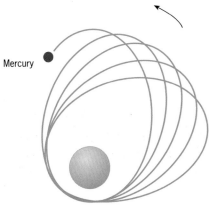

Fig 25.18 **Spacetime is curved...**
Q: What mysterious force draws the asteroid towards the Sun?
A (Newton): Gravity – but I can't explain it.
A (Einstein): No force needed. It's just geometry and I can explain it – even if you can't understand it!

Einstein showed that a lump of matter caused the spacetime around it to be curved. Two masses travelling near each other, such as the Earth and the Moon, will move in 'straight lines through spacetime' – *but we see them as moving in curved paths*. The Earth has more mass, so its spacetime curving effect is greater. Both bodies orbit around the same point, but the Moon orbits in a larger, less curved path.

The geometry works – and the theory predicts the orbits of planets more accurately than Newton's theory of gravity. The elliptical orbit of the planet Mercury swings around the Sun in a way that only Einstein's theory could explain (Fig 25.19).

The general theory of relativity is now the basis of large-scale cosmological theory.

Fig 25.19 **The orbit of Mercury swings ('precesses') around the Sun. Newton's theory predicted this but at the wrong rate. General relativity gets it right**

SUMMARY

In this chapter you have learned how a simple principle – **the constancy of the speed of light in a vacuum for all observers moving at a steady speed** – leads to some remarkable consequences about the nature of time, space (distance), mass and energy. In particular you should now:

■ Understand the principle of the Michelson –Morley experiment and the significance of its null result.

■ Know the two hypotheses of Einstein's theory of special relativity which explained the null result in the Michelson–Morley experiment.

■ Understand the concept of a frame of reference.

■ Understand how Einstein's hypotheses lead to the Lorentz factor.

■ Understand and be able to use the formulae for how the Lorentz factor affects time, distance and mass:

$$t' = \gamma t \qquad x' = \gamma x \qquad m = \gamma m_0$$

■ Understand the equivalence of mass and energy and $E = mc^2$.

■ Know about relativistic energy, and that momentum is linked by the energy and momentum–energy equations:

$$E = \gamma m_0 c^2 = E_K + m_0 c$$
$$E^2 = p^2 c^2 + (m_0 c^2)^2$$

(The chapter ended with a very simple look at general relativity. The derivations of formulae are not usually required for A-level examinations.)

QUESTIONS

For speeds fairly close to the speed of light, use:

$$\gamma = \frac{1}{\sqrt{1 - \dfrac{v^2}{c^2}}}$$

For speeds much smaller than the speed of light, you can use the approximation:

$$\gamma = 1 + \frac{v^2}{2c^2}$$

Take $c = 3 \times 10^8$ m s^{-1} where necessary.

1 Explain why the fundamental measurement of distance requires a measurement of time.

2 A distant quasar is moving away from us at half the speed of light.

a) At what speed does light from the quasar reach Earth?

b) What effect, in general, does the quasar's relative speed have on the light we receive?

3

a) At what relative speed must a clock travel to make it run at half the rate of a clock at rest?

b) At this speed, how long would a metre stick appear to be to **(i)** a traveller with the stick, **(ii)** an observer at rest?

4 The data at the start of this Questions section give two versions of a formula for the Lorentz factor γ. How much of an error is introduced in a simple calculation by using the approximate formula for speeds up to $0.2c$?

5 Muons are created in the upper atmosphere by cosmic rays. They are short-lived particles which decay into an electron and two neutrinos. The graph in Fig 25.Q5 shows how the number N of muons changes with time t.

For every 1000 muons detected at a height of 2000 m, 700 are detected at sea level.

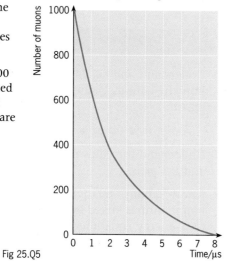

Fig 25.Q5

a) Use the graph to estimate how long it would take 1000 stationary muons to decay to 700.

b) How far would a light photon moving through the atmosphere travel in this time?

c) Muons produced by cosmic radiation travel at a speed of 99.8 per cent of the speed of light. Use the theory of special relativity to explain why such a high percentage of the muons produced by cosmic radiation reach sea level. (A quantitative answer is expected.)

[Adapted with permission from NEAB: 1996, Specimen question Paper 2, Section 2, Option C, Q5]

6 A moving clock always runs slower than a stationary clock. Does this mean that the moving clock has a mechanical fault of some kind? Explain your answer.

7 The cosmic rays with the highest energies are protons with kinetic energy 10^{10} GeV. (1 GeV = 10^9 eV.) This energy is much greater than the proton's rest mass energy.

a) Show that the value of γ for such a proton is about 10^{10}.

b) Write down the speed of this proton to 1 significant figure.

c) Our galaxy is 10^5 light years in diameter. How wide is it from the point of view of the proton?

d) How long would the proton take to cross the galaxy **(i)** as observed by us, **(ii)** as measured in the proton's reference frame?

8 The density of a moving object is bound to be greater than its density at rest for *two* reasons. What are they?

9 An electron has a rest mass of 9.11×10^{-31} kg. Calculate:

a) the relativistic mass,

b) the momentum of an electron travelling at a speed of $0.9c$.

10 Outline the physical reasons why it is impossible to make an object with rest mass travel at the speed of light.

11 Electrons in a TV tube are accelerated by a potential difference of 25 kV.

a) How much work in joules is done in accelerating an electron?

b) How much kinetic energy does the electron gain?

c) The tube works to a tolerance of 0.2 per cent. Does the tube design need to take account of a relativistic increase in mass of the electrons? If so, what aspect of the design would be affected?

12 The total energy E of a body of rest mass m_0 is given by the formula $E = \gamma m_0 c^2$. The total energy is the sum of the body's rest mass energy and its kinetic energy. Find **(i)** the total energy and **(ii)** the kinetic energy of:

a) a proton travelling at $0.995c$,

b) an electron travelling at $0.995c$.

Comment on the ratio of rest mass energy to kinetic energy in each case. (Proton rest mass 1.67×10^{-27} kg, electron rest mass 9.11×10^{-31} kg.)

SPACETIME PHYSICS

There are some quite difficult concepts in this chapter which you may need to go over more than once. Use the Chapter Map below to identify them and to help you see how the concepts are associated.

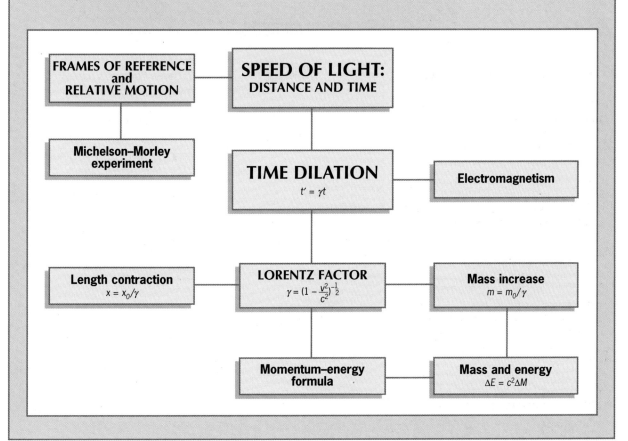

FRAMES OF REFERENCE
and
RELATIVE MOTION

SPEED OF LIGHT:
DISTANCE AND TIME

Michelson–Morley
experiment

TIME DILATION

$t' = \gamma t$

Electromagnetism

Length contraction

$x = x_0 / \gamma$

LORENTZ FACTOR

$\gamma = (1 - \frac{v^2}{c^2})^{-\frac{1}{2}}$

Mass increase

$m = m_0 / \gamma$

Momentum–energy
formula

Mass and energy

$\Delta E = c^2 \Delta M$

26 Deep matter

Warning to children

Children, if you dare to think
of the greatness, rareness, muchness,
fewness of this precious only
endless world in which you say
you live, you think of things like this:

Blocks of slate enclosing dappled
red and green, enclosing tawny
yellow nets, enclosing white
and black acres of dominoes,
where a neat brown paper parcel
tempts you to untie the string.
In the parcel a small island,
on the island a large tree,
on the tree a husky fruit.
Strip the husk and pare the rind off:
In the kernel you will see
blocks of slate enclosed by dappled
red and green, enclosed by tawny
yellow nets, enclosed by white
and black acres of dominoes,
where the same brown paper parcel –

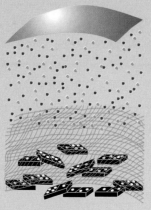

Children, leave the string untied!
For who dares undo the parcel
finds himself at once inside it,
on the island, in the fruit,
blocks of slate about his head,
finds himself enclosed by dappled
green and red, enclosed by yellow
tawny nets, enclosed by black
and white acres of dominoes,
with the same brown paper parcel
still untied upon his knee.

And if he then should dare to think
of the fewness, muchness, rareness
greatness of this endless only
precious world in which he says
he lives – he then unties the string.

Robert Graves (1895–1985)

SCIENTISTS ONCE THOUGHT that atoms were indivisible – the word 'atom' means 'cannot be cut'. They looked more closely, and found electrons, protons and neutrons. Again, they thought these were the smallest particles. Then physicists found they were not, but were divisible into a whole array of even smaller particles.

Where does this subdivision end, and will we ever know we have found *truly* fundamental particles? These are questions to which there is so far no definite answer.

The strange world of fundamental particles

In a study of the world of subatomic particles you are entering a very strange world indeed. What we see about us in everyday life is, on the whole, *stable* matter.

The theory of how the world of stable matter came about is called the **standard model**. This model says that about 15 billion years ago there was a 'singularity', a highly unstable point which was the entire Universe. Then, time and space began in the **Big Bang**. Chapter 28 deals in more detail with this stage in the development of the material of the Universe.

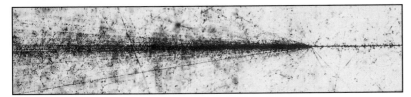

Fig 26.1 **Left: A very high cosmic ray iron nucleus collides with a silver or bromine nucleus in photographic emulsion, and produces a jet of particles (about 850 mesons)**

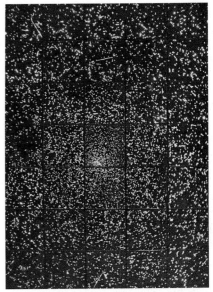

Today, researchers into the fundamental particles of matter use very large **particle accelerators** at international laboratories such as CERN (Centre Européen de Recherche Nucléaire) and other large international and national laboratories.

However, the most highly energetic particles come free: they are the **cosmic rays** that arrive from outer space, particles (most of them protons) which have been accelerated in the magnetic fields of exploding stars to speeds approaching the speed of light.

The practical application of particle physics is not as remote from everyday life as you might imagine. For instance, particle accelerators and detectors are routinely used in hospitals to diagnose and treat diseases (see Chapter 24). Archaeologists and art researchers do detection work with these expensive pieces of equipment. They have also been used to discover and improve new materials for industry and the home, such as superconducting materials, semiconductors and video screen surfaces.

Fig 26.1 **Above: One particle from space travelling close to the speed of light collides with an atomic nucleus in the atmosphere. The kinetic energy is converted to new particles. More collisions create more particles, giving the 'cosmic shower' of about 2×10^5 particles in 5 m × 7 m as recorded by discharge chambers**

Below: A cosmic ray sulpher nucleus collides with a nucleus in photographic emulsion. Green: fluorine nucleus, blue: other nuclear fragments, yellow: 16 pions

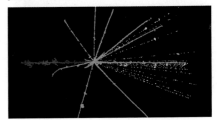

WHAT YOU SHOULD KNOW BEFORE TACKLING THIS CHAPTER

The basis of atomic and nuclear physics is covered in Chapters 17 and 18. You should know the following things about the size and nature of a **typical atom**:

- It is about 10^{-10} m across and consists of a small central **nucleus** just 10^{-14} m across surrounded by a 'cloud of wavy particles', the **electrons**.

- The nucleus contains practically all the mass of an atom and (except for hydrogen) is made up of two types of **nucleon**: **protons** and **neutrons.** Protons are stable, but free neutrons (not in an atom) decay as shown in Fig 26.2. This can also happen inside a nucleus which has too many neutrons.

You should also know from Chapter 18 that three kinds of force are needed to explain what happens in an atom.

- An **electromagnetic** force binds electrons to the nucleus.

- A **strong nuclear** force binds nucleons together inside the nucleus.

- There is also a rather more mysterious **weak nuclear** force that is (for example) involved in beta decay (see page 401).

This chapter has many terms and ideas that may be unfamiliar to you. Take it step by step and check your understanding before moving on. It may help you to refer now and then to Table 26.15 on page 594: it lists all the particles mentioned and their properties, and you can gradually build up a picture of them all.

Fig 26.2 **A neutron is stable only inside a nucleus. Free neutrons decay with a half-life of 900 s. Charge, momentum and mass-energy are conserved**

?

A Outline the arguments to support the ideas that:

(a) the electromagnetic force is stronger than the gravitational force,

(b) the strong force is stronger than the electromagnetic force.

(You may need to refer back to Chapter 18.)

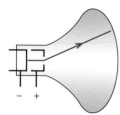

name cathode ray tube
particles electrons
energy 1 keV (1×10^3 V)
discoveries mass and charge of electron

name van de Graaff generator
particles protons/small ions
energy 10 MeV (10×10^6 V)
discoveries structure of nuclei

name electron microscope
particles electrons
energy 25 keV (25×10^3 V)
discoveries structure of crystals, viruses and cells

name cyclotron
particles protons/small ions
energy 25 MeV (25×10^6 V)
discoveries structure of nuclei and transmutation of elements

name linear accelerator
particles electrons/positrons/protons
energy 50 GeV (50×10^9 V)
discoveries evidence for quarks, tau lepton

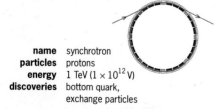

name synchrotron
particles protons
energy 1 TeV (1×10^{12} V)
discoveries bottom quark, exchange particles

Fig 26.3 **Particle accelerators have been used to produce the range of particles described in this chapter**

1 THE SIMPLE ATOM GETS COMPLICATED

The simple Rutherford–Bohr model of an atom had three particles: neutron, proton and electron. But the observed energy spectrum of beta particles emitted in radioactive decay showed the need for another particle, the **neutrino,** which carries off both energy and momentum. This is a very small, uncharged particle, with a mass of zero or close to zero (see page 402).

Then the British theoretical physicist Paul Dirac said in 1930 that, mathematically, each charged particle had to have a matching **antiparticle**, of identical mass but of opposite charge. So there had to be an electron with a positive charge. This anti-electron or **positron** was first observed experimentally in 1932.

Again because theory suggested it, the Japanese physicist Hideki Yukawa proposed in 1935 that another particle ought to exist, with a mass between that of an electron and a nucleon. This was called a **meson** (the Greek word for 'middleweight'). In fact, Yukawa and his colleagues found *two* 'mesons'. One had the properties predicted by Yukawa and is now called a **pion** or **pi-meson**. The other turned out to be a kind of giant electron now called a **muon**.

Too many particles?

More and more 'mesons' – middleweight particles – were discovered in the next thirty years, together with other particles which were more massive than the nucleons. This was all very confusing, and rather like the state chemistry was in before the idea of a Periodic Table (Table 18.2) and the theory of electron shells. The equivalent theory in particle physics is the standard model, involving the particles discussed in this chapter.

The first new particles discovered were thought to be the building blocks of ordinary matter and were called fundamental particles. But by the 1960s there were even more of these 'fundamental' particles than there were elements in the Periodic Table! Most existed only for extremely short periods of time. To understand and categorise them, physicists needed to identify their properties and compare them. We will look at this shortly, but first we look at how particles are discovered.

2 DISCOVERING PARTICLES

The electron was the first subatomic particle to be discovered, in 1897. The cathode ray tube technique used then (see page 252) is used in **particle accelerators** today. In the cyclotron, for example, charged particles are accelerated in a magnetic field at right angles to their direction of motion, which makes them move in the arc of a circle. In a given magnetic field, the radius of the circle depends on the charge, mass and speed of the particle.

These accelerated particles are then used in collisions to produce other subatomic particles, rather like the formation of particles by cosmic rays (Fig 26.1). **Particle detectors** enable the newly formed particles to be located and identified. It is not these particles that are detected, but their tracks formed as ionised particles of the material they pass through. As we shall see, unionised particles leave no tracks themselves, but other clues of their presence can be detected.

Energy and matter

Accelerated (highly energetic) particles are useful for the investigation of new particles because of a consequence of the **special principle of relativity** – that energy and mass are interconvertible (see page 567).

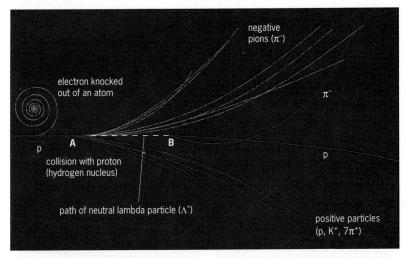

Fig 26.4(a) **A high-energy proton collides with a stationary proton, producing 18 new particles**

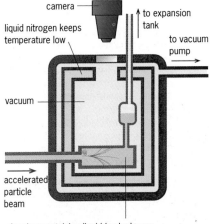

Fig 26.4(b) **A bubble chamber. The beam enters the chamber, and the liquid pressure is momentarily reduced at the expansion tank. This allows liquid in the chamber to boil only at points where ions are present. So tracks of bubbles appear where there are charged particles at that moment**

Fig 26.4(b) shows a **bubble chamber**, a detector of charged particles. An incoming proton travels at a speed very close to the speed of light. It has very high kinetic energy. As it collides at **A** with a stationary proton (a hydrogen nucleus), this energy creates all the new particles and provides them with the kinetic energy to move away. Following the Einstein relation $E = mc^2$:

mass-energy of incoming proton → masses plus kinetic energies
of new particles produced

Fig 26.4(a) shows an example of such a collision. The bubble tracks of some particles arc downwards, and others arc upwards. The small tightly curved spiral is from an electron knocked violently out of an atom by the incoming proton, showing that particles curving upwards are negatively charged, while those curving downwards are positively charged.

At **B**, two particles appear as if from nothing. They are a newly created proton and a negative meson known as a **pion**, symbol π. We have to guess that these two particles were formed at **B** by a decay or a collision involving a neutral particle that left no track as it travelled from the first collision at **A** to the start of the two new tracks at **B**.

Some collisions and decays

Particles that are observed in a detector, such as a bubble or cloud chamber, tend to do one of two things:

● In simple collisions, they collide with a nucleus already there, and then may possibly break up the nucleus and/or create new particles.

● They self-destruct by 'decaying' into other particles or by colliding with an anti-particle.

We now look at both kinds of events in more detail.

Simple collisions

Look at Fig 26.5(b), a **cloud chamber**, and Fig 26.5(a) which shows collisions between alpha particles and the nuclei of hydrogen, helium and nitrogen in the cloud chamber.

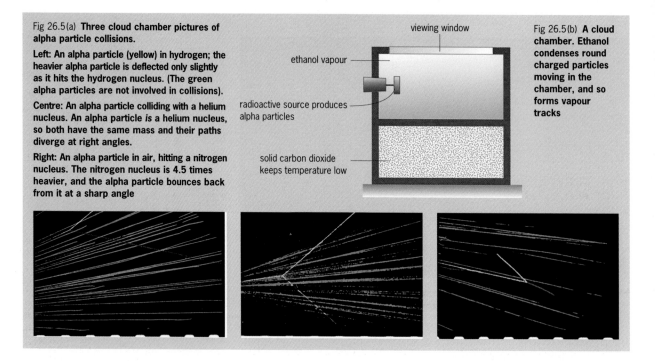

Fig 26.5(a) **Three cloud chamber pictures of alpha particle collisions.**

Left: An alpha particle (yellow) in hydrogen; the heavier alpha particle is deflected only slightly as it hits the hydrogen nucleus. (The green alpha particles are not involved in collisions).

Centre: An alpha particle colliding with a helium nucleus. An alpha particle *is* a helium nucleus, so both have the same mass and their paths diverge at right angles.

Right: An alpha particle in air, hitting a nitrogen nucleus. The nitrogen nucleus is 4.5 times heavier, and the alpha particle bounces back from it at a sharp angle

viewing window

ethanol vapour

radioactive source produces alpha particles

solid carbon dioxide keeps temperature low

Fig 26.5(b) **A cloud chamber. Ethanol condenses round charged particles moving in the chamber, and so forms vapour tracks**

The pictures show that the angle between the tracks of the alpha particle and the nucleus after collision is decided simply by the ratio of the mass of the alpha particle to that of the nucleus it hits.

For another simple collision that indicates different masses, look back at Fig 26.1(c) in which a heavy cosmic ray particle, in this case a sulphur nucleus, collides with a nucleus in the photographic emulsion. The heavy blue tracks are nuclei, the thin yellow ones are the tracks of smaller, less ionising particles identified as pions.

Particles that decay

Unstable particles are like unstable (radioactive) nuclei: for a particular particle we can predict *what* will happen, but not *when*. The newly created particles will decay, eventually, into the **stable particles of matter**, which are protons, electrons and neutrinos. Occasionally, the electromagnetic particle the **photon** is emitted, or the decay is completely to photons, leaving no matter behind. As with radioactive nuclei, we can be certain that half the number of any type of unstable particle will decay in a particular time, so we use the idea of the *half-life* of the particle.

We now know that every known particle has its antiparticle, equal and opposite in its properties. For example, the positron is an antiparticle, exactly the same as an electron but opposite in charge. Particles and antiparticles make up what we call **matter** and **antimatter**.

When an antiparticle collides with its ordinary-matter particle pair, they annihilate each other. Their mass-energies are converted to new particles, namely photons and massive particles, that appear in the detector.

3 THE SEARCH FOR ORDER: CLASSIFYING PARTICLES

The problem facing researchers was how to put all these new particles into some sort of order, and perhaps find out *why* they existed. Researchers started off by naming them with Greek letters, more or less randomly as they turned up. These are a few examples:

μ	mu 'mesons' or muons	Ξ	xi particles
π	pi mesons or pions	Σ	sigma particles
K	kappa mesons or kaons	Ω	omega particles
Λ	lambda particles		

Charge

Charge is either positive or negative. It is also 'quantised': both positive and negative charge come in simple multiples of the **elementary charge,** $-e$ or $+e$, of value $\pm 1.6 \times 10^{-19}$ C). But charge is no help in *classifying* so many particles.

Mass

Mass seemed a good basis for classification, and particles discovered in the 1940s and 1950s were put into three groups mainly by mass, as in Table 26.1.

There were problems with this grouping: a muon is nearly as massive as a pion, for example. Also, new particles were discovered, such as some new leptons which were much more massive than the lightest meson, and new mesons were heavier than the proton, say. But as we shall see next, the groups had properties other than mass that made them fundamentally different.

Table 26.1 **An attempt to classify on the basis of mass (not very useful)**

Name	Mass	Examples
mesons	medium	pions, kaons
baryons	heavy	protons, neutrons, lambda, sigma, xi
leptons	light	electrons, muons, neutrinos

■ See questions 1 to 6.

?

B Show that the energy equivalent of the mass of a proton is about 1 GeV.

Units of mass and energy in particle physics

Obeying the laws of relativity, the mass of a particle varies with its speed. This means that we have to use the value of its mass when it is not moving, its **rest mass** (see page 566). The principle of relativity also requires that mass and energy are equivalent, linked by the Einstein formula $E = mc^2$. In fact, particle physicists *measure mass in energy units.*

Investigations usually involve charged particles which are accelerated by high voltages and gain mass-energy as they accelerate. It is convenient for particle physicists to *measure energy and therefore mass in terms of an electrical unit,* namely the **electronvolt, eV.**

One electronvolt, 1 eV, is the energy gained by a particle carrying the electronic charge $e = 1.6 \times 10^{-19}$ C when it moves through a potential difference of 1 V.

$$1 \text{ eV} = 1.6 \times 10^{-19} \text{ C} \times 1 \text{ JC}^{-1}$$

$$= 1.6 \times 10^{-19} \text{ J}$$

and: $\qquad 1 \text{ J} = 6.25 \times 10^{18} \text{ eV}$

The mass of a proton is 1.6726×10^{-27} kg. In energy units, this is equivalent to:

$$E = m_p c^2 = 1.6726 \times 10^{-27} \times (3 \times 10^8)^2 \text{ J}$$

$$= 1.5053 \times 10^{-10} \text{ J}$$

Rest mass in GeV/c^2

The mass-energy of a proton is about 10^9 eV – or 1 GeV, a nice easy number to work with. We need to be careful with this unfamiliar set of units. For example, from the Einstein formula we have:

$$m = E/c^2$$

$$\frac{\text{energy in eV}}{(\text{light speed})^2} = \frac{\text{GeV}}{c^2}$$

Thus the rest mass of a proton m_p is written as 1, with units **GeV/c^2**, or:

$$m_p = 1 \text{ GeV } c^{-2}$$

The masses listed in the tables in this chapter are given in units of GeV/c^2. Examples of calculations based on this unit are on page 603.

New fundamental properties

Charge and mass were not enough to make the distinctions between all the particles found, so workers identified other common properties. These properties are just as fundamental and mysterious as charge and mass, and physicists have given them weird names including *charm, bottomness, baryon number* and *strangeness*. Using these properties, physicists have been able to classify the numerous particles, as we shall see later.

Matter and forces: hadrons and leptons

We are familiar with the fact that charge is acted upon by electromagnetic forces, and that mass is acted upon by gravity. A key to identifying the differences between particles was the way they were affected by these and other forces.

Protons in a nucleus are affected by electromagnetic forces and by gravity. But they stay inside the nucleus, in spite of the strong electric force of repulsion due to their positive charges, because there is another, stronger force holding them together. This is the **strong nuclear force** (see page 577).

All particles that 'feel' the strong nuclear force are called **hadrons**, subdivided into **mesons** and **baryons**. Most particles that do not 'feel' the strong force are called **leptons**. This distinction neatly separated most particles into two significant groups, as in Table 26.2: pions are hadrons, and 'feel' the strong nuclear force, while electrons and muons are leptons and are unaffected by the strong force. (This broad classification could be developed further, whereas the grouping Table 26.1 could not.)

Table 26.2 **Examples of hadrons which feel the strong nuclear force, and leptons which do not. Note that gravitons (if they exist) and photons are non-hadrons**

Hadrons		Leptons
mesons, eg pions, kaons, eta mesons		include electrons muons,neutrinos
baryons, eg protons, neutrons, omega, sigma and lambda particles		photons, gravitons (if they exist)

The lepton family is described in more detail on page 583.

Another important difference between hadrons and leptons is that *leptons have no substructure* – they are (as far as we know) *fundamental* particles – while hadrons are built from even smaller (fundamental) particles called **quarks** (see page 588).

Most hadrons and leptons are *unstable*: they very quickly decay into other particles, and finally end up as the only **stable particles** in the Universe – protons, electrons and neutrinos. Even the neutron is unstable, since outside a nucleus it decays with a half-life of about 11 minutes to a proton, an electron and an antineutrino:

$$n \rightarrow p + e^- + \bar{v}$$

Particles and antiparticles

Each hadron has its *antiparticle*, for example:

proton and antiproton	p^+ and p^-
neutron and antineutron	n and $\bar{n}$
kappa meson (or kaon)	K^+ and K^-

Antiparticles are often represented by using a bar over the symbol, as for the antineutron, especially when they have no opposite charge.

Forces and interactions

Forces always exist as pairs: when two particles exert any kind of force on each other, the force on one is the same as the force on the other, each acting in the opposite direction. This effect is called interaction between the particles. We commonly refer to the four *fundamental* interactions: electromagnetic, weak, strong and gravitational. As we shall see, when particles interact, there is always another particle that 'carries the force' – it 'mediates the interaction'. The mediating force-carrying particles known as **exchange particles** are in a group called **gauge bosons**, see Table 26.3.

Table 26.3 **Gauge bosons**

Interaction	Particle	Relative strength
strong	gluon (g)	1
electromagnetic	photon (γ)	10^{-2}
weak	W, Z bosons	10^{-7}
gravity	graviton (not yet found)	10^{-36}

Modern theory suggests that at very high energies all these forces merge into just one. There is experimental evidence that at very high energies the electromagnetic and weak forces are unified – they become the electroweak force.

Time scales of interactions

Each interaction has a characteristic time scale (which is quoted as a half-life, see page 605. The stronger the force, the quicker things happen, so when two hadrons interact due to the strong force, events happen on a time scale shorter than about 10^{-10} s. This also applies to a hadron's internal change (an 'interaction' involving one particle), which we observe as a spontaneous decay.

In the weak interaction, things happen slowly. Normal weak interactions produce decays in about a millionth of a second – very slow by particle interaction standards.

The time scale of an interaction or change can be measured in detectors. The time gives valuable clues about the event involved, and hence the type of particle produced, even though it is not 'seen' by the detector.

See question 7. ■

The weak nuclear force

As well as being affected by the electromagnetic force, charged hadrons are affected by the *weak nuclear force* and, since they all have mass, they are affected by *gravity*.

The weak force seems the most mysterious of all the forces. Like the strong force, it acts over a very short range. However, modern theory shows that when particles have a high enough energy, the weak force and the electric force become the same **electroweak force**. We will see on page 593 that the weak force seems to work as one quark changes into another.

Leptons

Leptons are generally much less massive than hadrons (Greek: hadron = heavy, lepton = small). Leptons are affected by three of the four fundamental interactions (forces): the electromagnetic (if charged), the weak and gravity.

There are three kinds of lepton: **electrons e, muons μ,** and **tauons τ.** Each of these is associated with its own kind of **neutrino ν.**

Electrons and neutrinos are stable particles, but both muons and tauons decay. Their decay is due to the weak force, so lifetimes are much longer than for the unstable hadrons. The tauon is much more massive than the other leptons and has enough mass-energy to decay into hadrons. Table 26.4 shows the main leptons and their neutrinos.

	Leptons			Antileptons		
	Electron	Muon	Tauon	Positron	Muon plus	Tauon plus
	e^-	μ^-	τ^-	e^+	μ^+	τ^+
Neutrinos	ν_e	ν_μ	ν_τ	$\bar{\nu}_e$	$\bar{\nu}_\mu$	$\bar{\nu}_\tau$
Lepton no.	1	1	1	−1	−1	−1

Table 26.4 **Leptons and their neutrinos. Each associated neutrino is subtly different from the others**

Matter and antimatter

When a particle meets its antiparticle there is instant annihilation of mass. The mass-energy of the pair appears as new particles (supplying both mass and kinetic energy) and/or radiation. The most commonly observed matter–antimatter collision is that for electrons:

electron + positron → 2 photons

$$e^- + e^+ \rightarrow 2\gamma$$

Why *two* photons? This is because photons carry not only energy but momentum, as explained below. Momentum has to be conserved. If we think of the two particles meeting head on with the same speed, the original momentum adds to zero. If the result was a single photon, the momentum would not be zero. The momentum can only be the same before and after the collision with two photons carrying *equal quantities* of momentum travelling in *opposite* directions, as in Fig 26.6.

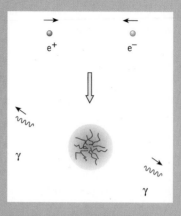

Fig 26.6 **Annihilation of an electron–positron pair converts their mass-energy into photons. Momentum is conserved by the photons of equal mass-energy moving away in exactly opposite directions. Note also that charge is conserved**

Pair production

Just as an electron–positron pair can annihilate to produce photons, so photons with enough energy can 'decay' into these and even more massive particles. Fig 26.7 shows such an event seen in a cloud chamber; it is called **pair production**. The event occurs only in a strong electric field – near a charged nucleus for example.

Fig 26.7 **Gamma ray photons (not detected) enter the bubble chamber from the left. One photon decays to an electron–positron pair, with the electron spiralling down and the positron up. (The straighter track is for an electron displaced by the same photon from an atom in the chamber.)**

On the right is the decay of a second photon to an electron–positron pair

4 LOOKING FOR PATTERNS OF BEHAVIOUR

So far, we have seen how particles are classified in terms of well-known properties, like mass and charge, and their response to forces. But the key to the fundamental nature of particles was in investigating what happens *and what does not happen* when particles interact or otherwise change, either naturally by spontaneous decay or when they interact at high energies.

The rules of behaviour

Fundamental particles must obey the basic ground rules of physics, which should be well known to you. These are:

- **Momentum conservation.** In any interaction between particles in a system, the total momentum must stay constant.
- **Mass-energy conservation.** In any interaction between particles in a system, mass-energy must neither be created nor destroyed.
- **Charge conservation.** In any interaction between particles in a system, the total electrical charge in the system must not change.

The following examples illustrate these rules.

Example 1. Neutron decay

$$n \rightarrow p + e^- + \bar{v}_e$$

The neutron has mass-energy of 939.6 MeV, the proton has 938.3 MeV and the electron 0.5 MeV. The neutron can thus change into the particles on the right with an energy of about 0.8 MeV left over to be shared by the kinetic energy of the products and the creation of the antineutrino. The neutron has zero charge and so the sum of the charges on the products is also zero.

Example 2. Proton–antiproton annihilation

$$p + \bar{p} \rightarrow \pi^+ + \pi^- + \pi^0$$

This reaction clearly conserves charge. The combined mass-energy of the protons is 1876 MeV. The mass-energy of each charged pion is 140 MeV, with 135 MeV for the neutral pion. Since the mass-energy of the protons is more than enough to produce the total mass of the pions (415 MeV), they fly off at very high speeds, in the directions that conserve momentum.

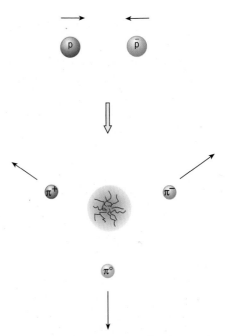

Fig 26.8 **Proton–antiproton annihilation**

Example 3. When the expected doesn't happen

Particle physicists soon realised that there were some possible changes that did obey the three basic conservation laws but which never actually occurred, even when enough energy was supplied. For example, a proton did not spontaneously decay into a positron and a neutral pion, even though charge, energy and momentum could all be conserved.

$$p \not\rightarrow e^+ + \pi^0$$

This is just as well, otherwise all protons would eventually decay and the atomic nucleus – hence atoms, chemistry, life, stars – could not exist!

> **?**
>
> **D** Use the charge conservation rule to check whether or not any of the following reactions can occur, at least in principle:
>
> **(a)** $K^+ + n \rightarrow K^0 + \Sigma^+$
>
> **(b)** $p + n \rightarrow p + n + \pi^-$
>
> **(c)** $K^+ + n \rightarrow \pi^0 + \Delta^+$

Baryon number explains why some changes could not happen

Many other proposed changes were never observed in practice. So physicists decided that there must be some sort of 'charge' that would not be conserved by these changes – the equations would not balance. A theory was proposed that particles possess (or lack) another property that must be conserved. It was a quality that, for example, a proton possessed but that neither electrons nor pions did.

Evidence built up that this new kind of charge was possessed only by the heavier particles, the baryons. Unlike electric charge, there was no *force* associated with it. But like electric charge it came in same-size 'chunks'. It is called **baryon number**, B.

The proton and neutron were given baryon number 1, and electrons, neutrinos, positrons and pions had baryon number 0.

Thus, proton decay could not occur because baryon number would not be conserved: the two sides in example 3 did not balance:

baryon equation: $\qquad\qquad 1 \neq 0$

But when a neutron decays into a proton (example 1), baryon number *is* conserved because the neutron also has a baryon number of 1.

Table 26.5 **Some particles and their baryon numbers**

Particle	Baryon no. B
baryons eg proton, neutron, sigma	1
baryon antiparticles eg antiproton	−1
mesons, eg pion, kaon; leptons eg electron, positron	0

E Look at Table 26.5 for the baryon numbers of the particles to check which of the following reactions would **not** occur because baryon number is not conserved. (All equations conserve charge.)

(a) $\pi^+ + n \rightarrow K^0 + \Sigma^+$

(b) $\pi^+ + p \rightarrow K^+ + \Sigma^+$

(c) $K^+ + n \rightarrow \pi^0 + \pi^+$

✔

As shown in Table 26.4, there are three pairs of leptons and each pair is associated with its own lepton number. This complicates things but does allow useful predictions about those ghostly leptons, the neutrinos.

See questions 9 – 11. ■

F Calculate how long it takes for a disturbance to travel across a nucleus (10^{-14} m) at light speed (3×10^8 m s^{-1}).

G First estimate the size of a cup. The speed of sound in a ceramic is about 5000 m s^{-1}. Show that a cup takes less than a millisecond to break.

Now look back at the reaction in example 2, in which two baryons (proton and antiproton) are converted to pions. Here the right side is entirely of pions, and so has *zero* baryon number. Thus the left side must also have zero baryon number. This can be true only if the antiproton has a *negative* baryon number (B = −1).

Note in Table 26.2 (page 582) that mesons and baryons are grouped together as hadrons. But note that mesons have a baryon number of 0. Leptons also have baryon number 0 but have their own conserved property, **lepton number** L: as Table 26.4 shows, lepton number may be 1 or −1. It can also be 0. To conserve lepton number, when leptons are made out of non-leptons they are made in pairs:

$$\pi^- \rightarrow \mu^- + \overline{\nu_\mu}$$

Pions are not leptons (Table 26.2), so the left side of the equation has a lepton number 0. On the right side is a muon of lepton number +1 and a muon antineutrino with a lepton number −1.

More non-events lead to more new properties

Identifying reactions that seemed never to happen became an important tool in classifying particles. This approach led to **strange particles** – particles with the conserved property of **strangeness**. This was to be explained by the theory of quarks (see page 588), which dramatically simplified the task of classifying particles.

Strangeness and time

Reactions that involve the strong nuclear force seemed to happen on a very short time scale, typically 10^{-23} s. This is the time it takes for the fastest possible effect – the strong interaction – to cross a nucleus at the speed of light.

A typical reaction involving strongly interacting particles takes this very short time. For example:

$$p + \pi^- \rightarrow K^0 + \Lambda \qquad \text{Time} = 10^{-23} \text{ s}$$

The Λ (lambda) particle is unstable and breaks up – but oddly, it does so in a time of about 3×10^{-10} s.

Let us put this into an everyday time scale. When you drop a china cup on to a hard floor, it breaks up in less than a millisecond. (For a rough estimate, consider how long it takes a shock wave travelling at the speed of sound to cross a cup. The cup usually breaks into several pieces, all produced at much the same time, all of random sizes.)

Both of the times for the particle reactions are very short, but one is 10^{13} times longer than the other. On the cup scale of time, it was as if you dropped a cup, which immediately (at least, in 10^{-4} s) broke into two large pieces. Then, *three years later*, these two pieces broke up into lots of smaller pieces.

The new exceptionally long-lived particles were not clearly identified by mass, charge and baryon number, so they were simply called **strange particles** and labelled as K (kappa – a kaon), Λ (lambda), Σ (sigma), Ξ (xi) and Ω (omega), see Table 26.6.

Fig 26.9 **A cup cannot break up faster than it takes for the shock wave to pass through it**

Their behaviour was explained by yet another fundamental 'charge' or property which, was called **strangeness**, symbol S. Particles can have strangeness numbers from +3 to –3. This introduces the idea of 'threeness' which we return to on page 588.)

The strangeness of interactions is conserved only for reactions in which the strong force is important. For example, we can produce the larger baryons, called **hyperons**, by colliding an electron and a positron at very high speeds. The collision has enough energy to make a large number of particles. When this happens, if hyperons are produced, they have to be as *pairs* with equal but opposite strangeness:

$$e^- + e^+ + \text{a lot of mass-energy} \rightarrow \Lambda^0 + \overline{\Lambda^0} \text{ plus other particles}$$

The ordinary baryons (protons and neutrons) have zero strangeness. Strange particles Λ and Σ have *negative* strangeness, and their antiparticles have *positive* strangeness.

Table 26.6 **Some strange particles and their strangeness values**

Particle	Strangeness S
K^+, K^0 (mesons)	+1
K^-, K^0 (mesons)	–1
Λ (baryon)	–1
Σ (baryon)	–1
Ξ (baryon)	–2
Ω (baryon)	–3

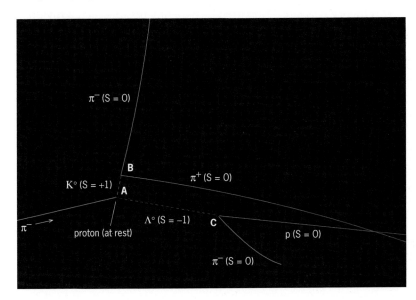

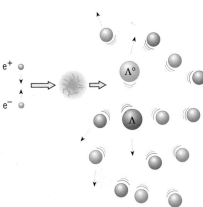

Fig 26.10 **Electrons and positrons collide at high energies to produce a lambda and an antilambda particle (both strange) plus lots of other particles, mostly mesons**

Fig 26.11 **A pion interacts with a proton to produce two strange particles (K^0, Λ^0), which leave no tracks. These then decay as shown.**

Strangeness is conserved at A (strong interaction), but changes by ±1 at B and C

Breaking the law of strangeness conservation

Some conservation rules work only when the strong force is involved, others only for the weak force. One rule is the conservation rule for strangeness. It has been found that strangeness-carrying particles can decay into particles with zero strangeness, *thus breaking the conservation law*. But they can only do this via the weak interaction, and when they do, they can only change the total strangeness number by just one step at a time.

Thus a **xi hyperon** (also called a cascade hyperon) first decays into a lambda hyperon in a weak interaction, losing a strangeness of 1. Then the lambda hyperon changes into hadrons (particles responding to the strong force) of strangeness 0, again with a strangeness change of unity.

Step 1: $\Xi^- \rightarrow \Lambda^0 + \pi^-$
  $S = -2$ $S = -1$ $S = 0$ $\Delta S = +1$

Step 2: $\Lambda^0 \rightarrow \pi^- + p^+$
  $S = -1$ $S = 0$ $S = 0$ $\Delta S = +1$

?

H Test the following reactions for conservation of strangeness:

(a) $\pi^- + p \rightarrow K^+ + \Sigma^-$

(b) $K^- + p \rightarrow K^+ + \Xi^-$

(c) $\pi^- + p \rightarrow K^- + \Sigma^+$

(d) $\pi^+ + n \rightarrow \pi^+ + \Lambda$

Table 26.15 may be useful.

See questions 12 and 13.

5 QUARKS BRING ORDER OUT OF CONFUSION

The ways described above for classifying particles were immensely helpful – interactions could be predicted. But like the Periodic Table in Mendeleev's day, there was no underpinning *theory* or *model* of particles which explained *why* the patterns existed.

Then a revolutionary new theory was suggested to explain the 'threeness' referred to on page 587: the heavy particles (baryons) might in fact be simple combinations of *three smaller particles*. This theory required a startling idea, that *electric charge has to be subdivided into thirds*: the new particles could have charges of 1/3 or 2/3 of the electronic charge.

At the time, very few physicists believed in the *real* existence of these hypothetical particles, at first not even the American physicist Murray Gell-Mann (born 1929), who suggested the new particles and named them **quarks**.

This theory was later supported by experiments in which individual protons were bombarded by very energetic electrons. Just as the Geiger–Marsden alpha particle bombardment experiment showed that the atom has a small central mass, so these experiments showed that the density of a proton was not uniform, suggesting that it was made up of even smaller masses. These turned out to be the quarks.

The quark model of hadrons

The original simple quark model said that there were three quarks. These are now called the **up**, **down** and **strange** quarks. They are summarised in Table 26.7.

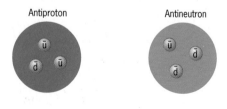

Proton　　　　　Neutron

(d) down quark: charge = $-\frac{1}{3}$

(u) up quark: charge = $+\frac{2}{3}$

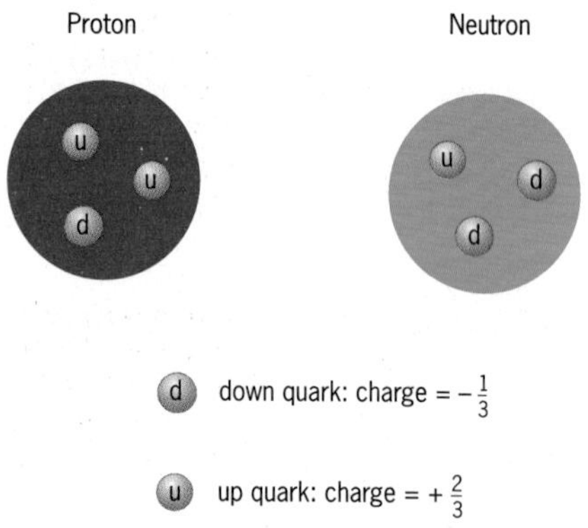

Fig 26.12 **The quark structure of the proton and the neutron**

Quark	Symbol	Charge/e	Baryon no.	Strangeness
up	u	+2/3	1/3	0
down	d	–1/3	1/3	0
strange	s	–1/3	1/3	–1

Table 26.7 **The first quarks**

Table 26.8 and Fig 26.12 show how these quarks are arranged in the most common baryons: protons and neutrons.

Table 26.8 **Quarks in protons and neutrons**

	Proton	u	u	d	Neutron	u	d	d
Charge	1	2/3	+2/3	–1/3	0	2/3	–1/3	–1/3
Baryon no.	1	1/3	+1/3	+1/3	1	1/3	+1/3	+1/3
Strangeness	0	0	0	0	0	0	0	0

Antiquarks

But of course, quarks must have **antiquarks**, and these were needed to explain the existence of mesons (the middleweight hadrons). Table 26.9 lists the three original antiquarks, and Table 26.11 shows how they combine with quarks to form some mesons. In antiquarks, charge and strangeness have swapped over.

Antiproton　　　　　Antineutron

(ū) up antiquark: charge = –2/3

(d̄) down antiquark: charge = +1/3

Fig 26.13 **The quark structure of the antiproton and the antineutron**

Table 26.9 **The three basic antiquarks**

Antiquark	Symbol	Charge/e	pin	Baryon no.	Strangeness
up	ū	–2/3	1/2	–1/3	0
down	d̄	+1/3	1/2	–1/3	0
strange	s̄	+1/3	1/2	–1/3	+1

Quarks in baryons

Table 26.10 lists the quark content of some baryons. Note that all baryons are made of quarks and all antibaryons are made of anti-quarks. The baryon number of a quark is +1/3 and of the antiquark −1/3, which explains why the baryon numbers are either +1 or −1 for these particles.

Particle	Proton	Antiproton	Neutron	Antineutron	Lambda	Antilambda
quarks	uud	$\bar{u}\bar{u}\bar{d}$	udd	$\bar{u}\bar{d}\bar{d}$	uds	$\bar{u}\bar{d}\bar{s}$
charge	1	−1	0	0	0	0
strangeness	0	0	0	0	−1	1

Table 26.10 **Quarks in some baryons**

?

I Use Table 26.10 to explain why:

(a) the antiproton has negative charge and zero strangeness,

(b) the lambda has zero charge and unit negative strangeness.

Quarks and antiquarks make mesons

The significant difference between mesons and baryons (both are types of hadrons) is that mesons have baryon number 0. This is because mesons are made up of a quark and an antiquark, whose baryon contribution always cancels out. Table 26.11 lists some common mesons and their antiparticles.

Meson	π^+	π^-	π^0	K^+	K^-	K^0	K^0 antikaon	ϕ phi
Quarks	$u\bar{d}$	$\bar{u}d$	$u\bar{u}$	$u\bar{s}$	$\bar{u}s$	$d\bar{s}$	$\bar{d}s$	$s\bar{s}$

Table 26.11 **The quarks in some mesons**

Table 26.11 clearly shows that the π^- is the antiparticle of the π^+: what was a quark in the particle has become an antiquark in the antiparticle, and vice versa. It also shows that some particles are their own antiparticles – changing an *up quark* for the *up antiquark* and vice versa makes no difference to the π^0.

Why single quarks don't exist

Quarks are 'invisible': they never appear on their own. Whenever there is enough energy to pull apart a pair of quarks in a pion, the energy is always enough to *create two more quarks*, which combine to form another pion (Fig 26.14). So all we see – and all we get – are pions.

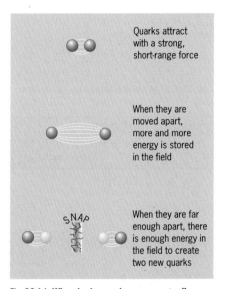

Quarks attract with a strong, short-range force

When they are moved apart, more and more energy is stored in the field

SNAP

When they are far enough apart, there is enough energy in the field to create two new quarks

Fig 26.14 **Why single quarks never actually appear**

More quarks

The three quark–antiquark pairs listed in Tables 26.7 and 26.9 are insufficient to describe the structure of all the known hadrons. Table 26.12 is an up-to-date list of quarks. How they were predicted and discovered is explained next.

Table 26.12 **All the quarks and their numbers**

Name	Symbol	Charge	Baryon no.	Flavour			
				Strangeness	Charm	Topness	Bottomness
Quarks							
up	u	+2/3	1/3	0	0	0	0
down	d	−1/3	1/3	0	0	0	0
strange	s	−1/3	1/3	−1	0	0	0
top	t	+2/3	1/3	0	0	1	0
bottom	b	−1/3	1/3	0	0	0	1
charm	c	+2/3	1/3	0	1	0	0
Antiquarks							
up	$\bar{u}$	−2/3	−1/3	0	0	0	0
down	$\bar{d}$	+1/3	−1/3	0	0	0	0
strange	$\bar{s}$	+1/3	−1/3	1	0	0	0
top	$\bar{t}$	−2/3	−1/3	0	0	−1	0
bottom	$\bar{b}$	+1/3	−1/3	0	0	0	−1
charm	$\bar{c}$	−2/3	−1/3	0	−1	0	0

?

J Find the quark structure of the following hadrons, using Table 26.12 and the following data for the particles:

Particle	Charge	Baryon no	Strangeness
antiproton	−1	−1	0
sigma minus1	−1	+1	–
xi minus	−1	+1	−2
kaon minus	−1	0	−1

6 THE REALLY FUNDAMENTAL PATTERN (POSSIBLY)

Table 26.13 **Lepton and quark families known about in 1974**

Leptons		Quarks		
$Q = -1$ e^- μ^-		u ?		$Q = +2/3$
$Q = 0$ v_e v_μ		d s		$Q = -1/3$

After accepting the quark model in the late 1960s, physicists noticed a new kind of symmetry – this time between *leptons* and quarks. Leptons contain no quarks (they are indivisible), and everything else does, so it makes sense to think of leptons and quarks as the basic, *really* fundamental building blocks of matter. Table 26.13 shows the pattern of leptons and quarks known by 1974 (omitting antiparticles).

The pattern suggested that a quark should exist in the gap above the strange (s) quark. It was expected to have a charge of +2/3, and theory predicted that hadrons containing this quark should have masses of about 3 GeV/c^2. Late in 1974, two laboratories announced the discovery of two new heavy hadrons, the J and psi (Ψ) particles. They turned out to be the *same* particle, with a mass of 3.1 GeV/c^2. Confusingly but appropriately, it is called the J/Ψ.

Charm enters physics

Physicists then realised that the new particle was a heavy meson containing the predicted missing quark, carrying yet another conserved quantity called **charm**. The J/Ψ consists of two *charmed quarks*, one the antiquark of the other ($c\bar{c}$).

The new J/Ψ particle was predicted, but in the search for it, evidence was found for a particle that was not predicted at all. This was a *new lepton* – but an amazingly massive one, twice as heavy as a proton. This *third* kind of lepton was named the **tau,** τ.

This discovery ruined the existing pattern between leptons and quarks and if the lepton–quark pattern was to be preserved it meant that yet another pair of quarks should exist (see Table 26.14).

Table 26.14 **Lepton and quarks known about in 1975**

Leptons			Quarks			
$Q = -1$ e^- μ^- τ			u c top?			$Q = +2/3$
$Q = 0$ v_e v_μ v_τ			d s bottom?			$Q = -1/3$

Quarks are found indirectly – by finding particles with properties and interactions that could only exist if they carry the properties of their component quarks. Physicists named the new quarks **top** and **bottom** from their positions in the table, and had to invent two new conserved quantities, **topness** and **bottomness**.

Top and bottom quarks were predicted to be much more massive than other quarks (Fig 26.15), and so should produce equally massive combinations. Such particles need a great deal of energy to be created and the new proton synchrotron opened in 1975 at Columbia University (New York State) with an energy of 30 GeV very soon found a new massive particle that was a combination of a bottom quark and an antibottom quark. This b meson was named the **upsilon,** Y. It has a mass of 9.5 GeV/c^2.

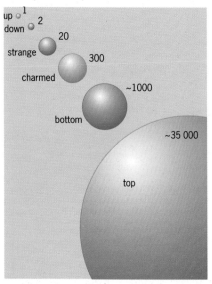

Fig 26.15 **Quarks come in different sizes. The top quark is 10 000 times as massive as the up quark. Top quarks are hard to make**

The search for the top quark

The missing piece of the jigsaw was the top quark. It was predicted to have a mass of more than 90 GeV/c^2, twenty times as massive as a bottom quark. A particle containing a top quark would have a mass greater than this.

The energy needed to produce such a massive particle is available when *protons* are made to collide with *antiprotons* in powerful synchrotrons. This was possible in the new 1800 GeV Tevatron collider at Fermilab (Chicago, USA), and in April 1994 evidence for the existence of the top quark was announced. Experiments in March 1995 matched a mass of about 175 GeV/c^2 for the top quark.

The uncertainty principle

At the heart of particle physics is the idea of an uncertainty throughout nature, though not the everyday physical uncertainty caused by imprecise measurements or the use of techniques that cannot measure every factor in a complex situation.

In 1927, the German physicist Werner Heisenberg (1901–1976) showed that there was an, inbuilt uncertainty in the ability to measure the state of any small particle, such as an electron or a photon, however accurate the instrumentation. His **uncertainty principle** has since been developed beyond the simple problem of measurement into a statement about the fundamental nature of the Universe.

Heisenberg imagined an experiment to measure the momentum and position of, say, an electron. One way of doing this would be to use an 'imaginary microscope' that could see the electron. This could only happen if a photon hit the electron and bounced back into the eyepiece. But the photon carries momentum and by its interaction with the electron would exchange some (unknown) fraction of its momentum, so that the electron would no longer have its momentum.

You could try to make this uncertainty in the momentum very small by using a photon of very small momentum. But a photon is also a wave that extends over space, having a characteristic wavelength λ. The position of the electron would not be determined to an accuracy better than the value of λ, and the smaller its momentum the larger is the wavelength of the photon. This means that by using a low momentum (low energy) photon we improve our knowledge of the electron's *momentum*, but lose accuracy in determining its *position*.

Fig 26.16 **Heisenberg's imaginary experiment. You try to 'see' an electron using a photon microscope. But the photon has momentum and collision moves the electron, changing its position *and* momentum. Thus, we cannot be certain of its 'observed' velocity or position**

1   2

The incoming photon has momentum h/λ, where λ is its wavelength and h is the Planck constant, value 6.6×10^{-34} J s (see page 363). The photon *could* transfer all this momentum to the electron. Thus the collision has produced an uncertainty in the electron's *momentum* of:

$$\Delta p = h/\lambda$$

As explained above, the uncertainty Δx in the *position* of the electron is of the order of the light wavelength, so we can put $\Delta x = \lambda$.

Multiplying both these uncertainties gives:

$$\Delta p \Delta x = \left(\frac{h}{\lambda}\right)\lambda = h$$

This represents the best possible accuracy. In practice, we must accept that the uncertainty is always greater, so we write:

$\Delta p \Delta x \geq h$ The momentum–position formula   [1]

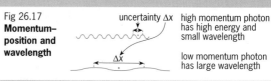

Fig 26.17 **Momentum–position and wavelength**   uncertainty Δx — high momentum photon has high energy and small wavelength

Δx — low momentum photon has large wavelength

(This argument, originally Heisenberg, is too simple and can be misleading. The uncertainty is *not* due to any method of measurement, but is in the nature of the moving whether a photon, an electron, a spaceship or anything else.)

Uncertainty in energy and time

There is also an uncertainty in the *energy* of a photon or an electron; both have a wave and a particle aspect. The energy E of a photon of frequency f, for example, is given by $E = hf$.

Now think about trying to measure this energy by measuring the frequency. Suppose our frequency measurer is able to identify one wave (as above) and so measure to an accuracy of 1 Hz, and we try to measure the frequency of a 1000 Hz wave.

In 1 second we can say $f = (1000 \pm 1)$ Hz. The uncertainty in the result is $\Delta f = 1$.

We can do better by taking a reading for a longer time, say 20 s, so that we measure 20 000 wavelengths.

Then the result could be put as $(20\ 000 \pm 1)/20$. This gives an uncertainty $\Delta f = 1/20$. So for $\Delta t = 1$, $\Delta f = 1$. For $\Delta t = 20$, $\Delta f = 1/20$.

In both cases: $\Delta f\, \Delta t = 1$ [2]

This is true, however accurate the measurement.

Now, both photons and particles are wavelike, with energy and frequency linked by $E = hf$. So we can write: $\Delta f = \Delta E/h$

and from equation 2: $\Delta f\, \Delta t = \Delta E \Delta t / h = 1$

giving: $\Delta E \Delta t = h$

Again, this is the best possible result, so in general we have an *uncertainty principle* involving energy and time that says:

$\Delta E \Delta t \geq h$ This is the energy-time formula   [3]

This means that if we wish to measure energy accurately we must take a long time to do the measuring.

7 THE AMAZING IMPLICATIONS OF UNCERTAINTY

The uncertainty principle outlined in the Extension box has some amazing consequences for particle physics. It implies that a quantum of energy can exist for a very short time, *provided that the product of energy and time is less than the value of the Planck constant*, that is, if $\Delta E \Delta t < h$.

This also applies to matter, since energy and matter are equivalent. So particles could also exist for very small times. These particles, called **virtual particles**, can be produced without breaking the law of conservation of mass-energy.

The significance of this feature of the uncertainty principle is seen as fundamental to a description of the nature of matter, as it exists now and in the early stages of the Universe.

?

K An electron has a mass-energy of 8.2×10^{-14} J. For how long, in principle, could such a particle exist without the Universe noticing? ($h = 6.6 \times 10^{-34}$ J s)

8 BOSONS, THE CARRIERS OF THE INTERACTIONS

It is the possibility of virtual particles that underlies the idea of gauge bosons as particles that 'mediate the interaction' between particles.

Newton was worried by his theory of gravity. He thought it unscientific because the force of gravity was having to act through *empty* space, so there was nothing to transmit the force. Modern physics agrees with Newton's worry, and says that the forces between particles are carried by *other particles*. These are the **exchange particles** or **bosons**. Table 26.3 on page 583 lists the bosons needed to explain the four fundamental interactions.

Photons

The most familiar boson is the **photon**, which carries the electro-magnetic force (interaction) between charged particles.

Fig 26.18 models what is believed to happen when an electron gets close to another electron and repels it. It is called a Feynman diagram, named after Richard Feynman who devised such diagrams to help illustrate exchange processes. As the electrons get closer, each emits a photon towards the other. Photons carry momentum and so the momentum of each electron changes. When we observe a change in momentum we explain it as the result of a force. At this level the concept of force is not very useful; it is better to work purely in terms of momentum changes.

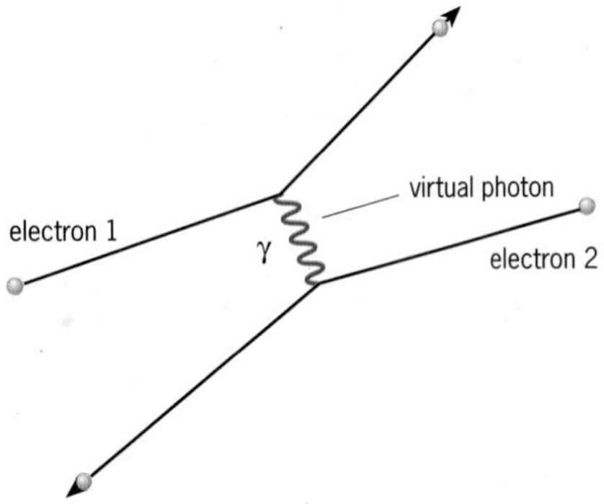

The diagram is not a picture of what happens; it just reminds us of the particles involved. It needs a very complex wave-quantum calculation to describe an interaction between particles completely.

Surprisingly, the photon is never actually observed. It is a ghost, a **virtual photon**, something perfectly acceptable on the uncertainty principle. Virtual particles are illustrated by lines which begin and end on the interacting particles. The 'real' particles are shown as labelled lines entering and leaving the diagram.

In this model, a charged particle is surrounded by a cloud of virtual photons. When the particle nears another, it exchanges virtual photons. There is an associated momentum exchange that we observe as a force causing a change in the velocity of both particles. The force mediated by photons can pull as well as push.

Fig 26.18 **A Feynman diagram of an electron repelling another electron via the exchange of a virtual photon. This diagram is not realistic: it summarises a continuous and complicated process**

W and Z particles

The *weak* interaction can be mediated by two W particles and a Z particle. Fig 26.19 shows what happens when an electron with enough energy interacts with a proton and produces a neutron and a neutrino.

The carriers of the weak force have been detected outside a nucleus. Unlike photons, they have mass, of about 90 GeV/c^2.

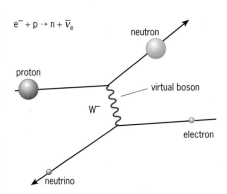

$$e^- + p \to n + \bar{\nu}_e$$

Fig 26.19 **A high-energy electron becomes a neutrino as it converts a proton to a neutron. The interaction is mediated by a virtual boson (W⁻) which exists long enough to do its job ($\Delta t = h/\Delta E$). The diagram can be read backwards and then it describes neutron decay**

Gluons

Gluons carry the *strong* force in interactions between quarks. The force holding nucleons inside a nucleus are the result of gluons leaking temporarily for quantum reasons between one quark system (a proton, say) and another. Gluons are massless.

Fig 26.20(a)–(c) show three Feynman diagrams for interactions between quarks. Bear in mind that gluons always *pull*, whilst photons can both pull and push.

■ See questions 14–17.

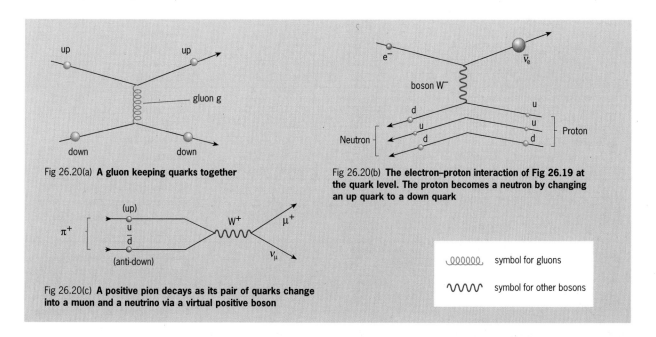

Fig 26.20(a) **A gluon keeping quarks together**

Fig 26.20(b) **The electron–proton interaction of Fig 26.19 at the quark level. The proton becomes a neutron by changing an up quark to a down quark**

Fig 26.20(c) **A positive pion decays as its pair of quarks change into a muon and a neutrino via a virtual positive boson**

⟨000000⟩	symbol for gluons
ᏙᏙᏙ	symbol for other bosons

The decay of quarks

Mesons are short lived: they contain a particle and an antiparticle. When a pion decays it forms a muon and a muon neutrino. Neither muons nor neutrinos contain quarks – they are leptons. 'Quarkness' is conserved, since a pion contains a quark and an antiquark. The decay is mediated by a W⁺ particle. This is shown in the Feynman diagram of Fig 26.20(c).

A neutron also decays through the weak interaction, so needs the mediation of a W particle, in this case a W⁻. Figs 26.19 and 26.20(b) show two versions of this – if you read them backwards! Fig 26.20(b) shows that a down quark has changed to an up quark. Again 'quarkness' is conserved, and so is baryon number.

?

L A positron–electron pair can interact to produce a pair of muons.

(a) Write down a particle equation, bearing in mind that charge must be conserved. (There are two possible answers.)

(b) Draw a Feynman diagram for the reaction, assuming that the reaction is mediated by a virtual Z boson.

Virtual particles that decay

Many new particles are produced by the collision of high-energy electrons and positrons in ring accelerators. Their production always involves the production of one or more virtual particles whose temporary and unseen existence can be inferred from the results.

Table 26.15 **A summary of the particles**

CLASS	Name and Symbol		Particles				Antiparticles				Mass (GeV/c^2)	Half-life/s	Discovered, source
		Properties	Q	B	S	Quark content	Q	B	S	Quark content			
HADRONS — Baryons	proton	p	+1	1	0	uud	−1	−1	0	$\bar{u}\bar{u}\bar{d}$	0.983	stable	1911–19 alpha scattering
	neutron	n	0	1	0	udd	0	−1	0	$\bar{u}\bar{d}\bar{d}$	0.940	900	1932 alpha bombardment of beryllium
	lambda	Λ	0	1	−1	uds	0	−1	+1	$\bar{u}\bar{d}\bar{s}$	1.115	2.6×10^{-10}	1951 cosmic rays
	sigma plus	Σ^+	+1	1	−1	uus	−1	−1	+1	$\bar{u}\bar{u}\bar{s}$	1.189	0.8×10^{-10}	1953 cosmic rays
	sigma minus	Σ^-	−1	1	−1	dds	+1	−1	+1	$\bar{d}\bar{d}\bar{s}$	1.197	1.5×10^{-10}	1953 accelerator
	sigma zero	Σ^0	0	1	−1	uds	0	−1	+1	$\bar{u}\bar{d}\bar{s}$	1.192	6×10^{-20}	1956 accelerator
	xi minus	Ξ^-	−1	1	−2	dss	+1	−1	+2	$\bar{d}\bar{s}\bar{s}$	1.321	1.6×10^{-10}	1952 cosmic rays
	xi zero	Ξ^0	0	1	−2	uss	0	−1	+2	$\bar{u}\bar{s}\bar{s}$	1.315	3×10^{-10}	1959 accelerator
	omega minus	Ω^-	−1	1	−3	sss	+1	−1	+3	$\bar{s}\bar{s}\bar{s}$	1.672	0.8×10^{-10}	1964 accelerator
	charmed lambda	Λ_c	1	1	0	udc	−1	−1	0	$\bar{u}\bar{d}\bar{c}$	2.28	2×10^{-13}	1975 accelerator
HADRONS — Mesons	pi zero	π^0	0	0	0	$u\bar{u}$ or $d\bar{d}$	0	0	0	$\bar{u}u$ or $d\bar{d}$	0.135	0.8×10^{-16}	1949 accelerator
	pion	π^+, π^-	+1	0	0	$u\bar{d}$	−1	0	0	$\bar{u}d$	0.140	2.8×10^{-8}	1947 cosmic rays
	K zero	K^0	0	0	+1	$d\bar{s}$	0	0	−1	$\bar{d}s$	0.498	$5 \times 10^{-8}, 1 \times 10^{-10}$	1947 cosmic rays
	kaon	K^+, K^-	+1	0	+1	$u\bar{s}$	−1	0	−1	$\bar{u}s$	0.494	1.2×10^{-8}	1947 cosmic rays
	J/Psi	J/Ψ	0	0	0	$c\bar{c}$	0	0	0	$\bar{c}c$	3.1	10^{-20}	1974 accelerator
	D zero	D^0	0	0	0	$c\bar{u}$	0	0	0	$\bar{c}u$	1.87	10^{-17}	1976 accelerator
	D plus	D^+	+1	0	0	$c\bar{d}$	0	0	0	$\bar{c}d$	1.87	4×10^{-13}	1976 accelerator
	upsilon	Y	0	0	0	$b\bar{b}$	0	0	0	$\bar{b}b$	9.46	10^{-20}	1977 accelerator
LEPTONS	electron, positron, e⁻, e⁺		−1	0	0	—	+1	0	0	—	0.00051	stable	e⁻1897 cathode rays e⁺ 1932 cosmic rays
	muon	μ	−1	0	0	—	+1	0	0	—	0.1056	2×10^{-6}	1937 cosmic rays
	tauon	τ	−1	0	0	—	+1	0	0	—	1.784	3×10^{-13}	1975 accelerator
	electron neutrino	ν_e	0	0	0	—	0	0	0	—	<50 eV?	stable ?	1956 nuclear reactor
	muon neutrino	ν_μ	0	0	0	—	0	0	0	—	<0.5 MeV?	stable ?	1962 accelerator
	tau neutrino	ν_τ	0	0	0	—	0	0	0	—	<70 MeV?	stable ?	not yet observed
GAUGE BOSONS	photon	γ	**Charge** 0								0	stable	1923 X-rays scattered from electrons in atoms
	W	W^+, W^-	+1, −1								83	10^{-25}	1983 proton-antiproton annihilation (CERN)
	Z	Z	0								93	10^{-25}	1983 proton-antiproton annihilation (CERN)
	gluon	g	0								0	stable	1979 electron-positron annihilation (DESY)
QUARKS	up	u	**Charge** +2/3				**Baryon number** +1/3				~0.005		1964 (theory. Observed 1968-72 electron scattering (Stanford), neutrino scattering (CERN)
	down	d	−1/3				+1/3				~0.01		
	strange	s	−1/3				+1/3				~0.1		
	charmed	c	+2/3				+1/3				~1.5		1974 inferred from existence of J/psi particle etc.
	bottom	b	−1/3				+1/3				~4.7		1977 inferred from discovery of upsilon particle
	top	t	+2/3				+1/3				~175		1994/95

See questions 18–21. ■

9 DETECTING SYSTEMS AND TECHNIQUES FOR FINDING SUBATOMIC PARTICLES

?

N Sketch the directions of magnetic field, electron velocity and the resulting force on the electron, which together make it move in a circle in the magnetic field.

The most common techniques for finding the mass and charge of a particle of matter are based on the following: a charged particle moving at right angles to the direction of a magnetic field B experiences a force which acts perpendicularly to its direction of motion and the direction of the magnetic field. Remember that a force that acts at right angles to the direction of motion of a body makes it move in a circular path.

Quarks and the weak interaction

To follow this section, you need to refer to the quark content of particles given in Table 26.15.

When a reaction is mediated by the weak interaction, one type of quark is *always* changed into another. The weak interaction is the only interaction that can do this. In all other cases, new quarks may be formed, but are formed in quark–antiquark pairs so that the total quantity of 'quarkness' remains constant. This is another way of saying that baryon number is conserved: a quark has baryon number +1/3 and its antiquark baryon number is –1/3.

For example, in the following *strong* reaction:

$$\pi^- + p \rightarrow n + \pi^+ + \pi^- + \pi^0$$

the quarks on the left side are ($\bar{u}$d) and (uud). On the right we have (udd), (u$\bar{d}$), ($\bar{u}$d) and (u$\bar{u}$). The extra quarks on the right side are a u$\bar{u}$ pair and a d$\bar{d}$ pair. In other words, if we count a quark as +1 and an antiquark as –1, the total quarkness on the left (3) equals the sum on the right.

The best known **weak reaction** is neutron decay:

$$n \rightarrow p + e^- + \bar{\nu}_e$$

On the left the quarks are (udd). On the right only the proton contains quarks (uud). Quarkness (baryon number) is conserved, but a down quark has turned into an up quark. This seems a minor result, but consider the reaction:

$$K^+ \rightarrow \pi^0 + \pi^+$$

On the left, the K^+ has quarks (u$\bar{s}$). On the right, the π^0 can be either a (u$\bar{u}$) or a (d$\bar{d}$) and the π^+ is (u$\bar{d}$). Quarkwise, the π^0 is self-cancelling. But to make the π^+, a *strange* antiquark has become a *down* antiquark. *This is the reason why strangeness is conserved in strong reactions, but may not be conserved in weak reactions.*

Whenever a quark changes type in a reaction, we can point to the weak interaction as its cause. Check that the following decays are due to weak interactions:

$$\Omega^- \rightarrow \Xi^0 + \pi^-$$

$$\Omega^- \rightarrow \Lambda + K^-$$

The radius r of the circle can be used to calculate the charge-to-mass ratio Q/m of the particle – provided its speed v is known:

$$Q/m = v/Br$$

Fig 26.21 shows the paths made by many particles as they move through a detecting device, in this case a **bubble chamber** (see page 596). The paths are curved, and particles carrying opposite charge curve in opposite directions – one clockwise, the other anti-clockwise.

Nowadays, in large physics laboratories, interpreting the images is done by computers. The tracks of the particles are detected electrically and the data fed directly into the computer. No human eye actually sees the particle tracks, let alone the particles.

Early particle detectors

The **cloud chamber** was one of the earliest detectors. You may have one at school or college (see page 393 for more details). Another cheap detector is the simple **photographic plate**. These were used a great deal in the 1940s and 1950s to investigate the cheapest source of unusual or fast-moving particles – cosmic rays.

The **Geiger–Müller detector** is described on page 394. An array of GM detectors can track the path of a particle in three dimensions, and the same principle is used in the most modern detector system, the **multiwire array** described on page 596.

M Check that the quark total in the following reaction is the same on both sides:

$$K^- + p \rightarrow \Sigma^- + \pi^+$$

Is strangeness conserved in this reaction?

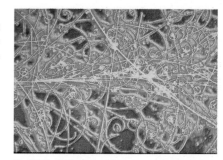

Fig 26.21 **Bubble chamber photo with tracks of numerous particles, curving when the particles are charged**

O Give the basic physical reasons why **(a)** more massive particles, **(b)** less highly charged particles move in curves of larger radii in a magnetic field

Modern particle detectors

1 The bubble chamber

A bubble chamber, the main tool for particle studies in the 1950s and 60s, is a tank filled with 'superheated' liquid hydrogen. The liquid hydrogen is under pressure to keep it just on the point of boiling without actually doing so. When the pressure is slightly reduced, bubbles may form – think of opening a can of fizzy drink.

But the bubbles need something to trigger them, just like the water drops in a cloud chamber. Any particle moving through the liquid creates ions, round which tiny bubbles form and show the particle's path. The tank is in a strong magnetic field, so the charged particles follow curved paths. Fig 26.21 shows some paths of particles in a bubble chamber.

The drawback of the bubble chamber is that it only produces a bubble track when the pressure on the liquid is slightly reduced. This has to be timed just *before* the particle enters it – and as its arrival is unpredictable, it is easy to miss a particle of interest. Then, the pressure has to be increased to restore the bubble chamber to a bubble-free state. The cycle takes about a second, which is a slow process by modern standards.

Speeding up the detecting process

Spark detectors were designed to avoid the problem of the random picture-taking and long recycling time of bubble chambers. The detectors are arranged as a set of thin metal plates spaced closely together in an inert gas, forming a **spark chamber**, as in Fig 26.22(a). When a charged particle passes through, the gas along its track is ionised.

On each side are separate Geiger–Müller tubes which check that a particle of interest has passed into the spark chamber. If so, they immediately trigger a high voltage between the plates. The ions of the particle track are still there and they 'cascade' to produce a spark. At the same time, a camera shutter opens to photograph the spark track, as in Fig 26.22(b). The sparking clears the ions away very quickly when the voltage is switched off, so that the chamber is ready again in a fraction of a second.

2 Multiwire chambers

Multiwire chambers are a development of the spark chamber in which the plates are replaced by an array of many thin wires. A particle's trail is revealed by a pulse of the ions it produces moving on to the nearest charged wire. The wires are very close together and

Fig 26.22(a) **A spark chamber. The GM tubes act as 'coincidence' detectors. A particle has to pass through one in each row to trigger the operation of the spark chamber**

Fig 26.22(b) **The spark chamber at CERN in operation**

See question 22. ■

Fig 26.23 **Detection parts of a multiwire chamber. The detectors are stacked either in planes or in cylinders with wires between, to give a three-dimensional track trace**

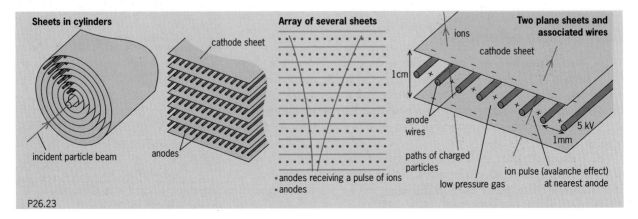

are connected separately to a computer recording system. They act like very many Geiger–Müller detectors in one large tube.

An improvement on the simple multiwire chamber is the **drift chamber** (Fig 26.24). Sensing is by separate wires: the ions drift onto them under the action of an electric field produced by the other wires in the array. The drift speed of the ions is known, and a computer calculates the origin of the ions from the time they take to drift to the sensor.

3 Total detection systems

The most modern detection system is a LEP detector, which uses a **L**arge **E**lectron–**P**ositron accelerator. Several different kinds of particle detectors surround the place in the accelerator in which particle-producing collisions occur. The detectors feed data into a computer which produces a picture of the paths, and can do this from any angle of view. The system also compares the energy and path curvature with those of known particles, so it detect any *new* particle not in its memory. A typical experiment with a LEP detector, with a collision event lasting a millionth of a second or so, might produce millions of particles, so the system needs to be automated.

Fig 26.25(b) shows the general arrangement of detectors round the collision chamber at the European Particle Physics Laboratory at CERN, Geneva, where particles are produced.

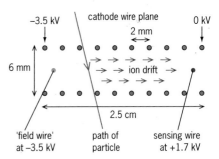

Fig 26.24 **One of the many sensing elements in a drift chamber. The varying voltages on the cathode wires cause ions to drift towards the sensing wire at a steady speed. The arrival of a particle triggers a timing circuit and the times taken for ions to arrive at the sensing wires are used to calculate its track**

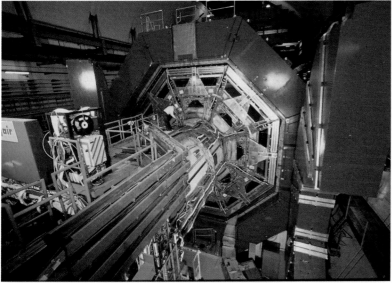

Fig 26.25 **The beam tube from CERN's 27 km LEP accelerator ring enters the L3 particle detector. (The doors are open for maintenance.) See photo 26.26(b) for a view of the other side of a detector array**

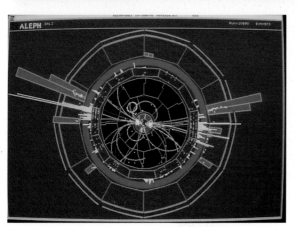

Fig 26.26(a) **An image produced by the ALEPH detector computer, generated from the data reaching the detector systems: the decay of a Z⁰ boson (a carrier of the electroweak force), via a quark–antiquark pair**

Fig 26.26(b) **The ALEPH detector at CERN with its cover removed. Several types of practice detector surround the central area in which particle-producing collisions occur**

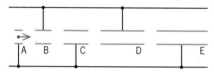

0 V   −V (V)

12 cm

charge⁺ Q

electrodes: hollow copper cylinders

Fig 26.27(a) **Part of a linear accelerator, like that at Stanford Linear Accelerator Center. The charged particle is accelerated by the field between the electrodes. It gains kinetic energy equal to the electrical potential energy transferred, QV**

Fig 26.27(b) **The particle moves from one electrode to the next in the linear accelerator: the same voltage difference is switched from A to B, B to C, and so on, so that the particle is always being accelerated. Why must each electrode be longer than the preceding one?**

Fig 26.28 **The 3 km building housing the accelerator at Stanford Linear Accelerator Centre, California**

10 MODERN PARTICLE ACCELERATORS

Modern particle accelerators are of three main kinds: linear accelerators (linacs), cyclotrons and synchrotrons.

1 Linear accelerators

In a **linear accelerator**, charged particles are accelerated along a straight evacuated tunnel by a set of switched anodes which produce a moving accelerating field, as in Figs 26.27(a) and (b). They are cheaper than accelerators with circular tracks because they don't need bending magnets, and cost depends mostly on that of the coils and special steel required for the magnets.

The most powerful linear accelerator (linac), at the Stanford Linear Accelerator Center, California, is 3 km long and accelerates electrons up an energy of 50 GeV. The electrons come from a heated filament, just as in a TV set. Then they pass through a chain of 100 000 electrodes. The electrons leave one cylinder and are pulled towards the next by a comparatively small difference in voltage set up between the pair of electrodes at just the right time. The electrons are made to move close to the speed of light by very rapid voltage switching. This is done by feeding a high frequency radio signal to each electrode at just the right (but always changing) frequency to make the electrode more positive than the one before it. In effect, the electrons are surfing along this radio wave.

Linear accelerators (linacs) are both cheaper than circular accelerators, and much more efficient at feeding energy to electrons. This is because when charged particles are accelerated they emit radio waves and so lose energy, and when the particles move in a circular path they are always being accelerated towards the centre of the circle, even if they move at a constant speed round it.

The radiation emitted by accelerating particles is called *synchrotron radiation*. Since electrons have such a large charge compared to their mass, they lose a large fraction of their kinetic energy moving at high speeds in a circle. The effect is less for more massive particles such as protons.

2 Circular accelerators – cyclotrons

In **cyclotrons**, a magnetic field forces charged particles to move in a circular path (Fig 26.29). The voltage on the D-shaped parts alternates so that the particles accelerate from one D to the other.

Fig 26.29 **The cyclotron. The potential of each D-shaped part alternates, so the positive particle is attracted from one part to the other, accelerating and spiralling out from the centre**

poles of strong electromagnet above and below D-shaped part

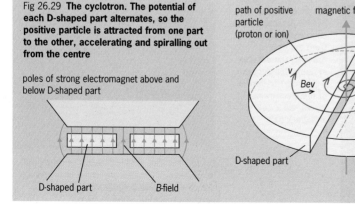

D-shaped part   B-field

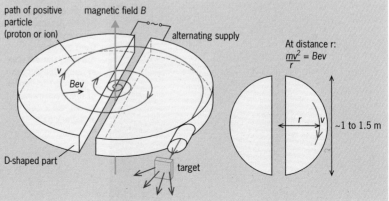

path of positive particle (proton or ion)

magnetic field B

alternating supply

At distance r:
$$\frac{mv^2}{r} = Bev$$

D-shaped part   target

~1 to 1.5 m

The particles gain speed, so the path is actually an outward spiral. The maximum energy depends on how many times they can spiral before leaving the machine. Cyclotrons provide energetic particles for medical treatment and for the testing of advanced industrial materials.

3 Synchrotrons

At very high energies, particles become more massive because of relativistic effects near light speed. So in a cyclotron they take longer to complete a circuit and then the alternating voltage that accelerates them fails to synchronise with them.

Synchrotrons overcome this problem: the frequency of the alternating voltage changes to match the increasing flight time of the particles. But at high energies the particles are moving at speeds so close to that of light that they actually make a circuit in a constant time: instead of getting faster, they just gain more momentum via an increase in mass. This effect underlies the design of synchrotrons.

They are very large machines. Particles are accelerated by passing through hollow anodes as in linear accelerators, but the track is circular. With a curved track, the particles can keep being accelerated without running out of track (as they eventually do in a linear accelerator) and in modern synchrotrons they keep moving in the circular path indefinitely, in *storage rings*.

Particles are first accelerated by smaller linacs or cyclotrons, then enter the synchrotron already travelling very close to light speed. They are held in a path of constant radius by strong magnetic fields, which are made even stronger as the particles gain more kinetic energy.

Most modern discoveries about fundamental particles come from **super-synchrotrons**. Their radius is very large, and superconducting coils are used to activate the electromagnets. Very large currents in superconducting coils avoid the heating effects that ordinary conductors would produce.

■ See question 24.

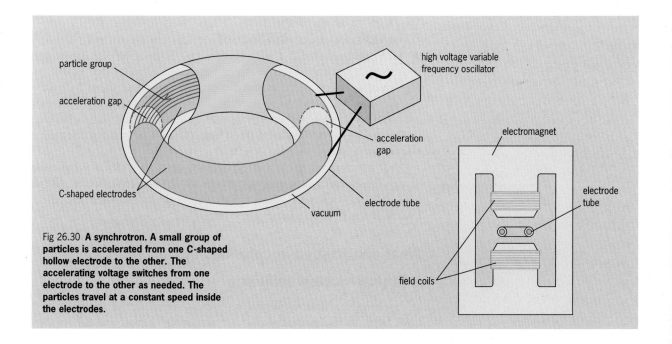

Fig 26.30 **A synchrotron. A small group of particles is accelerated from one C-shaped hollow electrode to the other. The accelerating voltage switches from one electrode to the other as needed. The particles travel at a constant speed inside the electrodes.**

11 TARGETS AND COLLISIONS

Fixed and moving targets

High energies are required to investigate subatomic particles. In a collision, the 'spare' energy appears as a flux of particles. The larger the mass of a newly created particle, the greater the energy needed, which depends on the total accelerating voltage available.

In the late 1970s, the physicists at CERN came up with a brilliant idea. If you could fire two particles in opposite directions and make them collide, you would double the available energy at a stroke, without increasing the maximum kinetic energies of the particles. This is the principle used in the **large electron–positron collider** (LEP collider), seen in Fig 26.31(a), and in the proton–antiproton **super proton synchrotron** at CERN.

The forces used to accelerate and bend particles and antiparticles have the same effect on both, except that particles move and curve in one direction and antiparticles in the other.

Fig 26.31(a) **The arrangement of magnets and accelerating electrodes for the LEP collider at the European Particle Physics Laboratory at CERN, Geneva**

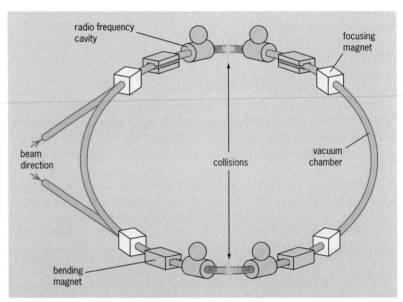

Fig 26.31(b) **Aerial view showing where the CERN rings run underground. The LEP collider is 27 km in circumference. The smaller circle marks the route of the super proton synchrotion**

The LEP accelerates each beam of electrons and positrons to about 90 GeV, giving a total available energy of 180 GeV. In the super proton synchrotron, each beam of particles is accelerated and kept in a storage ring 6.9 km in circumference at an energy of 315 GeV per particle. The beams are then made to cross so that the particles collide with a combined energy of 630 GeV. To get the same effective energy with protons hitting a fixed target would need an accelerating voltage of 150 000 GeV!

Only a small fraction of the particles actually collide when the beams cross. But as the beam contains 10^{14} particles per cm^3, there are enough collisions. The particles that do not collide are then available for the next pass.

What can happen in collisions at high energies

1 Proton–proton collisions

Hit an atom hard enough and it emits photons as the excited atom goes back to its ground state. Hit a proton (a lot harder!): it also becomes excited, and a *very* short time later emits **pions**. Unlike photons, a pion is unstable and decays very quickly, in 10^{-8} s, into a muon and a neutrino. The muon (a kind of giant electron) decays more slowly, in 2×10^{-6} s, into an electron and a neutrino. The Universe is surprisingly full of neutrinos – which don't decay: about a thousand billion (10^{12}) pass through your body every second, at the speed of light.

Fig 26.32 illustrates a simple and fairly low energy collision between two protons. Fig 26.33 shows a more complex event in which collision produces a starburst of tracks, showing the production of pairs of pions.

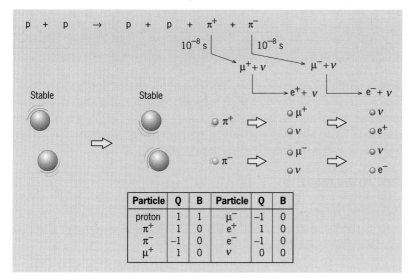

Particle	Q	B	Particle	Q	B
proton	1	1	μ^-	-1	0
π^+	1	0	e^+	1	0
π^-	-1	0	e^-	-1	0
μ^+	1	0	v	0	0

> **?**
> **O** Use the data in the table in Fig 26.32 to check that:
> **(a)** charge is conserved,
> **(b)** baryon number is conserved.

Fig 23.32 **A simple collision with just enough spare energy to create two new particles**

Fig 26.33 **Bubble chamber particle tracks. Pion tracks fan out, oppositely curved for each in a pion pair. Spirals are due to electrons and positrons**

2 Proton–antiproton collisions

When protons and antiprotons collide, they *annihilate* each other. The energy available is therefore not just their acquired kinetic energy but also the energy due to their rest masses. The rest mass energy is small, however, compared with the kinetic energy for particles travelling close to the speed of light – and is usually ignored (see below).

3 Decays

The only **stable particles** we meet in particle physics are the proton, electron and neutrino (and their antiparticles). When a particle decays it must obey the conservation laws. Charge, mass-energy, momentum and baryon number are always strictly conserved. Strangeness can be 'lost' if the weak interaction is involved.

Mass-energy
In any decay, the new particles must always have less total rest mass than the original particle that decays. Any mass-energy left over will appear as kinetic energy in the new particles.

Momentum
In simple decays, we assume that the original particle is at rest and so has zero momentum. This means that the particles must move off in the particular directions and at the particular speeds which combine to produce zero momentum.

Why stable particles are stable

The stable particles are stable because there is no particle with smaller mass that they can decay into without breaking a conservation law. *There is no baryon with less mass than a proton*, so it is stable. It could decay into pions and still conserve energy, charge and momentum – but this would not conserve baryon number. Pions are mesons with zero baryon number and so the decay is forbidden. Similarly, electrons could decay into neutrinos but for the fact that charge has to be conserved. (Refer back to Table 26.15 for categories of particles.)

See question 23. ■

13 MASS, ENERGY AND TIME AT RELATIVISTIC SPEEDS

When particles approach the speed of light, their effective (inertial) mass increases. Thus both their kinetic energy and momentum can increase without much change in their speeds. This fact is a consequence of the **special principle of relativity**, discussed on page 557. Here we will simply apply the results.

Relativistic mass

The mass of a particle at rest is its rest mass, m_0. This is the value quoted in reference books. For instance, the mass of an electron is 9.11×10^{-31} kg. The inertial mass of the particle moving at a speed v is however given by the formula:

P Use the mass–speed relationship and a calculator to find how many times more massive a particle becomes when it is moving at 0.99 c.

$$m = \frac{m_0}{\sqrt{1 - \frac{v^2}{c^2}}} = m_0\left(1 - \frac{v^2}{c^2}\right)^{-\frac{1}{2}} = m_0\gamma$$

γ is the Lorentz factor, symbol γ (see page 564).

Energy and momentum at (almost) the speed of light

In high-energy physics, energy calculations are simplest when we use the particle's **momentum**, p. The **total energy** of the particle, in relativity theory, is always and exactly $E = mc^2$ where m is the relativistic mass: $m = E/c^2$. For a particle moving at speed v, its momentum p is mv, hence:

$$p = mv = Ev/c^2$$

or: $$E = pc^2/v \qquad [1]$$

Now, for particles with very high energies, $v \approx c$, so:

$$E = pc \quad \text{and} \quad p = E/c \qquad [2]$$

Momentum units

Equation 2 means that:

$$\text{momentum units} = \text{energy units}/c$$

Just as particle physicists measure *mass* in units of GeV/c^2, so they measure *momentum* in terms of units GeV/c.

Equation 2 ignores the rest mass energy m_0c^2 of the particle. This means it also applies to photons, which have a rest mass of zero.

EXAMPLE

Q In the electron–positron collider at CERN, the particles are accelerated to energies of 50 GeV. What percentage error is introduced into calculations by assuming that the rest masses are zero?

A The rest mass energy of an electron is:

$$\text{mass of electron} \times (\text{speed of light})^2$$
$$= 9.1 \times 10^{-31} \text{ kg} \times (3 \times 10^8)^2 \text{ m}^2 \text{ s}^{-2}$$
$$= 8.2 \times 10^{-14} \text{ J}$$

An electronvolt is 1.6×10^{-19} J, so the rest mass energy:

$$= (8.2 \times 10^{-12})/(1.6 \times 10^{-19})$$
$$= 5.1 \times 10^5 \text{ eV} = 0.51 \text{ MeV} = 0.000 \, 51 \text{ GeV}$$

Thus: percentage error $= 0.000 \, 51/50 \times 100 = 0.001\%$

Massive particles at speeds close to the speed of light: the momentum-energy formula

For particles large enough for their rest mass to be significant, we need to use a more complete expression relating the conservation of total energy E and momentum p. This is:

$$E^2 = p^2 c^2 + m_0^2 c^4 \qquad [3]$$

> ✔
>
> Formula 3 is also relativistic, and its proof is straightforward – if you need to know it. Question 25 leads you through the proof.

Decays

You need to use equation 3 to calculate the mechanical outcome of particle decays.

EXAMPLE

Kaon decay

Q A positive kaon decays in about 10^{-8} s as follows:

$$K^+ \rightarrow \pi^+ + \pi^0$$

With what energy and at what speed do the pions move away?

A Suppose the kaon was at rest when the decay occurred. The rest mass of the kaon is 0.497 GeV/c^2. The total energy of the kaon has to supply the total energy of the pions, that is, both rest mass (of 0.140 GeV/c^2) and kinetic energy.

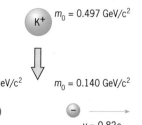

Fig 26.34 **A kaon decaying into two pions. Charge, energy and momentum must be conserved**

$m_0 = 0.497$ GeV/c^2

$m_0 = 0.140$ GeV/c^2 $m_0 = 0.140$ GeV/c^2

$v = -0.82c$ $v = 0.82c$

π^+ π^-

The pions have equal mass and share the kaon energy between them. Momentum is conserved, so the combined momentum of the pions must be zero: the momentum of the pions must be equal and opposite, and they must move apart with the same speed and the same kinetic energy.

Total energy for each pion $= 0.497/2 = 0.249$ GeV
We now use the momentum-energy formula to calculate the momentum p of a pion:

$$E^2 = p^2 c^2 + m_0^2 c^4$$

Rearranging: $\quad p^2 c^2 = E^2 - m_0^2 c^4$

So: $\quad p^2 = E^2/c^2 - m_0^2 c^2$

The rest mass m_0 of each pion is 0.140 GeV/c^2.

So: $p^2 = [(0.249 \text{ GeV})^2/c^2] - [(0.140 \text{ GeV}/c^2)^2 \times c^2]$
$= (0.249^2)(\text{GeV}^2/c^2) - (0.140^2)(\text{GeV}^2/c^4) \times c^2$

and: $\quad p^2 = 0.0422 \text{ (GeV}/c)^2$

giving: $\quad p = 0.205 \text{ GeV}/c$

TIP! Note that care needs to be taken with units – it is best to 'work them out' fully, as shown in the Example. If you find yourself multiplying or dividing by 9×10^{16} you are probably getting it wrong!

We can see how useful the chosen units are by finding the speed of each pion using equation 1 on page 602:

$$E = \frac{pc^2}{v}$$

that is: $\quad v = \frac{pc^2}{E} = 0.205 \times \frac{\text{GeV}}{c} \times c^2 \times \frac{1}{0.249 \text{GeV}}$

$$= \frac{0.205c}{0.249} = 0.82c$$

Collisions

In a collision, any new particles carry away kinetic energy *and* their rest mass energy. Both energies are provided by the total energy of the colliding particles.

We shall consider two situations involving a proton and an antiproton. In the first, the moving antiproton hits the proton which is a stationary target. In the second, both particles collide while moving with equal energies towards each other (as in a collider). The total energy of each moving particle is 2 GeV. Suppose that in each case a single particle is produced as a result of the collision. What total energy will this particle have?

1 A collision with a stationary target

The total energy is 2 GeV for the moving antiproton plus the rest mass energy of 1 GeV of the target proton, total 3 GeV (see Fig 26.35).

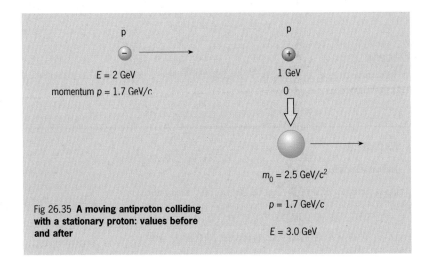

p

$E = 2$ GeV

momentum $p = 1.7$ GeV/c

p

1 GeV

0

$m_0 = 2.5$ GeV/c^2

$p = 1.7$ GeV/c

$E = 3.0$ GeV

Fig 26.35 **A moving antiproton colliding with a stationary proton: values before and after**

The total momentum is due to the incoming particle only, and is calculated (as above) from:

$$p^2 = \frac{E^2}{c^2} - m_0^2 c^2$$

which gives:

$$p^2 = 3\left(\frac{\text{GeV}}{c}\right)^2$$

and:

$$p = 1.7 \text{ GeV/}c$$

The new particle must carry away exactly this quantity of momentum. To agree with equation 3 on page 603, its rest mass m_{new} must be given by:

$$m_{\text{new}}^2 = \frac{E^2}{c^4} - \frac{p^2}{c^2}$$

$$= 3^2\left(\frac{\text{GeV}}{c^2}\right) - 1.7^2\left(\frac{\text{GeV}}{c^2}\right)^2$$

so that:

$$m_{\text{new}} = 2.5 \text{ GeV/}c^2$$

See questions 26–29. ■ This is in fact a Z particle.

Collisions in a collider

As in Fig 26.36, the energy for each colliding particle is 2 GeV, totalling 4 GeV. The total momentum before collision is zero, since each particle is moving with the same speed but in opposite directions. The momentum of the new particle must also be zero. This means that its *kinetic energy is zero*, so all the energy must appear as the rest mass of the new particle: which is thus 4 GeV/c^2.

This shows that the collider system is able to create particles of greater rest mass than the stationary target system is able to.

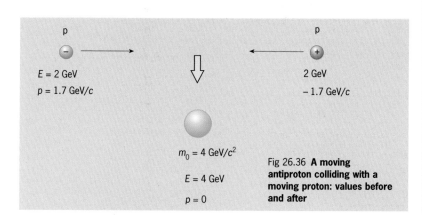

p — $E = 2$ GeV, $p = 1.7$ GeV/c

p — 2 GeV, -1.7 GeV/c

$m_0 = 4$ GeV/c^2
$E = 4$ GeV
$p = 0$

Fig 26.36 **A moving antiproton colliding with a moving proton: values before and after**

Key formulae for relativistic motion

Lorentz factor $\quad \gamma = \left(1 - \dfrac{v^2}{c^2}\right)^{-\frac{1}{2}}$

Relativistic mass $m = \gamma m_0$
Total energy = kinetic energy + rest energy
$$E = E_k + E_0$$
Momentum $p = mv = Ev/c^2$
Total energy $E = mc^2$
Energy-momentum formula:
$E^2 = p^2 c^2 + m_0^2 c^4$
Time dilation: $\Delta t^0 = \gamma \Delta t$

14 TIME DILATION

A muon has a half-life of about 2×10^{-6} s. This means that in a beam of muons, half have decayed in this short time, and in about 10^{-5} s, most will have decayed.

A typical muon that can be produced by a cosmic ray or a collision in an accelerator may be travelling close to light speed. Such a muon should travel a distance of about $c\Delta t$ in that time:

$$\text{distance before decay} = 2.2 \times 10^{-6} \text{ s} \times 3 \times 10^8 \text{ m}$$

$$\approx 700 \text{ m}$$

R A flux of 80 muons is produced in a cosmic ray shower. How many will have (should have!) decayed after four half-lives have elapsed?

'Long-life' muons

Cosmic ray experiments show that many muons are observed at ground level when collisions occur at a height of over 10 000 m in the atmosphere. Why did these muons live so long?

The extra lifetime is a consequence of the special principle of relativity. To a fixed observer, the observed time Δt_0 of a process occurring in a time Δt in a moving system is longer:

$$\Delta t_0 = \frac{\Delta t}{\sqrt{1 - \dfrac{v^2}{c^2}}} = \gamma \Delta t$$

As a result, a muon travelling at $0.99c$ would actually have a mean lifetime as observed by us on Earth of 1.6×10^{-5} s, that is, seven times longer. This why a significant fraction would be undecayed after travelling from high in the atmosphere.

It is this time dilation effect which allows experimenters to observe particles with much shorter lifetimes (10^{-10} to 10^{-20} s) in detectors which are several metres from the collision site (Fig 26.37).

S Check the value of 1.6×10^{-5} s for the lifetime of a muon travelling at $0.99c$.

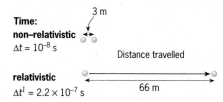

Time:
non–relativistic
$\Delta t = 10^{-8}$ s
3 m
Distance travelled

relativistic
$\Delta t^l = 2.2 \times 10^{-7}$ s
66 m

Fig 26.37 **The effect of time dilation at 0.999c. A particle with a lifetime of 10^{-8} s should travel 3 m at about the speed of light. Relativistic time dilation increases the observed lifetime to 2.2×10^{-7} s when the particle is travelling at 0.999c. Thus the distance actually travelled is 66 m**

Matter as we know it

Table 26.16 shows the two families, the quarks and the leptons, and their three generations. Each particle has its antiparticle. They make up all the matter and antimatter that has so far been discovered. But most of the Universe may consist of **dark matter**, which has not yet been identified.

getting heavier →

QUARKS	up	charmed	top
	down	strange	bottom
LEPTONS	electron	muon	tauon
	electron neutrino	muon neutrino	tauon neutrino

make up hadrons and feel the strong nuclear force

stand alone and don't feel nuclear force

⇧ These make up ordinary matter

⇧ Why do these exist?

Table 26.16 **The quarks and the leptons**

And finally...

Atoms and nuclei exist because they are made of particles *which cannot have the same set of quantum numbers while part of the same system.* The reason why is outside the scope of A-level physics.

This Chapter has shown that the rules which define the behaviour of particles are due to the fundamental properties of quarks. *But no one knows why quarks have these properties.*

The rich variety of particles exists – and just for very short times – only in conditions of high energy such as during collisions in accelerators, in the interiors of stars and also at the time of the Big Bang. Most of the Universe makes do with protons, neutrons and electrons.

The next main goal of particle physicists is to find a particle called the **Higgs boson** which is important because its interaction with other particles is believed to be the source of their **mass**. Like all bosons, it is a 'field particle', which means that it carries a force of some kind.

We observe mass as something that makes it hard to accelerate something. The harder it is to accelerate, the greater the mass of the object. But imagine trying to push a sheet of aluminium in a strong magnetic field (see page 270). You feel a force opposing your efforts, which we explain as being due to eddy currents excited in the aluminium by its motion through the magnetic field. If you didn't know about magnetism and electricity, you might confuse this effect with the 'inertial force' that seems to stop you giving instantaneous velocity to a mass. In a similar way, the **Higgs field** opposes motion by exchanging virtual bosons – just as the exchange of photons mediates electromagnetic forces. The effect is to produce an interaction of some kind that appears as inertial resistance to motion. You can

Fig 23.38 **Peter Higgs, whose theory aims to unify electromagnetism, the weak and strong nuclear forces**

think of the Higgs boson as somehow making space 'sticky'.

The theory put forward by Peter Higgs of Edinburgh University as long ago as 1964 is now seen to be very useful, as it links with the strange 'theories of everything' – superstring theory, supersymmetry and supergravity – now being explored by the 'super' theorists. The mass of the Higgs boson is predicted to be several hundred GeV/c^2, which is too massive to be revealed by current accelerators. But it will be the prime target of the new large hadron collider (LHC) at CERN.

Physics would quite like to know what causes mass – and be able to link quantum physics with Einstein's gravity theories.

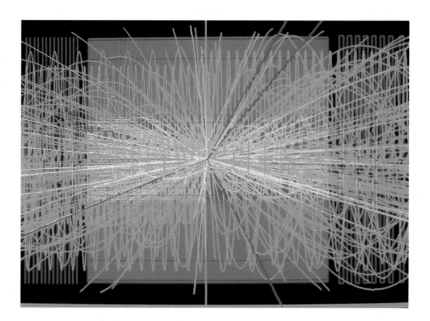

Fig 26.39 **A computer simulation of an event expected by the Higgs theory. (No accelerator is yet powerful enough to produce it.) After a proton–proton collision, a Higgs boson decays to produce four muons – straight red tracks – and many other particles**

SUMMARY

After studying this chapter, you should know and understand the following:

■ Three particles – the neutron, proton and electron – are able to explain the main behaviours of stable atoms.

■ The four types of interaction between particles require additional particles – bosons – which mediate the interactions (they 'carry the forces').

■ Many new particles have been discovered, initially from the collisions of cosmic rays (high speed protons, electrons and atomic nuclei) and later from accelerator experiments.

■ Particles and their antiparticles are classified by rest mass (baryon, meson, lepton) and by other properties (baryon number, strangeness) and their response to forces.

■ These properties were discovered by applying conservation rules in reactions.

■ The behaviour, interactions and nature of particles may be explained more simply using the quark model.

■ There are three lepton families and three quark families.

■ There are several types of accelerator, each based on a physical principle. Accelerators give particles high energies at which they interact.

■ There are several types of particle detector, each design based on a particular underlying physical principle.

■ For simple reactions there are calculations and decays involving the masses, energies and momenta of high speed, relativistic particles.

■ Conservation rules enable particle interactions to be analysed. Some use the quark model and Feynman diagrams.

QUESTIONS

Data
Electronic charge $e = 1.6 \times 10^{-19}$ C
Electron rest mass $= 9.1 \times 10^{-31}$ kg
Light speed $= 3.0 \times 10^8$ m s^{-1}
See Table 26.15, page 594, for data on particles.

1 Explain why the electronvolt is a unit of *energy*.

2 An electron is accelerated through a potential difference of 1.5×10^6 volts. Calculate the energy it acquires in **a)** joules, **b)** MeV.

3 Calculate the rest *energy* of a neutron (ie the energy equivalent of its rest mass of 1.675×10^{-27} kg) in **a)** joules, **b)** GeV.

4 Calculate the rest *mass* of an electron (9.11×10^{-31} kg) in units of **a)** MeV/c^2, **b)** GeV/c^2.

5
a) Estimate your own rest energy in **(i)** joules, **(ii)** GeV.
b) What is your rest mass in units GeV/c^2?

6 The highest energy produced in an accelerator for a particle is about 1800 GeV. Would this energy be enough to boil a kettle of water?

7 Name the four *fundamental interactions*. Experiments to find the graviton have so far been unsuccessful. Suggest a reason for this.

8 Which of the following particle reactions *fail* to conserve charge?
a) $\pi^- + p \rightarrow \Sigma^- + K^+$
b) $\pi^- + p \rightarrow n + \pi^+$
c) $n + \pi^+ \rightarrow \Lambda + K^0$
d) $\pi^- + p \rightarrow \Lambda + K^+$
e) $K^- + p \rightarrow \Omega^- + K^+ + K^0$

9 What are the main differences between **a)** leptons and hadrons, **b)** mesons and baryons?

10 In **a)** an atom of oxygen, **b)** an oxygen nucleus, how many **(i)** leptons, **(ii)** baryons are there?

11 Which of the following reactions *fail* to conserve baryon number?
a) $\pi^- + p \rightarrow K^- + \pi^+$
b) $K^- + p \rightarrow \Sigma^- + \pi^+$
c) $\Sigma^- \rightarrow n + \pi^-$
d) $K^+ + (?) p \rightarrow K^+ + \pi^+$
e) $K^- + p \rightarrow n + \Sigma^+ + K^-$

12 Each of the following reactions is forbidden. State the conservation law or laws that would be broken in each case if they were to occur.
a) $p + \bar{p} \nrightarrow \mu^+ + e$
b) $\gamma + p \nrightarrow n + \pi^0$
c) $p + p \nrightarrow p + \pi^+$
d) $p + p \nrightarrow p + \pi^+ + \pi^0 + \pi^-$

13 In which of the following is the law of conservation of strangeness broken?
a) $\Lambda \rightarrow p + \pi^-$
b) $\pi^- + p \rightarrow \Lambda + K^0$
c) $p + \bar{p} \rightarrow \bar{\Lambda} + \Lambda$
d) $\Xi^- \rightarrow \Lambda + \pi^-$
e) $\pi^- + p \rightarrow \pi^- + \Sigma^+$
f) $\Xi^0 \rightarrow p + \pi$

14 What is the main difference between strong and weak reactions with regard to **a)** time, **b)** the numbers and types of quark before and after a reaction, **c)** the conservation of strangeness?

15 All of the following reactions have been observed. Which of them is due to the strong interaction and which the weak?
a) $\pi^+ + p \rightarrow \Sigma^+ + K^+$
b) $\Omega^- \rightarrow \Xi^- + \pi^0$
c) $\Omega^- \rightarrow \Lambda + K^-$
d) $K^- + p \rightarrow n + K^0$

16 Explain what the following Feynman diagrams represent and write down an equation for any reaction:

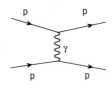

(a)

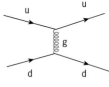

(b)

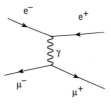

(c)

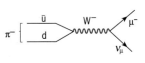

(d)

Fig 26.Q16

17 Draw Feynman diagrams to illustrate the following reactions/interactions. Where appropriate, show the quark status.

a) $e^+ + e^+ \rightarrow e^+ + e^+$

b) $\pi^+ + p \rightarrow \Sigma^+ + K^+$

c) $K^+ \rightarrow \pi^0 + \pi^+$

d) $n \rightarrow p + e^- + \bar{v}_e$

18 What are the main differences between cyclotrons and synchrotrons? Explain the reasons for these differences.

19 What are the differences in quark content between baryons, mesons and leptons?

20 Use the quark model to explain why a) antibaryons have negative baryon number, b) mesons have zero baryon number.

21 The following *strong* reactions have been observed.

$$\pi^- + p \rightarrow n + \pi^+ + \pi^- + \pi^0$$
$$K^- + p \rightarrow \Sigma^- + \pi^+$$

a) Use the data in Table 26.15 (page **XXX**) to list the quark content of each of the particles involved for each reaction.

b) Count a quark as +1 and an antiquark as –1. Make up a rule about the total 'quark content' on either side of each reaction.

c) Use the rule you have made in b) to explain why the following reaction does not occur:

$$\pi^+ + p \rightarrow \Sigma^+ + K^+ + \Lambda$$

22 Neutral particles, such as neutrons and lambdas, do not make tracks in detectors such as bubble chambers. Explain a) why this is so, b) how bubble chambers may nevertheless provide evidence for their existence.

23 There is a certain kind of particle – called a resonance particle – which decays in about 10^{-23} s.

a) What interaction (force) is likely to be responsible for the decay?

b) Are these particles likely to be hadrons or leptons? Give a reason for your answer.

24 The two particles in question 20 enter a magnetic field at the speeds that you have calculated. The field is at right angles to their motion and they begin to move in a circle. Calculate the radii of the two circles for a field of 1 mT (10^{-3} tesla).

25 This question leads you through a simple proof of the general equation which allows accurate calculation of the momentum, rest mass and total energy of a particle moving at relativistic speeds:

$$E^2 = p^2 c^2 + m^2{}_0 c^4$$

a) The total energy of a particle of mass m moving at speed v is given by $E = mc^2$. The particle has a rest mass m_0. Write down the value of E in terms of m_0 and the Lorentz factor (involving v and c).

b) Square the relationship in a) and multiply out the brackets.

c) You should have an equation with E^2 on its own and also as part of another expression. Replace the second E^2 with its equivalent: mc^2.

d) You should now have a simple equation involving m, m_0, v and c. All you have to do now to produce the formula is to use the fact that momentum $p = mv$.

Use the data given in Table 26.15, page 594, to help you answer questions 5, 6, 7 and 8.

26 An elephant on roller skates collides with a table tennis ball travelling at 50 m s^{-1}.

a) Describe the effect on (i) the elephant, (ii) the table tennis ball.

b) Estimate without detailed calculation the change in momentum of the table tennis ball (mass 5 g) and the elephant (mass 10 tonnes).

c) Choose the correct answer: The kinetic energy of the table tennis ball changes (i) a lot, (ii) hardly at all.

d) An electron (mass 0.0005 u) travelling at 10^5 m s^{-1} collides with a proton (mass 1 u) and bounces off. Does the kinetic energy of the electron change (i) a lot, (ii) hardly at all?

e) When a high speed electron ($0.99c$) hits a proton in an accelerator, it may lose a great deal of kinetic energy. Suggest some reasons why.

27 An alpha particle (mass number 4) travelling at a speed of 2×10^6 m s^{-1} collides with an oxygen nucleus (mass number 16) in a cloud chamber. The oxygen nucleus was originally at rest. Assume that the particles move in the same straight line before and after the collision.

a) Use the units given to write down the momentum of the alpha particle before the collision.

b) What is the total momentum of the particles after the collision?

c) Write down an expression linking the separate momenta of the particles after the collision. This should involve two unknown quantities.

d) Repeat the instructions in a), b) and c), this time for the kinetic energies of the particles before and after the collision. This should also produce an expression with two unknowns.

e) Solve the simultaneous equations resulting from your answers to c) and d) to find the speeds at which the alpha particle and the radon nucleus move off after the collision.

28 In a colliding beam experiment, an electron with total energy 4.7 GeV collides with a positron of the same total energy. A single new particle is produced.

a) What is the momentum of the new particle?

b) What is its rest mass energy?

c) What is its rest mass?

d) Is the new particle a lepton or a baryon?

29 Calculate the speed of a) an electron, b) a proton that have been accelerated through a potential difference of 500 V.

DEEP MATTER

This chapter map contains the main concepts of particle physics contained in this chapter. Also the important instruments that have been used to discover and build up an understanding of fundamental particles. Refer to the syllabus you are following to check the areas you need to know, and use the Chapter Map to understand the ways that the ideas interlink.

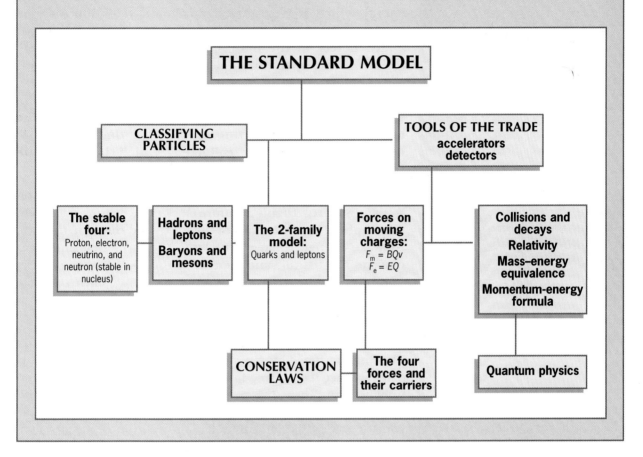

THE STANDARD MODEL

CLASSIFYING PARTICLES

TOOLS OF THE TRADE
accelerators
detectors

The stable four:
Proton, electron, neutrino, and neutron (stable in nucleus)

Hadrons and leptons
Baryons and mesons

The 2-family model:
Quarks and leptons

Forces on moving charges:
$F_m = BQv$
$F_e = EQ$

Collisions and decays
Relativity
Mass–energy equivalence
Momentum-energy formula

CONSERVATION LAWS

The four forces and their carriers

Quantum physics

27 Astrophysics

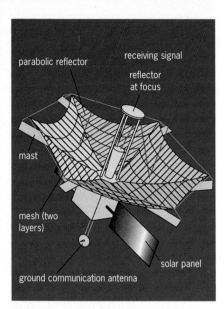

The radio telescope with a dish of Kevlar cables which orbits the Earth

ALL WE KNOW about stars is based on tiny quantities of energy which arrives as electromagnetic radiation carrying information about them. This is collected by telescopes which detect in the optical, radio, microwave and X-ray regions of the spectrum. Telescopes have to be as large as astronomers can afford to make them. The larger they are, the more energy they collect and the more detail there is in the images they make. The longer the wavelength of the radiation, the harder it is to see fine detail, so more information has to be collected, which is why radio telescopes are usually very much bigger than optical ones.

But several radio telescopes can combine to collect radiation from one source. Highly accurate clocks synchronise data from telescopes at opposite ends of the Earth and combine the data to make a single 'image'. The largest telescope on Earth has one end in Australia and another in the USA, forming a Very Long Baseline Interferometry system (VLBI). Each collector can be quite small, perhaps ten metres or so across.

In early 1997, the very largest VLBI started operating. It has a baseline *twice the diameter of Earth*, because it is in Earth orbit. The dish, designed by Japanese radio astronomers, weighs just 226 kg. Since it is in free fall, and is 'weightless', its dish need only be very flimsy. It is made of very strong Kevlar wire and looks like a string potholder. To fit into the launch rocket nose cone, it folded like an umbrella, and then in orbit, its radial arms extended, pulling the fine wires into a dish shape 8.4 m across. It sends data at a rate of 130 megabits per second to ground stations in Japan, USA, Spain and Australia. Jodrell Bank in England is involved in processing the data.

Gas clouds in the Eagle nebula, taken by the Hubble Space Telescope

In this chapter, we look first at the main features of the Solar System – the Sun and its planets, asteroids and comets and how they were formed. We then study the Sun as a typical star, the only star close enough to study in detail.

Our knowledge and understanding of the nearby Universe is based largely on the information of different forms of electromagnetic radiation. Though the nearest stars are mostly much larger and brighter than the Sun, they are so far away that they appear as tiny pinpoints of light, even in a large telescope. Yet we are able to gain a surprising amount of information from the tiny amount electromagnetic radiation reaching us.

Astronomers have fitted the different kinds of star into a pattern that links their sizes, brightness and ages. This pattern is the **Hertzsprung-Russell diagram**. The brightest stars we see are relatively close and lie in our own branch of our galaxy, which is called the **Galaxy** and the **Milky Way** and contains 100 billion stars. Galaxies in general will be dealt with in the next chapter, 28 Cosmology.

The physics of stars is mostly to do with their sources of internal energy, and how they reach equilibrium by matching energy emission with the rate of energy production from nuclear and gravitational processes.

Finally, we consider what happens when a star's supply of energy is running out, and its steady state turns into the dramatic processes that produce red giants, white dwarfs, supernovae, neutron stars and black holes.

1 THE SOLAR SYSTEM

The planets move round the Sun in elliptical orbits, held in their paths by the gravitational force between them and the Sun. The Solar System also includes many smaller rocky objects called **asteroids** which form a belt between the orbits of Mars and Jupiter.

Periodic comets

There are also **periodic comets,** such as Halley's comet. Comets are much smaller than planets and are made of dust and the 'ice' of water and other gaseous elements (Figs 27.1 and 2). Periodic comets move in very large orbits which extend far beyond the orbit of Pluto, the furthest planet. Comets may have their origin in a region filled with a large number of objects (estimated to be 10^{12}) called the **Oort Cloud** (see Fig 4.1, page 56). The Oort Cloud is more than a thousand times more distant than Pluto, at the outer limit of the Solar System.

The total mass of the Solar System is about 2.0×10^{30} kg. Some 99.9 per cent of this is concentrated in the Sun. The largest planet, Jupiter, for example, has a mass of 1.899×10^{27} kg – less than 0.1 per cent of the System's mass.

The numbers are astronomical

Astronomy requires very large numbers when distances are measured in standard SI units like metres and kilometres, so astronomers also use two other units of distance – the astronomical unit and the parsec. Popular astronomy also uses the light-year.

The **astronomical unit**, AU, is the mean distance of the Earth from the Sun, 150×10^9 m. On this scale, Jupiter is just over 5 AU from the Sun and Pluto about 40 AU. The AU is a convenient unit for astronomers as it is the basis of a method for *measuring* astronomical distances using parallax, described on page 12. The AU gives rise to the second unit, the **parsec**, pc. Parsec is short for 'parallax second' and is the distance of a star which appears to shift its position in the sky by 1 second of arc as the Earth moves from the end of one diameter of its orbit to the other. It has a value of 2.06×10^5 AU (3.1×10^{16} m).

The **light year**, ly, is the distance travelled by light in 1 standard year, 9.5×10^{15} m. 1 pc is about 3.3 ly.

To get an idea of the scales measured by these units: the *nearest* star to Earth is Proxima Centauri, a very faint companion star of the brightest star in the constellation of the Centaur. It is at a distance of 4×10^{16} m, which is the same as 4.2 ly or 1.3 pc.

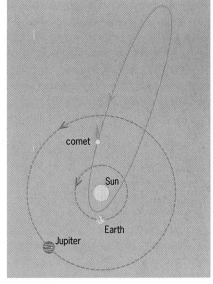

Fig 27.1 **A typical orbit of a comet is a highly eccentric ellipse**

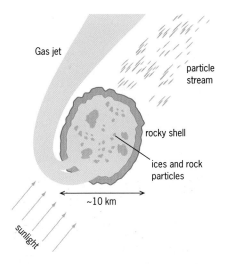

Fig 27.2 **The structure of a comet head and the effect of sunlight**

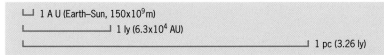

Fig 27.3 **Astronomical distance units (not drawn to scale)**

See questions 1 to 4. ■

?

A An astronomy book states: The orbital distance of the planets from the Sun follows a regular spacing: roughly, each planet lies twice as far out as the preceding one.
Use Table 27.1 to check this statement.

2 THE MAIN FEATURES OF THE SOLAR SYSTEM

Table 27.1 gives the main physical data of the Sun and planets. The planets form two groups, the **inner planets** and the **outer planets**. It is not easy to show them all to the same scale on a diagram, but with some artistic licence Fig 27.4 illustrates their size. More information is given in the Feature box.

Table 27.1 **Data on the planets**

Planet	Mean dist. from Sun (10^9m)	Mean dist. in A (Earth=1)	Period (Earth years)	Rotation period (hours)	Inclination of equator to orbital plane (degrees)	Surface temp (°C)	Density (kg m^{-3})	Mass (Earth = 1)	Surface g-field (N kg^{-1})	Number of satellites	Name of largest satellite	Mass of largest satellite (kg)	Mean density of satellite/s (kg m^{-3})
Mercury	58	0.4	0.24	1416	0	350	5400	0.056	3.7	0			
Venus	108	0.7	0.61	5832	117	460	5300	0.815	8.9	0			
Earth	150	1.0	1	24	23	288	5500	1	9.8	1	Moon	7.2×10^{22}	3340
Mars	228	1.5	2	25	25	20	4000	0.11	3.8	2	Phobos	9.6×10^{15}	2200
Jupiter	778	5.2	12	10	3	−23	1300	317.9	24.9	16	Ganymede	1.5×10^{23}	2000
Saturn	1427	9.5	29	10	27	120	700	95.1	10.5	17	Titan	1.3×10^{23}	1800
Uranus	2870	19.1	84	11	98	−180	1600	14.56	8.8	15	Oberon	6.0×10^{21}	1600
Neptune	4497	30.0	165	16	30	−220	2300	17.24	11.2	8	Triton	2.2×10^{22}	2000
Pluto	5900	39.3	248	154	98	−230	2000	0.0018	0.6	1	Charon	2.2×10^{21}	2400

Earth data: Mass = 5.97×10^{24} kg   Radius = 6.38×10^6 m

THE PLANETS IN THE SOLAR SYSTEM

SPACE PROBES AND SATELLITES are providing remarkable pictures of the planets and information about their composition. The following are brief descriptions of them.

As the nearest planet to the Sun, **MERCURY** travels very fast round it, completing an orbit in 88 Earth-days. It is affected by strong tidal gravitational forces from the Sun which gives it a rotation (spin) period two-thirds of its orbital period. Its surface is Moon-like, with many impact craters and a temperature ranging from a very cold 100 K at night to a very hot 700 K during the day. It has a very thin atmosphere, mainly of hydrogen.

EARTH the only planet with life and abundance of (liquid) water, facts which are connected. See Chapter 23 for more about this interesting planet.

VENUS is almost the same size as Earth and not much closer to the Sun. It would be quite Earth-like if its atmosphere was cool enough for rain! As a dry planet, it kept its carbon dioxide (98 per cent). The greenhouse effect helps to produce a mean surface temperature of 700 K. It has an atmospheric pressure 95 times that of Earth. Clouds of concentrated sulphuric acid hide the surface, but it has been photographed by satellite landers (short lived!) and by radar.

MARS is much smaller than Earth with a low surface gravity field strength and a correspondingly small escape speed. Hence, it cannot hold a dense atmosphere. Its surface is marked not only by impact craters but also by water channels. The little water that still exists forms polar icecaps. There may be larger quantities underground. Mars is the last of the terrestrial (Earth-like) planets, and the next most likely to support life.

Fig 27.4 **The Solar System planets**

JUPITER is the largest planet and is typical of the Jupiter group of planets formed of 'gas' as opposed to 'rock'. Its surface is likely to be liquid hydrogen. It is hidden by a very active banded atmosphere with the famous Red Spot, a storm 40 000 km long. The outer layers of metallic and liquid hydrogen are insulating, and so Jupiter may have a hot solid core (20 times the mass of Earth), made of silicates. Jupiter is just too small to have become a star and joined the Sun as a binary star system.

SATURN is the planet with the large ring system easily visible from Earth. It is the second planet of the Jupiter group, and is very similar, being slightly smaller but with a very low density (less dense than water). Saturn's rings are very thin – just a few kilometres – but extend to 140 000 km from the surface. They are made of small particles of water ice or ice-covered rocky grains. The Voyager spacecraft discovered new rings round Saturn, and also discovered rings round Jupiter, Uranus and Neptune.

URANUS was first discovered in 1781 by William Herschel who was making bigger and better telescopes. It is likely to have an Earth-sized rocky core surrounded by ice-rock slush and a dense atmosphere of hydrogen and helium with extra water, ammonia and methane. Its axis of rotation is unusual, being along the line Uranus-Sun.

NEPTUNE is the first planet discovered in 1846 when Newton's laws were applied to finding why the orbit of Uranus wobbles. Little is known about Neptune. It has an atmosphere of hydrogen and helium with a large permanent storm system called the Great Dark Spot (above white spot). It emits three times as much energy as it gets from the Sun, so it must have a hot core.

PLUTO was not discovered until 1930, and its satellite or twin-planet **CHARON** in 1978. Both are rocky planets covered with frozen methane. Its orbit is eccentric (strongly elliptical) and at times it is closer to the Sun than Neptune.

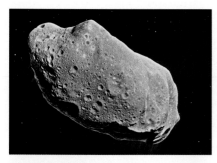

Fig 27.4 **Asteroid Ida, one of the bodies in the belt beyond the inner planets. It is 52 km long and is thought to have originated from a larger body that broke up. The picture was taken by spacecraft Galileo while on its way to Jupiter**

The inner planets

The four inner planets are Mercury, Venus, Earth and Mars, which all lie within 1.5 AU of the Sun. Then there is a belt of 'minor planets' and asteroids (Fig 27.4) at an average distance of 2.8 AU, containing an estimated 50 000 objects. The inner planets and the asteroids are *rocky bodies* made of materials similar to the material on and in the Earth. Their cores are iron and nickel and they have lighter elements (mainly in compounds) such as oxygen, silicon, aluminium, magnesium with smaller quantities of potassium, sodium and calcium.

Fig 27.5 **Plots for the planets: (a) Left: masses (log–log), (b) Right: densities, against distance (log–linear) from the Sun**

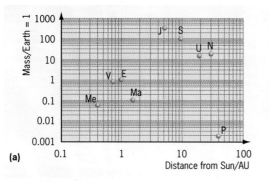

(a)

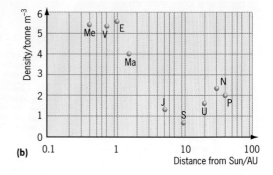

(b)

?

B Put the data in Table 27.1 into a spreadsheet. Use the data to investigate how such things as temperature, orbital period and rotational period vary with distance from the Sun for all the planets. Comment on any patterns you find.

C Give **(a)** two main chemical properties, **(b)** two main physical properties of the planets in the Solar System.

See question 5. ■

The outer planets

The outer planets, called the Jupiter group, are Jupiter, Saturn, Uranus and Neptune, plus an oddity, the 'double planet' Pluto-Charon. These planets are much further away from the Sun than the inner planets and are much larger, both in size and in mass. They are also much less dense, suggesting that their composition is different from the inner planets' composition. Their bulk must be mostly the light elements hydrogen and helium, which are gases at the planets' surfaces, but liquid or even solid at depth.

Pluto was discovered only in 1930: it is very small and is likely to be a rocky planet.

Planetary motion

All the planets are in orbit round the Sun and move in the same direction – anticlockwise when viewed from our North Pole. Their paths all lie close to the same plane.

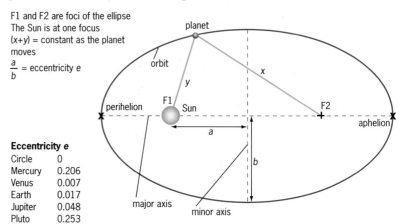

F1 and F2 are foci of the ellipse
The Sun is at one focus
$(x+y)$ = constant as the planet moves
$\frac{a}{b}$ = eccentricity e

Eccentricity e	
Circle	0
Mercury	0.206
Venus	0.007
Earth	0.017
Jupiter	0.048
Pluto	0.253

Fig 27.6 **Planetary orbits are ellipses. In this diagram, eccentricity e is exaggerated**

The orbits are elliptical, which means that sometimes they are further from the Sun than at other times. Books of data tend to give the *mean* distance of the planet from the Sun. For most planets this distance is usually quite close to the actual distance at any time, but Mercury and Pluto have more elliptical orbits. In fact, Pluto sometimes gets closer to the Sun than the next inner planet, Neptune.

The Sun is at one **focus** of the planet's ellipse (Fig 27.6). Accurate prediction of a planet's motion requires very careful mathematics – not only is the mathematics of elliptical motion more complicated than that of circular motion, but each planet is affected by the gravitational forces between itself and other planets. However, the orbits of most planets are so nearly circular that we can assume circular motion and get good enough results for our purposes.

The laws of planetary motion

The gravitational force acting on the planet produces an inward acceleration v^2/r where v is the orbital speed of the planet at a distance r from the Sun. Newton's gravitation law (see page 47) tells us the size of the force:

$$GMm/r^2$$

where G is the universal gravitational constant, M the mass of the Sun and m the mass of the planet. So, by applying Newton's second law of motion, force = mass × acceleration, we have:

$$G\frac{Mm}{r^2} = m\frac{v^2}{r}$$

This equation simplifies to:

$$v = \sqrt{\frac{GM}{r}} \qquad [1]$$

which tells us that the speed of a planet in its orbit does not depend on its mass but only on its distance from the Sun.

This is a very powerful result. For example, it allows us to calculate the mass of the Sun from the orbital speed of any planet. We also have to know the value of the gravitational constant G, a value which was not known in Newton's day and remains one of the most difficult of fundamental physical constants to measure accurately (see page 47).

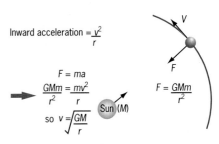

$$\text{Inward acceleration} = \frac{v^2}{r}$$

$$F = ma$$
$$\frac{GMm}{r^2} = \frac{mv^2}{r}$$
$$\text{so } v = \sqrt{\frac{GM}{r}}$$

$$F = \frac{GMm}{r^2}$$

Fig 27.7 **Speed in orbit**

?

D The Earth's orbit is almost circular with a radius of 1.5×10^{11} m. It takes 1 year (3.16×10^7 s) to complete its orbit.
(a) Show that its orbital speed is about 3×10^4 m s^{-1}.
(b) G is 6.7×10^{-11} N m^2 kg^{-2}. Calculate the mass of the Sun.

Kepler's laws of planetary motion, and Newton's explanations

Newton produced his famous laws of motion in order to explain the motion of the planets. He used these, and the idea of gravity as a force, to deduce some laws of planetary motion already discovered by years of patient observation, particularly by the astronomer Johannes Kepler (Austria/Bohemia 1571–1630) on the orbit of Mars. Three of Kepler's laws proved to be the most useful:

1 The law of ellipses. The orbit of each planet is an ellipse, with the Sun at one focus.

2 The law of equal areas. A line drawn from a planet to the Sun sweeps out equal areas in equal times, as in Fig 27.8(a).

3 The harmonic law. The *square* of the orbital period T of a planet is directly proportional to the *cube* of its average distance R from the Sun. This is usually written as:

$$\frac{T^2}{R^3} = \text{constant} \qquad [1]$$

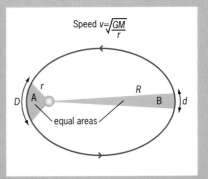

$$\text{Speed } v = \sqrt{\frac{GM}{r}}$$

Fig 27.8(a) **By Kepler's second law of planetary motion, a planet covers the distances D and d in equal times. It travels faster when close to the Sun, so D > d. Kepler found that the areas 'swept out' by the radius line in equal times were also equal: A = B**

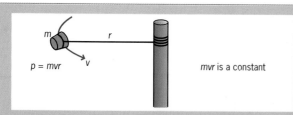

Fig 27.8(b) **Conservation of angular momentum. When you spin an object (here a rubber bung) round a rod, as the string winds up, *r* gets less but *v* increases to compensate**

Newton gave a physical explanation and proof of Kepler's laws, using his laws of motion and the idea of gravity as a force obeying an inverse square law.

Kepler's law 1

Newton was able to prove that when a body moves under the action of a centripetal force (ie with a direction always along the line joining the mass with a particular point), its path must be a circle, an ellipse or a hyperbola. A hyperbola, unlike the other two, is an open-ended path; a comet, for example, may follow a hyperbola, and then it will never return.

Kepler's law 2

Equation 1 on page 617 gives us a qualitative idea of Kepler's law 2. As *r* decreases (as in Fig 27.8(a)), the planet moves faster (*v* increases). In a given time, the line joining it to the Sun will sweep out an area such as *A*. When it is further from the Sun the planet will move more slowly, so that the radius line will sweep out an area such as *B*. Areas *A* and *B* are equal if *r* increases enough to compensate for the decrease in *v*.

The simplest mathematical proof of Kepler's law 2 uses the fact that when a planet moves in its orbit its angular momentum is conserved:

angular momentum $p = mvr$ = constant   [2]

Look at Fig 27.9.

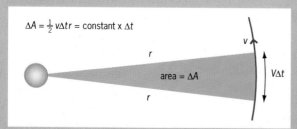

Fig 27.9 **Diagram to illustrate proof of Kepler's law 2**

When the planet moves for a very short time Δt its radius line of length *r* sweeps out a small area ΔA. ΔA is (approximately) the area of a triangle of base $v\Delta t$ and height *r*. So:

$$\Delta A = \tfrac{1}{2}v\Delta t r = vr \times \tfrac{1}{2}\Delta t$$

But since *m* is a constant for any planet, from equation 2, *vr* is a constant. Thus the area ΔA is the same for any value of *r* when the planet moves for equal times Δt.

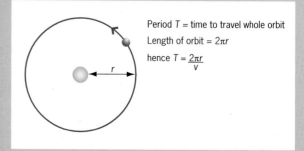

Fig 27.10 **Obtaining the period of orbit from speed and radius**

Kepler's law 3

Consider a planet moving at a speed *v* in a circle of radius *r*. The orbital period *T* is the time taken to complete one orbit, a circumference distance $2\pi r$. So:

$$T = \text{distance/speed} = 2\pi r/v$$

From equation 2 above:

$$v = \sqrt{\frac{GM}{r}}$$

and substituting for *v* gives:

$$T = 2\pi r \times \sqrt{\frac{r}{GM}}$$

When we square both sides we get:

$$T^2 = 4\pi^2 r^2 \times \frac{r}{GM}$$

which we can simplify to:

$$T^2 = \frac{4\pi^2}{GM} \times r^3$$

That is, the square of the orbital period is directly proportional to the cube of the average distance between Sun and planet. This is normally given as Kepler's law 3 in the form:

$$\frac{T^2}{r^3} = \frac{4\pi^2}{GM} = \text{constant}$$

for all planets. It applies to all bodies orbiting in gravitational fields, and may be used to help measure the masses of stars (see page 635).

?

E Use the data in Table 27.1 to check that the Kepler law 3 is true for the planets Venus and Mars. (You could use a spreadsheet to tackle all the planets.)

See question 6. ■

3 THE FORMATION AND EVOLUTION OF THE SOLAR SYSTEM

Dark matter

The Milky Way contains a great deal of '**dark matter**', often detected because it cuts out light. It may be lit up by nearby stars, as in Fig 27.11(a). This shows a region of dark matter in the constellation of Orion, an example of a **nebula** made of gas and dust lit up by stars hidden behind it with others in front. The gas is identified from both its visible spectrum and radio spectrum as being mostly hydrogen. The interaction of the dust with light suggest that it is made of small grains of 'dirty ice', as shown in Fig 27.12.

Fig 27.11(a) **An image in the visible spectrum of the nebula in Orion, 1500 light years from Earth. The bright central region of hydrogen gas hides a cluster of very hot stars called Trapezium which ionise the hydrogen with their ultraviolet radiation. Notice the two bright blue stars in the centre and below the centre on the right. Their colour indicates that they are very new stars, as are many others in this nebula**

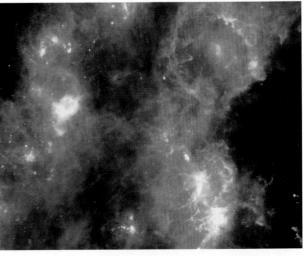

Fig 27.11(b) **As shown by this infrared image of the Orion and Monoceros constellations, there are areas which are strong emitters of infrared radiation. One is the nebula in Orion shown in Fig 27.11(a), the lowest bright patch seen here at bottom right**

The account that follows is thought to be the way that the Sun and the planets of the Solar System were formed, as well as broadly the way in which other planetary systems in the Universe were formed.

Formation of a star

To produce a star the size of the Sun would require a (roughly spherical) cloud of dust and gas a few light years in diameter. In the star's formation, much of the material is lost and doesn't end up in the star. The cloud may be stable for billions of years. During this time, the gravitational forces that tend to make it clump together are offset by the random thermal motion of the particles. Then the cloud starts a **gravitational collapse**. This may be triggered by a disturbance which makes a part of the cloud more dense than the rest.

?

F (a) Suggest why hot gas clouds are less likely to collapse than cold ones.
(b) How could a 'chance disturbance' trigger a collapse in a cloud too hot to collapse of its own accord?

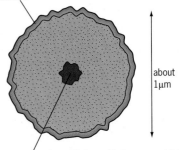

Mantle ('dirty ice': water, carbon dioxide, methane, ammonia)

about 1μm

core (possibly iron, silicates or graphite)

Fig 27.12 **A grain of interstellar dust. The core is thought to be of the same solids we find in meteorites. Like the mantle, the heads of comets are made of 'dirty ice', and the gases stream off to form the comet tails**

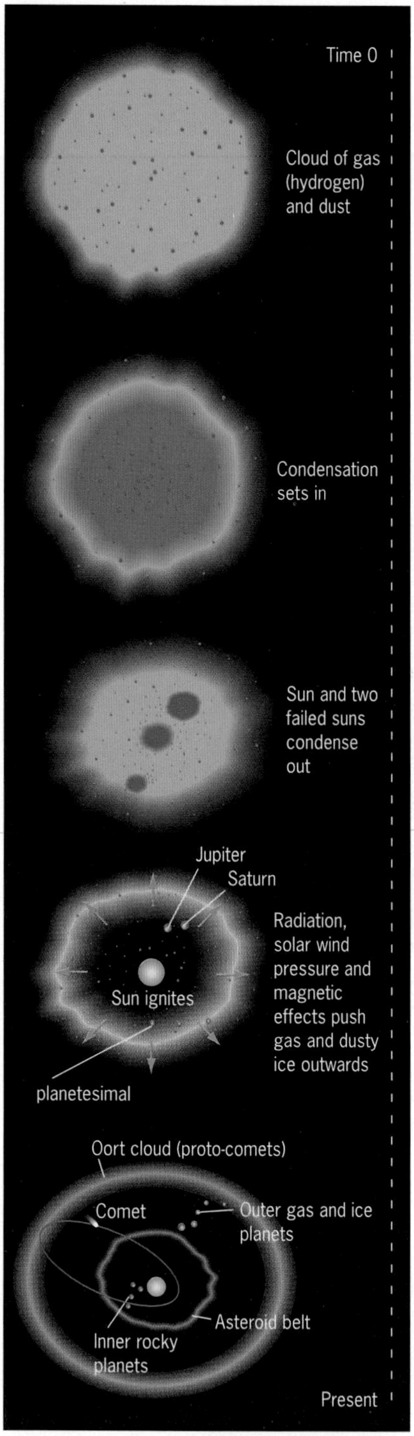
Time 0
Cloud of gas (hydrogen) and dust
Condensation sets in
Sun and two failed suns condense out
Jupiter
Saturn
Radiation, solar wind pressure and magnetic effects push gas and dusty ice outwards
Sun ignites
planetesimal
Oort cloud (proto-comets)
Comet
Outer gas and ice planets
Asteroid belt
Inner rocky planets
Present

Fig 27.13 **The standard model of the formation of the Solar System from a gas and dust cloud**

?

G Explain why high temperatures are needed before nuclear fusion can begin.

Phase 1: gravitational contraction of the gas cloud

When the cloud starts to contract under gravity, gas and dust fall towards the centre of mass, gaining speed as potential energy is converted to kinetic energy. Collisions make the extra motion random. The gas gets hotter. Most energy remains trapped, but the cloud begins to lose energy as infrared radiation. At this point, we could detect the cloud from Earth as an infrared source.

Eventually, the core in the contracting cloud becomes white hot, but we cannot see it because the cooler outer layers absorb the radiation. Because infrared radiation is less easily absorbed, we observe the large dark cloud as a strong source of radiation, an infrared 'hot spot'. You can see several hot spots in Fig 27.11(b).

Phase 2: a protostar is born

From the start of collapse, it takes about a million years for a mass of gas to reach the stage of being a **protostar**. The core is several times larger than the star it will eventually become, and its heat generates an outflow of particles which blows away the gas and dust in the cloud that still surrounds it.

The hot core is now visible as a **pre-main sequence star** (see page 638). It continues to gain energy by a slow gravitational contraction. The material in the core is now a plasma, with all the atoms (mostly protons from hydrogen) stripped of their electrons.

Phase 3: a real star at last

After another 50 million years or so, the core is suddenly at a high enough density and temperature – about 10 million kelvin – to tap a new source of energy: **nuclear fusion**. Some protons now fuse together, eventually forming helium, in the **proton–proton chain**, a process called 'hydrogen burning'. This is shown in Fig. 27.45 on page 642. Vast quantities of energy are suddenly released as the core becomes a nuclear fusion reactor (see page 624).The energy increases the temperature even further and hydrogen burning extends to more of the core.

The star reaches equilibrium

The cloud stops collapsing when the kinetic energy of the particles at all levels is great enough to balance the inward pull of gravity. Compare this with what happens in the Earth's atmosphere: the thermal energy of the air molecules makes them move fast enough not to be pulled to the surface by the Earth's gravity field.

4 THE FORMATION OF PLANETS, AND THE INEVITABILITY OF SPIN

But while these dramatic events are going on in the core, the gas cloud itself is also changing.

Making a disk

All objects in the Milky Way – indeed in all galaxies – rotate about the centre of mass of the galaxy. Think about our original dusty nebula such as the one in Fig 27.11(a). It gradually changes from a large mass to one with an axis. The outer edge moves faster than the inner edge, so the nebula spins very slowly on its axis: it has its own *angular momentum* about this axis (see page 174).

Now think of a mass of atoms, molecules and dust at a distance r from the centre of the nebula: it is rotating about the centre of the nebula at a very low speed v, and has angular momentum mvr. As the nebula contracts, this angular momentum is conserved, that is, mvr stays constant. Its mass m stays the same, but the mean value of r gets less, so *speed v must increase*. The result is that the shrinking cloud spins faster and faster.

Some of the gravitational force pulling the material inwards provides the centripetal force for the spinning mass. Less force is available to pull the particles inwards, so the rate of contraction is less than it would be if the cloud were not spinning. But this effect applies only to the direction at *right angles* to the axis of spin. There is no spinning *along* the spin axis, so the cloud condenses more rapidly in this direction, as shown in Fig 27.14. In **condensation**, small dust particles form when atoms and molecules clump together. The cloud eventually flattens into a pre-planetary disc.

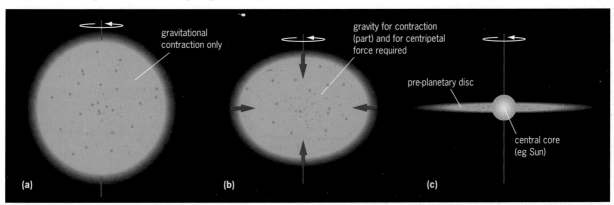

(a)   (b)   (c)

We now have another possible equilibrium, when the rotational speed of dust and gas particles is just large enough to keep them in orbit. The gravitational pull of the new central star at the distance of the particles matches the centripetal force needed at that speed.

Fig 27.14 **Why spherical spinning clouds go flat when they contract**

■ See questions 7, 8 and 9.

Making planets

The dust grains are not moving in neatly ordered orbits. They have a large random motion and collisions still happen. The particles stick together to form clumps, a process called **accretion**. As the clumps get larger they are more likely to bump into others, so they get even larger. At a particular size a clump is large enough for its gravitational attraction to become significant – it has become a **planetesimal,** able to attract smaller particles to make an even bigger object.

As time goes on, the larger planetesimals grow even larger by **accretion** of smaller ones, eventually undergoing **contraction** and reaching the size of a **planet**. This process is shown in Fig 27.15.

Fig 27.15 **The planetesimal theory**

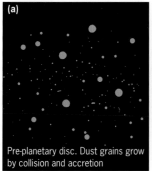

(a)

Pre-planetary disc. Dust grains grow by collision and accretion

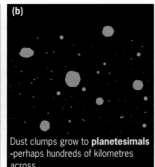

(b)

Dust clumps grow to **planetesimals** -perhaps hundreds of kilometres across

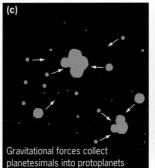

(c)

Gravitational forces collect planetesimals into protoplanets

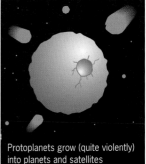

Protoplanets grow (quite violently) into planets and satellites

With the formation of new planets, there is intense heating when bodies make inelastic collisions with each other, and their kinetic energy becomes the internal energy of random motion of their particles. In the Solar System it is likely that at least the rocky planets became hot enough for their cores to be molten. This caused the separation of the planetary material into layers of different chemical structure and density (see Chapter 23, which deals with the structure of the Earth).

The Sun and planets were formed roughly at the same time. The Sun is estimated to be about 5 billion years old (5×10^9 y). Radioactive dating shows that the oldest rocks of the Earth and Moon are about 4.6 billion years old.

Evidence for the accretion model

There is not much evidence for the model of planetary formation just described, but all the rocky planets and many satellites have large impact craters which were made after the surface had cooled down and solidified (Fig 27.16). Such craters are harder to find on Earth because of its active geology (plate tectonics), its eroding atmosphere and the fact that two-thirds of its surface is covered by water. But there is still evidence of impacts on Earth (see page 57).

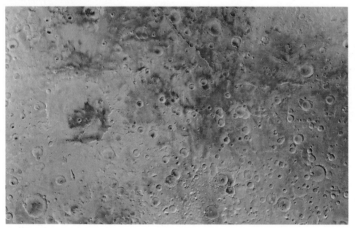

Fig 27.16(a) **Left: Craters on Mars, from images taken by Viking I and II space probes. The Huygens crater, centre left, is about 500 km in diameter**

Fig 27.16(b) **Right: Craters on the Moon, taken by the Galileo spacecraft**

Why there are two main groups of planets in the Solar System

Now think about the early stage of condensation, accretion and contraction. The new Sun created a large temperature gradient in the nearby surrounding cloud. Near the Sun, only those materials with a high melting point would solidify, such as rocks and metals with melting points above 400 or 500 K. Further out it would be cold enough for water ice to form – and even further out it would be cold enough to freeze compounds like methane and elements like hydrogen, oxygen and argon.

Gas particles are light and easy to move. The Sun emits not only radiation but also a stream of particles called the **solar wind**, containing protons and electrons. Both the radiation and the solar wind exert forces on particles near the Sun, and tends to push the lighter ones furthest.

It is assumed that this process began while the Sun was a protostar. The more massive particles of rock and metal were separated from the gases as rocks and metals formed planetesimals close to the Sun, whilst gases were forced further out and became 'icy' planetesimals made of frozen gases.

?

H Outline the evidence for the formation of planets and planetary satellites by accretion.

5 STARS: THE SUN AS A STAR

The Sun is an average kind of star and its main physical characteristics are shown in Table 27.2.

We can measure the Sun's surface temperature in two ways:

● Its spectrum can be plotted and fitted to the *black body curve* (see page 361).
● The value of its energy output can be fitted into the Stefan–Boltzmann equation for radiation (see page 362).

The Sun's surface temperature of 5780 K provides a black body (temperature) radiation which is mostly in the infrared, but peaks at a wavelength of 500 nm which we see as a yellowish-green.

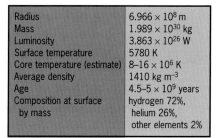

Radius	6.966×10^8 m
Mass	1.989×10^{30} kg
Luminosity	3.863×10^{26} W
Surface temperature	5780 K
Core temperature (estimate)	$8–16 \times 10^6$ K
Average density	1410 kg m^{-3}
Age	$4.5–5 \times 10^9$ years
Composition at surface by mass	hydrogen 72%, helium 26%, other elements 2%

Table 27.2 **Data about the Sun**

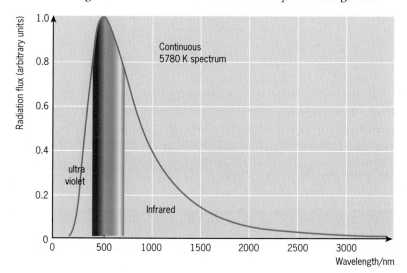

Fig 27.17 **The continuous spectrum of the Sun**

As described on page 618, the *mass* of an astronomical object can be calculated from the period of any object that is in orbit round it. The *diameter* of the Sun can be measured from its angular diameter and the value of the astronomical unit (AU). This measurement gives us a solar radius of the visible sphere, the **photosphere**, of 7.0×10^8 m.

EXAMPLE

Q Find the average density of the Sun.

A The Sun has an average radius of 7×10^8 m, so its volume is:
$v = 4/3\pi r^3$
$= 4/3 \times \pi \times (7.0 \times 10^8)^3$
$= 1.4 \times 10^{27}$ m^3

Its mass is 2.0×10^{30} kg, so its density is:
ρ = mass/volume
$= 1.4 \times 10^3$ kg m^{-3}

This is a little greater than the density of water.

?

I Use the data in Table 27.2 to calculate the surface area of the Sun.

The luminosity of the Sun and the solar constant

Luminosity *L* is a measure of the total radiation energy emitted by a star. At the distance of Earth's orbit, we measure the incident solar radiation energy to be 1.370 J per square metre per second (or W m^{-2}). This quantity is called the **solar constant *S***. Some energy is absorbed in the atmosphere, so less than this quantity actually reaches the Earth's surface.

The Sun radiates evenly in all directions and so the radiation obeys an **inverse square law**. Its **intensity *I*** is:

$$I \propto \frac{1}{r^2}$$

The total radiation energy emitted by the Sun passes through a sphere of radius at the distance of the Earth (1 AU) at the rate of 1.370×10^3 W m^{-2}. See Fig 27.18. We can therefore calculate the total radiation energy emitted by the Sun as $4\pi R^3 S$ where *R* is 1 AU (1.496×10^{11} m).

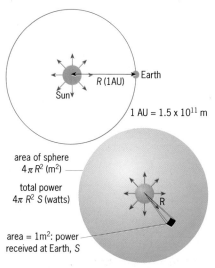

Fig 27.18 **Measuring the Sun's luminosity**

?

J Use the data in the text to show that the luminosity of the Sun is 3.90×10^{26} W.

K Use the data in the text to estimate the surface temperature of the Sun.

This value of luminosity may be used to calculate the Sun's surface temperature T, assuming that it obeys the Stefan-Boltzmann radiation law:

$$L = \sigma A T^4 = 4\pi R^2 \sigma T^4$$

where A is the surface area of the Sun and σ is the Stefan-Boltzmann constant, 5.57×10^{-8} W m^{-2} K^{-4}.

6 THE SUN

The composition of the Sun

The emission spectrum of the Sun has a number of dark lines in it (Fig 27.20). The spectrum of dark lines is due to absorption by cooler gases just above the hot visible surface that we see. The lines are called **Fraunhofer lines** after Josef von Fraunhofer (Bavaria, 1787–1826) who developed the spectroscope and discovered the dark lines. They indicate the elements that are present at the Sun's surface.

The *intensity* of the absorption lines for an element can tell us how much of the element is present: the more of the element that is at the surface, the more absorption takes place and the darker the line (Figs. 27.19 and 20). Such measurements show that the Sun's atmosphere consists of 72 per cent hydrogen, 26 per cent helium and 2 per cent of what astronomers call *heavy* elements, those with more than 2 protons in the nucleus!

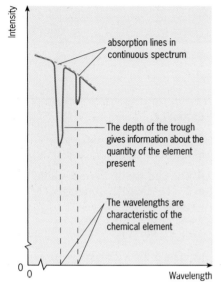

Fig 27.19 **Absorption lines (Fraunhofer lines) in the solar spectrum**

?

L (a) What are the two main features of the spectrum of radiation emitted by the Sun?
(b) Which feature **(i)** allows a measurement of its surface temperature, **(ii)** can be used to deduce the chemical composition of the solar atmosphere?

See question 10. ■

Fig 27.20 **The absorption spectrum of the Sun, expanded to show many of the dark Fraunhofer lines used to identify elements in the gases at the Sun's surface**

The Sun as an energy source

The Sun obtains its energy from **nuclear fusion**. In this process, positively charged nuclei collide with each other with enough kinetic energy to overcome the energy barrier produced by the electric *repulsion* forces between them. For this to happen, the particles have to be travelling at very high speeds, which are reached by a tiny fraction of the particles at any moment only when the temperature is more than about 10^7 K. The Sun's core is estimated to have temperatures in the range 0.8 to 1.6×10^7 K.

Nuclei are very small, and fusion reactions will be a continuous source of energy only when a plasma is at a density that is high enough for a sufficiently high rate of collisions. The **plasma** in the active core of the Sun is at high pressure and has a density 160 times that of water, 1.6×10^5 kg m^{-3}. This is high enough to maintain the reactions.

The solar fusion reactions

The process that converts hydrogen to helium in the Sun is the **proton–proton** chain. This is shown in Fig 27.21.

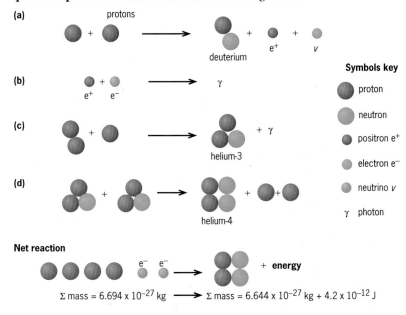

(a)
protons

deuterium

(b)

(c)
helium-3

(d)
helium-4

Symbols key

proton

neutron

positron e^+

electron e^-

neutrino v

γ photon

Net reaction

e^- e^- + **energy**

Σ mass = 6.694 x 10^{-27} kg $\longrightarrow$ Σ mass = 6.644 x 10^{-27} kg + 4.2 x 10^{-12} J

Fig 27.21 **The proton–proton chain**

?

M Use the Einstein relation $E = mc^2$ to show that to provide its luminosity of 3.9×10^{26} W, the Sun must lose mass at the rate of more than 4 billion kilograms per second.

■ See questions 11 and 12.

The Sun in equilibrium

The Sun is in equilibrium, and any changes are self-correcting. If the fusion reaction slowed down for some reason, the core would cool slightly and its particles would exert a smaller pressure. The force of gravity would cause a small collapse, and this would increase the temperature as gravitational potential energy becomes random kinetic energy. In turn, the rate of energy production by nuclear reactions would increase, the pressure of the hotter plasma would also increase, and so the balance would be restored.

The internal structure of the Sun

We can only observe the visible surface of the Sun (its photosphere) and have to deduce its internal structure by computer modelling using observations and calculations. The model in Fig 27.22 on page 626 matches what we actually observe at the surface.

The Sun's core

The **core** is the site of the nuclear fusion reactions which provide the energy of the Sun. This energy travels to the surface first as radiation and then as convection currents in the cooler gases of the outer layers.

Radiation from the core consists of photons of very high energy, mostly as X-rays. The photons travel at the speed of light but don't get very far. They collide continuously with particles in the plasma and move about randomly, travelling a distance of about 1 millimetre between collisions and taking millions of years to reach the outer layers. This means that the core itself is opaque to our observation.

The convection zone

Nuclear fusion ceases at temperatures less than about 6×10^6 K, and there is a rapid fall in temperature at the Sun's fusion boundary to about 10^6 K. High energy photons emerging from the core first

Fig 27.22 **The structure of the Sun**

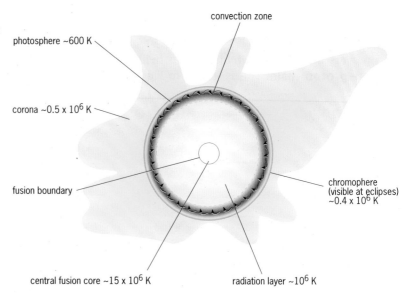

convection zone

photosphere ~600 K

corona ~0.5 x 10^6 K

fusion boundary

chromophere (visible at eclipses) ~0.4 x 10^6 K

central fusion core ~15 x 10^6 K

radiation layer ~10^6 K

?

N The region above the Sun's core is opaque to radiation, and efficient energy flow can only occur via convection. Explain why this is so. Relate your answer to what happens when heating water in a microwave oven.

supply energy to the gas of the **radiation layer**, making it hotter, but losing energy themselves as they do so. The heated gas expands and rises to form convection currents, just like water in a pan on a hot stove, in the **convection zone**.

This zone has a temperature which is low enough for hydrogen and helium *atoms* to form. Photons are still there as carriers of energy, but energy can be transported more quickly by convection than by photons. This is because the photons leaving the core are easily absorbed, by exciting or ionising the atoms. The convection zone also is therefore opaque.

The photosphere

Convection ceases where the Sun's atmosphere becomes too thin. Energy now leaves the Sun as radiation, which means that it is the top of the convection zone that we actually see from Earth – the **photosphere**. Fig 27.23 shows what it looks like: it has a granular structure of light and dark light areas when photographed using a filter which passes only the light from very hot hydrogen.

The chromosphere and the solar corona

The Sun has two regions outside the photosphere, which can be seen only during an eclipse of the Sun. Next to the photosphere is the **chromosphere**, a very thin region of low density. It is seen as a flash of bright pink hydrogen light at the start of a solar eclipse. Astronomers have deduced from this that the chromosphere is only a few thousand kilometres thick, consisting mostly of hydrogen at about a thousandth of the density of the photosphere gases. But the particles in this zone are, on average, travelling much faster, because the chromosphere is at a much higher temperature – up to 400 000 K. Particles (mostly protons) emerging from this region at high speed provide the solar wind.

At total eclipse, we can see yet another zone of light-emitting gas, the **corona** (Fig 27.24). It is at a temperature of 500 000 K or so, and emits bright lines from highly ionised atoms – neon, calcium, iron and nickel, in addition to hydrogen. The shape of the corona suggests that magnetic fields are involved, and these may be the mechanism by which energy gets into the very low density gases in the corona. The activity associated with **sunspots** (Fig 27.25) is also largely magnetic.

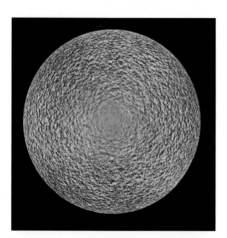

Fig 27.23 **The convection currents at the outer limit of the Sun's convection zone. The image was taken by the SOHO spacecraft using the Doppler effect: areas of dark colour are hot gases welling up towards the observer and those of lighter colour are cooler gases moving down**

Fig 27.24 **The Sun's corona during an eclipse**

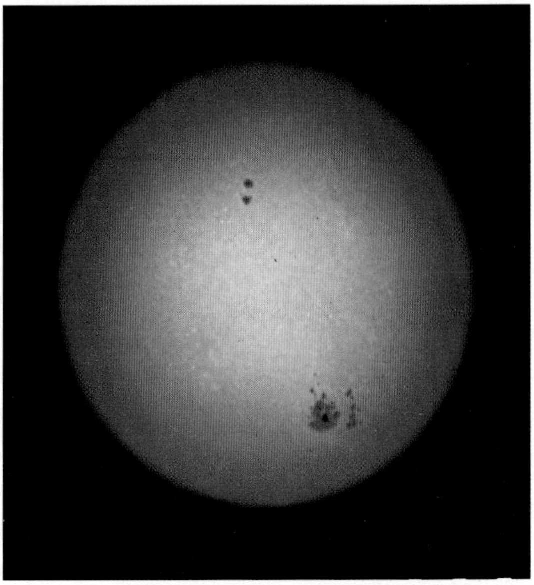

Fig 27.25 **Sunspots are dark regions of the photosphere that are cooler than their surroundings. They are the sites of intense magnetic storms and emit hot gases as solar flares. A typical flare emits energy of 10^{25} J, equivalent to a nuclear explosion of 2 billion megatonnes. Particles from flares cause disruption to radio communications. Flares produce other atmospheric disturbances, and so may have an effect on global weather patterns**

■ See questions 13 and 14.

How long will the Sun last?

The core of the Sun is believed to consist of helium and hydrogen in the ratio 60 per cent helium to 40 per cent hydrogen. Only about 10 per cent of the Sun's total mass of hydrogen is in the core. These are computer estimates as it is impossible to *measure* the amounts of hydrogen and helium at different levels in the Sun's interior. When the available hydrogen of the core is converted into helium, the proton–proton reaction will stop. The Sun as we know it will cease to exist – but see page 641 for what may happen next.

Computer models estimate that about 1×10^{29} kg of hydrogen is present in the core for conversion to helium. But only 0.7 per cent of the hydrogen's mass can be converted to *radiation energy* by way of the fusion reaction; the rest remains as helium. This means that the mass actually available for energy to keep the Sun going is reduced to 7×10^{26} kg.

We know that the Sun emits energy at the rate of 3.9×10^{26} W, or 1.2×10^{34} J per year. This is equivalent to a mass loss of:

$$\Delta m = \frac{E}{c^2} = \frac{1.2 \times 10^{34}}{9 \times 10^{16}} = 1.33 \times 10^{17} \text{ kg per year}$$

Thus, at a rough estimate, the Sun will use up its core hydrogen in $(7 \times 10^{26})/(1.33 \times 10^{17}) = 5 \times 10^9$ years.

7 THE EVIDENCE

We gain evidence about the nature of the Solar System and the wider Universe of stars, nebulae and galaxies by collecting information from the electromagnetic spectrum that reaches us from space. Until recently, all our telescopes have been ground-based, so this information has been filtered by the Earth's atmosphere. The atmosphere is transparent only to certain parts of the spectrum, absorbing almost all infrared, as well as X- and ultraviolet radiation. It also adds unwanted signals called **'noise'** to other parts of the spectrum.

Turbulence, twinkling and 'seeing'

Turbulence in pockets of air in the upper atmosphere affects light from stars, so that they seem to 'twinkle'. This effect stops even powerful telescopes from producing fine detail, a defect astronomers call poor **'seeing'**. It explains why all large observatories are built on high mountains, as far up into the atmosphere as possible.

Seeing problems can be best (but expensively) avoided by using Earth satellites as platforms for instruments capable of detecting radiations at all frequencies. The most powerful instrument of this kind is the Hubble Space Telescope (HST), launched in 1990, which has added greatly to our knowledge of stars and understanding of the Universe at large.

The basic physics of telescopes has been explained and discussed in Chapter 19.

See questions 15 and 16. ■

8 TELESCOPES IN MODERN ASTRONOMY

The simple Newtonian reflector is the most commonly used telescope (look back to Fig 19.37). The image is formed on a photographic plate or (more often nowadays) by an electronic detector. The image is never perfect. There are three main problems:

- **Diffraction** produces a circular pattern rather than a point image of a star (or other object).
- **'Seeing'** gives loss of detail when air currents in the atmosphere cause random motion of the image that reaches the detector.
- **'Grain'** results because the detecting device has a lower limit to the size of object it can detect. This can be because of the size of the grains of light-sensitive chemical in a photographic plate, or the pixel size in a charge-coupled detector.

As a general rule, *seeing* is the main problem for large ground-based telescopes. The most modern telescopes improve seeing by using **active optics**: a rapid acting light sensor coupled to a computer monitors the image and adjusts the shape of a small, flexible *correcting mirror* to make the image as small and fixed in position as possible. This method is used in the European Space Observatory's flexible 3.5 metre reflecting telescope at La Silla in Chile, shown in Fig 27.26, as well as in the largest reflecting telescope in the world, the Keck telescope, at the Mauna Kea Observatory in Hawaii (look back to Fig 19.38).

Fig 27.26(a) **Left: The European Southern Observatory at La Silla in Chile. The buildings house different telescopes**

Fig 27.26(b) **Right: ESO's 3.5-metre New Technology Telescope. It uses computer-operated active optics to correct images. This is particularly useful for getting good images of faint distant galaxies**

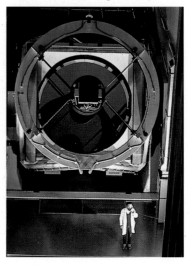

Radio telescopes

There are two main types of *radio* telescope, those with **dish aerials** and those with **linear arrays** of aerials.

Dish telescopes work in much the same way as reflecting telescopes, in that they gather as much of the signal as possible at as great a resolution as possible. They work at wavelengths which are very much longer than light, and the detector is a tuned circuit, as opposed to the photographic plate of an optical telescope.

As Fig 27.27 shows, the largest dish aerial is in a natural hollow in the ground at Arecibo in Puerto Rico. It has a diameter of 305 m. This telescope can only detect signals that enter it from overhead, so can only record whatever is above it as the Earth rotates. Steerable telescopes can point to any object above their horizon but have to be smaller. The largest is a 100 m diameter dish at the Max Planck Institute near Bonn in Germany.

Fig 27.27 **The Arecibo radio observatory in Puerto Rico. The dish reflects radio waves up to the receiving antenna suspended 130 m above the dish**

Diffraction

Diffraction limits the accuracy with which any telescope can determine the position of an object. A dish of 100 wavelengths diameter can position an object within a region 1 degree wide, which is very much larger than the angular size of a star. Radio astronomy often uses the radio signal from molecular hydrogen at a wavelength of 0.21 m. This means that the early radio telescopes with dishes about 25 m or even 50 m wide could not pinpoint the positions of **radio stars** – stars emitting radio waves – accurately enough to match them with likely visible sources. A group of stars may be as close as 0.5 seconds of arc, and to separate their images would require a dish 50 000 wavelengths in diameter.

Receiving power

The larger the dish, the more energy the telescope can receive per second. This receiving power is simply proportional to the area of the dish:

$$\text{receiving power} \propto (\text{dish diameter})^2$$

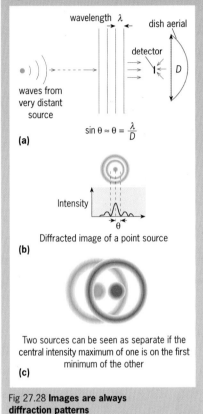

The larger the wavelength of a signal, the larger is the diffraction pattern produced. The critical value is the angular radius θ of the circular central peak of the diffraction pattern, as in Fig 27.28.

wavelength λ   dish aerial

detector

waves from very distant source

$$\sin\theta \approx \theta = \frac{\lambda}{D}$$

(a)

Intensity

θ

Diffracted image of a point source

(b)

Two sources can be seen as separate if the central intensity maximum of one is on the first minimum of the other

(c)

Fig 27.28 **Images are always diffraction patterns**

■ See question 17.

Q (a) Estimate the diameter of a radio telescope operating at a wavelength of 0.21 m that could resolve as separate objects two radio sources with an angular separation of 0.5 arcseconds.
(1 second of arc = 1/3600 of a degree, or 4.85×10^{-6} radians.)
(b) Suggest **(i)** two possible sources of unwanted signals ('noise') in a radio telescope, and **(ii)** two ways of reducing these signals.

■ See question 18.

Fig 27.29 **The array of aerials at the Mills Cross radio telescope at Molongo in Canberra Australia**

Reflector smoothness

The dish of a radio telescope is made of metal which reflects radio signals in much the same way that a silvered surface reflects light. The surface need not be solid: in fact, it is usually a wire mesh, which is as good as a continuous metal surface provided that the gaps between the wires are less than a twentieth of the shortest wavelength detected. (See also the Opener to this chapter.)

Line aerials

A simple **aerial** is a metal wire in which a **signal voltage** is induced when electromagnetic waves pass it, due to the electric component of the wave (look back at Fig 16.3). The signal is strongest when the length of the aerial is matched to the length of the electromagnetic wave. Then, there is **resonance** between the signal voltage and the wave that is being received. Such an aerial is 'tuned'. Everyday examples are the dipole aerials used for VHF radio and UHF TV reception.

Modern computer programs improve the radio image by adding up signals obtained at different times, enhancing the true signal and cancelling out noise.

(a) Tuned dipole aerial

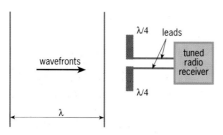

(b) Using an array improves *directionality and gain*. Polar diagrams illustrate how *sensitive* the array is in different directions

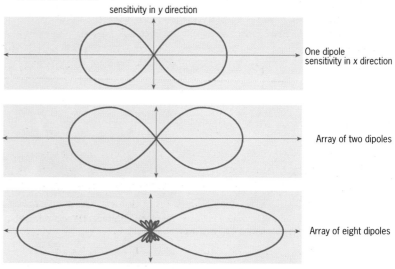

(c) An array of two aerials. They are spaced half a wavelength apart

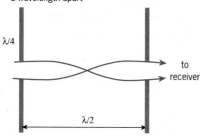

Fig 27.30 **Line aerials. A dipole aerial, see** (a), **is directional. The response to an incoming signal is shown by polar diagrams, see** (b). **The directionality and the efficiency of the aerial can be improved by using more than one dipole in an array, see** (c)

Fig 27.31 **An artist's impression of ROSAT in orbit. ROSAT, launched in 1990, carries an X-ray and far ultraviolet telescope, and is expected to detect over 100 000 new X-ray sources. Four concentric grazing-incidence surfaces of focal length 2.4 m focus radiation to the detector**

Specialist telescopes: gamma ray, X-ray, ultraviolet and infrared

Atmospheric absorption severely limits the use of ultraviolet, X-rays and infrared for observation, so balloons, high-flying aircraft, space laboratories (Space Lab) and satellites have been used to collect data in these regions of the spectrum.

X-ray telescopes have been carried in the Uhuru satellite (launched in 1970, the first to carry an X-ray telescope), the ROSAT satellite in 1990, and several others. It is difficult to focus X-rays as they tend to go straight through materials or get absorbed in them. An ordinary mirror reflector is as useless as a lens would be. Instead, they are focused to a detector by a set of slightly angled cylindrical surfaces which they reach at grazing angles. As explained on page 648, X-ray sources are closely linked with detecting the existence of **black holes**, described on page 647.

Gamma rays are more penetrating than X-rays and reach ground level. There is a gamma ray telescope 10 m in diameter at the Whipple Observatory in Arizona, and NASA used the space shuttle to launch the Gamma Ray Observatory in 1991.

Hot stars – those with surface temperatures greater than 10 000 K – emit most of their energy as **ultraviolet** radiation. This region gives the most useful spectral lines for studying the composition of very hot stars and regions of space where new stars are being formed (Fig 27.32). Ultraviolet is strongly absorbed in the atmosphere, so most research uses satellite-borne telescopes such as the International Ultraviolet Explorer. The Hubble Space Telescope also contains an ultraviolet instrument.

Fig 27.32 **Taken by the Ultraviolet Imaging Telescope launched in 1990, this picture is the ultraviolet image of spiral galaxy Messier 81 in the Great Bear constellation. It is about 9 million light years away. The bright spots show where new stars are being formed**

Ground-based **infrared astronomy** uses the narrow bands of wavelengths not absorbed by the atmosphere's carbon dioxide and water vapour. The electronic devices used to detect infrared have to be cooled close to absolute zero (about 2.5 K).

In 1983, IRAS, the Infrared Astronomical Satellite, mapped the whole sky at wavelengths between 12 and 57 μm. It detected some 250 000 infrared sources, identified as stars, galaxies and gas clouds. Five comets were also discovered. Some strong infrared sources are believed to be regions of space rich in gas and dust in which young stars are forming. The gravitational collapse of the cloud causes it to heat up (see page 620).

Cosmic background radiation

The most spectacular and significant event of infrared astronomy was the discovery of the **cosmic background radiation**, found to be uniform from all points of the Universe. This black body radiation is characteristic of a temperature of 2.7 K and is believed to be the remnant of the Big Bang with which the Universe began. (Though often called 'microwave', the wavelengths of background radiation are millimetres rather than micrometres.) It was discovered in 1964 as 'noise' while investigators were testing communications receivers for possible use at such short wavelengths. There is more about this in Chapter 28.

■ See questions 19, 20 and 21.

> **?**
>
> **P (a)** What are the two main factors which determine the clarity of an image made by an astronomical telescope? Suggest two ways for improving images.
> **(b) (i)** Do image detectors or recorders rely on the wave or particle nature of light? **(ii)** What regions of the electromagnetic spectrum are observed by telescopes whose image detectors rely on their wave aspects?
> **(c)** Why are large telescopes placed either on high, remote mountain peaks or in space?

9 USING THE EVIDENCE

Measuring astronomical distances

Astronomical distances are large and very difficult to imagine. Yet the grand theories of cosmology (see Chapter 28) depend on knowing as accurately as possible the distances to the furthest parts of the Universe. Measurement started with the distance of the Sun from the Earth.

Measuring the Astronomical Unit (AU)

Distances are now measured scientifically in *light-seconds*, but converted for everyday use to familiar units like metres. All this is described in Chapter 2. Astronomical distances are measured in

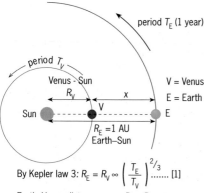

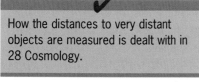

By Kepler law 3: $R_E = R_V \infty \left(\dfrac{T_E}{T_V} \right)^{2/3}$ ...... [1]

Earth–Venus distance = $x = R_E - R_V$

so $\quad R_V = R_E - x$

and equation 1 becomes:

$R_E = (R_E - x) \left(\dfrac{T_E}{T_V} \right)^{2/3}$

As x, T_E and T_V are known, R_E (= 1 AU)
can be calculated

Fig 27.33 **Measuring the astronomical unit**

How the distances to very distant
objects are measured is dealt with in
28 Cosmology.

terms of the Earth–Sun distance or astronomical unit (AU) and the parsec (pc), a unit derived from the AU (look back at Fig 27.3).

Kepler's third law can give us an accurate measurement of the **ratios** of distances in the Solar System. It states that for any planet, the ratio $T^2/R^3 = $ constant, where T is the orbital period of the planet and R is its mean distance from the Sun. This is a consequence of Newton's laws of gravitation (see page 618). So, for the Earth and Venus we can write:

$$\frac{T_E{}^2}{R_E{}^3} = \frac{T_V{}^2}{R_V{}^3}$$

or, rearranging:

$$R_E{}^3 = 1 \text{ AU} = \left(\frac{T_E}{T_V} \right)^2 \times R_V{}^3$$

The orbital periods of the planets are known very accurately, and we can find the value of 1 AU in metres by measuring the distance of Venus from the Sun. This used to be very difficult (see page 11). The distance from Earth to Venus has now been measured very accurately by radar. The rest is simple geometry, as shown in Fig 27.33.

Measuring the distance to a star

Distances to the nearer stars are measured using **parallax**, as described on page 12. What is measured is the apparent change in the angle of the 'line of sight' to a star, measured against the background of very distant stars, as the Earth moves from one end of the diameter of its orbit to the other. This gives a base line 2 AU long, as Fig 27.34 shows. The Universe is so large that this method can only be used for stars that are relatively near to Earth.

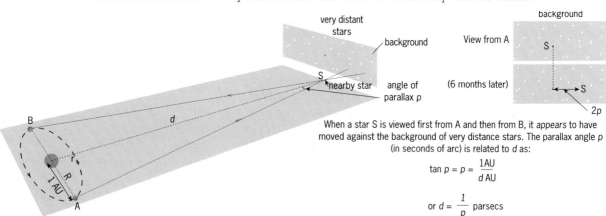

When a star S is viewed first from A and then from B, it *appears* to have moved against the background of very distance stars. The parallax angle p (in seconds of arc) is related to d as:

$$\tan p = p = \frac{1 \text{AU}}{d \text{ AU}}$$

or $d = \dfrac{1}{p}$ parsecs

Fig 27.34 **Parallax and the parsec**

The parallax angle p is very small: the nearest star to Earth has a parallax angle of only 0.772 seconds of arc, or **arcseconds**. This is about 0.2 thousandths of a degree. If p is measured in radians, then we can write:

$$p \text{ radians} = R/d = 1/d \text{ AU} \quad \text{since } R = 1 \text{ AU}$$

So:

$$\text{distance } d \text{ of a star} = 1/p \text{ AU} \quad \text{when } p \text{ is in radians}$$

But astronomers always measure parallax in *seconds of arc*: a second is a sixtieth of a minute of arc which is in turn a sixtieth of a degree. Thus they use the distance unit called the **parsec**, pc, in a more convenient formulation:

$$d = 1/p \text{ parsecs} \quad \text{when } p \text{ is measured in arcseconds}$$

A parsec is more convenient because it comes directly from the measurement. It is also so much bigger than an AU that it is more appropriate for astronomical distances.

1 parsec = 3.09×10^{16} m = 2.06×10^5 AU = 3.26 light years

Measuring parallax requires very accurate instruments. Sirius is considered a very close star at a distance of 0.38 pc. Measuring the parallax angle of 0.38 arcseconds is like measuring the diameter of a 10p piece at a distance of five kilometres! The smallest angle that can be measured with suitable accuracy using a telescope is about 0.001 arcsecond. This allows measurement only of stars quite close to us.

The Sun is a member of the **Milky Way** galaxy. This has a diameter of 25 000 pc, so most of the stars we can see are too far away to measure using parallax.

?

Q Show that a distance of 1 parsec is equivalent to 2.06×10^5 AU.

R (a) How far away is a star with a parallax of 0.001 arcsec in **(i)** pc, **(ii)** AU?
(b) Why is there an upper limit to the distances astronomers can measure using the parallax technique?

■ See question 22.

Temperature

The temperature of a star is its *surface temperature*, the temperature of its photosphere. Most stars produce a continuous spectrum of radiation that follows Planck's rule for *black body radiation* (see pages 363 and 623).

Astronomers produce the spectrum of a star and measure the intensity over a range of wavelengths using a spectrometer (once, photographs were produced, but now recording is electronic). The observed spectrum is then matched to the black-body (Planck) curve to estimate the surface temperature T of the star (look back to Fig 17.1). The simplest method is to measure the wavelength of the maximum intensity in the spectrum, λ_{max}, and then use Wien's displacement law (see page 361):

$$\lambda_{max}T = 2.898 \times 10^{-3} \text{ m K}$$

Astronomical objects range in temperature from just above absolute zero to 10^5 K. This means that the maximum wavelengths in their spectra range from radio to X-rays.

Measuring luminosity, the energy output of a star

The **luminosity L** of a star is a measure of the *total* radiant energy it emits per second. Thus L is measured in watts. The radiation follows an *inverse square law* so that the energy received per square metre per second, the **flux density F** at the Earth is:

$$F = \frac{L}{4\pi d^2}$$

where d is the distance of the star in metres. F is a very small quantity and is the sum for radiation at all frequencies. This is difficult to measure on Earth, since most stars emit most of their radiation in the ultraviolet, which is absorbed in the Earth's atmosphere. Of course, a star's energy may also be absorbed by interstellar clouds as well, making measurement even more difficult.

■ See question 23.

Measuring the masses of stars

It is surprising enough that astronomers can measure the masses of stars many light years away. It is even more surprising that they can measure the masses of stars that they know are there but cannot see! The methods start with measurements of double or **binary stars**, pairs of stars close enough to orbit round each other.

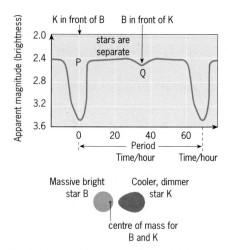

Fig 27.35 **Graph for the variations in brightness of Algol, a binary star system in the Perseus constellation**

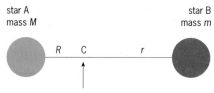

Fig 27.36 **C is the centre of mass of the binary star system. Imagine two weights balancing on a lever: *MR = mr***

See question 24. ■

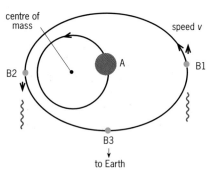

$\Delta\lambda$ = maximum difference from standard wavelength

$$v = \frac{\Delta\lambda}{\lambda} \times c$$

$$r = \sim \frac{vT}{2\pi}$$

Spectra obtained at different times – so period T can also be found

increasing wavelength

Relative to Earth:

B1: Star moving away, λ increases

B2: Star approaching, λ decreases

B3: No relative line of light motion, no change in λ

λ

$\Delta\lambda$

Fig 27.37 **Using the Doppler shift to find the speed of a component of a spectroscopic binary**

See questions 25 and 26. ■

Eclipsing binary star systems

The brightness of the star Algol varies on a very short time scale: it goes from maximum to minimum brightness and back again in just a few hours (see Fig 27.35). This variation is because Algol is a pair of stars, one brighter than the other, that orbit around each other. So, seen from Earth, they pass in front of each other. We see them at maximum brightness when both stars are visible. When dim star K passes in front of bright star B, it 'eclipses' it and we get minimum brightness – P on the graph. There is a smaller dip in brightness at Q when star B cuts out some light from star K as it eclipses it.

Most stars are found in small groups, and often contain pairs of stars close enough together to orbit around their common centre of mass. Fig 27.36 shows the principle of a common centre of mass.

If both stars are visible, we can observe their motions and measure the period of each star. The smaller star is further away from the centre of mass and so moves more slowly with a longer orbital period. However, many binary stars are close enough for the period to be measured in hours rather than days or years.

The period of rotation is determined by the distance between the stars and their masses, and these are related by Kepler's third law (page 618), the law of moments (for centre of mass) and Newton's laws of motion.

When the stars are near enough to Earth to be seen separately in our telescopes, we can measure both the radii of their orbits and their orbital periods separately. The next Extension box gives the mathematics involved.

Spectroscopic binary star systems

Many binary star systems are either so close together or so far away from Earth that telescopes cannot separate them as two distinct objects. But when we study the spectrum of light from the stars we can detect *two sets of absorption lines*, one for each star. The absorption lines arise because elements in the outer, cooler atmosphere of a star absorb light from the star at particular wavelengths (see page 624).

We can also detect a **Doppler shift** for each set of lines from a binary star system. When one of the stars is moving towards us, the frequency of the light it emits increases (wavelength decreases). When it moves away from us, frequency decreases (wavelength increases). With c the speed of light, the change in wavelength $\Delta\lambda$ compared with the unaltered wavelength λ of a spectral line is proportional to the speed v of the star:

$$\frac{\Delta\lambda}{\lambda} = \frac{v}{c}$$

This allows us to measure the speed of each star in a binary system and also their orbital periods. This data can be used to measure their separation and hence the masses of each star.

The simple treatment given here and in the next Extension box assumes that the plane of the stars' orbits are circles and that their planes are neatly edge-on to Earth. In fact, this is unlikely, and more complex calculations are required.

?

S Why are stars more likely to form as binary or multiple star systems than as single stars, like the Sun? Why don't we notice this when we look into the night sky?

Measuring the masses of stars

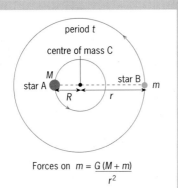

Fig 27.38 **Diagram of orbiting binary stars. Larger star A has mass _M_ and star B has mass _m_. The diagram is drawn for the reference frame of C. (Note that C moves as the stars move)**

period t

centre of mass C

star A

M

R r

star B

m

Forces on $m = \dfrac{G(M + m)}{r^2}$

Fig 27.38 shows the quantities for measuring star masses. The effective mass of the system acts at the centre of mass of the two stars and has a value of $M + m$. Considering the movement of smaller star B and assuming that it is moving in a circle of radius r about the centre of mass of the system, we can apply Newton's laws:

centripetal force due to gravity = mass × inward acceleration

$$\frac{G(M + m)m}{r^2} = \frac{mv^2}{r}$$

which leads to: $$v^2 = \frac{G(M + m)}{r} \qquad [1]$$

So, if the stars are close enough to Earth, and their distance is known, we can obtain a value for r by measuring angles and get v from either spectroscopic data (the Doppler effect) or from a knowledge of r and the period t:

$$v = \frac{2\pi r}{t} \qquad [2]$$

Thus by applying these values into equation 1 we obtain the sum of the masses:

$$(M + m) = \frac{v^2 r}{G} = \frac{4\pi^2 r^3}{Gt^2} \qquad [3]$$

For nearby binaries, the distances from the centre of mass for both the small and the large star, r and R, can be measured. Now look back at Fig 27.36: the law of moments tells us:

$$mr = MR \qquad [4]$$

For example, if $r = 16$ AU and $R = 4$ AU, then star A is 4 times the mass of star B, and it is simple to use equation 3 to find the individual masses of A (M) and B (m).

In practice, it is easier to make calculations in terms of the astronomical unit (AU), the Earth year

and the solar mass M_0. This means that we need to obtain a value for G in terms of these quantities.

Considering the **Sun–Earth system**:
$r = 1$ AU, Earth's period $T = 1$ year, Earth mass = M_E. The combined mass is $(M_0 + M_E)$ which we can approximate to M_0 since the Earth's mass is negligible compared to that of the Sun.

So we can use equation 3 above to write:

$$M_0 = \frac{4\pi^2 r^3}{GT^2}$$

But $r = 1$ AU and $T = 1$ year. So, using these units we have:

$$M_0 = 4\pi^2/G$$

giving: $$G = 4\pi^2/M_0 \qquad [5]$$

Suppose we have a binary star system with values for r and R of 16 AU and 4 AU. We measure the period t of the smaller star as 30 years. What is its mass?

From equation 4 we can write:

$$M = 4m \quad \text{so that} \quad (M + m) = 5m$$

From equation 3, and substituting for G from equation 5, we can write:

$$(M + m) = 5m = \frac{4\pi^2 r^3}{Gt^2} = \frac{4\pi^2 r^3}{t^2} \times \frac{M_0}{4\pi^2} = \frac{r^3}{t^2} M_0 \qquad [6]$$

Thus: $$5m = \frac{16^3}{30^2} M_0$$

giving the mass of the smaller star as 0.91 solar masses. The larger star then has mass $3.7\ M_0$.

You might recognise the term r^2/t^3 in equation 6 from Kepler's third law (page 618). The simple result of equation 5 could have been derived directly from this law. Here, it is derived from first principles to remind you of the Newtonian physics involved.

Spectroscopic binaries

Spectroscopic binaries are too far away for either r or R to be measured, but we can measure the *orbital speed v* and the *periods* of the orbits. So we can put $r = vt/2\pi$ (from equation 2 above) into equation 6 to get:

$$(M + m) = \frac{v^3 t}{2\pi} M_0 \qquad [7]$$

As shown above, we can then continue to find both M and m in terms of the mass of the Sun.

?

T Derive equation 7 in the Extension box from equations 2 and 6.

U What is the difference between an *eclipsing* and a *spectroscopic* binary system? What main astronomical measurement can data from such systems provide?

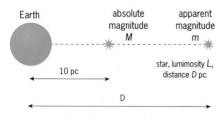

Actual distance *D* pc is standardised to 10 pc

Earth — absolute magnitude *M* — apparent magnitude *m*

star, lumimosity *L*, distance *D* pc

10 pc

D

Power (radiation) actually detected from star is:

$$P = \frac{L}{4\pi D^2} \text{ per m}^2$$

Power that would be detected if star was at 10 pc is:

$$P = \frac{L}{4\pi (10)^2} \text{ per m}^2$$

Thus

$$\frac{\text{'absolute' power/m}^2}{\text{observed power/m}^2} = \frac{D^2}{100}$$

In terms of the (logarithmic) magnitude scale, this transforms to:

$$M = m + 5 - 5 \log D$$

Fig 27.39 **Apparent and absolute magnitude**

10 STAR TYPES

The Sun is a typical star since it emits both visible and invisible radiation as a result of *thermonuclear processes* in its core. Thermonuclear processes begin only when the density and temperature at the core are high enough, and this happens only when the mass of the gas cloud that forms the star is large enough for a big enough energy gain from its gravitational collapse. If the mass is too small, the body will be a *warm* star with a core temperature less than 10^7 K which emits mostly infrared radiation. Such stars are called **brown dwarf** stars.

Using the most powerful telescopes, the only differences we can see between stars is in their *colour* and *brightness*. These characteristics are the basis of what is still the most useful way of classifying stars.

Colour

The colour of a star is due to the spectrum of the visible light it emits. As explained above, the emission spectrum can be used to measure *surface temperature*. The total energy that a star emits is called its **luminosity**, and this depends not only on its temperature but also on its size.

Apparent and absolute magnitude

The modern definition of **apparent visual magnitude** is now precisely logarithmic and is based on instrumental measurements of energy received in the visible part of the spectrum. The magnitude numbers have been chosen to fit as closely as possible to the historical naked eye observations. The brightness of a star (intensity of the starlight received on Earth) is measured using photographic plates or photoelectric sensors. This gives the **apparent magnitude** *m* of a star.

The eye perceives the brightest visible star to be roughly 100 times brighter than the faintest star. The scale of magnitude *m* is chosen so that stars *increase* in measured brightness by equal amounts as we go from *m* = 6 to *m* =1. So the lower the number the brighter the star, and for the brightest we can see, *m* = 1. The scale chosen means that a star of magnitude 2 is 2.51 times as bright as a star of magnitude 3. There are 5 steps between 6 (faintest) and 1 (brightest), so a star of magnitude 1 is brighter than a star of magnitude 6 by the ratio given by five steps:

$2.51 \times 2.51 \times 2.51 \times 2.51 \times 2.51 = (2.51)^5 = 100$

Instruments now allow astronomers to detect stars as faint as *m* = 23 or so. Table 27.3 shows the numbers of stars in the sky brighter than selected apparent visual magnitudes.

Magnitude	Number of stars brighter than the magnitude listed
5	1620
10	324 000
15	32 000 000
20	1 000 000 000 (10^9)

Table 27.3

In Britain it is quite difficult to see stars as faint as magnitude 5.

Absolute magnitude

The brightness a star would have if placed 10 parsecs from Earth is its **absolute magnitude**. The brightest star we see, of apparent magnitude 1, is more than 10 parsecs away and would appear much brighter if brought to this standard distance. If it appeared to be, say, 2.51 times brighter (ie *one magnitude brighter*), its apparent magnitude would become *zero*. Remember that brighter stars have smaller numbers. If it became 4 magnitudes brighter it would move from a magnitude of 1 to a magnitude of −3. All this works mathematically since we use a logarithmic scale.

We can't actually move stars around like this, but we can calculate how bright they would be 10 parsecs away. The absolute magnitude *M* of a star whose actual distance from Earth is *D* parsecs is related to its apparent magnitude *m* by the formula:

$$M = m + 5 - 5\log D$$

Brightness: apparent and absolute magnitude

In 102 BC, the Greek astronomer Hipparchus made the first *measurement* of star characteristics, classifying them into six levels of *brightness*, or **magnitude**. Stars of the *first magnitude* were the brightest, and stars of the *sixth magnitude* were the dimmest, just visible to the eye on a very clear night.

Astronomers still use this idea of stellar magnitude, but they distinguish between **apparent magnitude** and **absolute magnitude** (see the Extension box on page 636). A star may look bright because, though quite small, it is very close to Earth; or it may be a more luminous object much further away.

The light emitted from a star obeys the inverse square law of radiation, so the quantity of energy received per unit area at a particular position is inversely proportional to its distance from the star. To find the star's 'real' brightness, we have to allow for this distance effect, as follows. We calculate how bright the star would look at a **standard distance of 10 parsecs** (32.6 ly). At this distance, the Sun would be classified as a rather faint star of magnitude 4.9.

?

V For the star Spica, $D = 67$ pc, $m = 1.0$. Use the formula in the text to show that the absolute magnitude of Spica is −3.1.

W What is the difference between *apparent* and *absolute* visual magnitude? Explain the reason for having an absolute magnitude.

■ See questions 27. 28 and 29.

11 THE HERTZSPRUNG–RUSSELL DIAGRAM

In 1913, Henry Norris Russell (USA, 1877–1957) had the ingenious idea of plotting stars according to two properties: their absolute magnitude M and their **spectral class** (or spectral type). Hot blue stars are labelled spectral class O, less hot and less blue stars are B type, yellowish-white stars like the Sun are labelled G, and so on.

Russell's plot looked like Fig 27.40(a). Note that he plotted the stars with the bluer stars on the left. As explained above, colour is related closely to the temperature of the star. This was realised by Ejnar Hertzsprung (Denmark, 1873–1967) and the plot is now called a **Hertzsprung–Russell diagram**, abbreviated to H–R diagram.

Modern astronomers use *luminosities* instead of the absolute magnitude values, and temperatures instead of spectral class, as shown in Fig 27.40(b). Being pretty independent in their ways, astronomers break the convention of labelling axes by scaling the x-axis with temperature *decreasing* as it goes from left to right.

Fig 27.40(a) **Left: The absolute magnitude of the brightest and nearest stars according to their spectral class**

Fig 27.40(b) **Right: The Hertzsprung–Russell diagram for the brightest stars in the sky**

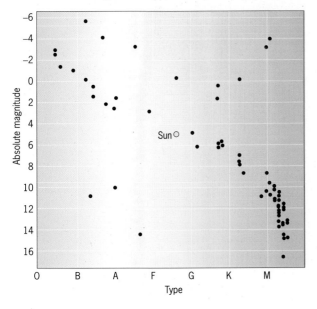

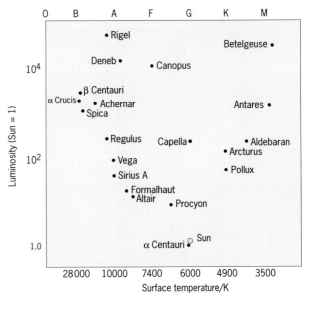

Colour	Blue-white	Blue-white	Blue-white	Bluish-white to white	White to yellowish-white	Yellow-orange	Reddish-orange
Spectral class	O	B	A	F	G	K	M
Surface temperature	30 000	11 000–30 000	7500–11000	6000–7500	5000–6000	3500–5000	about 3500

Table 27.4 **Relating spectral class to the surface temperatures of stars**

?

X (a) Why do we suppose that blue stars are hotter than red stars?
(b) How is it possible for some red stars to be as bright (even absolutely!) as blue stars?

See questions 30 and 31. ■

You might expect that the hotter the star, the brighter it will be, and in general this is true. But Hertzsprung found that some stars were exceptions to this rule. Some were very bright but reddish in colour, so must be quite cool. He reasoned that they were bright because they were very large: they emitted radiation at a fairly low intensity per unit area but looked bright because they have a very large surface area.

A typical star of this kind is the red star in the constellation of Orion, called Betelgeuse. This star has a surface temperature of only 3100 K but is 20 thousand times brighter than the Sun. It is so large that if it was in the Solar System its radius would extend to the orbit of Mars. Such a star is called a **red giant**.

There are also blue-white stars with a low luminosity: this means they must be small hot stars, and are named **white dwarf** stars.

Star groups in the H–R diagram

The stars on the H–R diagram of Fig 27.41 fall into three main groups. At the top of the diagram we have bright but rather cool stars. They are bright because they are large, as explained above, and are the **supergiant** and **giant** groups. At the very bottom of the diagram are the **white dwarf** stars, which are much fainter, but they range in temperature from about 6000 K to 20 000 K. Many stars plot on to a curving line, going from quite small but very bright and so very hot stars (blue white) to small faint cool stars (reddish white). These stars form the **main sequence** stars, and are, technically, also dwarf stars. The Sun is a typical main sequence star.

Fig 27.41 **Star groups on the Hertzsprung–Russell diagram**

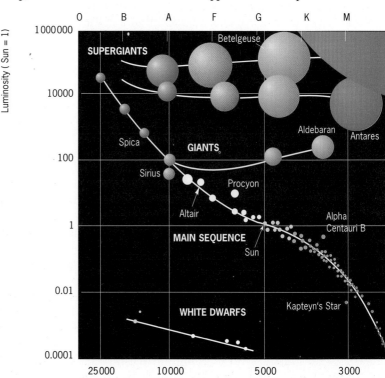

All the main sequence stars are small compared with the giants, but their masses vary from about a fifth of the Sun's mass to about 40 solar masses. A star's position on the main sequence is almost entirely decided by its mass and its age, as will be explained soon. Table 27.5 shows the main properties of some typical main sequence stars.

Table 27.5 **Properties of some main sequence stars**

Spectral class	Example	Surface temperature /K	Mass (Sun = 1)	Luminosity (Sun = 1)	Radius (Sun = 1)	Expected lifetime /10^6 years
O	Naos	40 000	40	500 000	20	1
B	Spica	15 000	7	800	4	80
A	Sirius	11 000	2	26	2	2 500
F	Procyon	6 600	1.3	2.5	1.2	5 000
G	Sun	5 800	1	1	1	10 000
K	Aldebaran	4 300	0.78	0.16	0.7	20 000
M	few named	3 300	0.21	0.01	0.3	50 000

Why large hot stars are rare

Star catalogues contain data on many thousands of stars. Most are dwarf stars that occupy the cooler regions of the main sequence, and all are comparatively close to Earth. Large hot stars are rare. This is because the hotter the star, the shorter its lifetime (See Table 27.5).

Any very hot stars that formed in our neighbourhood at the same time as the Sun (about 5 billion years ago) have stopped being very luminous. The reason is that increasing the mass of a star increases the density and temperature of its core. In fact, doubling a star's mass increases its energy production by nuclear fusion by more than ten times. So it uses up its nuclear fuel ten times as quickly. To explain this fully, we need to think about how a star is formed and what happens to it as time goes on. So now we look at **star formation** and **stellar evolution**.

12 LIFE AND DEATH OF STARS

Star formation – the cloud condensation model revisited

It is believed that all stars form in the way described for the Sun, that is, by the gravitational collapse of a large, very massive cloud of gas with a very tiny fraction of 'dust' (small solid particles). Large clouds may be a few light years in diameter with enough mass to produce stars which are much more massive than the Sun.

Large clouds tend to produce several stars rather than just one. Binary (double) stars are very common in the Milky Way; for example, Sirius, Algol, Capella, Alpha Centauri and Castor (Alpha Geminorum) are double stars, orbiting about a common centre of mass.

We also find many quite large groups of stars that are about the same age and probably formed from the same very large gas cloud at the same time. A typical example is the small constellation the **Pleiades** (Fig 27.42) which contains six very bright, young stars that can be seen with the naked eye. A telescope shows that Pleiades has about a hundred stars. Such groups are called **open clusters**, and some are close enough to Earth to make a splendid sight through good binoculars. (Look back also to Fig 27.11(b), the infrared image of a star-forming gas cloud.)

Fig 27.42 **The main stars in the Pleiades, an open cluster in the constellation of Taurus. They are relatively young – about 50 million years old – and surrounding them is the cloud of cold gas and interstellar dust left over from the cluster formation. Light is seen reflected by the cloud**

Reaching the main sequence

A new hot *protostar* gets its energy from gravitational collapse. When nuclear fusion begins, it joins the main sequence at a position that depends on its luminosity and temperature – it is now a *zero-age main sequence star*. Stars on the main sequence have reached a stable balance between their rate of energy production and energy emission, and are no longer collapsing under gravity forces. But how long does a star remain stable and so remain on the main sequence?

The mass–luminosity connection

Compared with the Sun, a bright A-type star like Sirius has twice the mass and so twice the available hydrogen as a source of energy. But Sirius has a shorter expectation of life than the Sun because it emits radiation at much more than twice the rate of the Sun. In general, the luminosity L of a star is approximately proportional to the third power of its mass:

$$L \propto M^3$$

With *available* energy E proportional to mass M, we can write:

$$E = \text{constant} \times M = k_1 M \qquad [1]$$

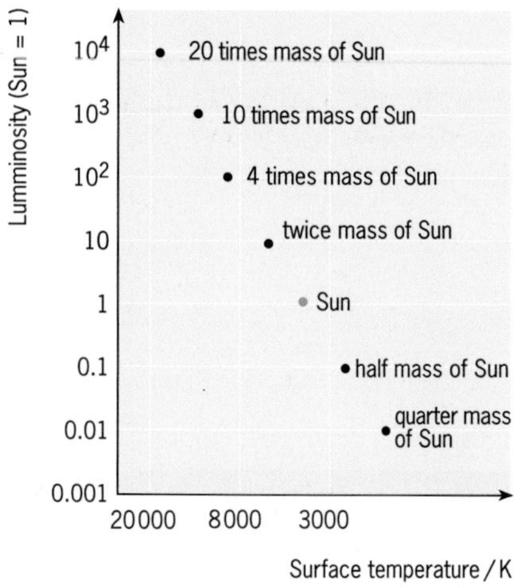

The rate of loss of energy (luminosity) $\Delta E/\Delta t$ is proportional to M^3, so:

$$\Delta E/\Delta t = k_2 M^3 \qquad [2]$$

where k_2 is some other constant. Stars tend to be in equilibrium for most of their lifetimes, so they emit energy at a steady rate. Thus, when the energy change ΔE equals all the available energy E then Δt represents the star's lifetime, T. Thus:

$$E/T = k_2 M^3 \quad \text{so that} \quad E = k_2 M^3 \times T \qquad [3]$$

So, combining equations 1 and 3 we can write:

$$E = k_1 M = k_1 k_2 M^3 \times T$$

or rearranging:

$$T = K/M^2 \quad \text{where } K \text{ is a new constant} \qquad [4]$$

Thus, a star of double the Sun's mass should have a quarter of its lifetime. The most massive stars may have 50 solar masses, with a life expectation of 1/2500th that of the Sun. On the other hand, a star of 0.1 solar mass will have a lifetime 100 times greater.

As a result, it is not surprising that the Milky Way has more old, small, dull stars than large, hot, bright ones. A more detailed theory predicts the lifetimes of stars of different masses, and matches the observed age of the Earth for example (about 5×10^9 years) and the Big Bang model for the age of the Universe.

Fig 27.43 **How mass, luminosity and temperature are linked for stars on the main sequence**

?

Y Check that the lifetimes of the star types listed in Table 27.5 are in general agreement with their mass and luminosity data.

See questions 33 and 34. ■

The deaths of stars

When main sequence stars eventually run out of their main nuclear fuel, hydrogen, what happens next depends on the pressure and temperature in the core. This in turn depends on the star's mass. First let us consider the fate of a star the size of the Sun.

13 THE DEATH OF A SMALL, SUN-SIZED STAR

Hydrogen fusion stops first in the central core, which then starts to cool down. But round the core is a spherical shell quite rich in hydrogen where fusion still occurs. As the inner core cools, its pressure decreases and gravity wins the long battle to make everything collapse. The hot hydrogen-fusing shell falls inwards and heats up even more as gravitational potential energy is lost.

The birth of a red giant

The rate of fusion in the collapsing shell increases, emitting more strong radiation, so that the next layer out in the star actually becomes hotter. This is the *convection zone* (see page 626), and it expands as it gets hotter. In fact, it expands so much that its outer surface, the *photosphere*, gets *cooler*. At this point, a star like the Sun would expand to reach as far as the orbit of the Earth.

The photosphere is now at about 3000 K instead of the 6000 K it was during the main hydrogen-burning period. But the increased surface area means that more energy is emitted and the star becomes much brighter – its luminosity has increased by a thousand times. The star is now a **red giant**. Looking at the H–R diagram of Fig 27.44, note that by now the star has moved away from the main sequence and up towards the top right.

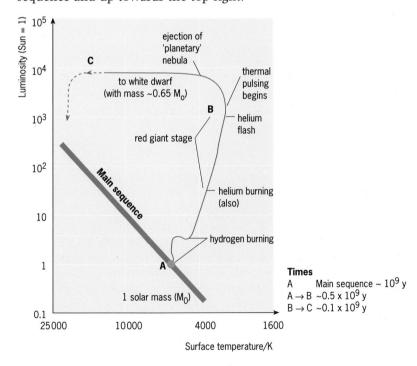

Fig 27.44 **Final stages in the evolution of a Sun-sized main sequence star**

Even red giants don't last for ever

The story doesn't end there. The core is still collapsing gravitationally, and still getting hotter. It is now also very dense with all its particles very close together. Most of the hydrogen is used up and now there is helium and electrons.

When the core is hot enough, another nuclear fusion process begins: helium is converted to carbon by the **triple alpha process**. Two helium-4 nuclei first combine to form beryllium-8; then this combines with another helium-4 nucleus to form carbon-12.

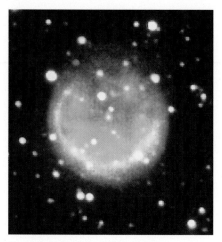

Fig 27.45 **A planetary nebula in the constellation of Aquila. You can see the shell of gas emitted from the blue-ringed dying star in the centre**

This new energy-producing process in the superdense core starts up very quickly – in only a few minutes – and may generate enough energy to heat up the adjacent layer which is now also rich in helium. This may then trigger explosive reactions, and matter in the next outer layer of the red giant may be blasted away – in a 'superwind' lasting for some tens of years. Then, as it cools, the star settles down again. Every few thousand years this 'helium burning flash' occurs again. A star like this seems to be pulsating, its size and luminosity varying over a long time scale.

Old red giants may become planetary nebulae

This time scale is too long to have been observed. The theory has been developed to account for objects we *can* see, called **planetary nebulae**. They are falsely called 'planetary' because early observers with small telescopes thought that they were planets. A planetary nebula is formed when the red giant has ejected enough material to leave the hot superdense core bare and in view. It illuminates the expanding shell of gas which now surrounds it. We can see many objects like this: one is shown in Fig 27.45.

White dwarf stars

The dense carbon core is not hot enough for any more fusion reactions and is now a **white dwarf**. As time goes on, it cools and becomes invisible as a **black dwarf** star.

Nuclear reactions in stars: the proton–proton chain

The first reaction to occur in the hot dense core of a star is the conversion of hydrogen to helium. This is the **proton–proton chain** (or pp chain). Look back at Fig 27.21. It begins when the temperature is at about 10^7 K, but works best at 10^8 K. At this temperature, the protons are moving fast enough on average for the speedier ones to overcome the electric repulsion force between the protons. When they get close enough, the strong nuclear force takes over and two protons bind together. They don't form a kind of helium, as you might expect, since the double-proton nucleus is unstable. A proton is converted into a neutron and the combination becomes deuterium, an isotope of hydrogen:

$$^1_1H + ^1_1H \rightarrow ^2_1H + ^0_1e^+ + \nu \qquad [1]$$

When a deuterium nucleus collides with a proton, it form an isotope of helium:

$$^2_1H + ^1_1H \rightarrow ^3_2He + \gamma \qquad [2]$$

The chain is completed when two helium-3 nuclei combine to form helium-4 (ordinary helium and also the alpha particle) and two loose protons:

$$^3_2He + ^3_2He \rightarrow ^4_2He + ^1_1H + ^1_1H \qquad [3]$$

The positive beta particle (positron) emitted in stage 1 soon meets an ordinary electron, its antiparticle, and they annihilate each other to produce a high-energy photon:

$$e^+ + e^- \rightarrow \gamma$$

The total energy produced in this reaction is due to the mass difference between the combining particles (4 protons plus the 2 electrons that have disappeared) and the final helium-4 nucleus.

For every helium-4 nucleus formed, 4.2×10^{-12} J is produced and is the main source of energy in ordinary stars. The neutrinos do not interact with matter and leave the core at almost the speed of light, carrying some energy with them. As described on page 641 and above, helium burning occurs in the triple alpha process when hydrogen is used up, producing carbon.

Further nuclear fusion reactions are possible which would build up carbon into larger nuclei – in theory nuclei up to iron. All these reactions are *exothermic*, converting mass to energy. But reactions making larger nuclei than iron are *endothermic* and take energy from the system (see page 384). Then, energy is converted to mass, and if this happens the core cools down very quickly – with possibly catastrophic results.

14 FATE OF A STAR MORE MASSIVE THAN THE SUN

In an ordinary smallish star like the Sun, the triple alpha reaction would be the end of the story. When all the helium is converted to carbon, the star's temperature is too low to trigger further fusion reactions and there is not enough gravitational potential energy available to raise the temperature by further collapse.

But stars more massive than the Sun may continue to create heavier nuclei *and* gain energy. For example, carbon-12 and helium-4 combine to form oxygen-16, oxygen-16 and helium-4 combine to form neon-20, and so on. The theory supporting all this is backed up by laboratory experiments on Earth, and the theory predicts quite accurately the proportions of elements detected in the atmospheres of stars.

A five solar mass star

A main sequence star of five solar masses is a bright star, not only larger than the Sun but with a surface temperature of over 20 000 K. It gains its energy by converting its hydrogen to helium, but by a faster process than the simple proton–proton chain. The temperature in its core is more than double that in the Sun – about 3.4×10^7 K. Such a star emits 500 times as much energy per second as the Sun and, as explained above, it will spend less time on the main sequence. Its lifetime in this phase is only 400 million years, compared to 10 billion years for the Sun.

The first phase after hydrogen burning is much the same as for a Sun-sized star: burning helium, the star becomes a **red giant**.

As before, the energy released blasts away the outer gas atmosphere and it changes from a red giant to a smaller white star. The star is now a helium star as all its hydrogen has been used up or blown away. The triple alpha process in the core then provides its energy for about 10 million years.

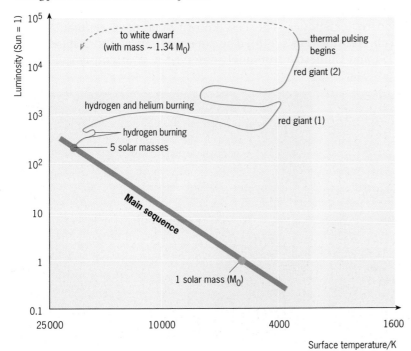

Fig 27.46 **The evolution of a star of five solar masses. Stars larger than five solar masses follow a similar track but they may become supernovae that explode and end up as neutron stars**

The death of the five solar mass star

When the helium in the core is all converted to carbon, it ceases to produce energy. Then the star contracts again. There are two possibilities for what happens next.

A degenerate black dwarf. . .

One is that the star continues to lose mass by its stellar wind and quietly becomes a white dwarf star, entirely of carbon and about 1.3 to 1.5 times the Sun's mass. With this mass the particles in the core becomes so packed together that gravitational forces cannot make the star any smaller. In this state, the matter is said to be *degenerate*. There is no energy to be had from gravitational contraction. The core gradually cools down and the end point is a **black dwarf** star.

. . . or a supernova?

Alternatively, the star might still be massive enough for its outer layers *not* to be degenerate. The star can still gain energy from gravitational contraction and so the carbon core can become hot enough for carbon burning to take place. The core is practically solid carbon, with nuclei and electrons very close together. Carbon nuclei convert to larger nuclei almost instantaneously across the entire core of perhaps 1.4 solar masses. The energy released in a few milliseconds is immense: *the rate of energy emission is equivalent to the power emitted by a whole galaxy of ten billion stars.* What we observe is a new *very* bright star appearing in the sky, a **supernova**. This a **type I** supernova.

All the outer layers of the star are ejected at very high speeds – up to 10^7 m s^{-1}. The supernova reaches its peak of visibility after about 40 days when the hot material has expanded to the size of the Solar System.

Fig 27.47 shows what is thought to be a supernova remnant. We see the red filaments where particles of the remnant are colliding with the low density gases (mostly hydrogen) that fill the space between stars.

In a typical galaxy, supernovae occur about once every 50 years. This hasn't happened nearby recently, so we haven't been able to study this spectacular astronomical sight at close quarters with modern techniques. The best known supernova remnant is the **Crab Nebula** (Fig 27.48). The supernova explosion happened in 1054 and Chinese astronomers observed it, courteously calling it a 'guest star'.

Fig 27.47 **Left: The Vela supernova remnant of a star that exploded about 12 000 years ago**

Fig 27.48 **Right: The Crab nebula. The gas blown away from the star is green, yellow and red, and the blue glow is of energetic electrons spiralling through a magnetic field round the star's remnant, which is a pulsar – a rotating neutron star. (The nebula is 10 light years across, and about 7000 light years from Earth)**

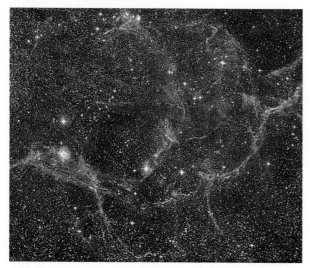

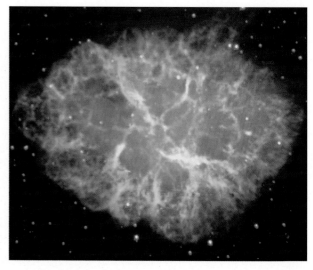

Very massive stars become neutron stars

Stars with masses greater than about 8 solar masses develop central cores hot enough to trigger carbon burning in a steady non-explosive way. Nuclear fusion reactions gradually make successively larger nuclei until the central core is made entirely of iron nuclei – the largest nucleus which can be made with a net *release* of energy. Outside this iron-building core are layers of gradually decreasing temperature in which other reactions take place, each producing as heavy a nucleus as it can for the ambient temperature, see Fig 27.49.

Eventually, iron-building stops and, as we have seen before, the core stops producing energy and begins to contract under its own huge gravitational forces. The inner core gets hotter and hotter until at a temperature of 5 billion kelvin (5×10^9 K) the laws of physics do the unexpected.

The outer layers of the star, and even the core itself, is partly supported by the *radiation pressure* of the photons in hot plasma. At a temperature over 5 billion kelvin, the photons have enough energy to break down the iron nuclei into helium nuclei. But in doing this, the photons lose a great deal of energy, so their radiation pressure suddenly drops and there is a rapid gravitational collapse.

The outer layers fall catastrophically onto the core, which is still mostly iron. The core, originally about the size of the Earth, is compressed to a small sphere with a radius of just 50 km. This produces a concentration of matter which is denser than an atomic nucleus, a density at which electrons and the helium protons are forced together to become neutrons and neutrinos.

Meanwhile, the much greater mass of stellar material round the small core is collapsing inwards while undergoing different kinds of nuclear fusion. The temperature of this mass increases tremendously as its potential energy is lost, making the rate of the nuclear reactions increase explosively. The exploding matter falls onto the neutron core (Fig 27.50), bounces off it and creates a shock wave which blows all the outer material into space at very high speed. What we observe is another supernova explosion, called a **type II**. This leaves behind the hot superdense core, a **neutron star**, which has a very strong magnetic field.

Pulsars

A **pulsar** is a rotating neutron star that emits short bursts (pulses) of radiation at radio wavelengths and at very regular intervals. See Figs 27.48 and 27.52.

The first pulsar was discovered in 1967 by 22-year-old research student Jocelyn Bell (now Burnell) and Professor Anthony Hewish whose team worked at the Mullard Radio Astronomy Observatory in Cambridge, England.

Jocelyn Bell noticed that one of the many radio sources detected by a very simple radio telescope produced sharp pulses at a very steady rate of just under one a second. The team wondered whether this could be a signal from some kind of extra-terrestrial civilisation.

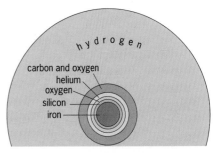

Fig 27.49 **Late in the life of a very massive star, there are different types of nuclear fusion in the layers. When the iron core reaches a critically high temperature, its nuclei disintegrate into helium. The core then cools, collapses and triggers a supernova explosion**

outward rebound blast → infalling material ←
flux of neutrinos due to disintegration of iron nuclei →

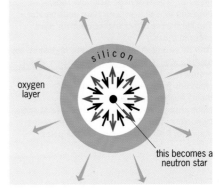

Fig 27.50 **The massive star's core of neutrons collapses, and the surrounding material (in which reactions continue) falls into it and bounces off**

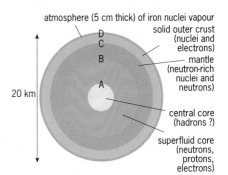

Point	Distance from centre/km	Density/kg m^{-3}
A	1	7×10^{17}
B	7	2.4×10^{17}
C	9	4×10^{14}
D	10	7×10^9

Mass ≳ 3 solar masses

Fig 27.51 **The structure of a neutron star, based on theories of gravitational attraction and nuclear particles**

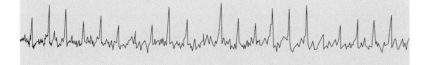

Fig 27.52 **The signals from pulsar CPO 328, recorded with the radio telescope at Nançay in France. The pulses last 7 milliseconds and occur at intervals of precisely 0.714 518 603 seconds**

But more observations revealed similar signals from different parts of the sky. It was clear that this was a new type of astronomical object – a pulsating radio star or **pulsar**. Professor Hewish was awarded the Nobel Prize for the discovery in 1974.

The nature of pulsars was worked out by Thomas Gold (Austria and USA). In 1968 he suggested that they were in fact rotating neutron stars. A neutron star, with its strong magnetic field, has condensed from a much larger star which was rotating. So it will still have much the same angular momentum and, being very small, it should rotate very rapidly. A rotating magnetic field will emit radio waves continually, and it will do so along the line defined by the most strong concentration of lines of force (Fig 27.53).

As the neutron star rotates, its radio signal is detected by a radio telescope on Earth every time the magnetic poles face Earth. This is the *lighthouse model* of pulsars: the radio signal is seen at intervals, just as we see a flash when the rotating shield moves away from the light in a lighthouse.

Fig 27.53 **The magnetic field and radio emission of a pulsar, a rotating neutron star. The lighthouse effect makes neutron stars seem to pulse.**
With its strong magnetic field and high spin, the neutron star is a dynamo. The dynamo effect produces a strong electric field which pulls electrons from the surface and accelerates them. Accelerated electrons emit electromagnetic radiation, in this case at radio frequency. The effect is strongest where the magnetic field is strongest – at the poles

Pulsar: a rotating neutron star

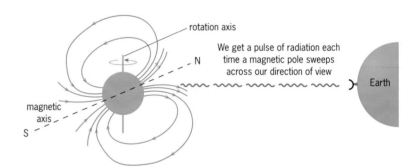

The signal from the Crab pulsar (Fig 27.48) reaches Earth every 0.033 s, which means it rotates 30 times a second. It emits radiation at all wavelengths from radio to X-radiation, and its total pulse radiation loss is 10^{28} W. This is 100 times the total radiation emitted by the Sun. Earth also receives non-pulsed radiation from it, and its total luminosity, of 10^{31} W, is 10 000 times greater than the Sun's.

See questions 35 and 36. ■

Supernovae and nucleosynthesis

The elements nickel, copper, gold, tin, platinum, lead, uranium, thorium – in fact, all elements with nuclei more massive than the iron nucleus – were formed in a type II supernova explosion. When the core collapses to form neutrons, many are ejected into the surrounding gas cloud that contains a range of nuclei. The neutrons penetrate nuclei freely and convert them to heavier and heavier nuclei. These then undergo radioactive decay (beta decay) to form stable elements.

The ejected gas cloud from a supernova explosion is later recycled to form other stars and planets. Apart from hydrogen, all the elements in your body were made in the core or atmosphere of an exploding star.

Black holes

It has been calculated that when a supernova remnant has a mass greater than about 2.5 solar masses, the neutron star formed is so dense and massive that no radiation can possibly escape from it. Then, the neutron star is a **black hole**.

Electromagnetic radiation is a form of energy, and according to relativity theory, the energy has an associated mass given by the Einstein formula $E = mc^2$. A photon of energy E (ie hf) has a mass:

$$m = E/c^2 = hf/c^2$$

The gravitational field of the neutron star acts on this effective mass. Its effect is not to *slow down* light leaving the star since this is forbidden by the laws of relativity. But the energy of the photons gets less until they reach zero, that is, until they cease to exist. Another way of looking at it is to consider the idea of *escape speed* (see page 66). An object is a black hole if the escape speed from its surface is greater than the speed of light.

The theory of black holes was first worked out by Karl Schwarzschild (Germany 1873–1916). The largest radius that an object can have and still be a black hole is called the *Schwarzschild radius*, and depends on the mass of the object. For an object to be a black hole, it has to be both very massive and very small. These conditions are satisfied only when matter is crushed to a density much greater than that of a neutron star.

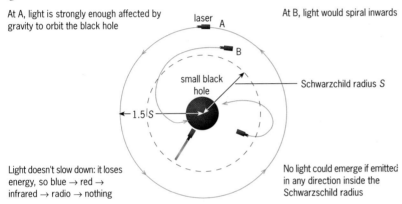

At A, light is strongly enough affected by gravity to orbit the black hole

At B, light would spiral inwards

laser A

B

small black hole

Schwarzschild radius S

1.5 S

Light doesn't slow down: it loses energy, so blue → red → infrared → radio → nothing

No light could emerge if emitted in any direction inside the Schwarzschild radius

Fig 27.54 **Light and the black hole**

?

Z The escape speed from a black hole is greater than the speed of light. Describe what happens to a photon emitted from inside the hole along a radius, bearing in mind that light cannot slow down.

How big would the Sun be if it became a black hole?

The escape speed from a spherical mass is given by the formula:

$$v = \sqrt{\frac{2GM}{r}}$$

where M is the mass and r is its radius. G is the universal gravitational constant, 6.67×10^{-11} N m^2 kg^{-2} (or m^3 kg^{-1} s^{-2}) (see page 47). Thus for light speed (3×10^8 m s^{-1}) and the Sun's mass (2×10^{30} kg) we can calculate the Schwarzschild radius for the Sun as:

$$r = 2GM/c^2$$

$$= \frac{2 \times 6\cdot67 \times 10^{-11} \times 2 \times 10^{30}}{9 \times 10^{16}} \text{ m}$$

$$= 3000 \text{ m}$$

The Schwarzschild radius is simply proportional to mass M. So, to be a black hole, the Earth, which has a mass about a three hundred thousandths of the Sun's mass, would have a Schwarzschild radius of just 1 cm!

■ See question 37.

On the edge of the event horizon

The matter could be so dense in a black hole that its actual radius is smaller than its Schwarzschild radius. Then, the Schwarzschild radius defines a sphere called the **event horizon**. Nothing can escape from any point inside the event horizon.

Detecting black holes

As neither radiation nor matter can leave a black hole, how do astronomers ever find one?

Imagine a situation in which some material, such as gas or dust particles, is falling into a black hole. As it gets closer and closer, the material moves faster and faster. Collisions cause ionisation and the accelerated ions emit electromagnetic radiation in a very intense, directional polarised beam called *synchrotron radiation*.

The acceleration of particles close to (but outside) the event horizon of a black hole will be so great that the radiation is extremely energetic and will consist of X-ray photons. This suggests that where astronomers detect X-rays, they might find black holes.

Many stars are binary stars. Such stars orbit round a common centre of mass. Quite often, one star is large, massive and bright, whilst its companion is too faint to see. There is a star in the constellation Cygnus named Cygnus X–1 which orbits an invisible companion once every 5.6 days. We know this because its light has a Doppler shift with this period. The star has a high luminosity: it must be a very hot supergiant with a mass at least 15 times that of the Sun and a surface temperature of over 30 000 K. For a star of this mass to move at such a high speed of rotation, its invisible companion must also be very massive. It could, of course, be just a very large dead star, but theory suggests that very large dead stars should turn into black holes.

Another clue is that, as well as having the ordinary radiation expected from a large bright star, Cygnus X–1 also emits vast quantities of X-radiation – in fact at a rate of about 4×10^{30} W, which is 10 000 times the total power radiated by the Sun.

Fig 27.55 **The black hole binary model for X-ray stars. Matter from the supergiant star is pulled away by the intense gravity field of the black hole. By the time it reaches the accretion disc, it is at a temperature of 10^6 K – hot enough to emit X-rays**

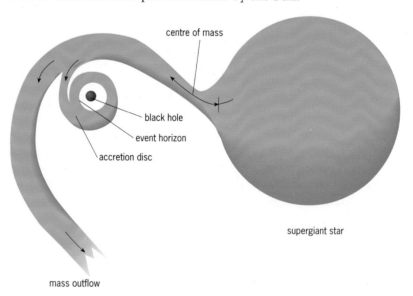

To explain these facts, astronomers have suggested the arrangement shown in Fig 27.55. The supergiant star is close enough to a black hole for its matter to be pulled away to form a disc round the black hole. Some of the material spirals into the black hole and some is propelled away by the gravitational slingshot effect. The accelerated matter moves fast enough to be ionised and so become the source of very intense X-ray photons.

14 THE WIDER UNIVERSE

This chapter has dealt with the basic physics of stars, and of the only planetary system we know about – our own Solar System. (In 1996 evidence was found for the existence of large planets round a few nearby stars.)

Astronomers can study in detail only the nearest stars. These are relatively *very* near, being our closest neighbours amongst the estimated 100 billion stars in the Milky Way, which is 100 000 light years wide. And there are billions of galaxies in the observed Universe.

What exactly a galaxy is, and what other objects exist even further away than this is the topic of the next chapter, where we move on to even larger scales of time, energy and distance.

SUMMARY

Having studied this chapter, you should:

■ Know about the main features of the solar planetary system.

■ Know that the motion of planets is described by Kepler's three laws, and how they are related to Newton's laws of motion and gravitation.

■ Know the current model of the Sun, its source of energy and the formation of the Sun and planets by the gravitational collapse of a large cloud of gas and dust.

■ Understand the nature and formation of the inner and outer groups of planets and know the differences between the two groups.

■ Understand the origin and nature of comets and meteorites.

■ Know how we obtain information about the Sun from a spectroscopic study of its radiation.

■ Know the distance units used in astronomy (astronomical unit, parsec and light year).

■ Know how it is possible to measure the mass, temperature and energy output (luminosity) of a star from measurements of its distance and a study of its spectrum.

■ Understand how binary stars are used to measure stellar masses and so relate mass and luminosity for various star types.

■ Know how stars are classified according to their spectra, and how this classification is related to surface temperature and luminosity, as illustrated by of the Hertzsprung–Russell (H–R) diagram.

■ Know that stars are also classified according to their brightness, measured as apparent magnitude and absolute magnitude.

■ Understand the formation and evolution of stars of different masses, leading to such objects and events as red giant stars, white dwarf stars, planetary nebulae, supernova explosions, neutron stars, pulsars and black holes.

QUESTIONS

The Solar System, and the Sun as a star

1 Explain the meaning of the following:
(a) Oort Cloud, **(b)** satellite, **(c)** planetesimal,
(d) parsec, **(e)** light year.

2 Explain qualitatively (ie without using mathematical formulae) the physical reason why planets further from the Sun move more slowly in their orbits than planets closer to the Sun.

3 Explain the meaning of *parallax*. How can a measurement of parallax be used to determine the distance of an object you cannot reach, like a star or a ship at sea?

4 Use the data on the planets in Table 27.1, page 614, to answer the following:
a)   **(i)** Calculate the orbital speed of Mars.
   (ii) Check that Kepler's third law applies to the planets Mercury and Mars.
b) Use the value of orbital speed you calculated in **a) (i)** to obtain a value for the mass of the Sun.
$$ (v = \sqrt{\frac{GM}{r}}; \text{take } G = 6.67 \times 11^{-11} \text{ m}^3 \text{ kg}^{-1} \text{ s}^{-2}). $$

5 The planets in the Solar System can be placed into two main groups.
a) Describe the main planetary features that distinguish between the groups.
b) Suggest how the way that the Solar System was formed may have produced these differences.

6 The table gives some data for the four largest satellites of Jupiter.

Satellite	Distance from Jupiter/10^3 km	Orbital period/days
Io	422	1.8
Europa	671	3.6
Ganymede	1070	
Callisto	1880	

Table 27.Q6

a) Complete the table, calculating the periods of Ganymede and Callisto. (Use Kepler's third law.)
b) show that the mass of Jupiter is 1.90×10^{27} kg. (Take $G = 6.67 \times 10^{-11}$ m^3 kg^{-1} s^{-2})

7 How does the condensation model of the formation of the Solar System explain:
a) how a spherical gas cloud becomes a flattened disc,
b) why the disk is rotating?

8 Describe the observational evidence for the theory that stars form from a cloud of gas and dust which collapses and at the same time gradually heats up.

9 Suggest why most stars occur in small groups that are quite close together and roughly the same age. (Remember that these groups are *not* constellations.)

10 Explain how the spectrum of a star can give information about:
a) its surface temperature,
b) the elements in its atmosphere,
c) its speed relative to Earth.

11
a) Explain briefly the meaning of **(i)** plasma, **(ii)** nuclear fusion.
b) How can nuclear fusion be a source of energy? How does this energy make the centre of the Sun *hot*?

12
a)   **(i)** What are the three stages in the formation of helium from hydrogen in the interior of a star like the Sun?
   (ii) Why is this process a source of energy?
b) The net result of the fusion process in the Sun (the *proton–proton chain*) is summarised as follows:
$$ 4^1\text{H} \rightarrow {}^4\text{He} + 2e^+ + 2v + 2\gamma + 24.68 \text{ MeV} $$
   (i) Explain the meaning of each of the symbols used in the reaction above.
   (ii) Calculate the energy in joules produced by the process. (Electronic charge = 1.60×10^{-19} C.)

13 Describe briefly how energy produced in the core of the Sun reaches the surface to be emitted as radiation.

14 Explain the meaning of the following terms:
a) photosphere, **b)** convective zone, **c)** corona, **d)** Fraunhofer line.

15
a) What is meant by the *aperture* of a telescope?
b) Give two reasons why it is an advantage for telescopes to have as large an aperture as possible.

16 What is meant by the *magnifying power* of a telescope? What determines how big the magnifying power is?

17 Why does the dish of a radio telescope have to be much larger than the objective mirror of an optical telescope?

18
a) Compare the resolving powers of **(i)** a radio telescope of dish diameter 50 m, used to receive radiation of frequency 150 MHz, **(ii)** an optical telescope with a mirror diameter of 4 m.
b) Explain why it is difficult using such telescopes to identify a radio source with any particular optical source, such as a star or a galaxy.

19 Give two reasons why a space telescope, such as the Hubble Space telescope, is likely to produce better information from astronomical objects than ground-based telescopes.

20 Describe and explain any technique you know of that enhances the image quality produced by a ground-based telescope without the need to make a single very large dish or receiver.

21 Describe the special problems for astronomers who want to investigate astronomical objects using **a)** X-rays, **b)** ultraviolet radiation, **c)** infrared radiation.

22 When the star Alpha Centauri is viewed from Earth on two occasions six months apart, the star appears to have changed its position relative to some very distant stars by 0.750 arcsec. Calculate the distance of Alpha Centauri from Earth in: **a)** parsecs, **b)** light years.

23 The following measurements were made for a particular star:
A. the wavelength at which its radiation intensity was a maximum.
B. the distance of the star from Earth.
C. the total radiation energy per unit area per second reaching the Earth.
Explain how these measurements can be used to estimate:
a) the surface temperature,
b) the luminosity of the star.

24 A binary system has two stars of mass 3 and 8 solar masses, separated by a distance of 120 AU.
a) How far from the larger star is the centre of mass of the system?
b) What is the period (in Earth years) of the smaller star round the common centre of mass?

25 Explain what is meant by the terms: **a)** eclipsing binary, **b)** spectroscopic binary stars.

26 Explain how spectrographic measurements from a binary star system could be used to calculate:
a) their (apparent) orbital speeds,
b) their orbital periods.

27 Explain the difference between the *apparent* and *absolute* magnitudes of a star.

28 The star Spica has an apparent (visual) magnitude of +1. On a particular night the planet Mars has an apparent visual magnitude of +2.
a) Which would look brighter?
b) How much brighter?

29 The relation between apparent magnitude m and absolute visual magnitude M for a star is given by $M = m + 5 - 5\log D$, where D is the distance of the star from Earth in parsecs.
Calculate the absolute visual magnitudes of the following stars:

Star	Distance/pc	Apparent visual magnitude	
Capella	13	+0.08	Table 27.Q29
Rigel	430	+0.12	
Procyon	3.4	+0.38	

30 Explain the meanings of **a)** spectral class, **b)** luminosity, **c)** Hertzsprung–Russell diagram.

31 What, in terms of the H–R diagram, is meant by **a)** main sequence, **b)** white dwarf, **c)** giant star?

32 Explain how **a)** the spectral class of a star is related to its surface temperature, **b)** the luminosity of a star is related to its mass.

33 Why do large massive stars tend be very hot and have comparatively short lifetimes?

34 Compare the evolution of a star of one solar mass and that of a more massive star of, say, 5 solar masses. Illustrate your answer with a diagram showing evolutionary tracks on a Hertzsprung–Russell diagram.

35 It is likely that pulsars are actually rapidly rotating neutron stars. What is the evidence for this?

36 When first observed, the radiation from a pulsar has a pulsing rate (frequency) of 0.200 000 per second. It emits radiation energy at a rate of 5.00×10^{30} W.
a) Calculate how quickly the pulsar rotates in:
  (i) rotations per second, **(ii)** radians per second.
b) The pulsar is observed for 4 years (1.26230×10^8 s) and it is found that the pulsing rate has decreased in that time to 0.199 800 per second.
  (i) How much energy has the pulsar lost via radiation in the four years?
  (ii) Calculate its new angular velocity in radians per second.
c) Suppose the lost radiation energy was supplied by a reduction in the kinetic energy of rotation of the pulsar. The kinetic energy of a spinning body is given by $E_K = \frac{1}{2}I\omega^2$,
where ω is the angular velocity and I is a constant which depends upon the mass of the body, its shape and its radius. Estimate the value of I.
d) The simplest model for a rotating neutron star estimates that it has a radius of from 10 to 20 km and a uniform density. This particular neutron star has a radius of 15 km, and the value of I for a sphere of mass M and radius r is $0.2\,Mr^2$. Estimate the mass of the neutron star.

37 Describe and explain what should in theory happen to a beam of light which might possibly be emitted from a body inside a gravitational black hole.

See page 667 for the CHAPTER MAP that covers Chapters 27 and 28

28 Cosmology

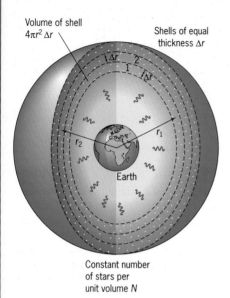

Volume of shell $4\pi r^2 \Delta r$

Shells of equal thickness Δr

Earth

Constant number of stars per unit volume N

Number of stars per unit volume of space = N

Volume of a spherical shell centred on Earth = $4\pi r^2 \Delta r$

So number of stars in each shell is proportional to r^2

Assume a constant mix of star types, on average; then light emitted from a shell is also proportional to r^2

But light obeys the inverse square law, getting weaker with distance, in proportion to $1/r^2$

So each shell produces the same intensity of light at the Earth. If there are an infinite number of shells, the night sky should be infinitely bright!

Olbers' paradox. The intensity of light from each shell of stars decreases as r increases. But the number of stars in each shell increases as r increases – just enough to compensate

1 parsec = 3.26 light years.

WITH SO MANY STARS in the sky, why is it dark at night? This question puzzled astronomers in the seventeenth century and came to be known as Olbers' paradox, after Heinrich Olbers (Germany, 1758–1840) who posed it.

'The Newtonian Universe is infinite in size and contains an infinite number of stars', astronomers thought, 'So everywhere you look, your line of sight should end on the surface of a star, and the whole night sky should be as bright as the surface of an average star.' This is clearly not the case. Even the greatest astronomers could find no flaw with the argument, which is why it is called a paradox. The margin gives the argument mathematically.

But Olbers' paradox is no longer a puzzle. The Universe is still infinite and may well contain an infinite number of stars, but it is no longer a static Newtonian universe. Because of the way it is expanding, light from the very distant stars will never reach us: they are receding so rapidly that the energy of the light they emit is effectively reduced to zero. This conveniently leaves us with darkness at night.

Cosmology – the study of the Universe

The Universe contains not only the very large – such as the huge groups of stars called **galaxies**. It also contains the very small – the particles of matter dealt with in Chapter 26. You will come across both in this chapter, which deals with the main features of the Universe, its age and size, and links the current theory of the Universe in the **hot Big Bang model**.

1 THE MAIN FEATURES OF THE UNIVERSE

Beyond the Solar System

The stars we see with the unaided eye are the nearest stars to Earth, at distances ranging from just over 1 parsec (the star Alpha Centauri) to 500 or so parsecs.

We see the Milky Way as a broad, dispersed band of light in the night sky. A pair of binoculars shows that it is composed of a host of faint stars, all very close together in the field of view.

There are an estimated hundred billion stars in the Milky Way system. A collection of stars this size is called a **galaxy**. Plotted in the three dimensions, the stars form the patterns shown in Figs 28.2(a) and (b).

Most of the stars lie in a narrow **disc** with a central globular **nucleus**. Surrounding the disk is a larger elliptical region, the **halo**, which contains lone stars at a lower density, and a number of

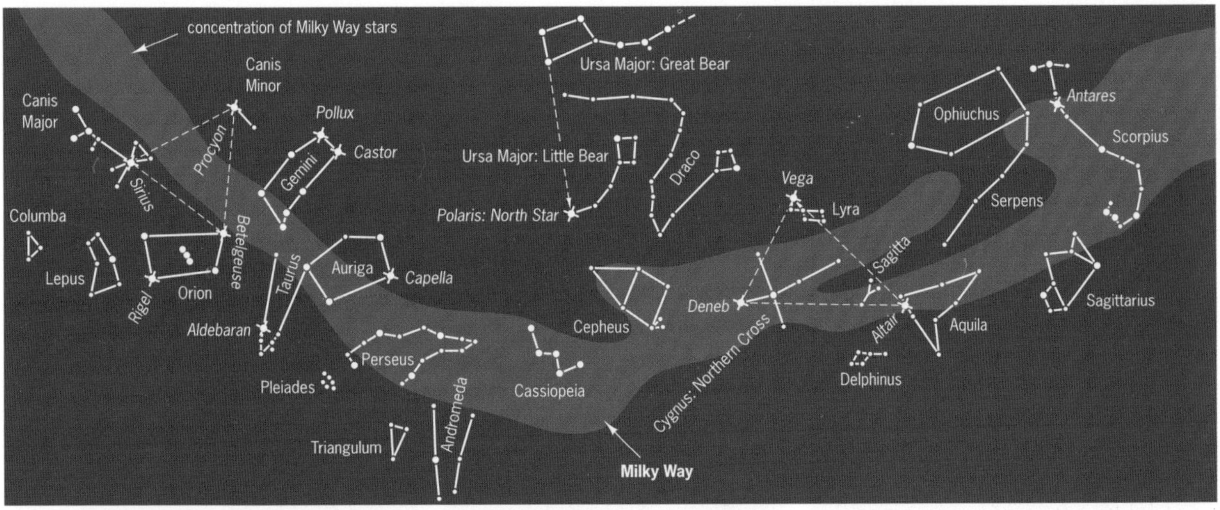

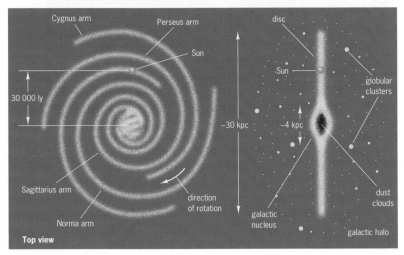

Fig 28.1 **Constellations in the Milky Way seen in the northern sky**

Fig 28.2(a) **Left: Top view of the Milky Way galaxy. The spiral arms of our galaxy have been mapped using radio waves emitted by the gas clouds they contain**

Fig 28.2(b) **Right: Schematic model of the Milky Way galaxy. The Sun is at the edge of a spiral arm, about 9 kpc from the galactic centre**

globular star clusters. Each cluster is 2 to 100 pc across, containing about a thousand to a million Sun-sized stars of roughly the same age, packed closely together.

As Fig 28.2 shows, the stars are arranged in long arms that spiral out from the nucleus, forming a disc 30 kpc in diameter. The Sun is a star of slightly below average size on the edge of a spiral arm, about two-thirds of the way out from the galactic centre.

Fig 28.3 **The Milky Way in a northern summer sky, seen as a band of stars and illuminated gas nebulae. Dust clouds block out the light of some stars**

Beyond the Milky Way

Early astronomers discovered many bright objects larger than stars and looking like clouds of bright gas and dust, so they were called **nebulae** after the Latin for 'clouds'.

Charles Messier (France, 1730–1817) first drew up a list of nebulae, stellar clusters and galaxies, which was published as the *Messier Catalogue*. Most of the bright nebulae in the northern half of the sky are still labelled by Messier numbers. For example, the Andromeda Galaxy (Fig 28.4) is Messier 31, or M31 for short. In another system, used for numbering nebulae, galaxies and clusters, the letters NGC prefix the number.

Later, astronomers found that many of these nebulae were not clouds of gas but collections of very many billions of stars, very much further away than the ordinary stars in the Milky Way. They were called **extra-galactic** objects.

Fig 28.4 **The Andromeda Galaxy, the only galaxy we can see without a telescope. It is about twice as large as our own galaxy and is 770 kpc distant. (The two other bright objects are also galaxies)**

■ See questions 1 and 2.

Fig 28.6 **The Southern Pinwheel spiral galaxy M83, some 30 000 ly in diameter and 10 million ly distant in the constellation of Hydra**

Fig 28.7 **The elliptical galaxy NGC 1199**

Fig 28.8 **The irregular galaxies the Large and Small Magellanic Clouds. They are the Milky Way's nearest-neighbour galaxies, in the Local Group, at a distance of about two Milky Way diameters**

See question3. ■

2 CLASSES OF GALAXIES

Galaxies fall into three main classes (see Figs 28.5 and 6):
● Spiral galaxies, rather like the Milky Way.
● Elliptical galaxies, with no apparent structure.
● Irregular galaxies, with no definite shape or structure.

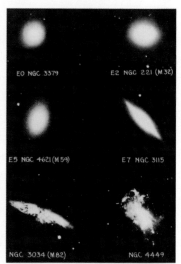

Fig 28.5 **In 1925, the American astronomer Edwin Hubble classified galaxies according to their appearance, in a system which has been widely accepted with minor revisions. These two pictures show the three main classes. Left: kinds of *spiral galaxies* that differ in clarity of nucleus, shape of arms and whether or not a dark central bar is present. Right: The first four are *elliptical galaxies*, and the bottom two are *irregular galaxies***

Spiral galaxies

A spiral galaxy is typically 30 kpc in diameter. Typically, it is a bright object in the sky, with many young hot stars in the arms, and with about 10^{11} stars producing a total luminosity 10^{10} times that of the Sun. It spins about its centre of mass, as in Fig 28.2(a).

The spectra of the stars in the arms show that they contain heavy elements (metals) which can only have been made in a supernova explosion (see page 644), so these stars are recycling material that has formed at least one star already. A spiral galaxy also contains a lot of hydrogen as **interstellar gas** at a density of about 6 atoms per cubic metre of space between stars.

We detect the gas from the radio waves it emits at a wavelength of 21 cm. Astronomers have worked out the positions of the spiral arms in our Galaxy by plotting the Doppler change to this wavelength caused by their rotation.

The separate arms of a spiral galaxy are assumed to have been caused by huge compressive waves passing through the cloud of gas and dust from which the galaxy was formed. These waves compress the cloud and trigger the gravitational collapse that eventually leads to star formation.

Elliptical galaxies

Elliptical galaxies contain mostly old stars. They are not very bright and rarely contain traces of heavy elements (metals). On average, they are smaller than spiral galaxies, at 10 kpc or so, and much less luminous, at about 10^8 solar luminosity. Elliptical galaxies have little gas and dust, so it is likely that star formation has ceased.

Irregular galaxies

Irregular galaxies are smaller than the first two classes, at an average size of 7 kpc. They have a mass of about 10^8 solar masses but at 10^9 solar luminosities they are quite bright for their mass. They contain young, metal-poor stars but are rich in gas and dust, so they have regions of active star formation.

3 MEASURING THE SIZES, MASSES AND DISTANCES OF GALAXIES

Table 28.1 lists some of the distances of objects in the Universe.

Table 28.1

Object		Distance	Measurement method/s
Sirius	a nearby star	2.6 pc	parallax (trigonometry)
Large Magellanic Cloud	small spiral galaxy close to Milky Way Galaxy	50 kpc	Cepheid variables
Andromeda Galaxy M31	largest galaxy (spiral) in our Local Cluster	770 kpc	Cepheid variables
Virgo Cluster	nearest large cluster of galaxies in our local supercluster	20 Mpc	Cepheid variables, novas (exploding stars), supergiant stars
Hercules Supercluster	nearest supercluster to ours	200–300 Mpc	supernovas, brightest galaxies in a cluster
Hydra Cluster	a large cluster of galaxies	1200 Mpc	brightest galaxies in a cluster
3C–324	furthest galaxy so far measured (radio)	2500 Mpc	red shift assuming $H = 85$ km s^{-1} Mpc^{-1}
Q0000-263	furthest quasar so far measured	3600 Mpc	red shift, as above

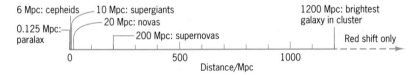

Fig 28.9 **Measuring astronomical objects**

The distances to the nearest stars can be found by triangulation, the method of **parallax** described on page 632. But extra-galactic objects such as galaxies are too far away to show parallax.

The nearest large galaxy to Earth is the Andromeda Galaxy (M31), shown in Fig 28.4. In 1923, it was the first to have its distance measured. Astronomers assumed that it was twice as massive as the Milky Way and so should have twice its overall luminosity. The light output received on Earth was measured and related to its distance d as described for stars on page 633.

$$\text{Flux density of light received } F = \frac{L}{4\pi d^2}$$

This simple, rather inaccurate measurement gave a distance at 450 kiloparsec which placed M31 well beyond the furthest stars in the Milky Way. The corrected distance is 770 ±30 kpc.

Distance, brightness and cepheid variables

Edwin Hubble (1889–1953) was the first astronomer to study galaxies and their distances in detail, observing with the then largest telescope in the world, the 2.5 m diameter Newtonian reflector at Mount Wilson in California.

He also used a very good method for measuring the distances to galaxies based on an odd kind of star called a **cepheid variable**, or cepheid for short. Delta Cephei in the constellation Cepheus was the first discovered, and later, cepheid variables were detected in M31.

These stars vary in their brightness very regularly over time, as shown in Fig 28.10(a). The period of the variation is related to the star's intrinsic brightness – its absolute magnitude, see Fig 28.10(b).

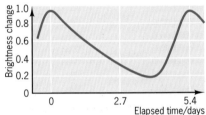

Fig 28.10(a) **Brightness curve for the cepheid variable star Delta Cephei. The period of the variation in brightness is regular, with a peak-to-peak time of about 5.4 days**

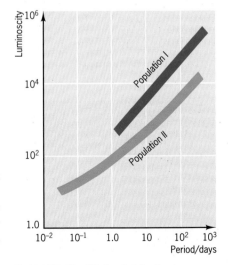

Fig 28.10(b) **Simplified period–luminosity relation for cepheids, showing the linear relationship for absolute magnitude and the log of the period**

?

A A Cepheid variable of known absolute magnitude –2.5 is measured from Earth to have an apparent magnitude of 14.5. Show that the star is about 25 kpc away.

See questions 4, 5 and 6 and the first ■ Assignment at the end of this chapter.

Unfortunately for Hubble, there are two kinds of Cepheid star, and he used data for the wrong kind, so his measured distances for galaxies containing cepheids were all too low. For example, for M31 he obtained a distance of 330 kpc, but it was later found to be at 770 kpc.

The Cepheid relationship gives the absolute magnitude of the variable star, and so the distance D (in parsecs) of a star can be calculated from its apparent magnitude m and its absolute magnitude M as follows:

$$M = m + 5 - 5\log D$$

Cepheids are brighter than the Sun, but even so, when they are further away than about 6 Mpc they get too faint to be useful for calculating differences.

Using very bright objects

The same magnitude–distance formula is used to measure distances further out, using very bright stars of known luminosity, such as **supergiant** stars to 10 Mpc, and even **supernovas** (to 200 Mpc). The same method can be used to 1200 Mpc by assuming that the brightest galaxies all have much the same size and so have similar luminosity.

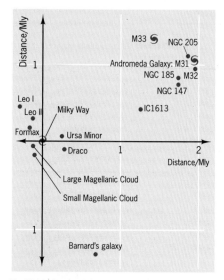

Fig 28.11 **Some of the main galaxies in the Local Group, from a view above the Milky Way. Most of these galaxies are elliptical or irregular, M31 and the Milky Way being the largest. The Local Group (centre) occupies about a cubic megaparsec of space (fig 28.13)**

Fig 28.12 **Elliptical and spiral galaxies in the Virgo cluster which contains several thousand galaxies. At a distance of 19 mpc, it is the nearest main cluster to the Local Group**

4 CLUSTERS AND SUPERCLUSTERS OF GALAXIES

The Milky Way and the Andromeda Galaxy are in fact the two largest galaxies in a **cluster** of 30 galaxies – the **Local Group** – occupying a region of space about 1 Mpc across and affecting each other gravitationally. The largest two, The Milky Way and the Andromeda Galaxy, orbit each other and have the greatest gravitational effect on the whole group.

The **Virgo cluster** is our nearest-neighbour cluster, 20 Mpc away in the direction of the constellation Virgo. A small part is shown in Fig 28.12. It is a huge cluster, about 3 Mpc across, spanning a region of the sky 14 times as large as the Moon but too faint to be seen with the naked eye. It contains several thousand galaxies and its gravitational field is large enough to affect the movement of our own Local Group.

Distance measurements show that nearly all galaxies are grouped in such **clusters**. Clusters range in size from the very small with just an isolated pair of galaxies, to the very large which may have widely spread galaxies, or close enough to be tightly bound by a common gravitational field in which individual galaxies orbit.

Clusters of galaxies are themselves in larger groupings called **superclusters**. Our Local Supercluster is a roughly disc-shaped collection of 382 galaxy clusters about 700 Mpc across (Fig 28.13), with the Local Group at the centre, and the Virgo cluster at the edge.

There is then a 100 Mpc region of empty space before we reach the next supercluster (in the constellation of Hercules), as in Fig 28.14. This pattern of superclusters and voids continues on an even larger scale, as shown in Fig 28.15. This may resemble the way that matter arranged itself in the first moments of the early Universe.

Fig 28.13 **Left: The Local Supercluster with the Local Group at the centre. Each sphere represents a cluster of galaxies, and its size is proportional to the number of galaxies in it. Above disc: red; below disc: yellow**

Fig 28.14 **Centre: Between the Local Supercluster and a neighbouring one, the Hercules Supercluster, there is a void with very few galaxies in it**

Fig 28.15 **Right: With the Milky Way at the centre, this is a two-dimensional picture of nearby clusters and superclusters, showing the voids between. In three dimensions, the clusters and superclusters look like fragmented bubbles, with the voids being the cavities in the bubbles. Each dot represents one or more galaxies**

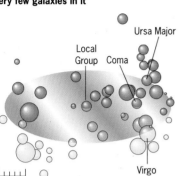

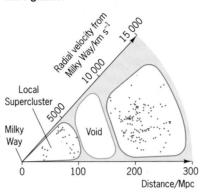

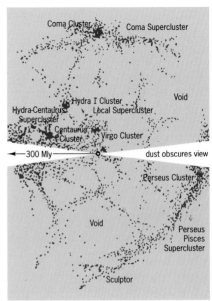

5 THE EXPANDING UNIVERSE

In 1929, Hubble used the Doppler effect to measure the speeds of 24 bright galaxies at different distances from Earth. He found that all of them were moving away from us, and the further away the galaxy was, the faster it was moving. He found a simple relation, called **Hubble's law**, between the distance D and the speed of recession v:

$$v = HD$$

where H is the **Hubble constant**. The speed of recession v is obtained by measuring the **red shift** of a known spectral line. Red shifts for galaxies in five comparatively close clusters are seen in Fig 28.16, and Fig 28.17 shows these shifts plotted on a graph to calculate v.

Fig 28.16 **Left: For each of the named galaxies, the H and K lines in the spectrum of ionised calcium shift towards the longer wavelength (red) end compared with a laboratory reference spectrum. The speed of recession is then calculated using the Doppler formula**

Fig 28.17 **Right: The data obtained from the spectra in Fig 28.16 is plotted here, showing a simple relationship between speed of recession and distance**

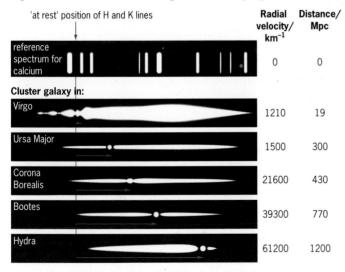

Cluster galaxy in:	Radial velocity/ km^{-1}	Distance/ Mpc
reference spectrum for calcium	0	0
Virgo	1210	19
Ursa Major	1500	300
Corona Borealis	21600	430
Bootes	39300	770
Hydra	61200	1200

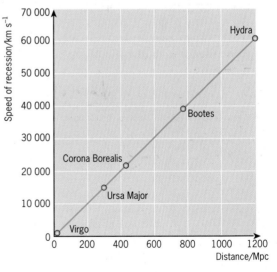

The Doppler effect is a shift in frequency due to the relative motion between source and observer. It is usually quantified in terms of a wavelength change $\Delta\lambda$. So, for a relative speed v we have:

$$\frac{\Delta\lambda}{\lambda_e} = \frac{v}{c} \qquad \text{or:} \qquad \frac{\lambda_0 - \lambda_e}{\lambda_e} = \frac{\lambda_0}{\lambda_e} - 1 = \frac{v}{c}$$

where λ_e is the wavelength of the radiation when *emitted*, λ_o is the wavelength *observed* on Earth, and c is the speed of light. In cosmology

See Appendix 4 for a mathematical description of the Doppler effect.

we are usually interested in light from objects moving away from us, so that λ_o is greater than λ_e, that is, the wavelength has 'shifted to the red' end of the spectrum.

The z factor

Astronomers define a quantity z, the **red shift**, as:

$$z = \frac{\lambda_o}{\lambda_e} - 1 = \frac{v}{c}$$

or, incorporating the Hubble law:

$$z = \frac{HD}{c}$$

Red shift

Fig 28.18

Imagine a still and a moving source of light, S_{still} and S_{mov}, as in Fig 28.18. Each emits light of frequency f from the same kind of excited atoms. Light speed is c, and the moving source is travelling at speed v away from the observer at O.

One second after the start of emission, f waves from source S_{still} occupy a distance of c metres. However, the same number of waves from the moving source have to spread out over a greater distance $(c + v)$. Remember that light travels at the same speed from both a fixed and a moving source (special relativity theory).

The wavelength of the second's-worth of light from S_{still} is observed by O to have a wavelength λ_e, which is the same for both emitter S_{still} and observer. This we can calculate as:

$$\lambda_e = \frac{\text{distance}}{\text{number of waves}} = \frac{c}{f}$$

But the observed wavelength of the light from the moving source is greater: the same number of waves (f) are spread over a greater distance $(c + v)$. Hence for this source the observed wavelength is λ_o where:

$$\lambda_o = \frac{c + v}{f}$$

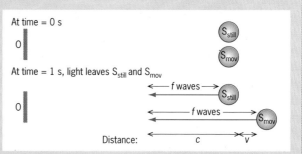

At time = 0 s

At time = 1 s, light leaves S_{still} and S_{mov}

Distance:

Thus, the ratio of observed wavelength to emitted wavelength is:

$$\frac{\lambda_o}{\lambda_e} = \frac{c + v}{c} = 1 + \frac{v}{c}$$

This is often rearranged as follows:

$$\frac{\lambda_o}{\lambda_e} - 1 = \frac{\lambda_o - \lambda_e}{\lambda_e} = \frac{\Delta\lambda}{\lambda_e} = \frac{v}{c}$$

where $\Delta\lambda$ is the change in wavelength due to the motion of the source.

In cosmology we have the red shift factor z defined as:

$$\frac{\text{change in wavelength}}{\text{original wavelength}},$$

That is:

$$z = \frac{v}{c}$$

■ See question 7.

Everything is moving away from everything else

It may seem that somehow the Earth is at the centre of a Universe that is doing its best to recede as far from it as possible. But this is an illusion. *Every large scale feature* of the Universe is moving away from every other one. This is explained by general relativity theory as an expansion of space itself, carrying with it the matter it creates and is created by.

Fig 28.19 shows a simple two-dimensional analogy. When a marked balloon is blown up, all the markings move apart. The rate of separation is the **speed of recession** and depends upon the separation, exactly as in the Hubble law.

Fig 28.19 **A model for Hubble expansion. When a balloon is blown up, every mark on it moves further away from every other mark. (One mark could be the Milky Way.) Each pair of marks move apart at a rate proportional to their 'original' separation**

The Hubble constant

The **Hubble constant** H is hard to measure accurately because distance measurements are so inexact for the furthest galaxies. The units of the constant are usually given as:

$$\frac{\text{speed in km s}^{-1}}{\text{distance in Mpc}}$$

A generally accepted value for H is 85 ± 30 km s^{-1} Mpc^{-1}. Fig 28.20 shows how the value as measured by astronomers has changed since Hubble's day. This is not because H is changing physically but because it is very hard to measure H accurately.

Hubble constant H and the age of the Universe

The Hubble constant gives us a rough measure of the age of the Universe. First assume that H has actually been constant since the Universe formed, a time t_0 ago. In that time, any two points have moved apart by a distance in Mpc of D, at a steady speed v, so:

$$v = D/t_0$$

But also: $v = HD$

So we have: $H = 1/t_0$

This gives the age of the Universe as:

$$t_0 = 1/H = \frac{1 \text{ Mpc}}{85 \text{ km s}^{-1}}$$

A megaparsec is 3.1×10^{19} km, so t_0 comes to about 1.2×10^{10} years – 12 billion years (12 Gy or 12 *aeons*).

Hubble's own value for H was ten times larger than the accepted value above. Because he got his distances wrong, his estimate was 800 km s^{-1} Mpc^{-1}, making the Universe ten times younger, at 1.2 Gy, less than the age of the oldest rocks on Earth!

Recent measurements have decreased the value of H to 50 km s^{-1} Mpc^{-1}. This gives an estimated age of 20 Gy, and avoids the paradox that some stars would be older than the Universe. The values for t_0 assume a constant H, but theory suggests that the Universe expanded more quickly when it was younger, which would drive the estimated age lower (Fig 28.21). The whole story is yet to emerge.

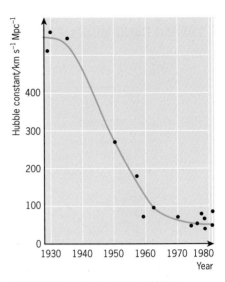

Fig 28.20 **The calculated value of H has gradually become more reliable, and smaller. H has not changed, but the techniques for measuring distance have improved since Hubble's day**

(a)

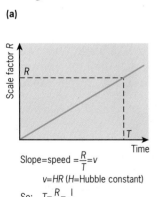

Slope=speed $=\dfrac{R}{T}=v$

$v=HR$ (H=Hubble constant)

So: $T=\dfrac{R}{v}=\dfrac{1}{H}$

(b)

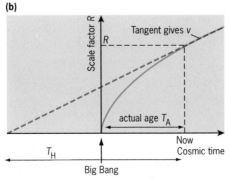

Rate of change of R = v(a recession speed)

$v = HR$

As in (a), $T_H = \dfrac{R}{v} = \dfrac{1}{H}$

Hubble time $T_H > T_A$

Fig 28.21 **The Hubble constant H only gives a correct age for the Universe if it has expanded uniformly, that is, if H is constant over time, as in** (a). **But if expansion is as in** (b), **then the Universe is younger than it looks**

B Check the three quoted values for the age of the Universe in years using the three values of H given.

■ See questions 8, 9 and 10.

THE BIG BANG MODEL OF THE UNIVERSE

Fig 28.22 **The standard Big Bang model for the Universe**

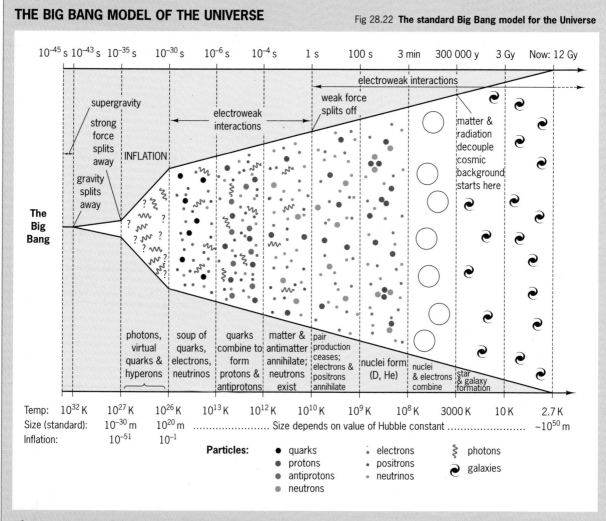

The current model for the origin and nature of the Universe begins with a seeming impossibility and ends in uncertainty. The impossibility is that at the beginning everything must have been a single point – a **singularity**. This is not only difficult to imagine but it also means that no known laws of physics could apply.

The success of the model is that once we are past that first impossible moment, the laws of physics are able to explain most of the Universe's observable features. Of course, there are still many gaps in the theories and many observations that don't fit very well. There are also missing observations: no one has detected gravitational waves, let alone gravitons, the particles that theory predicts carry the gravitational force. There is also the puzzle that most of the mass that must be in the Universe to explain gravitational effects seems to be undetectable.

There follows a summary of the evolution of the Universe as described by the standard **Big Bang** model, shown in Fig 28.22, together with some of the main evidence in favour of the model. As you read, you may find it helpful to refer back to Chapter 26 for descriptions of sub-atomic particles.

During the first three minutes

If we run the film of the expanding Universe backwards, we reach a point where everything – spacetime, radiation and matter – is compressed into dimensions incredibly smaller than the size of a proton. At this very earliest stage, the four fundamental forces (interactions) of nature formed one single unified force, often called **supergravity**.

At 10^{-43} seconds

The micro-Universe has expanded to be about 10^{-35} m across at a temperature of 10^{32} K, with a mass-energy density of 10^{97} kg m^{-3}. Nothing surrounds it, not even a *space* or 'vacuum' since the Universe is all there is. At this time, ordinary gravity splits away (decouples) from the original single force.

Expansion continues, and the tightly packed highly energetic photons convert some of their energy into a sea of *virtual quarks* which combine to form larger virtual particles and antiparticles. Virtual particles of matter exist for only a very short time. This is allowed by the Heisenberg uncertainty principle, provided that their mass-energy and lifetimes multiply to a smaller value than the Planck constant h.

After 10^{-35} seconds: a sudden rapid inflation
The Universe has been expanding steadily, while cooling. But at about 10^{-35} s, the *strong* force separates out, leaving the *electromagnetic* and the *weak* forces coupled together as the *electroweak* force. With the separation of the strong forces, there is a sudden increase in the rate of expansion, an **inflation**, which lasts until 10^{-32} s have elapsed. In this time, the Universe has expanded by a factor of 10^{50}, to become about the size of an orange.

By the end of this period of inflation, the Universe has cooled from 10^{32} K to about 10^{26} K. *Real* particles can now be formed from photons which have enough energy to produce particle-antiparticle pairs: *quarks* and *antiquarks, electrons* and *positrons*. So photons become fewer and less energetic. As expansion and cooling continue, the quarks combine to form very massive but short-lived particles (hyperons), and eventually the stable *protons* and *antiprotons*.

At one microsecond
A microsecond is a very long time in the early Universe. At this time it has reached the size of the Solar System and its temperature has dropped to 10^{13} K, too low for the creation of any more heavy particles from photons. In fact, the quantity of matter is now decreasing rapidly as protons combine with antiprotons, annihilating each other.

If protons and antiprotons had been created in exactly equal quantities, the Universe would now consist entirely of radiation, all at too low an energy to create more particles. Fortunately, there was an excess of protons over antiprotons in the ratio 10 000 000 001 to 10 000 000 000. This left enough protons over to make all the stars and galaxies that now exist.

At one second
The temperature has fallen to 10^{10} K, at which photons no longer have enough energy to create electron–positron pairs. There is a second great matter–antimatter annihilation as those that existed meet. This creates more photons. But again, there is an excess of matter (electrons) over antimatter (positrons) which is just enough to match the protons' charge – so that the total electric charge in the Universe is zero.

At this stage the Universe is a hot plasma consisting of photons, electrons and protons in equal proportions. There is also a very large but unknown number of neutrinos.

Neutrons are produced by collisions between protons and electrons. But neutrons are unstable, and another equilibrium develops between creation and decay which eventually produces two neutrons for every 14 protons. (The significance of these numbers is explained in question 11.)

Later, at a slightly lower temperature, protons and neutrons can combine on collision to form the nuclei of deuterium (hydrogen with 1 proton, 1 neutron) and then helium. (The ratio of neutrons to protons, predicted by nuclear and particle theory, should have produced a hydrogen–helium ratio of 75% to 25% by mass. This ratio is confirmed by present-day observation.) Any larger nuclei are broken up by collision until the plasma cools down further.

At three minutes
Eventually, during the next three minutes, the stable isotopes of the light nuclei lithium (^{7}Li) and beryllium (^{9}Be) are formed – but in very tiny amounts. By this time, the average distance between particles is so large that they are extremely unlikely to meet and form larger nuclei. This process has to wait until the nucleosynthesis in stars (see page 646).

During the next 300 000 years
In its first 300 000 years, the Universe is first a plasma consisting of electrons, protons and helium nuclei. All atoms are ionised. Photons are scattered by the charged particles and cannot travel without losing energy by collisions. We say the Universe at this stage is 'opaque' to radiation.

Eventually, the temperature drops to about 3000 K and nuclei rapidly capture electrons to form neutral atoms which photon energies are too low to break up. The Universe becomes transparent to its own radiation.

When the temperature falls a lot more, the maximum wavelength of the radiation is about 10^{-3} m, in the near infrared. It is from this stage of the Universe's evolution that a relic of radiation at that time, **cosmic background radiation**, is still detectable. This radiation is roughly the same from all directions and shows a pattern of intensity at a range of wavelengths characteristic of black body radiation (pages 305 and 361) of a Universe which has cooled to about 3 K as it expanded.

■ See question 12.

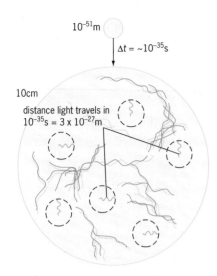

10^{-51}m

$\Delta t = \sim 10^{-35}$s

10cm

distance light travels in
10^{-35}s = 3×10^{-27}m

Fig 28.23 **It is thought that during and after expansion, any inhomogeneities had no time to smooth themselves out because they were 'frozen' into the Universe by rapid inflation**

The four fundamental interactions are gravitation, the weak, the electromagnetic and the strong.

The first 3 billion years: a puzzle

On the whole, we are more certain of what happened in the first three minutes of the Universe than in the first 3 billion years. Modern particle theory strongly supports the story given in the last Extension box. But on the simple Big Bang model, the Universe should be full of galaxies spaced roughly equal distances apart, since originally the particles should have been spaced evenly apart. It would not be too difficult to model some instabilities which started the process of condensation into stars and galaxies. But we don't know how matter could have arranged itself into the galactic superclusters we observe and that give the structure of matter in the Universe the bubble-like *non-homogeneous* (uneven) appearance seen in Fig 28.15.

GUT and Guth

A way out of this problem was proposed in 1980 by the American theoretical physicist Alan Guth. He developed the supergravity theory, that at high enough energies the fundamental interactions (forces) in nature become one. This is the **grand unification theory** or **GUT**, which predicts the very rapid inflation described in the Extension box. Any small random inhomogeneities (unevennesses) in the Universe just before the inflation are magnified as small bubbles of slightly differing density. Then, as time goes on, the Universe grows too fast to smooth itself out again, and the inhomogeneities remain as superclusters and clusters of galaxies (Fig 28.23).

THE COSMIC BACKGROUND RADIATION

THE EXISTENCE OF a low temperature radiation that fills all space was discovered by accident in 1965. It had been predicted as early as 1948 by George Gamow, Ralph Alpher and Robert Hermann, but later forgotten.

Arno Penzias and Robert Wilson, physicists working for the Bell Telephone Company, were testing a small, very sensitive radio telescope they had designed and built to detect radiation of 7 cm wavelength. They found a high level of 'noise' in the system which they couldn't eliminate. They also found that, unlike most noise in radio telescopes, it didn't vary with time of day or with the direction the telescope pointed. Eventually, they accepted that the signal wasn't noise in the machine but *radiation of cosmological origin*.

At a nearby university, physicists were reinventing the theories of Gamow, Alpher and Hermann, and the report of Penzias' and Wilson's discovery seemed convincing evidence in favour of the Big Bang model, which at the time had little evidence to support it. Since then, many very accurate measurements have been made on the background radiation and have confirmed it as a relic of the primal universal radiation (Fig 28.24).

One problem with the radiation was that it was too uniform to be true! The variation and clustering of mass in the Universe was expected be matched by a similar – if not so dramatic – variation in radiation density: whatever had caused matter not to be distributed uniformly should also have affected the radiation. Then, several years of very careful measurements using a satellite, the Cosmic Background Explorer (COBE), the radiation was shown in 1992 to be 'lumpy'.

Fig 28.25 shows the map of the Universe made by the background radiation. The different shades of colour show differences of just 0.01 per cent in the temperature of the radiation. It is effectively a snapshot of the Universe taken when it was just 300 000 years old.

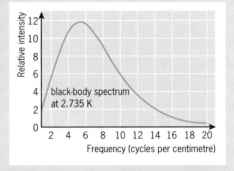

Fig 28.24 **The spectrum of microwave radiation measured by the Cosmic Background Explorer satellite (COBE). Note its similarity to the black body curves in Fig 14.4**

black-body spectrum at 2.735 K

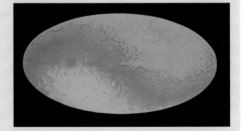

Fig 28.25 **Microwave map of the whole sky taken by the Cosmic Background Explorer satellite at 5.7×10^{-3} m, showing that the radiation is not absolutely smooth but is slightly lumpy. (The pink–blue colour change is due to the motion of the Solar System through the Universe, and the horizontal band to emission from the Milky Way)**

6 THE FUTURE OF THE UNIVERSE

You need Einstein's theory of general relativity for a full explanation of an expanding Universe. But the Newtonian model is good enough to explain some important features.

Think of a galaxy at the edge of a large spherical volume of space inside the Universe, as in Fig 28.26. It contains clusters of galaxies, all moving apart from each other. Newtonian theory says that only galaxies *inside* the spherical volume will affect the motion of the galaxy. The effects of galaxies outside the sphere cancel each other out. Our chosen galaxy (and all others, because we can choose any sphere as long as it is large enough) will have a force on it tending to slow down its outward movement.

What eventually happens to the galaxy depends on a balance between its kinetic energy and its gravitational potential energy. The galaxy has total energy E, composed of kinetic energy E_K and potential energy E_P:

$$E = E_K + E_P$$

Bear in mind that E_K is always positive, but E_P is negative, becoming zero at infinity. There are three possible scenarios, as shown in Fig 28.27:

1. *E* is zero

This means that at the start the kinetic energy is equal in size to the (negative) potential energy. Thus the galaxy moves away, gradually slowing down as kinetic energy is converted to potential energy. After an infinite time our galaxy will reach infinity. The galaxy was originally given its 'escape speed' – but only just.

2. *E* is negative

In this case, E_P is numerically larger than E_K. There is not enough kinetic energy for the galaxy to escape the pull of gravity on it. Eventually, it will stop and fall back, like a ball thrown into the air on Earth. This looks forward to an eventual Big Crunch as everything smashes together again.

3. *E* is positive

In this scenario, there is enough kinetic energy for the galaxy to move to infinity more quickly than in scenario 1. This last scenario is almost meaningless in a Newtonian universe but makes sense in general relativity where gravitation does not really exist as a force. What we have called E is related to a quantity which decides the degree of curvature of spacetime. Both theories agree that what decides the force-curvature involved is the quantity of matter in the Universe.

To summarise, if there is too little matter we have scenarios 1 and 3, if enough, then scenario 2 will operate.

Ninety per cent of the Universe is missing

Estimates of the density of matter in the Universe that are based on the matter we can actually see give very low values – about 4×10^{-28} kg m^{-3}, or one hydrogen atom to every 10 m^3. To keep the Universe expanding steadily, as in scenario 1, we need about 70 atoms in this volume of space. However, when astronomers measure the gravitational pull of the material in a galaxy, or even a supercluster of galaxies, they find that the visible material contributes only about 10 per cent of the gravitational mass.

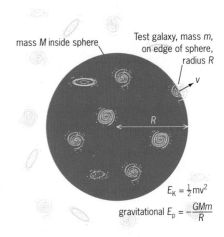

Fig 28.26

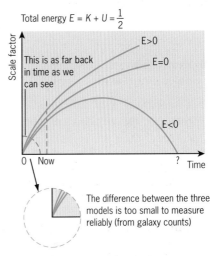

Fig 28.27 **The three possible fates of the Universe. The scale factor of the *y*-axis is any very large distance, undistorted by local gravitation**

■ See question 13.

■ See the second Assignment at the end of this chapter.

Radius today	13×10^9 ly
	1.2×10^{23} m
	4×10^9 Mpc
Hubble constant *H*	85 km s^{-1} Mpc^{-1}
Age	10 Gy
Hubble time *1/H*	12 Gy
Mass	5.7×10^{53} kg
Density	1.5×10^{-26} kg m^{-3}
Number of stars (Sun-sized)	2.9×10^{23}
Number of galaxies	1.6×10^{12}

Table 28.2 **Data about the Universe**

The invisible material is not expected to affect the light output much, so is unlikely to be clouds of dust. (Even gas clouds form only 5 to 15 per cent of the mass of a galaxy.) The search is now on for the missing material. It could exist in the form of dead *dark stars*, unknown forms of matter made of *weakly interacting massive particles* (WIMPs), *neutrinos* (which are extremely numerous and, with a mass of about 2.4 eV, they may have enough total mass), or even *black holes* (just small enough at 10^6 solar masses not to produce dramatic effects on galaxies) or some combination of all these.

Table 28.2 summarises data about aspects of the Universe described in this chapter.

SUMMARY

After studying this chapter, you should:

■ Understand why Olbers' Paradox suggests that a static, infinite Universe is an impossibility.

■ Know that the Universe consists of stars and interstellar material grouped in units: galaxies, clusters of galaxies and superclusters.

■ Know how the distances to stars and distant objects can be measured.

■ Understand how the Doppler effect is used to measure the velocities of galaxies.

■ Know about the observed expansion of the Universe on a large scale, red shift, the Hubble law and the Hubble constant.

■ Know the standard Big Bang model of the Universe, the importance of particle physics in describing the first minutes of its evolution and be able to give evidence in support of the model (microwave background radiation, cosmic abundance of elements).

■ Know the age and possible fates of the Universe, linked to its density and the value of the Hubble constant.

QUESTIONS

Useful formulae: $M = m + 5 - 5\log \Delta v = HD$

$$z = \frac{\Delta\lambda}{\lambda_e} = \frac{v}{c}$$

1 Some nebulae are nearby gas clouds, others are collections of billions of stars. Why were early nineteenth-century astronomers unable to tell the difference between these kinds of nebulae?

2 What are the differences between galaxies and globular clusters?

3 Draw a diagram outlining the main structural features of a spiral galaxy.

4 A typical supergiant star has an absolute magnitude $M = -8.0$. It is observed from Earth to have an apparent magnitude of $m = 18$. How far away is this star (in megaparsecs)?

5 Write an essay of 1000 words on: The problems of measuring astronomical distances.

6 Assuming we can measure the distance of a galaxy, how can we estimate the number of stars it contains? You should be able to describe two methods, both based on the characteristics of an average star.

7 The mean red shift for a number of galaxies in a distant cluster of galaxies is $z = 0.85$.
a) Estimate the distance of the cluster in Mpc.
b) Why is it important to obtain a *mean* value for the red shift?

8 Edwin Hubble measured the distance to comparatively nearby galaxies to be much less than they actually are. Explain why this error meant that:
a) his value of the Hubble constant was too large.
b) his estimate of the age of the Universe was too small.

9 Use the Hubble law $v = HD$ to calculate the speeds of recession of galaxies at distances of:
(i) 0.05 Mpc, **(ii)** 0.5 Mpc, **(iii)** 1 Mpc, **(iv)** 100 Mpc.
(Take $H = 85$ km s^{-1} Mpc^{-1}.)

10 The Hubble constant is not known to any great degree of accuracy.
a) What are the main problems in measuring the value of this constant?
b) The current value is about 85 km s^{-1} Mpc^{-1}. Discuss why astronomers would be worried if an accurate and reliable determination gave a result of twice this value.

11 The best current measurements of the percentage abundances of different kind of nuclei by mass in the Universe are:

hydrogen	helium	everything else
73	25	2

After the annihilation of matter with antimatter, the Universe was hot enough for electrons to combine with protons to form neutrons. Neutrons have a half-life of about 11 minutes. The standard model uses knowledge of the behaviour of subatomic particles to predict that in the next three minutes there was a stable ratio of about two neutrons for every 14 protons in the Universe. These would then combine to form as much helium as possible.

a) How many helium nuclei ($^4_2$He) would be formed from these 16 nucleons?

b) How many protons would be left?
c) What would be percentage of helium mass to total mass for this sample?
d) What percentage by mass should be hydrogen? Suggest why this value is not exactly the same as the value above.

12 Why is a knowledge of particle physics essential for an understanding of what happened in the first few minutes of the Universe?

13 What physical factors decide whether the Universe will keep expanding for ever, or will eventually stop expanding and start to contract again?

Assignment

CEPHEID VARIABLES

A Cepheid variable is a type of star which varies in its brightness over a precise, fixed period of time. The period of variation is related to the star's intrinsic brightness – its absolute magnitude (that is, the star's brightness observed at a distance of 10 parsecs). So the longer the period, the greater the absolute magnitude. This has enabled astronomers to calculate the distance of Cepheid variables and hence the distance of the remote galaxies in which many of them are found.

Table 28.A1 shows the period in days and the absolute magnitude M of some cepheid variable stars in our own Galaxy. The stars were close enough for the distances D (in parsecs) to be measured, so that their absolute magnitudes could be calculated from their apparent magnitudes m using the formula:

$$m - M = 5\log (D/10)$$

Period days	Absolute magnitude
2.0	−2.4
3.1	−2.6
4.9	−3.1
5.4	−3.5
5.5	−3.6
6.3	−3.8
6.7	−3.9
8.0	−3.9
9.8	−4.1
10.8	−3.6
10.9	−4.4
13.6	−4.8
18.9	−5.1
27	−5.6
41.4	−5.8
45	−6.2

Table 28.A1 **The periods and magnitudes of some Cepheid variables**

1
a) Draw a graph of absolute magnitude against period for this set of variable stars. Fit the best curve to this data, by eye.
b) Describe the trend shown by the graph.
c) Comment on the importance of Cepheid variables as distance indicators in astronomy.

2
a) A cepheid variable in a distant galaxy is measured to have a period of 12 days. **(i)** Estimate the absolute magnitude of this star. **(ii)** What is the likely percentage error in this value?
b) Repeat part **a)** for a star which has a period of 42 days.
c) The star with a period of 42 days has a mean apparent magnitude of +12.5. Use the formula given above to calculate its distance, with an estimate of the error in this calculation.
d) The star with a period of 42 days has a mean apparent magnitude of +19. How far away is it?

3 A good telescope system may soon be able to measure stars down to an apparent magnitude of 29. The brightest cepheid has an absolute magnitude of −6.4. How far out would this allow Cepheid variables to be used to standardise a distance scale?

Assignment

MISSING MATTER

Newton's law of gravitation allows us to calculate the mass M of any large massive body with two simple measurements:

The orbital speed v of any object moving round the mass.

The distance R of the body from the centre of mass.

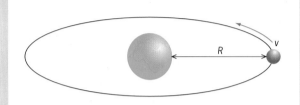

Fig 28.A1

Equating the required **centripetal force** to the **gravitational force** on the orbiting object we have:

$v = \sqrt{(GM/R)}$

This formula can be used to calculate the mass of the Sun from the orbital speed and distance of a planet. On a much larger scale, we can obtain a value for the mass of a galaxy. Stars are in orbit around their galactic centre: the Sun for example is about 9 kpc from the centre of the Milky Way and has an orbital speed of about 210 km s⁻¹. The gravitational force making the Sun orbit is due to the mass of all stars *nearer* the centre than the Sun.

1 Use this information and data at the end of this assignment to estimate the mass of the Milky Way. (You need to assume that it is a sphere with the Sun at its edge.)

Astronomers can also estimate the mass of a galaxy by measuring its **luminosity**. A galaxy with a fairly normal mix of stars of different types should have an absolute luminosity which depends simply on the number of stars it contains. This is expressed in the **mass to luminosity ratio**. Table 28.A2 gives some values in terms of this ratio for the Sun (= 1).

Object	Mass/luminosity
Sun	1
Nucleus of a spiral galaxy	1 to 3
Disc of a spiral galaxy	2 to 5

Table 28.A2

One of the major cosmological problems now facing astronomers is that these two estimates give widely different results for the masses not only of galaxies but clusters and superclusters of galaxies. The gravitational mass for all large objects in the Universe seems to be about ten times as great as the luminous mass. In other

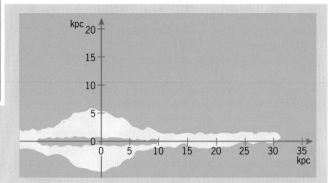

Fig 28.A2

Table 28.A3

Distance d/kpc	Red shift z/10⁻⁴
0	0
3	3.0
5	5.22
6	6.33
7	6.33
10	6.16
12	6.00
14	6.16
17	6.40
20	6.33
22	6.40
25	6.26
27	6.50
30	6.67
35	6.67

words, 90 per cent of the Universe is invisible. It doesn't even shield light, so the prime suspects – gas and dust clouds – are innocent.

Table 28.A3 gives data about galaxy NGC 2998, seen in Fig 28.A2. It is in terms of the red shift z of light from stars at different distances d from the galactic centre. The values have been corrected to deduct the red shift caused by the whole galaxy moving away from Earth.

2
a) Convert the z values in the table to orbital speeds in km s⁻¹ using the relationship $z = v/c$.
b) Plot a graph of this orbital speed against distance d from galactic centre.

Most of the light of the galaxy comes from the galactic centre – out to about 10 kpc in this example. If most of the mass were concentrated in this central region, the movement of stars in the spiral arms would be different.

3 Assume that most of the mass of the galaxy is concentrated in the galactic centre – say, out to 10 kpc. You can produce values for the expected speed at different distances by using a modified form of the formula to make calculation easier:

$v = K/\sqrt{R}$

where K is a constant, v is in km s⁻¹, and R is in kpc.
a) Use the data for stars at 10 kpc to estimate a value for K, then calculate the expected speeds at distances of 15, 20, 25, 30 and 35 kpc
b) Plot these values on the same axes as your graph in 2.
c) Comment on the difference between the two graphs. What can you deduce about the mass distribution in the galaxy? (Hint: Think about what the speed variation would be like if M increased steadily with R.) Sketch a graph of a likely distribution.

Data: 1 kpc = 3.09×10^{19} m, c = 3.00×10^8 m s⁻¹, G = 6.67×10^{-11} m³ kg⁻¹ s⁻²)

ASTROPHYSICS AND COSMOLOGY

This Chapter Map brings together the main concepts in **27 Astrophysics** and **28 Cosmology**, linking them with the ideas and tools common to both areas. Refer to the syllabus you are following to identify the topics you need to study.

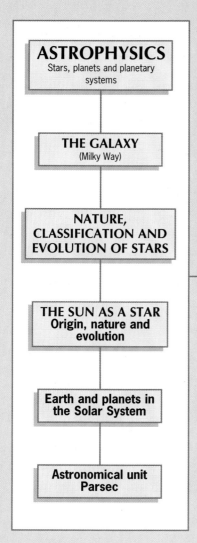

ASTROPHYSICS
Stars, planets and planetary systems

THE GALAXY
(Milky Way)

NATURE, CLASSIFICATION AND EVOLUTION OF STARS

THE SUN AS A STAR
Origin, nature and evolution

Earth and planets in the Solar System

Astronomical unit
Parsec

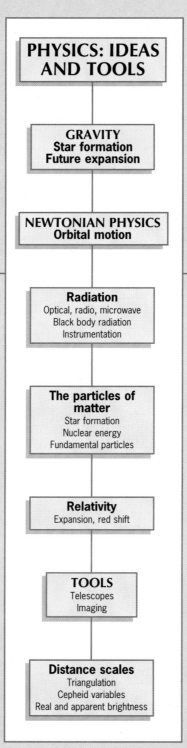

PHYSICS: IDEAS AND TOOLS

GRAVITY
Star formation
Future expansion

NEWTONIAN PHYSICS
Orbital motion

Radiation
Optical, radio, microwave
Black body radiation
Instrumentation

The particles of matter
Star formation
Nuclear energy
Fundamental particles

Relativity
Expansion, red shift

TOOLS
Telescopes
Imaging

Distance scales
Triangulation
Cepheid variables
Real and apparent brightness

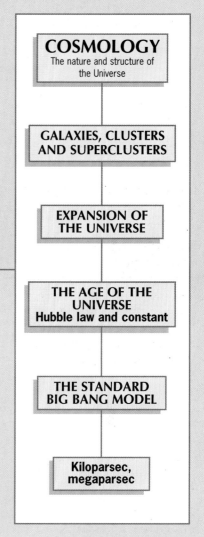

COSMOLOGY
The nature and structure of the Universe

GALAXIES, CLUSTERS AND SUPERCLUSTERS

EXPANSION OF THE UNIVERSE

THE AGE OF THE UNIVERSE
Hubble law and constant

THE STANDARD BIG BANG MODEL

Kiloparsec, megaparsec

APPENDIX 1: DATA AND RELATIONSHIPS

A Physical constants and symbols

1. Fundamental physical constants

Constant	Preferred value	Units
Avogadro constant N_A	$6.022\ 136\ 7 \times 10^{23}$	mol^{-1}
Boltzmann constant k	$1.380\ 658 \times 10^{-23}$	$J\ K^{-1}$
electron charge e	$1.602\ 177\ 33 \times 10^{-19}$	C
electron rest mass m_e	$9.109\ 389\ 7 \times 10^{-31}$	kg
Faraday constant F	$9.648\ 530\ 9 \times 10^{4}$	$C\ mol^{-1}$
gravitational constant G	$6.672\ 59 \times 10^{-11}$	$N\ m^2\ kg^{-2}$
light speed in a vacuum c	$2.997\ 924\ 58 \times 10^{8}$	$m\ s^{-1}$
molar gas constant R	$8.314\ 510$	$J\ K^{-1}\ mol^{-1}$
neutron rest mass m_n	$1.674\ 928\ 6 \times 10^{-27}$	kg
permeability of a vacuum μ_0	$4\pi \times 10^{-7}$	$H\ m^{-1}$
permittivity of a vacuum ε_0	$8.854\ 187\ 817 \times 10^{-12}$	$F\ m^{-1}$
Planck constant h	$6.626\ 075\ 5 \times 10^{-34}$	$J\ s$
proton rest mass m_p	$1.672\ 623\ 1 \times 10^{-27}$	kg
Stefan–Boltzmann constant σ	$5670\ 51 \times 10^{-8}$	$W\ m^{-2}\ K^{-4}$
unified atomic mass constant u	$1.660\ 540\ 2 \times 10^{-27}$	kg

NOTE In most calculations at A-level, the above constants are used rounded off to 2 significant figures only.

2. Other useful constants and values in physics

(a) General
Electric force constant $1/4\pi\varepsilon_0$ — $9.0 \times 10^{9}\ m\ F^{-1}$
Volume of 1 mole of gas at s.t.p. — $22.4 \times 10^{-2}\ m^3$

(b) Earth data
Earth mass — $5.98 \times 10^{24}\ kg$
age of Earth — $\sim 4.5 \times 10^{9}\ y$
Earth radius (mean) — $6.37 \times 10^{6}\ m$
distance from Sun — $1.50 \times 10^{11}\ m$
mean gravitational field strength g — $9.81\ N\ kg^{-1}$
acceleration of free fall — $9.81\ m\ s^{-2}$
Earth constant in gravitation GM — $4.0 \times 10^{14}\ N\ m^2\ kg^{-1}$
mean atmospheric pressure — $1.01 \times 10^{5}\ Pa$
escape speed — $1.1 \times 10^{4}\ m\ s^{-1}$
solar constant — $1.37 \times 10^{3}\ W\ m^{-2}$

(c) Astronomical data
Hubble constant (disputed) — $50\text{–}85\ km\ s^{-1}\ Mpc^{-1}$
astronomical unit AU — $1.5 \times 10^{11}\ m$
light year ly — $9.46 \times 10^{15}\ m = 6.32 \times 10^{4}\ AU = 0.31\ pc$
parsec pc — $3.09 \times 10^{14}\ m = 2.06 \times 10^{5}\ AU = 3.26\ ly$
age of Sun — $5 \times 10^{9}\ y$
luminosity of Sun — $3.90 \times 10^{26}\ W$
mass of Sun — $2.00 \times 10^{30}\ kg$
sidereal year — $3.16 \times 10^{7}\ s$
mass of Moon — $7.35 \times 10^{22}\ kg$
radius of Moon — $1.74 \times 10^{6}\ m$
Earth–Moon distance (mean) — $3.84 \times 10^{8}\ m$
mean density of Universe — $10^{-31}\ kg\ m^{-3}$

3. Standard symbols for quantities

A	displacement, amplitude, area
a	acceleration
B	flux density
C	capacitance
c	specific heat capacity
d	separation
E	energy, e.m.f.
F	force
f	frequency
I	current, intensity
k	any constant
l	length
m, M	mass
n, N	number (no dimensions)
P	power
p	momentum, pressure
Q	charge, energy transferred thermally
R	resistance: electrical, thermal
r	radius, separation
s	distance, displacement
T	kelvin temperature
t	time or Celsius temperature
U	internal energy
V	voltage, p.d., potential
v	speed, velocity, volume
W	work
x	distance, displacement, extension
Δ	change in quantity
θ	Celsius temperature, angle
λ	wavelength, decay constant
ρ	density, resistivity
σ	electric conductivity
ϕ	flux
ω	angular velocity/frequency
ε	electromotive force (e.m.f.)

B Formulae and relationships

Refer to Section A for the identities of most symbols in the formulae and relationships in Section B. See brackets below the formulae for other symbols.

1. Motion and forces

linear momentum $p = mv$

final speed $v_1 = v_0 + at$

final speed $v_1^2 = v_0^2 + 2ax$

distance $x = v_0 t + \frac{1}{2}at^2$

force = rate of change of momentum = ma (mass constant)

impulse = $F\Delta t$

kinetic energy = $\frac{1}{2}mv^2$

gravitational potential energy difference = $mg\Delta h$

energy transferred (work) = force component × displacement

components of force in 2 perpendicular directions

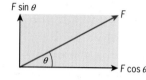

moment of force about a point = force × perpendicular distance from point to line of action of force

circular motion $a = v^2/r$

centripetal force $F = mv^2/r$

2. Electricity

current $I = AvnQ$

charge $\Delta Q = I\Delta t$

resistance $R = V/I$, $R = \rho l/A$

resistance in series $R = R_1 + R_2 + ...$

resistance in parallel $1/R = 1/R_1 + 1/R_2 + ...$

conductance = 1/resistance

capacitance $C = Q/V$

capacitance in series $1/C = 1/C_1 + 1/C_2 + ...$

capacitance in parallel $C = C_1 + C_2 + ...$

energy stored = $\frac{1}{2}QV = \frac{1}{2}CV^2$

discharge of capacitor $Q = Q_0 e^{-t/RC}$
  (RC = time constant)

power $P = IV$

3. Fields and potential

(a) All fields:

field strength $E = -dV/dr \approx -\Delta V/\Delta r$

(b) Electrical:

electric field $E = F/Q$

uniform field between parallel plates $E = V/d$, $E = \sigma/\varepsilon_0$

capacitance of parallel plate capacitor $C = \varepsilon_0 \varepsilon_1 A/d$
  (ε_r = relative permittivity)

point charge $F = \dfrac{1}{4\pi\varepsilon_0} \dfrac{Q_1 Q_2}{r^2}$

electric field strength $E = \dfrac{1}{4\pi\varepsilon_0} \dfrac{Q}{r^2}$

electric potential $V = \dfrac{1}{4\pi\varepsilon_0} \dfrac{Q}{r}$

(c) Gravitational:

field strength $g = F/m = -GM/r^2$
  (M = mass of Earth or other body)

force $F = -Gm_1 m_2/r^2$
  (r = separation of centres)

gravitational potential energy change $\Delta E = GM(1/r_1 - 1/r_2)$

4. Matter

density = mass/volume

(a) Solids: material in tension

Hooke's law: tension $F = kx$
  (k = spring constant)

stress = tension/cross-sectional area

strain = extension/original length

Young modulus = stress/strain

elastic strain energy = stress/strain

elastic strain energy = $\frac{1}{2}kx^2$

elastic strain energy per unit volume = $\frac{1}{2}$stress × strain

(b) Liquids

liquid (fluid) pressure $p = \rho gh$

Bernouilli's equation: $p = \frac{1}{2}\rho v^2 + \rho gh = \text{constant}$

power available from fluid flow $P = \frac{1}{2}A\rho v^2$

Stokes' law: $F = 6\pi\eta rv$
  (η = viscosity coefficient)

Reynolds number $R_e = \rho vr/\eta$

pressure drag force for object moving in a fluid $F = A\rho v^2$

drag force in turbulent flow $F = Br^2 \rho v^2$

Archimedes' principle: upthrust on a body wholly or partly immersed in a fluid = weight of fluid displaced

(c) Gases

($\overline{c^2}$ = mean square speed, m = mass of molecule, N = molecules, n = moles, R = gas constant)

ideal gas equation: $pV = nRT$

kinetic theory of gases: $pV = \frac{1}{3}Nm\overline{c^2}$

pressure $p = \frac{1}{3}\rho \overline{c^2}$

for 1 mole of an ideal gas: mean kinetic translational energy = $\frac{3}{2}RT$

5. Thermal physics

rate of thermal energy flow (conduction):
  power $P = k(A\Delta t)/l$
  (k = thermal conductivity, A = cross-sectional area, l = length, $\Delta\theta$ = temperature difference)

thermal energy ('heat') transfer $\Delta Q = mc\Delta\theta$

thermal energy $\Delta Q = \Delta U - \Delta W$
  (W = work done on system)

work done by a gas $\Delta W = -p\Delta V$

kinetic energy of a monatomic molecule $E_K = \frac{3}{2}kT$

change of state $\Delta Q = mL$
  (L = specific latent heat)

efficiency of a heat engine $\eta = (T_{hot} - T_{cold})/T_{hot}$

entropy change $\Delta S = \Delta Q/T = k\Delta \ln W$
  (k = Boltzmann constant)

6. Oscillations and waves

(a) Simple harmonic motion

equation of motion $a = -(k/m)x$
  (k = force per unit displacement)

Displacement–time relationships:

displacement $x = A \cos \omega t$

angular frequency^2 $\omega^2 = k/m$

periodic time $T = \dfrac{2\pi}{\omega} = 2\pi\sqrt{\dfrac{m}{k}}$

frequency $f = \dfrac{1}{T} = \dfrac{\omega}{2\pi} = \dfrac{1}{2\pi}\sqrt{\dfrac{k}{m}}$

maximum velocity $v_{max} = \omega A$

maximum acceleration $a_{max} = \omega v_{max} = \omega^2 A$

total energy $= \frac{1}{2}kA^2$

(b) Wave speeds

general wave equation: $= A\sin(kx - \omega t)$

all waves: speed $c = f\lambda$

compression wave in mass–spring system $c = x\sqrt{(k/m)}$
(x = spacing, k = force per unit displacement)

transverse wave on string $c = \sqrt{(t/\mu)}$
(T = tension, μ = mass per unit length)

electromagnetic waves in free space $c = 1/\sqrt{(\varepsilon_0\mu_0)}$

(c) Radiation and light

Wien displacement equation: $\lambda_{max}T$ = constant

Stefan–Boltzmann equation: $P = \sigma A T^4$

Doppler effect $\Delta\lambda/\lambda_e = \Delta f/f = v/c$
(λ_e = wavelength emitted)

(d) Diffraction and Young's slits

through narrow slit: $n\lambda = b\sin\theta$
(n = order, b = slit width, θ = angles of minima)

Rayleigh criterion: $\theta \geq \lambda/D$

diffraction grating: $n\lambda = d\sin\theta$
(d = grating spacing, θ = angles of maxima)

Young's double slits: $\Delta x = \lambda D/s$
(Δx = fringe separation, s = slit spacing)

8. Electromagnetism

(a) Magnetic fields

force on current-carrying conductor $F = BIl$

force on moving charge $F = BQv$
(v = velocity perpendicular to field)

Flux density:
– inside a long solenoid $B = \mu_0 NI/l$
(N = turns)

– near a long straight wire $B = \mu_0 I/2\pi r$
(r = radial distance)

– at centre of circular coil $B = \mu_0 NI/2r$
(r = radius of coil)

reluctance $= l/\mu_0\mu_r A$
(μ_r = relative permeability, A = area)

flux = current turns/reluctance

(b) Induction

induced e.m.f. = rate of change of flux linked

induced e.m.f. $\epsilon_i = Nd\phi/dt$
(N = linked turns, $d\phi/dt$ = rate of change of flux)

e.m.f. in secondary $\epsilon_s = MdI/dt$
(M = mutual inductance)

p.d. across coil $V = LdI/dt$
(L = self inductance)

(c) Transformers

e.m.f. in secondary $\epsilon_s = V_P N_S/N_P$
(V_P = p.d. across primary, N_S and N_P = turns)

in transformer: $I_P V_P > I_S \epsilon_S$

8. A.c. circuits

r.m.s. current $I_{r.m.s.} = I_0/\sqrt{2}$
(I_0 = peak current)

r.m.s. voltage $V = V_0/\sqrt{2}$
(V_0 = peak voltage)

inductive reactance $X_L = \omega L$
(L = self inductance)

capacitative reactance $X_C = 1/\omega C$

impedance $Z = \sqrt{(R^2 + X^2)}$

9. Atomic, nuclear and particle physics

radioactive decay $dN/dt = -\lambda N$
(λ = decay constant)

particles: number remaining $N = N_0 e^{-\lambda t}$
(N_0 = initial number)

half-life $t_{\frac{1}{2}} = \ln 2/\lambda \approx 0.693/\lambda$

mass-energy relationship $\Delta E = c^2\,\Delta m$

photons: energy–frequency relationship $E = hf$

particles: wavelength–momentum relationship $\lambda = h/p$
(p = momentum)

radius of a nucleus $r \propto A^{1/3}$
(A = mass number)

momentum-energy relationship for relativistic particles $E^2 = p^2 c^2 + m_0^2 c^4$
(p = momentum, m_0 = rest mass)

10. Special relativity

relativistic constant $\gamma = \left|1 - \dfrac{v^2}{c^2}\right|^{-\frac{1}{2}} = \dfrac{1}{\sqrt{1 - \dfrac{v^2}{c^2}}}$

time dilation $\Delta t = \gamma\Delta t_0$

Lorentz contraction $t = t_0/\gamma$

relativistic mass $m = \gamma m_0$

11. Astronomy and astrophysics

Kepler's laws:

$R^2\omega$ = constant

$GmT^2 = 4\pi R^3$
(T = orbital period: year, R distance from Sun)

luminosity of star $L = 4\pi R^2 \sigma T^4$
(σ = Stefan–Boltzmann constant)

apparent magnitude $m = -2.5\log I$ + constant
(I = intensity)

distance of star $d = m - M + 5\log d$
(m, M = apparent and absolute magnitudes)

Hubble law $v = HD$
(H = Hubble constant)

age of Universe $= 1/H$

red shift $z = (\lambda_0/\lambda_e) - 1 = v/c$

APPENDIX 2: MATHEMATICS FOR PHYSICS

At A-level you are using mathematics in physics in four main roles:

Role 1: computation – to get the right (and *suitably accurate*) answers in straightforward routine problems and calculations;

Role 2: to **model** what might be sets of complicated data and relationships in a way that makes them easier to understand;

Role 3: to **prove** some of the formulae and relationships that make up the models in 2;

Role 4: to **explore and investigate** the possible consequences of models when data, factors or even the models themselves are changed: *What will happen if …?*

Role 1: Computation

Only rarely will a physics activity not involve **Role 1**, and every examination will test your skills in this area. This means that you will need to become fluent and confident in:

- using simple arithmetic – and powers of 10
- using electronic calculators
- your choice and use of significant figures
- setting up, rearranging and solving linear equations
- drawing and interpreting graphs

Role 2 : Modelling with mathematics

Role 2 will also be an important feature of your physics studies. Many definitions, laws and rules in physics are expressed mathematically. We have simple **linear relationships** such as:

distance covered = speed × time $x = vt$

hydrostatic pressure = density × gravitational field strength × depth $p = \rho g h$

force stretching a spring = spring constant × extension $F = kx$

Next, we have relationships in which one variable is raised to a *power*, that is, it is squared or cubed:

kinetic energy = constant × speed squared
$$E_K = 1/2mv^2$$

volume of a sphere = 4/3π × radius^3
$$= \tfrac{4}{3}\pi r^3$$

power generated in a resistor
= current^2 × resistance $P = I^2R$

distance fallen from rest
= constant × (time of fall)^2 $x = \tfrac{1}{2}gt^2$

Inverse relationships are a special case of a 'power law', with a quantity raised to a power of −1. The classic example is Boyle's law:

volume of a fixed mass of = constant/pressure of gas at constant temperature

$$V = k/p = kp^{-1} \text{ or } pV = k$$

TIP: Note the third version of the formula (in **bold**): it gives us a quick way to check if any relationship is an inverse one like Boyle's law.

Inverse square relationships describe some of the most important laws in physics: Newton's law of gravitation and Coulomb's law of force between charges, for example.

$$F = G\frac{Mm}{r^2} \qquad F = k\frac{Q_1Q_2}{r^2}$$

where G is the universal gravitational constant and k the electrical constant (usually written as $1/4\pi\varepsilon_0$). Inverse square relationships in these two examples are linked to the fact that space is three-dimensional – see Chapters 3 and 10 for more about this.

Equally important, but a little more difficult to handle, are **exponential** relationships. These are quite difficult ideas and are best understood after you have studied the relevant physics. They are likely to appear in two forms.

1. Relationships involving the number e
$$N = N_0e^{-kt} \text{ (radioactive decay)}$$
$$Q = Q_0e^{-t/RC} \text{ (decay of charge on a capacitor)}$$

2. Relationships in a logarithmic form:
$$ln(N/N_0) = -kT$$
where ln is logarithm to base e, also written $\log_e$.

noise level in dB = 10 log (I/I_0)
where log is logarithm to base 10, or $\log_{10}$.

3. Relationships in a form based on the defining physical process
(change in a short time) proportional to (how much there is):
$$\Delta N = \pm kN\Delta t \qquad dN/N = \pm kdt$$

Handling the formulae
You should be able to use any of these relationships (formulae, equations) to calculate the value of an unknown quantity, given the values of the others involved. This may mean rearranging the equation to make the unknown quantity the subject of the equation (*algebra*). You will need to insert the right numbers in the right places and carry out the necessary *arithmetic*, probably using an electronic calculator (see *Hints and tips for computation* on

page 675). You then clearly state the result with the correct units and to the appropriate number of significant figures (*physics*).

You should also be able to analyse two sets of related numerical data to test for the relationship between them. In simple cases you can do this 'by inspection' – does one variable double when the other one does, for example. If so, the relationship is *linear*:

$$y = \text{constant} \times x$$

Simple calculation might show that when you multiply a pair of related numbers together you always get much the same answer – showing an *inverse* relationship as in Boyle's law:

$$xy = k \quad \text{so} \quad y = k/x$$

A simple test for an exponential relationship

Table A2.1 shows some data which follow (approximately) an exponential relationship. You can check this by looking for the 'equal ratio property': as one quantity increases by equal amounts the other changes, so that dividing any value by the next one gives the same result. The y values increase successively by a factor of about 1.4 – a near enough agreement for such a simple test.

Table A2.1 **The equal ratio test for exponential dependence**

x	2	4	6	8	10
y	0.8	1.2	1.6	2.2	3
ratios of successive y values		1.5	1.3	1.4	1.4

Graphing relationships

Graphing results or data should make it obvious whether they follow a pattern or not – and which pattern they follow. Fig A2.1 shows the types of graph given by various mathematical relationships.

As a general rule, plot the **independent** variable on the horizontal (x) axis and the **dependent** variable on the vertical (y) axis. For example, the *speed* of an accelerating vehicle changes with the *time* during which it is accelerating. We usually define time as the independent variable, speed the dependent variable.

But note that in physics it is not always sensible to obey this rule. For example, an elastic object stretches when we apply a force to it. The stretch – its extension – is the variable that depends on the force applied. But when we plot a force–extension graph we always plot the dependent variable *extension* on the x-axis and the independent variable *force* on the y-axis. This is because the equation linking force and extension is:

$$\text{force} = \text{constant} \times \text{extension}$$

and the constant is the *slope* of the graph, as explained next.

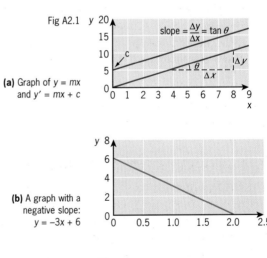

Fig A2.1

(a) Graph of $y = mx$ and $y' = mx + c$

(b) A graph with a negative slope: $y = -3x + 6$

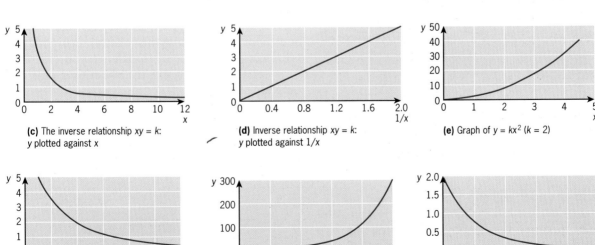

(c) The inverse relationship $xy = k$: y plotted against x

(d) Inverse relationship $xy = k$: y plotted against $1/x$

(e) Graph of $y = kx^2$ ($k = 2$)

(f) Inverse square relationship $y = k/x^2$

(g) Exponential growth: $y = ke^x$

(h) Exponential decay: $y = 2e^{-x}$

Slopes and tangents

The rate at which some quantity changes with respect to another variable is given by the **slope** (or tangent, gradient) of the graph line. Mathematically, the formula that gives a straight-line graph is of the form:

$$y = mx + c$$

where m is the slope of the graph and c its intercept on the y-axis. The intercept occurs when $x = 0$, so here $y = c$.

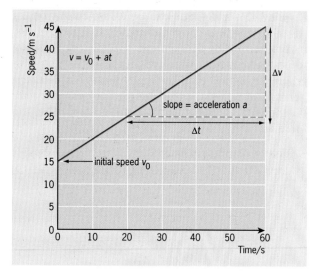

Fig A2.2 **Slope and intercept: graph of accelerated motion**

This idea is illustrated in Fig A2.2: speed is plotted on the y-axis and time on the x-axis. The slope of the line is the *constant* rate of change of speed, that is, the acceleration. The intercept on the y-axis is the starting speed – the speed when $t = 0$.

Fig A2.3 shows how to find the slope of a non-linear graph at a point by drawing a tangent.

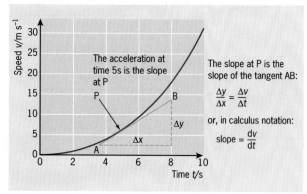

Fig A2.3 **Drawing a tangent to a graph**

The area under a graph

The area enclosed by ('under') a graph is a measure of the quantity you get by multiplying the two variables together. This idea works even when they are varying continuously. This area is also the result of using calculus to get the *integral* of whatever

relationship links the two variables. This is shown in Fig A2.4.

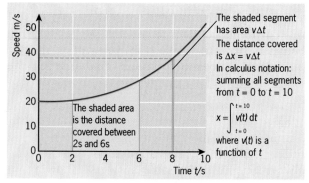

Fig A2.4 **The area under a graph, with calculus**

Turning curves into straight-line graphs

Plotting y against x in a relationship such as $y = ax^2$ will produce a curve. Plotting y against x^2 will produce a straight line-graph of slope a. This is a useful way of both checking that data are related by a power relationship – and for finding the value of the linking constant a.

If you don't know how one set of values (y) is related to another set (x) but suspect that it has the form: $y = ax^n$, a plot of log y against log x will help. The plotted relationship would have the form:

$$\log y = \log a + n\log x$$

and will be a straight-line graph of slope n and intercept log a.

Scattergraphs

Data do not always follow a neatly obvious mathematical relationship. When one variable is plotted against another the points may be 'scattered' around a straight line or a curve. In this case we can think of the **line of best fit** as a kind of average of the way the system described by the data behaves. If the data is too scattered it may be wrong to say that it supports a simple law or relationship. This is usually more of problem in subjects like biology, and there are well-known statistical techniques for estimating the degree of confidence you can claim for a relationship. Most spreadsheet programs give help for doing this.

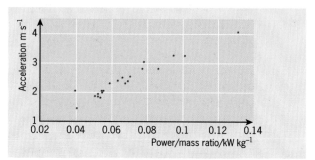

Fig A2.5 **Graph showing a scatter: How 21 types of car obey Newton's laws**

Role 3: Mathematical proofs

You should be able to follow the proofs of formulae and relationships which are derived from basic definitions, such as:

(a) centripetal acceleration v^2/r from the basic definition of acceleration as rate of change of velocity,

(b) the gravitational potential energy formula $-GM/r$ from the definitions of work and the inverse square law,

(c) the energy stored in a charged capacitor $E = \frac{1}{2}QV = \frac{1}{2}CV^2$ by considering the area under a voltage–charge graph,

(d) the kinetic theory formula $pV = \frac{1}{3}Nm\overline{c^2}$ from conservation of momentum and the definition of pressure.

Very few A-level syllabuses expect you to reproduce such proofs in an examination – but check your own syllabus about this.

Role 4: Exploring and investigating with mathematics: modelling

Traditionally physicists have modelled 'real life' by making simplifying assumptions – such as zero friction, perfect spheres, uniform density etc. – and used some quite advanced mathematics to describe what might happen. The result is a model of reality – which with any luck will be close enough to the real thing for useful predictions to be made – and a better understanding of what is going on.

The most useful kind of mathematics for this has been the **calculus** – which was in fact invented by Isaac Newton (and others) to help solve the problems that arose when he was producing his models ('laws') of motion and of gravity. Calculus remains one of the most useful tools for mathematical modelling. See page 677 for a list of the main results in calculus useful at A-level.

But increasingly both working physicists and students are using more basic, numerical methods to make mathematical models – using **computer spreadsheet** programs or programmable graphical calculators. These methods allow you to change the conditions – the values of constants and variables – and calculate the effects very easily and quickly, as the electronics does all the hard work. The spreadsheet programs have most features in common, but have slightly different ways for setting up equations. It will be very worthwhile learning how to use these aids, and using them to investigate mathematical models in physics (without having to do too much maths!).

There are three main ways in which you should find spreadsheets useful:

1. Entering and storing experimental data and plotting graphs. Example: investigating data about car performance, experimental results.

2. Using previously derived formulae with real or made-up data and constants to model what happens when data and/or constants change. Essentially all the mathematical physics has been done by whoever derived the formula. Examples: using derived equations of motion, modelling projectile motion with different speeds and air resistance, investigating simple harmonic motion, investigating the addition of sound waves of different frequencies.

3. Using numerical methods based on simple fundamental definitions to make models. This relies on repeated calculations in small steps (iterations), and no derivation of possible advanced equations is needed. Examples: modelling radioactive decay, capacitor discharge, gravitational potential near a planet, rocket motion. (See Assignments to Chapters 2 and 4 for example.)

See *Spreadsheet tasks* on page 678 for suggested models that you might investigate.

Hints and tips for computation

Powers and indices

If $x = a^y$ then $y = \log_a x$ eg $x = 10^3$, $\log_{10} x = 3$

$$x = e^4 \quad \log_e x = \ln x = 4$$

Table A2.2 **Powers, indices and prefixes**

Sub-multiple	Prefix	Symbol		Multiple	Prefix	Symbol
10^{-1}	deci	d		10^1	deca	da*
10^{-2}	centi	c		10^2	hecto	h*
10^{-3}	milli	m		10^3	kilo	k
10^{-6}	micro	μ		10^6	mega	M
10^{-9}	nano	n		10^9	giga	G
10^{-12}	pico	p		10^{12}	tera	T
10^{-15}	femto	f		10^{15}	peta	P*
10^{-18}	atto	a		10^{18}	exa	E*

*rarely used

Scientific (or standard) notation uses powers of ten, eg:

542.3 is written 5.423×10^2

0.05423 is written 5.423×10^{-2}

The ^{2} and the $^{-2}$ are also called **indices**.

Computers and calculators may present standard form coded with the letter E standing for *exponent* (index), eg:

5.423E2   or   5.423E–2

Always give final answers and results in standard form. *Take care when plugging numbers into a calculator – what you press is not what you see!* For example, 5.423×10^2 is entered into a calculator as:

step 1 **5.432**
step 2 **EXP** (the exponent key)
step 3 **2**

In other words, follow the notation above. If you add × 10 between steps 1 and 2 the answer will be 10 times too big. Try it and see!

When you *multiply* numbers given in powers of 10, **add** the indices:

$3 \times 10^4 \times 2 \times 10^5 = 6 \times 10^9$
$3 \times 10^4 \times 2 \times 10^{-5} = 6 \times 10^{-1}$

In general, $a^5 \times a^2 = a^7 \quad a^m \times a^n = a^{(m+n)}$

When *dividing*, **subtract** the indices:

$3 \times 10^4 \div 2 \times 10^5 = 1.5 \times 10^{-1}$
$3 \times 10^4 \div 2 \times 10^{-5} = 1.5 \times 10^9$

You should work out that:

$10^2 = 100$, $10^1 = 10$, $10^{-1} = 1/10 = 0.1$ and $10^{-2} = 1/10^2 = 1/100 = 0.01$.

The missing element in this sequence is 10^0. You would be correct in guessing that $10^0 = 1$. In fact, anything raised to power zero is 1: $a^0 = 1$. Try multiplying $a^2 \times a^{-2}$!

Algebra: rearranging and solving equations

Linear equations have no variables ('unknowns') raised to a power higher than 1, eg:

$v = u + at \qquad V = IR \qquad F = ma$

To calculate a value for one unknown you need to have the values of all the others: eg you need to know that $u = 0$ m s^{-1}, $a = 9.8$ m s^{-2} and $t = 10$ s to find a value for speed:

$v = 0 + 9 \times 10 = 98$ m s^{-1}

To find, for example, how long it will take for an object falling from rest to reach a speed of 98 m s^{-1} you first rearrange the formula to make t the subject. You can do this in steps:

1. move u across and change its sign: $v - u = at$
2. which is the same as: $at = v - u$
3. then divide both sides by a: $t = (v - u)/a$
4. and insert the known values: $t = (98 - 0)/9.8$
   $t = 10$ s

Quadratic equations involve quantities raised to the power 2: Simple examples include:

$E = \frac{1}{2}mv^2$
$v^2 = u^2 + 2as$

You can solve these types using exactly the same technique as above for linear equations. Sometimes

(but quite rarely at A-level) you may have to solve a 'proper' quadratic equation of the type:

$ax^2 + bx + c = 0$

Here a and b are constants (numbers) and x is the unknown. In general, such an equation will have two solutions (ie give two values for x). Elementary maths courses give lots of tricks for solving such equations, but brute force uses the general solution:

$$x = \frac{-b \pm \sqrt{b^2 - 4ac}}{2a}$$

EXAMPLE

Q An object is thrown straight up into the air with an initial speed of 20 m s^{-1}. Use the formula $s = ut + \frac{1}{2}at^2$ to find how long it will take to reach a height of 10 m. Note that $a = -9.8$ m s^{-2}.

A Insert the numbers into the equation:

$10 = 20t + 0.5 \times (-9.8)t^2$

Tidy up and rearrange: $4.9t^2 - 20t + 10 = 0$
Use the solution formula:

$x = \dfrac{+20 \pm \sqrt{400 - (4 \times 4.9 \times 10)}}{2 \times 4.9}$

$= \dfrac{20 \pm \sqrt{204}}{9.8}$

which gives two values for x: 3.5 s and 0.58 s. The object is 10 m high going up (after 0.58 s) *and* coming down (after 3.5 s).

Trigonometry

Fig A2.6 shows the three main trigonometric ratios, based upon the right-angled triangle: sine, cosine and tangent.

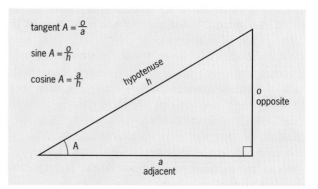

tangent $A = \frac{o}{a}$
sine $A = \frac{o}{h}$
cosine $A = \frac{a}{h}$
hypotenuse h
o opposite
A
a adjacent

Fig A2.6 **Trigonometric ratios**

You will occasionally come across the inverse of these ratios, the cosecant, secant and cotangent:

cosec $A = 1/\sin A$ sec $A = 1/\cos A$ cot $A = 1/\tan A$

Do not confuse these with the following terminology (as used on calculators):

$\sin^{-1} n$ meaning *the angle whose sine is n*
$\cos^{-1} n$ meaning *the angle whose cosine is n*
$\tan^{-1} n$ meaning *the angle whose tangent is n*

Looking at the triangle in Fig A2.6, you should be able to recognise that:

$\sin A = \cos (90 - A)$
$\cos A = \sin (90 - A)$
$\cot A = \tan (90 - A)$

As the triangle is right angled, Pythagoras' theorem gives:

$a^2 + b^2 = c^2$

From this it is easy to show that:

$\sin^2 A + \cos^2 A = 1$

Other trigonometrical identities that you might find useful are:

$\sin 2A = 2 \sin A \cos A$
$\cos 2A = \cos^2 A - \sin^2 A = 2 \cos^2 A - 1 = 1 - 2 \sin^2 A$
$\sin(A \pm B) = \sin A \cos B \pm \cos A \sin B$
$\cos(A \pm B) = \cos A \cos B \mp \sin A \sin B$

Radians and small angle approximations

The radian is a measure of angle based on a circle such that:

angle in radians = length of arc subtended divided by the radius of the circle

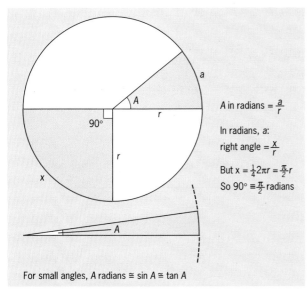

A in radians $= \frac{a}{r}$

In radians, a:

right angle $= \frac{x}{r}$

But $x = \frac{1}{4}2\pi r = \frac{\pi}{2}r$

So $90° = \frac{\pi}{2}$ radians

For small angles, A radians $\cong$ sin A $\cong$ tan A

Fig A2.7 **Angles in radians**

When the angles involved are small, the segment of the circle is very close to a right-angled triangle, and we can put:

angle A in radians = $\sin A$ = $\tan A$

This approximation is correct to two significant figures for angles less than 5 degrees.

TIP: Calculators can work in degrees or radians (or even the unusual American system of grads) – so when using a calculator to work out trigonometrical expressions check that it is operating in the correct mode.

Useful calculus formulae

Derivatives are formed by **differentiating** a function.

Table A2.3 **Integrals and differentials**

y	x^n	sin x	cos x	tan x	cot x	cosec x	e^{kx}	ln x	ln kx
dy/dx	nx^{n-1}	cos x	$-\sin x$	$\sec^2 x$	$-\csc^2 x$	$-\csc x \cot x$	ke^{kx}	$1/x$	$1/x$

Integrals

These are indefinite integrals: an arbitrary constant should be added to the right side in each case.

f(x) (function of x)	$\int f(x)\, dx$
x^n	$\frac{x^{n+1}}{n+1}$ unless $n = 1$
$\frac{1}{x}$	$\ln x$
e^x	e^x
a^x	$\frac{a^x}{\ln a}$
$\frac{1}{\sqrt{a^2 - x^2}}$	$\sin^{-1}\frac{x}{a}$

Differentiating and integrating combinations of functions of x

Note: We use the ' notation to show a derivative:

$u = f(x)$ $u' = du/dx$
$v = f(x)$ $v' = dv/dx$

Using this notation we have the following results:

$$(uv)' = u'v + uv'$$

$$\left(\frac{u}{v}\right)' = \frac{u'v - uv'}{v^2}$$

$$\int uv'\, dx = uv - \int u'v\, dx$$

Using spreadsheets in physics

Spreadsheets are computer programs that can handle large amounts of data and plot graphs of various kinds. They have built-in mathematical and statistical formulae and you can write your own to create mathematical models of a physical system. Such a model may contain several variables and constants and can be set up quite easily.

Once the formulae are put in, the spreadsheet can perform hundreds of calculations in a few seconds and also plot several graphs to illustrate the model. You can change some of the constants to see the consequences, which appear very quickly on the graphs.

Data is inserted in the spreadsheet in cells, labelled by their position in a grid which has letters along the top and numbers down the side.

Nowadays most common spreadsheets use the same symbols and conventions for formulae and operators. The most commonly used in physics modelling are as follows. As printed on your spreadsheets, upright lettering is given here for variables, D for Δ and PI or PI() for π.

= introduces a formula you want to construct, eg: = B6 + C6 is the formula telling the spreadsheet to add the contents of cell B6 to the contents of cell C6

* means multiply, so 23×42 is written as 23*42

^ shows an exponent, so 3^2 is written as 3^2

E denotes powers of 10, so 5×10^4 is written as 5E4 or as 5*10^4)

/ means divide by, so $4 \div 3$ is written as 4/3

Spreadsheet tasks

A. Data handling
This includes recording experimental data, sorting data into an order, plotting graphs and charts, taking averages, finding correlations and patterns.

B. Modelling
You can either use standard equations and formulae (eg from textbooks) to find out what happens when a key numerical factor or quantity is changed, or, more fundamentally, to build up a model from the basic definitions of quantities.

In the first case, somebody has already done the clever work of using algebra or geometry to analyse the situation. In the second case the spreadsheet does the work by making lots of small calculations as quantities vary by a small amount each time – the human user does not need to be expert at advanced mathematics! Examples of both kinds are listed below, with hints for the formulae or basic definitions involved.

Model	Hints and formulae
1. Accelerated motion	v=v+Dv, Dv=vDt, s=s+Ds, Ds=vDt
2. Motion in a uniform gravity field	g=9.8, Dv=gDt, v=v+Dv, average velocity=0.5(v'+v)
3. Projectile motion with air resistance	Modify Model 2. Set a value for angle A, initial speed V; vertical velocity = Vsin(A), horizontal = Vcos(A), vertical acceleration is –g. Calculate vertical and horizontal coordinates (height, distance). Plot appropriately
4. Simple harmonic motion (horizontal spring)	You need: spring constant k, mass m, initial disturbance x; speed v=v+Dv, a=–kx/m, Dv=aDt, x=x+vDt
5. Beats	Plot sum of waves at two frequencies defined by sin (2PIft), eg 4Hz, 6Hz with time increasing by 0.005s.
7. Radioactive decay	You need: decay constant k, initial number of nuclei N, sensible time interval. N=N+DN, DN=–kNDt
8. Radioactive decay with daughter products	As above, plus daughter B (zero at start) with decay constant b, DB=+DN–bBDt
9. Rockets	You need: initial mass of rocket M, initial rocket speed V, flow rate of ejected gases R in kg/s, speed of ejected gases v M=M–DM, DM=RDt, V=V+DV, DV=vDM/M
10. Charge and discharge of a capacitor	
11. Analysing planetary data	Get the data from any astronomy source book; look for patterns
12. Radiations from hot objects: the Planck function	Calculate the main constant, use exponential formula in spreadsheet. Find the effect of changing temperature, check Wien's law, etc

APPENDIX 3: EXPERIMENTS AND INVESTIGATIONS

Why do experiments?

There are a number of good reasons for doing practical laboratory work in A-level physics. They can be summarised by the following four objectives:

1. Practicals help you understand physics better – by making abstract ideas more real.

2. They help you learn the skills of handling equipment safely and …

3. …well enough to get accurate and reliable measurements.

4. They help you develop the more general skills of designing, carrying out and evaluating an investigation.

Doing practical work helps you understand physics and should ensure that you get better grades in examinations. The second objective is fundamental to any laboratory work: safety is an obvious necessity, and for every practical procedure you should make a *risk assessment* – see your teacher about this. Objective 3 is also fundamental: poor results tend to give rise to confused ideas, and poor techniques may lead to unsafe practice.

The fourth objective combines the other three – and more: most importantly, it tests your understanding of a theory or model by making you predict an outcome based upon that theory, that is, you form a reasoned prediction or **hypothesis**.

Next, you need to know enough about the use and availability of equipment to *design* a workable plan and then use it to take *appropriate, relevant and reliable measurements*. You then go back to the model and compare it with the outcomes: Do they support the hypothesis?

The list of possible laboratory tasks on pages 682–684 are divided into three main categories:

A. Illustrative or learning tasks – to make abstract ideas more real and understandable.

B. Measurement tasks – which may achieve the above as well as teach you the use of basic measuring and data-gathering instruments and techniques, and give you an idea of the scale of important quantities in physics (the value of g, the Young modulus for steel, the specific heat capacity of copper, for example).

C. Investigations – which bring together the skills and ideas you have learnt in designing and carrying out the test of your own hypothesis or model.

Appropriate accuracy

In all experiments, there is a trade-off between obtaining accurate and reliable measurements and the time and effort it would require to get them. The accuracy aimed at should match the needs of the task, which means that when planning an experiment or investigation you must think about:

- the range of measurements of a variable (enough to match an aim or test a prediction).
- the number of readings you need to take (enough to ensure reliability within the time available).
- the accuracy aimed at for each reading (appropriate to the aim).
- the choice of measuring equipment (matching the desired accuracy and range).
- whether or not some measurements or sets of measurements need to be repeated.

All measurements are inaccurate – to some extent

We can measure or estimate the accuracy of a measurement in terms of either a 'percentage error' or an 'uncertainty' in its value. For example:

voltage across resistor $R = 11.0$ V $\pm$ 2%
voltage across resistor $R = (12.0 \pm 0.2)$ V

The value you quote should have enough significant figures to match the uncertainty. It would be wrong to quote the reading of a digital voltmeter used in this case, say, as 11.025 V, even though that is what it says. But how we do we know how accurate a measurement is?

Estimating accuracy

Measurements can be uncertain for a number of reasons:

Limitations due to the scale markings of instruments (eg a millimetre rule can only measure to say 0.2 mm – depending on how good your eyesight is, the thickness of a meter pointer and the size of the smallest scale divisions).

They might be inherently variable (eg the speed of sound in the open air, the water pressure in a tap – even the mains voltage).

Human error: response time (eg timing the fall of mass or the period of an oscillation with a stop-clock); reading a scale at an angle and so introducing a parallax error.

Interference by other events or factors – 'noise in the system' which produces an effect similar in size to the one you are trying to measure (eg the gravitational force between two masses, the Hall voltage in a slab of semiconductor).

Calibration errors – a faulty scale, often because the zero value is incorrect (eg thermometers whose 0 °C marking is not at the ice point).

Of these, you can usually make an estimate only of the first cause – just by looking at the instrument you intend to use. The uncertainty in the others will be revealed when you have collected the data and find that there is a range of values in a final result, or when you plot a graph and find that the points match an expected line or curve only approximately. As a rule, more than one source of uncertainty will be present.

The effect is shown in Fig A3.1, where error bars (uncertainties) have been plotted to indicate the estimated (predicted) errors in each measurement of force and extension for a stretched wire. The plot is expected to be a straight line, but the values do not fall exactly on any single line: the one plotted is a 'line of best fit'. This has been drawn by eye – but a good spreadsheet will plot an accurate best fit of data if needed.

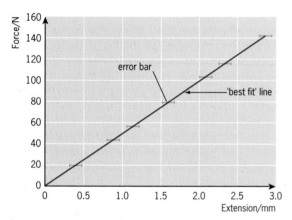

Fig P.1 **Graph of force–extension for a wire, with error bars**

Combining uncertainties

The graph of force and extension in Fig A3.1 may be used to measure the Young modulus for the metal of the wire. The formula is:

$$E = \frac{\text{stress}}{\text{strain}} = \frac{\Delta F}{A}\frac{L}{\Delta x} = \frac{L}{\pi r^2}\frac{\Delta F}{\Delta L}$$

Here, A is the cross-sectional area of the wire (πr^2 where r is the radius). Radius r is measured with a micrometer screw gauge to be (0.32 ± 0.01) mm. l is the unstretched length of wire, measured to be (1.225 ± 0.002) m. The slope of the graph force–extension graph, $\Delta F/\Delta L$, is measured as:

$100N/2.0$ mm $= 5.0 \times 10^4$ N m^{-1}

The uncertainty in the slope is due to the uncertainty in the instruments used *plus* the inherent uncertainty in the way the wire behaves, and perhaps

some slippage in the system when heavy weights are added. We can find a reasonable estimate of the uncertainty by calculating the slopes of the most extreme lines that we could draw through the error bars. This works out to be:

slope $= (5.0 \pm 0.4) \times 10^4$ N m^{-1}

When we calculate E from these values, *the percentage uncertainty of the result is the sum of the percentage uncertainties in the values used to make the calculation,* see Table A3.1.

Table A3.1

Quantity:	l	r	r	$\Delta F/\Delta l$
value	1.25 m	0.32 mm	0.32 mm	5.0×10^4 N m^{-1}
% uncertainty	0.2	3	3	4

The total percentage uncertainty in the final value of E is therefore about 10 per cent. Note that r appears twice because $r^2 = r \times r$.

We calculate E from the data as 1.943×10^{-11} Pa – to 4 significant figures. But how many are justified? A 10 per cent uncertainty is about 0.2×10^{-11} Pa, so we should quote the result of the experiment as $E = (1.9 \pm 0.2) \times 10^{-11}$ Pa.

Look for the critical measurement

When you become familiar with this approach to estimating likely errors and uncertainties in results based on laboratory measurements, you will be able to spot the most critical measurements. These are the ones that introduce the greatest uncertainty.

In the example above, it is the measurement of the radius r of the wire. Not only is it hard to measure but its effect is doubled because the calculation involves r^2. This means that if you want to improve the final accuracy you should look for ways of improving this measurement.

Improve accuracy by repeating readings

You can improve accuracy with an instrument of limited sensitivity (ability to make fine-scale measurements) by repeating the measurements. As a general rule, taking N measurements of the same quantity, each having an uncertainty of x per cent, reduces the error in the mean of the measurements by a factor $\sqrt{N}$, that is, the error in the mean reduces from x per cent to $x/\sqrt{N}$ per cent.

NOTE: When you obtain a value by adding two quantities, you do not add the percentage errors but take the uncertainty in the final value to be that of the least certain value.

Units and dimensions

Quantities in physics may or may not have units attached to them. Those that don't may be *simple numbers*, like the number of atoms in a sample; or they may be **ratios** of quantities which have the same units, such as strain, the ratio of an extension (in m) to an original length (also in m).

Some ratios are easy to spot because they are named as relative quantities, like *relative density* (the ratio of the density of a material to the density of water) or *relative permittivity* (the ratio of an electric force constant for a material compared with that of a vacuum). These are called **dimensionless** numbers – the word 'dimensions' is used with a special meaning in physics to denote the physical factors linked together in the quantity.

The most basic dimensions are length (l), mass (m) and time (t). There is more about these in the section *Using the method of dimensions*.

The basic units in physics are defined by the *Système Internationale* or **SI** system. For example, the dimension of length is measured in the base unit the metre and the dimension of mass has the kilogram as its associated unit. All physical quantities can be measured in terms of just seven **base units** and two **supplementary units**, listed in Table A3.2.

Table A3.2 **SI quantities and units**

Quantity	Name of unit	Unit symbol
Base unit:		
length	metre	m
mass	kilogram	kg
time	second	s
electric current	ampere	A
thermodynamic temperature	kelvin	K
amount of substance	mole	mol
luminous intensity	candela	cd
Supplementary quantities:		
plane angle	radian	rad
solid angle	steradian	sr

Derived quantities

Other quantities are derived from the base quantities via their definitions. Table A3.3 shows examples.

Table A3.3

Quantity	Units
area = (distance)^2	m^2
speed = distance/time	m s^{-1}
acceleration = speed/time	m s^{-2}
force = mass × acceleration	kg m s^{-2}

Some of these quantities are important enough to have named units of their own, for example, force, where the unit combination kg m s^{-2} is also called a newton (N). This makes life simpler! Again, the ratio of voltage to current is called resistance, defined as $R = V/I$ and measured in ohms (Ω). In

base units we would have to work out the units of voltage from its definition as joules per coulomb, where a joule is defined as a force times a distance:

joule = force × distance
has units kg m s^{-2} × m or kg m^2 s^{-2}

But a coulomb is a derived unit too:

coulomb = current × time
and has base units A s

So the base units of voltage are:

$$\frac{\text{joules}}{\text{coulombs}} = \frac{\text{kg m}^2}{\text{A s}}$$

which simplifies to kg m^2 A s^{-3}; and the base units of resistance are:

$$\text{resistance} = \text{voltage/current} = \frac{\text{kg m}^2 \text{ A s}^{-3}}{\text{A}}$$
$$= \text{kg m}^2 \text{ s}^{-3}$$

This explains why so many common quantities have their own named unit!

See question 1.

Using the method of dimensions

Dimensions and units can be used to check the validity of a formula or equation – and even to derive the form of an equation. In some books you will see the basic dimensions labelled as length l, mass m, time t, current I, etc.

We can derive the form of an equation as follows. We might guess that the period T of a simple pendulum depends on its mass m, its length l and the strength of the gravitational field (or acceleration of free fall) g.

Let us say we decide T = a function of (m,l,g). On the left side T has units (dimensions) of time only, in s. We observe that the longer the pendulum the slower it oscillates, and guess that a more massive pendulum swings faster and also that the stronger the gravity field the faster it will swing. We could write down the right side of the equation as a combination of the quantities as follows: length/(mass × acceleration due to gravity). Inserting units gives:

$$\frac{\text{m}}{\text{kg} \times \text{m s}^{-2}} = \frac{\text{s}^2}{\text{kg}}$$

This is clearly wrong: there is no mass unit on the right side, and time is measured in seconds not (seconds)^2. This means that the period does not depend on mass m. We know it depends on length l and the only combination of l and g that fits is $\sqrt{(l/g)}$. Check this! You will find that the units are now matching on both sides of the equation. But there may of course be a dimensionless factor involved. A proper physical treatment gives us the result $T = 2\pi\sqrt{(l/g)}$.

Using *l*, *m* and *t*

The left side of the equation has dimension T. The right side of the equation might involve length (l), g ($l\,T^{-2}$) and mass (m):

$$T = kl^x g^y m^z \text{ (where } k \text{ is a constant)}$$

The possible relationship between dimensions is:

$$T = (l)^x \times (lT^{-2})^y \times (m)^z = (l)^{(x+y)} \times (T)^{-2y} \times m^z$$

We see immediately that $z = 0$, as m is not involved. Looking at the indices we can write an equation:

For T: $1 = -2y$, giving $y = -\frac{1}{2}$
For l: $0 = x + y$

and using our knowledge of y we get $x = -y = \frac{1}{2}$.

This means that the formula for the period of the pendulum must be:

$$T = kl^{\frac{1}{2}} g^{-\frac{1}{2}} = k\sqrt{(l/g)}$$

See question 2.

Experiments and learning activities

Below is a list of practical activities, mostly laboratory-based, that you may carry out during your course. Some are essential – see your syllabus for what is expected of you. The list is a long one and you probably won't have time to tackle them all. Some may be demonstrated by your teacher or fellow-students.

Using basic instruments

You should be familiar with the use of the following measuring instruments and other equipment used in A-level physics:

(a) General

length measurers (metre and millimetre rules, micrometer screw gauge, travelling microscope, vernier scales)
stopwatch
electronic timer
speed measurers (ticker timers and tapes and/or electronic systems)
force measurers (eg newton meters)
weighing devices (eg electronic balance)
thermometers

(b) Electrical

ammeters and voltmeters (analogue and/or digital)
multimeter
cathode ray oscilloscope
resistance substitution box
capacitance substitution box
electrometer or other device for measuring small currents and charges
electrical power supplies (low voltage, high voltage, e.h.t.)

radioactivity detectors – ratemeters and/or scalers
signal generators
data loggers (used with a computer)
3 cm wave apparatus

A. Illustrative experiments (learning experiments)

Matter and its properties

Modelling metal structures with a bubble raft
Observing what happens when materials (eg copper, polythene) stretch and break
Comparing the effect of forces on different materials: paper, string, plastics, metals
Comparing composite with non-composite materials
Forces in liquids: pressure and depth; surface tension
Liquid flow and viscosity
Measuring acceleration
Collisions – elastic and inelastic
Investigating angular velocity, acceleration and conservation of angular momentum
Boyle's law
The gas laws (how pressure and volume vary with temperature)
Investigating the effects of fluid flow (streamline and turbulent flow, Stokes' law and terminal speed, Bernouilli effects, aerofoils)

Thermal physics

Investigating natural cooling
Observing infrared radiation from hot objects
Plotting temperature against time for a change of state
Calibrating a thermistor

Electricity and magnetism

The motion of ions in a liquid
Resistors in series and parallel
Action of a potential divider
The relationship between voltage and charge for a capacitor
Charging and discharging a capacitor via resistors
Capacitors in series and parallel
Voltage–current relationships for thermistors, diodes, lamps and other devices (eg using a data logger)
Electric field patterns
Equipotentials around conductors
Looking at magnetic domain structure (eg in garnet film)
Magnetising and demagnetising materials
The magnetic fields near current-carrying conductors: single wire; coils
Force on a wire in a magnetic field
The behaviour and efficiency of small electric motors
The Hall effect: use of Hall probe to investigate magnetic fields
Induced voltages and the laws of electromagnetic induction

The behaviour of an inductor: time delay in reaching steady current; variation of current and reactance with frequency

Measurements in a.c. circuits with L, C and R

Phase differences in an LCR circuit

Microelectronics

Behaviour of logic gates, truth tables

Solving problems using logic gates

Making astable and bistable circuits

Waves and oscillations (including light and optics)

Observing simple oscillating systems (mass on a spring, simple pendulum)

Resonance effects (eg Barton's pendulum)

Coupled pendulums/springs – beats and resonance

Investigating what affects the speed of a wave in a coil spring

Longitudinal and transverse waves in a spring

Overtones in a vibrating string

Reflection of waves

Standing (sound) waves in a closed tube

Looking at diffraction and interference effects in a ripple tank

The diffraction and interference of light: Young's fringes, air wedge fringes, Lloyd's mirror fringes

The properties of a diffraction grating

Diffraction and interference of microwaves

Observing polarisation of light

Effect of temperature changes on the colour spectrum of a filament lamp

Observing spectra – emission and absorption

Investigating the inverse square law for light

Investigating refraction of light, including total internal reflection, dispersion, action of an optical fibre

Investigating the action of lenses in producing an image

Investigating the working of simple optical instruments and devices

Atomic and simple particle physics

Observing cathode rays: the effects of E and B fields

Identifying the charge on electrons

Observing electron diffraction

Observing the photoelectric effect

Looking at emission and absorption spectra

Observing alpha particles in a cloud chamber

Observing alpha particle scattering (plus analogue experiment with a 'potential hill')

The response of a GM tube as applied anode voltage varies

Measuring background radiation

Investigating the range and penetration of ionising radiations

Verifying the inverse square law for gamma rays

Astronomy and astrophysics

Observing and identifying stars and planets

Oberving the colour changes and spectrum of a heated filament

Observing Rayleigh scattering

Using telescopes and assessing their properties (field of view, magnifying power)

Observing the emission spectra of elements

Magnifying power of a telescope

B. Measuring constants and properties

You should be able to describe or comment on experiments to measure the following quantities.

Matter and mechanics

Densities of solids and liquids

Spring constants

Young modulus

The acceleration of free fall, g

Rotational inertia (moment of inertia) of a flywheel

Thermal physics

Specific thermal capacities: solids, liquids

The 2 specific heats of a gas; C_p, C_v

Specific latent heats

Thermal conductivity of a good and a bad conductor

Electricity and magnetism

Resistance of a conductor

Resistivity of a metal and/or a semiconductor

Electric current

Charge

Power delivered or developed

Conductivity of an ionic solution

The Coulomb constant $1/4\pi\varepsilon_0$

Capacitance

Self-inductance

Reactance in a.c. circuits

Magnetic field strength

Electric field strength

Waves and oscillations (including light and optics)

Speed of sound in: air; carbon dioxide; a metal

Focal lengths of lenses and curved mirrors

Speed of light (in air or in an optical fibre)

The wavelength of light: Young's fringes and/or diffraction grating

Atomic and particle physics

The specific charge of the electron

Charge of an electron (Millikan's experiment)

The Planck constant

Half-life (also analogue experiments)

The activity of an alpha source

C. Investigations

You should be familiar with the four main stages involved in scientific investigation:

Planning and designing: make predictions or hypotheses when appropriate, identify key variables, choose equipment and the range of measurements required, safety implications.

Making and recording observations: work safely and use equipment skilfully to obtain relevant data of appropriate accuracy, using correct units, clearly recorded; use tables/charts; modify procedures in the light of experience.

Analysing data: use graphs and appropriate mathematical procedures, estimate accuracy, identify trends, match outcomes with predictions, present conclusions clearly in good English with correct units and sensible significant figures.

Evaluating procedures and data: comment on the validity, accuracy and reliability of conclusions or final values, suggest improvements to strategies, techniques and instrumentation.

There is no room in this book to list more than a very small sample of interesting and successful investigations that you could carry out as an A-level student. What follows is meant simply to indicate the range.

How air resistance affects the motion of a thrown ball.
How temperature affects the bounce of a squash ball – and why.
What happens when copper is 'heat treated'?
The length of a flash from a camera.
The efficiency of a bicycle.
The stretching of polythene.
The chaotic behaviour of a dripping tap.
Light variations from a variable star.
The strength of concrete.
The pressure–volume relation for a rubber balloon.
Air flow in a heated room
Thermal conduction across a window or wall.
Diffusion of inks in water.

The flight of model gliders.
The speed of ripples on water.
What affects the resolution of the human eye?
The scattering of beta particles by different materials.
Radon gas in houses.
The lifetimes of electric batteries and their rate of use.
The polarisation of reflected (or scattered) light.
Streamline and turbulent flow.
Thermal energy losses from the human body.
Constructing and using a resistance thermometer.
The mechanical (or thermal) properties of new
  packaging materials.
The efficiency of a small transformer.
Glass springs

Simulations

Bear in mind that some of the most difficult and/or dangerous experiments in physics can be simulated using a computer (eg in astronomy, nuclear physics). See your teacher for some up-to-date examples. You can also create your own simulations using a spreadsheet, as described in Appendix 2, but this is not a substitute for trying to do them 'for real'. Use computers to extend the laboratory work that you do, such as by changing the values of variables and constants.

Questions

1 Work out the base units involved in

  (a) pressure,
  (b) volume,
  (c) angular velocity,
  (d) density,
  (e) the watt,
  (f) the gravitational constant G.

2 Check that the following formulae are dimensionally matched, that is, the base units are the same on both sides of the equation:

  (a) power $P = VIt$,
  (b) centripetal acceleration $a = v^2/r$,
  (c) speed of a wave on a string $v = \sqrt{(T/\mu)}$ where T is the tension force and μ is the mass per unit length of the string.

APPENDIX 4: THE DOPPLER EFFECT

The Doppler effect is the change in frequency produced when a source of waves (eg sound, light) moves relative to the observer. For example, you hear a change in the pitch (frequency) of the sound of a car horn or a police or ambulance siren as it passes you. The frequency increases when the vehicle is moving towards you, and decreases as it moves away. The effect is widely used to measure vehicle speeds, and also to measure the relative motion of distant stars, galaxies and other objects.

Remember:
period is the inverse of frequency $T = 1/f$
wave speed = wavelength × frequency $v = f\lambda$

The Doppler effect for sound waves

Fig A4.1(a) shows the sound waves emitted by a source S at a speed v and a frequency f_S – which is the frequency heard by the observer O when both are at rest. In this treatment we also assume that the medium is air and that it is not moving.

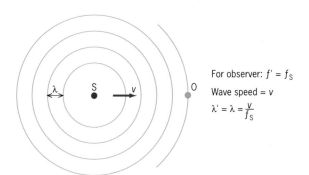

For observer: $f' = f_S$
Wave speed = v
$\lambda' = \lambda = \dfrac{v}{f_S}$

Fig A4.1(a) **Source and observer are stationary**

1. Stationary source and a moving observer

Consider a stationary source with an *observer* moving towards it at speed v_0, as in Fig A4.1(b).

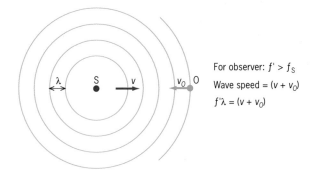

For observer: $f' > f_S$
Wave speed = $(v + v_0)$
$f'\lambda = (v + v_0)$

Fig A4.1(b) **A stationary source and a moving observer**

The wavelength of the sound is not affected by the movement of the observer. But the observer crosses more wave crests per second than a stationary observer would. The speed of the waves relative to the observer has increased and so has the observed frequency. The relative speed of the waves is now $v + v_0$ and the frequency heard is f'.

The wave formula gives us:

$$v + v_0 = f'\lambda \quad \text{or} \quad f' = (v + v_0)/\lambda$$

The frequency as emitted by the source is $f = v/\lambda$, so the ratio of frequencies is:

$$\frac{\text{observed frequency}}{\text{source frequency}} = \frac{f'}{f} = \frac{v + v_0}{v} = 1 + \frac{v_0}{v} \quad [1]$$

If the observer is moving away from the source the frequency decreases – fewer crests pass the observer per second – and the formula becomes:

$$\frac{f'}{f} = 1 - \frac{v_0}{v} \quad [2]$$

These formulae are often combined as:

$$f' = f\left(1 \pm \frac{v_0}{v}\right) \quad [3]$$

2. Stationary observer and a moving source

Fig A4.2 shows what happens when the *source* is moving towards a stationary observer at a steady speed v_S.

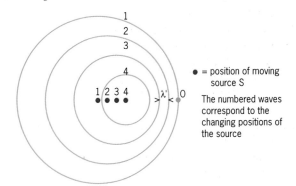

● = position of moving source S

The numbered waves correspond to the changing positions of the source

Fig A4.2 **Moving source with stationary observer**

The effect is to *reduce* the wavelength, because in a given time more waves fill the space between source and observer than when both are at rest, as in Fig A4.1(a). The observed frequency thus increases – the observer is passed by more crests in that time. But the waves pass at the same speed, so the observed frequency is f' such that $v = f'l'$, or:

$$\lambda' = \frac{v}{f'}$$

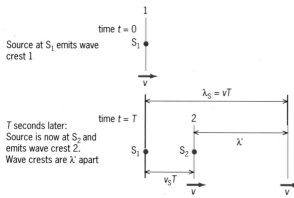

Fig A4.3 **Doppler effect due to a moving source. Note that wave crests only are shown**

Fig A4.3 helps us to find the value of the new wavelength. In the time t, where t is the period of the oscillation, the wave crest moves a distance vt. But in that time, the source has moved a distance $v_S t$, and so the next crest is just λ' behind its predecessor, and as the diagram shows:

$$l' = vt - v_S t$$

Now, period $T = 1/f$, and $\lambda' = v/f'$, so that we can rewrite this as:

$$\frac{v}{f'} = \frac{v}{f} - \frac{v_S}{f} = \frac{1}{f}(v - v_S)$$

which may be rewritten as:

$$f' = f\left(\frac{v}{v - v_S}\right)$$

or, as usually quoted:

$$f' = f\left|\frac{1}{1 - \frac{v_S}{v}}\right| \qquad [4]$$

This confirms that when the source moves towards the observer (v_S is positive) the observed frequency *increases*. When the source is moving away the formula becomes:

$$f' = f\left|\frac{1}{1 + \frac{v_S}{v}}\right| \qquad [5]$$

and f' is less than f.

3. Both source and observer moving

When both source and observer are moving the observed frequency is given by:

$$f' = f\left(\frac{v \pm v_0}{v \mp v_S}\right) \qquad [6]$$

+v_0: observer approaching source
−v_0: observer receding from source
−v_s: source approaching observer
+v_s: source receding from observer

Check that this formula reduces to equations 2, 3, 4 and 5 for appropriate values of v_0 and v_S. When

using it, bear in mind the importance of the *signs* + and −. It is easy to get this right if you remember that relative motion *towards* each other *increases* frequency, motion *away decreases* frequency.

B. The Doppler effect with electromagnetic waves

It is the fundamental idea in the *special theory of relativity* that for two observers moving at a steady speed relative to each other it is impossible to say which of them is moving – or if both are moving. Therefore we have the same formula for the case 'moving observer–fixed source' as for the case 'fixed observer–moving source'. It must also be true that both observer and source would measure the speed of light to be the same, however fast either of them moved.

The full derivation of the relativistic Doppler effect formula is beyond A-level, but when source and/or observer are moving at speeds much less than the speed of light the formula is simply:

$$f' = f\left(1 + \frac{v}{c}\right) \qquad [7]$$

This is sometimes given as the ratio of wavelengths. Since $v = f'\lambda' = f\lambda$, then

$$\frac{\lambda'}{\lambda} = \frac{f}{f'} = \frac{1}{\left(1 + \frac{v}{c}\right)} = \frac{c}{c + v} \qquad [8]$$

In many radar applications the *difference* Δf between emitted and received frequencies is measured, so that equation 7 is used in the form:

$$\frac{\Delta f}{f} = \frac{v}{c} \qquad [9]$$

Doppler shift due to a moving reflector
Radar speed measurements of moving vehicles rely on the reflection of waves. Equation 9 then becomes

$$\frac{\Delta f}{f} = \frac{2v}{c} \qquad [10]$$

The factor 2 arises because when a plane mirror moves, the images in it move twice as fast.

Student tasks

Look at Fig 4.A3. Draw a clear labelled diagram showing the positions of three wave crests at time $t = 2T$. Show that λ^1 stays the same.

Try to manipulate equation 7 algebraically to derive formula 9.

ACKNOWLEDGEMENTS

The authors and publisher are most grateful for the kind assistance of the following:

We wish to thank Dr J.J. Wellington, Reader in the Division of Education at the University of Sheffield, for his careful reading of the text for accuracy of ideas and in particular for helping to make the reading level of the text more accessible to students.

We also thank Phillip Henry for his painstaking review of the mathematics and algebraic expressions, to ensure that they are correct.

Any errors that remain are the responsibility of the authors and publisher.

Poem
The poem by Robert Graves that appears on page 576 is reproduced by kind permission of Carcanet Press. It appeared in *Robert Graves: Poems selected by himself*, published by Penguin Books Ltd in 1957.

Examination questions
The publisher thanks examination boards for their permission to reproduce examination questions. Questions are acknowledged as follows:

AEB   The Associated Examining Board
NEAB   Northern Examinations and Assessment Board
OCSEB   Oxford and Cambridge Schools Examining Board
UCLES   University of Cambridge Local Examinations Syndicate
ULEAC   University of London Examinations and Assessment Council
WJEC   Welsh Joint Education Committee

Index
Thanks are due to Julie Rimington for compiling the index.

Every effort has been made to trace the holders of copyright. If any have been overlooked, the publisher will be pleased to make the necessary arrangements at the earliest opportunity.

Photographs
We are grateful to the following for permission to reproduce photographs.

AEA Technology, 24.25; Allsport UK, page 30 (Hewitt), 3.12 (Vandystadt/Klein), 5.31 (Cannon), 5.37 (Squire), 6.14(b) (Delecour), 7.5 (Rondeau); Ardea London/Weisser, 18.42; Aviation Picture Library, 3.4, 8.29, 8.30 (right), 16.6; Biophoto Associates, page 360 above; John Birdsall Photography, page 302, 18.40, 19.13, 19.23, 19.26; British Telecom, 21.11; CERN Media Service, 26.26; Collections, 2.14 (Shout), 5.22 (Shuel), 6.15 (Sieveking); Corbis UK/Bettmann, 6.17, 25.5 (UPI); English Electric Valve Company 14.3, 16.7; European Space Agency, 4.17(b), 4.23(b), 4.24, 4.29; Mary Evans Picture Library, page 49; Genesis Space Picture Library, page 612; Geological Survey of Canada, 4.3, 23.22; Geoscience Features, 3.A1, 5.41; David Grace, 9.13, 9.20, 9.24, 9.26, 9.28, 10.12 above and centre, page 464; Ronald Grant Archive, page 556; Enrico Gratton, Universities of Urbana-Champaign and Pennsylvania/Derek Armstrong, University of Toronto, page 536 below; Peter Gould, 3.25, 5.20 right, 7.3(a), 16.15(b), 16.16, 16.23, 17.7 centre, 20.16, 20.21(a), 20.23(a); Dr.A.L. Hodson, University of Leeds, 26.1 above right; IBM UK, page 360 below, page 446; ICCE Photo Library/Boulton, 8.46 below; Instron, 5.33; LaCoste and Romberg Gravity Meters Inc, 23.24(b); Andrew Lambert Photographs, 3.6, 3.16, 3.22, 5.20 left, 5.42, 6.20, 7.16 below, 8.1(b), 8.17, 8.28, 8.34, page 230 below, 10.12 below, 15.17, 16.2, 16.8, 16.33, 16.37, 17.12, 19.31(a), 20.11(b), 20.23(b), 22.5, 22.17, 22.39; Frank Lane Picture Agency/Hosking, page 40; London Features International, 16.9(a) (Mazur), page 486 below (Fisher); Microscopix/Syred, 5.39, 19.50; Dr.D.S. Moore, 5.49; NASA, page 56, 4.25; National Grid Company, page 282; National Medical Slidebank, 24.24(b) left and right; Nuffield Chelsea Curriculum Trust/*RNAP SG1* (fig A53), 7.17; Oxford Scientific Films/Bernard, page 327; Panos Pictures/Hartley, 21.9; Performing Arts Library/Barda, 6.26(a); *Peterborough Evening Telegraph*, 4.4; MRP Photography, page 376 below; Polaroid International, 16.38; Powerstock, 2.7 (Nelson), 5.10; QA/Eurotunnel, 5.1 left; Quadrant Picture Library, 5.9, 5.51(b), 8.1(a), 8.7, 8.41 above, page 260 (Hoare); Rapho/Chaverou, 2.2; Redferns/Landy, page 486 above; Rex/Game, page 82; Ann Ronan at Image Select, 2.1, 4.9, 7.1; Science Photo Library, page 5 (Sauze), 2.3 (NASA), 2.4 (NOAO), 4.2 (Nunuk), 4.15 (NASA), 4.17(a), 4.22 (Migdale), 5.43 (Muller/Struers), 5.50 (Aprahamian/Sharples Stress Engineers), 114 (Menzel), 6.1 (Stevenson), 6.26(b) (Watts), 7.2 (Parker), page 168 (Sauze), 8.6 (Burgess), 8.18, 8.22 and 26.5 (Megna/Fundamental Photos), 8.30 left (Burgess), 8.40 (Takara), 8.46 above (Mead), page 199 (Maisonneuve, Publiphoto Division), page 200 (Morgan), 9.12(b) (Fielding/Johnson Matthey), 10.10(b) (Garradd), page 246 above (Fermi National Accelerator Laboratory/Parker), page 246 below (FNAL), 11.A4 (Parviainen), page 262 (NASA), page 301 (Bond), 14.1, page 318 (Parker), 15.13 (Fletcher), page 337 (Plailly),

688 ■ Acknowledgments

16.9(b) (Meadows/Peter Arnold Inc), 16.10(a) (Evelegh), 16.11(b) left (Royal Greenwich Observatory), 16.11(b) right (Max Plank Inst. For Radio Astronomy), 16.28(a) (NASA), 16.28(b) (Space Telescope Science Inst.), 16.32 centre and below (Physics Department, Imperial College), page 360 centre (Schild/Smithsonian Astrophyical Observatory), 17.7 above (Physics Department, Imperial College), 17.7 below (Imperial College), page 376 (Bartel), 18.19 (Bond), 18.25, 18.26 (Wilson), 18.28(b) (Parker), 18.43 (King-Holmes), page 412 (Nurisany and Perennou), 19.38 left and 19.38 right (Ressmeyer/Starlight), 19.47 (Syred), 19.48 (Parker), 19.49 (Schild), 19.51 (NRSC), 19.52 (Schleichkorn/Custom Medical Stock), 19.53 (CNRI), 19.55 (University of Medicine and Dentistry of New Jersey), 19.56 left (BM), 19.56 above (Plailly), page 445 (Biggs), 20.11(a) (Morgan), 21.21(b) (Rain), 21.24 left (Hart-Davis), 21.24 centre (Kulyk), 21.24 right (McIntyre), page 507 (CNES 1986 Distribution Spot Image), page 508 (van Sant /Geosphere Project), 23.37 (NOAA), 23.39 left (Worldsat Productions), 23.39 right (NRSC), 23.40 (ESA Eurimage), 23.41 (NOAA), 23.42 (Feldman/NASA GSFC), page 536 above (CNRI), 24.4 (Meadows), 24.14 (Kulyk), 24.15 (Tompkinson), 24.17 (Schiller), 24.19 (Parker), 24.20(b) (Dohrn), 24.22(b) (Fraser/Med. Physics, RVI, Newcastle upon Tyne), 24.22(c) (CNRI), 24.23 (Sikora), 24.24(a) (McIntyre), page 555 (Parker), 26.1 right (Fowler, Bristol University), 26.1 below (Powell, Fowler, Perkins), 26.5(a) right,

26.21 (Loiez), 26.22(b) (CERN), 26.25 (CERN), 26.28 (Parker), 26.31(b) (CERN), 26.33 (Loiez/CERN), 26.38 (Parker), 26.39 (Parker), page 614 above left (Butler), page 614 below left (NASA), page 614 above right (ESA), page 614 below right (USGS), page 615 above left (NASA), page 615 below left (NASA), page 615 above right (NASA), page 615 centre right (NASA), page 615 below right (Hardy), 27.4, 27.11(a) (Royal Observatory, Edinburgh), 27.11(b) (NASA), 27.16(a) (USGS), 27.16(b) (NASA), 27.20 (NOAO), 27.23 (ESA), 27.25 (NASA), 27.26(a) (Ressmeyer/Starlight), 27.26(b) (Ressmeyer/Starlight), 27.27 (Ressmeyer/Starlight), 27.29 (Scagell), 27.31 (Dornier Space), 27.32 (NASA), 27.42 (Royal Observatory, Edinburgh), 27.44 (Schild/Smithsonian Astrophysical Laboratory), 27.48 (Hester and Scowen, Arizona State University), 28.1 (Schad), 28.4 (Hallas), 28.5 left, 28.5 right (Hencoup and Marten), 28.12 (Royal Observatory, Edinburgh), 28.25 (NASA); Science and Society Picture Library, 5.17, 16.10(b); Starland Picture Library, 27.24 (NOAO), 27.47 (ESO), 28.6, 28.7, 28.8 (NOAO); M.R. Stringer, Medical Physics Department, University of Leeds, Leeds General Infirmary, page 338; SD Pictures, 5.1 right; Volvo Cars UK, 8.14.

The cover photograph of a bubble chamber image is by Patrice Loiez, CERN, from Science Photo Library.

ANSWERS TO NUMERICAL QUESTIONS

CHAPTER 2

Self-test questions
A 2.2×10^6 y
B 1.7×10^{-4} s
H 5 s
I **(a)** 96 m, **(b)** 2.7 m
K **(a)** 6 s
M **(a)** 0.8 m, **(b)** 100 m
P 510 km h^{-1} at 80° N of E

End-of-chapter questions
1 (a) 500 km, **(b)** 261 s
3 (b) 1 ns
4 Sirius 8.63 ly, Alpha Centauri 4.38 ly
5 120 m s^{-1}
8 110 m
9 (b) −0.453 m s^{-2}
10 Note: distance to Mars is 8×10^{10} m, not as in question data. **(a)** 7920 km, **(b)** 6 h, **(c)** 21.5 year, **(d)(i)** 1400 s (1360 s) **(ii)** 1.36×10^7 m, **(iii)** 46 days
11 (b) 150 m
12 38 m
14 380 km h^{-1}
15 5.7 knots 30° N of E

CHAPTER 3

Self-test questions
E **(a)** 2.1 N s, **(b)** 2.1 N s, **(c)(i)** 14 N, **(ii)** 14 N
F 5 N s
G 2.4 kg
H **(a)** 3 kg, **(b)** 29 N
K **(a)** 0.25 m
M 5.3 s
R **(a)** 12 m s^{-1}, **(c)** a close call! (Why?)
S 16 m s^{-1}
U 112 km

End-of-chapter questions
1 (a) 29 m s^{-1}, **(b)** 88 m
7 (a) 28.0 N, **(b)** 5.83 m s^{-2} **(c)** 5.83 m s^{-2}
10 14 m s^{-1}
11 (b) 1.2 s, **(c)** 7.3 m
14 (a) 28 m s^{-1}, **(b)** 870 m s^{-2}
15 (a) 10 s **(b)** 44.5 m **(c)(i)** 30 m s^{-1}, **(ii)** 3 s
16 (a)(i) 240 N, **(ii)** 240 N, **(b)** 0.8 m s^{-1}
18 (a) 3.1×10^{11} m s^{-2}, **(b)** 2.8×10^{-19} N
21 (a) 2×10^{20} N, **(b)** 2.7×10^{-3} m s^{-2}

CHAPTER 4

Self-test questions
C **(a)** 196 J
D **(a)** 18.5 J, **(c)** 6 m s^{-1}
F **(b)** 5050 kJ
J **(b)** 7.9 km s^{-1}
K **(a)** 7.7 km s^{-1}, **(b)** 93 min, **(c)** 15.5
L 1.3×10^8 m (tricky – a Moonday is 27.3 Earthdays)

End-of-chapter questions
2 (a) 1000 N
4 (a) 10 m s^{-1}, **(b)** 18 J or 36%
6 (a) 0, **(b)** 6 kJ
7 0.18 J
8 (b) ~10^7 J kg^{-1}, **(c)** escape speed: 5 km s^{-1}, **(i)** 8.862×10^4 s, **(ii)** 2.04×10^7 m
9 1.5 km
11 (a)(i) 1.4×10^3 m s^{-1}, **(ii)** 6.9×10^3 m s^{-1}, **(b)** 2.2×10^{12} s
13 15 kN

CHAPTER 5

Self-test questions
B 8.5 units, 4.7° E of N
D 15(.2) m
I 210 N
J **(a)** 2.5 mm, **(b)** 21 mm

End-of-chapter questions
1 (a) 1.39 kN tension, 1.60 kN compression
2 (a) 0.5 m, **(b)** 1 m
3 (a) 3.0×10^8 N m^{-2}
4 (a) 0.25 mm, **(b)** 22.5 kN
5 (a)(i) 50 kN, **(ii)** 10^7 N m^{-2}, **(iii)** 5×10^{-5}
6 (c) 14, **(d)** 1.3 J, **(e)** 1.4×10^3 N m^{-1}
10 170 J
16 1.7 m^3

CHAPTER 6

Self-test questions
A 800 Hz
H 1.2 Hz, 0.83 s
I 1.6 cm
J **(a)** 700 Hz, **(b)** 1400 Hz
L **(a)** 32 cm, **(b)** 16 cm
M 258.5 Hz, 253.5 Hz

End-of-chapter questions
5 1.3 km s^{-1}, 1.0×10^{16} m s^{-1}
6 8.2 cm
7 30 m s^{-1}
9 (a) 288 Hz, **(b)** 30 N, **(c)** 8 Hz
10 (c) 28 Hz

CHAPTER 7

Self-test questions
B 0.148 or 1:6.8
C **(a)** 6.8 MJ, **(b)** 14p
F $\pi/4{:}\pi/2\sqrt{3}$ or 1:1.15
G 4
J **(a)** 900 m s^{-1}, **(b)** 0.4, **(c)** 0.04 (approx. values)

End-of-chapter questions
1 (b) 7.3% approx, **(c)** 5×10^{-21} J
2 (b)(ii) 2.6×10^{-10} m, 3.7×10^{-20} J
3 (b)(i) 1.35×10^{-20} J, **(ii)** 3.1 kJ, **(iii)** 1.28×10^{-20} J, **(c)** ~2×10^{18} m^{-2}, **(d)** between 10^{-9} m and 10^{-10} m

6 26.6 lb in^{-2} ($\equiv 1.83 \times 10^5$ Pa)
7 **(b)(ii)** 418 m s^{-1}, **(iii)** 0.93 × 10^5 Pa, **(c)(i)** √10
8 **(a)** 430 m s^{-1}, **(b)** 11 kJ
9 **(b)** 480 m s^{-1}, **(d)** 1.7 × 10^{-10} Pa, **(c)** 3.1 h (10^4 s)

CHAPTER 8

Self-test questions
A Are you flat-footed? About 10^5 N m^{-2} (Pa)
B 7.8 m s^{-2} (it's always μg!)
D **(a)** π radians s^{-1},   **I** 1.2 kPa
  (b) 47 m s^{-1} **N** 100 m s^{-1}
E 375 kg m^2 **O** about 400 km! (So what has
H 3 × 10^4 Pa   gone wrong?)

End-of-chapter questions
3 **(a)** 1.4 × 10^4 N, **(c)** 7 m s^{-2} (0.7 g), **(d)** 4.4 s
4 **(a)** 0.67, **(b)** ~50 m
5 **(a)** 2.4 rad s^{-1}, **(b)** 7.3 rad s^{-1}, **(c)** 3 times
7 **(a)** 375 m s^{-1}
14 8 MPa
15 **(a)** 15 kPa, **(b)** 45 kN, **(c)** 27 kN
17 825 kg m^{-3}
18 **(b)(iii)** 0.038 J s^{-1}, **(iv)** 8.5 × 10^{-2} N s m^{-2}
19 **(b)(i)** 1 kN, **(ii)** 40 kW

CHAPTER 9

Self-test questions
C 6.2 × 10^{18}
D **(a)** 0.5 A, **(b)** 2 mA, **(c)** 40 nA
E **(a)** 8 × 10^4 C, **(b)** 4320 C, **(c)** 100 µC
F ratio 1 (drift velocities are same)
H 100
I resistances are the same
J 12 V bulb filament dia. 20 times larger than for 240 V bulb
K iron:copper resistance 5.2:1, in series iron wire hotter, in parallel, copper hotter
L 3240 J   **M** 1 MΩ   **N**   0.42 A

End-of-chapter questions
3 **(a)** 2 Ω, **(b)** 1.44 J s^{-1}
4 **(a)** 6 V, **(b)(ii)** 4.8 V, **(c)** 4.8 V
5 **(a)** 11 kΩ, **(b)** 0.27 mA, **(c)** 3/4
6 **(a)(ii)** near full scale, **(b)(ii)** near full scale
7 **(a)** 1 J, **(b)** 1000 W, **(c)** 100 A
8 **(a)** 5000 µC, **(b)** $\Delta T/RC = 0.1$, $\Delta Q = 500$ C, **(c)** 4500 µC, **(d)** 450 µC, left 4050 µC
10 **(a)** and **(b)**
11 **(a)** 0.6 mA, **(b)(i)** ≅12.5 squares = 0.13 mC, **(ii)** 20 µC
12 **(a)** 20 mA, **(c)** 10 µC, **(d)** 0.58 ms

CHAPTER 10

Self-test questions
A 10 V
C **(a)** 1.6 × 10^{-19} J, 1.6 × 10^{-16} J, 1.6 × 10^{-13} J, **(b)** 1 eV, 1000 eV (1 keV), 1 000 000 eV (1 MeV)

E **(a)** 180 V
F **(a)** ~110 m^2, **(b)** ~19 m^2
H 2.3 × 10^{-8} N, about 10^{18} times the weight of a hydrogen atom

End-of-chapter questions
1 **(a)** 250 N C^{-1}, **(b)** 2.5 ×10^3 N C^{-1}, **(c)** 4 × 10^3 N C^{-1}
2 **(a)** 10 N, **(b)** 1.0 × 10^{-4} N, **(c)** 2.5 × 10^{-4} N, **(d)** 1.0 × 10^{-7} N, **(e)** 2.5 × 10^{-10} N
3 **(a)** 5000 V m^{-1}, **(b)** 5000 V m^{-1}, **(c)** 5000 V m^{-1}
4 2.5 × 10^{-4} N
5 0.28 µC
6 1.6 × 10^7 m s^{-1}
7 **(a)** ×2 **(b)** no difference, **(c)** ×2
8 **(a)** 7.7 g; 4, 12, 8, 5, 10 and 7, **(b)** 5, 7, 3, 11, 9 and 8.
9 **(a)** 200 V m^{-1}, **(b)** 200 N C^{-1}
10 **(a)** area 1.13 × 10^{-2} m^2, length 0.57 m, **(b)** 10 nC, **(d)** half needed
11 **(a)(i)** 27 pF, **(ii)** 1.3 × 10^{-7} J, **(b)(i)** 1.4 × 10^{-5} N
12 4.8 µC
13 2.3 × 10^{-22} N, force ratio 1.24 × 10^{36}
14 2 m, 67 nC
15 4.6 × 10^{-14} m

CHAPTER 11

Self-test questions
A 0.05 N
E **(a)** 1.6 × 10^{-3} m s^{-1}, **(b)** 3.1 × 10^{-5} V. A 1 mm misalignment → resistance 2.7 × 10^{-5} Ω in direction of current. 5A would cause p.d of 130 µV

End-of-chapter questions
2 **(c)(i)** 1 × 10^{-2} N, **(ii)** 0.1 N m
3 **(a)** 18.8 m, **(b)(i)** 0.57 N, **(ii)** 0.028 mm
5 **(b)(i)** 1.1 × 10^{-7} s
7 **(e)** $e/m = 1.7 \times 10^{11}$ C kg^{-1}
8 **(c)(i)** 0.21 m s^{-1}, **(ii)** 3.2 × 10^{-5} m^3 s^{-1}

CHAPTER 12

Self-test questions
A 0.5 V   **J** 0.5 H

End-of-chapter questions
1 0.26 V
4 **(a)** 7.5 × 10^{-2} Wb, (b) 0.75 A
5 D
6 flux and e.m.f. **(a)** ×2, **(b)** ×1/2, **(c)** ×1, **(d)** ×2
7 **(a)** resistance ×1/2, **(b)** reluctance ×1/2
8 **(c)** same, **(d)** initial 0.5 A s^{-1}, final 0.67 A
9 **(a)** 20:1 (step up), **(b)** 0.15A

CHAPTER 13

Self-test questions
A 322 V   **B** 11.4 V
C **(a)** 1:10 (step up), **(b)** Current reduced 10, energy losses reduced ×100

D **(a)** 159 Ω, **(b)** 1.6 Ω, **(c)** 0.16 Ω
E **(a)** 6.3×10^{-3} Ω, **(b)** 0.63 Ω, **(c)** 6.3 Ω
F **(a)** 101 Ω, 100 Ω, **(b)** 100.2 Ω, 7.9 kΩ
G 225 kHz

End-of-chapter questions

2 **(a)** 20 V, **(b)** 14.1 V, **(c)** 100 Hz
3 53 Ω, r.m.s.current 4.4 A
4 **(a)** 4 V, **(b)** 1.7 kHz, **(c)** 28 mA
5 **(a)** 14.4 V, **(b)** 20.3 V, **(c)** 18.9 V, **(d)** 100 Hz,
    (e) (i) RC from 1 s to 0.01 s = ripple period, o/p
    voltage not smoothed, **(ii)** as **(i)**
6 **(a)** 265 μF, **(c)** 4.2 mH, **(d) (i)** 11 Ω, **(ii)** 13 Ω
7 **(b)** 2 μF, 20 mH, **(c) (i)** 8 Ω, 16 Ω, **(ii)** 1.3 kΩ, 630 Ω,
    (d) about 1 Ω (small compared to speaker impedance),
    typically L_2 = 200 μH, C_2 = 200 μF
8 **(a)** 2.5 Hz, **(b)** 8.6 H

CHAPTER 14

Self-test questions
A 184 J, 184 J   **B** 81.6 °C, 354.8 K

End-of-chapter questions
1 **(a) (i)** 1.51 kJ, **(ii)** 0.187 kJ kg^{-1} K^{-1}
2 **(a)** 540 K m^{-1}, 1.9 W, **(b)** 190 K m^{-1}, 0.66 W, **(c)** 1.2 W
3 **(b) (i)** 12.16 °C, 3.84 °C
4 **(a)** 18.8 W, **(b)** 12.6 °C, **(c)** 9.7 W
5 **(a)** 13 W m^{-2} K^{-1}, 12.5 W m^{-2} K^{-1}, **(c)** 31 W m^{-2} K^{-1}

CHAPTER 15

Self-test questions
G 61%   **H** 1500% (approx. only)

End-of-chapter questions
1 (1.75) (12 cm)   **6** 416 K
2 8.6 kJ   **7** 14 kJ, 10.5 kJ
4 150 K   **9** 4 J K^{-1}
5 **(b)** 0.08 mole,
    (c) (i) 400 kPa, **(ii)** 600 kPa,
    (iii) 300 kPa, **(iv)** 200 kPa

CHAPTER 16

Self-test questions
E **(b)** 3 cm or more
G 2 million times
H 6×10^{-8} radians

End-of-chapter questions
3 **(b)**, **(c)**
6 **(b)** ~10
9 **(b)** ~10
10 23°, 51°
11 **(a)** 3.6×10^{-4} radians, **(b)** only just!
12 1.3
13 1.8×10^{-4} radians, **(b) (i)** 1/16, **(ii)** 4, **(c)** 3°

CHAPTER 17

Self-test questions
A 160 W m^{-2}, 2×10^{16} W m^{-2}

End-of-chapter questions
2 **(a)** 9×10^{-6} m, **(b)** 1.1 kW
3 **(a)** 12 000 K, **(b)** 9.7×10^{-7} m
4 5200 K
6 **(a) (i)** 1×10^{-18} J, **(ii)** 3×10^{-19} J, **(iii)** 1×10^{-19} J,
    (iv) 2×10^{-23} J
7 ~4×10^{18}
9 **(b)** 10^{15} Hz, **(d)** 1.06×10^{-19} J
13 **(c)** 5.7°, **(d) (i)** 2.7×10^{-11} m,
    (ii) 5×10^{-6} rad (3×10^{-4} degrees)
14 **(a)** 1.03×10^{-11} m, **(b)** 4×10^{-34} J s
16 **(a)** 4.6×10^{-24} N s, **(b)** 1.4×10^{-10} m
18 **(a)** 2×10^{-10} m, **(b)** 3.3×10^{-24} N s, **(c)** 6×10^{-18} J,
    (d) 5×10^{-18} J

CHAPTER 18

Self-test questions
E 7×10^{-15} m   **I** 250 T
F 5×10^{-14} m   **K** 7.2×10^{10} J
H **(a) (i)** 3.2×10^{-19} J, **(ii)** 6.4×10^{-13} J,   **M** about 12 000
    (b) (i) 6.8 MeV, **(ii)** 1.6×10^{20} eV

End-of-chapter questions
5 **(a)** 3.0523×10^{-30} kg, **(b) (i)** 2.75×10^{-13} J, **(ii)** 1.72 MeV
6 194 MeV
8 **(a)** 2 mT, **(b)** Ne_{20} 10.4 m, Ne_{21}10.9 m, Ne_{22} 11.4 m
10 4.5×10^{-15} m
11 1.5×10^{17} kg m^{-3}
15 **(a) (i)** 650 s^{-1} **(ii)** 0.6 s^{-1}
16 **(a)** 1.5×10^{12} s^{-1}, **(b)** 280 s
19 **(b)** 1.8×10^{-5} s^{-1}, **(c)** 1.8×10^{-5} s^{-1}, **(d)** 3.9×10^{4} s
21 **(a)** ~125 s^{-1}, **(b)** 8×10^{-4} kg, **(c)** 3×10^{-12} s^{-1},
    (d) 6.5×10^{13}, **(e)** 1.45×10^{-12} kg,
    (f) 6.9×10^{11} (C_{12} to C_{14})

CHAPTER 19

Self-test questions
B **(a)** 6×10^{-7} m, **(b) (i)** 4×10^{-7} m, **(ii)** 1.33
E **(a)** −14.5 dioptre, +34.5 dioptre, **(b)** +20 dioptre,
    (c) 50 mm
G 1.3 m, 0.93 m
I 2.8 m, 1.9 m × 2.5 m
K ×30
P **(a)** about 2.5 cm
S **(b) (i)** 6.6×10^{-23} N s, **(ii)** 2.4×10^{-15} J, **(c)** 15 kV

End-of-chapter questions
5 **(a)** 50.21 mm, **(b)** 8.6 m
8 **(b) (i)** 7.4 mm, **(ii)** 29×10^{-3} rad, **(iii)** 3.2,
    (c) (i) 0.85 m, **(ii)** 16, **(iii)** 0.15 rad
10 **(a)** 24, **(b)** 13
11 0.32 rad

CHAPTER 20

Self-test questions
A 8 m, 4 m, 2.67 m, 2.5 m
C 10^6 s, 10^{11} m
D 1.60×10^{-19} J
E 4×10^{-19}, 6.5×10^{-10}
N 63 mm

End-of-chapter questions
1 1.1×10^{-4} m s^{-1}
4 3.4×10^{21} m^{-3}
6 0.32 nC, 0.64 nC
7 **(a)** 1.1×10^{-4},
   (b) 13 nC,
   (c) 2.4 pF, 5.5 kV

CHAPTER 21

Self-test questions
A **(a)** 1500 m, 300 m, 3 m, 0.6 m,
   3 cm,
   (b) 1.5 Mhz, 3.75 Mhz,
   150 Mhz, 1.5 Ghz, 30 Ghz
B 1.001 Mhz, 0.999 MHz
C 1.1 MHz
D 110 stations
E 133 stations
F 5 bits, 6 bits, 7 bits
I 4.3×10^9 quantization levels

End-of-chapter questions
2 **(a)** 2 kHz,
   (b)(i) 300 Hz,1 Mhz,
   0.9997 Mhz, 1.0003 Mhz,
   (ii) 3.4 kHz,1Mhz,
   0.9966 Mhz, 1.0034 Mhz,
   (d) 6.8 kHz,
   (e) about 58 stations,
   (f) 10 stations at most
3 **(c)(i)** 20 stations

CHAPTER 22

Self-test questions
H 50 mV
I 10^9 times
J +10
K 90 kΩ
L 0.18 V
N (difficult to get one value !!)
   0.5 Mhz: $L = \sim200$ μH,
   frequency = 1.5 Mhz: $L = \sim60$ μH

End-of-chapter questions
1 **(a)** 2-input NAND, **(b)** 3-input AND,
(c) 2-input exclusive-OR, **(d)** 2-input
parity gate or comparator
2 **(a)**

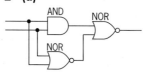

(b) 5 NOR, two used as NOT, **(c)** 6 gates

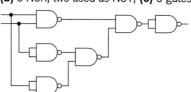

8

Time /s	0	1	2	3	4
V_B/V	0	0	6	0.7	0

9 **(a)**

Amplifier	Input at non-inverting input
A	4 V
B	3 V
C	2 V
D	1 V

3 **(a)(i)** 2-input parity gate or
   comparator, **(ii)** o/p remains 'high',
   (b) same as **Q2(c)**,
   without the last (RHS) gate
4 2-input AND, 2-input exclusive-OR.
5

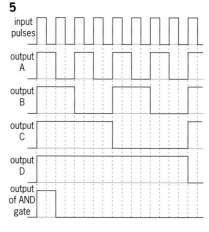

6 **(a)** 91 nA, 91 μV,
   (b)(i) 0V (virtual earth), **(ii)** 10 nA,
   (iii) –1.0 mV,
   (c) add another inverter (o/p at F
   should be +1 mV)
7 **(a)** +3 V,
   (b) +11 V,
   (c) –2 V

5

time/ms	0	1	2	3	4	5	6	7
binary code	1011	1101	1000	0001	0100	0010	1010	1111
digital output								

6 sample rate 1 kHz, max frequency 500 Hz, **(a)** 125 μs, **(d)** 2 samples per cycle
7 **(a)** 256, **(b)** 125 μs
8 **(a)** 16, **(b)(i)** 2^{16} = 65536 levels
9 **(a)** 8000 samples s^{-1}, **(b)** 64 kbit s^{-1}, **(c)** 2048 kbit s^{-1} = 2 Mbit s^{-1},
   (d) 2048, **(e)** 131 072 000, **(f)** over 78 000
10 **(a)** 256 levels, **(c)** 125 μs

(b)

Analogue voltage	Amplifier outputs			
	A	B	C	D
0	1	1	1	1
1	0	1	1	1
2	0	0	1	1
3	0	0	0	1
4	0	0	0	0

(c) 1.5 V output is 0111, 4.5 V output is 0000

10 (a)

High input	Voltage output
A	2.50V
B	1.25 V
C	0.63 V
D	0.31 V

(b)

Binary input	Analogue voltage output/volts
0000	0.00
0001	0.31
0010	0.63
0011	0.94
0100	1.25
0101	1.56
0110	1.88
0111	2.19
1000	2.50
1001	2.81
1010	3.13
1011	3.44
1100	3.75
1101	4.06
1110	4.37
1111	4.69

(c) X 1001, Y 0110, Z 1101
(d) 32 levels needs a 5-bit digital signal, add 5th input resistor value $32R$

CHAPTER 23

Self-test questions
G **(a)** 4×10^{13} J s^{-1}, **(b)** ~1/4
H **(a)** 463.84 m s^{-1}, **(b)** 0.033731 N, **(c)** 3.4396×10^{-3}
I 0.4

End-of-chapter questions
1 **(b)** 8×10^4 K
2 **(a)** 36 000, **(b)** 7.6×10^8 y
7 **(c)** 33 km
12 1.7 hours

CHAPTER 24

Self-test questions
B 15.4 MHz   **D** 0.15

End-of-chapter questions
3 75 mm s^{-1} **6** 8.25×10^{-12} m
4 ~0.15 mm   **9** **(a)** 32.4 m^{-1}, **(b)** 71 mm

CHAPTER 25

Self-test questions
B 1.3
C **(b)(i)** 600 m, **(ii)** 9 km
F you get ~22 times more massive

End-of-chapter questions
2 (a) 3×10^8 m s^{-1}
3 (a) 2.6×10^8 m s^{-1}, **(b)** 0.5 m
5 (a) 0.7 μs, **(b)** 210 m
7 (b) 3×10^8 m s^{-1}, **(c)** 10^{-5} ly (~10^{11} m), **(d)(i)** 10^5 years, **(ii)** 10^{-5} years (about 5 minutes)
9 (a) 2.09×10^{-30} kg, **(b)** 5.64×10^{-22} N s
11 (a) 4×10^{-15} J, **(b)** 4×10^{-15} J
12 (a)(i) 1.51×10^{-8} J, **(ii)** 1.49×10^{-8} J, **(b)(i)** 8.20×10^{-12} J, **(ii)** 8.12×10^{-12} J

CHAPTER 26

Self-test questions
F 3×10^{-23} s   **P** 50 times
K 8×10^{-21} s   **R** 75

End-of-chapter questions
10 (a) 2.4×10^{-13} J, **(b)** 1.5 MeV
11 (a) 1.5×10^{-10} J, **(b)** 0.9 GeV
12 (a) 0.51 MeV/c^2, **(b)** 0.000 51 GeV/c^2
19 (a) 0, **(b)** 9.4 GeV, **(c)** 9.4 GeV/c^2
20 (a) 1.3×10^7 m s^{-1}, **(b)** 3.1×10^5 m s^{-1}
21 (a) 74 mm, **(b)** 3.2 m

CHAPTER 27

Self-test questions
D **(b)** 2×10^{30} kg   **O** ~100 km
I 6.1×10^{18} m^2 **R** **(a)(i)** 1 kpc, **(ii)** 2×10^7 AU
K 5800 K

End-of-chapter questions
6 Ganymede 7.2 days, Callisto 17 days
18 (a) ratio 2.7×10^5
22 (a) 1.3 pc, **(b)** 4.3 ly
24 (a) 32.7 AU, **(b)** 246 years
29 Capella −0.49, Rigel −8.0, Procyon +2.7
36 (a)(i) 0.200 000 times a second, **(ii)** 0.62 831 853 rad s^{-1}, **(b)(i)** 6.31×10^{38} J, **(ii)** 0.627 690 21 rad s^{-1}, **(c)** $I = 1.6 \times 10^{42}$ kg m^2, **(d)** 4×10^{34} kg

CHAPTER 28

End-of-chapter questions
9 (i) 4.25 km s^{-1}, **(ii)** 43 km s^{-1}, **(iii)** 85 km s^{-1}, **(iv)** 8.5×10^3 km s^{-1}
4 1.6 Mpc
7 (a) 3 Mpc
11 (a) 1, **(b)** 12, **(c)** 25%, **(d)** 75%

INDEX